Lecture Notes in Computer Science 1974

Edited by G. Goos, J. Hartmanis and J. van Leeuwen

Springer
*Berlin
Heidelberg
New York
Barcelona
Hong Kong
London
Milan
Paris
Singapore
Tokyo*

Sanjiv Kapoor   Sanjiva Prasad (Eds.)

FST TCS 2000:
Foundations of
Software Technology
and Theoretical
Computer Science

20th Conference
New Delhi, India, December 13-15, 2000
Proceedings

 Springer

Series Editors

Gerhard Goos, Karlsruhe University, Germany
Juris Hartmanis, Cornell University, NY, USA
Jan van Leeuwen, Utrecht University, The Netherlands

Volume Editors

Sanjiv Kapoor
Sanjiva Prasad
Indian Institute of Technology, Delhi
Department of Computer Science and Engineering
Hauz Khas, New Delhi 110016, India
E-mail:{skapoor/sprasad}@cse.iitd.ernet.in

Cataloging-in-Publication Data applied for

Die Deutsche Bibliothek - CIP-Einheitsaufnahme

FST TCS 2000: Foundations of software technology and theoretical
computer science : 20th conference ; proceedings, New Delhi, India,
December 13 - 15, 2000. Sanijv Kapoor ; Sanijva Prasad (ed.). – Berlin ;
Heidelberg ; New York ; Barcelona ; Hong Kong ; London ; Milan ;
Paris ; Singapore ; Tokyo : Springer, 2000
    (Lecture notes in computer science ; Vol. 1974)
    ISBN 3-540-41413-4

CR Subject Classification (1998): F.3, D.3, F.4, F.2, F.1, G.2

ISSN 0302-9743
ISBN 3-540-41413-4 Springer-Verlag Berlin Heidelberg New York

Springer-Verlag Berlin Heidelberg New York
a member of BertelsmannSpringer Science+Business Media GmbH
© Springer-Verlag Berlin Heidelberg 2000
Printed in Germany

Typesetting: Camera-ready by author, data conversion by Boller Mediendesign
Printed on acid-free paper   SPIN 10781381   06/3142   5 4 3 2 1 0

Preface

The Foundations of Software Technology and Theoretical Computer Science conference (FST TCS) is a well-established annual event in the theoretical computer science community. The conference provides a forum for researchers to present interesting new results in several areas of theoretical computer science. The conference is now in its twentieth year and has continued to attract high-quality submissions and reputed invited speakers.

This year's conference attracted 141 submissions (of which 5 were withdrawn) from over 25 countries. Each submission was reviewed by at least three referees, with most receiving more than four reviews. The Program Committee met in New Delhi on 5 and 6 August 2000, with many members participating electronically over the Internet. The discussions continued over the Internet for several days and we finally selected 36 papers for presentation at the conference and inclusion in the proceedings. We thank the Program Committee for their superlative efforts in finding top quality reviewers and working extremely hard to ensure the quality of the conference. We also thank all our reviewers for providing detailed and informative feedback about the papers. Rich Gerber's START program greatly simplified managing the submissions and the PC work.

We are grateful to our six invited speakers, Peter Buneman, Bernard Chazelle, Allen Emerson, Martin Grötschel, José Meseguer, and Philip Wadler for agreeing to speak at the conference and for providing written contributions that are included in the proceedings.

With the main conference this year, there are two satellite workshops — on *Recent Advances in Programming Languages* and on *Computational Geometry*.

FST TCS is being hosted this year by the Indian Institute of Technology, Delhi, after a hiatus of eight years. We thank the Organizing Committee for their efforts and several others who have been generous with their time and energy in assisting us — Mohammed Sarwat, Kunal Talwar, Surender Baswana, Rohit Khandekar, and Harsh Nanda, in particular.

We express our gratitude for the financial and other support from the various sponsors, IBM India Research Laboratory, New Delhi, in particular. We also thank TCS (TRDDC), Tata Infotech, Silicon Automation Systems, Cadence Design Systems, IIT Delhi, and others.

We also thank the staff at Springer-Verlag, especially Alfred Hofmann, for making the production of these proceedings flow very smoothly.

December 2000   Sanjiv Kapoor
                                                                      Sanjiva Prasad

Organization

FST TCS 2000 is organized by the Department of Computer Science, IIT Delhi, under the aegis of the Indian Association for Research in Computer Science (IARCS).

Program Committee

Pankaj Agarwal *(Duke)*
Manindra Agrawal *(IIT, Kanpur)*
Tetsuo Asano *(JAIST)*
Vijay Chandru *(IISc, Bangalore)*
Rance Cleaveland *(Stony Brook)*
Anuj Dawar *(Cambridge)*
Sampath Kannan *(U Penn)*
Sanjiv Kapoor *(IIT, Delhi)* **(Co-chair)**
Kamal Lodaya *(IMSc, Chennai)*
Madhavan Mukund *(CMI, Chennai)*
Gopalan Nadathur *(Minnesota)*
Seffi Naor *(Bell Labs and Technion)*
Tobias Nipkow *(TU Munich)*
Luke Ong *(Oxford)*
C. Pandu Rangan *(IIT, Chennai)*
Paritosh Pandya *(TIFR)*
Benjamin Pierce *(U Penn)*
Sanjiva Prasad *(IIT, Delhi)* **(Co-chair)**
Sridhar Rajagopalan *(IBM, Almaden)*
Abhiram Ranade *(IIT, Bombay)*
Dave Sands *(Chalmers)*
A. Prasad Sistla *(U Illinois, Chicago)*
Michiel Smid *(Magdeburg)*
Mandayam K. Srivas *(SRI)*

Organizing Committee

Sandeep Sen *(IIT, Delhi)* **(Chair)**
Naveen Garg *(IIT, Delhi)* **(Treasurer)**
S.N. Maheshwari *(IIT, Delhi)*

Referees

Mark Aagaard
Parosh Abdulla
Andreas Abel
Luca Aceto
Bharat Adsul
Salvador Lucas Alba
Eric Allender
Rajeev Alur
Roberto Amadio
Henrik R. Andersen
Andre Arnold
Anish Arora
S. Arun-Kumar
Amitabha Bagchi
Paolo Baldan
Chitta Baral
Franco Barbanera
Adi Ben-Israel
Stefan Berghofer
Karen Bernstein Jeffrey
Pushpak Bhattacharya
Ingrid Biehl
Somenath Biswas
Marcello Bonsangue
V. S. Borkar
E. Boros
Ahmed Bouajjani
Julian Bradfield
Roberto Bruni
Didier Caucal
Supratik Chakraborty
Manuel Chakravarty
Bernadette Charron
Andrea Corradini
Flavio Corradini
Jean-Michel Couvreur
Mary Cryan
Geir Dahl
Mads Dam
Vincent Danos
Alex Dekhtyar
Rocco De Nicola
Deepak Dhar
Catalin Dima

Juergen Dingel
Gilles Dowek
Deepak D'Souza
Phan Minh Dung
Bruno Dutertre
Sandro Etalle
Leonidas Fegaras
Sandor Fekete
Amy Felty
Gerard Ferrand
Rudolf Fleischer
Wan Fokkink
Phyllis Frankl
Lars-Ake Fredlund
Ari Freund
Daniel Fridlender
Jan Friso Groote
Philippa Gardner
Naveen Garg
Paul Gastin
Simon Gay
Yael Gertner
F. Geurts
Neil Ghani
John Glauert
Ganesh Gopalakrishnan
Andy Gordon
R. Govindarajan
Bernd Grobauer
Gopal Gupta
Peter Habermehl
Thomas Hallgren
Mikael Hammar
Ramesh Hariharan
Tero Harju
James Harland
Sariel Har-Peled
Reinhold Heckmann
Nevin Heintze
Fergus Henderson
Rolf Hennicker
Jesper G. Henriksen
Andreas Herzig
Michael Hicks

Pat Hill
Yoram Hirshfeld
Joshua Hodas
Haruo Hosoya
Alan Hu
Michaela Huhn
Graham Hutton
Costas Iliopoulos
Anna Ingolfsdottir
Mathew Jacob
Radha Jagadeesan
R. Jagannathan
David Janin
Alan Jeffrey
Thierry Jeron
Somesh Jha
Felix Joachimski
Bengt Jonsson
Charanjit Jutla
Kyriakos Kalorkoti
Ravi Kannan
Deepak Kapur
S. Sathya Keerthi
Ravindra Keskar
Sanjeev Khanna
Moonjoo Kim
Kamala Kirthivasan
Barbara König
Jochen Konemann
Goran Konjevod
Guy Kortsarz
Marc van Kreveld
Ajay Kshemakalyani
Herbert Kuchen
V. S. Anil Kumar
K. Narayan Kumar
S. Ravi Kumar
Vijay Kumar
Orna Kupferman
Jim Laird
Charles Lakos
Leslie Lamport
Cosimo Laneve
Kung-Kiu Lau

Bill Leal
Xavier Leroy
Christos Levcopoulos
Jean-Jacques Levy
Patrick Lincoln
C. E. Veni Madhavan
P. Madhusudan
Meena Mahajan
Anil Maheshwari
Jean-Yves Marion
Narciso Mart-Oliet
John Matthews
Guy McCusker
B. Meenakshi
Ron van der Meyden
Adam Meyerson
Tiusanen Mikko
Jon Millen
Dale Miller
Swarup Mohalik
Faron Moller
Remi Morin
Ben Moszkowski
Shin-Cheng Mu
Supratik Mukhopadhyay
Ketan Mulmuley
Andrzej Murawski
Anca Muscholl
Alan Mycroft
Lee Naish
Kedar Namjoshi
Y. Narahari
Giri Narasimhan
Tom Newcomb
Joachim Niehren
Martin Odersky
David von Oheimb
Atsushi Ohori
Friedrich Otto
Linda Pagli
Catuscia Palamidessi
Prakash Panangaden
Marina Papatriantafilou
Michel Parigot
Lawrence C. Paulson

Francois Pessaux
Antoine Petit
Ion Petre
Frank Pfenning
Jean-Eric Pin
Amir Pnueli
Bernard Pope
Ernesto Posse
K. V. S. Prasad
T. K. Prasad
Femke van Raamsdonk
J. Radhakrishnan
N. Raja
Sriram Rajamani
C. R. Ramakrishnan
Rajeev Raman
R. Ramanujam
K. Rangarajan
S. Srinivasa Rao
Julian Rathke
Laurent Regnier
Jakob Rehof
M. Rodriguez-Artalejo
W.-P. de Roever
Suman Roy
Abhik Roychoudhury
Oliver Ruething
Jan Rutten
Mark Ryan
Peter Ryan
Andrei Sabelfeld
Thomas Santen
A. Sanyal
Andre Schiper
Thomas Schreiber
Friedrich Schroeer
Nicole Schweikardt
Jonathan P. Seldin
Peter Selinger
Sandeep Sen
Sanjit Seshia
Anil Seth
Peter Sewell
J. Shahabuddin
Natarajan Shankar

Priti Shankar
Anil Shende
Bruce Shepherd
Rajeev Shorey
Amir Shpilka
R. K. Shyamasundar
Joseph Sifakis
Robert de Simone
Arindama Singh
Gurdip Singh
D. Sivakumar
G. Sivakumar
Milind Sohoni
Oleg Sokolsky
S. Sridharan
Y. N. Srikant
Aravind Srinivasan
Mark-Oliver Stehr
Andrew Stevens
Charles Stewart
Colin Stirling
Chris Stone
K. G. Subramaninan
K. V. Subrahmanyam
V. R. Sule
Eijiro Sumii
S. P. Suresh
Carolyn Talcott
Prasad Tetali
Hendrik Tews
Denis Therien
P. S. Thiagarajan
Hayo Thielecke
Henk van Tilborg
Mikko Tiusanen
Philippas Tsigas
John Tucker
Frits Vaandrager
Kasturi Varadarajan
Moshe Vardi
Vasco T. Vasconcelos
Helmut Veith
Santosh Vempala
S. Venkatesh
R. Venugopal

Oleg Verbitsky
V. Vinay
Sundar Vishwanathan
Mahesh Viswanathan
Walter Vogler

Igor Walukiewicz
Pascal Weil
Herbert Wiklicky
Thomas Wilke
Harro Wimmel

James Worrell
Eric van Wyk
Kwangkeun Yi
Clement Yu
Shiyu Zhou

Table of Contents

Invited Presentations

Contributions

Model Checking:
Theory into Practice

E. Allen Emerson*

Department of Computer Sciences and Computer Engineering Research Center
The University of Texas at Austin, Austin TX-78712, USA
emerson@cs.utexas.edu
http://www.cs.utexas.edu/users/emerson/

Abstract. Model checking is an automatic method for verifying correctness of reactive programs. Originally proposed as part of the dissertation work of the author, model checking is based on efficient algorithms searching for the presence or absence of temporal patterns. In fact, model checking rests on a theoretical foundation of basic principles from modal logic, lattice theory, as well as automata theory that permits program reasoning to be completely automated in principle and highly automated in practice. Because of this automation, the practice of model checking is nowadays well-developed, and the range of successful applications is growing. Model checking is used by most major hardware manufacturers to verify microprocessor circuits, while there have been promising advances in its use in software verification as well. The key obstacle to applicability of model checking is, of course, the state explosion problem. This paper discusses part of our ongoing research program to limit state explosion. The relation of theory to practice is also discussed.

1 Introduction

There is a chronic need for more effective methods of constructing correct and reliable computer software as well as hardware. This need is especially pressing for concurrent, distributed, real-time, or, more generally, *reactive systems*, which can exhibit, unpredictably, any one of an immense set of possible ongoing, ideally infinite, behaviors. Many safety critical and economically essential real-world applications are reactive systems. Examples include: computer operating systems, computer network communication protocols, on-board avionics control systems, microprocessors, and even the internet.

The traditional approach to program verification involves the use of axioms, and inference rules, together with hand proofs of correctness. Because of the inherent difficulty and sheer tediousness of manual proof construction, this manual, *proof-theoretic* strategy has, by-and-large, fallen from favor. While this approach has introduced fundamental principles such as coherent design and compositional proofs, the evidence suggests that it does not scale to real systems.

---

* This work was supported in part by NSF grant CCR-980-4736 and TARP project 003658-0650-1999.

S. Kapoor and S. Prasad (Eds.): FST TCS 2000, LNCS 1974, pp. 1–10, 2000.

We have developed an alternative *model-theoretic* strategy for program reasoning (cf. [7]). The informal motivation is that, while manual proofs are not feasible, it ought to be possible to devise fully automated design and reasoning methods using the basic model theory of modal temporal logic. Temporal logic, a tensed form of modal logic, has been shown to provide a most useful approach to specifying and reasoning about the complex and ongoing behavior reactive systems [27]. The linear temporal logic LTL provides modalities corresponding to natural language tenses such as Fp (eventually p), Gp (henceforth p), Xp (nexttime p), and pUq (eventually q and p until then) that are well-suited to describing ongoing behavior along a (discrete) timeline corresponding to an execution sequence of a reactive program. Additional expressiveness is gained through use of the path quantifiers A (for all futures) and E (for some future) in the branching time logics CTL and CTL* (cf. [9]).

The automation of model-theoretic reasoning is permitted by the fact that core decision problems for temporal and classical modal logic are decidable. This provides a means, in principle, of letting programs reason about programs. Thus, it is significant that the following key problems associated with (propositional) temporal logic are decidable, in some cases efficiently. First, there is the *satisfiability* problem: given temporal logic specification f, does it have a model M? Second, there is the *model checking* problem: Is candidate Kripke structure model M a genuine model of temporal specification f?. Satisfiability asks if f is realizable in *some* model, and is useful in automatic program synthesis. Model checking decides if f is true in a *particular* model, and caters for automatic program verification.

We introduced model checking as an algorithmic method of verifying correctness of finite state reactive programs [7] (cf. [4], [6], [30]). Model checking has turned out to be one of the more useful approaches to verification of reactive systems, encompassing not only finite state but some finitely represented, infinite state systems as well. Our original approach is based on fixpoint computation algorithms to efficiently determine if a given finite state transition graph defines a model of correctness specification given in the temporal logic CTL. An advantage of this algorithmic approach is that it caters for both verification and debugging. The latter is particularly valuable in the early stages of systems development when errors predominate and is widely used in industrial applications. Symbolic model checking, which equals our original fixpoint based model checking algorithm plus the data structures for symbolic state graph representation using binary decision diagrams [3] (cf. [24]), is now a standard industrial tool for hardware verification.

The remainder of the paper is organized as follows. In section 2 we describe some recent technical advances in limiting state explosion. In section 3 we discuss factors relevant to the transition of model checking from theory into practice. Some closing remarks are given in section 4.

2 Limiting State Explosion

The most common calls from industrial users of formal verification tools are for (a) increased capacity of the tool to handle really large programs/designs, and for (b) increased automation. Model checking is especially popular in industrial usage because it is fully automated in principle. However, model checking's capacity is still limited. Model checking's capacity and utility are limited primarily by space complexity, and secondarily by time complexity. It is important that (the representation of) the state graph of the program being verified fit within the main memory of the computer on which the model checking tool is running, so as to avoid the slow down of many orders of magnitude which occurs when paging to disk. Of course, the underlying theoretical vexation is the well-known problem of combinatorial *state explosion*: e.g., if each sequential process of a parallel system has just 10 local states, a system with 1000 processes, could have as many as 10^{1000} global system states. A better handle on state explosion is essential to increasing the capacity of model checking tools. Thus, a good deal of effort has been devoted to limiting state explosion and making model checking more space efficient.

The technique of *abstraction* is central to the effective management of state explosion. In general terms, abstraction means the replacement of an intractably large system by a much smaller, abstract system through suppression of inessential detail and elimination of redundant information. The abstracted system should still preserve relevant information about the original system, and a desirable attribute is that the abstracted system should be equivalent in some appropriate sense to the original system. In the sequel, we will describe some of our work on abstraction utilizing the regularity of structure present in many reactive systems composed of multiple similar subcomponents, and mention some related open problems.

2.1 Symmetry

Symmetry Quotient Reduction. One useful approach to abstraction exploits the symmetry inherent in many concurrent systems composed of multiple interchangeable subprocesses or subcomponents. Using symmetry we have been able to verify a resource controller system with 150 processes and about 10^{47} states in roughly an hour on a Sparc. We developed a "group-theoretic" approach to symmetry reduction in [14]. The global state transition graph M of such systems often exhibits a great deal of symmetry, characterized by the the group of graph automorphisms of M. The basic idea is to reduce model checking over the original, intractably large structure M to model checking over the smaller quotient structure $\overline{M}$, where symmetric states are identified. The quotient graph can be exponentially smaller. Technically, let $\mathcal{G}$ be any group contained in *Aut M* $\cap$ *Auto f*, where *Aut M* is the group of permutations of process indices defining automorphisms of M and *Auto f* is the group of permutations that leave f and crucial subformulas thereof invariant. We define an equivalence relation on states of M so that $s \equiv_{\mathcal{G}} t$ iff $t = \pi(s)$ for some permutation $\pi \in \mathcal{G}$.

The quotient $\overline{M}$ is obtained by identifying states that are $\equiv_{\mathcal{G}}$-equivalent. Then, because $\mathcal{G}$ respects both the symmetry of the structure M and the ("internal") symmetry of the CTL* specification formula f, we can show $M, s \models f$ iff $\overline{M}, \bar{s} \models f$, where $\bar{s}$ represents the equivalence class $[s]$ of states symmetric to s. Since it turns out, in many practical cases, to be both easy and efficient to construct the quotient, we have reduced the problem of model checking CTL* properties over an intractably large structure to model checking over a relatively small structure, with the proviso that the symmetry of the structure is appropriately respected by the specification. Other work on symmetry may be found in, e.g., [21], [19], [5].

Annotated Symmetry Quotient Reductions. One can also "trade group theory for automata theory". In [15] we introduced a powerful alternative method. This method is also more uniform in that it permits use of a *single* annotated quotient $\overline{M} = M/Aut\ M$ for model checking for *any* specification f, without computing and intersecting with $Auto\ f$. The idea is to augment the quotient with "guides", indicating how coordinates are permuted from one state to the next in the quotient. An automaton for f designed to run over paths through M, can be modified into another automaton run over $\overline{M}$ using the guides to keep track of shifting coordinates. This automata-theoretic method is much more more expressive than the purely "group-theoretic approach" above. In particular, we can show that the automata-theoretic approach makes it possible to efficiently handle reasoning under fairness assumption, unlike the purely "group-theoretic" approach above. Another augmented "model-theoretic" approach is introduced in [16], accommodating both fairness and discrete real-time in the uniform framework of the Mu-calculus.

Open problems related to quotient symmetry reduction. Despite the expressive power of annotated symmetry quotient reduction, efficiently handling of general specifications in which all process indices participate along a path still seems a difficult one. It would be interesting, and useful for hardware verification, to see if the work on fair scheduling (e.g., $GF\,ex_1 \wedge \ldots GF\,ex_n$) could be adapted to efficiently handle, e.g., the tighter requirement of round-robin scheduling $((ex_1; ex_2; \ldots ex_n)^\omega)$) of an otherwise fully symmetric system. Such a composite seems to possess only cyclic symmetry, thereby limiting compression.

Simple Symmetry. Another important type of syntactic reasoning based on symmetry we call *simple symmetry* (originally dubbed state symmetry in [14]). It simplifies reasoning based on the symmetry of the specification, the structure, and individual states, and does not entail calculating a quotient structure. For instance, with an appropriately symmetric start state and with appropriate global symmetry, an *individual* process (or component) is correct iff *all* processes are correct. It permits "quantifier inflation" and "quantifier reduction": $M, s \models f_1$ iff $M, s \models \bigwedge_i f_i$, provided all f_j are identical up to re-indexing. Quantifier inflation via simple symmetry was used in [26] to facilitate verification of memory arrays, and more recently to radically reduce case analysis in [25]. In [16] we provided some powerful instances of the application of simple symmetry. Model checking

$E(FP_1 \wedge \ldots \wedge FP_n)$ is an NP-complete problem, apparently requiring time exponential in n in the worst case. We use simple symmetry to show that for a system that (a) is fully symmetric; (b) has fully symmetric start state s_0; and (c) is already known to be *re-settable* so that $AGEFs_0$ holds, the above property then amounts to EFP_1, which is in polynomial time. This can be quite useful, since many systems do possess such symmetry and re-settability is a common requirement for many hardware circuits and embedded systems anyway. Many more useful applications of simple symmetry are possible.

Approximate Symmetry. In conversations we have had with industrial hardware engineers, it comes out that while symmetry reduction is often applicable due to the presence of many similar subcomponents, there are also many instances where it is not — quite — applicable. That is, the systems are not genuinely symmetric but "approximately" symmetric, for example, because of one different component or slight differences among all components. This limits the scope of utility of symmetry reduction techniques.

In [17] and [10] we proposed and formalized three progressively "looser" notions of approximate symmetry: *near* symmetry, *rough* symmetry, and, finally, *virtually* symmetry. Each can be applied to do group-theoretic quotient reduction to various asymmetric systems. The correspondence established in each case between the original large structure and the small quotient structure is exact, a bisimulation (up to permutation), in fact. Near symmetry can permit symmetry reduction on systems comprised of 2 similar but not identical processes. Rough symmetry can accommodate, e.g., for fixed $k \geq 2$, systems with k similar but not identical process, prioritized statically. Rough symmetry can handle the readers writers problem for k readers and ℓ writers. Virtual symmetry is yet more general and can accommodate dynamically varying priorities.

Additional open problems related to symmetry. There are number of important unsolved problems here. One is to broaden the scope of approximate symmetry reduction as much as possible. Obviously, a system that is "absolutely" asymmetric is going to lack sufficient regularity and redundancy to permit identification of approximate symmetries for reduction. Finding a sufficiently broad notion of approximate symmetry that is flexible enough to cover every useful applications is open, and perhaps an inherently ill-posed problem. But a practical solution is highly desirable. One possibility is to formulate a very general notion of approximate symmetry parameterized by degree of divergence from actual symmetry. We might look for a notion that is universally applicable in principle, but which (a) may in general result in a collapsed graph that is only loosely corresponds, say by a conservative or liberal abstraction, to the original structure; but (b) which progresses toward exact abstraction as the system approaches genuine symmetry. Our previous work with virtual general symmetry suggests a measure based on the number of "missing" arcs. Another technical problem that would have significant practical ramifications would be to extend approximate symmetry to annotated quotient structures (cf. [15], [16]).

2.2 Parameterized Verification

Symmetry quotient reductions, and many other abstraction methods, address reasoning about systems with k processes for a possibly large natural number constant k. It is often more desirable to reason about systems comprised of n homogeneous processes, where n is a natural number parameter. This gives rise to the *Parameterized Model Checking Problem (PMCP)*: decide whether a temporal property is true for all (sufficiently large) size instances n of a given system. The advantage is that one application of PMCP settles all cases of interest; there is then no need to be concerned whether a system known to be correct for 100 processes might fail for 101 processes or 1000 processes. In general, PMCP is undecidable (cf. [AK86]). But because of its practical importance many researchers have addressed this problem, obtaining interesting results and partial solutions (cf., e.g., [23], [2], [18], [22]). But most of these have potentially serious limitations, e.g., they require human assistance ("process invariants", etc.), are only partially automated (may not terminate), are sound but not complete, or lack a well-defined domain of applicability.

Parameterized Synchronous Systems. In [12] we formulated an algorithm for determining if a parameterized synchronous system satisfies a specification. The method is based on forming a single finite *abstract graph* (cf. [Lu84]) which encodes the behavior of concrete systems of all sizes n. An abstract state $\overline{s}$ records for a state s, which process locations are occupied. The theory we developed in [12] we later applied in [13] to verify correctness of the Society of Automotive Engineers SAE-J1850 protocol. This protocol operates along a single wire bus in (Ford) automobiles, and coordinates the interactions of (Motorola) microcontrollers distributed among the brake units, the airbags, the engine, etc. The general goal is to verify parameterized correctness so that the bus operates correctly no matter how many units are installed on the bus. (Thus, the same bus architecture could be used in small cars and large trucks.) The specific property we verified, by using a meta-tool built on-top of SMV, was that higher priority messages could not be overtaken by lower priority messages.

Open problem related to parameterized synchronous systems. A restriction on the mathematical model used in [12] is that conventional mutual exclusion algorithms cannot be implemented in it (cf. [18]), roughly because a system with 1 processes in a local state is, by definition of the abstract graph, indistinguishable from one with strictly more than 1 processes in that local state. It would be desirable to overcome this limitation. One possibility is to refine the abstract graph to distinguish between exactly 1 vs. 2 vs. 3 or more processes in an abstract state, to accommodate mutual exclusion. Or perhaps this method could be adjoined with use of a signal token as in [11], possession of which arbitrates among competing processes.

Reducing Many Processes to Few. In recent work [EK00] we give a fully automatic (algorithmic), sound and complete solution to PMCP in a rather broad framework. We consider asynchronous systems comprised of many homogeneous copies of a generic process template. The process template is represented as a synchronization skeleton [8], where enabling guards have a special character, either disjunctive or conjunctive. Correctness properties are expressed using CTL*\X (CTL* minus the nexttime X) We reduce model checking for systems of arbitrary size n to model checking for systems of size (up to) a small *cutoff* size c. This establishes decidability of PMCP as it is only necessary to model check a finite number of relatively small systems. In a number of interesting cases, we can establish polynomial time decidability. For example, we can reduce the problem of checking mutual exclusion for a critical section protocol to reasoning about systems with 2 processes. We emphasize this *algorithmically* establishes correctness for systems of *all* sizes n. This method generalizes and has been applied to systems comprised of multiple heterogeneous classes of processes, e.g., m readers and n writers.

3 Theory and Practice in Model Checking

In this section we discuss factors related to the usefulness of and use of model checking. The core theoretical principles underlying model checking rest on a few basic ideas including, e.g. some from modal logic (the finite model theorem), lattice theory (the Tarski-Knaster theorem permitting branching time modalities to be calculated easily), and automata theory (the language containment approach). These ideas are well-known. It has also become increasingly clear that model checking is useful in practice. Model checking is employed by most major hardware companies, e.g., Cadence, IBM, Intel, and Motorola, for verification and debugging of microprocessors circuits. Model checking is showing promise for the verification of software, and is being used or is under investigation by, e.g., Lucent, Microsoft, NASA. We would now like to offer an account of why it is that model checking turns out to be useful in practice. It involves the following factors.

Model checking = search $\wedge$ efficiency $\wedge$ expressiveness. Plainly, efficiency is key. Impressive progress has been made with symbolic data structures such as BDDs, and significant advances with various forms of abstraction are ongoing. But expressiveness is also key. An efficient automated verification method that could not express most of the important correctness properties of interest would be of little use, as would a method that lacked the modeling power to capture reactive systems of interest. It seems model checking fares well on these fronts.

Temporal logic provides a powerful and flexible language for specification. Overall, it is quite adequate to the task of reasoning about reactive systems. This is perhaps a bit surprising when we recall that we are using *propositional* temporal logic. But the work of Kamp [20], showing that LTL is essentially equivalent to the First Order Language of Linear Order, does provide a measure of expressive completeness. It is certainly the case that there are properties such as "P holds at every even moment, and we don't care about the odd moments" which are not expressible in LTL. However, they are expressible in the framework of ω-fsa's which means that our technical model checking machinery is still applicable. The author, in fact, views automata as just generalized formulae of temporal logic. Automata and temporal logic, broadly speaking, are the same thing. The advantage of temporal logic, which is sometimes important in practice, is the close connection with tensed natural language.

A finite state framework suffices. At least it does in practice for most of the applications most of the time. This is genuinely surprising, but, there are several reasons for it. First, from the standpoint of specifications, most all propositional modal and temporal logics, including LTL and CTL, have the (bounded) *finite model property*. If a formula f in such a logic is satisfiable in any model, then it is satisfiable in a finite model (of size bounded by some function of the length of f, e.g., $exp(|f|)$). If a system can be specified in propositional temporal logic, it can be realized by a finite state program. Two decades of experience shows that many (of at least the crucial parts of) reactive systems can be specified in propositional temporal logic, and hence should be realized by a finite state program. Second, from the standpoint of the reactive programs themselves, we see that most solutions to synchronization problems presented in the literature are, indeed, finite state. Typically, in a concurrent program, we can cleanly separate out the finite state synchronization skeletons [7] of the individual processes, which synchronize and coordinate the processes, from the sequential code. For example, in the synchronization skeleton for the solution to the mutual exclusion problem, we abstract out the details of the code manipulating the critical section and obtain just a single node. In the field we are starting to see a trend in model checking of software of abstracting out the irrelevant sequential parts and boiling down the program to a system of finite state synchronization skeletons.

Model checking is highly automated. It is fully automated in principle and highly automated in practice. Human intervention is sometimes required in practice for such things as determining a good variable ordering when doing BDD-based symbolic model checking. It can also be required in doing certain abstractions to get the model to be checked. For instance, in the case where one is verifying a module in isolation, typically a human must understand the environment in which the module operates and provide an abstraction of the environment for the module to interact with. Still, model checking is highly automated even in practice. Model checking seems to be more popular in industrial usage than theorem proving because of the high degree of automation. Theorem provers, while having in principle unbounded capacity, require human expertise to supply key lemmas, and, a skilled "operator" of the tool. In practice, the operator is usually someone with a Ph.D. in CS, EE, or, quite often, Mathematics. For this reason, deployment of theorem provers may be hindered in an industrial setting. Due to the automation of model checkers, they can successfully be used by engineers and programmers at the M.S. or B.S. level. In management's view, this facilitates the wide-scale deployment of verification technology in the organization.

4   Conclusion

There is nowadays widespread interest in *Computer Aided Verification* of reactive systems, as evidenced by the attention paid to the topics at FST-TCS, as

well as such conferences as CAV, TACAS, CONCUR, FMCAD, etc. The reason
is that such automated techniques as model checking as well as partially automated theorem proving have by now been shown to actually work on a variety
of "industrial strength" examples. Pnueli [28] argues that, due to the success of
techniques such as model checking on actual applications, we are on the verge of
an era of *Verification Engineering*. Of course, there is still a gulf between what
we need to do and what we currently have the capacity to do. Basic advances
as well as concerted engineering efforts are called for. One popular, and valuable, idea is the integration of theorem provers with model checkers; a number
of researchers are pursuing this topic. For the present, however, this researcher's
primary interest is still to try to push the idea of model-theoretic automation as
far as possible, aspiring to *Completely Automated Verification*.

References

1. K. Apt and D. Kozen. Limits for automatic verification of finite-state concurrent
   systems. *Information Processing Letters*, 15, pages 307-309, 1986.
2. M.C. Browne, E.M. Clarke and O. Grumberg. Reasoning about Networks with
   Many Identical Finite State Processes. *Information and Control*, 81(1), pages 13-
   31, April 1989.
3. R. E. Bryant, "Graph-Based Algorithms for Boolean Function Manipulation" *IEEE
   Transactions on Computers*, 35(8): 677-691 (1986).
4. E. M. Clarke and E. A. Emerson, "Design and Synthesis of Synchronization Skeletons using Branching Time Temporal Logic", *Logics of Programs Workshop*, IBM
   Yorktown Heights, New York, Springer LNCS no. 131, pp. 52-71, 1981.
5. E. M. Clarke, R. Enders, T. Filkorn, S. Jha, "Exploiting Symmetry In Temporal
   Logic Model Checking", *Formal Methods in System Design*, vol. 9, no. 1/2, pp.
   77-104, Aug. 96.
6. E. M. Clarke, E. A. Emerson, and A. P. Sistla, "Automatic Verification of Finite-
   State Concurrent Systems Using Temporal Logic Specifications", *ACM Trans. on
   Prog. Lang. and Sys (TOPLAS)* 8(2): 244-263 (1986).
7. E. A. Emerson, *Branching Time Temporal Logic and the Design of Correct Concurrent Programs*, Ph.D. Dissertation, Harvard University, 1981.
8. E. A. Emerson, E M. Clarke, "Using Branching Time Temporal Logic to Synthe-
   size Synchronization Skeletons", *Science of Computer Programming*, 2(3): 241-266
   (1982)
9. E. A. Emerson, J. Y. Halpern: " 'Sometimes' and 'Not Never' Revisited: On
   Branching versus Linear Time Temporal Logic", *Journal of the Assoc. Comp.
   Mach. (JACM)*, 33(1): 151-178 (1986).
10. E. A. Emerson, J. Havlicek, and R. J. Trefler, "Virtual Symmetry", LICS'00, pp.
    121-132.
11. E.A. Emerson and K.S. Namjoshi. Reasoning about Rings. In *Conference Record
    of POPL '95: 22nd ACM SIGPLAN-SIGACT Symposium on Principles of Pro-
    gramming Languages*, pages 85-94, 1995.
12. E.A. Emerson and K.S. Namjoshi. Automatic Verification of Parameterized Syn-
    chronous Systems. In *Computer Aided Verification, Proceedings of the 8th Inter-
    national Conference*. LNCS , Springer-Verlag, 1996.
13. E. A. Emerson and K. S. Namjoshi, "Verification of Parameterized Bus Arbitration
    Protocol", Conference on Computer Aided Verification (CAV), pp. 452–463, 1998.

14. E.A. Emerson and A.P. Sistla,. Symmetry and Model Checking. In *Formal Methods in System Design*, vol. 9, no. 1/2, pp. 105-131, Aug. 96.

15. E. A. Emerson and A. P. Sistla, "Utilizing Symmetry when Model-Checking under Fairness Assumptions: An Automata-Theoretic Approach", *ACM Trans. on Prog. Lang. and Systems (TOPLAS)*, pp. 617–638, vol. 19, no. 4, July 1997.

16. E. A. Emerson and R. J. Trefler, "Model Checking Real-Time Properties of Symmetric Systems", MFCS 1998: 427-436.

17. E. A. Emerson, R. J. Trefler, "From Asymmetry to Full Symmetry: New Techniques for Symmetry Reduction in Model Checking", CHARME 1999: 142-156.

18. S.M. German and A.P. Sistla. Reasoning about Systems with Many Processes. *J. ACM*,39(3), July 1992.

19. C. Ip and D. Dill. Better verification through symmetry. In *Formal Methods in System Design*, vol. 9, no. 1/2, pp. 41-76, Aug. 1996.

20. Kamp, J. A. W., "Tense Logic and the Theory of Linear Order", Ph.D. thesis, University of California, Los Angeles, 19868.

21. R. P. Kurshan, *Computer Aided Verification*, Princeton Univ. Press, 1994.

22. R.P. Kurshan and K. McMillan. A Structural Induction Theorem for Processes. In *Proceedings of the Eight Annual ACM Symposium on Principles of Distributed Computing*, pages 239-247, 1989.

23. B. Lubachevsky. An Approach to Automating the Verification of Compact Parallel Coordination Programs I.*Acta Informatica 21*, 1984.

24. K. McMillan, *Symbolic Model Checking*, Ph.D. Dissertation, CMU, 1992.

25. K. McMillan, Verification of Infinite State Systems by Compositional Model Checking, CHARME'99.

26. M Pandey and R. E. Bryant, "Exploiting Symmetry When Verifying Transitor-Level Circuits by Symbolic Trajectory Evaluation", CAV 1997: 244-255.

27. A. Pnueli. The Temporal Logic of Programs. In *Proceedings of the eighteenth Symposium on Foundations of Computer Science*. 1977.

28. A. Pnueli, "Verification Engineering: A Future Profession" (A. M. Turing Award Lecture), Sixteenth Annual ACM Symposium on Principles of Distributed Computing (PODC 1990), San Diego, August, 1997;
http://www.wisdom.weizmann.ac.il/~amir/turing97.ps.gz

29. F. Pong and M. Dubois. A New Approach for the Verification of Cache Coherence Protocols. *IEEE Transactions on Parallel and Distributed Systems*, August 1995.

30. J-P. Queille and J. Sifakis, "Specification and Verification of Concurrent Systems in CESAR", *International Symposium on Programming*, Springer LNCS no. 137, pp 337-351, 1982.

31. A. P. Sistla, Parameterized Verification of Linear Networks Using Automata as Invariants, CAV, 1997, 412-423.

32. I. Vernier. Specification and Verification of Parameterized Parallel Programs. In *Proceedings of the 8th International Symposium on Computer and Information Sciences*, Istanbul, Turkey, pages 622-625,1993.

33. P. Wolper and V. Lovinfosse. Verifying Properties of Large Sets of Processes with Network Invariants. In J. Sifakis(ed) *Automatic Verification Metods for Finite State Systems*, Springer-Verlag, LNCS 407, 1989.

An Algebra for XML Query

Mary Fernandez[1], Jerome Simeon[2], and Philip Wadler[3]

[1] ATT Labs, mff@research.att.com
[2] Bell Labs, Lucent Technologies, simeon@research.bell-labs.com
[3] Avaya Labs, wadler@avaya.com

Abstract. This document proposes an algebra for XML Query. The algebra has been submitted to the W3C XML Query Working Group. A novel feature of the algebra is the use of regular-expression types, similar in power to DTDs or XML Schemas, and closely related to Hasoya, Pierce, and Vouillon's work on Xduce. The iteration construct involves novel typing rules not encountered elsewhere (even in Xduce).

1 Introduction

This document proposes an algebra for XML Query.

This work builds on long standing traditions in the database community. In particular, we have been inspired by systems such as SQL, OQL, and nested relational algebra (NRA). We have also been inspired by systems such as Quilt, UnQL, XDuce, XML-QL, XPath, XQL, and YATL. We give citations for all these systems below.

In the database world, it is common to translate a query language into an algebra; this happens in SQL, OQL, and NRA, among others. The purpose of the algebra is twofold. First, the algebra is used to give a semantics for the query language, so the operations of the algebra should be well-defined. Second, the algebra is used to support query optimization, so the algebra should possess a rich set of laws. Our algebra is powerful enough to capture the semantics of many XML query languages, and the laws we give include analogues of most of the laws of relational algebra.

In the database world, it is common for a query language to exploit schemas or types; this happens in SQL, OQL, and NRA, among others. The purpose of types is twofold. Types can be used to detect certain kinds of errors at compile time and to support query optimization. DTDs and XML Schema can be thought of as providing something like types for XML. Our algebra uses a simple type system that captures the essence of XML Schema [35]. The type system is close to that used in XDuce [19]. Our type system can detect common type errors and support optimization. A novel aspect of the type system (not found in Xduce) is the description of projection in terms of iteration, and the typing rules for iteration that make this viable.

The best way to learn any language is to use it. To better familiarize readers with the algebra, we have implemented a type checker and an interpreter for the algebra in OCaml[24]. A demonstration version of the system is available at

S. Kapoor and S. Prasad (Eds.): FST TCS 2000, LNCS 1974, pp. 11–45, 2000.

http://www.cs.bell-labs.com/~wadler/topics/xml.html#xalgebra

The demo system allows you to type in your own queries to be type checked and evaluated. All the examples in this paper can be executed by the demo system.

This paper describes the key features of the algebra. For simplicity, we restrict our attention to only three scalar types (strings, integers, and booleans), but we believe the system will smoothly extend to cover the continuum of scalar types found in XML Schema. Other important features that we do not tackle include attributes, namespaces, element identity, collation, and key constraints, among others. Again, we believe they can be added within the framework given here.

The paper is organized as follows. A tutorial introduction is presented in Section 2. Section 3 explains key aspects of projection and iteration. A summary of the algebra's operators and type system is given in Section 4. We present some equivalence and optimization laws of the algebra in Section 5. Finally, we give the static typing rules for the algebra in Section 6. Section 7 discusses open issues and problems.

Cited literature includes: SQL [16], OQL [4,5,13], NRA [8,15,21,22], Quilt [11], UnQL [3], XDuce [19], XML Query [33,34], XML Schema [35,36], XML-QL [17], XPath [32], XQL [25], and YaTL [14].

2 The Algebra by Example

This section introduces the main features of the algebra, using familiar examples based on accessing a database of books.

2.1 Data and Types

Consider the following sample data:

```
<bib>
  <book>
    <title>Data on the Web</title>
    <year>1999</year>
    <author>Abiteboul</author>
    <author>Buneman</author>
    <author>Suciu</author>
  </book>
  <book>
    <title>XML Query</title>
    <year>2001</year>
    <author>Fernandez</author>
    <author>Suciu</author>
  </book>
</bib>
```

Here is a fragment of a XML Schema for such data.

```
<xsd:group name="Bib">
  <xsd:element name="bib">
    <xsd:complexType>
      <xsd:group ref="Book"
          minOccurs="0" maxOccurs="unbounded"/>
    </xsd:complexType>
  </xsd:element>
</xsd:group>

<xsd:group name="Book">
  <xsd:element name="book">
    <xsd:complexType>
      <xsd:element name="title" type="xsd:string"/>
      <xsd:element name="year" type="xsd:integer"/>
      <xsd:element name="author" type="xsd:integer"
          minOccurs="1" maxOccurs="unbounded"/>
    </xsd:complexType>
  </xsd:element>
</xsd:group>
```

This data and schema is represented in our algebra as follows:

```
type Bib =
  bib [ Book* ]
type Book =
  book [
    title  [ String ],
    year   [ Integer ],
    author [ String ]+
  ]
let bib0 : Bib =
  bib [
    book [
      title  [ "Data on the Web" ],
      year   [ 1999 ],
      author [ "Abiteboul" ],
      author [ "Buneman" ],
      author [ "Suciu" ]
    ],
    book [
      title  [ "XML Query" ],
      year   [ 2001 ],
      author [ "Fernandez" ],
      author [ "Suciu" ]
    ]
  ]
```

The expression above defines two types, Bib and Book, and defines one global variable, bib0.

The Bib type consists of a bib element containing zero or more value of type Book. The Book type consists of a book element containing a title element (which contains a string), a year element (which contains an integer), and one or more author elements (which contain strings).

The Bib type corresponds to a single bib element, which contains a *forest* of zero or more Book elements. We use the term forest to refer to a sequence of (zero or more) elements. Every element can be viewed as a forest of length one.

The Book type corresponds to a single book element, which contains one title element, followed by one year element, followed by one or more author elements. A title or author element contains a string value and a year element contains an integer.

The variable bib0 is bound to a literal XML value, which is the data model representation of the earlier XML document. The bib element contains two book elements.

The algebra is a strongly typed language, therefore the value of bib0 must be an instance of its declared type, or the expression is ill-typed. Here the value of bib0 is an instance of the Bib type, because it contains one bib element, which contains two book elements, each of which contain a string-valued title, an integer-valued year, and one or more string-valued author elements.

For convenience, we define a second global variable book0, also bound to a literal value, which is equivalent to the first book in bib0.

```
let book0 : Book =
  book [
    title  [ "Data on the Web" ],
    year   [ 1999 ],
    author [ "Abiteboul" ],
    author [ "Buneman" ],
    author [ "Suciu" ]
  ]
```

2.2 Projection

The simplest operation is projection. The algebra uses a notation similar in appearance and meaning to path navigation in XPath.

The following expression returns all author elements contained in book0:

```
    book0/author
==> author [ "Abiteboul" ],
    author [ "Buneman" ],
    author [ "Suciu" ]
 :  author [ String ]+
```

The above example and the ones that follow have three parts. First is an expression in the algebra. Second, following the ==>, is the value of this expression.

Third, following the :, is the type of the expression, which is (of course) also a legal type for the value.

The following expression returns all `author` elements contained in `book` elements contained in `bib0`:

```
    bib0/book/author
==> author [ "Abiteboul" ],
    author [ "Buneman" ],
    author [ "Suciu" ],
    author [ "Fernandez" ],
    author [ "Suciu" ]
  : author [ String ]*
```

Note that in the result, the document order of `author` elements is preserved and that duplicate elements are also preserved.

It may be unclear why the type of `bib0/book/author` contains *zero* or more authors, even though the type of a `book` element contains *one* or more authors. Let's look at the derivation of the result type by looking at the type of each sub-expression:

```
    bib0              : Bib
    bib0/book         : Book*
    bib0/book/author  : author [ String ]*
```

Recall that `Bib`, the type of `bib0`, may contain *zero* or more `Book` elements, therefore the expression `bib0/book` might contain zero `book` elements, in which case, `bib0/book/author` would contain no authors.

This illustrates an important feature of the type system: the type of an expression depends only on the type of its sub-expressions. It also illustrates the difference between an expression's run-time value and its compile-time type. Since the type of `bib0` is `Bib`, the best type for `bib0/book/author` is one listing zero or more authors, even though for the given value of `bib0` the expression will always contain exactly five authors.

2.3   Iteration

Another common operation is to iterate over elements in a document so that their content can be transformed into new content. Here is an example of how to process each book to list the authors before the title, and remove the year.

```
    for b in bib0/book do
      book [ b/author, b/title ]
==> book [
      author [ "Abiteboul" ],
      author [ "Buneman" ],
      author [ "Suciu" ],
      title  [ "Data on the Web" ]
    ],
```

```
    book [
      author [ "Fernandez" ],
      author [ "Suciu" ],
      title  [ "XML Query" ]
    ]
  : book [
      author[ String ]+,
      title[ String ]
    ]*
```

The for expression iterates over all book elements in bib0 and binds the variable b to each such element. For each element bound to b, the inner expression constructs a new book element containing the book's authors followed by its title. The transformed elements appear in the same order as they occur in bib0.

In the result type, a book element is guaranteed to contain one or more authors followed by one title. Let's look at the derivation of the result type to see why:

```
  bib0/book        : Book*
  b                : Book
  b/author         : author [ String ]+
  b/title          : title  [ String ]
```

The type system can determine that b is always Book, therefore the type of b/author is author[String]+ and the type of b/title is title[String].

In general, the value of a for loop is a forest. If the body of the loop itself yields a forest, then all of the forests are concatenated together. For instance, the expression:

```
  for b in bib0/book do
    b/author
```

is exactly equivalent to the expression bib0/book/author.

Here we have explained the typing of for loops by example. In fact, the typing rules are rather subtle, and one of the more interesting aspects of the algebra, and will be explained further below.

2.4   Selection

Projection and for loops can serve as the basis for many interesting queries. The next three sections show how they provide the power for selection, quantification, join, and regrouping.

To select values that satisfy some predicate, we use the where expression. For example, the following expression selects all book elements in bib0 that were published before 2000.

```
   for b in bib0/book do
     where value(b/year) <= 2000 do
         b
==> book [
     title   [ "Data on the Web" ],
     year    [ 1999 ],
     author [ "Abiteboul" ],
     author [ "Buneman" ],
     author [ "Suciu" ]
     ]
:    Book*
```

The value operator returns the scalar (i.e., string, integer, or boolean) content of an element.

An expression of the form

```
where e₁ do e₂
```

is just syntactic sugar for

```
if e₁ then e₂ else ()
```

where e_1 and e_2 are expressions. Here () is an expression that stands for the empty sequence, a forest that contains no elements. We also write () for the type of the empty sequence.

According to this rule, the expression above translates to

```
for b <- bib0/book in
   if value(b/year) < 2000 then b else ()
```

and this has the same value and the same type as the preceding expression.

2.5 Quantification

The following expression selects all book elements in bib0 that have *some* author named "Buneman".

```
   for b in bib0/book do
     for a in b/author do
       where value(a) = "Buneman" do
           b
==> book [
     title   [ "Data on the Web" ],
     year    [ 1999 ],
     author [ "Abiteboul" ],
     author [ "Buneman" ],
     author [ "Suciu" ]
     ]
:    Book*
```

In contrast, we can use the `empty` operator to find all books that have *no* author whose name is Buneman:

```
for b in bib0/book do
    where empty(for a in b/author do
                    where value(a) = "Buneman" do
                    a) do
        b
==> book [
    title   [ "XML Query" ],
    year    [ 2001 ],
    author  [ "Fernandez" ],
    author  [ "Suciu" ]
    ]
:     Book*
```

The `empty` expression checks that its argument is the empty sequence `()`.

We can also use the `empty` operator to find all books where all the authors are Buneman, by checking that there are no authors that are not Buneman:

```
for b in bib0/book do
    where empty(for a in b/author do
                    where value(a) <> "Buneman" do
                    a) do
        b
==> ()
:     Book*
```

There are no such books, so the result is the empty sequence. Appropriate use of `empty` (possibly combined with `not`) can express universally or existentially quantified expressions.

Here is a good place to introduce the `let` expression, which binds a local variable to a value. Introducing local variables may improve readability. For example, the following expression is exactly equivalent to the previous one.

```
for b in bib0/book do
    let nonbunemans = (for a in b/author do
                          where value(a) <> "Buneman" do
                          a) do
        where empty(nonbunemans) do
            b
```

Local variables can also be used to avoid repetition when the same subexpression appears more than once in a query.

2.6   Join

Another common operation is to *join* values from one or more documents. To illustrate joins, we give a second data source that defines book reviews:

```
type Reviews  =
  reviews [
    book [
      title  [ String ],
      review [ String ]
    ]*
  ]
let review0 : Reviews =
  reviews [
    book [
      title  [ "XML Query" ],
      review [ "A darn fine book." ]
    ],
    book [
      title  [ "Data on the Web" ],
      review [ "This is great!" ]
    ]
  ]
```

The Reviews type contains one reviews element, which contains zero or more book elements; each book contains a title and review.

We can use nested for loops to join the two sources review0 and bib0 on title values. The result combines the title, authors, and reviews for each book.

```
for b in bib0/book do
  for r in review0/book do
    where value(b/title) = value(r/title) do
      book [ b/title, b/author, r/review ]
==>
  book [
    title  [ "Data on the Web" ],
    author [ "Abiteboul" ],
    author [ "Buneman" ],
    author [ "Suciu" ]
    review [ "A darn fine book." ]
  ],
  book [
    title  [ "XML Query" ],
    author [ "Fernandez" ],
    author [ "Suciu" ]
    review [ "This is great!" ]
  ]
: book [
    title [ String ],
    author [ String ]+
    review [ String ]
  ]*
```

Note that the outer-most `for` expression determines the order of the result. Readers familiar with optimization of relational join queries know that relational joins commute, i.e., they can be evaluated in any order. This is not true for the XML algebra: changing the order of the first two `for` expressions would produce different output. In Section 7, we discuss extending the algebra to support unordered forests, which would permit commutable joins.

2.7   Restructuring

Often it is useful to regroup elements in an XML document. For example, each `book` element in `bib0` groups one title with multiple authors. This expression regroups each author with the titles of his/her publications.

```
for a in distinct(bib0/book/author) do
  biblio [
    a,
    for b in bib0/book do
      for a2 in b/author do
        where value(a) = value(a2) do
          b/title
  ]
==> biblio [
      author [ "Abiteboul" ],
      title  [ "Data on the Web" ]
    ],
    biblio [
      author [ "Buneman" ],
      title  [ "Data on the Web" ]
    ],
    biblio [
      author [ "Suciu" ],
      title  [ "Data on the Web" ],
      title  [ "XML Query" ]
    ],
    biblio [
      author [ "Fernandez" ],
      title  [ "XML Query" ]
    ]
  : biblio [
      author [ String ],
      title  [ String ]*
    ]*
```

Readers may recognize this expression as a self-join of books on authors. The expression `distinct(bib0/book/author)` produces a forest of author elements with no duplicates. The outer `for` expression binds `a` to each author element, and the inner for expression selects the title of each book that has some author equal to a.

Here distinct is an example of a built-in function. It takes a forest of elements and removes duplicates.

The type of the result expression may seem surprising: each biblio element may contain *zero* or more title elements, even though in bib0, every author co-occurs with a title. Recognizing such a constraint is outside the scope of the type system, so the resulting type is not as precise as we might like.

2.8   Aggregation

We have already seen several several built-in functions, such as children, distinct, and value. In addition to these, the algebra has five built-in aggregation functions: avg, count, max, min and sum.

This expression selects books that have more than two authors:

```
for b in bib0/book do
    where count(b/author) > 2 do
        b
==> book [
    title   [ "Data on the Web" ],
    year    [ 1999 ],
    author [ "Abiteboul" ],
    author [ "Buneman" ],
    author [ "Suciu" ]
    ]
:   Book*
```

All the aggregation functions take a forest with repetition type and return an integer value; count returns the number of elements in the forest.

2.9   Functions

Functions can make queries more modular and concise. Recall that we used the following query to find all books that do not have "Buneman" as an author.

```
for b in bib0/book do
    where empty(for a in b/author do
                    where value(a) = "Buneman" do
                    a) do
        b
==> book [
    title   [ "XML Query" ],
    year    [ 2001 ],
    author [ "Fernandez" ],
    author [ "Suciu" ]
    ]
:   Book*
```

A different way to formulate this query is to first define a function that takes a string s and a book b as arguments, and returns true if book b does not have an author with name s.

```
fun notauthor (s : String; b : Book) : Boolean =
   empty(for a in b/author do
            where value(a) = s do
               a)
```

The query can then be re-expressed as follows.

```
for b in bib0/book do
   where notauthor("Buneman"; b) do
      b
==> book [
      title   [ "XML Query" ],
      year    [ 2001 ],
      author [ "Fernandez" ],
      author [ "Suciu" ]
   ]
:   Book*
```

We use semicolon rather than comma to separate function arguments, since comma is used to concatenate forests.

Note that a function declaration includes the types of all its arguments and the type of its result. This is necessary for the type system to guarantee that applications of functions are type correct.

In general, any number of functions may be declared at the top-level. The order of function declarations does not matter, and each function may refer to any other function. Among other things, this allows functions to be recursive (or mutually recursive), which supports structural recursion, the subject of the next section.

2.10   Structural Recursion

XML documents can be recursive in structure, for example, it is possible to define a **part** element that directly or indirectly contains other **part** elements. In the algebra, we use recursive types to define documents with a recursive structure, and we use recursive functions to process such documents. (We can also use mutual recursion for more complex recursive structures.)

For instance, here is a recursive type defining a part hierarchy.

```
type Part =
   Basic | Composite
type Basic =
   basic [
      cost [ Integer ]
   ]
```

```
type Composite =
  composite [
    assembly_cost [ Integer ],
    subparts [ Part+ ]
  ]
```

And here is some sample data.

```
let part0 : Part =
  composite [
    assembly_cost [ 12 ],
    subparts [
      composite [
        assembly_cost [ 22 ],
        subparts [
          basic [ cost [ 33 ] ]
        ]
      ],
      basic [ cost [ 7 ] ]
    ]
  ]
```

Here vertical bar (|) is used to indicate a choice between types: each part is either basic (no subparts), and has a cost, or is composite, and includes an assembly cost and subparts.

We might want to translate to a second form, where every part has a total cost and a list of subparts (for a basic part, the list of subparts is empty).

```
type Part2 =
  part [
    total_cost [ Integer ],
    subparts [ Part2* ]
  ]
```

Here is a recursive function that performs the desired transformation. It uses a new construct, the case expression.

```
fun convert(p : Part) : Part2 =
  case p of
    b : basic =>
      part[
        total_cost[ value(b/cost) ],
        subparts[]
      ]
  | c : composite =>
      let s = (for q in children(c/subparts) do convert(q)) in
        part[
          total_cost[
```

```
        value(c/assembly_cost) +
          sum(for t in s/total_cost do value(t))
        ],
        subparts[ s ]
      ]
  end
```

Each branch of the case is labeled with an element name, `basic` or `composite`, and with a corresponding variable, `b` or `c`. The `case` expression checks whether the value of p is a `basic` or `composite` element, and evaluates the corresponding branch. If the first branch is taken then b is bound to the value of p, and the branch retuns a new part with total cost the same as the cost of b, and with no subparts. If the second branch is taken then c is bound to the value of p. The function is recursively applied to each of the subparts of c, giving a list of new subparts s. The branch returns a new part with total cost computed by adding the assembly cost of c to the sum of the total cost of each subpart in s, and with subparts s.

One might wonder why b and c are required, since they have the same value as p. The reason why is that p, b, and c have different types.

```
p : Part
b : Basic
c : Composite
```

The types of b and c are more precise than the type of p, because which branch is taken depends upon the type of value in p.

Applying the query to the given data gives the following result.

```
    convert(part0)
==> part [
      total_cost [ 74 ],
      subparts [
        part [
          total_cost [ 55 ],
          subparts [
            part [
              total_cost [ 33 ],
              subparts []
            ]
          ]
        ],
        part [
          total_cost [ 7 ],
          subparts []
        ]
      ]
    ]
  :   Part2
```

Of course, a `case` expression may be used in any query, not just in a recursive one.

2.11 Processing Any Well-Formed Document

Recursive types allow us to define a type that matches any well-formed XML document. This type is called `UrTree`:

```
type UrTree  =
    UrScalar
  | ~ [ UrTree* ]
```

Here `UrScalar` is a built-in scalar type. It stands for the most general scalar type, and all other scalar types (like `Integer` or `String`) are subtypes of it. The tilde (`~`) is used to indicate a wild-card type. In general, `~[t]` indicates the type of elements that may have any tag, but must have children of type t. So an `UrTree` is either an `UrScalar` or a wildcard element with zero or more children, each of which is itself an `UrTree`. In other words, any single element or scalar has type `UrTree`.

The use of `UrScalar` is a small, but necessary, extension to XML Schema, since XML Schema provides no most general scalar type. In contrast, the use of tilde is a significant extension to XML Schema, because XML Schema has no type corresponding to `~[t]`, where t is some type other than `UrTree*`. It is not clear that this extension is necessary, since the more restrictive expressiveness of XML Schema wildcards may be adequate. Also, note that `UrTree*` is equivalent to the `UrType` in XML Schema.

In particular, our earlier data also has type `UrTree`.

```
      book0 : UrTree
==> book [
       title  [ "Data on the Web" ],
       year   [ 1999 ],
       author [ "Abiteboul" ],
       author [ "Buneman" ],
       author [ "Suciu" ]
     ]
   :   UrTree
```

A specific type can be indicated for any expression in the query language, by writing a colon and the type after the expression.

As an example, we define a recursive function that converts any XML data into HTML. We first give a simplified definition of HTML.

```
type HTML =
  ( UrScalar
  | b [ HTML ]
  | ul [ (li [ HTML ])* ]
  )*
```

An HTML body consists of a sequence of zero or more items, each of which is either: a scalar; or a b element (boldface) with HTML content; or a ul element (unordered list), where the children are li elements (list item), each of which has HTML content.

Now, here is the function that performs the conversion.

```
fun html_of_xml( t : UrTree ) : HTML =
  case t of
    s : UrScalar =>
        s
    | e =>
        b [ name(e) ],
        ul [ for c in children(e) do li [ html_of_xml(c) ] ]
  end
```

The case expression checks whether the value of x is a subtype of UrScalar or otherwise, and evaluates the corresponding branch. If the first branch is taken, then s is bound to the value of t, which must be a scalar, and the branch returns the scalar. If the second branch is taken, then e is bound to the value of t, which must not be a scalar, and hence must be an element. The branch returns the name of the element in boldface, followed by a list containing one item for each child of the element. The function is recursively applied to get the content of each list item.

Applying the query to the book element above gives the following result.

```
    html_of_xml(book0)
==> b [ "book" ],
    ul [
      li [ b [ "title" ],  ul [ li [ "Data on the Web" ] ] ],
      li [ b [ "year" ],   ul [ li [ 1999 ] ] ],
      li [ b [ "author" ], ul [ li [ "Abiteboul" ] ] ],
      li [ b [ "author" ], ul [ li [ "Buneman" ] ] ],
      li [ b [ "author" ], ul [ li [ "Suciu" ] ] ]
    ]
:   Html_Body
```

2.12 Top-Level Queries

A query consists of a sequence of top-level expressions, or *query items*, where each query item is either a type declaration, a function declaration, a global variable declaration, or a query expression. The order of query items is immaterial; all type, function, and global variable declarations may be mutually recursive.

A query can be evaluated by the query interpreter. Each query expression is evaluated in the environment specified by all of the declarations. (Typically, all of the declarations will precede all of the query expressions, but this is not required.) We have already seen examples of type, function, and global variable declarations. An example of a query expression is:

```
query html_of_xml(book0)
```

To transform any expression into a top-level query, we simply precede the expression by the `query` keyword.

3 Projection and Iteration

This section describes key aspects of projection and iteration.

3.1 Relating Projection to Iteration

The previous examples use the / operator liberally, but in fact we use / as a convenient abbreviation for expressions built from lower-level operators: `for` expressions, the `children` function, and `case` expressions.

For example, the expression:

```
book0/author
```

is equivalent to the expression:

```
for c in children(book0) do
  case c of
    a : author => a
  | b => ()
  end
```

Here the `children` function returns a forest consisting of the children of the element book0, namely, a title element, a year element, and three author elements (the order is preserved). The `for` expression binds the variable v successively to each of these elements. Then the `case` expression selects a branch based on the value of v. If it is an `author` element then the first branch is evaluated, otherwise the second branch. If the first branch is evaluated, the variable a is bound to the same value as x, then the branch returns the value of a. If the second branch is evaluated, the variable b is bound to the same value as x, then then branch returns (), the empty sequence.

To compose several expressions using /, we again use `for` expressions. For example, the expression:

```
bib0/book/author
```

is equivalent to the expression:

```
for c in children(bib0) do
  case c of
    b : book =>
      for d in children(b) do
        case d of
          a : author => d
```

```
        | e => ()
        end
  | f => ()
  end
```

The `for` expression iterates over all `book` elements in `bib0` and binds the variable `b` to each such element. For each element bound to `b`, the inner expression returns all the `author` elements in `b`, and the resulting forests are concatenated together in order.

In general, an expression of the form $e \ / \ a$ is converted to the form

```
for v₁ in e do
  for v₂ in children(v₁) do
    case v₂ of
      v₃ : a => v₃
    | v₄ => ()
    end
```

where e is an expression, a is a tag, and v_1, v_2, v_3, v_4 are fresh variables (ones that do not appear in the expression being converted).

According to this rule, the expression `bib0/book` translates to

```
for v1 in bib0 do
  for v2 in children(v1) do
    case v2 of
      v3 : book => v3
    | v4 => ()
    end
```

In Section 5 we introduce laws of the algebra, which allow us to simplify this to the previous expression

```
for v2 in children(bib0) do
  case v2 of
    v3 : book => v3
  | v4 => ()
  end
```

Similarly, the expression `bib0/book/author` translates to

```
for v5 in (for v2 in children(bib0) do
             case v2 of
               v3 : book => v3
             | v4 => ()
             end) do
  for v6 in children(v5) do
    case v6 of
      v7 : author => v7
    | v8 => ()
    end
```

Again, the laws will allow us to simplify this to the previous expression

```
for v2 in children(bib0) do
  case v2 of
    v3 : book =>
      for v6 in children(v3) do
        case c of
          v7 : author => d
        | v8 => ()
        end
  | v4 => ()
  end
```

These examples illustrate an important feature of the algebra: high-level operators may be defined in terms of low-level operators, and the low-level operators
may be subject to algebraic laws that can be used to further simplify the expression.

3.2   Typing Iteration

The typing of for loops is rather subtle. We give an intuitive explanation here,
and cover the detailed typing rules in Section 6.

A *unit* type is either an element type $a[t]$, a wildcard type $\tilde{}[t]$, or a scalar
type s. A for loop

```
for v in e₁ do e₂
```

is typed as follows. First, one finds the type of expression e_1. Next, for each unit
type in this type one assumes the variable v has the unit type and one types
the body e_2. Note that this means we may type the body of e_2 several times,
once for each unit type in the type of e_1. Finally, the types of the body e_2 are
combined, according to how the types were combined in e_1. That is, if the type
of e_1 is formed with sequencing, then sequencing is used to combine the types
of e_2, and similarly for choice or repetition.

For example, consider the following expression, which selects all author elements from a book.

```
for c in children(book0) do
  case c of
    a : author => a
  | b => ()
  end
```

The type of children(book0) is

```
title[String], year[Integer], author[String]+
```

This is composed of three unit types, and so the body is typed three times.

assuming `c` has type `title[String]` the body has type `()`
        "   `year[Integer]`   "   `()`
        "   `author[String]`   "   `author[String]`

The three result types are then combined in the same way the original unit types were, using sequencing and iteration. This yields

`(), (), author[String]+`

as the type of the iteration, and simplifying yields

`author[String]+`

as the final type.

As a second example, consider the following expression, which selects all `title` and `author` elements from a book, and renames them.

```
for c in children(book0) do
  case c of
    t : title  => titl [ value(t) ]
  | y : year   => ()
  | a : author => auth [ value(a) ]
  end
```

Again, the type of `children(book0)` is

`title[String], year[Integer], author[String]+`

This is composed of three unit types, and so the body is typed three times.

assuming `c` has type `title[String]` the body has type `titl[String]`
        "   `year[Integer]`   "   `()`
        "   `author[String]`   "   `auth[String]`

The three result types are then combined in the same way the original unit types were, using sequencing and iteration. This yields

`titl[String], (), auth[String]+`

as the type of the iteration, and simplifying yields

`titl[String], auth[String]+`

as the final type. Note that the title occurs just once and the author occurs one or more times, as one would expect.

As a third example, consider the following expression, which selects all basic parts from a sequence of parts.

```
for p in children(part0/subparts) do
  case p of
    b : basic      =>  b
  | c : composite =>  ()
  end
```

The type of `children(part0/subparts)` is

    (Basic | Composite)+

This is composed of two unit types, and so the body is typed two times.

    assuming p has type   Basic   the body has type Basic
              "   Composite   "   ()

The two result types are then combined in the same way the original unit types were, using sequencing and iteration. This yields

    (Basic | ())+

as the type of the iteration, and simplifying yields

    Basic*

as the final type. Note that although the original type involves repetition one or more times, the final result is a repetition zero or more times. This is what one would expect, since if all the parts are composite the final result will be an empty sequence.

In this way, we see that `for` loops can be combined with `case` expressions to select and rename elements from a sequence, and that the result is given a sensible type.

In order for this approach to typing to be sensible, it is necessary that the unit types can be uniquely identified. However, the type system given here satisfies the following law.

$$a[t_1 \mid t_2] = a[t_1] \mid a[t_2]$$

This has one unit type on the left, but two distinct unit types on the right, and so might cause trouble. Fortunately, our type system inherits an additional restriction from XML Schema: we insist that the regular expressions can be recognized by a top-down deterministic automaton. In that case, the regular expression must have the form on the left, the form on the right is outlawed because it requires a non-deterministic recognizer. With this additional restriction, there is no problem.

4   Summary of the Algebra

In this section, we summarize the algebra and present the grammars for expressions and types.

4.1 Expressions

Figure 1 contains the grammar for the algebra, i.e., the convenient concrete syntax in which a user may write a query. A few of these expressions can be rewritten as other expressions in a smaller *core* algebra; such reducible expressions are labeled with "*". We define the algebra's typing rules on the smaller core algebra. In Section 5, we give the laws that relate a user expression with its equivalent expression in the core algebra. Typing rules for the core algebra are defined in Section 6.

We have seen examples of most of the expressions, so we will only point out details here. We define a subset of expressions that correspond to *data values*. An expression is a data value if it consists only of scalar constant, element, sequence, and empty sequence expressions.

We have not defined the semantics of the binary operators in the algebra. It might be useful to define more than one type of equality over scalar and element values. We leave that to future work.

4.2 Types

Figure 2 contains the grammar for the algebra's type system. We have already seen many examples of types. Here, we point out some details.

Our algebra uses a simple type system that captures the essence of XML Schema [35]. The type system is close to that used in XDuce [19].

In the type system of Figure 2, a scalar type may be a `UrScalar`, `Boolean`, `Integer`, or `String`. In XML Schema, a scalar type is defined by one of fourteen primitive datatypes and a list of facets. A type hierarchy is induced between scalar types by containment of facets. The algebra's type system can be generalized to support these types without much increase in its complexity. We added `UrScalar`, because XML Schema does not support a most general scalar type.

A type is either: a type variable; a scalar type; an element type with literal tag a and content type t; a *wildcard* type with an unknown tag and content type t; a sequence of two types, a choice of two types; a repetition type; the empty sequence type; or the empty choice type.

The algebra's external type system, that is, the type definitions associated with input and output documents, is XML Schema. The internal types are in some ways more expressive than XML Schema, for example, XML Schema has no type corresponding to `Integer*` (which is required as the type of the argument to an aggregation operator like `sum` or `min` or `max`), or corresponding to ~`[t]` where t is some type other than `UrTree*`. In general, mapping XML Schema types into internal types will not lose information, however, mapping internal types into XML Schema may lose information.

4.3 Relating Values to Types

Recall that *data* is the subset of expressions that consists only of scalar constant, element, sequence, and empty sequence expressions. We write $\vdash d : t$ if data d has type t. The following type rules define this relation.

tag	a			
function	f			
variable	v			
integer	c_{int}	$::=$ $\cdots \mid -1 \mid 0 \mid 1 \mid \cdots$		
string	c_{str}	$::=$ `""` $\mid$ `"a"` $\mid$ `"b"` $\mid \cdots \mid$ `"aa"` $\mid \cdots$		
boolean	c_{bool}	$::=$ **false** $\mid$ **true**		
constant	c	$::=$ $c_{\text{int}} \mid c_{\text{str}} \mid c_{\text{bool}}$		
operator	op	$::=$ **+** $\mid$ **-** $\mid$ **and** $\mid$ **or**		
		$\mid$ **=** $\mid$ **!=** $\mid$ **<** $\mid$ **<=** $\mid$ **>=** $\mid$ **>**		
expression	e	$::=$ c	scalar constant	
		$\mid$ v	variable	
		$\mid$ $a[e]$	element	
		$\mid$ $\tilde{\ }e[e]$	computed element	
		$\mid$ e , e	sequence	
		$\mid$ $()$	empty sequence	
		$\mid$ **if** e **then** e **else** e	conditional	
		$\mid$ **let** v **=** e **do** e	local binding	
		$\mid$ **for** v **in** e **do** e	iteration	
		$\mid$ **case** e **of** $v{:}p$ **=>** e $\mid$ v **=>** e **end**	case	
		$\mid$ $f(e;\ldots;e)$	function application	
		$\mid$ $e : t$	explicit type	
		$\mid$ **empty**(e)	emptiness predicate	
		$\mid$ **error**	error	
		$\mid$ e **+** e	plus	
		$\mid$ e **=** e	equal	
		$\mid$ **children**(e)	children	
		$\mid$ **name**(e)	element name	
		$\mid$ e **/** a	projection	*
		$\mid$ **where** e **then** e	conditional	*
		$\mid$ **value**(e)	scalar content	*
		$\mid$ **let** $v : t$ **=** e **do** e	local binding	*
pattern	p	$::=$ a	element	
		$\mid$ $\tilde{\ }$	wildcard	
		$\mid$ s	scalar	
query item	q	$::=$ **type** x **=** t	type declaration	
		$\mid$ **fun** $f(v{:}t;\ldots;v{:}t){:}t$ **=** e	function declaration	
		$\mid$ **let** $v : t$ **=** e	global declaration	
		$\mid$ **query** e	query expression	
data	d	$::=$ c	scalar constant	
		$\mid$ $a[d]$	element	
		$\mid$ d , d	sequence	
		$\mid$ $()$	empty sequence	

Fig. 1. Algebra

```
tag          a
type name    x
scalar type  s ::= Integer
                 | String
                 | Boolean
                 | UrScalar
type         t ::= x           type name
                 | s           scalar type
                 | a[t]        element
                 | ~[t]        wildcard
                 | t , t       sequence
                 | t | t       choice
                 | t*          repetition
                 | ()          empty sequence
                 | ∅           empty choice
unit type    u ::= a[t]        element
                 | ~[t]        wildcard
                 | s           scalar type
```

Fig. 2. Type System

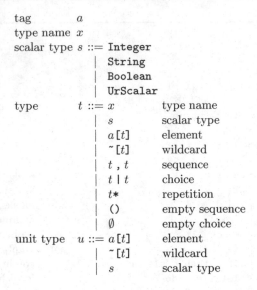

$$\overline{\vdash c_{\text{int}} \;:\; \texttt{Integer}}$$

$$\overline{\vdash c_{\text{str}} \;:\; \texttt{String}}$$

$$\overline{\vdash c_{\text{bool}} \;:\; \texttt{Boolean}}$$

$$\overline{\vdash c \;:\; \texttt{UrScalar}}$$

$$\frac{\vdash d \;:\; t}{\vdash a[d] \;:\; a[t]}$$

$$\frac{\vdash d \;:\; t}{\vdash a[d] \;:\; \text{\textasciitilde}[t]}$$

$$\frac{\vdash d_1 \;:\; t_1 \qquad \vdash d_2 \;:\; t_2}{\vdash d_1 , d_2 \;:\; t_1 , t_2}$$

$$\overline{\vdash () \;:\; ()}$$

$$\frac{\vdash d \; : \; t_1}{\vdash d \; : \; t_1 \mid t_2}$$

$$\frac{\vdash d \; : \; t_2}{\vdash d \; : \; (t_1 \mid t_2)}$$

$$\frac{\vdash d_1 \; : \; t \qquad \vdash d_2 \; : \; t*}{\vdash (d_1, d_2) \; : \; t*}$$

$$\overline{\vdash () \; : \; t*}$$

We write $t_1 <: t_2$ if for every data d such that $\vdash d \; : \; t_1$ it is also the case that $\vdash d \; : \; t_2$, that is t_1 is a subtype of t_2. It is easy to see that $<:$ is a partial order, that is it is reflexive, $t <: t$, and it is transitive, if $t_1 <: t_2$ and $t_2 <: t_3$ then $t_1 <: t_3$. We also have that $\emptyset <: t$ for any type t, and $a[t] <: {}^\sim[t]$. We have $s <: \mathtt{UrScalar}$ for every scalar type s. We have $t_1 <: (t_1 \mid t_2)$ and $t_2 <: (t_1 \mid t_2)$ for any t_1 and t_2. If $t <: t'$, then $a[t] <: a[t']$ and $t* <: t'*$. And if $t_1 <: t_1'$ and $t_2 <: t_2'$ then $t_1, t_2 <: t_1', t_2'$.

We write $t_1 = t_2$ if $t_1 <: t_2$ and $t_2 <: t_1$. Here are some of the equations that hold.

$$
\begin{aligned}
\mathtt{UrScalar} &= \mathtt{Integer} \mid \mathtt{String} \mid \mathtt{Boolean} \\
(t_1, t_2), t_3 &= t_1, (t_2, t_3) \\
t, () &= t \\
(), t &= t \\
t_1 \mid t_2 &= t_2 \mid t_1 \\
(t_1 \mid t_2) \mid t_3 &= t_1 \mid (t_2 \mid t_3) \\
t \mid \emptyset &= t \\
\emptyset \mid t &= t \\
t_1, (t_2 \mid t_3) &= (t_1, t_2) \mid (t_1, t_3) \\
(t_1 \mid t_2), t_3 &= (t_1, t_3) \mid (t_2, t_3) \\
t, \emptyset &= \emptyset \\
\emptyset, t &= \emptyset \\
a[t] \mid {}^\sim[t] &= {}^\sim[t] \\
t* &= () \mid t, t*
\end{aligned}
$$

We also have that $t_1 <: t_2$ if and only iff $t_1 \mid t_2 = t_2$.

We define $t?$ and $t+$ as abbreviations, by the following equivalences.

$$
\begin{aligned}
t? &= () \mid t \\
t+ &= t, t*
\end{aligned}
$$

$$e/a$$
$$\Rightarrow \texttt{for } v_1 \texttt{ in } e \texttt{ do} \qquad\qquad (1)$$
$$\qquad \texttt{for } v_2 \texttt{ in children}(v_1) \texttt{ do}$$
$$\qquad\quad \texttt{case } v_2 \texttt{ of}$$
$$\qquad\qquad v_3 \; : \; a \; \texttt{=>} \; v_3$$
$$\qquad\qquad | \; v_4 \; \texttt{=>} \; \texttt{()}$$

$$\texttt{where } e_1 \texttt{ then } e_2$$
$$\Rightarrow \texttt{if } e_1 \texttt{ then } e_2 \texttt{ else ()} \qquad (2)$$

$$\texttt{value}(e)$$
$$\Rightarrow \texttt{case children}(e) \texttt{ of} \qquad\qquad (3)$$
$$\qquad v_1 \; : \; \texttt{UrScalar => } v_1$$
$$\qquad | \; v_2 \; \texttt{=>} \; v_2 : \emptyset$$

$$\texttt{let } v : t \texttt{ = } e_1 \texttt{ do } e_2$$
$$\qquad \texttt{let } v \texttt{ = } (e_1 : t) \texttt{ do } e_2 \qquad (4)$$

Fig. 3. Definitions

5   Equivalences and Optimization

5.1   Equivalences

Figure 3 contains the laws that relate the reducible expressions (i.e., those labeled with "*" in Figure 1) to equivalent expressions. In these definitions, $e1\{e2/v\}$ denotes the expression $e1$ in which all occurrences of v are replaced by $e2$.

In Rule 1, the projection expression e/a is rewritten as described previously. Rule 2 rewrites a **where** expression as a conditional, as described previously. Rule 3 rewrites $\texttt{value}(e)$ as a **case** expression which checks whether the content of e is a scalar value, and if so, returns it. If e is not scalar value, its value is returned with the empty choice type, which may indicate an error. Rule 4 rewrites the **let** expression with a type as a **let** expression without a type by moving the type constraint into the expression.

5.2   Optimizations

Figure 4 contains a dozen algebraic simplification laws. In a relational query engine, algebraic simplifications are often applied by a query optimizer before a physical execution plan is generated; algebraic simplification can often reduce the size of the intermediate results computed by a query interpreter. The purpose of our laws is similar – they eliminate unnecessary **for** or **case** expressions, or they enable other optimizations by reordering or distributing computations. The set of laws given is suggestive, rather than complete.

$$E ::= \text{if } [] \text{ then } e_1 \text{ else } e_2$$
$$| \quad \text{let } v = [] \text{ do } e$$
$$| \quad \text{for } v \text{ in } [] \text{ do } e$$
$$| \quad \text{case } [] \text{ of } v_1 : p \Rightarrow e_1 \mid v_2 \Rightarrow e_2 \text{ end}$$

$$\text{for } v \text{ in } () \text{ do } e \Rightarrow () \tag{5}$$

$$\text{for } v \text{ in } (e_1 , e_2) \text{ do } e_3$$
$$\Rightarrow (\text{for } v \text{ in } e_1 \text{ do } e_3) , (\text{for } v \text{ in } e_2 \text{ do } e_3) \tag{6}$$

$$\text{for } v \text{ in } e_1 \text{ do } e_2$$
$$\Rightarrow e_2 \{ e_1 / v \}, \quad \text{if } e : u \tag{7}$$

$$\text{case } a[e_0] \text{ of } v_1 : a \Rightarrow e_1 \mid v_2 \Rightarrow e_2 \text{ end}$$
$$\Rightarrow e_1 \{ a[e_0] / v_1 \} \tag{8}$$

$$\text{case } a'[e_0] \text{ of } v_1 : a \Rightarrow e_1 \mid v_2 \Rightarrow e_2 \text{ end}$$
$$\Rightarrow e_2 \{ a'[e_0] / v_2 \}, \quad \text{if } a \neq a' \tag{9}$$

$$\text{for } v \text{ in } e \text{ do } v \Rightarrow e \tag{10}$$

$$E[\text{if } e_1 \text{ then } e_2 \text{ else } e_3]$$
$$\Rightarrow \text{if } e_1 \text{ then } E[e_2] \text{ else } E[e_3] \tag{11}$$

$$E[\text{let } v = e_1 \text{ do } e_2]$$
$$\Rightarrow \text{let } v = e_1 \text{ do } E[e_2] \tag{12}$$

$$E[\text{for } v \text{ in } e_1 \text{ do } e_2]$$
$$\Rightarrow \text{for } v \text{ in } e_1 \text{ do } E[e_2] \tag{13}$$

$$E[\text{case } c_0 \text{ of } v_1 : p \Rightarrow e_1 \mid v_2 \Rightarrow e_2 \text{ end}]$$
$$\Rightarrow \text{case } e_0 \text{ of } v_1 : p \Rightarrow E[e_1] \mid v_2 \Rightarrow E[e_2] \text{ end} \tag{14}$$

Fig. 4. Optimization Laws

Rules 5, 6, and 7 simplify iterations. Rule 5 rewrites an iteration over the empty sequence as the empty sequence. Rule 6 distributes iteration through sequence: iterating over the sequence e_1 , e_2 is equivalent to the sequence of two iterations, one over e_1 and one over e_2. Rule 7 eliminates an iteration over a single element or scalar. If e_1 is a unit type, then e_1 can be substituted for occurrences of v in e_2.

Rules 8 and 9 eliminate trivial case expressions.

Rule 10 eliminates an iteration when the result expression is simply the iteration variable v.

Rules 11–16 commute expressions. Each rule actually abbreviates a number of other rules, since the *context variable* E stands for a number of different expressions. The notation $E[e]$ stands for one of the six expressions given with expression e replacing the hole $[]$ that appears in each of the alternatives. For instance, one of the expansions of Rule 13 is the following, when E is taken to be for v in $[]$ do e.

$$\text{for } v_2 \text{ in (for } v_1 \text{ in } e_1 \text{ do } e_2) \text{ do } e_3$$
$$\Rightarrow \text{for } v_1 \text{ in } e_1 \text{ do (for } v_2 \text{ in } e_2 \text{ do } e_3)$$

Rules 7 and 10 together with the above expansion of Rule 13 are exactly analogous to the three monad laws used with list, bag, and set comprehensions in nested relational algebra [6,8,22,21] algebra, and derived from a similar use in functional programming [28]. In effect, these three laws show that the for loop introduced here is the analogue of a monad for semi-structured data.

Note that the sophisticated type rule for for loops ensures that the left side of Rule 10 is well typed whenever the right side is. (Originally, a less sophisticated type rule was used, for which this is not the case.)

In Section 3.1 we claimed that the expression bib0/book translates to

```
for v1 in bib0 do
  for v2 in children(v1) do
    case v2 of
      v3 : book => v3
    | v4 => ()
    end
```

and that this simplifies to

```
for v2 in children(bib0) do
  case v2 of
    v3 : book => v3
  | v4 => ()
  end
```

We can now see that the translation happens via Rule 1, and the simplification happens via Rule 7.

In that Section, we also claimed that the expression bib0/book/author translates to

```
for v5 in (for v2 in children(bib0) do
             case v2 of
               v3 : book => v3
             | v4 => ()
             end) do
  for v6 in children(v5) do
    case v6 of
      v7 : author => v7
```

```
   | v8 => ()
   end
```

and that this simplifies to

```
for v2 in children(bib0) do
  case v2 of
    v3 : book =>
      for v6 in children(v3) do
        case c of
          v7 : author => d
        | v8 => ()
        end
  | v4 => ()
  end
```

We can now see that the translation happens via two applications of Rule 1, and the simplification happens via Rule 7 and the above instance of Rule 13.

To reiterate, these examples illustrate an important feature of the algebra: high-level operators may be defined in terms of low-level operators, and the low-level operators may be subject to algebraic laws that can be used to further simplify the expression.

6 Type Rules

We explain our type system in the form commonly used in the programming languages community. For a textbook introduction to type systems, see, for example, Mitchell [23].

6.1 Environments

The type rules make use of an environment that specifies the types of variables and functions. The type environment is denoted by Γ, and is composed of a comma-separated list of variable types, $v : t$ or function types, $f : (t_1; \ldots; t_n) \to t$. We retrieve type information from the environment by writing $(v : t) \in \Gamma$ to look up a variable, or by writing $(f : (t_1; \ldots; t_n) \to t) \in \Gamma$ to look up a function.

The type checking starts with an environment that contains all the types declared for functions and global variables. For instance, before typing the first query of Section 2.2, the environment contains: $\Gamma = \mathtt{bib0} : \mathtt{Bib}, \mathtt{book0} : \mathtt{Book}$. While doing the type-checking, new variables will be added in the environment. For instance, when typing the query of section 2.3, variable b will be typed with Book, and added in the environment. This will result in a new environment $\Gamma' = \Gamma, \mathtt{b} : \mathtt{Book}$.

6.2   Type Rules

We write $\Gamma \vdash e \; : \; t$ if in environment Γ the expression e has type t.

The definition of `for` uses an auxiliary type judgement, given below, and the definition of `case` uses an auxiliary function, given below.

$$\frac{}{\Gamma \vdash c_{\text{int}} \; : \; \texttt{Integer}}$$

$$\frac{}{\Gamma \vdash c_{\text{str}} \; : \; \texttt{String}}$$

$$\frac{}{\Gamma \vdash c_{\text{bool}} \; : \; \texttt{Boolean}}$$

$$\frac{(v : t) \in \Gamma}{\Gamma \vdash v \; : \; t}$$

$$\frac{\Gamma \vdash e \; : \; t}{\Gamma \vdash a[e] \; : \; a[t]}$$

$$\frac{\Gamma \vdash e_1 \; : \; \texttt{String} \qquad \Gamma \vdash e_2 \; : \; t}{\Gamma \vdash \; {}^{\sim}e_1[e_2] \; : \; {}^{\sim}[t]}$$

$$\frac{\Gamma \vdash e_1 \; : \; t_1 \qquad \Gamma \vdash e_2 \; : \; t_2}{\Gamma \vdash e_1 \, , e_2 \; : \; t_1 \, , t_2}$$

$$\frac{}{\Gamma \vdash () \; : \; ()}$$

$$\frac{\Gamma \vdash e_1 \; : \; \texttt{Boolean} \qquad \Gamma \vdash e_2 \; : \; t_2 \qquad \Gamma \vdash e_3 \; : \; t_3}{\Gamma \vdash \texttt{if } e_1 \texttt{ then } e_2 \texttt{ else } e_3 \; : \; (t_2 \mid t_3)}$$

$$\frac{\Gamma \vdash e_1 \; : \; t_1 \qquad \Gamma, v : t_1 \vdash e_2 \; : \; t_2}{\Gamma \vdash \texttt{let } v = e_1 \texttt{ do } e_2 \; : \; t_2}$$

$$\frac{\Gamma \vdash e_1 \; : \; t_1 \qquad \Gamma; \texttt{for } v : t_1 \vdash e_2 : t_2}{\Gamma \vdash \texttt{for } v \texttt{ in } e_1 \texttt{ do } e_2 \; : \; t_2}$$

$$\frac{\Gamma \vdash e_0 \; : \; u \qquad u' \mid t' = split^p(u) \qquad \Gamma, v_1 : u' \vdash e_1 \; : \; t_1 \qquad \Gamma, v_2 : t' \vdash e_2 \; : \; t_2}{\Gamma \vdash \texttt{case } e_0 \texttt{ of } v_1 : p \texttt{ => } e_2 \mid v_2 \texttt{ => } e_3 \texttt{ end} \; : \; (t_1 \text{ if } u' \neq \emptyset) \mid (t_2 \text{ if } t' \neq \emptyset)}$$

$$\frac{\begin{array}{c} (f : (t_1; \ldots; t_n) \to t) \in \Gamma \\ \Gamma \vdash e_1 \; : \; t_1' \qquad t_1' <: t_1 \\ \cdots \\ \Gamma \vdash e_n \; : \; t_n' \qquad t_n' <: t_n \end{array}}{\Gamma \vdash f(e_1; \ldots; e_n) \; : \; t}$$

$$\frac{\Gamma \vdash e : t}{\Gamma \vdash \texttt{empty}(e) \;:\; \texttt{Boolean}}$$

$$\frac{}{\Gamma \vdash \texttt{error} \;:\; \emptyset}$$

$$\frac{\Gamma \vdash e \;:\; t' \qquad t' <: t}{\Gamma \vdash (e : t) \;:\; t}$$

$$\frac{\Gamma \vdash e_1 \;:\; \texttt{Integer} \qquad \Gamma \vdash e_2 \;:\; \texttt{Integer}}{\Gamma \vdash e_1 + e_2 \;:\; \texttt{Integer}}$$

$$\frac{\Gamma \vdash e_1 \;:\; t_1 \qquad \Gamma \vdash e_2 \;:\; t_2}{\Gamma \vdash e_1 = e_2 \;:\; \texttt{Boolean}}$$

$$\frac{\Gamma \vdash e \;:\; \texttt{Integer*}}{\Gamma \vdash \texttt{sum}\; e \;:\; \texttt{Integer}}$$

$$\frac{\Gamma \vdash e \;:\; t}{\Gamma \vdash \texttt{count}\; e \;:\; \texttt{Integer}}$$

$$\frac{}{\Gamma \vdash \texttt{error} \;:\; \emptyset}$$

The definition of for uses the following auxiliary judgement. We write $\Gamma \vdash v \;:\; t e t'$ if in environment Γ where the bound variable of an iteration v has type t_1 that the body e of the iteration hast type t_2.

$$\frac{\Gamma, v : u \vdash e \;:\; t'}{\Gamma; \texttt{for}\; v : u \vdash e : t'}$$

$$\frac{}{\Gamma; \texttt{for}\; v \;:\; () \vdash e : ()}$$

$$\frac{\Gamma; \texttt{for}\; v : t_1 \vdash e : t_1' \qquad \Gamma; \texttt{for}\; v : t_2 \vdash e : t_2'}{\Gamma; \texttt{for}\; v : t_1 \,,\, t_2 \vdash e : t_1' \,,\, t_2'}$$

$$\frac{}{\Gamma; \texttt{for}\; v : \emptyset \vdash e : \emptyset}$$

$$\frac{\Gamma; \texttt{for}\; v : t_1 \vdash e : t_1' \qquad \Gamma; \texttt{for}\; v : t_2 \vdash e : t_2'}{\Gamma; \texttt{for}\; v : t_1 \mid t_2 \vdash e : t_1' \mid t_2'}$$

$$\frac{\Gamma; \texttt{for}\; v : t \vdash e : t'}{\Gamma; \texttt{for}\; v : t* \vdash e : t'*}$$

To determine the types in a **case** expression, we use the function $split^p(t)$, where p is a pattern (either an element a, or a wildcard $\tilde{\ }$, or a scalar s) and t is a type. For mnemonic convenience we write $a[t'] \mid t'' = split^a(t)$ or $\tilde{\ }[t'] \mid t'' = split^{\tilde{\ }}(t)$ or $s' <: s \mid t' = split^s(t)$ but one should think of the function as returning a pair consisting of two types t and t', or in the last instance a scalar type s' and a type t'. The function $split^p(t)$ is undefined if type t involves sequencing, since a **case** expression acts on elements or scalars, not sequences.

$$
\begin{aligned}
split^a(s) &= a[\emptyset] \mid s \\
split^a(a[t]) &= a[t] \mid \emptyset \\
split^a(a'[t]) &= a[\emptyset] \mid a'[t] && \text{if } a \neq a' \\
split^a(\tilde{\ }[t]) &= a[t] \mid a[t] \\
split^a(t_1 \mid t_2) &= a[t_1' \mid t_2'] \mid (t_1'' \mid t_2'') && \text{where } a[t_i'] \mid t_i'' = split^a(t_i) \\
split^a(\emptyset) &= a[\emptyset] \mid \emptyset
\end{aligned}
$$

$$
\begin{aligned}
split^{\tilde{\ }}(s) &= \tilde{\ }[\emptyset] \mid s \\
split^{\tilde{\ }}(a[t]) &= \tilde{\ }[t] \mid \emptyset \\
split^{\tilde{\ }}(\tilde{\ }[t]) &= \tilde{\ }[t] \mid \emptyset \\
split^{\tilde{\ }}(t_1 \mid t_2) &= \tilde{\ }[t_1' \mid t_2'] \mid (t_1'' \mid t_2'') && \text{where } \tilde{\ }[t_i'] \mid t_i'' = split^{\tilde{\ }}(t_i) \\
split^{\tilde{\ }}(\emptyset) &= \tilde{\ }[\emptyset] \mid \emptyset
\end{aligned}
$$

$$
\begin{aligned}
split^s(s') &= s' <: s \mid \emptyset && \text{if } s' <: s \\
&= \emptyset <: s \mid s' && \text{otherwise} \\
split^s(a[t]) &= \emptyset <: s \mid a[t] \\
split^s(\tilde{\ }[t]) &= \emptyset <: s \mid \tilde{\ }[t] \\
split^s(t_1 \mid t_2) &= (s_1 \mid s_2) <: s \mid (t_1' \mid t_2') && \text{where } s_i <: s \mid t_i' = split^s(t_i) \\
split^s(\emptyset) &= \emptyset <: s \mid \emptyset
\end{aligned}
$$

6.3   Top-Level Expressions

We write $\Gamma \vdash q$ if in environment Γ the query item q is well-typed.

$$
\frac{}{\Gamma \vdash \textbf{type } x = t}
$$

$$
\frac{\Gamma, v_1 : t_1, \ldots, v_n : t_n \vdash e : t' \qquad t' <: t}{\Gamma \vdash f(v_1 : t_1 ; \ldots ; v_n : t_n) : t \ = \ e}
$$

$$
\frac{\Gamma \vdash e : t' \qquad t' <: t}{\Gamma \vdash \textbf{let } v : t = e}
$$

$$
\frac{\Gamma \vdash e : t}{\Gamma \vdash \textbf{query } e}
$$

We extract the relevant component of a type environment from a query item q with the function $environment(q)$.

$$
\begin{aligned}
environment(\textbf{type}\ x\ =\ t) &= ()\\
environment(\textbf{fun}\ f(v_1\!:\!t_1;\ \ldots;\ v_n\!:\!t_n)\!:\!t) &= f:(t_1;\ \ldots;\ t_n) \rightarrow t\\
environment(\textbf{let}\ v\ :\ t\ =\ e) &= v:t
\end{aligned}
$$

We write $\vdash q_1 \ldots q_n$ if the sequence of query items $q_1 \ldots q_n$ is well typed.

$$
\frac{\Gamma = environment(q_1),\ \ldots,\ environment(q_n) \qquad \Gamma \vdash q_1 \quad \cdots \quad \Gamma \vdash q_n}{\vdash q_1 \ldots q_n}
$$

7   Discussion

The algebra has several important characteristics: its operators are orthogonal, strongly typed, and they obey laws of equivalence and optimization.

There are many issues to resolve in the completion of the algebra. We enumerate some of these here.

Data Model. Currently, all forests in the data model are ordered. It may be useful to have unordered forests. The distinct operator, for example, produces an inherently unordered forest. Unordered forests can benefit from many optimizations for the relational algebra, such as commutable joins.

The data model and algebra do not define a global order on documents. Querying global order is often required in document-oriented queries.

Currently, the algebra does not support reference values, which are defined in the XML Query Data Model. The algebra's type system should be extended to support reference types and the data model operators ref and deref should be supported.

Type System. As discussed, the algebra's internal type system is closely related to the type system of XDuce. A potentially significant problem is that the algebra's types may lose information when converted into XML Schema types, for example, when a result is serialized into an XML document and XML Schema.

The type system is currently first order: it does not support function types nor higher-order functions. Higher-order functions are useful for specifying, for example, sorting and grouping operators, which take other functions as arguments.

The type system is currently monomorphic: it does not permit the definition of a function over generalized types. Polymorphic functions are useful for factoring equivalent functions, each of which operate on a fixed type. The lack of polymorphism is one of the principal weaknesses of the type system.

Operators. We intentionally did not define equality or relational operators on element and scalar types undefined. These operators should be defined by consensus.

It may be useful to add a fixed-point operator, which can be used in lieu of recursive functions to compute, for example, the transitive closure of a collection.

Functions. There is no explicit support for externally defined functions.

The set of builtin functions may be extended to support other important operators.

Recursion. Currently, the algebra does not guarantee termination of recursive expressions. In order to ensure termination, we might require that a recursive function take one argument that is a singleton element, and any recursive invocation should be on a descendant of that element; since any element has a finite number of descendants, this avoids infinite regress. (Ideally, we should have a simple syntactic rule that enforces this restriction, but we have not yet devised such a rule.)

References

1. S. Abiteboul, R. Hull, V. Vianu. *Foundations of Databases*. Addison Wesley, 1995.
2. Richard Bird. *Introduction to Functional Programming using Haskell*. Prentice Hall, 1998.
3. P. Buneman, M. Fernandez, D. Suciu. UnQL: A query language and algebra for semistructured data based on structural recursion. *VLDB Journal*, to appear.
4. Catriel Beeri and Yoram Kornatzky. Algebraic Optimization of Object-Oriented Query Languages. *Theoretical Computer Science* 116(1&2):59–94, August 1993.
5. Francois Bancilhon, Paris Kanellakis, Claude Delobel. *Building an Object-Oriented Database System*. Morgan Kaufmann, 1990.
6. Peter Buneman, Leonid Libkin, Dan Suciu, Van Tannen, and Limsoon Wong. Comprehension Syntax. *SIGMOD Record*, 23:87–96, 1994.
7. David Beech, Ashok Malhotra, Michael Rys. A Formal Data Model and Algebra for XML. W3C XML Query working group note, September 1999.
8. Peter Buneman, Shamim Naqvi, Val Tannen, Limsoon Wong. Principles of programming with complex object and collection types. *Theoretical Computer Science* 149(1):3–48, 1995.
9. Catriel Beeri and Yariv Tzaban, SAL: An Algebra for Semistructured Data and XML, *International Workshop on the Web and Databases (WebDB'99)*, Philadelphia, Pennsylvania, June 1999.
10. R. G. Cattell. *The Object Database Standard: ODMG 2.0*. Morgan Kaufmann, 1997.
11. Don Chamberlin, Jonathan Robie, and Daniela Florescu. Quilt: An XML Query Language for Heterogeneous Data Sources. *International Workshop on the Web and Databases (WebDB'2000)*, Dallas, Texas, May 2000.
12. Vassilis Christophides and Sophie Cluet and Jérôme Siméon. On Wrapping Query Languages and Efficient XML Integration. *Proceedings of ACM SIGMOD Conference on Management of Data*, Dallas, Texas, May 2000.
13. S. Cluet and G. Moerkotte. Nested queries in object bases. *Workshop on Database Programming Languages*, pages 226–242, New York, August 1993.
14. S. Cluet, S. Jacqmin and J. Siméon The New YAT$_L$: Design and Specifications. *Technical Report*, INRIA, 1999.
15. L. S. Colby. A recursive algebra for nested relations. *Information Systems* 15(5):567–582, 1990.

16. Hugh Darwen (Contributor) and Chris Date. *Guide to the SQL Standard: A User's Guide to the Standard Database Language SQL* Addison-Wesley, 1997.

17. A. Deutsch, M. Fernandez, D. Florescu, A. Levy, and D. Suciu. A query language for XML. In *International World Wide Web Conference*, 1999. http://www.research.att.com/~mff/files/final.html

18. J. A. Goguen, J. W. Thatcher, E. G. Wagner. An initial algebra approach to the specification, correctness, and implementation of abstract data types. In *Current Trends in Programming Methodology*, pages 80–149, Prentice Hall, 1978.

19. Haruio Hosoya, Benjamin Pierce, XDuce : A Typed XML Processing Language (Preliminary Report) *WebDB Workshop* 2000.

20. M. Kifer, W. Kim, and Y. Sagiv. Querying object-oriented databases. *Proceedings of ACM SIGMOD Conference on Management of Data*, pages 393–402, San Diego, California, June 1992.

21. Leonid Libkin and Limsoon Wong. Query languages for bags and aggregate functions. *Journal of Computer and Systems Sciences*, 55(2):241–272, October 1997.

22. Leonid Libkin, Rona Machlin, and Limsoon Wong. A query language for multi-dimensional arrays: Design, implementation, and optimization techniques. *SIGMOD* 1996.

23. John C. Mitchell *Foundations for Programming Languages*. MIT Press, 1998.

24. The Caml Language. http://pauillac.inria.fr/caml/.

25. J. Robie, editor. XQL '99 Proposal, 1999. http://metalab.unc.edu/xql/xql-proposal.html.

26. H.-J. Schek and M. H. Scholl. The relational model with relational-valued attributes. *Information Systems* 11(2):137–147, 1986.

27. S. J. Thomas and P. C. Fischer. Nested Relational Structures. In *Advances in Computing Research: The Theory of Databases*, JAI Press, London, 1986.

28. Philip Wadler. Comprehending monads. *Mathematical Structures in Computer Science*, 2:461-493, 1992.

29. Philip Wadler. A formal semantics of patterns in XSLT. Markup Technologies, Philadelphia, December 1999.

30. Limsoon Wong. An introduction to the Kleisli query system and a commentary on the influence of functional programming on its implementation. *Journal of Functional Programming*, to appear.

31. World-Wide Web Consortium XML Query Data Model, Working Draft, May 2000. http://www.w3.org/TR/query-datamodel.

32. World-Wide Web Consortium, XML Path Language (XPath): Version 1.0. November, 1999. /www.w3.org/TR/xpath.html

33. World-Wide Web Consortium, XML Query: Requirements, Working Draft. August 2000. http://www.w3.org/TR/xmlquery-req

34. World-Wide Web Consortium, XML Query: Data Model, Working Draft. May 2000. http://www.w3.org/TR/query-datamodel/

35. World-Wide Web Consortium, XML Schema Part 1: Structures, Working Draft. April 2000. http://www.w3.org/TR/xmlschema-1

36. World-Wide Web Consortium, XML Schema Part 2: Datatypes, Working Draft, April 2000. http://www.w3.org/TR/xmlschema-2.

37. World-Wide Web Consortium, XSL Transformations (XSLT), Version 1.0. W3C Recommendation, November 1999. http://www.w3.org/TR/xslt.

Irregularities of Distribution, Derandomization, and Complexity Theory*

Bernard Chazelle

Department of Computer Science,
Princeton University, and NEC Research Institute
chazelle@cs.princeton.edu

Abstract. In 1935, van der Corput asked the following question: Given an infinite sequence of reals in $[0, 1]$, define

$$D(n) = \sup_{0 \le x \le 1} \Big| \, |S_n \cap [0, x]| - nx \, \Big|,$$

where S_n consists of the first n elements in the sequence. Is it possible for $D(n)$ to stay in $O(1)$? Many years later, Schmidt proved that $D(n)$ can never be in $o(\log n)$. In other words, there are limitations on how well the discrete distribution, $x \mapsto |S_n \cap [0, x]|$, can simulate the continuous one, $x \mapsto nx$. The study of this intriguing phenomenon and its numerous variants related to the irregularities of distributions has given rise to *discrepancy theory*. The relevance of the subject to complexity theory is most evident in the study of probabilistic algorithms. Suppose that we feed a probabilistic algorithm not with a perfectly random sequence of bits (as is usually required) but one that is only pseudorandom or even deterministic. Should performance necessarily suffer? In particular, suppose that one could trade an exponential-size probability space for one of polynomial size without letting the algorithm realize the change. This form of derandomization can be expressed by saying that a very large distribution can be simulated by a small one for the purpose of the algorithm. Put differently, there exists a measure with respect to which the two distributions have low discrepancy. The study of discrepancy theory predates complexity theory and a wealth of mathematical techniques can be brought to bear to prove nontrivial derandomization results. The pipeline of ideas that flows from discrepancy theory to complexity theory constitutes the *discrepancy method*. We give a few examples in this survey. A more thorough treatment is given in our book [15]. We also briefly discuss the relevance of the discrepancy method to complexity lower bounds.

1 Facts from Discrepancy Theory

Let $(V, \mathcal{S})$ be a set system, where $V = \{v_1, \ldots, v_n\}$ is the ground set and $\mathcal{S} = \{S_1, \ldots, S_m\}$, with $S_i \subseteq V$. We wish to color the elements of V red and blue so

* Proceedings of FSTTCS-2000. This work was supported in part by NSF Grant CCR-96-23768, ARO Grant DAAH04-96-1-0181, and NEC Research Institute.

that, within each S_i, no color greatly outnumbers the other one. To do that, we choose a function χ that maps each $v_j \in V$ to an element in $\{-1, 1\}$, and we define the *discrepancy* of the set S_i to be

$$\chi(S_i) = \sum_{v_j \in S_i} \chi(v_j).$$

The maximum value of $|\chi(S_i)|$, over all $S_i \in \mathcal{S}$, is the discrepancy of the set system under the given coloring. The *discrepancy* of the set system itself, denoted by $D_\infty(\mathcal{S})$, refers to its minimum discrepancy under all possible colorings. The L^2 norm creates an easier environment to work with, and so we also define

$$D_2(\mathcal{S}) \overset{\text{def}}{=} \min_{\chi} \sqrt{\chi(S_1)^2 + \cdots + \chi(S_m)^2},$$

where the minimum is taken over all colorings $\chi : V \mapsto \{-1, 1\}$. The discrepancy can be characterized by using matrices, which is sometimes more convenient. Let A be the *incidence matrix* of the set system $(V, \mathcal{S})$: the n columns are indexed by the elements of V and the m rows are the characteristic vectors of the sets S_i, so that A_{ij} is 1 if $v_j \in S_i$ and 0 otherwise. The discrepancy of the set system, also denoted by $D_\infty(A)$, can be expressed as the L^∞ norm of a column vector. Generally, for any $p \in \{1, 2, \ldots, \infty\}$, we have $D_p(A) - \min_{x \in \{-1, 1\}^n} \|Ax\|_p$. The following result of Spencer [44] is tight.

Theorem 1. *Any set system $(V, \mathcal{S})$ such that $|V| = |\mathcal{S}| = n$ has $O(\sqrt{n})$ discrepancy.*

For general set systems with m sets, the bound becomes $O(\sqrt{n \ln(2m/n)})$. A simple, elegant result concerns the case of small-degree set systems. The degree refers to the maximum number of sets containing a given element. The classical Beck-Fiala theorem [7] states that:

Theorem 2. *The discrepancy of a set system of degree at most t is less than $2t$.*

Techniques for proving lower bounds often involve spectral arguments and, in particular, harmonic analysis. The latter comes from the fact that set systems are often defined by using a convolution operator, which the Fourier transform diagonalizes. Bounding the eigenvalues gives us a handle on the L^2-norm discrepancy. Perhaps the simplest result obtained in this manner is Roth's $\frac{1}{4}$-Theorem [40].

Theorem 3. *Any two-coloring of the integers $\{1, \ldots, n\}$ contains an arithmetic progression whose discrepancy is $\Omega(n^{1/4})$.*

There exists a wealth of techniques and results for geometric set systems. In such cases, it is useful to define the notion of volume discrepancy. Consider the problem of placing a set P of n points in the unit cube $[0, 1]^d$ to minimize the discrepancy with respect to axis-parallel boxes. The (volume) discrepancy of a box $B = \prod_{k=1}^{d} [p_k, q_k)$ is defined as

$$D(B) \overset{\text{def}}{=} n \cdot \text{vol}(B) - |P \cap B|.$$

Theorem 4. *There is a set of n points in $[0,1]^d$ such that the volume discrepancy of any box in $[0,1]^d$ is $O(\log n)^{d-1}$ in absolute value.*

Here is a construction in two dimensions [46,47]. Given a nonnegative integer m, let $\sum_{i \geq 0} b_1(i)\, 2^i$ be its binary decomposition, and let

$$x_1(m) = \sum_{i \geq 0} \frac{b_1(i)}{2^{i+1}} \in [0,1).$$

The numbers $x_1(m)$, for $0 \leq m < n$, form the classical *van der Corput* sequence. We can use it to define the *bit-reversal* point set:

$$\left\{ (x_1(m), m/n) \;\middle|\; 0 \leq m < n \right\}.$$

This easily generalizes to d dimensions. Choose $d-1$ relatively prime numbers: $2 = p_1, p_2, \ldots, p_{d-1}$. The integer m has a unique decomposition in base p_k, $m = \sum_{i \geq 0} b_k(i) p_k^i$, so we can define

$$x_k(m) = \sum_{i \geq 0} \frac{b_k(i)}{p_k^{i+1}}.$$

The point set

$$P = \left\{ \left(x_1(m), \ldots, x_{d-1}(m), \frac{m}{n} \right) : 0 \leq m < n \right\}$$

is called *Halton-Hammersley* [25] and satisfies Theorem 4.

What about the L^2 norm? Let P be a set of n points in the unit square. Given a box B_q of the form $[0, q_1) \times [0, q_2)$, where $q = (q_1, q_2)$, the discrepancy of B_q is

$$D(B_q) = n \cdot \text{area}\,(B_q) - |P \cap B_q|.$$

We define the L^2-norm discrepancy of P as

$$D_2(P) \stackrel{\text{def}}{=} \sqrt{\int_{[0,1]^2} D(B_q)^2 \, dq}.$$

The following result is by Davenport [22].

Theorem 5. *It is possible to find a set P of n points in $[0,1]^2$ such that $D_2(P) = O(\sqrt{\log n})$.*

We forsake the Halton-Hammersley construction and, instead, turn to a construction based on irrational lattices. Take the set of $n = 2k - 1$ points of the form

$$\left(\{j\varphi\}, \frac{|j|}{n} \right),$$

for all j ($|j| < k$), where $\{x\} \overset{\text{def}}{=} x \pmod 1$ is the fractional part of x and $\varphi = \frac{1}{2}(\sqrt{5}+1)$ is the golden ratio. The only property we use about the golden ratio is the size of the partial quotients of its continued fraction expansion, so many other choices exist for Theorem 5.

We generalize the discrepancy to $\mathbf{R}^d$ in the obvious manner. Given a point $q = (q_1, \ldots, q_d)$ in the unit cube $[0,1]^d$, let B_q denote the box $[0, q_1) \times \cdots \times [0, q_d)$. Fix a set P of n points in $[0,1]^d$, and as usual define the volume discrepancy $D(B_q)$ at a point $q \in [0,1]^d$ as $D(B_q) = nq_1 \cdots q_d - |P \cap B_q|$. We write $D_2(P) = \sqrt{\int_{[0,1]^d} D(B_q)^2 \, dq}$. The following bound is due to Roth [39], and shows the optimality of Theorem 5.

Theorem 6. *Given a set P of n points in $[0,1]^d$, the mean-square discrepancy for axis-parallel boxes satisfies*

$$D_2(P) > c(\log n)^{(d-1)/2},$$

for some constant $c = c(d) > 0$.

In two dimensions, we have this interesting lower bound by Schmidt [41], which shows a rare divergence between L^2 and L^∞ behaviors.

Theorem 7. *Given n points in $[0,1]^2$, there exists a box B such that $|D(B)| = \Omega(\log n)$.*

We now consider rotated boxes. Given a set P of n points in $[0,1]^2$, the discrepancy of a (rotated) box R is defined naturally as $D(R) = n \cdot \text{area}\,(R \cap [0,1]^2) - |P \cap R|$. By rotated box, we mean any rectangle not necessarily parallel to the axes. The following upper bound was established by Beck; see Beck and Chen's book [6].

Theorem 8. *It is possible to place n points in the unit square $[0,1]^2$, so that any (rotated) box R satisfies $|D(R)| = O(n^{1/4}\sqrt{\log n}\,)$.*

A quasi-matching lower bound was first proven by Beck [5], using his beautiful *Fourier transform method* (other proof techniques exist).

Theorem 9. *Given n points in the unit square $[0,1]^2$, there exists a rotated box R such that $|D(R)| = \Omega(n^{1/4})$.*

The same bound holds for disks as well. The proof, by Montgomery [34,35], also uses harmonic analysis.

Theorem 10. *Given n points in the unit square $[0,1]^2$, there exists a disk K such that $|D(K)| = \Omega(n^{1/4})$.*

2   Sampling

The red-blue discrepancy of a set system tells us how well we can sample its ground set by choosing about half of its elements. What about different sample sizes? For example, given a collection of n points in the plane, is it possible to choose a subset of constant size, such that any disk that encloses at least one percent of the points also includes at least one sample point? Surprisingly, the answer is yes. The surprise is that the sample size can be kept independent of n. The magic lies in the notion of VC dimension.

Let $(V, \mathcal{S})$ be a (finite or infinite) set system. Given $Y \subseteq V$, let $(Y, \mathcal{S}|_Y)$ denote the set system *induced* by Y, ie, $\{Y \cap S \mid S \in \mathcal{S}\}$. A subset Y of V is said to be *shattered* (by $\mathcal{S}$) if $\mathcal{S}|_Y = 2^Y$, ie, every subset of Y (including the empty set) is of the form $Y \cap S$, for some $S \in \mathcal{S}$. The supremum of all sizes of finite shattered subsets of X is called the *Vapnik-Chervonenkis dimension* (or *VC-dimension* for short) of the set system.

Let $(V, \mathcal{S})$ be a finite set system, where $|V| = n$ and $|\mathcal{S}| = m$. Given any $0 < \varepsilon < 1$, a set $N \subseteq V$ is called an ε-*net for* $(V, \mathcal{S})$ if $N \cap S \neq \emptyset$, for any $S \in \mathcal{S}$ with $|S|/|V| > \varepsilon$. A set $A \subseteq V$ is called an ε-*approximation* for $(V, \mathcal{S})$ if, for any $S \in \mathcal{S}$,

$$\left| \frac{|S|}{|V|} - \frac{|A \cap S|}{|A|} \right| \leq \varepsilon.$$

Equivalently, given a random v uniformly distributed in V, for each $S \in \mathcal{S}$,

$$\Big| \text{Prob}[v \in S] - \text{Prob}[v \in S \mid v \in A] \Big| \leq \varepsilon.$$

The following was proven by Chazelle and Matoušek [18], building on the foundational work in [16,26,29,48].

Theorem 11. *Let $(V, \mathcal{S})$ be a set system of VC-dimension d. Given any $r \geq 2$, a $(1/r)$-approximation for $(V, \mathcal{S})$ of size $O(dr^2 \log dr)$ can be computed in time $O(d)^{3d}(r^2 \log dr)^d |V|$.*

Theorem 12. *Let $(V, \mathcal{S})$ be a set system of VC-dimension d. Given any $r \geq 2$, a $(1/r)$-net for $(V, \mathcal{S})$ of size $O(dr \log dr)$ can be computed in time $O(d)^{3d}(r^2 \log dr)^d |V|$.*

Note that the set systems are usually understood as members of an infinite family; for example the set of all points in $\mathbf{R}^2$ and the set of all disks. The term *range space* is often used in the literature to refer to such a family.

3   Geometric Algorithms

Suppose that we are given a set H of n hyperplanes in $\mathbf{R}^d$. We wish to subdivide $\mathbf{R}^d$ into a small number of simplices, so that none of them is cut by too many hyperplanes. Given a parameter $\varepsilon > 0$, a collection $\mathcal{C}$ of closed full-dimensional simplices is called an ε-*cutting* if: (i) their interiors are pairwise disjoint, and together they cover $\mathbf{R}^d$; and (ii) the interior of any simplex of $\mathcal{C}$ is intersected by at most εn hyperplanes of H.

Cuttings are among the most useful, versatile tools in computational geometry, as they lay the grounds for efficient divide-and-conquer [1,2,20,26,27]. Using some of the sampling technology for finite VC dimension discussed earlier, Chazelle [11] proved the following:

Theorem 13. *Given a collection H of n hyperplanes in $\mathbf{R}^d$, for any $r > 0$, there exists a $(1/r)$-cutting for H of optimal size $O(r^d)$. A full description of the cutting, including the list of hyperplanes intersecting the interior of each simplex, can be found deterministically in $O(nr^{d-1})$ time.*

Here are some direct applications of cuttings: Point location is understood here as the problem of preprocessing an arrangement of n hyperplanes in $\mathbf{R}^d$ so that, given a query point, the face of the arrangement that contains the point can be found quickly. Simplex range searching is the problem of preprocessing n points in $\mathbf{R}^d$ so that given a query simplex the points inside it can be counted quickly.

Theorem 14. *Point location among n hyperplanes in $\mathbf{R}^d$ can be done in $O(\log n)$ query time, using $O(n^d)$ preprocessing.*

Theorem 15. *To decide whether n points and n lines in the plane are free of any incidence can be done in $n^{4/3} \cdot O(\log n)^{1/3}$ time.*

Theorem 16. *Given n points in $\mathbf{R}^d$, there exists a data structure of size m (for any $n \leq m \leq n^d$), which allows simplex range searching to be done in time $O(n^{1+\varepsilon}/m^{1/d})$ per query, for any fixed $\varepsilon > 0$.*

A far more involved application of cuttings and the discrepancy method gives the following result (and its corollary), which was proven by Chazelle [12]. The complexity is tight in the worst case.

Theorem 17. *The convex hull of a set of n points in $\mathbf{R}^d$ can be computed deterministically in $O(n \log n + n^{\lfloor d/2 \rfloor})$ time, for any fixed $d > 1$.*

Theorem 18. *The Voronoi diagram of a set of n points in $\mathbf{E}^d$ can be computed deterministically in $O(n \log n + n^{\lceil d/2 \rceil})$ time, for any fixed $d > 1$.*

Applications to linear and quadratic programming include the following results by Chazelle and Matoušek [18].

Theorem 19. *The ellipsoid of minimum volume that encloses a set of n points in $\mathbf{R}^d$ can be computed in time $d^{O(d^2)}n$.*

Theorem 20. *Linear programming with n constraints and d variables can be solved in $d^{O(d)} n$ time.*

These last two results build on important previous work. In particular, we mention the general formalism for linear programming developed by Sharir and Welzl [43], known as *LP-type*. The first algorithm for linear programming with a running time linear in the number of constraints was found by Megiddo [32,33]. Subsequent improvements were found in [19,21,23,24,42].

4   Linear Circuit Complexity

Let A be an n-by-n matrix with 0/1 elements. Consider the task of assembling A by forming a sequence of column vectors $U_1, \ldots, U_s \in \mathbf{Z}^n$, where $s \geq n$ and (i) $(U_1, \ldots, U_n)$ is the n-by-n identity matrix; (ii) $A = (U_{s-n+1}, \ldots, U_s)$; and (iii) for any $i = n+1, \ldots, s$, there exist $j, k < i$ and $\alpha_i, \beta_i \in \mathbf{Z}$, such that $U_i = \alpha_i U_j + \beta_i U_k$. The minimum length s of any sequence that satisfies these three conditions is called the *complexity* of A. It is easy to see that all 0/1 matrices have complexity $O(n^2)$ and that a random one has complexity $\Omega(n^2/\log n)$.

The complexity of A is the same as the linear circuit complexity of computing $A^T x$. (A circuit consists of gates that can add linear forms.) For the case where $|\alpha_i|, |\beta_i| = O(1)$ (which is to be understood from now on), Chazelle's *spectral lemma* [14] gives us a line of attack:

Lemma 1. *The complexity of an n-by-n 0/1 matrix A is $\Omega(\max_k k \log \lambda_k)$, where λ_k is the k-th largest eigenvalue of $A^T A$.*

Of course, the same lemma applies to the circuit complexity as well. A recent variant by Chazelle and Lvov [17] gives us another powerful tool which bypasses the need to bound individual eigenvalues.

Lemma 2. *The complexity of an n-by-n 0/1 matrix A is*

$$\Omega_\varepsilon\left(n \log\left(\operatorname{tr} M/n - \varepsilon\sqrt{\operatorname{tr} M^2/n}\right)\right),$$

where $M = A^T A$ and $\varepsilon > 0$ is an arbitrarily small constant.

The complexity of range searching relates to the complexity of certain geometric matrices. A box matrix refers to a set system formed by points and axis-parallel boxes. Simplex matrices, on the other hand, denote the incidence matrices of set systems formed by points and simplices in $\mathbf{R}^d$. The following results, by Chazelle [9,10,13,14], make heavy use of the discrepancy method.

Theorem 21. *There are n-by-n box matrices of circuit complexity $\Omega(n \log \log n)$ in $\mathbf{R}^2$ and monotone circuit complexity $\Omega(n(\log n/\log\log n)^{d-1})$ in $\mathbf{R}^d$.*

Theorem 22. *There are n-by-n simplex matrices of circuit complexity $\Omega(n \log n)$ and monotone circuit complexity $\Omega(n^{4/3})$ in $\mathbf{R}^2$.*

Recall that the monotone circuit model disallows the use of subtraction. While the monotone complexity of these problems is essentially resolved (there are quasi-matching upper bounds), the nonmonotone case is still wide open.

References

1. Agarwal, P.K. Partitioning arrangements of lines II: Applications, *Disc. Comput. Geom.* 5 (1990), 533–573.
2. Agarwal, P.K. Geometric partitioning and its applications, in *Computational Geometry: Papers from the DIMACS Special Year,* eds., Goodman, J.E., Pollack, R., Steiger, W., Amer. Math. Soc., 1991.
3. Agarwal, P.K., Erickson, J. Geometric range searching and its relatives, in *Advances in Discrete and Computational Geometry,* eds. Chazelle, B., Goodman, J.E., Pollack, R., *Contemporary Mathematics* 223, Amer. Math. Soc., 1999, pp. 1–56.
4. Alon, N., Spencer, J.H. *The Probabilistic Method,* Wiley-Interscience, 1992.
5. Beck, J. Irregularities of distribution, I, *Acta Math.* 159 (1987), 1–49.
6. Beck, J., Chen, W.W.L. *Irregularities of Distribution,* Cambridge Tracts in Mathematics, 89, Cambridge University Press, 1987.
7. Beck, J., Fiala, T. "Integer-making" theorems, *Discrete Applied Mathematics* 3 (1981), 1–8.
8. Beck, J., Sós, V.T. Discrepancy theory, in *Handbook of Combinatorics,* Chap. 26, eds., Graham, R.L., Grötschel, M., Lovász, L., North-Holland, 1995, pp. 1405–1446.
9. Chazelle, B. Lower bounds on the complexity of polytope range searching, *J. Amer. Math. Soc.* 2 (1989), 637–666.
10. Chazelle, B. Lower bounds for orthogonal range searching: II. The arithmetic model, *J. ACM* 37 (1990), 439–463.
11. Chazelle, B. Cutting hyperplanes for divide-and-conquer, *Disc. Comput. Geom.* 9 (1993), 145–158.
12. Chazelle, B. An optimal convex hull algorithm in any fixed dimension, *Disc. Comput. Geom.* 10 (1993), 377–409.
13. Chazelle, B. Lower bounds for off-line range searching, *Disc. Comput. Geom.* 17 (1997), 53–65.
14. Chazelle, B. A spectral approach to lower bounds with applications to geometric searching, *SIAM J. Comput.* 27 (1998), 545–556.
15. Chazelle, B. *The Discrepancy Method: Randomness and Complexity,* Cambridge University Press, 2000.
16. Chazelle, B., Friedman, J. A deterministic view of random sampling and its use in geometry, *Combinatorica* 10 (1990), 229–249.
17. Chazelle, B., Lvov, A. A trace bound for the hereditary discrepancy, *Proc. 16th Annual ACM Symp. Comput. Geom.* (2000), 64–69. To appear in Disc. Comput. Geom.
18. Chazelle, B., Matoušek, J. On linear-time deterministic algorithms for optimization problems in fixed dimension, *J. Algorithms* 21 (1996), 579–597.
19. Clarkson, K.L. Linear programming in $O(n \times 3^{d^2})$ time, *Inform. Process. Lett.* 22 (1986), 21–24.
20. Clarkson, K.L. New applications of random sampling in computational geometry, *Disc. Comput. Geom.* 2 (1987), 195–222.
21. Clarkson, K.L. Las Vegas algorithms for linear and integer programming when the dimension is small, *J. ACM* 42 (1995), 488–499.
22. Davenport, H. Note on irregularities of distribution, *Mathematika* 3 (1956), 131–135.
23. Dyer, M.E. On a multidimensional search technique and its application to the Euclidean one-centre problem, *SIAM J. Comput.* 15 (1986), 725–738.

24. Dyer, M.E., Frieze, A.M. A randomized algorithm for fixed-dimensional linear programming, *Mathematical Programming* 44 (1989), 203–212.
25. Hammersley, J.M. Monte Carlo methods for solving multivariable problems, *Ann. New York Acad. Sci.* 86 (1960), 844-874.
26. Haussler, D., Welzl, E. ε-nets and simplex range queries, *Disc. Comput. Geom.* 2 (1987), 127–151.
27. Matoušek, J. Construction of ε-nets, *Disc. Comput. Geom.* 5 (1990), 427–448.
28. Matoušek, J. Geometric range searching, *ACM Comput. Surv.* 26 (1994), 421–461.
29. Matoušek, J. Approximations and optimal geometric divide-and-conquer, *J. Comput. Syst. Sci.* 50 (1995), 203–208.
30. Matoušek, J. Derandomization in computational geometry, *J. Algorithms* 20 (1996), 545–580.
31. Matoušek, J. *Geometric Discrepancy: An Illustrated Guide*, Algorithms and Combinatorics, 18, Springer, 1999.
32. Megiddo, N. Linear-time algorithms for linear programming in R^3 and related problems, *SIAM J. Comput.* 12 (1983), 759–776.
33. Megiddo, N. Linear programming in linear time when the dimension is fixed, *J. ACM* 31 (1984), 114–127.
34. Montgomery, H.L. On irregularities of distribution, in *Congress of Number Theory* (Zarautz, 1984), Universidad del País Vasco, Bilbao, 1989, pp. 11–27.
35. Montgomery, H.L. *Ten Lectures on the Interface Between Analytic Number Theory and Harmonic Analysis*, CBMS Regional Conference Series in Mathematics, No. 84, Amer. Math. Soc., Providence, 1994.
36. Motwani, R., Raghavan, P. *Randomized Algorithms*, Cambridge University Press, 1995.
37. Niederreiter, H. *Random Number Generation and Quasi-Monte Carlo Methods*, CBMS-NSF, SIAM, Philadelphia, PA, 1992.
38. Pach, J., Agarwal, P.K. *Combinatorial Geometry*, Wiley-Interscience Series in Discrete Mathematics and Optimization, John Wiley & Sons, Inc., 1995.
39. Roth, K.F. On irregularities of distribution, *Mathematika* 1 (1954), 73–79.
40. Roth, K.F. Remark concerning integer sequences, *Acta Arithmetica* 9 (1964), 257–260.
41. Schmidt, W.M. Irregularities of distribution, VII, *Acta Arithmetica* 21 (1972), 45–50.
42. Seidel, R. Small-dimensional linear programming and convex hulls made easy, *Disc. Comput. Geom.* 6 (1991), 423–434.
43. Sharir, M., Welzl, E. A combinatorial bound for linear programming and related problems, *Proc. 9th Annual Symp. Theoret. Aspects Comput. Sci.*, LNCS, 577, Springer-Verlag, 1992, pp. 569–579.
44. Spencer, J. Six standard deviations suffice, *Trans. Amer. Math. Soc.* 289 (1985), 679–706.
45. Spencer, J. *Ten Lectures on the Probabilistic Method*, CBMS-NSF, SIAM, 1987.
46. van der Corput, J.G. Verteilungsfunktionen I. *Proc. Nederl. Akad. Wetensch.* 38 (1935), 813–821.
47. van der Corput, J.G. Verteilungsfunktionen II. *Proc. Nederl. Akad. Wetensch.* 38 (1935), 1058–1066.
48. Vapnik, V.N., Chervonenkis, A.Ya. On the uniform convergence of relative frequencies of events to their probabilities, *Theory of Probability and its Applications* 16 (1971), 264–280.

Rewriting Logic as a Metalogical Framework

David Basin[1], Manuel Clavel[2], and José Meseguer[3]

[1] Institut für Informatik, Universität Freiburg, Germany
[2] Department of Philosophy, University of Navarre, Spain
[3] Computer Science Laboratory, SRI International, USA

Abstract. A metalogical framework is a logic with an associated methodology that is used to represent other logics and to reason about their metalogical properties. We propose that logical frameworks can be good metalogical frameworks when their logics support reflective reasoning and their theories always have initial models.

We present a concrete realization of this idea in rewriting logic. Theories in rewriting logic always have initial models and this logic supports reflective reasoning. This implies that inductive reasoning is valid when proving properties about the initial models of theories in rewriting logic, and that we can use reflection to reason at the metalevel about these properties. In fact, we can uniformly reflect induction principles for proving metatheorems about rewriting logic theories and their parameterized extensions. We show that this reflective methodology provides an effective framework for different, non-trivial, kinds of formal metatheoretic reasoning; one can, for example, prove metatheorems that relate theories or establish properties of parameterized classes of theories. Finally, we report on the implementation of an inductive theorem prover in the Maude system, whose design is based on the results presented in this paper.

1 Introduction

A logical framework is a logic with an associated methodology that is employed for representing and using other logics, theories, and, more generally, formal systems. A number of logical frameworks have been proposed and to compare them and analyze their effectiveness, it is helpful to distinguish between their intended applications. In particular, we can distinguish between *logical frameworks*, where the emphasis is on reasoning *in* a logic, in the sense of simulating its derivations in the framework logic, and *metalogical frameworks*, where the emphasis is on reasoning *about* logics and even about relationships *between* logics. Metalogical frameworks are more powerful, as they include the ability to reason about a logic's entailment relation, as opposed to merely being adequate to simulate entailment.

Induction plays a central rôle in distinguishing logical frameworks from their metalogical counterparts. In a logical framework, representations of proof rules are used to construct derivations of (object logic) entailments. This approach is taken in logical frameworks like Isabelle [34] and the Edinburgh LF [22]. There,

S. Kapoor and S. Prasad (Eds.): FST TCS 2000, LNCS 1974, pp. 55–80, 2000.

one may formalize logics and theories where induction is present *within* the theory (e.g., Peano Arithmetic), but induction is not present *over* the theories. That is, the framework does not support induction over the terms and proofs of a theory. In contrast, in a metalogical framework, it is essential to have induction over theories. Standard proof-theoretic arguments usually require induction over the formulae or derivations of the object theory. Induction is essential too for computer science applications, like reasoning about operational semantics.

1.1   Reflective Metalogical Frameworks

In this paper, we propose a new approach to metalogical frameworks motivated by the following observation. A logic's syntax and proofs can be viewed as algebras, whose carrier sets are inductively built from syntax and proof constructors. A logical framework and a metalogical framework can share these as a common basis. However, whereas for a logical framework the application of these constructors suffices to simulate derivations of the object logic, for a metalogical framework, our representation must additionally preserve the inductive nature of these algebras. That is, a formalization in the metalogic should have an *initial model* corresponding to the syntax and proofs of the formalized object logic.

Our proposal is that for some logical frameworks—namely, those that are reflective and whose theories have initial models—we can take the step from a logical framework to a metalogical framework by reflecting at the metalevel the induction principles for the formalized logics. We sum this up with the slogan *"logical frameworks with reflection and initiality are metalogical frameworks"*.

After making this idea precise, we give a concrete realization of it using rewriting logic and present an example. Our example is a standard one in metareasoning: the deduction theorem for *minimal logic (of implication)*. Rewriting logic is not the only candidate for a reflective metalogical framework, but we believe it is a good one. Rewriting logic has been demonstrated to be a good logical framework [11,23,24,30,36,37] and it is balanced on a point where it is expressive enough to naturally formalize different entailment systems, but it is weak enough so that its theories always have initial models. This means that there are sound induction principles for reasoning with respect to these models. To prove metatheorems about theories in rewriting logic and their parameterized extensions, the key is to *reflect* these reasoning principles at the metalevel.

Overall, we see our contributions as both theoretical and practical. Theoretically, our work contributes to answering the question *"what is a metalogical framework?"* by proposing reflective logical frameworks, whose theories have initial models, as a possible answer. Moreover, it illuminates the interrelationship between logical and metalogical frameworks, and the rôle of reflection as a key ingredient for turning a logical framework with initial models into a metalogical one. Practically, we provide evidence that rewriting logic, combined with reflection, is an effective metalogical framework that can be used for nontrivial kinds of metatheoretic reasoning.

1.2   Related Work

Various approaches have been considered in the past to strengthen logical frameworks so that they can function as metalogical frameworks. All of these differ significantly from our proposal both in their logical basis and in the rôle of reflection in metareasoning.

One approach is to formalize theories in a framework logic supporting some notion of module, where each module is explicitly equipped with its own induction principle. For example, in [3], theories were formalized by collections of parameterized modules (Σ-types) within the Nuprl type theory (a constructive, higher-order logic), and each module included its own induction principle for reasoning about terms or proofs. This approach is powerful and can be used, for example, to relate different theories formalized in this way.

An alternative approach is to formalize theories directly using inductive definitions in a framework logic or framework theory that is strong enough to formalize the corresponding induction principles. A simple example of this is the first-order theory FS_0 of [19], which has been used by [25] to carry out experiments in formal metatheory. In FS_0, inductive definitions are terms in the framework theory, which has an induction rule for reasoning about such terms.

Another common choice is to formalize theories as inductive definitions in strong "foundational" framework logics such as higher-order logic or set-theory [21,33], or in a type theory like the calculus of constructions with inductive definitions [32]. In higher-order logic and set theory one can internally develop a theory of inductive definitions, where inductive definitions correspond to terms in the metatheory (e.g., formalized as the least fixedpoint of a monotonic function) and, from the definition, induction principles are formally derived within the framework logic. Alternatively, in the calculus of constructions, given an inductive definition, induction principles are simply added, soundly, to the metalogic. Current research in this area focuses on appropriate induction principles for logics that support higher-order abstract syntax [17,26,35].

Organization

The remainder of our paper is organized as follows. In Section 2 we present the idea of a reflective metalogical framework and abstractly formalize our requirements for such a metalogic. In Section 3 we present background material on rewriting logic, membership equational logic, and the Maude language. In Section 4 we discuss induction principles for membership equational theories and present a simple notion of parameterized membership equational theory. In Section 5 we discuss how rewriting logic can be used as a logical framework, and in Sections 6 and 7 we show how to combine initiality and reflection to use rewriting logic as a metalogical framework. In particular, we show how to reflect, in a uniform way, induction principles for reasoning, at the metalevel, about theories and their parameterized extensions. After this, we present in Section 8 an example of formal metareasoning using rewriting logic as a metalogical framework, namely the proof of the deduction theorem, and we draw conclusions in Section 9.

2   Reflective Metalogical Frameworks

In this section we begin by defining reflective logics. Based on this, we then describe properties sufficient for a reflective logical framework to function as a reflective metalogical framework.

2.1   Reflective Logics

Intuitively, a reflective logic is a logic in which important aspects of its metatheory, such as theories and entailment, can be represented and reasoned about *in* the logic. A general axiomatic notion of reflective logic was recently proposed in [7,14]. The notion is itself expressed in terms of the more general axiomatic notion of an *entailment system* [27], which captures the entailment relation of a logic. For our purposes here, an entailment system $\mathcal{E}$ consists of the following:

1. a class *Sign* of *signatures*, where each signature $\Sigma \in Sign$ specifies the syntax of a language;
2. a function *sen* assigning to each signature $\Sigma \in Sign$ a set $sen(\Sigma)$ of its sentences;
3. for each signature $\Sigma \in Sign$, an *entailment relation* $\vdash_\Sigma$, where $\vdash_\Sigma \subseteq \mathcal{P}(sen(\Sigma)) \times sen(\Sigma)$ and $\vdash_\Sigma$ satisfies the properties of reflexivity, monotonicity, and transitivity (or cut); in what follows, we omit the subscript of $\vdash_\Sigma$ when Σ is clear from the context.

A *theory* in $\mathcal{E} = (Sign, sen, \vdash)$ is then a pair $T = (\Sigma, \Gamma)$ consisting of a signature $\Sigma \in Sign$ and a set of sentences $\Gamma \subseteq sen(\Sigma)$. We can extend the entailment relation to theories in the obvious way by defining $(\Sigma, \Gamma) \vdash \varphi$ iff $\Gamma \vdash \varphi$, for $\varphi \in sen(\Sigma)$.

Definition 1 *Given an entailment system $\mathcal{E}$ and a nonempty set of theories $\mathcal{C}$ in it, a theory U is $\mathcal{C}$-universal if there is a function, called a* representation function,

$$\overline{(\_\vdash\_)} : \bigcup_{T \in \mathcal{C}} (\{T\} \times sen(T)) \longrightarrow sen(U),$$

such that for each $T \in \mathcal{C}, \varphi \in sen(T)$,

$$T \vdash \varphi \ \text{ iff } \ U \vdash \overline{T \vdash \varphi}. \tag{1}$$

If, in addition, $U \in \mathcal{C}$, then the entailment system $\mathcal{E}$ is called $\mathcal{C}$-reflective. Finally, a reflective logic *is a logic whose entailment system is $\mathcal{C}$-reflective for $\mathcal{C}$, the class of all finitely presentable theories in the logic.*

2.2   Requirements for a Reflective Metalogical Framework

We now consider what we require from a logical framework so that it can function as a metalogical framework. As indicated in Section 1.2, various approaches to formal metareasoning have been proposed in the past. Our approach is based on reflective reasoning and initiality, and here we present, abstractly our requirements for this. They are:

1. the logical framework is *weak* enough so that there are valid induction principles for reasoning about all its theories,
2. the logical framework is *expressive* enough so that it really is a viable logical framework, and
3. the logical framework is *reflective*.

Note that 1 specifies a requirement on the framework logic and can be alternatively formulated in an abstract and logic-independent way. If the framework logic is such that its theories have initial models, then an appropriate form of inductive reasoning is always valid when proving sentences with respect to the initial models of its theories. This method is very general; for example, for equational logic, induction and initiality are equivalent concepts [31].[1]

We now explain why the requirements listed above are sufficient for turning a logical framework into a metalogical framework. If requirement 2 is satisfied, then logics and their entailment relations can be represented as theories in the logical framework, and if requirement 1 is also satisfied, then these representations can preserve the inductive nature of the algebras characterizing the syntax and proofs of the logics that they represent. As a consequence, proof-theoretic arguments requiring induction over the formulae or over the derivations of a logic can be applied in the framework logic. This is enough when proving theorems about a logic. These theorems can be formalized as sentences about the initial model of the theory representing the object logic under consideration, and can be proved by induction.

However, when dealing with metatheorems we often require something more. Metatheorems may relate *different* logics in a family of logics. Consider, for example, the deduction theorem for minimal logic (of implication). This is actually a metatheorem not about a particular deduction system, but rather a metatheorem that relates different deduction systems: one in which $A \to B$ is proved and a second (which is obtained from the first by adding the axiom A) in which B is proved. In our setting, this means that sentences formalizing metatheorems should relate initial models of different theories. Here is where reflection plays a decisive rôle. Namely, if the logical framework satisfies requirement 3, then: (3a)

---

[1] Since the notion of initiality is very general, the corresponding inductive reasoning principles may in each case take different forms. For example, in an equational logic allowing infinitary operations of arity smaller than a given regular cardinal α, the inductive principles will be transfinite. We are mainly interested in logical frameworks suitable for representing *finitary* logics. Therefore, we will in practice be interested in a finitary framework logic whose theories have initial models and whose induction principles are also finitary.

it contains a universal theory where the metalevel of its logic can be reflected, and (3b) the universal theory is itself a theory in the logical framework. By 3b and 1 the universal theory has an initial model. The key then is to exploit 3a in order to formalize relationships between the initial models of object theories as theorems about the initial model of the universal theory, and turn, by reflection, induction principles for reasoning about the initial models of object theories into induction principles for reasoning about the initial model of the universal theory.

In the following sections, we will give a concrete instance of these ideas for the case of rewriting logic. In particular, we will show that for a certain class of rewriting logic theories the induction principles for reasoning, at the metalevel, about these theories correspond, in a simple way, to the induction principles for reasoning about the inductive properties of the theories. Moreover, by reasoning by induction in the universal theory we can inductively reason about properties satisfied by families of theories. This provides us with capabilities analogous to what is possible in metalogical frameworks based on parameterized inductive definitions.

3   Background

In this section we provide background material on rewriting logic, membership equational logic, and the Maude language. The material presented here is standard. We postpone discussion of the reflective aspects to Section 6.

3.1   Rewriting Logic

Rewriting logic [28] is a simple logic whose sentences are sequents of the form $t \longrightarrow t'$, with t and t' Ω-terms on a given signature Ω. Theories in rewriting logic are triples (Ω, E, R), with Ω a signature of operators, E a set of Ω-equations, and R a collection of (possibly conditional [28]) Ω-rewrite rules.

The inference rules of rewriting logic [28] allow the derivation of all rewrites possible in a given theory. Thus, from the logical point of view, we can think of rewriting logic as a framework logic in which formulae are formalized as elements of the initial model of an equational theory (Ω, E) and an inference system is formalized by expressing each inference rule as a (possibly conditional) rewrite rule. Rewriting is understood *modulo* the equations E. This supports a flexible and abstract kind of inference where the equations can take care of structural bookkeeping. For example, when formalizing sequent calculi, structural rules for sequents can be "internalized" by rewriting modulo appropriate equational axioms such as associativity, associativity-commutativity, and so on.

Since a rewrite theory (Ω, E, R) has an underlying equational theory (Ω, E), rewriting logic is parameterized by the choice of the equational logic. An attractive choice in terms of expressiveness is *membership equational logic* [29], a logic that has sorts, subsorts, overloading of function symbols, and is capable of expressing partiality using equational conditions. Since we can view an equational theory (Ω, E) as a rewrite theory $(\Omega, E, \emptyset)$, there is an obvious sublogic inclusion, $MEqtl \subseteq RWLogic$, from membership equational logic into rewriting logic. Both membership equational logic and rewriting logic have initial models [28,29], which provide the basis for reasoning by induction.

3.2 Membership Equational Logic

Membership equational logic is an expressive version of equational logic. A full account of the syntax and semantics of membership equational logic can be found in [6,29]. Here we define the basic notions needed in this paper.

A *signature* in membership equational logic is a triple $\Omega = (K, \Sigma, S)$ with K a set of *kinds*, Σ a K-kinded signature $\Sigma = \{\Sigma_{w,k}\}_{(w,k) \in K^* \times K}$, and $S = \{S_k\}_{k \in K}$ a pairwise disjoint K-kinded family of sets. We call S_k the set of *sorts* of kind k. The pair (K, Σ) is what is usually called a many-sorted signature of function symbols; however we call the elements of K kinds because each kind k now has a set S_k of associated *sorts*, which in the models will be interpreted as subsets of the carrier for the kind. Also, as usual, we denote by T_Σ the K-kinded algebra of ground Σ-terms, and by $T_\Sigma(X)$ the algebra of Σ-terms on the K-kinded set of variables X.

The atomic formulae of membership equational logic are either equations $t = t'$, where t and t' are Σ-terms of the same kind, or *membership assertions* of the form $t : s$, where the term t has kind k and $s \in S_k$. Sentences are Horn clauses on these atomic formulae, i.e., sentences of the form

$$\forall(x_1, \ldots, x_m).\, A_1 \wedge \ldots \wedge A_n \Rightarrow A_0 \,,$$

where each A_i is either an equation or a membership assertion, and each x_j is a K-kinded variable. For example, Figure 1 gives a set of membership equational axioms specifying minimal logic of implication, where SentConstant, Formula, and Theorem are sorts formalizing sentential constants, formulae, and theorems, respectively.[2] A theory in membership equational logic is a pair (Ω, E), where E is a finite set of sentences in membership equational logic over the signature Ω. The way in which partiality is expressed in membership equational logic is by the fact that terms always have a kind, but may not have a sort. Terms for which a sort cannot be established from the axioms E correspond to *undefined* or *error* elements.

We employ standard semantic concepts from many-sorted logic. Given a signature $\Omega = (K, \Sigma, S)$, an Ω-*algebra* is a many-kinded Σ-algebra (that is, a K-indexed-set $A = \{A_k\}_{k \in K}$ together with a collection of appropriately kinded functions interpreting the function symbols in Σ) together with an assignment to each sort $s \in S_k$ of a subset $A_s \subseteq A_k$. Hence, sorts can be thought of as unary predicates that semantically denote subsets of the appropriate kind. An algebra A and a (kind-respecting) valuation σ, assigning to variables of kind

---

[2] Note that we write the object logic connective $\rightarrow$ in infix. We will consider this example in more detail in Section 5.

$\forall A.\, A\!:\!\mathtt{SentConstant} \Rightarrow A\!:\!\mathtt{Formula},$
$\forall A.\, A\!:\!\mathtt{Theorem} \Rightarrow A\!:\!\mathtt{Formula},$
$\forall (A, B).\, A\!:\!\mathtt{Formula} \wedge B\!:\!\mathtt{Formula} \Rightarrow A{\rightarrow}B\!:\!\mathtt{Formula},$
$\forall (A, B).\, A\!:\!\mathtt{Formula} \wedge B\!:\!\mathtt{Formula} \Rightarrow A{\rightarrow}(B{\rightarrow}A)\!:\!\mathtt{Theorem},$
$\forall (A, B, C).\, A\!:\!\mathtt{Formula} \wedge B\!:\!\mathtt{Formula} \wedge C\!:\!\mathtt{Formula}$
$\quad \Rightarrow (A{\rightarrow}B){\rightarrow}((A{\rightarrow}(B{\rightarrow}C)){\rightarrow}(A{\rightarrow}C))\!:\!\mathtt{Theorem},$
$\forall (A, B).\, A\!:\!\mathtt{Formula} \wedge B\!:\!\mathtt{Formula} \wedge (A{\rightarrow}B)\!:\!\mathtt{Theorem} \wedge A\!:\!\mathtt{Theorem}$
$\quad \Rightarrow B\!:\!\mathtt{Theorem}$

Fig. 1. Membership equational axioms for minimal logic.

k values in A_k, satisfy an equation $t = t'$ iff $\sigma(t) = \sigma(t')$, where we overload notation by identifying σ with its unique homomorphic extension to Σ-terms. We write $A, \sigma \models t = t'$ to denote such a satisfaction. Similarly, $A, \sigma \models t\!:\!s$ holds iff $\sigma(t) \in A_s$.

Note that an Ω-algebra is nothing but a K-kinded first-order model with function symbols Σ and an alphabet of unary predicates $\{S_k\}_{k \in K}$. Therefore, the satisfaction relation can be extended to Horn and first-order formulae ϕ over these atomic formulae in the standard way. We write $A \models \phi$ when the formula ϕ is satisfied for all valuations σ, and then say that A is a model of ϕ. Similarly, a theory (Ω, E) in membership equational logic is simply a Horn theory for the associated signature, when Ω is viewed as first-order K-kinded signature. As usual, for ϕ a first-order sentence in the language of Ω, we write $(\Omega, E) \models \phi$ when all the models of the set E of sentences are also models of ϕ.

Theories in membership equational logic have initial models [29]. This provides the basis for reasoning by induction, as is explained in detail in Section 4.1. We write $(\Omega, E) \models \phi$ to denote that the initial model of the membership equational logic theory (Ω, E) is also a model of ϕ. Note that even though we restrict the axioms E to Horn clauses, we will employ first-order formulae ϕ to formalize properties satisfied by the initial model, that is, inductive properties.

3.3   The Maude System

The Maude system [9,13] implements rewriting logic and has been designed with the explicit aims of supporting executable specification and reflective computation. Theories are specified in Maude by modules, of which there are two kinds: *functional modules* and *system modules*. Maude's functional modules are theories in membership equational logic. Equations in Maude's functional modules are assumed to be Church-Rosser and terminating; they are executed by the Maude rewrite engine according to the rewriting techniques and operational semantics developed in [6]. Maude's system modules are rewrite theories. The rules in a system module are not necessarily Church-Rosser or terminating.

The semantics of a functional (respectively system) module is *initial*, i.e., such a module denotes the *initial model* in membership equational logic (respectively rewriting logic) of the theory thus specified. The syntax for functional modules

```
fmod MINIMAL is
sorts SentConstant Formula Theorem .
subsort SentConstant < Formula .
subsort Theorem < Formula .
op _→_ : Formula Formula -> Formula .
vars A B C : Formula .
mb A → (B → A) :  Theorem .
mb (A → B) → ((A →(B → C)) → (A → C)) : Theorem .
cmb B : Theorem if (A → B) : Theorem and  A : Theorem .
endfm
```

Fig. 2. The module MINIMAL.

is of the form fmod (Ω, E) endfm, with (Ω, E) a membership equational theory meeting the requirements mentioned above. Figure 2 gives an example of a functional module in Maude syntax, where LATEX symbols are used instead of ASCII characters to improve readability. Note that Maude's syntax for functional modules is syntactic sugar for introducing finite sets of membership axioms (Figure 2 is just the sugared version of Figure 1), and we will use it from now on to present membership equational theories. In particular, (possibly conditional) equations and membership axioms in Maude are Horn clauses in membership equational logic; any operation declaration op $f : s_1 \ldots s_n$ -> s corresponds to the Horn clause

$$\forall(x_1, \ldots, x_n). x_1 : s_1 \wedge \ \ldots \ \wedge x_n : s_n \Rightarrow f(x_1, \ldots, x_n) : s,$$

where x_i is a variable of kind k_i and $s_i \in S_{k_i}$, for $i \in \{1, \ldots, n\}$; also, any subsort declaration subsort s < s' can be reduced to the sentence

$$\forall x. x : s \Rightarrow x : s',$$

where x is a variable of kind k and $s, s' \in S_k$. Finally, kinds are not explicitly defined in Maude modules, but are instead inferred by the system as determined by the different connected components of the poset of sorts.

As additional syntactic sugar, we shall often write $\forall x : s. \phi(x)$ as shorthand for the formula $\forall x. x : s \Rightarrow \phi(x)$, for x a variable of kind k and $s \in S_k$. Moreover, for the formula $x : s \Rightarrow \phi(x)$, we will say that "$x$ is of sort s (in ϕ)."

4   Induction and Parameterization

In this section we introduce two concepts that play key rôles in the rest of the paper. We define an induction principle for membership equational theories and show how such theories can also be parameterized. We introduce these concepts in a simple setting that is adequate to illustrate the main ideas and carry out applications.

4.1   Induction Principles for Membership Equational Theories

Given that membership equational logic *is* a subset of equational Horn logic (indeed, they can be shown to be equivalent [29]) it follows immediately that any theory (Ω, E) has a unique (up to isomorphism) initial model [20]. The following is an induction principle for reasoning about properties of sorts, with respect to this model.

Definition 2 (Induction over sort definitions) *Let $T = (\Omega, E)$ be a theory in membership equational logic and let s be a sort in some S_k. Let $C_{[T,s]} = \{C_1, \ldots, C_n\}$ be those sentences in E that specify s, i.e., those C_i of the form*

$$\forall(x_1, \ldots, x_{p_i}). A_1 \wedge \ldots \wedge A_{q_i} \Rightarrow A_0 \,, \tag{2}$$

where, for some t of kind k and $s \in S_k$, A_0 is $t : s$.

For τ a first-order formula with free variable x of sort s over the signature Ω, an induction principle *for (Ω, E), with respect to $x : s$ and $\tau(x)$ is the formula*

$$\psi_1 \wedge \ldots \wedge \psi_n \Rightarrow \forall x : s. \, \tau(x) \tag{3}$$

where, for $1 \leq i \leq n$ and C_i of the form (2), ψ_i is

$$\forall(x_1, \ldots, x_{p_i}). [A_1]_\tau \wedge \ldots \wedge [A_{q_i}]_\tau \Rightarrow [A_0]_\tau \tag{4}$$

and, for $0 \leq j \leq q_i$,

$$[A_j]_\tau = \begin{cases} \tau(u) & \text{if } A_j = u : s, \text{ for } u \text{ of kind } k \\ A_j & \text{otherwise} \end{cases}$$

For a given membership equational theory (Ω, E), the above defines an induction schema (ind), given by (3), in many-kinded first-order logic over the signature Ω.[3] Note that for $q_i = 0$, the nullary conjunction in the antecedent of (4) is *true* and the implication can be replaced with the succedent $\tau(t)$.

In the initial model of a membership equational theory, sorts are interpreted as the smallest sets satisfying the axioms in the theory, and equality is interpreted as the smallest congruence satisfying those axioms. Alternatively, the sets interpreting sorts can be characterized as being inductively generated in stages. This corresponds to the fixedpoint characterization of the least Herbrand model of a collection of Horn clauses [38], and the induction principle we have given formalizes induction over the stages in which the set is inductively defined [1]. By induction over the stages of the inductive definition of a sort s, which amounts to an induction over the proof that some ground term of kind k is of sort s, we can establish that reasoning in the membership equational theory (Ω, E), augmented by (ind), is sound.

---

[3] This induction schema cannot be directly formalized in the sublogic membership equational logic, since it is not, in general, a sentence in membership equational logic. However, as we will later see, inference rules for (many-kinded) first-order theories—like this induction schema—can be encoded in rewriting logic and can be used to prove properties, at the metalevel, about membership equational theories.

Theorem 1 (Soundness) *Let* (Ω, E) *be a membership equational theory. If* $(\Omega, E \cup \{(ind)\}) \vdash \tau$, *then* $(\Omega, E) \models \tau$.

As an example, consider the membership equational theory for minimal logic previously given in Figure 1. Definition 2 gives rise to the following induction principle over the sort `Theorem`:

$[\forall(A, B). (A\!:\!\texttt{Formula} \wedge B\!:\!\texttt{Formula} \Rightarrow \tau(A{\rightarrow}(B{\rightarrow}A))) \wedge$
$\forall(A, B, C). (A\!:\!\texttt{Formula} \wedge B\!:\!\texttt{Formula} \wedge C\!:\!\texttt{Formula}$
$\quad \Rightarrow \tau((A{\rightarrow}B){\rightarrow}(A{\rightarrow}(B{\rightarrow}C)){\rightarrow}(A{\rightarrow}C))) \wedge$
$\forall(A, B). (A\!:\!\texttt{Formula} \wedge B\!:\!\texttt{Formula} \wedge \tau(A{\rightarrow}B) \wedge \tau(A) \Rightarrow \tau(B))]$
$\quad\quad \Rightarrow \forall A\!:\!\texttt{Theorem}.\, \tau(A)$

This axiom formalizes induction over the structure of proofs in minimal logic.

Note that other induction principles are possible. In particular *(ind)* takes *all* of the sentences that specify membership in *s* as constituting the inductive definition of *s*. In some cases, a subset of the sentences (sometimes called *generators* or *constructors*) is sufficient to characterize an inductive definition. Of course additional proof obligations then arise, e.g., sufficient completeness of the chosen subset (see [6,10]).

4.2   Parameterized Membership Equational Theories

When carrying out metalogical reasoning, we often reason not about a fixed theory, but about a *parameterized family* of theories. There are many different ways in which a theory in membership equational logic may be parameterized. For the purposes of this paper it will be enough to consider the notion of a *parameterized extension* of a given theory T which, intuitively, describes the extensions of T by a parametric set of new axioms.

Definition 3 *Let* $T = (\Omega, E)$ *be a theory in membership equational logic with* $\Omega = (K, \Sigma, S)$. *Then, a* parameterized extension *of* T *(by parameters* V, *and axioms* G*) is a membership equational theory* $T_G[V] = (\Omega[V], E \cup G)$, *with* $\Omega[V] = (K, \Sigma \cup V, S)$, *where the* K*-kinded signatures* Σ *and* V *are mutually disjoint and* V *consists only of constants. We call* $T_G[V]$ *a parameterized membership equational theory.*

Let $\beta : V \longrightarrow T_\Sigma$ *be a* K*-kinded function. Then,* $T_G[\beta] = (\Omega, E \cup \beta(G))$ *denotes an* instance *of* $T_G[V]$, *where* $\beta(G)$ *is the homomorphic extension of* β *to axioms.*

The substitution β is used to generate instances of a parameterized membership equational theory. Namely, the new axioms G are instantiated so that all instances of variables in V are replaced by ground terms. For example, if $G = \{f(v) : s\}$ and $\beta(v) = q(a, b)$, then the parametric axiom $f(v) : s$ is translated as $\beta(f(v) : s) = f(q(a, b)) : s$. The result is well-kinded under the (non-parameterized) signature Ω.

```
fmod MINIMAL Ξ[X] is
sorts SentConstant Formula Theorem .
subsort SentConstant < Formula .
subsort Theorem < Formula .
op  X : -> Formula .
op _→_ : Formula Formula -> Formula .
vars A B C : Formula .
mb  X :  Theorem .
mb A → (B → A) :  Theorem .
mb (A → B) → ((A →(B → C)) → (A → C)) : Theorem .
cmb B : Theorem if (A → B) : Theorem and  A : Theorem .
endfm
```

Fig. 3. The module $\text{MINIMAL}_\Xi[X]$.

Figure 3, provides an example of a parameterized module $\text{MINIMAL}_\Xi[X]$, with X a parameter of the kind of the sort `Formula`, and Ξ a set that contains only the parametric axiom X:`Theorem`. We will later see how this parameterized module can be used to formalize the deduction theorem.

5   Rewriting Logic as a Logical Framework

As we have already said, from the logical point of view we can think of rewriting logic as a framework logic in which an inference system can be formalized by expressing each inference rule as a (possibly conditional) rewrite rule in a rewrite theory (Ω, E, R). Note that rewriting logic is noncommittal about the structure and properties of the formulae expressed by Ω-terms. They are user-definable as an algebraic data type satisfying equational axioms, so that rewriting deduction takes place modulo such axioms. Because of this ecumenical neutrality and the simplicity of the rules of the logic, rewriting logic can be effectively applied as a logical framework. In [11,23,24,30,36,37], many examples of logic representations are given, including first-order linear logic, sequent presentations of modal and propositional logics, Horn logic with equality, the lambda calculus, and higher-order pure type systems, among others. In all such examples, the *representational distance* between the object logic and its representation in rewriting logic is virtually zero, that is, the representations are direct and reasoning with them faithfully simulates reasoning in the original logics.

In fact, there are several ways of conservatively representing a logic (with a finitary syntax and inference system) in rewriting logic. As mentioned before, a simple and direct way is to turn the inference rules into rewrite rules, which may be conditional if the inference rules have side conditions. Alternatively, we can use the underlying membership equational logic to represent theoremhood in a logic as a sort in a membership equational theory. Conditional membership axioms then directly support the representation of rules as schemas, which is typically used in presenting logics and formal systems. The module `MINIMAL`, presented previously in Figure 2, represents minimal logic in membership equational logic using this idea. A formula A is a theorem in minimal logic if and only if A is a term of sort Theorem in MINIMAL. Note that this representation preserves the inductive nature of the set of theorems and proofs in minimal logic.

Similarly, we can represent theoremhood in a parameterized family of logics as a sort in a parameterized membership equational theory.[4] As an example, consider the parameterized theory MINIMAL$_\Xi$[$\mathcal{X}$] in Figure 3, with $\mathcal{X}$ a parameter of the kind of the sort Formula, and Ξ a set that contains only the parametric axiom $\mathcal{X}$:Theorem. This parameterized theory represents the family of logics that includes any extension of minimal logic with a new axiom in the following sense: a formula B is a theorem in minimal logic extended with a new axiom A if and only if B is a term of sort Theorem in MINIMAL$_\Xi$[A], where MINIMAL$_\Xi$[A] is the instance MINIMAL$_\Xi$[β] of MINIMAL$_\Xi$[$\mathcal{X}$], with $\beta(\mathcal{X}) = A$.

The ability to represent parameterized families of logics is important for using rewriting logic also as a metalogical framework, and we will give an example of this in the experimental work reported on in Section 8.

6   Reflection in Rewriting Logic and Maude

In this section we explain how rewriting logic is reflective and how the Maude system implements reflective rewriting logic deduction. We also introduce a Boolean function that reflects at the metalevel the membership relation in membership equational logic, which will be used in later sections. Finally, we explain how to combine the use of rewriting logic as a logical framework and the reflective capabilities of Maude to build a theorem prover for carrying out inductive proofs.

6.1   Reflection in Rewriting Logic

Rewriting logic is reflective [7,15,16]. There is a universal theory UNIVERSAL, and a representation function $(\_ \vdash \_)$ encoding pairs consisting of a rewrite theory T and a sentence in it as sentences in UNIVERSAL. For any finitely presented rewrite theory T (including UNIVERSAL itself) and any terms t, t' in T, the representation function is defined by

$$\overline{T \vdash t \longrightarrow t'} = \langle \overline{T}, \overline{t} \rangle \longrightarrow \langle \overline{T}, \overline{t}' \rangle \,,$$

where $\overline{T}$, $\overline{t}$, $\overline{t}'$ are terms in UNIVERSAL. Then, the equivalence (1) in Section 2 holds for rewriting logic (as proved in [7,15,16]) and takes the form

$$T \vdash t \longrightarrow t' \text{ iff UNIVERSAL} \vdash \langle \overline{T}, \overline{t} \rangle \longrightarrow \langle \overline{T}, \overline{t}' \rangle \,.$$

---

[4] A sort in a parameterized membership equational theory can be used to represent theoremhood in a family of logics if and only if there is a one–to–one correspondence between logics in the family and instances of the parameterized membership equational logic theory, and this correspondence is such that theoremhood in a logic in the family can be represented as membership in this sort in the corresponding instance of the parameterized membership equational logic.

6.2   Reflection in Maude

Maude's language design and implementation make systematic use of the fact that rewriting logic is reflective to give the user a well-defined gateway to the metatheory of rewriting logic. This entry point is the predefined module META-LEVEL, which is a (partial) specification in Maude of UNIVERSAL [8]. In the module META-LEVEL, a Maude term t is reified as an element $\bar{t}$ of a data type Term of terms, and a Maude module T, i.e., a membership equational theory, is reified as a *ground term* $\overline{T}$ in a data type Module of modules. See [8] for a complete definition of the module META-LEVEL and of the metarepresentation map for theories and terms.

The metarepresentation of a parameterized membership equational theory $T_G[V]$ is similar to that of an unparameterized theory T, except that parameters are treated in a special way. The *parametric* nature of $T_G[V]$ is expressed in its metarepresentation $\overline{T_G[V]}$ by the fact that each parameter $v \in V$ is represented by a (meta-) variable $\bar{v}$ of sort Term.

Note that since, in Definition 3, a parameterized theory $T_G[V]$ is technically defined as an ordinary membership equational theory (plus some extra information), one *could* metarepresent $T_G[V]$ as an ordinary theory, and then one would get a ground term of sort Module, instead of a term with variables. Therefore, our notation $\overline{T_G[V]}$ is potentially confusing, since it depends on whether $T_G[V]$ is metarepresented as a parameterized entity or as an unparameterized one. Rather than introducing new notation, we have chosen to solve the possible ambiguities by the context in which they occur. In particular, having introduced $T_G[V]$ as a parameterized membership equational theory, $\overline{T_G[V]}$ will always denote the metarepresentation of $T_G[V]$ as a parameterized entity. The same rule applies when metarepresenting parameterized terms in parameterized membership equational theories.

To reason about metarepresented (parameterized) theories we have defined, in an extension META-IND of META-LEVEL, a Boolean function ($\_:\_in\_$) that reflects at the metalevel the membership relation in membership equational logic. (The theory is so named because it is in this theory that we will prove inductive metatheorems.) In particular, ($\bar{t}:\bar{s}$ in $\overline{T}$) checks, at the metalevel, whether the ground term t has the sort s in the functional module T. Specifically, this check of membership is based on the equivalence

$$T \vdash t:s \quad \Longleftrightarrow \quad \text{META-IND} \vdash \bar{t}:\bar{s} \text{ in } \overline{T} = \texttt{true}.$$

In [2] we give the full specification of ($\_:\_in\_$) for a restricted subclass $\mathcal{C}$ of modules. Members of $\mathcal{C}$ are modules that correspond, basically, to membership equational theories whose axioms are Horn clauses that only involve membership assertions (no equations). The Boolean function ($\_:\_in\_$) will play a key rôle in the rest of this paper.

In what follows, $\mathcal{C}$ will always denote the above mentioned subclass of modules. In addition, for any parameterized membership equational theory $T_G[V]$, we write $T_G[V] \in \mathcal{C}$ iff for any instance $T_G[\beta]$ of $T_G[V]$, $T_G[\beta] \in \mathcal{C}$. We also write

$\overline{V}$ for the set of variables of sort `Term` that metarepresent the parameters $v \in V$ in $T_G[V]$.

6.3 Building an Inductive Theorem Prover

In Section 4.1 we have presented induction principles for reasoning about first-order formulae over sorts defined in functional modules, i.e., membership equational theories. Here we explain how to combine the use of rewriting logic as a logical framework and the reflective capabilities of Maude to build a theorem prover for carrying out inductive proofs. The paper [10] provides further details on building theorem proving tools in Maude.

To build an inductive theorem prover, we use rewriting logic to specify its inference system (as explained in Section 5) and reflection to define strategies that control rule application. Strategies are needed here since inference rules will be specified as rewrite rules that are not necessarily Church-Rosser or terminating. Hence, it is important to have some way of controlling the application of these rewrite rules in order to drive rewriting in some desired direction. Maude users can control rewriting by specifying, at the metalevel, their own rewriting strategies. (See [7,9] for more details on defining strategies in Maude.)

Our inductive theorem prover—the ITP tool[5]—has a reflective design. The functional module T, about which we want to prove inductive theorems, is at the object level. An inference system $\mathcal{I}$ for inductive proofs uses T as data and therefore is specified as a system module ITP-RULES at the metalevel. In particular, ITP-RULES encodes syntax and proof rules for first-order logic as well as the induction over sort definitions introduced in Definition 2. Finally, different *proof tactics* to guide the application of the rewrite rules specifying the inference rules in $\mathcal{I}$ are strategies, which are defined at the meta-metalevel in a module ITP-TACTICS.

Operationally, to use the ITP tool, the user submits as an initial goal the pair formed by (the metarepresentation) of a functional module and (the representation of) the first-order sentence over its signature that is to be proved, and then this goal is successively transformed by rewriting—using the inference rules as rewrite rules—into different sets of subgoals, until (in the case of a successful proof) no subgoals are left. The application of the inference rules as rewrite rules is controlled by the user using strategies.

Finally, note that building the theorem prover using different levels of reflection results in a modular design with a clean separation between the logical and the control components. For example, we can simply extend the tool by specifying additional inference rules in the module ITP-RULES without changing the strategy language defined in the module ITP-TACTICS and vice versa.

---

[5] `http://sophia.unav.es/~clavel` contains the most recent version of this tool.

7 Rewriting Logic as a Metalogical Framework

We now show how the induction principles introduced in Section 4.1 for reasoning in membership equational theories can be uniformly reflected for reasoning, at the metalevel, about membership equational theories and their parameterized extensions. The results presented in this section provide the basis for using rewriting logic as a metalogical framework.

7.1 Inductive Theorems versus Inductive Metatheorems

The induction principle presented in Section 4.1 is well-suited for proving properties of ground terms of sort s in a given membership equational theory (Ω, E), with $\Omega = (K, \Sigma, S)$, and s a sort in some S_k, *when these properties are expressible as first-order formulae over the signature* Ω; when this is the case, a property P holds if the first-order formula that expresses P holds in the initial model of (Ω, E). Of course, there are many interesting properties satisfied by ground terms of sort s that cannot be expressed as first-order formulae over the signature Ω, despite the fact that they are typically proved by induction over the definition of the sort s, e.g., properties that *relate different membership equational theories*. Many of these properties can be naturally expressed, at the metalevel, as first-order formulae over the signature of META-IND. Consider, for example, the following property: let $T = (\Omega, E) \in \mathcal{C}$ and $T' = (\Omega', E') \in \mathcal{C}$ be membership equational theories; then the property

> **if** t *is a ground term of sort s in T,*
> **then** t *is also a ground term of sort s' in T'*

is not expressible as a first-order formula over either T or T'. Notice that, using the Boolean function (_:_in_), we can express this as a first-order formula at the metalevel, namely,

$$\forall x : \text{Term.} \, (x{:}\overline{s} \ \text{in} \ \overline{T} = \text{true} \Rightarrow x{:}\overline{s}' \ \text{in} \ \overline{T'} = \text{true}) . \tag{5}$$

The situation is similar when proving properties of ground terms of sort s for all instances $T_G[\beta]$ of a parameterized theory $T_G[V] \in \mathcal{C}$. Consider the following generalization of the above statement. Let $T_G[V] \in \mathcal{C}$ be a parameterized extension of $T = (\Omega, E)$, with $V = \{v_1, \dots, v_n\}$, and k_i the kind of the parameter v_i. Then we might formalize that for all instances $T_G[\beta]$ of $T_G[V]$

> **if** t *is a ground term of sort s in $T_G[\beta]$,*
> **then** t *is a ground term of sort s' in T'.*

Again, we cannot express this property as a first-order formula over the signature of any particular instance of $T_G[V]$. Notice, however, that, using the Boolean function (_:_in_), we can formalize this as

$$\forall(\overline{V} : \text{Term}, x : \text{Term}). \, ((\overline{v_1}{:}\overline{s}_1 \ \text{in} \ \overline{T} = \text{true} \wedge \dots \wedge \overline{v_n}{:}\overline{s}_n \ \text{in} \ \overline{T} = \text{true})$$
$$\Rightarrow (x{:}\overline{s} \ \text{in} \ \overline{T_G[V]} = \text{true} \Rightarrow x{:}\overline{s}' \ \text{in} \ \overline{T'} = \text{true})) , \tag{6}$$

where, for each parameter $v_i \in V$, s_i is a sort in S_{k_i}.

We claim that instances of both (5) and (6) can be proved by induction in a way that mirrors their expected inductive proofs (we will show this for a particular example in Section 8). To see this, the crucial observation—allowing us to mirror inductive reasoning over a sort s in a theory T (or in a parameterized extension $T_G[V]$) by inductive reasoning over META-IND—is the following. Let s be a sort in T defined by a set of Horn clauses $\{C_1, \dots, C_n\}$. By the definition of the Boolean function $(\_ : \_in\_)$, the set of ground terms u of sort Term such that

$$u : \overline{s} \text{ in } \overline{T} = \text{true} \tag{7}$$

is precisely the set of terms of the form $u = \overline{t}$, for t a ground term of sort s in T, and can be defined inductively by a set of Horn clauses $\{\overline{C}_1, \dots, \overline{C}_n\}$ that reflect, at the metalevel, the set of Horn clauses $\{C_1, \dots, C_n\}$. The idea is that we can then use $\{\overline{C}_1, \dots, \overline{C}_n\}$ to derive an induction rule $(\widetilde{ind})$ to prove metaproperties about the ground terms of sort s in T, in exactly the same way as we obtained the induction rule (ind) from $\{C_1, \dots, C_n\}$ in Definition 2. Since each $\overline{C}_i$ mirrors at the metalevel the corresponding C_i, inductive metareasoning with $(\widetilde{ind})$ also mirrors inductive reasoning with (ind). Notice that, when dealing with parameterized extensions, the resulting induction rule $(\widetilde{ind})$ will have to be universally quantified over the variables representing the parameters.

7.2   Metalevel Inference Rules for Parameterized Theories

We now formalize the above intuitions and introduce a new inference rule for proving a broad class of metatheorems about parameterized membership equational theories. These metatheorems correspond to inductive properties of the initial model of the module META-IND of the general form

$$\forall(\overline{V} : \text{Term}, x : \text{Term}). ((\overline{v}_1 : \overline{s}_1 \text{ in } \overline{T} = \text{true} \land \dots \land \overline{v}_m : \overline{s}_m \text{ in } \overline{T} = \text{true})$$
$$\Rightarrow ((x : \overline{s} \text{ in } \overline{T_G[V]} = \text{true} \land \Phi) \Rightarrow \phi)),$$

where ϕ and Φ range over first-order formulae over the signature of the module META-IND. The formulae (5) and (6) above are instances of this general form. In the case of (5), the nonparameterized theory T constitutes a trivial parameterized extension $T_\emptyset[\emptyset]$ of itself.

The soundness of the inference rule that we introduce is based on the fact that, for any instance $T_G[\beta]$ of $T_G[V]$, the set of terms of sort Term that metarepresent terms of sort s in $T_G[\beta]$ is inductively defined. In essence, the new inference rule reflects at the metalevel the induction principle defined in Definition 2 for reasoning over the terms of sort s in any instance of $T_G[V]$.

First, we define for any parameterized extension $T_G[V] = (\Omega[V], E \cup G) \in \mathcal{C}$, with $\Omega[V] = (K, \Sigma \cup V, S)$, and any sort s in some S_k, a set of clauses $\overline{C}_{[T_G[V], s]}$, that mirrors, at the metalevel, the set of Horn clauses $C_{[T_G[V], s]}$ that inductively define the terms of sort s in $T_G[V]$. (Recall that parameters in V are metarepresented as variables of sort Term.) Then, we define an inference rule for proving certain metatheorems about $T_G[V]$.

Definition 4 *Let $T_G[V] = (\Omega[V], E \cup G) \in \mathcal{C}$ be a parameterized extension, with $\Omega[V] = (K, \Sigma \cup V, S)$, and let s_0 be a sort in some S_k, such that $C_{[T_G[V],s_0]} = \{C_1, \ldots, C_n\}$ is the set of sentences of the form*

$$\forall(x_1, \ldots, x_{p_i}). \, t_1 : s_1 \wedge \ldots \wedge t_{q_i} : s_{q_i} \Rightarrow t_0 : s_0$$

that specify s_0 in $T_G[V]$.

We define $\overline{C}_{[T_G[V],s_0]} = \{\overline{C}_1, \ldots, \overline{C}_n\}$, where, for $1 \leq i \leq n$, $\overline{C}_i$ is

$$\begin{aligned}
\forall(x_1, \ldots, x_{p_i}).& \\
\hat{t}_1 : \overline{s}_1 \text{ in } & \overline{T_G[V]} = \texttt{true} \, \wedge \ldots \wedge \hat{t}_{q_i} : \overline{s}_{q_i} \text{ in } \overline{T_G[V]} = \texttt{true} \\
\Rightarrow \hat{t}_0 : \overline{s}_0 & \text{ in } \overline{T_G[V]} = \texttt{true},
\end{aligned}$$

where $\{x_1, \ldots, x_{p_i}\}$ are variables of the kind of the sort Term, *and, for $0 \leq i \leq q_i$, $\hat{t}_i$ is the metarepresentation of the term t_i, except that any variable x in t_i is not metarepresented but, instead, it is replaced by a (meta-) variable x of the kind of the sort* Term. *Note that, in general, some clauses in $\overline{C}_{[T_G[V],s_0]}$ may contain free variables.*

Definition 5 *Let $T_G[V] = (\Omega[V], E \cup G) \in \mathcal{C}$ be a parameterized extension, with $\Omega[V] = (K, \Sigma \cup V, S)$ and $V = \{v_1, \ldots, v_m\}$. Let s_0 be a sort in some S_k and let $C_{[T_G[V],s_0]} = \{C_1, \ldots, C_n\}$ be those sentences of the form*

$$\forall(x_1, \ldots, x_{p_i}). \, t_1 : s_1 \wedge \ldots \wedge t_{q_i} : s_{q_i} \Rightarrow t_0 : s_0, \tag{8}$$

that specify s_0 in $T_G[V]$. Finally, let τ be a first-order formula, with free variable x of the kind of the sort Term, *of the form*

$$\begin{aligned}
\forall \overline{V} : \texttt{Term.} \, ((\overline{v_1} : \overline{z}_1 \text{ in } \overline{T} = \texttt{true} \, \wedge \, \ldots \wedge \, \overline{v_m} : \overline{z}_m \text{ in } \overline{T} = \texttt{true}) \\
\Rightarrow ((x : \overline{s}_0 \text{ in } \overline{T_G[V]} = \texttt{true} \wedge \Phi) \Rightarrow \phi)),
\end{aligned}$$

where, for each parameter $v_i \in V$, $z_i \in S_{k_i}$, for k_i the kind of v_i.

An inductive *inference rule for* META-IND, *with respect to x and $\tau(x)$ is the formula*

$$\begin{aligned}
(\forall \overline{V} : \texttt{Term.} \, (\overline{v_1} : \overline{z}_1 \text{ in } \overline{T} = \texttt{true} \, \wedge \, \ldots \wedge \, \overline{v_m} : \overline{z}_m \text{ in } \overline{T} = \texttt{true}) \\
\Rightarrow [\overline{C}_1]_\tau \wedge \ldots \wedge [\overline{C}_n]_\tau) \\
\Rightarrow \forall x : \texttt{Term.} \, \tau(x),
\end{aligned} \tag{9}$$

where, for each C_i of the form (8), $\overline{C}_i$ is defined as in Definition 4, and $[\overline{C}_i]_\tau$ is the formula

$$\begin{aligned}
\forall(x_1, \ldots, x_{p_i}).& \\
[\hat{t}_1 : \overline{s}_1 \text{ in } & \overline{T_G[V]} = \texttt{true}]_\tau \wedge \ldots \wedge [\hat{t}_{q_i} : \overline{s}_{q_i} \text{ in } \overline{T_G[V]} = \texttt{true}]_\tau \\
\Rightarrow [\hat{t}_0 : \overline{s}_0 & \text{ in } \overline{T_G[V]} = \texttt{true}]_\tau
\end{aligned}$$

where, for $0 \leq j \leq q_i$,

$$[\hat{t}_j : \overline{s}_j \text{ in } \overline{T_G[V]} = \texttt{true}]_\tau = \begin{cases} (\Phi \Rightarrow \phi)(\hat{t}_j) & \text{if } s_j = s_0 \\ \hat{t}_j : \overline{s}_j \text{ in } \overline{T_G[V]} = \texttt{true} & \text{otherwise.} \end{cases}$$

The soundness of the inference rule (9) is proved in [2].

7.3   Building a Inductive Metatheorem Prover

In Section 6.3 we indicated how it is possible to use reflection in Maude to design modular, extensible, theorem proving tools. In particular, we explained the reflective design of the ITP tool and how we implemented the inference rule (3) for induction over sort definitions. For carrying out formal metatheory, we have also extended the ITP tool with the inductive inference rule (9). It is this extended version of the tool that we have used in the experimental work reported on in the next section.

8   An Example

In this section we give an example that illustrates how rewriting logic can be used as a reflective metalogical framework. Our example is a standard one in metareasoning, namely, the deduction theorem.

8.1   The Deduction Theorem for Minimal Logic

We present here the deduction theorem for minimal logic of implication. This theorem is interesting for several reasons. To begin with, it is a central metatheorem that holds for Hilbert systems for many logics and justifies proof under temporary assumption in the manner of a natural deduction system. Moreover, although relatively simple, it illustrates some subtle aspects of formal metareasoning. For example, it is actually a metatheorem not about a particular deduction system, but rather a metatheorem that relates different deduction systems: one in which $A \rightarrow B$ is proved, and a second (which is obtained from the first by adding the axiom A) in which B is proved. Indeed, since A is an arbitrary formula, the standard statement of the deduction theorem is actually a statement about the relationship between a *family* of pairs of deduction systems.

For A and B formulae, we write $\vdash_{\mathcal{M}} A$ to denote that A is a theorem in minimal logic, and $\vdash_{\mathcal{M}[A]} B$ to denote that if minimal logic is extended with the additional axiom A, then B belongs to the resulting set of theorems. The deduction theorem then states that, for any formulae A and B in minimal logic,

$$\text{if } \vdash_{\mathcal{M}[A]} B \quad \text{then} \quad \vdash_{\mathcal{M}} A \rightarrow B. \tag{10}$$

This metatheorem is proven by induction on the structure of derivations in minimal logic extended with the axiom A.

Formalization. Consider now the representation of minimal logic in rewriting logic provided by MINIMAL in Figure 2, and its parameterized extension $\text{MINIMAL}_{\subseteq}[\mathcal{X}]$ introduced in Figure 3. Recall that MINIMAL represents minimal logic in the sense that a formula A is a theorem in minimal logic if and only if A is a term of sort Theorem. We can rephrase the deduction theorem as follows: for any formulae A and B, if B is a term of sort Theorem in $\text{MINIMAL}_{\subseteq}[A]$, then $A \rightarrow B$ is a term of sort Theorem in MINIMAL.

Notice that this theorem states an implication between the truth of two membership assertions over two different membership equational theories (in fact, a whole family of such pairs, since A is a parameter). Hence, to formalize the deduction theorem, we must move up a level, to META-IND. We claim that the following formula formalizes the deduction theorem as a metatheorem about the initial model of META-IND:

$$\forall(\mathcal{X}, B) : \mathsf{Term}. \tag{11}$$
$$((\mathcal{X}:\overline{\mathsf{Formula}} \text{ in } \overline{\mathsf{MINIMAL}} = \mathsf{true} \;\wedge\; B:\overline{\mathsf{Formula}} \text{ in } \overline{\mathsf{MINIMAL}} = \mathsf{true})$$
$$\Rightarrow (B:\overline{\mathsf{Theorem}} \text{ in } \overline{\mathsf{MINIMAL}_{\equiv}[\mathcal{X}]} = \mathsf{true}$$
$$\Rightarrow (\overline{\mathcal{X}{\rightarrow}B}):\overline{\mathsf{Theorem}} \text{ in } \overline{\mathsf{MINIMAL}} = \mathsf{true})),$$

where in the term denoted by $\overline{\mathcal{X}{\rightarrow}B}$ the (meta-) variables $\mathcal{X}$ and B of sort Term are not metarepresented as if they were object level variables, but are instead preserved as (meta-) variables. From now on, we will follow the same convention for terms of this kind, i.e., terms that include elements of sort Term in META-IND.

When performing metareasoning we must reason about terms being well-sorted with respect to particular theories, in this case, the membership equational theory MINIMAL. For this reason, we have explicitly assumed in our formalization of the deduction theorem the well-typedness of $\mathcal{X}$ and B. Of course, standard textbook proofs also require this, but such well-formedness details are usually glossed over as trivial. The correctness of the formalization then follows from the definition of the Boolean function (_:_in_) and the fact that MINIMAL is a conservative representation of minimal logic.

Note, incidentally, that the requirement that B is a formula in minimal logic is actually superfluous and can be dropped (we will do so for the proof below). It is provable (again by induction) that any theorem in any extension of minimal logic with a new axiom is also a formula in minimal logic.

Proof of the Deduction Theorem. We show here how we prove (11). Note that our proof mirrors the standard proof of the deduction theorem.

To prove (11) in META-IND we apply the reflected version of the induction principle for the sort $\mathsf{Theorem}$ in the parameterized extension $\mathsf{MINIMAL}_{\equiv}[\mathcal{X}]$, that is, the corresponding instance of the inference rule (9). This reduces proving (11) to proving the formula given in Figure 4. Notice that the four conjuncts correspond to the cases involved in proving the deduction theorem by induction over the proof that B is a theorem in minimal logic extended with the axiom A. The first formalizes the case when B is $\mathcal{X}$. The next two conjuncts formalize the cases where B is either an instance of the K or S axiom schemata. The final conjunct formalizes the case of B being proved by an instance of *modus ponens*.

By the theorem of constants for membership equational logic [29], we can reduce proving this formula to proving the four conjuncts that result from replacing the variable $\mathcal{X}$ by a new constant symbol x of sort Term, under the assumption that

$$x:\overline{\mathsf{Formula}} \text{ in } \overline{\mathsf{MINIMAL}} = \mathsf{true}. \tag{12}$$

$\forall \mathcal{X}$: Term.$[$
  $\mathcal{X}$:$\overline{\text{Formula}}$ in $\overline{\text{MINIMAL}}$ = true
  $\Rightarrow$
  $[(\overline{\mathcal{X} \rightarrow \mathcal{X}})$:$\overline{\text{Theorem}}$ in $\overline{\text{MINIMAL}}$ = true
  $\wedge$
  $(\forall (A, B). (A$:$\overline{\text{Formula}}$ in $\overline{\text{MINIMAL}_{\text{S}}[\mathcal{X}]}$ = true $\wedge$
        B:$\overline{\text{Formula}}$ in $\overline{\text{MINIMAL}_{\text{S}}[\mathcal{X}]}$ = true$)$
    $\Rightarrow$ $(\overline{\mathcal{X} \rightarrow (A \rightarrow (B \rightarrow A))})$:$\overline{\text{Theorem}}$ in $\overline{\text{MINIMAL}}$ = true$)$
  $\wedge$
  $(\forall (A, B, C). (A$:$\overline{\text{Formula}}$ in $\overline{\text{MINIMAL}_{\text{S}}[\mathcal{X}]}$ = true $\wedge$
        B:$\overline{\text{Formula}}$ in $\overline{\text{MINIMAL}_{\text{S}}[\mathcal{X}]}$ = true $\wedge$
        C:$\overline{\text{Formula}}$ in $\overline{\text{MINIMAL}_{\text{S}}[\mathcal{X}]}$ = true$)$
    $\Rightarrow (\overline{\mathcal{X} \rightarrow ((A \rightarrow B) \rightarrow ((A \rightarrow (B \rightarrow C)) \rightarrow (A \rightarrow C)))})$:$\overline{\text{Theorem}}$ in $\overline{\text{MINIMAL}}$
    = true$)$
  $\wedge$
  $(\forall (A, B). (A$:$\overline{\text{Formula}}$ in $\overline{\text{MINIMAL}_{\text{S}}[\mathcal{X}]}$ = true $\wedge$
        B:$\overline{\text{Formula}}$ in $\overline{\text{MINIMAL}_{\text{S}}[\mathcal{X}]}$ = true $\wedge$
        $\overline{\mathcal{X} \rightarrow (A \rightarrow B)}$:$\overline{\text{Theorem}}$ in $\overline{\text{MINIMAL}}$ = true $\wedge$
        $\overline{\mathcal{X} \rightarrow A}$:$\overline{\text{Theorem}}$ in $\overline{\text{MINIMAL}}$ = true$)$
    $\Rightarrow (\overline{\mathcal{X} \rightarrow B})$:$\overline{\text{Theorem}}$ in $\overline{\text{MINIMAL}}$ = true$)]]$

Fig. 4. Goal resulting after induction

The proof of each of the resulting conjuncts mirrors the proof of the corresponding case in the standard inductive proof of the deduction theorem. In what follows, $\text{MINIMAL}_{\text{S}}[x]$ denotes the term of sort Module that results from $\text{MINIMAL}_{\text{S}}[\mathcal{X}]$ by replacing the free variable $\mathcal{X}$ of sort Term by the new constant symbol x. Consider, for example, how we prove the third conjunct:

$$\forall (A, B, C). \tag{13}$$
  A:$\overline{\text{Formula}}$ in $\overline{\text{MINIMAL}_{\text{S}}[x]}$ = true $\wedge$
  B:$\overline{\text{Formula}}$ in $\overline{\text{MINIMAL}_{\text{S}}[x]}$ = true $\wedge$
  C:$\overline{\text{Formula}}$ in $\overline{\text{MINIMAL}_{\text{S}}[x]}$ = true
  $\Rightarrow (\overline{x \rightarrow ((A \rightarrow B) \rightarrow ((A \rightarrow (B \rightarrow C)) \rightarrow (A \rightarrow C)))})$:$\overline{\text{Theorem}}$
    in $\overline{\text{MINIMAL}}$ = true$)$.

Using the theorem of constants again, we can reduce proving (13) to proving

$$\overline{x \rightarrow ((a \rightarrow b) \rightarrow ((a \rightarrow (b \rightarrow c)) \rightarrow (a \rightarrow c)))}$$:$\overline{\text{Theorem}}$

  in $\overline{\text{MINIMAL}}$ = true, $\quad$ (14)

under the assumptions that a, b, c are new constants of sort `Term` such that

$$a:\overline{\text{Formula}} \text{ in } \overline{\text{MINIMAL}_{\equiv}[x]} = \text{true} \wedge$$
$$b:\overline{\text{Formula}} \text{ in } \overline{\text{MINIMAL}_{\equiv}[x]} = \text{true} \wedge \qquad (15)$$
$$c:\overline{\text{Formula}} \text{ in } \overline{\text{MINIMAL}_{\equiv}[x]} = \text{true}.$$

Note that, from the assumptions (12) and (15), by using the fact (which must be proven separately) that any theorem in any extension of minimal logic with a new axiom is a well-formed formula in minimal logic, we can derive the formula

$$a:\overline{\text{Formula}} \text{ in } \overline{\text{MINIMAL}} = \text{true} \wedge$$
$$b:\overline{\text{Formula}} \text{ in } \overline{\text{MINIMAL}} = \text{true} \wedge$$
$$c:\overline{\text{Formula}} \text{ in } \overline{\text{MINIMAL}} = \text{true} \wedge \qquad (16)$$
$$x:\overline{\text{Formula}} \text{ in } \overline{\text{MINIMAL}} = \text{true}.$$

Finally, we prove (14) using the equations in `META-IND` and (16). This proof mirrors the proof that, for any formulae X, A, B, and C,

$$X \to ((A \to B) \to ((A \to (B \to C)) \to (A \to B))) \qquad (17)$$

is a theorem in minimal logic; in particular, this proof mirrors proving (17) by *modus ponens*, using the following instance of the S axiom

$$(A \to B) \to ((A \to (B \to C)) \to (A \to C))$$

and the following instance of the K axiom

$$[(A \to B) \to ((A \to (B \to C)) \to (A \to C))]$$
$$\to (X \to [(A \to B) \to ((A \to (B \to C)) \to (A \to C))]) .$$

8.2   Other Examples and Experience

We have used rewriting logic as a reflective metalogical framework to carry out a number of other proofs in formal metatheory based on more sophisticated versions of the deduction theorem for minimal logic. In particular, we have proved results similar to those of Basin and Matthews [4,5], who have shown how metatheorems that are parameterized by their scope of application can be proved using a theory of parameterized inductive definitions as a metatheory. For example, they present a generalized version of the deduction theorem that can be applied to all extensions of the language and axioms of minimal logic. From their theorem it follows that the deduction theorem holds for the minimal logic of implication and for any propositional extension of it, but not necessarily for extensions to modal logics (which would require adding new rules, as opposed to new axioms). Although rewriting logic is based on a rather different foundation than those considered in [5], our representation of the object logic is quite similar and—abstracting away from the details involved in moving between levels of representation—the basic structure of the proofs is also similar.

One promising area to apply our results is program transformation and metaprogramming. From a reflective declarative point of view, programs that transform other programs are first-order functions acting on terms that metarepresent theories, and the properties that they satisfy are metatheorems, as they are understood in this paper. This reflective declarative methodology has been used in [12] to specify polytypic programs like *map* and *cata* in Maude. Accordingly, polytypic programs are specified as metalevel functions that add to a module the equations defining the desired object function by structural induction over the sort definitions. Properties of polytypic programs, like the functoriality of *map*, are then metatheorems that can be proved, as it is showed in [12], using the corresponding induction rule (Definition 5).

Here we would also like to comment on our experience in proving these theorems and on the issue of managing proofs that combine reasoning at different levels. To the working logician or computer scientist, reflective metalogical frameworks may seem complicated and not particularly user-friendly since there is quite a bit of encoding involved in stating a metatheorem and in carrying out its proof. In particular, reasoning can involve three or more levels (object, meta, meta-meta, ...).[6]

In our case, we have been able to avoid many of the practical problems of working with a reflective hierarchy by exploiting the reflective capabilities of Maude to build tools and suitable interfaces that hide levels of reflection. As part of our work, we have built an interface—fully specified in Maude—to interact with the ITP inductive theorem prover described in Sections 6.3 and 7.3. As already explained, ITP automatically extracts from a theory the induction principles for reasoning over its sorts (Definition 2), and (in its metaprover extension) the induction rules that correspond to reflecting those induction principles at the metalevel when the task at hand is to prove a metatheorem (Definition 5). Proving an inductive theorem then amounts to computing a strategy at the meta-metalevel, or at the meta-meta-metalevel if the theorem is, as in the case of (11), a metatheorem about the initial model of META-IND. Fortunately, the interface we use hides all these levels of encoding from the user. Hence the user can actually abstract away many of the metarepresentation details and focus on the essential structure of proofs of theorems.

9   Conclusion

We have presented, both abstractly and concretely, a new approach to metatheoretic reasoning based on using reflective logical frameworks whose theories have initial models. Initial experiments demonstrate that the machinery for reflective deduction in membership equational logic provides a rich foundation for formalizing and proving metatheorems. Our experiments show, for example, that one

---

[6] Although note that reasoning about a logic encoded as an inductive definition in a logical framework like Isabelle also involves multiple levels, e.g., the framework's metalogic, the theory of inductive definitions, and the object logic. Moreover, there is often an additional language for writing tactics.

can prove metatheorems similar to those provable in logical frameworks based on parameterized inductive definitions, and that one has considerable flexibility in moving between theories and proving theorems that relate theories or establish properties of parameterized classes of theories. In essence, we can do this because the requirements that such metatheorems pose on the metatheory—namely, that one can build families of sets using parameterized inductive definitions and that one can reason about their elements by induction—are realizable in membership equational logic using reflection.

There are a number of directions for further work. One concerns generalizing our notion of a parameterized theory. Currently we can reason at the metalevel about families of theories that are parameterized by sets of new constants and new axioms, which may make use of the new constants. For proving other metatheorems it would be useful to develop a more general theory representation calculus where one could reason at the metalevel about families of theories that are parameterized by arbitrary sets of new sorts, operators, and axioms. In particular, this would allow us to prove metatheorems involving the more general *parameterized modules* of Full Maude [9,18].

Also, our example illustrates how it is possible to carry out proofs similar to those possible in stronger framework logics. However, it would be interesting to have a more formal comparison of the relative strengths of membership equational logic with reflection versus stronger metalogics like higher-order logic or set theory. Finally, related to this is the question of how easy it is to reflect induction principles other than structural induction, e.g., induction over an arbitrary, user-definable well-founded order.

Acknowledgments

This research was supported by DARPA through Rome Laboratories Contract F30602-C-0312, by DARPA and NASA through Contract NAS2-98073, by Office of Naval Research Contract N00014-99-C-0198, and by National Science Foundation Grant CCR-9900334. The authors also thank Narciso Martí-Oliet for his careful reading of a draft of this paper and his detailed suggestions for improving the exposition.

References

1. P. Aczel. An introduction to inductive definitions. In J. Barwise, editor, *Handbook of Mathematical Logic*, pages 739–782. North-Holland, Amsterdam, 1977.
2. D. Basin, M. Clavel, and J. Meseguer. Rewriting logic as a metalogical framework. Technical report, September 2000, http://maude.csl.sri.com.
3. D. Basin and R. Constable. Metalogical frameworks. In G. Huet and G. Plotkin, editors, *Logical Environments*, pages 1–29. Cambridge University Press, 1993.
4. D. Basin and S. Matthews. Scoped metatheorems. In *International Workshop on Rewriting Logic and its Applications*, volume 15, pages 1–12. Electronic Notes in Theoretical Computer Science (ENTCS), September 1998.

5. D. Basin and S. Matthews. Structuring metatheory on inductive definitions. *Information and Computation*, 2000. To appear.
6. A. Bouhoula, J.-P. Jouannaud, and J. Meseguer. Specification and proof in membership equational logic. *Theoretical Computer Science*, 236:35–132, 2000.
7. M. Clavel. *Reflection in General Logics and in Rewriting Logic with Applications to the Maude Language*. PhD thesis, University of Navarre, 1998.
8. M. Clavel, F. Durán, S. Eker, P. Lincoln, N. Martí-Oliet, and J. Meseguer. Metalevel computation in Maude. In C. Kirchner and H. Kirchner, editors, *Second International Workshop on Rewriting Logic and its Applications*, volume 15 of *Electronic Notes in Theoretical Computer Science*, pages 3–23, Pont-à-Mousson, France, September 1998. Elsevier.
9. M. Clavel, F. Durán, S. Eker, P. Lincoln, N. Martí-Oliet, J. Meseguer, and J. Quesada. Maude: Specification and programming in rewriting logic. SRI International, January 1999, http://maude.csl.sri.com.
10. M. Clavel, F. Durán, S. Eker, and J. Meseguer. Building equational proving tools by reflection in rewriting logic. In *Proceedings of the CafeOBJ Symposium '98, Numazu, Japan*. CafeOBJ Project, April 1998.
11. M. Clavel, F. Durán, S. Eker, J. Meseguer, and M.-O. Stehr. Maude as a formal meta-tool. In J. Wing and J. Woodcock, editors, *FM'99 — Formal Methods*, volume 1709 of *Lecture Notes in Computer Science*, pages 1684–1703. Springer-Verlag, 1999.
12. M. Clavel, F. Durán, and N. Martí-Oliet. Polytypic programming in Maude. To appear in Proc. WRLA 2000, ENTCS, Elsevier, 2000.
13. M. Clavel, S. Eker, P. Lincoln, and J. Meseguer. Principles of Maude. In J. Meseguer, editor, *First International Workshop on Rewriting Logic and its Applications*, volume 4 of *Electronic Notes in Theoretical Computer Science*, pages 65–89, Asilomar (California), September 1996. Elsevier.
14. M. Clavel and J. Meseguer. Axiomatizing reflective logics and languages. In G. Kiczales, editor, *Proceedings of Reflection'96*, pages 263–288, San Francisco (California), April 1996. Xerox PARC.
15. M. Clavel and J. Meseguer. Reflection and strategies in rewriting logic. In J. Meseguer, editor, *First International Workshop on Rewriting Logic and its Applications*, volume 4 of *Electronic Notes in Theoretical Computer Science*, pages 125–147, Asilomar (California), September 1996. Elsevier.
16. M. Clavel and J. Meseguer. Reflection in condition rewriting logic. Manuscript. Submitted for publication, 2000.
17. J. Despeyroux, F. Pfenning, and C. Schürmann. Primitive recursion for higher-order abstract syntax. In *Proceedings of the 3rd International Conference on Typed Lambda Calculi and Applications (TLCA'97)*, volume 1210 of *Lecture Notes in Computer Science*, Nancy, France, April 1997. Springer-Verlag.
18. F. Durán. *A Reflective Module Algebra with Applications to the Maude Language*. PhD thesis, University of Málaga, 1999.
19. S. Feferman. Finitary inductively presented logics. In *Logic Colloquium '88*. North-Holland, 1988.
20. J. Goguen and J. Meseguer. Models and equality for logical programming. In H. Ehrig, G. Levi, R. Kowalski, and U. Montanari, editors, *Proceedings TAPSOFT'87*, volume 250 of *Lecture Notes in Computer Science*, pages 1–22. Springer-Verlag, 1987.
21. M. Gordon and T. Melham. *Introduction to HOL: A Theorem Proving Environment for Higher Order Logic*. Cambridge University Press, 1993.

22. R. Harper, F. Honsell, and G. Plotkin. A framework for defining logics. *J. ACM*, 40(1):143–184, January 1993.
23. N. Martí-Oliet and J. Meseguer. Rewriting logic as a logical and semantic framework. Technical Report SRI-CSL-93-05, SRI International, Computer Science Laboratory, August 1993. To appear in D. Gabbay, ed., *Handbook of Philosophical Logic*, Kluwer Academic Publishers.
24. N. Martí-Oliet and J. Meseguer. General logics and logical frameworks. In D. Gabbay, editor, *What is a Logical System?*, pages 355–392. Oxford University Press, 1994.
25. S. Matthews, A. Smaill, and D. Basin. Experience with FS_0 as a framework theory. In G. Huet and G. Plotkin, editors, *Logical Environments*, pages 61–82. Cambridge University Press, 1993.
26. R. McDowell and D. Miller. A logic for reasoning with higher-order abstract syntax. In *Twelfth Annual IEEE Symposium on Logic in Computer Science*, June 1997.
27. J. Meseguer. General logics. In H.-D. Ebbinghaus et al., editor, *Logic Colloquium'87*, pages 275–329. North-Holland, 1989.
28. J. Meseguer. Conditional rewriting logic as a unified model of concurrency. *Theoretical Computer Science*, 96(1):73–155, 1992.
29. J. Meseguer. Membership algebra as a semantic framework for equational specification. In F. Parisi-Presicce, editor, *Proceedings of WADT'97*, volume 1376 of *Lecture Notes in Computer Science*, pages 18–61. Springer-Verlag, 1998.
30. J. Meseguer. Research directions in rewriting logic. In U. Berger and H. Schwichtenberg, editors, *Computational Logic, NATO Advanced Study Institute, Marktoberdorf, Germany, July 29 - August 6, 1997*. Springer-Verlag, 1998.
31. J. Meseguer and J. A. Goguen. Initiality, induction and computability. In M. Nivat and J. C. Reynolds, editors, *Algebraic Methods in Semantics*, pages 459–541. Cambridge University Press, 1985.
32. C. Paulin-Mohring. Inductive Definitions in the System Coq — Rules and Properties. In M. Bezem and J.-F. Groote, editors, *Proceedings of the conference Typed Lambda Calculi and Applications*, volume 664 of *Lecture Notes in Computer Science*, 1993.
33. L. C. Paulson. A fixedpoint approach to implementing (co)inductive definitions. In *Proceedings of the 12th International Conference on Automated Deduction (CADE-12)*, volume 814 of *Lecture Notes in Artificial Intelligence*, Nancy, France, June 1994. Springer-Verlag.
34. L. C. Paulson. *Isabelle : a generic theorem prover; with contributions by Tobias Nipkow*, volume 828 of *Lecture Notes in Computer Science*. Springer, Berlin, 1994.
35. C. Schürmann and F. Pfenning. Automated theorem proving in a simple metalogic for LF. In C. Kirchner and H. Kirchner, editors, *Proceedings of the 15th International Conference on Automated Deduction (CADE-15)*, volume 1421 of *Lecture Notes in Computer Science*, pages 286–300, Lindau, Germany, July 1998. Springer-Verlag.
36. M.-O. Stehr and J. Meseguer. Pure type systems in rewriting logic. In *Proc. of LFM'99: Workshop on Logical Frameworks and Meta-languages*, 1999. http://www.cs.bell-labs.com/~felty/LFM99/.
37. M.-O. Stehr, P. Naumov, and J. Meseguer. A proof-theoretic approach to the HOL-Nuprl connection with applications to proof translation. Manuscript, SRI International, http://www.csl.sri.com/~stehr/fi_eng.html, February 2000.
38. M. van Emden and R. Kowalski. The semantics of predicate logic as a programming language. *J. ACM*, 23:733–42, 1976.

Frequency Assignment in Mobile Phone Systems

Martin Grötschel

Konrad-Zuse-Zentrum für Informationstechnik and
Technische Universität Berlin

Wireless Communication and Frequencies

Wireless communication technology is the basis of radio and television broadcasting, it is used in satellite-based cellular telephone systems, in point-to-multipoint radio access systems, and in terrestrial mobile cellular networks, to mention a few such systems (see [5] for more detailed information).

Wireless communication networks employ radio frequencies to establish communication links. The available radio spectrum is very limited. To meet today's radio communication demand, this resource has to be administered and reused carefully in order to control mutual interference. The reuse can be organized via separation in space, time, or frequency, for example. The problem, therefore, arises to distribute frequencies to links in a "reasonable manner". This is the basic form of the frequency assignment problem. What "reasonable" means, how to quantify this measure of quality, which technical side constraints to consider cannot be answered in general. The exact specification of this task and its mathematical model depend heavily on the particular application considered.

Mobile Cellular Networks, GSM

I will concentrate here on terrestrial mobile cellular networks, an application that has revolutionized the telephone business in the recent years and is going to have further significant impact in the years to come. Even in this special application the frequency assignment problem has no universal mathematical model. I will focus on the GSM standard (GSM stands for "General System for Mobile Communication"), which has been in use since 1992. GSM is the basis of almost all cellular phone networks in Europe. It is employed in more than 100 countries serving several hundred million customers. The new worldwide standard UMTS (Universal Mobile Telecommunication System) is expected to become commercially available around 2002. It is frequently covered in the public press at present because of the enormous amounts of money telephone companies are paying in the national frequency auctions. UMTS handles frequency reuse in an even more intricate manner than GMS: frequency or time division are used in combination with code division multiple access (CDMA) technology.

Channel Spectrum

The typical situation in GSM frequency planning is as follows. A telephone company (let us call it the *operator*) has bought the right to use a certain spectrum of frequencies

S. Kapoor and S. Prasad (Eds.): FST TCS 2000, LNCS 1974, pp. 81–86, 2000.

$[f_{min}, f_{max}]$ in a particular geographical region, e.g., a country. The frequency band is – depending on the technology utilized – partitioned into a set of *channels*, all with the same bandwidth Δ. The available channels are usually denoted by $1, 2, \ldots, N$, where $N = (f_{max} - f_{min})/\Delta$. In Germany, for instance, an operator of a mobile phone network owns about 100 channels. On each channel available, one can communicate information from a transmitter to a receiver. For bidirectional traffic a second channel is needed. In fact, if an operator buys a spectrum $[f_{min}, f_{max}]$ he automatically obtains a paired spectrum of equal width for bidirectional communication. One of these spectra is used for mobile to base station (up-link), the other for base station to mobile (down-link) communication.

BTSs, TRXs, and Cells

To serve his customers an operator has to solve a number of nontrivial problems. In an intitial step the geographical distribution of the communication demand for the planning period is estimated. Based on these figures, a communication infrastructure has to be installed capable to serve the anticipated demand. The devices handling the radio communication with the mobile phones of the customers are called Base Transceiver Stations (BTS). They have radio transmission and reception equipment, including antennas and all necessary signal processing capabilities. An antenna of a BTS can be omni-directional or sectorized. The typical BTS used today operates three antennas each with an opening angle of 120 degrees. Each such antenna defines a *cell*. These cells are the basic planning units (and that is why mobile phone systems are also called *cellular phone systems*).

The capacity of a cell is defined by the number of transmitter/receiver units, called TRXs, installed for this antenna. The first TRX handles the signalling and offers capacity for up to six calls (by time division). Additional TRXs can typically handle 7 or 8 further calls – depending on the extra signalling load. No more than 12 TRXs can be installed for one antenna, i.e., the maximum capacity of a cell is in the range of 80 calls. That is why areas of heavy traffic (e.g., airports, business centers of big cities) have to be subdivided into many cells.

BSCs, MSCs, and the Core Network

In a next planning step, the operator has to locate and install the so called Base Station Controllers (BSCs). Each BTS has to be connected (in general via cable) to such a BSC, while a BSC operates several BTSs in parallel. A BSC is, e.g., in charge of the management of hand-overs.

Every BSC, in turn, is connected to a Mobile Service Switching Center (MSC). The MSCs are connected to each other through the so called *core network*, which has to carry the "backbone traffic". The location planning for BSCs and MSCs, the design of the topology of the core network, the optimization of the link capacities, routing, failure handling, etc., constitute major tasks an operator has to address. We do not intend to discuss here the roles of all the devices that make up a mobile phone network and their mutual interplay in detail. This brief sketch is just meant to indicate that telecommunication network planning is quite a complex task.

Channel Assignment, Hand-Over

We have seen that the TRXs are the devices that handle radio communication with the mobile phones of the customers. The operators in Germany maintain networks of about 5,000 to 15,000 TRXs and have around 100 channels available. Thus, the question arises how to best distribute the channels to the TRXs.

An operational mobile phone emits signals that allow the network to roughly keep track of where the mobile phone is currently located. This is done via so called control channels. Whenever a communication demand arises, the system decides which TRX is going to handle the communication. This decision is based on the signal strengths of the various TRXs that are able to communicate with the phone as well as on the current traffic. The mobile phone is tuned to the channel of the TRX that presently appears to serve the phone best. If the phone moves (e.g., in a car) the communication with its current TRX may become poor. The system monitors the reception quality and may decide to use a TRX from another cell. Such a switch is called hand-over.

This short discussion shows that a mobile phone typically is not only in one cell. In fact, some cells must overlap, otherwise hand-overs are not possible.

Interference

Whenever two cells overlap and use the same channel, interference (signal-to-noise ratio at the receiving end of a connection) occurs in the area of cell intersection. Moreover, antennas may cause interference far beyond their cell limits. The computation of the level of interference is a difficult task. It depends not only on the channels, the signals' strength and direction, but also on the shape of the environment, which strongly influences wave propagation. There are a number of theoretical methods and formulas with which interference can be quantified. Most mobile phone companies base their analysis of interference on some mathematical model taking transmitter power, distances, fading and filtering factors into account. The data for these models typically come from terrain and building data bases but may also include vegetation data. They combine this with practical experience and extensive measurements. The result is an interference prediction model with which the so called *co-channel interference* that occurs when two TRXs transmit on the same channel is quantified. There may also be *adjacent-channel interference* when two TRXs operate on channels that are adjacent (i.e., one TRX operates on channel i, the other on channel $i + 1$ or $i - 1$).

Reality is a bit more complicated than sketched before. Several TRXs (and not only two) operating on the same or adjacent channels may interfere with each other at the same time. And what really is the interference between two cells? It may be that two cells interfere only in 10% of their area but with high noise or that they interfere in 50% of their area with low noise. What if interference is high but almost no traffic is expected? How can a single "interference value" reflect such a difference in the interference behaviour? There is no clear answer.

The planners have to investigate such cases in detail and have to come up with a reasonable compromise. The result, in general, is a number, the *interference value*, which is usually normalized to be between 0 and 1. This number should – to the best of the knowledge of the planners – characterize the interference between two TRXs (in terms of the model, the technological assumptions, etc., used by the operator).

Separation and Blocked Channels

There are also hard constraints. If two or more TRXs are installed at the same location (or site), there are restrictions on how close their channels may be. For instance, if a TRX operates on channel i, a TRX at the same site is not allowed to operate on channels $i+1, i-1$. Such a restriction is called co-site *separation*. Separation requirements may even be tighter if two TRXs are not only co-site, but also serve the same cell. Separation requirements may apply also to TRXs that are in close proximity.

The situation is even more complex. Due to government regulations, agreements with operators in neighbouring regions, requirements from military forces, etc., an operator may not be allowed to use its whole spectrum of channels at every location. This means that, for each TRX, there may be a set of so called *blocked channels*.

Interference Graph

A feasible assignment of channels to TRXs clearly has to satisfy all separation constraints. Blocked channels must not be used. What should one do about interference?

On our way to an adequate mathematical representation of all technical constraints let us first introduce the *interference graph* $G = (V, E)$. G has a node for every TRX, two nodes are joined by an edge, if interference occurs when the associated TRXs operate on the same channel or on adjacent channels or if a separation constraint applies to the two TRXs. With each edge $vw \in E$, two interference values, denoted by $c^{co}(vw)$ and $c^{ad}(vw)$, are associated; the number $c^{co}(vw)$ is the co-channel interference that occurs when TRXs v and w operate on the same channel while $c^{ad}(vw)$ denotes the interference value coming up when v and w operate on adjacent channels. In general, $c^{co}(vw) \geq c^{ad}(vw)$. If a separation constraint applies to v and w then a suitable large number is allocated to $c^{co}(vw)$ and $c^{ad}(vw)$.

Two "Natural" Approaches

A first attempt to solve the frequency assignment problem is obvious. We try to find a spectrum, i.e., a number of channels $1, \dots, N$ such that the N available channels can be assigned to TRXs so that no interference occurs.

No interference is, of course, a good aim, but this task is unrealistic for several reasons. A mobile phone network is a "living system", i.e., new BTSs, antennas, etc., are installed regularly, old ones are replaced by new ones with different characteristics. It is impossible to change the spectrum each time the network changes. Moreover, the number of channels may be fixed or channels may only be available in bundles (i.e., one may buy 75, 100 or 125, but nothing else). Frequencies are expensive and cost reasons may require that some interference is tolerated. In fact, some interference may be unavoidable. We have data of mobile phone systems where the largest clique in the interference graph is about twice as large as the number of available channels and where the largest degree of a node is ten times as large as the number of available channels.

Another classic choice for the solution of the frequency assignment problem is the following. We choose a threshold value t and consider the graph $G^t = (V, E^t)$ where $E^t := \{ij \in E \mid c^{co}_{vw} \geq t\}$. Now we try to find the coloring number of G^t, or try to color G^t with N colors. In other words, we consider interference below t tolerable and try to find an assignment of channels to TRXs such that as few channels as possible are used (a color represents a channel, no two nodes with the same color are not allowed to be adjacent) or we try to use the available channels so that no "high interference" occurs. Of course, if for a given threshold t no feasible coloring can be found, one has to modify t and try again.

This approach is unable to handle separation constraints and ignores adjacent channel interference. It was the "standard approach" in the early days of the mobile phone era but did not prove efficient in the more complex environment we have today.

Minimizing Interference

There are several other ways of modelling the frequency assignment problem mathematically, see [5]. For reasons of brevity I will focus on an approach that was employed in a joint project of the Konrad-Zuse-Zentrum and E-PLUS, one of the four operators in Germany, and which has resulted in very satisfactory channel assignments.

Let $G = (V, E)$ be the interference graph introduced before. Let $C = [1, \ldots, N]$ be the set of available channels, and let, for each TRX $v \in V$, B_v denote the subset of channels blocked at node v. The values $c^{co}(vw)$, $c^{ad}(vw)$ denote, for each edge $vw \in E$, the co-channel and adjacent-channel interference arising when TRXs v and w operate on the same or on adjacent channels. Moreover, let $d(vw) \in \mathbb{Z}_+$ denote the separation necessary between the channels assigned to TRXs v and w. Thus, the input to a frequency assignment problem is a 7-tuple $(V, E, C, \{B_v\}_{v \in V}, d, c^{co}, c^{ad})$, briefly called *network* here. A *frequency assignment* for the network is a function $y : V \to C$. It is called *feasible* if it satisfies the following side constraints

$$y(v) \in C \backslash B_v \text{ for all } v \in V$$

$$|y(v) - y(w)| \geq d(vw) \text{ for all } vw \in E$$

The objective is to minimize the sum of co- and adjacent-channel interference, more formally:

$$\min_{y \text{ feasible}} \sum_{\substack{vw \in E: \\ y(v) = y(w)}} c^{co}(vw) \quad + \sum_{\substack{vw \in E: \\ |y(v) - y(w)| = 1}} c^{ad}(vw) \qquad \text{(FAP)}$$

This version of the frequency assignment problem is a generalization of list colorings in graph theory and it is related to the well known T-coloring problem.

There are several ways to reformulate (FAP) in terms of other standard models of combinatorial optimization, e.g., there is a stable set model and a so called orientation model which is related to linear ordering.

Several modifications of FAP have to be considered in practice. No operator wants to change all channel assignments whenever a new plan has to be computed. Some assignments have to stay fix (that is easy to achieve), sometimes one looks for the smallest number of channel adjustments within a certain range of interference, or one requires that, e.g., at most 100 of the assignments are changed.

FAP is difficult in terms of complexity theory. Deciding whether a TRX network allows a feasible assignment is $\mathcal{NP}$-complete; the optimization problem is strongly $\mathcal{NP}$-hard.

FAP is also difficult in practice. Nobody can solve realistic instances to optimality. Satisfactory lower bounds on the objective function value are very hard to obtain. All approaches based on polyhedral combinatorics and linear programming have failed so far. There is some hope to exploit semidefinite relaxations of FAP.

The whole available "zoo" of heuristics has been tried for the solution of FAPs. Considerable improvements over previous approaches can be achieved. There are some spectacular successes, but at present, the gap between lower and upper bounds – computable in practice – is still very large.

In my talk on this subject I will elaborate on the mathematical modelling of the FAP, on the development of heuristics and on the approaches with which lower bounds have been computed. I will present examples from practice that show what can be achieved today and how this mathematical approach compares to more traditional planning techniques.

This lecture is based on joint work of the telecommunications group at the Konrad-Zuse-Zentrum, particular on the work of Andreas Eisenblätter [3]. Further references are [1], [2], [4]. The FAP website [5] is another excellent source of information.

References

[1] R. Borndörfer, A. Eisenblätter, M. Grötschel, and A. Martin, "Frequency Assignment in Cellular Phone Networks", *Annals of Operations Research*, 76:73-93 (1998).
[2] R. Borndörfer, A. Eisenblätter, M. Grötschel, and A. Martin, "The orientation model for Frequency Assignment Problems", Technical Report TR 98-013, Konrad-Zuse-Zentrum für Informationstechnik Berlin, (1998a).
[3] A. Eisenblätter, "Frequency Assignment in GSM Networks: Models, Heuristics, and Lower Bounds", Ph.D. Thesis, TU Berlin 2000, to appear.
[4] A.M.C.A. Koster, "Frequency Assignment – Models and Algorithms", Ph.D. Thesis, Universiteit Maastricht, Maastricht, The Netherlands (1999).
[5] FAP Web – A Website about Frequency Assignment Problems, http://fap.zib.de

Data Provenance: Some Basic Issues

Peter Buneman, Sanjeev Khanna, and Wang-Chiew Tan

University of Pennsylvania

Abstract. The ease with which one can copy and transform data on
the Web, has made it increasingly difficult to determine the origins of a
piece of data. We use the term *data provenance* to refer to the process
of tracing and recording the origins of data and its movement between
databases. Provenance is now an acute issue in scientific databases where
it is central to the validation of data. In this paper we discuss some of
the technical issues that have emerged in an initial exploration of the
topic.

1   Introduction

When you find some data on the Web, do you have any information about how
it got there? It is quite possible that it was copied from somewhere else on the
Web, which, in turn may have also been copied; and in this process it may well
have been transformed and edited. Of course, when we are looking for a best
buy, a news story, or a movie rating, we know that what we are getting may be
inaccurate, and we have learned not to put too much faith in what we extract
from the Web. However, if you are a scientist, or any kind of scholar, you would
like to have confidence in the accuracy and timeliness of the data that you are
working with. In particular, you would like to know how it got there.

In its brief existence, the Web has completely changed the way in which data
is circulated. We have moved very rapidly from a world of paper documents
to a world of on-line documents and databases. In particular, this is having a
profound effect on how scientific research is conducted. Let us list some aspects
of this transformation:

- A paper document is essentially unmodifiable. To "change" it one issues a
  new edition, and this is a costly and slow process. On-line documents, by
  contrast, can be (and often are) frequently updated.
- On-line documents are often databases, which means that they have explicit
  structure. The development of XML has blurred the distinction between
  documents and databases.
- On-line documents/databases typically contain data extracted from other
  documents/databases through the use of query languages or "screen-scrap-
  ers".

Among the sciences, the field of Molecular Biology is possibly one of the
most sophisticated consumers of modern database technology and has generated

S. Kapoor and S. Prasad (Eds.): FST TCS 2000, LNCS 1974, pp. 87–93, 2000.

a wealth of new database issues [15]. A substantial fraction of research in genetics is conducted in "dry" laboratories using *in silico* experiments – analysis of data in the available databases. Figure 1 shows how data flows through a very small fraction of the available molecular biology databases[1]. In all but one case, there is a *Lit* – for literature – input to a database indicating that this is database is *curated*. The database is not simply obtained by a database query or by on-line submission, but involves human intervention in the form of additional classification, annotation and error correction. An interesting property of this flow diagram is that there is a cycle in it. This does not mean that there is perpetual loop of possibly inaccurate data flowing through the system (though this might happen); it means that the two databases overlap in some area and borrow on the expertise of their respective curators. The point is that it may now be very difficult to determine where a specific piece of data comes from. We use the term *data provenance* broadly to refer to a description of the origins of a piece of data and the process by which it arrived in a database. Most implementors and curators of scientific databases would like to record provenance, but current database technology does not provide much help in this process for databases are typically rather rigid structures and do not allow the kinds of *ad hoc* annotations that are often needed for recording provenance.

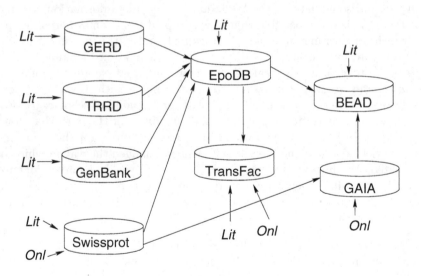

Fig. 1. The Flow of Data in Bioinformatics

The databases used in molecular biology form just one example of why data provenance is an important issue. There are other areas in which it is equally acute [5]. It is an issue that is certainly broader than computer science, with legal

---

[1] Thanks to Susan Davidson, Fidel Salas and Chris Stoeckert of the Bioinformatics Center at Penn for providing this information.

and ethical aspects. The question that computer scientists, especially theoretical computer scientists, may want to ask is what are the technical issues involved in the study of data provenance. As in most areas of computer science, the hard part is to formulate the problem in a concise and applicable fashion. Once that is done, it often happens that interesting technical problems emerge. This abstract reviews some of the technical issues that have emerged in an initial exploration.

2 Computing Provenance: Query Inversion

Perhaps the only area of data provenance to receive any substantial attention is that of provenance of data obtained via query operations on some input databases. Even in this restricted setting, a formalization of the notion of data provenance turns out to be a challenging problem. Specifically, given a tuple t in the output of a database query Q applied on some source data D, we want to understand which tuples in D contributed to the output tuple t, and if there is a compact mechanism for identifying these input tuples. A natural approach is to generate a new query Q', determined by Q, D and t, such that when the query Q' is applied to D, it generates a collection of input tuples that "contributed to" the output tuple t. In other words, we would like to identify the provenance by *inverting* the original query. Of course, we have to ask what we mean by contributed to? This problem has been studied under various names including "data pedigree" and "data lineage" in [1, 9, 7]. One way we might answer this question is to say that a tuple in the input database "contributes to" an output tuple if changing the input tuple causes the output tuple to change or to disappear from the output. This definition breaks down on the simplest queries (a projection or union). A better approach is to use a simple proof-theoretic definition. If we are dealing with queries that are expressible in positive relational algebra (SPJU) or more generally in positive datalog, we can say that an input tuple (a fact) "contributes to" an output tuple if it is used in *some* minimal derivation of that tuple. This simple definition works well, and has the expected properties: it is invariant under query rewriting, and it is compositional in the expected way. Unfortunately, these desirable properties break down in the presence of negation or any form of aggregation. To see this consider a simple SQL query:

```
SELECT name, telephone
FROM employee
WHERE salary > SELECT AVERAGE salary FROM employee
```

Here, modifying any tuple in the `employee` relation could affect the presence of any given output tuple. Indeed, for this query, the definition of "contributes to" given in [9] makes the whole of the `employee` relation contribute to each tuple in the output. While this is a perfectly reasonable definition, the properties of invariance under query rewriting and compositionality break down, indicating that a more sophisticated definition may be needed.

Before going further it is worth remarking that this characterization of provenance is related to the topics of truth maintenance [10] and view maintenance

[12]. The problem in view maintenance is as follows. Suppose a database (a view) is generated by an expensive query on some other database. When the source database changes, we would like to recompute the view without recomputing the whole query. Truth maintenance is the same problem in the terminology of deductive systems. What may make query inversion simpler is that we are only interested in what is in the database; we are not interested in updates that would add tuples to the database.

In [7] another notion of provenance is introduced. Consider the SQL query above, and suppose we see the tuple ("John Doe", 12345) in the output. What the previous discussion tells us is *why* that tuple is in the output. However, we might ask an apparently simpler question: given that the tuple appears in the output, *where* does the telephone number 12345 come from? The answer to this seems easy – from the "John Doe" tuple in the input. This seems to imply that as long as there is some means of identifying tuples in the employee relation, one can compute where-provenance by tracing the variable (that emits 12345) of the query. However, this intuition is fragile and a general characterization is not obvious; it is discussed in [7].

We remark that this second form of provenance, where-provenance, is also related to the view update problem [3]: if John Doe decides to change his telephone number at the view, which data should be modified in the employee relation? Again, where-provenance seems simpler because we are only interested in modifications to the existing view; we are not interested in insertions to the view.

Another issue in query inversion is to capture other query languages and other data models. For example, we would like to describe the problem in object-oriented [11] or semistructured data models [2] (XML). What makes these models interesting is that we are no longer operating at the fixed level of tuples in the relational model. We may want to ask for the why- or where-provenance of some deeply nested component of some structure. To this end, [7] studies the issue of data provenance in a "deterministic" model of semistructured data in which every element has a canonical path or identifier. Work on view maintainence based on this model has also been studied in [14]. This leads us to our next topics, those of citing and archiving data.

3   Data Citation

A digital library is typically a large and heterogeneous collection of on-line documents and databases with sophisticated software for exploring the collection [13]. However many digital libraries are also being organized so that they serve as scholarly resources. This being the case, how do we *cite* a component of a digital library. Surprisingly, this topic has received very little attention. There appear to be no generally useful standards for citations. Well organized databases are constructed with *keys* that allow us uniquely to identify a tuple in a relation. By giving the attribute name we can identify a component of a tuple, so there is usually a canonical path to any component of the database.

How we cite portions of documents, especially XML documents is not so clear. A URL provides us with a universal locator for a document, but how are we to proceed once we are inside the document? Page numbers and line numbers – if they exist – are friable, and we have to remember that an XML document may now represent a database for which the linear document structure is irrelevant. There are some initial notions of keys in the XML standard [4] and in the XML Schema proposals [16]. In the XML Document Type Descriptor (DTD) one can declare an ID attribute. Values for this attribute are to be unique in the document and can be used to locate elements of the document. However the ID attribute has nothing to do with the structure of the document – it is simply a user-defined identifier.

In XML-Schema the definition of a key relies on XPath [8], a path description language for XML. Roughly speaking a key consists of two paths through the data. The first is a path, for example `Department/Employee`, that describes the set of nodes upon which a key constraint is to be imposed. This is called the *target set*. The second is another path, for example `IdCard/Number` that uniquely identifies nodes in the target set. This second part is called the key path, and the rule is that two distinct nodes in the target set must have different values at the end of their key paths. Apart from some details and the fact that XPath is probably too complex a language for key specification, this definition is quite serviceable, but it does not take into account the hierarchical structure of keys that are common in well-organized databases and documents.

To give an example of what is needed, consider the problem of citing a part of a bible, organized by chapter, book and verse. We might start with the idea that books in the bible are keyed by name, so we use the pair of paths (`Bible/Book`, `Name`). We are assuming here that `Bible` is the unique root. Now we may want to indicate that chapters are specified by number, but it would be incorrect to write (`Bible/Book/Chapter`, `Number`) because this says that that chapter numbers are unique within the bible. Instead we need to specify a *relative key* which consists of a triple, (`Bible/Book`, `Chapter`, `Number`). What this means is that the (`Chapter`, `Number`) key is to hold at every node specified by by the path `Bible/Book`.

A more detailed description of relative keys is given in [6]. While some basic inference results are known, there is a litany of open questions surrounding them: What are appropriate path languages for the various components of a key? What inference results can be established for these languages? How do we specify foreign keys, and what results hold for them? What interactions are there between keys and DTDs. These are practical questions that will need to be answered if, as we do in databases, use keys as the basis for indexing and query optimization.

4   Archiving and Other Problems Associated with Provenance

Let us suppose that we have a good formulation, or even a standard, for data citation, and that document A cites a (component of a) document B. Whose responsibility is it to maintain the integrity of B? The owner of B may wish to update it, thereby invalidating the citation in A. This is a serious problem in scientific databases, and what is commonly done is to release successive versions of a database as separate documents. Since one version is – more or less – an extension the previous version, this is wasteful of space and the space overhead limits the rate at which one can release versions. Also, it is difficult when the history of a database is kept in this form to trace the history of components of the database as defined by the key structure. There are a number of open questions :

- Can we compress versions so that the history of A can be efficiently recorded?
- Should keeping the cited data be the responsibility of A rather than B?
- Should B figure out what is being cited and keep only those portions?

In this context it is worth noting that, when we cite a URL, we hardly ever give a date for the citation. If we did this, at least the person who follows the citation will know whether to question the validity of the citation by comparing it with the timestamp on the URL.

Again, let us suppose that we have an agreed standard for citations and that, rather than computing provenance by query inversion (which is only possible when the data of interest is created by a query,) we decide to annotate each element in the database with one or more citations that describes its provenance. What is the space overhead for doing this? Given that the citations have structure and that the structure of the data will, in part, be related to the structure of the data, one assumes that some form of compression is possible.

Finally, one is tempted to speculate that we may need a completely different model of data exchange and databases to characterize and to capture provenance. One could imagine that data is exchanged in packages that are "self aware"[2] and somehow contain a complete history of how they moved through the system of databases, of how they were constructed, and of how they were changed. The idea is obviously appealing, but whether it can be formulated clearly, let alone be implemented, is an open question.

References

[1] A. Woodruff and M. Stonebraker. Supporting fine-grained data lineage in a database visualization environment. In *ICDE*, pages 91–102, 1997.
[2] Serge Abiteboul, Peter Buneman, and Dan Suciu. *Data on the Web. From Relations to Semistructured Data and XML*. Morgan Kaufman, 2000.

---

[2] A term suggested by David Maier

[3] T. Barsalou, N. Siambela, A. Keller, and G Wiederhold. Updating relational databases through object-based views. In *Proceedings ACM SIGMOD*, May 1991.

[4] Tim Bray, Jean Paoli, and C. M. Sperberg-McQueen. *Extensible Markup Language (XML) 1.0*. World Wide Web Consortium (W3C), Feb 1998. http://www.w3.org/TR/REC-xml.

[5] P. Buneman, S. Davidson, M. Liberman, C. Overton, and V. Tannen. Data provenance. http://db.cis.upenn.edu/~wctan/DataProvenance/precis/index.html.

[6] Peter Buneman, Susan Davidson, Carmem Hara, Wenfei Fan, and Wang-Chiew Tan. Keys for XML. Technical report, University of Pennsylvania, 2000. http://db.cis.upenn.edu.

[7] Peter Buneman, Sanjeev Khanna, and Wang-Chiew Tan. Why and Where: A Characterization of Data Provenance. In *International Conference on Database Theory*, 2001. To appear, available at http://db.cis.upenn.edu.

[8] James Clark and Steve DeRose. *XML Path Language (XPath)*. W3C Working Draft, November 1999. http://www.w3.org/TR/xpath.

[9] Y. Cui and J. Widom. Practical lineage tracing in data warehouses. In *ICDE*, pages 367–378, 2000.

[10] Jon Doyle. A truth maintenance system. *Artificial Intelligence*, 12:231–272, 1979.

[11] R. G. G. Cattell et al, editor. *The Object Database Standard: Odmg 2.0*. Morgan Kaufmann, 1997.

[12] A. Gupta and I. Mumick. Maintenance of materialized views: Problems, techniques, and applications. IEEE Data Engineering Bulletin, Vol. 18, No. 2, June 1995., 1995.

[13] Michael Lesk. *Practical Digital Libraries: Books, Bytes and Bucks,*. Morgan Kaufmann, July 1997.

[14] Hartmut Liefke and Susan Davidson. View maintenance for hierarchical semistructured data. In *International Conference on Data Warehousing and Knowledge Discovery*, 2000.

[15] Susan Davidson and Chris Overton and Peter Buneman. Challenges in Integrating Biological Data Sources. *Journal of Computational Biology*, 2(4):557–572, Winter 1995.

[16] World Wide Web Consortium (W3C). *XML Schema Part 0: Primer*, 2000. http://www.w3.org/TR/xmlschema-0/ .

Fast On-Line/Off-Line Algorithms for Optimal Reinforcement of a Network and Its Connections with Principal Partition

Sachin B. Patkar[1] and H. Narayanan[2]

[1] Department of Mathematics, Indian Institute of Technology - Bombay,
Mumbai-400 076, India.
patkar@math.iitb.ernet.in
[2] Department of Electrical Engg., Indian Institute of Technology - Bombay,
Mumabi-400 076, India.
hn@ee.iitb.ernet.in

Abstract. The problem of computing the strength and performing optimal reinforcement for an edge-weighted graph $G(V, E, w)$ is well-studied [1,2,3,6,7,9]. In this paper, we present fast (sequential linear time and parallel logarithmic time) on-line algorithms for optimally reinforcing the graph when the reinforcement material is available continuosly on-line. These are first on-line algorithms for this problem. Although we invest some time in preprocessing the graph before the start of our algorithms, it is also shown that the output of our on-line algorithms is as good as that of the off-line algorithms, making our algorithms viable alternatives to the fastest off-line algorithms in situations when a sequence of more than $O(|V|)$ reinforcement problems need to be solved. In such a situation the time taken for preprocessing the graph is less that the time taken for all the invocations of the fastest off-line algorithms. Thus our algorithms are also efficient in the general sense. The key idea is to make use of the theory of Principal Partition of a Graph. Our results can be easily generalized to the general setting of principal partition of nondecreasing submodular functions.

1   Introduction

Let $G(V, E)$ denote a graph with V as the vertex set and E as the set of edges. We use $G(V, E, w)$ to denote a graph $G(V, E)$ with nonnegative edge-weights given by $w(.)$.

A fundamental problem concerning practical networks is that of making the connectivity reliable in the face of failure of individual edges. Cunningham [3] defined **strength** of a graph $G(V, E, w)$ as

$$\min_{\emptyset \neq X \subseteq E} \frac{w(X)}{\text{number of additional components created by destroying } X}$$

which is same as, $\min_{\emptyset \neq X \subseteq E} \frac{w(X)}{r(E) - r(E-X)}$, where $r(Z)$ denotes the rank of the subgraph on edge set Z, which is defined as the sum of the ranks of the connected

S. Kapoor and S. Prasad (Eds.): FST TCS 2000, LNCS 1974, pp. 94–105, 2000.

components of the subgraph on Z. Recall that the rank of a connected graph equals the number of vertices minus one. Thus the *strength* of a graph is a *measure of its invulnerability*.

In [3] Cunningham considered the problem of computing the strength of a graph $G(V, E, w)$, and optimal reinforcement of the edge-weights to raise the strength to a prescribed level.

In this paper we show the relation between these problems and the notion of the Principal Partition (**PP**) of a graph. We show that computing the PP of a graph beforehand allows us to solve the "successive (on-line) reinforcement" problem very efficiently. Furthermore, it also gives us a lot of information about the relationship between the amount of reinforcement material available and maximum strength realizable by utilizing it. Throughout, the algorithms based on PP are extremely simple and efficient.

The problem of computing optimal reinforcement to strengthen a graph $G(V, E, w)$ is as follows:

Problem 1. **Main Problem: Optimal Reinforcement of a Graph**
Given $G(V, E, w)$ and the required strength $\hat{\lambda}$, find a vector of increase (zero increase allowed) in the weights of the edges such that with the resulting edge-weights the graph has strength equal to $\hat{\lambda}$ and the total increase is minimum. (We are not permitted to introduce new edges.)

Using the well known ideas from "fractional programming" [3,4], one sees that it would suffice to consider the following family (parameterized by real values λ) of problems:

$$\min_{X \subseteq E} \{w(X) - \lambda * (r(E) - r(E - X))\}.$$

or equivalently,

$$\max_{X \subseteq E} \{w(X) - \lambda * r(X)\}.$$

This latter is the *Weighted Principal Partition Problem*
Given a graph $G(V, E)$ and a real positive weight assignment w to the edges, to find for each real λ, the collection of all subsets of E which maximize $w(.) - \lambda * r(.)$.

It turns out that it is sufficient to consider not more that $r(E)$ values of λ to solve this problem completely. Using these "critical values" the PP (the above collection of subsets) can be constructed and stored efficiently [6,16,17,19].

It may be remarked that efficient algorithms for computation of PP for the real weights case have already been given in [10,14,16,17] earlier. But to reveal the connection between the classical idea of PP and the "strength and reinforcement" problem, we present a new approach to computation of the required information from the PP, which runs in the same time as required by the previous best algorithms in [16,17].

The problem of *computing strength and optimal reinforcement* has been studied and solved efficiently by several researchers [1,2,3,6,7,9]. The best algorithms are due to Gabow [7] and Cheng and Cunningham [2]. They solve the problems of computing strength and min-cost optimal reinforcement each in time $O(|V|^2|E|\log(|V|^2/|E|))$. It may be remarked that, there are no known faster algorithms for the unit-cost optimal reinforcement problem. The ideas underlying the best algorithms also appeared in [1,3,9]. We may also mention that the many of the fast algorithms mentioned above are based on ideas from [8]. The connection of the problem of computing the strength with the classical problem of Principal Partition was brought out by Fujishige [6], who related the strength to the smallest critical value in a principal partition of a dual of the rank function of the given graph. We extend this connection of principal partition to optimal reinforcement, too, using it to build fast oracles suitable for on-line real-time computations.

We mainly consider the practical problem of doing optimal reinforcement. After computation of a certain skeleton of the principal partition, called Principal Sequence, we solve the problem of optimal reinforcement in fast time (sequential linear time and parallel logarithmic time) for every successive request of optimal reinforcement. Thus our algorithm is indeed on-line and real-time too. When the number of requests for successive optimal reinforcement is larger that $O(|V|)$, our algorithms turn out to be faster than the fastest off-line algorithms due to [7] [2], without any loss of quality of output.

We solve the following problems which relate to on-line version of the optimal reinforcement problem:

Problem 2. **Problem (P1):**
Given a graph $G(V, E, w)$, build an efficient oracle for the function $W(.)$ that maps the required strength λ to the minimum total amount of weight augmentation to be performed to increase the strength to λ.

The following is an "inverse" of the above problem:

Problem 3. **Problem (P2):**
Given a graph $G(V, E, w)$ and a specified amount W of total weight augmentation permitted, find the maximum strength achievable by using the amount W to augment weights of the edges. Note that we are not permitted to decrease any of the existing weights.

Now consider the following practical situation: Reinforcement material is made available in arbitrary quantities at arbitrary intervals. At every stage when the reinforcement arrives, we are required to utilize the reinforcement fully and optimally (that is, without saving some amount for future use, and making sure that the graph is strengthened to the best level using the current lot of reinforcement).

Furthermore, at any stage, the cumulative augmentation (or reinforcement) should look as if it were the optimal reinforcement done if the whole reinforcement were made available in one go and we were supposed to utilize it optimally.

We formally model the above scenario as the following problem and solve it in this paper:

Problem 4. **Problem (P3):**
Given a graph $G(V, E, w)$, build the oracles for the family of monotonic non-decreasing functions:

$$\{f_e : [\lambda_t, \infty] \to \mathbf{R}^+ \mid e \in E\},$$

which satisfy the following:
For $\lambda \geq \lambda_t$, $\{f_e(\lambda) \mid e \in E\}$ represent the weight augmentations to be carried out for each edge $e \in E$ such that the strength of the graph $G(V, E)$ goes up to λ. Furthermore, $\sum_{e \in E} f_e(\lambda)$ is required to be minimum.

Thus the decisions as to how much weight to be augmented for a given edge in the current stage would be taken by querying the above oracles. We would like these oracles to be efficient so that the above decisions could be taken **in real-time**, and we also want an efficient algorithm for building this family of oracles. It is clear that as a consequence of building the above oracles, we would have **on-line** and **real-time** algorithm to solve the problem of reinforcement of a graph optimally.

Now we review the literature for PP briefly: The PP of a graph for the case $\lambda = 2$ was constructed by Kishi and Kajitani (see [16] for details). For arbitrary real λ the problem was solved for matroids independently by Narayanan [13] and Tomizawa [19]. Extensive work has been done on these problems and their generalization to matroids, particularly in Japan [5,6,10,11,14,16,17,19]. Details may be found in [6,16]. Principal Partition problem also has many significant applications- to Electrical Network theory (see, for instance, [11,16]), Fault tolerant computing [12] and to Engineering Systems in general [11]. Using the Principal Partition of a graph, approximate algorithms were designed for an NP-hard problem of computing *Min-k-cut* of a graph (see [15]). Patkar and Narayanan [16,17] gave an $O(|E||V|^3 log|V|)$ algorithm for computing the whole Weighted Principal Partition of the graph.

Due to lack of space, certain details are omitted. Interested reader is referred to [18] for complete version of this paper.

2   Preliminaries and Notation

We deal throughout with finite sets. A function f on the subsets of S is said to be submodular if $f(A) + f(B) \geq f(A \cup B) + f(A \cap B) \ldots \forall A, B \subseteq S$. f is said to be supermodular iff $-f$ is submodular. We say that a function is normalized if takes value 0 on the emptyset. A function f is nondecreasing if $X \subseteq Y \Rightarrow f(X) \leq f(Y)$. A normalized, nondecreasing and submodular function is also called a polymatroid function.

$G(V, E) \times Z$ denotes the graph obtained by contracting $E - Z$ from $G(V, E)$. Note that we maintain $G \times Z$ as a multigraph, that means, every edge in $G \times Z$ corrsponds to some edge in the original graph G. $G(V, E) \bullet Z$ denotes the subgraph of $G(V, E)$ induced by Z. By abuse of notation $r(G)$ denotes the rank of the set of edges of the graph G. Note that $r(.)$ is normalized, nondecreasing and submodular function.

$\mathcal{V}(X)$ denotes the set of vertices spanned by the edge set X. $\mathcal{E}(U)$ denotes the set of edges having both the endpoints in U. The subgraph of G on $U \subseteq V$ is denoted by $(U, \mathcal{E}(U))$. Note that $w(.) - k * r(.)$ is a supermodular function when $k \geq 0$.

The following fact from the literature on the submodular functions [6,16,19] is very basic and useful.

Theorem 1. *[6,13,16,19] Let W be a set and $U \subseteq W$. Let $f : 2^W \to \mathbf{R}$ be a submodular (supermodular) function. Then the subsets which minimize (maximize) the function $f(.)$ over all those subsets of W which contain U, form a lattice under the usual operations of union and intersection. In particular there exist unique smallest and largest such sets.*

3   Some Relevant Properties of Principal Partition

Some properties of the Principal Partition which are relevant to this paper are as follows [16,19]:

1. There is a unique maximal set X^k and a unique minimal set X_k at which $w(.) - k * r(.)$ reaches the maximum. We call these sets *critical sets* in the Principal Partition of the graph.
2. If $k_1 \geq k_2$, it can be shown that $X^{k_1} \subseteq X_{k_2}$.
3. For finitely many values of k, $X_k \neq X^k$. Such values are called *critical values* in the Principal Partition of the graph.
4. There are at most $r(E)$ critical values in the Principal Partition of the graph.
5. Let $\lambda_1 > \lambda_2 > \lambda_3 \ldots > \lambda_t$ be the sequence of all critical values. The last critical value λ_t (see [6]) is equal to the **strength** of the given graph G under the edge-weights w. We also take $\lambda_0 = \infty$ as a convention.
   Furthermore, $X^{\lambda_i} = X_{\lambda_{i+1}}$ for $i = 1, 2, \ldots, t-1$, and $X_{\lambda_1} = \emptyset$, $X^{\lambda_t} = E$. The sequence $X_{\lambda_1} \subset X_{\lambda_2} \ldots X_{\lambda_t} \subset X^{\lambda_t}$ is called the **Principal Sequence**.

The following characterization of strength (thus that of smallest critical value) is implicit in [3].

Lemma 1. *Let $G(V, E, w)$ be the given graph with edge-weights given by $w(.)$. Let σ be such that $w(Z) - \sigma * r(Z) = w(E) - \sigma * r(E) = \max_{X \subseteq E}\{w(X) - \sigma * r(X)\}$ for a proper subset $Z \subset E$. Then σ is the **strength** of $G(V, E, w)$. The converse also holds.*

We state the following result about the computation of the Principal Sequence of $G(V, E, w)$ which will follow from the algorithm in Section 7.

Theorem 2. *The Principal Sequence of $G(V, E, w)$ can be computed using $O(|V|)$ invocations of the subroutine that computes the strength of a given subgraph of $G(V, E, w)$.*

4   An Algorithm Based on Principal Partition for Minimum-Weight Reinforcement

In this section we present an algorithm which will perform minimum-weight reinforcment using the Principal Sequence which is assumed to be available. The ideas from this algorithm will be used to build on-line algorithm (using the oracles as described in **Problem (P3)**) in the later section. We start with the following definition.

Definition 1. *Let $X_0 \subset X_1 \subset X_2 \ldots \subset X_t$ (with $X_0 = \emptyset$ and $X_t = E$) be the Principal Sequence of $G(V, E, w)$. Let $\lambda_1 > \lambda_2 \ldots > \lambda_t$ be the sequence of critical values. Let $E_i = X_i - X_{i-1}$, and $G_i = (G \bullet X_i) \times (X_i - X_{i-1})$ for $i = 1, 2, \ldots t$.*

Thus G_i is a minor of G obtained by first restricting it to X_i and then contracting out the subset X_{i-1}. By one of the properties of PP, λ_t is the strength of $G(V, E, w)$. We wish to increase the strength to $\hat{\lambda}$ by augmenting the weight function w to suitable $\hat{w}$. We also require that the total augmentation in the weights, that is $(\hat{w} - w)(E)$, is as small as possible.

4.1   Algorithm 1

Our algorithm is as follows:
Algorithm 1

- Let p be smallest index such that $\hat{\lambda} > \lambda_p$. Thus $\lambda_{p-1} \geq \hat{\lambda} > \lambda_p$.
- Let F_p be a subset of edges of E that forms a spanning forest of G_p. Recall that we maintain G_i's (graphs after contraction) as multigraphs, that means, every edge in G_i corrsponds to some edge in the original graph G.
- We add $\hat{\lambda} - \lambda_p$ to the weight of each of the edges in F_p.
- For each j from $p + 1$ to t we do something similar:
  - Let F_j be a subset of edges of E that forms a spanning forest of G_j.
  - We add $\hat{\lambda} - \lambda_j$ to the weight of each of the edges in F_j.

4.2   Proof of Correctness of Algorithm 1

We need to prove the following.

Theorem 3. *Algorithm 1 uses minimum weight augmentation in order to increase the strength of a graph $G(V, E, w)$ to a prescribed level. Furthermore, the minimum weight augmentation required to increase the strength from λ_t to $\hat{\lambda}$ equals $\sum_{j=p}^{t}(\hat{\lambda} - \lambda_j) * rank(G_j)$.*

To prove the theorem 3 we will use the following definitions and results.

Definition 2. *Let $\hat{G}(\hat{V}, \hat{E})$ be a graph with rank function $r(.)$ on the subset of edges. Let $\hat{w}(.)$ be a non-negative weight function on the edges of this graph and let λ be a non-negative real. We say that the graph is **molecular** with respect to $(\hat{w}, \lambda)$ if*

$$\hat{w}(\emptyset) - \lambda * r(\emptyset) = \hat{w}(\hat{E}) - \lambda * r(\hat{E}) = \max_{X \subseteq \hat{E}}\{\hat{w}(X) - \lambda * r(X)\} \qquad (1)$$

We need a few lemmas.

By the characterization of strength (lemma 1) it is clear that,

Lemma 2. *A graph is molecular w.r.t.* $(\hat{w}, \lambda)$, **if and only if** *its strength is equal to* λ *when the edge-weights are given by* $\hat{w}(.)$, *and the total edge-weight of such a graph is equal to* $(\lambda *$ *rank of the graph).*

The proofs of the following lemmas (lemma 3 and lemma 4) will easily follow from the algorithm given in a later section that constructs the PP of $G(V, E, w)$.

Lemma 3. *For each* $i = 1, 2, \ldots t$, *the strength of* $G \bullet (E_1 \cup E_2 \cup \ldots E_i)$ *equals* λ_i.

Lemma 4. *Let* G_i *be as defined before for* $i = 1, 2, \ldots t$. *Strength of* G_i *is equal to* λ_i *and it is molecular with respect to* (w, λ_i).

We state the following lemmas without proofs (see [18] for complete details).

Lemma 5. *Let* σ *be the strength of* $G(V, E, w)$. *Let* F *be any spanning forest of* $G(V, E)$. *If we increase the weights of each edge of* F *by* α, *then the strength of the weight-augmented graph is at least* $\sigma + \alpha$. *Furthermore, if* $G(V, E, w)$ *is molecular w.r.t.* (w, σ) *then the resulting strength is equal to* $\sigma + \alpha$, *and the resulting graph is molecular w.r.t.* $(w', \sigma + \alpha)$, *where* w' *denotes the augmented weights.*

Lemma 6. *Let* $G'(V', E')$ *be a graph with edge-weights* w'. *Let* σ *be the strength of* G'. *Let* $\emptyset \neq Z \subseteq E'$ *such that* $G' \bullet Z$ *has strength* σ_1 *that is at least as large as* σ. *Further suppose* $G' \times (E' - Z)$ *is molecular w.r.t.* (w', σ). *Let* $\sigma_1 \geq \lambda \geq \sigma$. *Then addition of weight* $\lambda - \sigma$ *to any spanning forest* F *of* $G' \times (E' - Z)$ *raises the strength of* G' *to* λ.

We state a simple lemma about strength after any contraction or augmentation operation. The proof follows immediately from the definitions.

Lemma 7. *Strength of a graph does not decrease after any augmentation or contraction.*

Now we are ready to prove theorem 3.

Proof of Theorem 3:

We first establish that the resulting weight function, provides the required strength, that is, $\hat{\lambda}$ *to the graph* $G(V, E)$.

We use the following notation: $\hat{G}_i = G \bullet (E_1 \cup \ldots E_i)$. G_i is as defined before, and let $r_i(.)$ denote the rank function of G_i for $i = 1, 2, \ldots, t$.

We prove by induction on $i = p, p+1, \ldots, t$, that after the augmentation step on G_i, $\hat{G}_i$ has strength $\hat{\lambda}$, but the strength of $\hat{G}_{i+1}$ remains at λ_{i+1}.

Induction base: Lemma 3 and lemma 4 tell us that prior to augmentation step the strength of G_p is λ_p, strength of $\hat{G}_{p-1}$ is λ_{p-1} and strength of $\hat{G}_p$ is λ_p.

If $p = 1$ the we make use of lemma 5 to conclude that strength of $\hat{G}_p$ becomes $\hat{\lambda}$. Otherwise, we apply lemma 6 to $\hat{G}_p$, $\hat{G}_{p-1}$ and G_p. This establishes that after the augmentation step on G_p, the strength of $\hat{G}_p$ becomes $\hat{\lambda}$. But, under the augmented weights, the strength of $\hat{G}_{p+1}$ remains at λ_{p+1} as

$$\text{strength of } \hat{G}_{p+1} \leq \text{strength of } G_{p+1} = \lambda_{p+1}.$$

and the strength of $\hat{G}_{p+1}$ would not decrease from its earlier value of λ_{p+1} after augmentation step (The inequality in the above follows from the lemma 7). Thus the induction base is proved:

The proof of the inductive hypothesis is along similar lines as above. In fact, the key idea is once again the use of lemma 6.

Now we establish that the weight added is minimum that is required to increase the strength to the prescribed level, $\hat{\lambda}$.

Towards this, one may look at the graph obtained from $G(V, E)$ by contracting out the set of edges $E_1 \cup E_2 \ldots E_{p-1}$. The resulting graph is on the set of edges $E_p \cup E_{p+1} \ldots E_t$. Let G' denote this resulting graph.

We once again make use of the lemma 7. For the strength of $G(V, E)$ to be $\hat{\lambda}$, it is required that the strength of G' should be at least $\hat{\lambda}$. Thus, G' must have at least $\hat{\lambda} * rank(G')$ as the total weight of the edges after the augmentation.

Now $rank(G') = \sum_{j=p}^{t} rank(G_j)$. Thus the weight of the the edges of G' is at least $\sum_{j=p}^{t} \hat{\lambda} * rank(G_j)$.

But original weight of the edges of G' was $\sum_{j=p}^{t} \lambda_j * rank(G_j)$. This follows from molecularity of G_j w.r.t. (w, λ_j) and lemma 2 .

Thus the minimum weight augmentation required to increase the strength from λ_t to $\hat{\lambda}$ equals

$$\sum_{j=p}^{t} (\hat{\lambda} - \lambda_j) * rank(G_j).$$

But then, our algorithm has used exactly the same (as above) amount of weight augmentation for increasing the strength, thus our algorithm has performed **minimum weight augmentation of edge weights to increase the strength from** λ_t **to** $\hat{\lambda}$. **q.e.d**

5   Minimum Weight Successive Augmentation to Increase the Strength of a Graph

In this section we show how the ideas underlying **Algorithm 1** can be used to solve **Problem (P3)**.

To show this, we make use of the spanning forests $F_1, F_2, \ldots F_t$ of the graphs $G_1, G_2, \ldots, G_t$, respectively.

Define for $\lambda > \lambda_t$,

$$f_e(\lambda) = \lambda - \lambda_j \text{ if } \quad e \in F_j \text{ and } \lambda > \lambda_j$$
$$= 0 \quad \text{otherwise} \tag{2}$$

Time required to build efficient oracles for the functions $f_e(.), e \in E$, is clearly dominated by the time required for finding the sets $E_1, E_2, \ldots, E_t$, which (by theorem 2) may be done by $O(|V|)$ invocations of Cheng and Cunningham's or Gabow's algorithm from [2,7].

After choosing the spanning forests $F_1, F_2, \ldots F_t$, we remember in each oracle for $f_e(.)$ the pair (j, λ_j) if $e \in F_j$ for some j, and if no such j exists the oracle returns 0.

Clearly, each oracle $f_e(.)$ answers in constant time.

Theorem 4. *If we use ideas in Algorithm 1, then the time required to solve* **Problem (P3)** *is* $(|V| * T_{strength})$, *where* $T_{strength}$ *denotes the time required to compute the strength of graph* $G(V, E, w)$. *The space complexity of the oracle for* $f_e(.)$ *for any edge* $e \in E$ *is also* $O(1)$. *Furthermore, each of the oracle of* **Problem (P3)** *requires constant time to provide answer, if it is fed with the input* λ.

From the above discussion and *Algorithm 1* we get the following lemma.

Lemma 8. *The function* $W(.)$ *that maps strength* λ *to the minimum total weight augmentation required to increase the strength to* λ *is given by the following piecewise linear function:* $W : [\lambda_t, \infty] \to \mathbf{R}_+$ *and it satisfies,*

- $W(\lambda_t) = 0$,
- *and the slope of* $W(.)$ *in the domain interval* $[\lambda_{i+1}, \lambda_i]$ *is equal to* $\sum_{j=i+1}^{t} r(G_j)$.

Thus an efficient oracle $W(.)$ as well as its inverse map could be built easily, after having computed the graphs $G_1, G_2, \ldots, G_t$ by $O(|V|)$ invocations of the algorithm that finds strength of a graph (on a sequence of smaller and smaller graphs).

Note that in then construction of the above oracle for $W(.)$, one could make use of the values $\lambda_1, \ldots, \lambda_t$ and the slopes of $W(.)$ in the domain intervals $[\lambda_{i+1}, \lambda_i]$ for $i = 0, 1, 2, \ldots, t - 1$.

With this information in the oracle, the oracle and its inverse can provide the answers in **time logarithmic in** $|V|$. Thus we have a good solution for problems **Problem (P1)** and **Problem (P2)**

6   Another Algorithm for Optimal Reinforcement of a Graph

We now present another, slightly modified, algorithm which would give a technique for a different approach for solving **Problem (P3)** (see [18] for details). **Algorithm 2**

- Let p be smallest index such that $\hat{\lambda} > \lambda_p$. Thus $\lambda_{p-1} \geq \hat{\lambda} > \lambda_p$.
- Let $\hat{F}_p$ be a subset of edges of E that forms a spanning forest of

$$G \times (E_p \cup E_{p+1} \cup \ldots E_t).$$

– We add $\hat{\lambda} - \lambda_p$ to the weight of each of the edges in $\hat{F}_p$.
– For each j from $p+1$ to t we do something similar:
  • Let $\hat{F}_j$ be a subset of edges of E that forms a spanning forest of

$$G \times (E_j \cup E_{j+1} \cup \ldots E_t).$$

  • We add $\hat{\lambda} - \lambda_j$ to the weight of each of the edges in $\hat{F}_j$.

Theorem 5. *The above algorithm computes a minimum weight augmentation to increase the strength from λ_t to $\hat{\lambda}$.*

7 An Algorithm to Compute Principal Sequence

In what follows, we describe a new approach based on Cunningham's algorithm [3] for computation of the Principal Sequence of a graph.

A simple modification of Cunningham's [2,3] algorithm computes σ and the smallest proper subset $Z \subset E$ such that

$$w(Z) - \sigma * r(Z) = w(E) - \sigma * r(E) = \max_{X \subseteq E}\{w(X) - \sigma * r(X)\} \qquad (3)$$

7.1 Computation of Principal Sequence Using the Subroutine to Compute the Strength

Let μ_1 denote the strength of $G(V, E, w)$. Let H_1 denote the largest subset of E such that

$$w(E - H_1) - \mu_1 * r(E - H_1) = w(E) - \mu_1 * r(E) = \max_{X \subseteq E}\{w(X) - \mu_1 * r(X)\}. \qquad (4)$$

Let μ_2, H_2 be obtained by performing the above procedure on the graph

$$G(V, E) \bullet (E - H_1).$$

Let μ_3, H_3 be obtained by performing the above procedure on the graph

$$G(V, E) \bullet (E - (H_1 \cup H_2))$$

and so on . . ..
In general, let μ_i, H_i be obtained by performing the above procedure on the graph

$$G(V, E) \bullet (E - (H_1 \cup H_2 \cup \ldots H_{i-1})).$$

We stop the above process when we find μ_t and H_t such that

$$E = H_1 \cup H_2 \cup \ldots \cup H_t.$$

Noting that the sequence of ranks of successive subgraphs is strictly decreasing, one sees that,

Lemma 9. $t \leq r(E)$.

7.2   Proof of Correctness of the Above Algorithm for Computation of the Principal Sequence

The discussion in this section will establish that the above algorithm has indeed built the Principal Sequence of the rank function of an edge-weighted graph.

Let $\mu_1, \mu_2, \ldots \mu_t$ and $H_1, H_2, \ldots H_t$ be as defined above, for an edge-weighted graph $G(V, E, w)$. Let us define $\lambda_i = \mu_{t+1-i}$, and $E_i = H_{t+1-i}$. We also follow the convention that $\lambda_0 = \infty$ and $E_0 = \emptyset$.

We make use of the following technical lemma (see [18] for proof).

Lemma 10. *Let σ be the strength of $G(V, E, w)$. Let Z be the smallest (proper) subset of E such that*

$$w(Z) - \sigma * r(Z) = w(E) - \sigma * r(E) = \max_{X \subseteq E}\{w(X) - \sigma * r(X)\} \qquad (5)$$

then,

1. *$G(V, E) \times (E - Z)$ is molecular w.r.t. (w, σ), and therefore has strength equal to σ.*
2. *If $Z \neq \emptyset$ then, $G(V, E) \bullet Z$ has strength strictly greater than σ.*

We will use the following well-known characterization of the Principal Sequence (see [16], *pp*.404).

Theorem 6. *[13,16,19] $Z_0 \subset Z_1 \subset \ldots \subset Z_l$, with $Z_0 = \emptyset$ and $Z_l = E$, is the Principal Sequence of $G(V, E, w)$ **if and only if** there exists $\gamma_1 > \gamma_2 \ldots > \gamma_l$ such that for each $i = 1, 2, \ldots, l$, $(G \bullet Z_i) \times (Z_i - Z_{i-1})$ is molecular w.r.t. (w, γ_i).*
Furthermore, the Principal Sequence exists and is unique.

Thus using theorem 6 and lemmas 2 3 4 it is clear that,

Theorem 7. *The algorithm to preprocess the graph has decomposed the edge set E into $E_1, E_2, \ldots, E_t$ such that G_i (as defined before) is molecular w.r.t. (w, λ_i) for $i = 1, 2, \ldots, t$. The nested sequence of sets*

$$\emptyset \subset E_1 \subset (E_1 \cup E_2) \ldots \subset (E_1 \cup E_2 \ldots E_t) \, (= E)$$

is the Principal Sequence of the graph $G(V, E, w)$.

Using lemma 9 and the above theorem we obtain theorem 2.

8   Acknowledgement

We thank the anonymous referees for pointing out a few discrepancies. We also thank DST-India and NRB-India, (grant no. 98DS018, 98NR002) for financial support.

References

1. Barahona, F.: Separating from the dominant of spanning tree polytope, *Oper. Res. Letters*, vol. 12, 1992, pp.201-203.
2. Cheng, E. and Cunningham, W.: A faster algorithm for computing the strength of a network, *Information Processing Letters*, vol. 49, 1994, pp.209-212.
3. Cunningham, W.: Optimal attack and reinforcement of a network, *JACM*, vol. 32, no. 3, 1985, pp.549-561.
4. Dinkelbach, W.: On nonlinear fractional programming, *Management Sci.*, vol. 13, 1967, pp.492-498.
5. Edmonds, J.: Submodular functions, matroids and certain polyhedra, *Proc. Calgary Intl. conference on Combinatorial Structures*, 1970, pp.69-87.
6. Fujishige, S.: *Submodular Functions and Optimization, Annals of Discrete Mathematics*, North Holland, 1991.
7. Gabow, H.N.: Algorithms for graphic polymatroids and parametric s-sets *J. Algorithms*, vol. 26, 1998, pp.48-86.
8. Gallo, G., Grigoriadis, M. and Tarjan, R.E.: A fast parametric network flow algorithm, *SIAM J. of Computing*, vol. 18, 1989, pp.30-55.
9. Gusfield, D.: Computing the strength of a graph, *SIAM J. of Computing*, vol. 20, 1991, pp.639-654.
10. Imai, H.: Network flow algorithms for lower truncated transversal polymatroids, *J. of the Op. Research Society of Japan*, vol. 26, 1983, pp. 186-210.
11. Iri, M. and Fujishige, S.: Use of matroid theory in operations research, circuits and systems theory, *Int. J. Systems Sci.*,vol. 12, no. 1, 1981, pp. 27-54.
12. Itai, A. and Rodeh, M.: The multi-tree approach to reliability in distributed networks, *in Proc. 25th ann. symp. FOCS*, 1984, pp. 137-147.
13. Narayanan, H.:*Theory of matroids and network analysis*, Ph.D. thesis, Department of Electrical Engineering, I.I.T. Bombay, 1974.
14. Narayanan, H.: The principal lattice of partitions of a submodular function, *Linear Algebra and its Applications*, 144, 1991, pp. 179-216.
15. Narayanan, H., Roy, S. and Patkar, S.B.: Approximation Algorithms for min-k-overlap Problems, using the Principal Lattice of Partitions Approach, *J. of Algorithms*, vol. 21, 1996, pp. 306-330.
16. Narayanan, H.: *Submodular Functions and Electrical Networks, Annals of Discrete Mathematics-54*, North Holland, 1997.
17. Patkar, S. and Narayanan, H. : Principal lattice of partitions of submodular functions on graphs: fast algorithms for principal partition and generic rigidity, *in Proc. of the 3rd ann. Int. Symp. on Algorithms and Computation, (ISAAC), LNCS-650*, Japan, 1992, pp. 41-50.
18. Patkar, S. and Narayanan, H. : Fast On-line/Off-line Algorithms for Optimal Reinforcement of a Network and its Connections with Principal Partition, *Technical Report, Industrial Mathematics Group, Department of Mathematics, IIT Bombay, available from authors via e-mail*, 2000.
19. Tomizawa, N.: Strongly irreducible matroids and principal partition of a matroid into strongly irreducible minors (in Japanese), *Transactions of the Institute of Electronics and Communication Engineers of Japan*, vol. J59A, 1976, pp. 83-91.

On-Line Edge-Coloring with a Fixed Number of Colors

Lene Monrad Favrholdt and Morten Nyhave Nielsen

Department of Mathematics and Computer Science
University of Southern Denmark, Odense
{lenem,nyhave}@imada.sdu.dk

Abstract We investigate a variant of on-line edge-coloring in which there is a fixed number of colors available and the aim is to color as many edges as possible. We prove upper and lower bounds on the performance of different classes of algorithms for the problem. Moreover, we determine the performance of two specific algorithms, *First-Fit* and *Next-Fit*.

1   Introduction

The Problem. In this paper we investigate the on-line problem EDGE-COLORING defined in the following way. A number k of colors is given. The algorithm is given the edges of a graph one by one, each one specified by its endpoints. For each edge, the algorithm must either color the edge with one of the k colors, or reject it, before seeing the next edge. Once an edge has been colored the color cannot be altered and a rejected edge cannot be colored later. The aim is to color as many edges as possible under the constraint that no two adjacent edges receive the same color.

Note that the problem investigated here is different from the classical version of the edge coloring problem, which is to color *all* edges with as *few* colors as possible. In [2] it is shown that, for the on-line version of the classical edge coloring problem, the greedy algorithm (the one that we call *First-Fit*) is optimal.

The Measures. To measure the quality of the algorithms, we use the competitive ratio which was introduced in [6] and has become a standard measure for on-line algorithms. For the problem EDGE-COLORING addressed in this paper, the competitive ratio of an algorithm A is the worst case ratio, over all possible input sequences, of the number of edges colored by A to the number of edges colored by an optimal off-line algorithm.

In some cases it may be realistic to assume that the input graphs are all k-colorable. Therefore, we also investigate the competitive ratio in the special case where it is known that the input graphs are k-colorable. This idea is similar to what was done in [1] and [3]. In these papers the competitive ratio is investigated on input sequences that can be fully accommodated by an optimal off-line algorithm with the resources available (in this paper the resource is, of course, the colors). Such sequences are called accommodating sequences. This

S. Kapoor and S. Prasad (Eds.): FST TCS 2000, LNCS 1974, pp. 106–116, 2000.

is generalized in [4], where the competitive ratio as a function of the amount of resources available is investigated.

This paper illustrates an advantage of analyzing accommodating sequences, apart from tailoring the measure to the type of input. A common technique when constructing a difficult proof is to start out investigating easier special cases. In our analysis of the general lower bound on the competitive ratio, the case of k-colorable input graphs was used as such a special case.

The Algorithms. We will mainly consider fair algorithms. A *fair* algorithm is an algorithm that never rejects an edge, unless it is not able to color it. Two natural fair algorithms are *Next-Fit* and *First-Fit* described in Sections 4 and 5 respectively.

The Graphs. The lower bounds on the competitive ratio proven in this paper are valid even if we allow multigraphs. The adversary graphs used for proving the upper bounds are all simple graphs. Thus, the upper bounds are valid even if we restrict ourselves to simple graphs. Furthermore, the adversary graphs are all bipartite except one which could easily be changed to a bipartite graph. Thus, the results are all valid for bipartite graphs too.

The Proofs. Due to space limitations we have omitted the details of some of the proofs. The full version can be found in [5].

2   Notation and Terminology

We label the colors $1, 2, \ldots, k$ and let $C_k = \{1, 2, \ldots, k\}$.

$K_{m,n}$ denotes the complete bipartite graph in which the two independent sets contain m and n vertices respectively.

The terms *fair^D*, *fair^R*, *on-line^D*, and *on-line^R* denote arbitrary on-line algorithms from the classes "fair deterministic", "fair randomized", "deterministic", and "randomized", respectively, for the EDGE-COLORING problem. The term *off-line* denotes an optimal off-line algorithm for the problem.

3   The Competitive Ratio

We begin this section with a formal definition of the competitive ratio for the problem EDGE-COLORING.

Definition 3.1. *For any algorithm $\mathbb{A}$ and any sequence S of edges, let $\mathbb{A}(S)$ be the number of edges colored by $\mathbb{A}$ and let $OPT(S)$ be the number of edges colored by an optimal off-line algorithm. Furthermore, let $0 \leq c \leq 1$.*

An on-line algorithm $\mathbb{A}$ is c-competitive if there exists a constant b such that $\mathbb{A}(S) \geq c \cdot OPT(S) - b$, for any sequence S of edges.

The competitive ratio *of $\mathbb{A}$ is $C_\mathbb{A} = \sup\{c \mid \mathbb{A} \text{ is } c\text{-competitive}\}$.*

3.1   A Tight Lower Bound for Fair Algorithms

In this section a tight lower bound on the competitive ratio for fair algorithms is given. Note that it is not possible to give a *general* lower bound greater than 0, since the algorithm that simply rejects all edges have a competitive ratio of 0.

Theorem 3.2. *For any fair on-line algorithm* $\mathbb{A}$ *for* EDGE-COLORING, $C_{\mathbb{A}} \geq 2\sqrt{3} - 3 \approx 0.4641$.

Proof. Let E_c denote the set of edges colored by *fair^R*, let E_u denote the set of edges colored by *off-line* and not by *fair^R*, and let E_d denote the set of edges colored by both *off-line* and *fair^R*. Thus, $E_u \cup E_d$ are the edges colored by *off-line*, and $E_d \subseteq E_c$. Similarly, for any vertex x, let $d_c(x)$, $d_u(x)$, and $d_d(x)$ denote the number of edges incident to x colored by *fair^R*, *not* colored by *fair^R*, and colored by both *fair^R* and *off-line* respectively. Let c be a constant such that $0 \leq c < \frac{1}{2}$. Then *fair^R* is c-competitive for any c such that $|E_c| \geq c(|E_d| + |E_u|)$, or $|E_c| - c|E_d| \geq c|E_u|$.

Now, the intuition is that, for each edge $e \in E_c$, *fair^R* earns one unit of some value. If *fair^R* can buy all edges in $E_u \cup E_d$ paying the fraction c of a unit for each, then $|E_c| \geq c(|E_d| + |E_u|)$. *fair^R* starts out buying all edges in E_d, paying c for each. The remaining value is distributed to the edges in E_u in two steps. In the first step, each vertex x receives the value $m(x) = \frac{1}{2}\big(d_c(x) - cd_d(x)\big)$. Note that $\sum_{x \in V} m(x) = |E_c| - c|E_d|$. In the next step, the value on each vertex is distributed equally among the edges in E_u incident to it. Thus, each vertex x with $d_u(x) \geq 1$ gives the value $m_u(x) = \frac{m(x)}{d_u(x)}$ to each edge in E_u incident to it. Note that $\sum_{(x,y) \in E_u} \big(m_u(x) + m_u(y)\big) \leq \sum_{(x,y) \in E_u} \big(m_u(x) + m_u(y)\big) + \sum_{d_u(x)=0} m(x) = \sum_{x \in V} m(x) = |E_c| - c|E_d|$. Thus, if $m_u(x) + m_u(y) \geq c$ for any edge $(x,y) \in E_u$, then $c|E_u| \leq \sum_{(x,y) \in E_u} \big(m_u(x) + m_u(y)\big) \leq |E_c| - c|E_d|$ and *fair^R* is c-competitive.

The inequalities below follow from two simple facts. (1) For any vertex $x \in V$, $d_d(x) + d_u(x) \leq k$, since *off-line* can color at most k edges incident to x. (2) For each edge $(x,y) \in E_u$, $d_c(x) + d_c(y) \geq k$, since *fair^R* is a fair algorithm. For any edge $(x,y) \in E_u$,

$$
\begin{aligned}
m_u(x) + m_u(y) &= \frac{1}{2}\left(\frac{d_c(x) - cd_d(x)}{d_u(x)} + \frac{d_c(y) - cd_d(y)}{d_u(y)}\right) \\
&\overset{(1)}{\geq} \frac{1}{2}\left(\frac{d_c(x) - cd_d(x)}{k - d_d(x)} + \frac{d_c(y) - cd_d(y)}{k - d_d(y)}\right) \\
&\overset{(2)}{\geq} \frac{1}{2}\left(\frac{d_c(x) - cd_d(x)}{k - d_d(x)} + \frac{k - d_c(x) - cd_d(y)}{k - d_d(y)}\right)
\end{aligned}
$$

Calculations show that this expression is greater than or equal to c as long as $c \leq 2\sqrt{3} - 3$. $\qquad\square$

In section 4 it is shown that values of k exist for which the competitive ratio of *Next-Fit* is arbitrarily close to $2\sqrt{3} - 3$. Thus, the result in Theorem 3.2 is tight. The next theorem in conjunction with Theorem 3.2, shows that all deterministic fair algorithms must have very similar competitive ratios.

3.2   An Upper Bound for Fair Deterministic Algorithms

Theorem 3.3. *No deterministic fair algorithm* $\mathbb{A}$ *for* EDGE-COLORING *is more than $\frac{1}{2}$-competitive.*

Proof. We construct a simple graph $G = (V_1 \cup V_2, E)$ in two phases. In Phase 1, only vertices in V_1 are connected. In Phase 2, vertices in V_2 are connected to vertices in V_1. Let $|V_1| = |V_2| = n$ for some large integer n.

In Phase 1, the adversary gives an edge between two unconnected vertices $x, y \in V_1$ with a common unused color. Since the edge can be colored, $fair^D$ will do so. This process is repeated until no two unconnected vertices with a common unused color can be found. At that point Phase 1 ends. For any vertex x, let $C_u(x)$ denote the set of colors not represented at x. At the end of Phase 1, the following holds true. For each color c and each vertex x such that $c \in C_u(x)$, x is already connected to all other vertices y with $c \in C_u(y)$. Since x can be connected to at most k other vertices, there are at most k vertices $y \neq x$ such that $c \in C_u(y)$. Thus, $\sum_{x \in V_1} C_u(x) \leq k(k+1)$.

The edges given in Phase 2 are the edges of a k-regular bipartite graph with V_1 and V_2 forming the two independent sets. Note that, by König's Theorem [7], such a graph can be k-colored.

From Phase 2, $fair^D$ gets at most $k(k+1)$ edges, but *off-line* rejects all edges from Phase 1 and accepts all edges from Phase 2, giving a performance ratio of at most $\frac{\frac{1}{2}(nk - k(k+1)) + k(k+1)}{nk} = \frac{nk + k(k+1)}{2nk} = \frac{1}{2} + \frac{k+1}{2n}$. If we allow n to be arbitrarily large, this can be arbitrarily close to $\frac{1}{2}$. $\qquad\square$

3.3   A General Upper Bound

Now follows an upper bound on the competitive ratio for any type of algorithm for EDGE-COLORING, fair or not fair, deterministic or randomized.

Theorem 3.4. *For any algorithm* $\mathbb{A}$ *for* EDGE-COLORING $C_{\mathbb{A}} \leq \frac{4}{7}$.

Proof. In Fig. 1, the structure of the adversary graph is depicted. Each box contains k vertices. When two boxes are connected, there are k^2 edges in a complete bipartite graph between the $2k$ vertices inside the boxes. Note that such a graph can be k-colored. The edges of the graph are divided into n levels, level $1, \ldots, n$. The adversary gives the edges, one level at a time, according to the numbering of the levels. The edges of level i are given in three consecutive phases:

1. H_i: Internal (horizontal) edges at level i. In total k^2 edges.
2. V_i: Internal (vertical) edges between level i and level $i+1$. In total $2k^2$ edges.
3. E_i: External edges at level i. In total $2k^2$ edges.

Let X_{H_i} be a random variable counting how many edges *on-line^R* will color from the set H_i, and let X_{V_i} and X_{E_i} count the colored edges from V_i and E_i respectively. For $i = 0, \ldots, n$, let EXT_i and INT_i be random variables counting

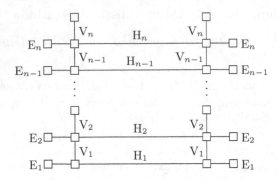

Fig.1. Structure of the adversary graph for the general upper bound on the competitive ratio.

the sum of all external and internal edges, respectively, colored by *on-line^R* after level i is given, i.e., $\text{EXT}_i = \sum_{j=1}^{i} X_{E_j}$ and $\text{INT}_i = \sum_{j=1}^{i}(X_{V_j} + X_{H_j})$. Note that $\text{EXT}_0 = \text{INT}_0 = 0$.

If the adversary stops giving edges after Phase 1 of level i, *off-line* will color $k^2(2i-1)$ edges in total, namely the edges in the sets $E_1, E_2, \dots, E_{i-1}$, and H_i. If the adversary stops giving edges after Phase 2 (or 3) of level i, *off-line* will color $2k^2i$ edges, namely the edges in the sets $E_1, E_2, \dots, E_{i-1}$, and V_i. The proof is divided into two cases.

Case 1: There exists a level $i \leq n$, where $E[\text{EXT}_i] > \frac{2}{7}k^2i$.

Let i denote the first level such that $E[\text{EXT}_i] > \frac{2}{7}k^2i$. Assume that the number of edges colored by *on-line^R* is at least $\frac{4}{7}$ of the number of edges colored by *off-line*. If the adversary stops the sequence after Phase 1 of level i, the following inequality must hold:

(1) $E[\text{INT}_{i-1}] + E[\text{EXT}_{i-1}] + E[X_{H_i}] \geq \frac{4}{7}k^2(2i-1)$.

If the adversary stops the sequence after Phase 2 of level i, the following inequality must hold:

(2) $E[\text{INT}_i] + E[\text{EXT}_{i-1}] \geq \frac{4}{7}k^2 2i$.

If *on-line^R* is $\frac{4}{7}$-competitive, both inequalities must hold. Adding inequalities (1) and (2) yields

(3) $2(E[\text{INT}_{i-1}] + E[\text{EXT}_{i-1}]) + 2E[X_{H_i}] + E[X_{V_i}] \geq \frac{16}{7}k^2i - \frac{4}{7}k^2$.

Now, $E[\text{INT}_{i-1}] \leq \frac{1}{2}(2k^2(i-1) - E[\text{EXT}_{i-1}] - E[X_{V_{i-1}}]) + E[X_{V_{i-1}}]$, $E[\text{EXT}_{i-1}] \leq \frac{2}{7}k^2(i-1)$, and $E[X_{V_{i-1}}] + 2E[X_{H_i}] + E[X_{V_i}] \leq 2k^2 - E[X_{E_i}] < \frac{12}{7}k^2$. Inserting these inequalities into (3) we arrive at a contradiction. Thus, in this case *on-line^R* is not $\frac{4}{7}$-competitive.

Case 2: For all $i \leq n$, $E[\text{EXT}_i] \leq \frac{2}{7}k^2i$.

The expected number of edges colored by *on-line^R* is
$E[\text{INT}_n] + E[\text{EXT}_n] \leq \frac{1}{2}(2k^2n - E[\text{EXT}_n] - E[X_{V_n}]) + E[X_{V_n}] + E[\text{EXT}_n] = k^2n + \frac{1}{2}(E[\text{EXT}_{n-1}] + E[X_{E_n}] + E[X_{V_n}]) \leq k^2n + \frac{1}{2}(\frac{2}{7}k^2(n-1) + 2k^2) = \frac{8}{7}k^2n + \frac{6}{7}k^2$.

Thus, we get an upper bound on the performance ratio of $\frac{\frac{8}{7}k^2 n + \frac{6}{7}k^2}{2nk^2} = \frac{4}{7} + \frac{3}{7n}$, which can be arbitrarily close to $\frac{4}{7}$, if we allow n to be arbitrarily large. $\square$

Thus, even if we allow probabilistic algorithms that are not necessarily fair, no algorithm is more than 0.11 apart from the worst fair algorithm when comparing competitive ratios.

4   The Algorithm *Next-Fit*

The algorithm *Next-Fit* (*NF*) is a fair algorithm that uses the colors in a cyclic order. *Next-Fit* colors the first edge with the color 1 and keeps track of the last used color c_{last}. When coloring an edge (u, v) it uses the first color in the sequence $\langle c_{\text{last}} + 1, c_{\text{last}} + 2, \ldots, k, 1, 2, \ldots, c_{\text{last}} \rangle$ that is not yet used on any edge incident to u or v, if any.

Intuitively, this is a poor strategy and it turns out that its competitive ratio matches the lower bound of section 3.1. Thus, this algorithm is mainly described here to show that the lower bound cannot be improved.

When proving upper bounds for *Next-Fit*, it is useful to note that any coloring in which each color is used on exactly n or $n + 1$ edges, for some $n \in \mathbb{N}$, can be produced by *Next-Fit*, for some ordering of the request sequence. The colors just need to be permuted so that the colors used on $n + 1$ edges are the lowest numbered colors.

Theorem 4.1. $\inf_{k \in \mathbb{N}} C_{NF}(k) = 2\sqrt{3} - 3 \approx 0.4641.$

Proof. The adversary constructs a graph G_{NF} in the following way. It chooses an $x \in C_k$ as close to $(\sqrt{3} - 1)k$ as possible and then constructs a $(k - x)$-regular bipartite graph $G_1 = (L_1 \cup R_1, E_1)$ with $|L_1| = |R_1| = k$ and a graph $G_2 = (L_2 \cup R_2, E_2)$ isomorphic to $K_{x,x}$. Now, each vertex in R_1 is connected to each vertex in L_2 and each vertex in R_2 is connected to each vertex in L_1. Call these extra edges E_{12}. The graph G_{NF} for $k = 4$ is depicted in Fig. 2.

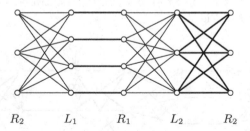

R_2 $\qquad$ L_1 $\qquad$ R_1 $\qquad$ L_2 $\qquad$ R_2

Fig. 2. The graph G_{NF} when $k = 4$, showing that $C_{NF}(4) \leq \frac{13}{28} \approx 0.4643$.

Assume first, that $k - x \leq 1$. In this case $|L_1| \leq |L_2| + 1$, so G_1 and G_2 can be colored by *Next-Fit* with C_{k-x} and $C_k \setminus C_{k-x}$ respectively. After this, *Next-Fit* will not be able to color any of the edges in E_{12}. It is possible however, to color all edges in $E_1 \cup E_{12}$ with k colors, because the subgraph of G_{NF} containing these edges is bipartite and has maximum degree k. Thus, for any $x \in C_k$, the competitive ratio of *Next-Fit* can be no more than $\frac{|E_1|+|E_2|}{|E_1|+|E_{12}|} = \frac{k(k-x)+x^2}{k(k-x)+2kx} = \frac{k^2-kx+x^2}{k^2+kx}$. This ratio attains its minimum value of $2\sqrt{3}-3$ when $x = (\sqrt{3}-1)k$. Thus, by allowing arbitrarily large values of k, it can be arbitrarily close to $2\sqrt{3}-3$.

If $k - x > 1$, then $|L_1| > |L_2|+1$ and thus it is not possible to make *Next-Fit* color all edges in G_1 using only C_{k-x}. In this case more copies of G_{NF} are needed. Let m be the smallest positive integer such that $m(k-x)$ is a multiple of k. Then mx is a multiple of k as well. In general, m copies of G_{NF}, $G_C^1, G_C^2, \ldots, G_C^m$, are used. A k-coloring of the m copies of G_{NF} in which each color is used the same number of times can be obtained in the following way. In G_C^i, G_1 is colored with the colors $(k-x)(i-1)+1 \bmod k$, $(k-x)(i-1)+2 \bmod k$, $\ldots$, $(k-x)i \bmod k$, and G_2 is colored with the remaining colors in C_k.   □

5 The Algorithm *First-Fit*

The algorithm *First-Fit* (*FF*) is a fair algorithm. For each edge e that it is able to color, it colors e with the lowest numbered color possible.

Theorem 5.1. $\inf\limits_{k \in \mathbb{N}} C_{FF}(k) \leq \frac{2}{9}(\sqrt{10}-1) \approx 0.4805.$

Proof. The adversary graph G_{FF} of this proof is inspired by the graph G_{NF}. It is not possible, though, to make *First-Fit* color the subgraph G_2 of G_{NF} with $C_k \setminus C_x$. Therefore, the graph is extended by the subgraph G_2' isomorphic to G_2. Each vertex in R_2 is connected to exactly $k - x$ vertices in L_2' and vice versa. Now, E_2 denotes the edges in G_2 and G_2' and the edges connecting them. Finally, $2kx$ new vertices are added, and each vertex in $R_2 \cup L_2'$ is connected to k of these vertices. Let E_3 denote the set of these extra edges. The graph G_{FF} for $k = 4$ is depicted in Fig. 3.

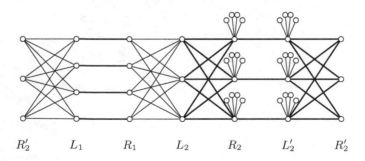

$R_2' \qquad L_1 \qquad R_1 \qquad L_2 \qquad R_2 \qquad L_2' \qquad R_2'$

Fig. 3. The graph G_{FF} when $k = 4$, showing that $C_{FF}(4) \leq \frac{25}{52} \approx 0.4808$.

If the edges in G_1 and the edges connecting G_2 and G_2' are given first (one perfect matching at a time), followed by the edges in G_2 and G_2' (one perfect matching at a time), it is obvious that *First-Fit* will color the edges in the desired way. After this, *First-Fit* will not be able to color any more edges of G_{FF}. On the other hand it is possible to k-color the set $E_1 \cup E_{12} \cup E_3$ of edges. Thus, the competitive ratio of *First-Fit* can be no more than $\frac{|E_1|+|E_2|}{|E_1|+|E_{12}|+|E_3|} = \frac{k(k-x)+2x^2+x(k-x)}{k(k-x)+2kx+2kx} = \frac{k^2+x^2}{k^2+3kx}$. This ratio attains its minimum value of $\frac{2}{9}(\sqrt{10} - 1)$, when $x = \frac{1}{3}(\sqrt{10} - 1)k$. Thus, for the graph G_{FF}, the optimal (from an adversary's point of view) value of x is an integer as close as possible to $\frac{1}{3}(\sqrt{10} - 1)k$, and by allowing arbitrarily large values of k, the ratio can be arbitrarily close to $\frac{2}{9}(\sqrt{10} - 1)$. □

6 k-Colorable Graphs

Now that we know that the competitive ratio cannot vary much between different kinds of algorithms for the EDGE-COLORING problem, it would be interesting to see what happens if we know something about the input graphs — for instance that they are all k-colorable. In this section we investigate the competitive ratio in the case where the input graphs are known to be k-colorable.

6.1   A Tight Lower Bound for Fair Algorithms

Theorem 6.1. *Any fair algorithm for* EDGE-COLORING *is* $\frac{1}{2}$-*competitive on k-colorable graphs.*

Sketch of the Proof. As in the proof of Theorem 3.2 the idea is that each colored edge is worth one unit of some value. The value of each colored edge e is distributed equally among its endpoints and, from there, redistributed to the uncolored edges adjacent to e. If each uncolored edge receives a total value of at least one, then there are at least as many colored edges as uncolored edges. Let d_c and d_u be defined as in the proof of Theorem 3.2. Then each uncolored edge (x, y) receives the value $\frac{1}{2}\left(\frac{d_c(x)}{d_u(x)} + \frac{d_c(y)}{d_u(y)}\right) \geq \frac{1}{2}\left(\frac{d_c(x)}{k-d_c(x)} + \frac{d_c(y)}{k-d_c(y)}\right) \geq \frac{1}{2}\left(\frac{d_c(x)}{k-d_c(x)} + \frac{k-d_c(x)}{d_c(x)}\right) \geq \frac{1}{2}\left(\frac{k/2}{k/2} + \frac{k/2}{k/2}\right) = 1$. The first inequality above follows from the fact that $d_c(x) + d_u(x) \leq k$, since the graph is k-colorable. The second inequality follows from the fact that $d_c(x) + d_c(y) \geq k$, since *fair*^{R} is fair. □

In Section 4 it is shown that, on k-colorable graphs, the competitive ratio of the algorithm *Next-Fit* is $\frac{1}{2}$ for all even k. Thus, the result in Theorem 6.1 is tight.

6.2   An Upper Bound for Deterministic Algorithms

Theorem 6.2. *For any deterministic algorithm $\mathbb{A}$ for* EDGE-COLORING, $C_{\mathbb{A}} \leq \frac{2}{3}$, *even on k-colorable graphs.*

Sketch of the Proof. The edges are given in two phases. In Phase 1, a large $\frac{k}{2}$-regular bipartite graph $G = (L \cup R, E)$ is given. After Phase 1, the vertex set L is divided in subsets according to the colors represented at each vertex. Vertices with the same color sets are put in the same subset. The same is done to R. Now, look at one such subset S. Assume that it has size n and that the number of colors represented at each vertex is d. In Phase 2, $\frac{n}{2}$ new vertices are added and connected to the vertices in S, creating a bipartite graph B in which the new vertices have degree k, and the vertices in S have degree $\frac{k}{2}$. Thus, looking at the whole graph, the total vertex degree of the vertices in B is $\frac{3}{2}nk$. *on-line^D* can color at most $k - d$ edges incident to each of the new vertices. Therefore, looking at the subgraph colored by *on-line^D*, the total degree of the vertices in B is at most $2 \cdot \frac{n}{2}(k - d) + nd = nk$. Since each of the vertices given in Phase 2 is connected to only one of the sets L and R, the whole graph is bipartite. Thus, *off-line* colors all of the edges.   □

6.3   The Algorithm *Next-Fit*

Theorem 6.3. *On k-colorable graphs,* $C_{NF}(k) \leq \begin{cases} \frac{1}{2}, & \text{if } k \text{ is even} \\ \frac{1}{2} + \frac{1}{2k^2}, & \text{if } k \text{ is odd} \end{cases}$

Proof. The adversary constructs a graph G_{NF} in the following way. First it constructs two complete bipartite graphs $G_1 = (L_1 \cup R_1, E_1)$ with $|L_1| = |R_1| = \lceil \frac{k}{2} \rceil$ and $G_2 = (L_2 \cup R_2, E_2)$ with $|L_2| = |R_2| = \lfloor \frac{k}{2} \rfloor$. G_1 can be colored with $\lceil \frac{k}{2} \rceil$ colors using each color $\lceil \frac{k}{2} \rceil$ times, and G_2 can be colored with $\lfloor \frac{k}{2} \rfloor$ colors using each color $\lfloor \frac{k}{2} \rfloor$ times. The edges in these two graphs are given in an order such that *Next-Fit* colors G_1 with $C_{\lceil \frac{k}{2} \rceil}$ and G_2 with $C_k \setminus C_{\lceil \frac{k}{2} \rceil}$. Now, each vertex in R_1 is connected to each vertex in L_2 and each vertex in R_2 is connected to each vertex in L_1. Let E_{12} denote these edges connecting G_1 and G_2. *Next-Fit* is not able to color any of the edges in E_{12}. It is, however, possible to color all edges in G_{NF} with C_k, since the graph is bipartite and has maximum degree k. Thus, in the case where the input graphs are all k-colorable, the competitive ratio of *Next-Fit* can be no more than $\frac{|E_1| + |E_2|}{|E_1| + |E_2| + |E_{12}|} = \frac{\lceil \frac{k}{2} \rceil^2 + \lfloor \frac{k}{2} \rfloor^2}{\lceil \frac{k}{2} \rceil^2 + \lfloor \frac{k}{2} \rfloor^2 + 2\lceil \frac{k}{2} \rceil \lfloor \frac{k}{2} \rfloor}$, which reduces to $\frac{1}{2}$ when k is even, and to $\frac{1}{2} + \frac{1}{2k^2}$ when k is odd.   □

6.4   The Algorithm *First-Fit*

The following theorem is an immediate consequence of Lemma 6.5 and Lemma 6.6.

Theorem 6.4. *On k-colorable graphs,* $C_{FF}(k) = \frac{k}{2k-1}$.

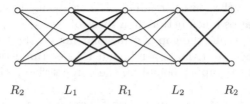

$$R_2 \qquad L_1 \qquad R_1 \qquad L_2 \qquad R_2$$

Fig. 4. The graph G_{NF} when $k = 5$

Thus, for small values of k, the competitive ratio of *First-Fit* on k-colorable graphs is significantly larger than that of *Next-Fit*, but the difference tends to zero as k approaches infinity.

Lemma 6.5. *On k-colorable graphs, $C_{FF}(k) \geq \frac{k}{2k-1}$.*

Proof. Let E be the edge set of an arbitrary k-colorable graph G. Assume that *First-Fit* is given the edges in E in some order. For $c \in C_k$, let E_c denote the set of edges that *First-Fit* colors with the color c. We will prove by induction on c that, for all $c \in C_k$, $\sum_{i=1}^{c} |E_i| \geq \frac{c}{2k-1} |E|$.

For the base case, consider $c = 1$. By the definition of *First-Fit*, each edge in $E \setminus E_1$ is adjacent to at least one edge in E_1. Furthermore, since G is k-colorable, each edge in E_1 is adjacent to at most $2(k - 1)$ other edges. Thus, $|E| \leq 2(k - 1)|E_1| + |E_1|$, or $|E_1| \geq \frac{1}{2k-1} |E|$.

For the induction step, let $c \in C_k$. Each edge in $E \setminus \cup_{i=1}^{c} E_i$ is adjacent to at least one edge in E_c. Moreover, since each edge in E_c is adjacent to at least $c - 1$ edges in $\cup_{i=1}^{c-1} E_i$, each edge in E_c is adjacent to at most $2(k-1)-(c-1) = 2k-c-1$ edges in $E \setminus \cup_{i=1}^{c} E_i$. Therefore, $|E_c| \geq \frac{1}{2k-c} |E \setminus \cup_{i=1}^{c-1} E_i|$. Thus, $\sum_{i=1}^{c} |E_i| \geq$
$$\sum_{i=1}^{c-1} |E_i| + \frac{|E|-\sum_{i=1}^{c-1}|E_i|}{2k-c} = \frac{|E|+(2k-c-1)\sum_{i=1}^{c-1}|E_i|}{2k-c} \geq \frac{|E|+(2k-c-1)\frac{c-1}{2k-1}|E|}{2k-c} = \frac{c}{2k-1}|E| \qquad \square$$

Lemma 6.6. *On k-colorable graphs, $C_{FF}(k) \leq \frac{k}{2k-1}$.*

Outline of the Proof. Inspired by the proof of Lemma 6.5, we construct a bipartite graph G and a *First-Fit* coloring of G such that all vertices have degree k and no edge is adjacent to more than one edge of each color. For such a graph the analysis in the proof of Lemma 6.5 is tight, meaning that *First-Fit* colors exactly $\frac{k}{2k-1}$ of the edges in G. Since G is bipartite, it can be k-colored off-line. $\square$

7 Conclusions

We have proven that the competitive ratios of algorithms for EDGE-COLORING can vary only between approximately 0.46 and 0.5 for deterministic algorithms and between 0.46 and 0.57 for probabilistic algorithms (it can, of course, be lower for algorithms that are not fair). Thus, we cannot hope for algorithms with competitive ratios much better than those of *Next-Fit* and *First-Fit*. In the case of k-colorable graphs the gap is somewhat larger: the (tight) lower bound for fair algoritms is $\frac{1}{2}$ and the upper bound for deterministic algorithms is $\frac{2}{3}$. In this case we have no upper bound on the competitive ratio for probabilistic algorithms.

We have shown that the performance of *Next-Fit* matches the lower bound on the competitive ratio in both the general case and in the special case of k-colorable graphs. Furthermore, we have found the exact competitive ratio of *First-Fit* on k-colorable graphs. For small values of k it is significantly better than that of *Next-Fit*, but for large values of k they can hardly be distinguished. In the general case, *First-Fit* is at most 0.016 better than *Next-Fit*. We believe that the competitive ratio of *First-Fit* is larger than that of *Next-Fit* but we have not proven it.

References

[1] Yossi Azar, Joan Boyar, Lene M. Favrholdt, Kim S. Larsen, and Morten N. Nielsen. Fair versus unrestricted bin packing. In *Seventh Scandinavian Workshop on Algorithm Theory*, volume 1851 of *Lecture Notes in Computer Science*, pages 200–213. Springer-Verlag, 2000.

[2] Amotz Bar-Noy, Rajeev Motwani, and Joseph Naor. The greedy algorithm is optimal for on-line edge coloring. *Information Processing Letters*, 44(5):251–253, 1992.

[3] J. Boyar and K. S. Larsen. The seat reservation problem. *Algorithmica*, 25(4):403–417, 1999.

[4] J. Boyar, K. S. Larsen, and M. N. Nielsen. The accommodating function — a generalization of the competitive ratio. In *Sixth International Workshop on Algorithms and Data Structures*, volume 1663 of *Lecture Notes in Computer Science*, pages 74–79. Springer-Verlag, 1999.

[5] Lene M. Favrholdt and Morten N. Nielsen. On-line edge-coloring with a fixed number of colors. Preprints 1999, no. 6, Department of Mathematics and Computer Science, University of Southern Denmark, Odense.

[6] Anna R. Karlin, Mark S. Manasse, Larry Rudolph, and Daniel D. Sleator. Competitive snoopy caching. *Algorithmica*, 3:79–119, 1988.

[7] Douglas B. West. *Introduction to Graph Theory*, page 209. Prentice Hall, Inc., 1996.

On Approximability of the Independent/Connected Edge Dominating Set Problems

Toshihiro Fujito

Department of Electronics, Nagoya University
Furo, Chikusa, Nagoya, 464-8603 Japan
fujito@nuee.nagoya-u.ac.jp

Abstract. We investigate polynomial-time approximability of the problems related to edge dominating sets of graphs. When edges are unit-weighted, the edge dominating set problem is polynomially equivalent to the minimum maximal matching problem, in either exact or approximate computation, and the former problem was recently found to be approximable within a factor of 2 even with arbitrary weights. It will be shown, in contrast with this, that the minimum weight maximal matching problem cannot be approximated within any polynomially computable factor unless $P=NP$.

The connected edge dominating set problem and the connected vertex cover problem also have the same approximability when edges/vertices are unit-weighted, and the former problem is known to be approximable, even with general edge weights, within a factor of 3.55. We will show that, when general weights are allowed, 1) the connected edge dominating set problem can be approximated within a factor of $3 + \epsilon$, and 2) the connected vertex cover problem is approximable within a factor of $\ln n + 3$ but cannot be within $(1 - \epsilon) \ln n$ for any $\epsilon > 0$ unless $NP \subset \text{DTIME}(n^{O(\log \log n)})$.

1 Introduction

In this paper we investigate polynomial-time approximability of the problems related to edge dominating sets of graphs. For two pairs of problems considered, it will be shown that, while both problems in each pair have the same approximability for the unweighted case, they have drastically different ones when optimized under general non-negative weights.

In an undirected graph an edge *dominates* all the edges adjacent to it, and an *edge dominating set (eds)* is a set of edges collectively dominating all the other edges in a graph. The problem EDS is then that of finding a smallest eds or, if edges are weighted, an eds of minimum total weight. Yannakakis and Gavril showed that EDS is *NP*-complete even when graphs are planar or bipartite of maximum degree 3 [24]. Horton and Kilakos extended this *NP*-completeness result to planar bipartite graphs, line and total graphs, perfect claw-free graphs, and planar cubic graphs [14]. A set of edges is called a *matching* (or *independent*)

S. Kapoor and S. Prasad (Eds.): FST TCS 2000, LNCS 1974, pp. 117–126, 2000.

if no two of them have a vertex in common, and a matching is *maximal* if no other matching properly contains it. Notice that any maximal matching is necessarily an eds, because an edge not in it must be adjacent to some in it, and for this reason it is also called an *independent edge dominating set*, and the problem IEDS asks for computing a minimum maximal matching in a given graph. Certainly, a smallest maximal matching cannot be smaller than a smallest eds. Interestingly, one can construct a maximal matching, from any eds, of no larger size in polynomial time [13], implying that the size of a smallest eds equals to that of a smallest maximal matching in any graph. Thus, EDS and IEDS are polynomially equivalent, in exact or approximate computation, when graphs are *unweighted*. Based on this and the fact that any maximal matching cannot be more than twice larger than another one, it has been long known that either problem, without weights, can be approximated within a factor of 2, and even with weights, EDS was very recently shown approximable within a factor of 2 [4,7]. We will present, in contrast with this, strong inapproximability results for weighted IEDS.

We next consider EDS with connectivity requirement, called the *connected edge dominating set (CEDS)* problem, where it is asked to compute a connected eds (ceds) of minimum weight in a given connected graph. Since it is always redundant to form a cycle in a ceds, the problem can be restated as that of going after a minimum tree whose vertices "covers" all the edges in a graph, and thus, it is also called *tree cover*. Although enforcing the independence property on EDS solutions does not alter (increase) their sizes as stated above, the connectivity condition certainly does (just consider a path of length 5). The *vertex cover (VC)* problem is another basic NP-complete graph problem [16], in which a minimum vertex set is sought in G s.t. every edge of G is incident to some vertex in the set, and when a vertex cover is additionally required to induce a connected subgraph in a given connected graph, the problem is called *connected vertex cover (CVC)* and known to be as hard to approximate as VC is [8]. These problems are closely related to EDS and CEDS in that an edge set F is an eds for G iff $V(F)$, the set of vertices touched by edges in F, is a vertex cover for G, and similarly, a tree F is a ceds iff $V(F)$ is a connected vertex cover. Since one can easily obtain a cvc of size $|F| + 1$ from a ceds F (tree), and conversely, a ceds of size $|C| - 1$ from a cvc C, these two problems have the same approximability for the unweighted case. The unweighted version of CEDS or CVC is also known to be approximable within a factor of 2 [22,2]. It is not known, however, if CEDS and CVC can be somehow related even if general weights are allowed, and the algorithm scheme of Arkin et al. for weighted CEDS gives its approximation factor in the form of $r_{\mathrm{St}} + r_{\mathrm{wvc}}(1 + 1/k)$, for any constant k, where $r_{\mathrm{St}}(r_{\mathrm{wvc}})$ is the performance ratio of any polynomial time algorithm for the Steiner tree (weighted vertex cover, resp.) problem [2]. By using the currently best algorithms for Steiner tree with $r_{\mathrm{St}} = 1 + \ln 3/2 \approx 1.55$ [21] and for weighted vertex cover with $r_{\mathrm{wvc}} = 2 - \log \log n / \log n$ [3] in their scheme, the bound for weighted CEDS is estimated at 3.55. After improving this bound to $3 + \epsilon$, we will show that weighted CVC is as hard to approximate as weighted set cover is, indicating that it is not approximable within a factor better than $(1-\epsilon)\ln n$ unless $NP \subset \mathrm{DTIME}(n^{O(\log\log n)})$ [6]. Lastly, we present an algorithm approximating weighted CVC within a factor of $r_{\mathrm{wvc}} + H(\Delta - 1) \leq \ln(\Delta - 1) + 3$, where $H(k)$ is the kth Harmonic number and Δ is the maximal vertex degree of a graph.

Since EDS is exactly the *(vertex) dominating set* problem on line graphs, it is worth comparing our results with those for independent/connected dominating set problems. The connected dominating set is as hard to approximate as set cover is, but can be approximated within a factor of $\ln n + 3$ for the unweighted case [9], and within $1.35 \ln n$ for the weighted case [10]. The *independent dominating set* problem (also called *minimum maximal independent set*), on the other hand, cannot be approximated, even for the unweighted case, within a factor of $n^{1-\epsilon}$ for any constant $\epsilon > 0$, unless *P=NP* [12].

2   Independent Edge Dominating Set

To show that it is extremely hard to approximate IEDS, let us first describe a general construction of graph G_ϕ for a given 3SAT instance (i.e., a CNF formula) ϕ, by adapting the one used in reducing SAT to minimum maximal independent set [15,12] to our case. For simplicity, every clause of ϕ is assumed w.l.o.g. to contain exactly three literals. Each variable x_i appearing in ϕ is represented in G_ϕ by two edges adjacent to each other, and the endvertices of such a path of length 2 are labeled x_i and $\bar{x}_i$; let E_v denote the set of these edges. Each clause c_j of ϕ is represented by a triangle (a cycle of length 3) C_j, and vertices of C_j are labeled distinctively by literals appearing in c_j; let E_c denote the set of edges in these disjoint triangles. The paths in E_v and triangles in E_c are connected together by having an edge between every vertex of each triangle and the endvertex of a path having the same label. The set of these edges lying between E_v and E_c is denoted by E_b. It is a simple matter to verify that, for a 3SAT instance ϕ with m variables and p clauses, G_ϕ constructed this way consists of $3(m+p)$ vertices and $2(m+3p)$ edges.

Lemma 1. *Let $M(G)$ denote a minimum maximal matching M in G. For any 3SAT instance ϕ with m variables and p clauses, and for any number t, there exists a graph G_ϕ on $3(m+p)$ vertices and $2(m+3p)$ edges, and a weight assignment $w : E \to \{1, t\}$ such that*

$$w(M(G_\phi)) = \begin{cases} \leq m+p & \text{if } \phi \text{ is satisfiable} \\ > t & \text{otherwise .} \end{cases}$$

Proof. Let $w(e) = 1$ if $e \in E_v \cup E_c$ and $w(e) = t$ if $e \in E_b$. Suppose that ϕ is satisfiable, and let τ be a particular truth assignment satisfying ϕ. Construct a matching M_τ in E_v by choosing, for each i, the edge with its endvertex labeled by x_i if $\tau(x_i)$ is true and the one having an endvertex labeled by $\bar{x}_i$ if $\tau(x_i)$ is false. Consider any triangle C_j in E_c. Since τ satisfies ϕ, at least one edge among those in E_b connecting C_j and E_v must be dominated by M_τ. This means that all the edges in E_b between C_j and E_v can be dominated by M_τ, plus one edge on C_j. Let M_c denote the set of such edges, each of which taken this way from each C_j. Then, $M_\tau \cup M_c$ is clearly a minimal matching since it dominates all the edges in G_ϕ. Since all the edges in $M_\tau \cup M_c$ are of weight 1, its weight is $|M_\tau \cup M_c| = m + p$. On the other hand, if ϕ is not satisfiable, there is no way to dominate all the edges in E_b only by any matching built inside $E_v \cup E_c$, and hence, any maximal matching in G_ϕ must incur a cost of more than t.   □

The computational hardness of approximating weighted IEDS easily follows from this lemma:

Theorem 1. *For any polynomial time computable function $\alpha(n)$, IEDS cannot be approximated on graphs with n vertices within a factor of $\alpha(n)$, unless P=NP.*

Proof. Given a 3SAT instance ϕ with m variables and p clauses, construct a graph G_ϕ and assign a weight $w(e) \in \{1, t\}$ to each edge $e \in E$, as in the proof of Lemma 1. Since G_ϕ consists of $3(m + p)$ vertices and $(m + p)\alpha(3(m + p))$ is computable in time polynomial in the length of ϕ, $m + 3p$, we can set $t = (m + p)\alpha(3(m + p)) = (m + p)\alpha(n)$. If a polynomial time algorithm A exists approximating IEDS within a factor of $\alpha(n)$, then, when applied to G_ϕ, A will output a number at most $(m + p)\alpha(n)$ if ϕ is satisfiable, and a number greater than $t = (m + p)\alpha(n)$ if ϕ is not satisfiable. Hence, A decides 3SAT in polynomial time.   □

It is additionally pointed out in this section that IEDS is complete for *exp-APX*, the class of *NP* optimization problems polynomially approximable within some exponential factors [1], by slightly modifying the construction used in Lemma 1 and allowing zero weight on edges. Previously, a more general problem, minimum weight maximal independent set, was shown to be *exp-APX*-complete [5].

Definition 1. Minimum Weighted Satisfiability (MinWSAT) *is the problem defined by*

Instance: *a CNF formula ϕ with nonnegative weights $w(x)$ on the variables appearing in ϕ.*
Solution: *truth assignment τ, either satisfying ϕ or setting all the variables to "true". The latter is called a* trivial *assignment.*
Objective: *minimize $w(\tau) = \sum_{\tau(x)=true} w(x)$.*

Theorem 2. *The weighted IEDS problem is complete for exp-APX.*

Proof. MinWSAT is known to be *exp-APX*-complete [19], and we reduce it to weighted IEDS. Given ϕ and weights w on its variables, define G_ϕ as before and edge weights w' such that

$$w'(e) = \begin{cases} w(x_i), & \text{if } e \in E_v \text{ and its endvertex is labeled by } x_i \\ \sum_i w(x_i), & \text{if } e \in E_b \\ 0, & \text{otherwise .} \end{cases}$$

If a maximal matching in G_ϕ contains no edges in E_b, it must have an edge in E_v labeled either x_i or $\bar{x}_i$ for each i. So, from an MinWSAT instance $\langle \phi, w \rangle$ and a maximal matching M in G_ϕ, a truth assignment τ for ϕ can be recovered in such a way that

$$\tau = \begin{cases} \text{truth assignment corresponding to } M \cap E_v, & \text{if } M \cap E_b = \emptyset \\ \text{trivial assignment,} & \text{otherwise .} \end{cases}$$

It is then straightforward to verify that any algorithm for IEDS with performance guarantee of α can be used to approximate MinWSAT within a factor at most α. □

3   Connected Edge Dominating Set

We first consider a restricted version of CEDS; for a designated vertex r called *root*, an r-ceds is a ceds touching r, and the problem r-CEDS is to compute an r-ceds of minimum weight. Given an undirected graph $G = (V, E)$ with edge weights $w : E \to \mathbb{Q}_+$, let $\boldsymbol{G} = (V, \boldsymbol{E})$ denote its directed version obtained by replacing each edge $\{u, v\}$ of G by two directed ones, (u, v) and (v, u), each of weight $w(\{u, v\})$. For the root r, a non-empty set $S \subseteq V - \{r\}$ is called *dependent* if S is not an independent set in G. Suppose $T \subseteq E$ is an r-ceds, and let $\boldsymbol{T}$ denote the directed counterpart obtained by choosing, for each pair of directed edges, the one directed away from the root to a leaf. Clearly, $w(T) = w(\boldsymbol{T})$. Moreover, let $\boldsymbol{T}$ be represented by its characteristic vector $x^{\boldsymbol{T}} \in \{0, 1\}^{\boldsymbol{E}}$, and, for any $x \in \mathbb{Q}^{\boldsymbol{E}}$ and $\boldsymbol{F} \subseteq \boldsymbol{E}$, let $x(\boldsymbol{F}) = \sum_{a \in \boldsymbol{F}} x_a$. Then, $x^{\boldsymbol{T}}$ satisfies the linear inequality $x(\delta^-(S)) \geq 1$ for all dependent sets $S \subseteq V$, where $\delta^-(S) = \{(u, v) \in \boldsymbol{E} \mid v \in S, u \notin S\}$, because, when an edge exists inside S, at least one arc of $\boldsymbol{T}$ must enter it. Thus, the following linear programming problem is a relaxation of r-CEDS:

$$Z_{\text{ceds}} = \min \sum_{a \in \boldsymbol{E}} w(a) x_a$$
$$\text{s.t.}$$
$$x(\delta^-(S)) \geq 1 \qquad \forall \text{ dependent set } S \subseteq V - \{r\} \qquad (1)$$
$$0 \leq x_a \leq 1 \qquad \forall a \in \boldsymbol{E}$$

Lemma 2. *For any feasible solution $x \in \mathbb{Q}^{\boldsymbol{E}}$ of (1), let $V_+(x) = \{u \in V \mid x(\delta^-(\{u\})) \geq 1/2\}$. Then, $V_+(x) \cup \{r\}$ is a vertex cover for G.*

Proof. Take any edge $e = \{u, v\} \in E$, and assume $r \notin e$. Then, $\{u, v\}$ is a dependent set, and $x(\delta^-(\{u, v\})) \geq 1$, which implies either $x(\delta^-(\{u\}))$ or $x(\delta^-(\{v\}))$ is at least $1/2$. Thus, $\{u, v\} \cap V_+(x) \neq \emptyset$. □

From this lemma it is clear that any tree $T \subseteq E$ containing all the vertices in $V_+(x) \cup \{r\}$ is an r-ceds for G, and, in searching for such T of small weight, it can be assumed w.l.o.g. that the edge weights satisfy the triangle inequality since any edge between two vertices can be replaced, if necessary, by the shortest path between them. Then, the problem of finding such a tree of minimum weight is called the *(metric) Steiner tree* problem: Given $G = (V, E)$ with edge weight $w : E \to \mathbb{Q}_+$ and a set $R \subseteq V$ of *required vertices* (or *terminals*), find a minimum weight tree containing all the required vertices and any others (called *Steiner vertices*). For this problem Rajagopalan and Vazirani considered the so called *bidirected cut relaxation* [20]:

$$Z_{\text{smt}} = \min \sum_{a \in \boldsymbol{E}} w(a) x_a$$

$$\text{s.t.}$$

$$x(\delta^-(S)) \geq 1 \qquad \forall \text{ valid set } S \subseteq V - \{r\} \qquad (2)$$
$$0 \leq x_a \leq 1 \qquad \forall a \in \boldsymbol{E}$$

where the root r is any required vertex and a set $S \subseteq V - \{r\}$ is *valid* if it contains a required vertex. Based on this relaxation, they designed a primal-dual approximation algorithm for metric Steiner tree and showed that it computes a Steiner tree of cost at most $(3/2 + \epsilon)Z_{\text{smt}}$ and that the integrality gap of (2) is bounded by $3/2$, when restricted to graphs in which Steiner vertices form independent sets (called *quasi-bipartite* graphs). Our algorithm for r-CEDS is now described as:

1. Compute an optimal solution x for (1).
2. Let $V_+(x) = \{u \in V \mid x(\delta^-(\{u\})) \geq 1/2\}$.
3. Compute a Steiner tree T with $R = V_+(x) \cup \{r\}$, the set of required vertices, by the algorithm of Rajagopalan and Vazirani.
4. Output T.

It is clear that this algorithm computes an r-ceds for G, except for one special case in which $R = \{r\}$ and so, $T = \emptyset$; but then, it is trivial to find an optimal r-ceds since G is a *star* centered at r. Not so clear from this description is polynomiality of its time complexity, and more specifically, that of Step 1. It can be polynomially implemented by applying the ellipsoid method to (1), if the separation problem for the polytope P_{ceds} corresponding to the feasible region of (1), is solved in polynomial time [11]. So, let y be a vector in $\mathbb{Q}^{\boldsymbol{E}}$. It is easily tested if $0 \leq y_a \leq 1$ for all $a \in \boldsymbol{E}$. To test whether $y(\delta^-(S)) \geq 1$ for every dependent set S, we consider y as a capacity function on the arcs of $\boldsymbol{G}$. For every arc a, not incident upon r, contract a by merging its two endvertices into a single vertex v_a, and determine an (r, v_a)-cut C_a of minimum capacity by, say the Ford-Fulkerson algorithm. It is then rather straightforward to see that

$$\min\{y(C_a) \mid a \in \boldsymbol{E} - \delta(\{r\})\} = \min\{y(\delta^-(S)) \mid S \subseteq V - \{r\} \text{ is dependent}\}$$

where $\delta(\{r\})$ is the set of arcs incident to r. So, by calculating $|E - \delta(\{r\})|$ minimum capacity (r, v_a)-cuts, we can find a dependent set S of minimum cut capacity $y(\delta^-(S))$. If $y(\delta^-(S)) \geq 1$, we thus conclude that $y \in P_{\text{ceds}}$, while, if not, the inequality $x(\delta^-(S)) \geq 1$ is violated by y and a separation hyperplane is found.

Notice that our graph G is quasi-bipartite when $V_+(x) \cup \{r\}$ is taken as the set of required vertices since it is a vertex cover for G, and for the approximation quality of solutions, we have[1]

Theorem 3. *The algorithm above computes an r-ceds of weight at most* $(3 + \epsilon)Z_{\text{ceds}}$.

Proof. Let $x \in \mathbb{Q}^E$ be an optimal solution of (1), and T be an r-ceds computed by the algorithm. As mentioned above, it was shown that $w(T) \leq (3/2 + \epsilon)Z_{\text{smt}}$ when graphs are quasi-bipartite [20]. So, it suffices to show that $2x$ is a feasible solution of (2) with $R = V_+(x) \cup \{r\}$ for then, $Z_{\text{smt}} \leq 2\sum_{a \in E} w(a)x_a = 2Z_{\text{ceds}}$, and hence, $w(T) \leq 2(3/2 + \epsilon)Z_{\text{ceds}}$. To this end, let $S \subseteq V - \{r\}$ be any valid set. If S is not an independent set in G, it is dependent, ensuring that $x(\delta^-(S)) \geq 1$. Suppose now S is an independent set. Since it is a valid set, S contains a vertex $u(\neq r)$ in $V_+(x)$. But then, $x(\delta^-(\{u\})) \geq 1/2$, and, since S is an independent set in G, $2x(\delta^-(S)) \geq 2x(\delta^-(\{u\})) \geq 1$. Thus, in either case, $2x$ satisfies all the linear constraints of (2). $\square$

Since the integrality gap of (2) is bounded by $3/2$ for quasi-bipartite graphs, we have

Corollary 1. *The integrality gap of (1) is bounded by 3.*

Lastly, since any ceds is an r-ceds for some $r \in V$, by applying the algorithm with $r = u$ for each $u \in V$ and taking the best one among all computed, CEDS can be approximated within a factor of $3 + \epsilon$.

4   Connected Vertex Cover

Savage showed that non-leaf vertices of any depth first search tree form a vertex cover of size at most twice the smallest size [22]. Since such a vertex cover clearly induces a connected subgraph, it actually means that a cvc of size no more than twice larger than the smallest vertex cover always exists and can be efficiently computed. When vertices are arbitrarily weighted, however, the weighted set cover problem can be reduced to it in an approximation preserving manner, as was done for node-weighted Steiner trees [18] and connected dominating sets [9]:

Theorem 4. *The weighted set cover problem can be approximated within the same factor as the one within which weighted CVC can be on bipartite graphs.*

Proof. From a set cover instance $(U, \mathcal{F})$ and $w : \mathcal{F} \rightarrow \mathbb{Q}_+$, where $\mathcal{F} \subseteq 2^U$ and $\cup_{S \in \mathcal{F}} S = U$, construct a bipartite graph G as a CVC instance, using a new vertex c, with vertex set $(U \cup \{c\}) \cup \mathcal{F}$ s.t. an edge exists between c and every $S \in \mathcal{F}$, and between $u \in U$ and $S \in \mathcal{F}$ iff $u \in S$. All the vertices in U and c are

---

[1] Independently of our work, Koenemann et al. recently obtained the same performance guarantee by the essentially same algorithm [17].

assigned with zero weights, while every vertex $S \in \mathcal{F}$ inherits $w(S)$, the weight of set S, from $(U, \mathcal{F})$.

For a vertex subset V' of G let $\Gamma(V')$ denote the set of vertices adjacent to a vertex in V'. Clearly, $\mathcal{F}' \subseteq \mathcal{F}$ is a set cover for $(U, \mathcal{F})$ iff $U \subseteq \Gamma(\mathcal{F}')$ in G, and moreover, for any set cover $\mathcal{F}'$, $\mathcal{F}' \cup U \cup \{c\}$ is a cvc of the same weight. On the other hand, for any cvc C for G, $U \subseteq \Gamma(C \cap \mathcal{F})$, i.e., $C \cap \mathcal{F}$ is a set cover, of the same weight because, if not and $u \notin \Gamma(C \cap \mathcal{F})$ for some $u \in U$, $\Gamma(\{u\}) \cap C = \emptyset$, and hence, there is no way to properly cover an edge incident to u by C. Thus, since it costs nothing to include c and vertices in U, any cvc for G can be assumed to be in the form of $\mathcal{F}' \cup U \cup \{c\}$ s.t. $\mathcal{F}'$ is a set cover for $(U, \mathcal{F})$, with its weight equaling to that of $\mathcal{F}'$. Therefore, any algorithm approximating CVC within a factor r can be used to compute a set cover of weight at most r times the optimal weight. $\square$

Due to the non-approximability of set cover [6], it follows that

Corollary 2. *The weighted CVC cannot be approximated in a factor better than* $(1 - \epsilon) \ln n$ *for any* $\epsilon > 0$, *unless* $NP \subset DTIME(n^{O(\log \log n)})$.

One simple strategy for approximating weighted CVC, which turns out to yield a nearly tight bound, is to compute first a vertex cover $C \subseteq V$ for $G = (V, E)$, and then to augment it to become connected by an additional vertex set $D \subseteq V - C$. While many good approximation algorithms are known for vertex cover, we also need to find such D of small weight. This problem is not exactly same as but not far from the weighted set cover, because it can be seen as a specialization of the *submodular set cover* problem [23], which in general can be stated simply as $\min_{D \subseteq N} \{w(D) \mid f(D) = f(N)\}$, given (N, f) where N is a finite set and $f : 2^N \to \mathbb{R}_+$ is a nondecreasing, submodular set function on N. For our case, take $N = V - C$, and $f(D) = \kappa(C) - \kappa(C \cup D)$ defined on $V - C$, where $\kappa(F)$ denotes the number of connected components in the subgraph $G[F]$ induced by F. Then, using the fact that $V - C$ is an independent set in G, it can be verified that f thus defined is indeed nondecreasing and submodular. Also notice that $G[C \cup D]$ is connected iff $f(D) = \kappa(C) - 1 = f(V - C)$. This way, the problem of computing minimum $D \subseteq V - C$ such that $G[C \cup D]$ is connected, is formulated exactly by the submodular set cover problem for $(V - C, f)$.

The greedy algorithm for submodular set cover, adapted to our case, is now described as:

1. Initialize $D \leftarrow \emptyset$.
2. Repeat until $G[C \cup D]$ becomes connected.
3.   Let u be a vertex minimizing $w(v)/(f(D \cup \{v\}) - f(D))$ among $v \in V - C$.
4.   Set $D \leftarrow D \cup \{u\}$.
5. Output D.

It was shown by Wolsey that the performance of the greedy algorithm for submodular set cover generalizes the one for set cover:

Theorem 5 ([23]). *The greedy algorithm for submodular set cover computes a solution of weight bounded by $H(\max_{j \in N} f(\{j\}))$ times the minimum weight.*

Since $\max_{j \in N} f(\{j\}) \leq \Delta - 1$, in our case, for a graph of maximal vertex degree Δ, the greedy heuristic works with an approximation factor bounded by $H(\Delta - 1) \leq 1 + \ln(\Delta - 1)$.

Theorem 6. *The algorithm above computes a cvc of weight at most $r_{\text{wvc}} + H(\Delta - 1) \leq \ln(\Delta - 1) + 3$ times the minimum weight.*

Proof. Let C^* be an optimal cvc and $C \cup D$ be the one computed by the algorithm above for G, where C is a vertex cover of weight at most twice that of the minimum vertex cover, and D is the greedy submodular set cover for $(V - C, f)$. Clearly, $w(C) \leq 2w(C^*)$. Observe that $G[C^* \cup C]$ remains connected because any superset of a cvc is still a cvc. But then, it means that $C^* - C \subseteq V - C$ is a submodular set cover for $(V - C, f)$. We thus conclude that

$$w(C \cup D) \leq 2w(C^*) + H(\Delta - 1)w(C^* - C) \leq (2 + H(\Delta - 1))w(C^*) .$$

$\square$

References

1. Ausiello, G., Crescenzi, P., Gambosi, G., Kann, V., Marchetti-Spaccamela, A., Protasi, M.: Complexity and Approximation: Combinatorial Optimization Problems and Their Approximability Properties. Springer-Verlag, Berlin Heidelberg New York (1999)
2. Arkin, E.M., Halldórsson, M.M., Hassin, R.: Approximating the tree and tour covers of a graph. IPL **47** (1993) 275–282
3. Bar-Yehuda, R., Even, S.: A local-ratio theorem for approximating the weighted vertex cover problem. In: Annals of Discrete Mathematics, Vol. 25. North-Holland (1985) 27–46
4. Carr, R., Fujito, T., Konjevod, G., Parekh, O.: A $2\frac{1}{10}$-approximation algorithm for a generalization of the weighted edge-dominating set problem. In: Proc. 8th ESA (to appear)
5. Crescenzi, P., Kann, V., Silvestri, R., Trevisan, L.: Structure in approximation classes. In: Proc. COCOON 95. Lecture Notes in Computer Science, Vol. 959. Springer-Verlag (1995) 539–548
6. Feige, U.: A threshold of $\ln n$ for approximating set cover. In: Proc. the Twenty-Eighth Annual ACM Symp. Theory of Computing. ACM (1996) 314–318
7. Fujito, T., Nagamochi, H.: Polyhedral characterizations and a 2-approximation algorithm for the edge dominating set problem. (submitted)
8. Garey, M.R., Johnson, D.S.: The rectilinear Steiner-tree problem is NP-complete. SIAM J. Applied Math. **32**(4) (1977) 826–834
9. Guha, S., Khuller, S.: Approximation algorithms for connected dominating sets. Algorithmica **20**(4) (1998) 374–387
10. Guha, S., Khuller, S.: Improved methods for approximating node weighted Steiner trees and connected dominating sets. Information and Computation (to appear)

11. Grötschel, M., Lovász, L., Schrijver, A.: Geometric Algorithms and Combinatorial Optimization. Springer-Verlag, Berlin (1988)
12. Halldórsson, M.M.: Approximating the minimum maximal independence number. IPL **46** (1993) 169–172
13. Harary, F.: Graph Theory. Addison-Wesley, Reading, MA (1969)
14. Horton, J.D., Kilakos, K.: Minimum edge dominating sets. SIAM J. Discrete Math. **6**(3) (1993) 375–387
15. Irving, R.W.: On approximating the minimum independent dominating set. IPL **37** (1991) 197–200
16. Karp, R.M.: Reducibility among combinatorial problems. In: Miller, R.E., Thatcher, J.W. (eds.): Complexity of Computer Computations. Plenum Press, New York (1972) 85–103
17. Koenemann, J., Konjevod, G., Parekh, O., Sinha, A.: Improved approximations for tour and tree covers. In: Proc. APPROX 2000 (to appear)
18. Klein, P., Ravi, R.: A nearly best-possible approximation algorithm for node-weighted Steiner trees. J. Algorithms **19**(1) (1995) 104–115
19. Orponen, P., Mannila, H.: On approximation preserving reductions: Complete problems and robust measures. Technical Report C-1987-28, Department of Computer Science, University of Helsinki (1987)
20. Rajagopalan, S., Vazirani, V.V.: On the bidirected cut relaxation for the metric Steiner tree problem. In: Proc. 10th Annual ACM-SIAM Symp. Discrete Algorithms. ACM-SIAM (1999) 742–751
21. Robins, G., Zelikovsky, A.: Improved Steiner tree approximation in graphs. In: Proc. 11th Annual ACM-SIAM Symp. Discrete Algorithms. ACM-SIAM (2000) 770–779
22. Savage, C.: Depth-first search and the vertex cover problem. IPL **14**(5) (1982) 233–235
23. Wolsey, L.A.: An analysis of the greedy algorithm for the submodular set covering problem. Combinatorica **4**(2) (1982) 385–393
24. Yannakakis, M., Gavril, F.: Edge dominating sets in graphs. SIAM J. Applied Math. **38**(3) (1980) 364–372

Model Checking CTL Properties of Pushdown Systems

Igor Walukiewicz*

Institute of Informatics, Warsaw University,
Banacha 2, 02-097 Warsaw, POLAND,
igw@mimuw.edu.pl

Abstract. A *pushdown system* is a graph $G(P)$ of configurations of a pushdown automaton P. The *model checking problem* for a logic L is: given a pushdown automaton P and a formula $\alpha \in L$ decide if α holds in the vertex of $G(P)$ which is the initial configuration of P. Computation Tree Logic (CTL) and its fragment EF are considered. The model checking problems for CTL and EF are shown to be EXPTIME-complete and PSPACE-complete, respectively.

1 Introduction

A *pushdown system* is a graph $G(P)$ of configurations of a pushdown automaton P. The edges in this graph correspond to single steps of computation of the automaton. The *pushdown model checking problem* (PMC problem) for a logic L is: given a pushdown automaton P and a formula $\alpha \in L$ decide if α holds in the vertex of $G(P)$ which is the initial configuration of P. This problem is a strict generalization of a more standard model checking problem where only finite graphs are considered.

In this paper we consider PMC problem for two logics: CTL and EF. CTL is the standard Computation Tree Logic [4, 5]. EF is a fragment of CTL containing only operators: exists a successor ($\exists \circ \alpha$), and exists a reachable state ($\exists F \alpha$). Moreover, EF is closed under conjunction and negation. We prove the following:

- The PMC problem for EF logic is PSPACE-complete.
- The PMC problem for CTL is EXPTIME-complete.

The research on the PMC problem continues for some time. The decidability of this problem for monadic second order logic (MSOL) follows from [8] (for a simpler argument see [2]). This implies decidability of the problem for all those logics which have effective translations to MSOL. Among them are the μ-calculus, CTL* as well as the logics considered here. This general result however gives only nonelementary upper bound on the complexity of PMC. In [9] an EXPTIME-completeness of PMC for the μ-calculus was proved. This result was slightly encouraging because the complexity is not that much bigger than the

---

* The author was supported by Polish KBN grant No. 8 T11C 027 16.

S. Kapoor and S. Prasad (Eds.): FST TCS 2000, LNCS 1974, pp. 127–138, 2000.

complexities of known algorithms for the model checking problem over finite graphs. In [1] it was shown that the PMC problems for LTL and linear time μ-calculus are EXPTIME-complete.

The PMC problem for EF was considered already in [1]. It was shown there that the problem is PSPACE-hard. Moreover it was argued the the general method of the paper gives a PSPACE algorithm for the problem. Later, a closer analysis showed that there is no obvious way of implementing the method in polynomially bounded space [6]. The algorithm presented here follows the idea used in [9] for the μ-calculus.

The EXPTIME hardness results for the alternation free μ-calculus and LTL show two different reasons for the hardness of the PMC problem. One is unbounded alternation, the other is the ability to compare two consecutive blocks of states on a path. The reachability problem for pushdown systems is of course solvable in polynomial time (see [7] for a recent paper on this problem). Over finite graphs the model checking problem for CTL reduces to a sequence of reachability tests. This suggested that PMC problem for CTL may be PSPACE-complete. In this light EXPTIME-hardness result is slightly surprising. The argument combines ideas from the hardness results for the μ-calculus and LTL. It essentially shows that $\exists G\alpha$ operator (there is a path on which α always holds) is enough to obtain EXPTIME-hardness. This result makes EF logic more interesting as it is a fragment of CTL that disallows $\exists G\alpha$ but still allows $\forall G\alpha$ (for all paths α always holds).

Next section gives definitions concerning logics and pushdown systems. Section 3 presents an assumption semantics of EF. This semantics allows to formulate the induction argument in the correctness proof of the model checking algorithm. The proof is described in Section 4. The final section presents EXPTIME-hardness result of the PMC problem for CTL.

2   Preliminaries

In this section we present CTL and EF logics. We define pushdown systems and the model checking problem.

CTL and EF logics Let *Prop* be a set of propositional letters; let $p, p', \ldots$ range over *Prop*.

The set of formulas of EF logic, $Form(EF)$, is given by the grammar: $\alpha ::= p \mid \neg\alpha \mid \alpha \wedge \beta \mid \exists \circ \alpha \mid \exists F\alpha$. For CTL the grammar is extended with the clauses: $\exists(\alpha_1 U \alpha_2) \mid \exists\neg(\alpha_1 U \alpha_2)$.

The models for the logic are labelled graphs $\langle V, E, \rho \rangle$; where V is the set of vertices, E is the edge relation and $\rho : Prop \to \mathcal{P}(V)$ is a labelling function assigning to each vertex a set of propositional letters. Such labelled graphs are called *transition systems* here. In this context vertices are also called states.

Let $M = \langle V, E, \rho \rangle$ be a transition system. The meaning of a formula α in a state v is defined by induction. Tha clauses for propositional letters, negation and conjunction are standard. For the other constructs we have:

- $M, v \vDash \exists\circ\alpha$ if there is a successor v' of v such that $M, v' \vDash \alpha$.
- $M, v \vDash \exists F\alpha$ if there is a path from v to v', s.t. $M, v' \vDash \alpha$.
- $M, v \vDash \exists(\alpha U\beta)$ if there is a path from v to v', s.t. $M, v' \vDash \beta$ and for all the verticies v'' on the path other than v' we have $M, v'' \vDash \alpha$.
- $M, v \vDash \exists\neg(\alpha U\beta)$ if there is a maximal (i.e., infinte or finite ending in a vertex without successors) path π from v s.t. for every vertex v' on π with $M, v' \vDash \beta$ there is an earlier vertex v'' on π with not $M, v'' \vDash \alpha$.

We will freely use abbreviations:

$$\alpha \vee \beta = \neg(\neg\alpha \wedge \neg\beta) \qquad \forall\circ\alpha = \neg\exists\circ\neg\alpha \qquad \forall G\alpha = \neg\exists F\neg\alpha$$

Using these one can convert every formula of EF logic to an equivalent *positive formula* where all the negations occur only before propositional letters.

Pushdown systems A *pushdown system* is a tuple $P = \langle Q, \Gamma, \Delta, q_0, \perp \rangle$ where Q is a finite set of *states*, Γ is a finite *stack alphabet* and $\Delta \subseteq (Q \times \Gamma) \times (Q \times \Gamma^*)$ is the set of *transition rules*. State $q_0 \in Q$ is the *initial state* and symbol $\perp \in \Gamma$ is the *initial stack symbol*.

We will use q, z, w to range over Q, Γ and Γ^* respectively. We will write $qz \rightarrowtail_\Delta q'w$ instead of $((q, z), (q', w)) \in \Delta$. We will omit subscript Δ if it is clear from the context.

In this paper we will restrict ourselves to pushdown systems with transition rules of the form $qz \rightarrowtail q'$ and $qz \rightarrowtail qz'z$. Operations pushing more elements on the stack can be simulated with only polynomial increase of the size of a pushdown system. We will also assume that $\perp$ is never taken from the stack, i.e., that there is no rule of the form $q\perp \rightarrowtail q'$ for some q, q'.

Let us now give the semantics of a pushdown system $P = \langle Q, \Gamma, \Delta, q_0, \perp \rangle$. A *configuration* of P is a word $qw \in Q \times \Gamma^*$. The configuration $q_0\perp$ is the *initial configuration*. A pushdown system P defines an infinite graph $G(P)$ which nodes are configurations and which edges are: $(qzw, q'w) \in E$ if $qz \rightarrowtail_\Delta q'$, and $(qzw, q'z'zw) \in E$ if $qz \rightarrowtail_\Delta q'z'z$; for arbitrary $w \in \Gamma^*$.

Given a valuation $\rho : Q \rightarrow \mathbb{P}(Prop)$ we can extend it to $Q \times \Gamma^*$ by putting $\rho(qw) = \rho(q)$. This way a pushdown system P and a finite valuation ρ define a, potentially infinite, transition system $M(P, \rho)$ which graph is $G(P)$ and which valuation is given by ρ as described above.

The *model checking problem* is:

given P, ρ and φ decide if $M(P, \rho), q_0\perp \vDash \varphi$

Please observe that the meaning of φ in the initial configuration $q_0\perp$ depends only on the part of $M(P, \rho)$ that is reachable from $q_0\perp$.

3   Assumption Semantics

For this section let us fix a pushdown system P and a valuation ρ. Let us abbreviate $M(P, \rho)$ by M.

We are going to present a modification of the semantics of EF-logic. This modified semantics is used as an induction assumption in the algorithm we are going to present later. From the definition of a transition system M it follows that there are no edges from vertices q, i.e., configurations with the empty stack. We will look at such vertices not as dead ends but as places where some parts of the structure where cut out. We will take a function $S : Q \to \mathbb{P}(Form(EF))$ and interpret $S(q)$ as an assumption that in the vertex q formulas $S(q)$ hold. This view leads to the following definition.

Definition 1. *Let* $S : Q \to \mathbb{P}(Form(EF))$ *be a function. For a vertex* v *of* $G(P)$ *and a formula* α *we define the relation* $M, v \vDash_S \alpha$ *as the least relation satisfying the following conditions:*

- $M, q \vDash_S \alpha$ *for every* $\alpha \in S(q)$.
- $M, v \vDash_S p$ *if* $p \in \rho(v)$.
- $M, v \vDash_S \alpha \wedge \beta$ *if* $M, v \vDash_S \alpha$ *and* $M, v \vDash_S \beta$.
- $M, v \vDash_S \alpha \vee \beta$ *if* $M, v \vDash_S \alpha$ *or* $M, v \vDash_s \beta$.
- $M, v \vDash_S \exists \circ \alpha$ *for* $v \notin Q$ *if there is a successor* v' *of* v *such that* $M, v' \vDash_S \alpha$.
- $M, v \vDash_S \forall \circ \alpha$ *for* $v \notin Q$ *if for every successor* v' *of* v *we have that* $M, v' \vDash_S \alpha$.
- $M, v \vDash_S \exists F \alpha$ *if there is a path from* v *to* v', *s.t.* $M, v' \vDash_S \alpha$ *or* $v' = q$ *for some state* $q \in Q$ *and* $\exists F \alpha \in S(q)$.
- $M, v \vDash_S \forall G \alpha$ *iff for every path* π *from* v *which is either infinite or finite ending in a vertex from* Q *we have that* $M, v' \vDash_S \alpha$ *for every vertex* v' *of* π *and moreover if* π *is finite and ends in a vertex* $q \in Q$ *then* $\forall G \alpha \in S(q)$.

Of course taking arbitrary S in the above semantics makes little sense. We need some consistency conditions as defined below.

Definition 2. *A set of formulas* B *is* saturated *if*

- *for every formula* α *either* $\alpha \in B$ *or* $\neg \alpha \in B$ *but not both;*
- *if* $\alpha \in B$ *and* $\beta \in B$ *then* $\alpha \wedge \beta \in B$;
- *if* $\alpha \in B$ *then* $\alpha \vee \beta \in B$ *and* $\beta \vee \alpha \in B$ *for arbitrary* β;
- *if* $\alpha \in B$ *then* $\exists F \alpha \in B$.

Definition 3 (Assumption function). *A function* $S : Q \to \mathbb{P}(Form(EF))$ *is* saturated *if* $S(q)$ *is saturated for every* $q \in Q$. *A function* S *is* consistent *with* ρ *if* $S(q) \cap Prop = \rho(q)$ *for all* $q \in Q$. *We will not mention* ρ *if it is clear from the context. We say that* S *is an* assumption function *(for* ρ*) if it is saturated and consistent.*

Lemma 1. *For every assumption function* S *and every vertex* v *of* M: $M, v \vDash_S$ α *iff not* $M, v \vDash_S \neg \alpha$.

The next lemma says that the truth of α depends only on assumptions about subformulas of α.

Definition 4. *For a formula α, let $cl(\alpha)$ be the set of subformulas of α and their negations.*

Lemma 2. *Let α be a formula. Let S, S' be two assumption functions such that $S(q) \cap cl(\alpha) = S'(q) \cap cl(\alpha)$ for all $q \in Q$. For every v we have that: $M, v \vDash_S \alpha$ iff $M, v \vDash_{S'} \alpha$.*

We have asumed that the initial stack symbol $\perp$ cannot be taken from the stack. Hence no state q is reachable from configuration $q_0\perp$. In this case our semantics is equivalent to the usual one:

Lemma 3. *For arbitrary S and α we have $M, q_0\perp \vDash_S \alpha$ iff $M, q_0\perp \vDash \alpha$.*

We finish this section with a composition lemma which is the main property of our semantics. We will use it in induction arguments.

Definition 5. *For a stack symbol z and an assumption function S we define the function $S \uparrow_z$ by: $S \uparrow_z (q') = \{\beta : M, q'z \vDash_S \beta\}$, for all $q' \in Q$.*

Lemma 4 (Composition Lemma). *Let α be a formula, z a stack symbol and S an assumption function. Then $S \uparrow_z$ is an assumption function and for every configuration qwz' reachable from qz' we have:*

$$M, qwz'z \vDash_S \alpha \quad iff \quad M, qwz' \vDash_{S\uparrow_z} \alpha$$

4   Model Checking EF

As in the previous section let us fix a pushdown system P and a valuation ρ. Let us write M instead of $M(P, \rho)$ for the transition system defined by P and ρ.

Instead of the model checking problem we will solve a more general problem of deciding if $M, qz \vDash_S \beta$ holds for given q, z, S and β. A small difficulty here is that S is an infinite object. Fortunately, by Lemma 2 to decide if $M, qz \vDash_S \beta$ holds it is enough to work with S restricted to subformulas of β, namely with $S|_\beta$ defined by $S|_\beta(q') = S(q') \cap cl(\beta)$ for all $q' \in Q$. In this case we will also say that S is *extending* $S|_\beta$.

Definition 6. *Let α be a formula, q a state, z a stack symbol, and $\overline{S} : Q \rightarrow \mathbb{P}(Form(EF))$ a function assigning to each state a subset of $cl(\alpha)$. We will say that a tuple $(\alpha, q, z, \overline{S})$ is good if there is an assumption function S such that $S|_\alpha = \overline{S}$ and $M, qz \vDash_S \alpha$.*

Below we describe a procedure which checks if a tuple $(\alpha, q, z, \overline{S})$ is good. It uses an auxiliary procedure Search(q, z, q') which checks whether there is a path from the configuration qz to the configuration q'.

– Check$(p, q, z, \overline{S}) = 1$ if $p \in \rho(q)$;
– Check$(\alpha \wedge \beta, q, z, \overline{S}) = 1$ if Check$(\alpha, q, z, \overline{S}) = 1$ and Check$(\beta, q, z, \overline{S}) = 1$;

- $\mathrm{Check}(\neg\alpha, q, z, \overline{S}) = 1$ if $\mathrm{Check}(\alpha, q, z, \overline{S}) = 0$;
- $\mathrm{Check}(\exists\circ\alpha, q, z, \overline{S}) = 1$ if either
  - there is $qz \rightarrowtail q'$ and $\alpha \in \overline{S}(q')$; or
  - there is $qz \rightarrowtail q'z'z$ and $\mathrm{Check}(\alpha, q', z', \overline{S}') = 1$, where $\overline{S}'$ is defined by:
    $\overline{S}'(q'') = \{\beta \in \mathrm{cl}(\alpha) : \mathrm{Check}(\beta, q'', z, \overline{S}) = 1\}$, for all $q'' \in Q$.
- $\mathrm{Check}(\exists F\alpha, q, z, \overline{S}) = 1$ if either
  - $\mathrm{Check}(\alpha, q, z, \overline{S}) = 1$; or
  - there is $qz \rightarrowtail q'$ and $\exists F\alpha \in \overline{S}(q')$; or
  - there is $qz \rightarrowtail q'z'z$ and $q'' \in Q$ for which $\mathrm{Search}(q', z', q'') = 1$ and $\mathrm{Check}(\exists F\alpha, q'', z, \overline{S}) = 1$; or
  - there is $qz \rightarrowtail q'z'z$ with $\mathrm{Check}(\exists F\alpha, q', z', \overline{S}') = 1$ for $\overline{S}'$ defined by:
    $\overline{S}'(q'') = \{\exists F\alpha : \mathrm{Check}(\alpha, q'', z, \overline{S}) = 1\} \cup \{\neg\exists F\alpha : \mathrm{Check}(\alpha, q'', z, \overline{S}) = 0\} \cup \{\beta \in \mathrm{cl}(\alpha) : \mathrm{Check}(\beta, q'', z, \overline{S}) = 1\}$, for all $q'' \in Q$.
- In other cases $\mathrm{Check}(\alpha, q, z, \overline{S}) = 0$.
- $\mathrm{Search}(q_1, z, q_2) = 1$ if either
  - there is $q_1 z \rightarrowtail q_2$; or
  - there is $q_1 z \rightarrowtail q_1' z' z$ and $q_2' \in Q$ for which $\mathrm{Search}(q_1', z', q_2') = 1$ and $\mathrm{Search}(q_2', z, q_2) = 1$.

Lemma 5. *We have* $\mathrm{Search}(q_1, z, q_2) = 1$ *iff there is a path from the configuration* $q_1 z$ *to the configuration* q_2. *The procedure can be implemented on a Turing machine working in* $\mathcal{O}(|Q|^2|\Gamma|)$ *time and space.*

Proof
The proof of the correctness of the procedure is easy. The procedure can be implemented using dynamic programming. The implementation can construct a table of all good values (q_1, z, q_2).   □

Lemma 6. *Procedure* $\mathrm{Check}(\alpha, q, z, \overline{S})$ *can be implemented on a Turing machine working in* $Sp(|\alpha|) = \mathcal{O}((|\alpha|\log(|Q|)|Q||\Gamma|)^2)$ *space.*

Proof
The proof is by induction on the size of α. All the cases except for $\alpha = \exists F\beta$ are straightforward.

For $\alpha = \exists F\beta$ consider the graph of exponential size which nodes are of the form $\mathrm{Check}(\exists F\beta, q, z, \overline{S})$ for arbitrary $q, z, \overline{S}$. The edges are given by the rules:

- $\mathrm{Check}(\exists F\beta, q_1, z, \overline{S}) \rightarrow \mathrm{Check}(\exists F\beta, q_2, z, \overline{S})$ whenever $q_1 z \rightarrowtail q_1' z' z$ and $\mathrm{Search}(q_1', z', q_2) = 1$;
- $\mathrm{Check}(\exists F\beta, q_1, z_1, \overline{S}_1) \rightarrow \mathrm{Check}(\exists F\beta, q_2, z_2, \overline{S}_2)$ if $q_1 z_1 \rightarrowtail q_2 z_2 z_1$ and $\overline{S}_2$ is defined by $\overline{S}_2(q'') = \{\beta \in \mathrm{cl}(\alpha) : \mathrm{Check}(\beta, q'', z_1, \overline{S}_1) = 1\} \cup \{\neg\exists F\alpha : \mathrm{Check}(\alpha, q'', z_1, \overline{S}_1) = 0\} \cup \{\exists F\alpha : \mathrm{Check}(\alpha, q'', z_1, \overline{S}_1) = 1\}$

Observe that by induction assumption we can calculate whether there is an edge between two nodes using space $Sp(|\beta|)$. A node Check($\exists F\beta, q, z, \overline{S}$) is *successful* if either Check($\beta, q, z, \overline{S}$) = 1 or there is $qz \rightarrowtail q'$ with $\exists F\beta \in \overline{S}(q')$.

It is easy to see that Check($\exists F\beta, q, z, \overline{S}$) = 1 iff in the graph described above there is a path from the node Check($\exists F\beta, q, z, \overline{S}$) to a successful node.

We need $\mathcal{O}(\log(|Q|)|\Gamma||Q||\beta|)$ space to store a node of the graph. So we need $\mathcal{O}((\log(|Q|)|\Gamma||Q||\beta|)^2)$ space to implement Savitch algorithm performing deterministic reachability test in this graph. We also need $S(|\beta|)$ space for an oracle to calculate edges and $\mathcal{O}(|Q|^2|\Gamma|)$ space for Search procedure. All this fits into $Sp(|\exists F\beta|)$ space. $\qquad\square$

Remark: It does not seem that this lemma follows from the fact that alternating machines with bounded alternation can be simulated by deterministic ones with small space overhead (c.f. the theorem attributed in [3] to a personal communication from A. Borodin).

Lemma 7. *A tuple* $(\alpha, q, z, \overline{S})$ *is good iff* Check($\alpha, q, z, \overline{S}$) = 1

Proof

The proof is by induction on the size of α. The case when α is a propositional letter is obvious. The case when $\alpha = \neg\beta$ follows from Lemma 1. The case for conjunction is easy using Lemma 2. We omit the case for $\alpha = \exists\circ\beta$ because the arguments is simpler than in the case of F operator.

Case $\alpha = \exists F\beta$. Suppose that $(\alpha, q, z, \overline{S})$ is good. This means that there is an assumption function S such that $S|_\alpha = \overline{S}$ and $M, qz \vDash_S \alpha$. By the definition of the semantic, there is a vertex v reachable from qz such that $M, v \vDash_S \beta$ or $v = q'$ and $\exists F\beta \in S(q')$. Suppose that v is such a vertex at the smallest distance from qz. We show that Check($\alpha, q, z, \overline{S}$) = 1 by induction on the distance to v.

If $v = qz$ then, as β is a subformula of α, we have by the main induction hypothesis that Check($\beta, q, z, \overline{S}$) = 1. So Check($\alpha, q, z, \overline{S}$) = 1. If $qz \rightarrowtail q'$ and $\exists F\beta \in S(q')$ then we also get Check($\alpha, q, z, \overline{S}$) = 1. Otherwise we have $qz \rightarrowtail q'z'z$ and $q'z'z$ is the first vertex on the shortest path to v.

Suppose that on the path to v there is a configuration of the form $q''z$ for some q''. Assume moreover that it is the first configuration of this form on the path. We have that Search(q', z', q'') = 1 and $M, q''z \vDash \exists F\beta$. As the distance to v from $q''z$ is smaller than from qz, we get Check($\exists F\beta, q'', z, \overline{S}$) = 1 by the induction hypothesis. Hence Check($\alpha, q, z, \overline{S}$) = 1.

Otherwise, i.e., when there is no configuration of the form $q''z$ on the path to v, we know that $v = q''wz'z$ for some $q'' \in Q$ and $w \in \Gamma^*$. Moreover we know that $q''wz'$ is reachable from $q'z'$. By Composition Lemma we have that $M, q''wz' \vDash_{S\uparrow_z} \beta$. Let $\overline{S}_1$ be a function defined by $\overline{S}_1(q_1) = (S\uparrow_z)|_\beta(q_1) \cup \{\neg\exists F\beta : \beta \notin S\uparrow_z(q_1)\}\cup\{\exists F\beta : \beta \in S\uparrow_z(q_1)\}$. It can be checked that $\overline{S}_1$ can be extended to an assumption function S_1. By Lemma 2 we have $M, q''wz' \vDash_{S_1} \beta$. Hence $M, q'z' \vDash_{S_1} \exists F\beta$. We have Check($\exists F\beta, q', z', \overline{S}_1$) = 1 from induction hypothesis. By definition of S_1 and the induction hypothesis we have that $S_1(q_1) = \{\gamma \in \mathrm{cl}(\beta) : $ Check(γ, q_1, z, S) = 1$\} \cup \{\neg\exists F\beta : $ Check(β, q_1, z, S) = 0$\} \cup \{\exists F\beta : $ Check(β, q_1, z, S) = 1$\}$. Which gives Check(α, q, z, S) = 1.

For the final case suppose that $\alpha = \exists F\beta$ and that $\text{Check}(\alpha, q, z, \overline{S}) = 1$. We want to show that $(\alpha, q, z, \overline{S})$ is good using additional induction on the length of the computation of $\text{Check}(\alpha, q, z, \overline{S})$. Let S be an assumption function such that $S|_\alpha = \overline{S}$.

Skiping a couple of easy cases suppose that there is $qz \rightarrowtail q'z'z$ and that we have $\text{Check}(\exists F\beta, q', z', \overline{S}') = 1$ for $\overline{S}'$ defined by $\overline{S}'(q'') = \{\gamma \in \text{cl}(\beta) : \text{Check}(\gamma, q'', z, S) = 1\} \cup \{\neg \exists F\beta\}$ or $\overline{S}'(q'') = \{\gamma \in \text{cl}(\beta) : \text{Check}(\gamma, q'', z, S) = 1\} \cup \{\exists F\beta\}$ depending on whether $\text{Check}(\beta, q'', z, S) = 0$ or not. By the induction hypothesis, $M, q'z' \vDash_{S'} \exists F\beta$ for an assumption function S' such that $S'|_\alpha = \overline{S}'$

Consider $S \uparrow_z$. We have that $S \uparrow_z |_\beta = S'|_\beta$ by the induction hypothesis. It is also the case that for every $q'' \in Q$, whenever $\exists F\beta \in S'(q'')$ then $\exists F\beta \in S \uparrow_z (q'')$. Hence, by Lemma 2 and the definition of our semantics, we have that $M, q'z' \vDash_{S\uparrow_z} \exists F\beta$. By Composition Lemma we have $M, q'z'z \vDash_S \exists F\beta$. Which gives $M, qz \vDash_S \exists F\beta$. So $(\alpha, q, z, \overline{S})$ is good. □

5   Model Checking CTL

In this section we show that the model checking problem for pushdown systems and CTL is EXPTIME hard. The problem can be solved in EXPTIME as there is a linear translation of CTL to the μ-calculus and the model checking for the later logic can be done in EXPTIME [9].

Let M be an alternating Turing machine using n tape cells on input of size n. For a given configuration c we will construct a pushdown system P_M^c, valuation ρ_M, and a CTL formula α_M such that: $M(P_M^c, \rho_M), q_0\bot \vDash \alpha_M$ iff M has an accepting computation from c. As P_M^c and α_M will be polynomial in the size of c this will show EXPTIME hardness of the model checking problem.

We will do the construction in two steps. First, we will code the acceptance problem into the reachability problem for a pushdown system extended with some test operations. Then, we will show how to simulate these tests in the model checking problem.

We assume that the nondeterminism of M is limited so that from every configuration M has at most two possible moves. A move is a pair $m = (a, d)$ where a is a letter to put and d is a direction for moving the head. We use $c \vdash_m c'$ to mean that c' is obtained from c by doing the move m. The transition function of M assigns to each pair (state,letter) a pair of moves of M. A computation of M can be represented as a tree of configurations. If the machine is in a universal state then the configuration has two sons corresponding to the two moves in the pair given by the transition function. If the machine is in an existential state then there is only one son for one of the moves from the pair.

An *extended pushdown system* is obtained by adding two kinds of test transitions. Formally each of the kinds of transitions depends on a parameter n which is a natural number. To make notation clearer we fix this number in advance. Transition $q \rightarrowtail^A q'$ checks whether the first n letters from the top of the stack form an accepting configuration of M. Transition $q \rightarrowtail^M q'$ checks, roughly, whether the first $2n$ letters from the top of the stack form two configurations such that the first is the successor of the second. A formal definition of these transitions is given below when we define a particular extended pushdown system.

Let us fix n as the size of input to our Turing machine. We define an extended pushdown system EP_M simulating computations of M on inputs of size n. The set of states of the system is $Q = \{q, q_M, q_A\}$. The stack alphabet is $\Gamma = \Gamma_M \cup Q_M \cup \text{Moves}_M \times \text{Moves}_M \cup \{E, L, R\}$; where Γ_M is the tape alphabet of M, Q_M is the set of states of M; Moves_M is the set of moves of M; and E, L, R are new special letters which stand for arbitrary, left and right element of a pair respectively. Before defining transitions of EP_M let us formalize the definition of $\rightarrowtail^A$ and $\rightarrowtail^M$ transitions. These transitions add the following edges in the graph of configurations of the system:

- For a transition $q \rightarrowtail^A q'$ and for an arbitrary $w \in \Gamma^*$ we have the edge $qcw \rightarrow q'cw$ if c is an accepting configuration of M.
- For a transition $q \rightarrowtail^M q'$, for an arbitrary $w \in \Gamma^*$ and a letter $? \in \{E, L, R\}$ we have the edge $qc'?(m_1, m_2)cw \rightarrow q'c'?(m_1, m_2)cw$ if (m_1, m_2) is the move form a configuration c and $c \vdash_m c'$ where $m = m_1$ if $? = L$; $m = m_2$ if $? = R$; and $m \in \{m_1, m_2\}$ if $? = E$.

Finally, we present the transition rules of EP_M. Below, a' stands for any letter other than E, L or R. We use c, c' to stand for a configuration of M, i.e., a string of length $n + 1$.

$$q \rightarrowtail q_A \qquad\qquad q_A c \rightarrowtail^A q$$
$$qa' \rightarrowtail q_M c' L(m_1, m_2)a' \qquad\qquad q_M \rightarrowtail^M q$$
$$qa' \rightarrowtail q_M c' E(m)a' \qquad\qquad qL(m_1, m_2) \rightarrowtail q_M c' R(m_1, m_2)$$
$$qR(m_1, m_2)c \rightarrowtail q \qquad\qquad qE(m)c \rightarrow q$$

It is easy to see that the transitions putting or taking a whole configuration from the stack can be simulated by a sequence of simple transitions working with one letter at the time. In the above, transition $q_A c \rightarrowtail^A q$ (which removes a configuration and at the same time checks whether it is accepting) is not exactly in the format we allow. Still it can be simulated by two transitions in our format. We use $G(EP_M)$ to denote the graph of configurations of EP_M, i.e., the graph which vertices are configurations and which edges correspond to one application of the transition rules.

The idea behind the construction of EP_M is described by the following lemma.

Lemma 8. *For every configuration c of M we have that: M accepts from c iff in the graph $G(EP_M)$ of configurations of EP_M configuration q is reachable from configuration qc.*

Proof

We present only a part of the argument for the left to right direction. The proof proceeds by induction on the height of the tree representing an accepting computation of M on c.

If c is an accepting configuration then we have a path $qc \to q_A c \to q$ in $G(EP_M)$.

Suppose now that the first move of M in its computation is (m_1, m_2) and it is an existential move. Then we have a path:

$$qc \to q_M c' E(m_1, m_2)c \to qc' E(m_1, m_2)c \to \cdots \to qE(m_1, m_2)c \to q$$

where the existence of a path $qc' E(m_1, m_2)c \to \cdots \to qE(m_1, m_2)c$ follows from the induction hypothesis.

Suppose now that the first move of M in an accepting computation from c is (m_1, m_2) and it is a universal move. We have a path:

$$qc \to q_M c' L(m_1, m_2)c \to qc' L(m_1, m_2)c \to \cdots \to qL(m_1, m_2)c \to$$
$$q_M c'' R(m_1, m_2)c \to qc'' R(m_1, m_2)c \to \cdots \to qR(m_1, m_2)c \to q.$$

Once again the existence of dotted out parts of the path follows from the induction hypothesis.

This completes the proof from the left to right direction. The opposite direction is analogous. $\qquad\square$

The next step in our proof is to code the above reachability problem into the model checking problem for a normal pushdown system. First, we change extended pushdown system EP_M into a normal pushdown system P_M. We add new states q_{TA}, q_{TM}, q_F and q_R^a for every letter a of the stack alphabet. The role of q_{TA} and q_{TM} is to initiate test performed originally by $\rightarrowtail^A$ and $\rightarrowtail^M$ transitions, respectively. State q_F is a terminal state signalling success. States q_R^a are used in the test. They take out all the letters from the stack and give information about what letters are taken out. In the rules below c, c' range over configurations; a, b over single letters; and a' over letters other than E, L or R.

$$
\begin{aligned}
qa' &\rightarrowtail q_A a' & q_A &\rightarrowtail q, q_{TA} \\
qa' &\rightarrowtail q_M c' L(m_1, m_2)a' & q_M &\rightarrowtail q, q_{TM} \\
qa' &\rightarrowtail q_M c' E(m_1, m_2)a' & qL(m_1, m_2) &\rightarrowtail q_M c' R(m_1, m_2) \\
qR(m_1, m_2)c &\rightarrowtail q & qE(m_1, m_2)c &\rightarrow q \\
q_{TA}a &\rightarrowtail q_R^a & q_{TM}a &\rightarrow q_R^a \\
q_R^a b &\rightarrowtail q_R^b & q\bot &\rightarrow q_F \bot
\end{aligned}
$$

Recall that $\bot$ is the initial stack symbol of a pushdown automaton. As before we use $G(P_M)$ to denote the graph of configurations of P_M.

To simplify matters we will use states also as names of propositions and take valuation ρ_M such that in a state q' exactly proposition q' holds, i.e., $\rho_M(q') = \{q'\}$.

First we take two EF formulas Accept and Move such that:

- $M(P_M, \rho_M), q_{TA}w \vDash$ Accept iff w starts with an accepting configuration of M.

– $M(P_M, \rho_M), q_{TM}w \vDash$ Move iff w is of the form $c'?(m_1, m_2)cw'$, (m_1, m_2) is the move of M, and $c \vdash_m c'$ where $m = m_1$ if $? = L$; $m = m_2$ if $? = R$; and $m \in \{m_1, m_2\}$ if $? = E$.

From states q_{TA} and q_{TM} the behaviour of P_M is deterministic. It only takes letters from the stack one by one. The formula Accept is $\bigvee_{i=1,\ldots,n+1} \exists \circ^i q_R^F$ where q_R^F signals an accepting state of M. The formula Move is slightly more complicated as it needs to code the behaviour of M. Still its construction is standard.

The formula we are interested in is:

$$\alpha = \exists \big[(q \vee q_A \vee q_M) \wedge (q_A \Rightarrow \exists \circ (q_{TA} \wedge \text{Accept})) \wedge$$

$$(q_M \Rightarrow \exists \circ (q_{TM} \wedge \text{Move})) \big] \, U \, q_F$$

It says that there is a path going only through states q, q_A or q_M and ending in a state q_F. Moreover, whenever there is a state q_A on the path then there is a turn to a configuration with a state q_{TA} from which Accept formula holds. Similarly for q_M.

Lemma 9. *For every word w over the stack alphabet: q is reachable from qw in $G(EP_M)$ iff $M(P_M, \rho_M), qw\bot \vDash \alpha$.*

Proof
The proof in both directions is by induction on the length of the path. We will only present a part of the proof for the direction from left to right.
If in $G(EP_M)$ the path is $qw \to q_A w \to q$ then in $G(P_M)$ we have:

$$qw\bot \longrightarrow q_a w\bot \longrightarrow q\bot$$
$$\searrow q_{TA}w\bot$$

The edge $q_A w \to q$ exists in $G(EP_M)$ only if w is an accepting configuration. Hence, we have that $M(P_M, \rho_M), q_{TA}w \vDash$ Accept and consequently we have the thesis of the lemma.
If the path is $qw \to q_M c'?(m_1, m_2)w \to qc'?(m_1, m_2)w \to \cdots$ then in $G(P_M)$ we have:

$$qw\bot \longrightarrow q_M c'?(m_1, m_2)w\bot \longrightarrow qc'?(m_1, m_2)w\bot \longrightarrow \cdots$$
$$\searrow q_{TM}c'?(m_1, m_2)w\bot$$

The edge $q_M c'?(m_1, m_2)w \to qc'?(m_1, m_2)w$ exists in $G(EP_M)$ only when the stack content $c'?(m_1, m_2)w$ satisfies the conditions of $\rightarrowtail^M$ transition. This means that $M(P_M, \rho_M), q_{TM}c'?(m_1, m_2)w\bot \vDash$ Move. From the induction assumption we have $M(P_M, \rho_M), qc'?(m_1, m_2)w \vDash \alpha$. Hence $M(P_M, \rho_M), qw \vDash \alpha$. $\square$

Theorem 1. *The model checking problem for pushdown systems and CTL is EXPTIME-complete*

Proof

The problem can be solved in EXPTIME as there is a linear translation of CTL to the μ-calculus and the model checking for the later logic can be done in EXPTIME [9].

To show hardness part let M be an alternating Turing machine as considered in this section. For an input word v of length n we construct in polynomial time a pushdown system P_M^v, valuation ρ_M and a formula α_M such that: v is accepted by M iff $M(P_M^v, \rho_M), q_0\bot \vDash \exists \circ^{n+1}\alpha$. Let c_0^v be the initial configuration of M on v. It has the length $n + 1$.

Valuation ρ_M and formula α_M are ρ and α as described before Lemma 9. The system P_M^v is such that started in $q_0\bot$ it first puts the initial configuration c_0^v on the stack and then behaves as the system P_M.

By Lemma 8 we have that M has an accepting computation from c_0^v iff there is a path from qc_0^v to q in $G(EP_M)$. By Lemma 9 this is equivalent to the fact that $M(P_M, \rho_M), qc_0^v\bot \vDash \alpha_M$. By the construction of P_M^v this the same as saying that $M(P_M^v, \rho_M), q_0\bot \vDash \exists \circ^{n+1}\alpha_M$. □

References

[1] A. Bouajjani, J. Esparza, and O. Maler. Reachability analysis of pushdown automata: Applications to model checking. In *CONCUR'97*, volume 1243 of *LNCS*, pages 135–150, 1997.

[2] D. Caucal. On infinite transition graphs a having decidable monadic theory. In *ICALP'96*, LNCS, 1996.

[3] A. K. Chandra, D. C. Kozen, and L. J. Stockmeyer. Alternation. *Journal of the ACM*, 28(1):114–133, 1981.

[4] E. Clarke and E. Emerson. Design and synthesis of synchronization skeletons using branching time temporal logic. In *Workshop on Logics of Programs*, volume 131 of *LNCS*, pages 52–71. Springer-Verlag, 1981.

[5] E. A. Emerson. Temporal and modal logic. In J. Leeuwen, editor, *Handbook of Theoretical Computer Science Vol.B*, pages 995–1072. Elsevier, 1990.

[6] J. Esparza. Private communication.

[7] J. Esparza, D. Hansel, and P. Rossmanith. Efficient algorithms for model checking pushdown systems. In *CAV '00*, LNCS, 2000. to appear.

[8] D. Muller and P. Schupp. The theory of ends, pushdown automata and second-order logic. *Theoretical Computer Science*, 37:51–75, 1985.

[9] I. Walukiewicz. Pushdown processes: Games and model checking. In *CAV'96*, volume 1102 of *LNCS*, pages 62–74, 1996. To appear in Information and Computation.

A Decidable Dense Branching-Time Temporal Logic*

Salvatore La Torre[1,2] and Margherita Napoli[2]

[1] University of Pennsylvania
[2] Università degli Studi di Salerno

Abstract. Timed computation tree logic (TCTL) extends CTL by allowing timing constraints on the temporal operators. The semantics of TCTL is defined on a dense tree. The satisfiability of TCTL-formulae is undecidable even if the structures are restricted to dense trees obtained from timed graphs. According to the known results there are two possible causes of such undecidability: the denseness of the underlying structure and the equality in the timing constraints. We prove that the second one is the only source of undecidability when the structures are defined by timed graphs. In fact, if the equality is not allowed in the timing constraints of TCTL-formulae then the finite satisfiability in TCTL is decidable. We show this result by reducing this problem to the emptiness problem of timed tree automata, so strengthening the already well-founded connections between finite automata and temporal logics.

1 Introduction

In 1977 Pnueli proposed Temporal Logic as a formalism to specify and verify computer programs [Pnu77]. This formalism turned out to be greatly useful for reactive systems [HP85], that is systems maintaining some interaction with their environment, such as operating systems and network communication protocols. Several temporal logics have been introduced and studied in literature, and now this formalism is widely accepted as specification language for reactive systems (see [Eme90] for a survey).

Temporal logic formulae allow to express temporal requirements on the occurrence of events. Typical temporal operators are "until", "next", "sometimes", "always", and a typical assertion is "p is true until q is true". These operators allow us only to express qualitative requirements, that is constraints on the temporal ordering of the events, but we cannot place bounds on the time a certain property must be true. As a consequence traditional temporal logics have been augmented by adding timing constraints to temporal operators, so that assertions such as "p is true until q is true within time 5" can be expressed. These logics, which are often referred to as real-time or quantitative temporal logics, are suitable when it is necessary to explicitly refer to time delays between events and then we want to check that some hard real-time constraints are satisfied.

---

* Work partially supported by M.U.R.S.T. grant TOSCA.

S. Kapoor and S. Prasad (Eds.): FST TCS 2000, LNCS 1974, pp. 139–150, 2000.

Besides the usual classification in linear and branching-time logics, real-time logics are classified according to the nature of the time model they use. Temporal logics based on discrete time models are presented in [EMSS90, JM86, Koy90, PH88]. An alternative approach is to model time as a dense domain. Temporal logics with this time model are MITL [AFH96], TCTL [ACD93], STCTL [LN97], and GCTL [PH88]. For more about real-time logics, see [AH93, Hen98].

In this paper we are interested in branching-time temporal logics which use a dense time domain and in particular we will consider the satisfiability problem in TCTL that was introduced by Alur et al. in [ACD93]. Given a formula φ we want to determine if there exists a structure M satisfying it. The syntax of TCTL is given by augmenting the temporal operators of CTL [CE81] (except for the "next" which is discarded since it does not have any meaning in a dense time domain) with a timing constraint of type $\approx c$, where $\approx$ is one among $<$, $\leq$, $>$, $\geq$, and $=$, and c is a rational number. The semantics of TCTL is given on a dense (or continuous) tree. It turns out that the satisfiability problem in TCTL is undecidable even if the semantics is restricted to dense trees obtained from timed graphs (*finite satisfiability* [ACD93]), that is, timed transition systems where the transitions depend also on the current value of a finite number of clock variables.

Another real-time branching-time temporal logic is STCTL [LN97] which is obtained by restricting both the semantics and the syntax of TCTL. Instead of a dense tree, a timed ω-tree is used to define the semantics of formulae, and the equality is not allowed in the timing constraints. With these restrictions the STCTL-satisfiability problem turns out to be decidable. This result is obtained by reducing the STCTL-satisfiability problem to the the emptiness problem of finite automata on timed ω-trees, which is shown to be decidable in [LN97]. Introducing the equality in the timing constraints causes the loss of the decidability. A similar kind of result was observed in MITL, where the decidability is lost when the restriction to non-singular intervals is relaxed [AFH96]. In this paper we prove that this indeed holds also for the finite satisfiability in TCTL. In particular, we reduce the finite satisfiability problem of TCTL-formulae without equality in the timing constraints, to the emptiness problem of timed tree automata, via translation to the satisfiability problem of TCTL-formulae with respect to a proper subclass of STCTL-structures. Restricting the class of STCTL-structures is necessary since along any path of a timed graph the truth assignments of the atomic propositions vary according to a sequence of left-closed right-open intervals, while in general in STCTL-structures the truth assignments change according to sequences of time intervals which are alternatively singular and opened. Having defined the language of a logic as the set of formulae which are satisfiable, as a consequence of the previous result we have that TCTL interpreted on timed graphs is language equivalent to a proper restriction of STCTL. Moreover, in this paper we also introduce a concept of a highly-deterministic timed tree automaton with the aim of matching the concept of regular tree in ω-tree languages. The use of the theory of timed tree automata to achieve the decidability of the TCTL finite satisfiability, strengthens the relationship between finite automata and temporal logics, also in the case of real-time logics. In a recent paper [DW99] an automata-theoretic approach to TCTL-model checking has been presented. There the authors introduced timed alternating tree automata and rephrased the model-checking problem as a particular word problem for these automata. For timed alternating tree automata, this decision problem is decidable while the emptiness problem is not decidable.

The rest of the paper is organized as follows. In section 2 we recall the main definitions and results from the theory of timed tree automata, and we introduce a concept of highly-deterministic timed tree automaton. In section 3 we recall the temporal logics TCTL and STCTL with the related decidability results. The main result of this paper is presented in section 4, where the finite satisfiability in TCTL is shown to be decidable via reduction to the emptiness problem of timed tree automata. Finally, we give our conclusions in section 5. Due to lack of space some proofs are omitted, for a full version of the paper see [URL].

2   Timed Tree Automata

In this section we recall some definitions and results concerning to timed automata [AD94, LN97], and introduce the concept of highly-deterministic timed tree automaton.

Let Σ be an alphabet and $dom(t)$ be a subset of $\{1, \ldots, k\}^*$, for an integer $k > 0$, such that (i) $\varepsilon \in dom(t)$, and (ii) if $v \in dom(t)$, then for some $j \in \{1, \ldots, k\}$, $vi \in dom(t)$ for any i such that $1 \leq i \leq j$ and $vi \notin dom(t)$ for any $i > j$. A Σ-valued ω-tree is a mapping $t : dom(t) \longrightarrow \Sigma$. For $v \in dom(t)$, we denote with $pre(v)$ the set of prefixes and with $deg(v)$ the arity of v. A path in t is a maximal subset of $dom(t)$ linearly ordered by the prefix relation. Often we will denote a path π with the ordered sequence of its nodes $v_0, v_1, v_2, \ldots$ where v_0 is ε. A timed Σ-valued ω-tree is a pair (t, τ) where t is a Σ-valued ω-tree and τ, called time tree, is a mapping from $dom(t)$ into the set of the nonnegative real numbers $\Re_+$ such that (i) $\tau(v) > 0$, for each $v \in dom(t) - \{\varepsilon\}$ (positiveness), and (ii) for each path π and for each $x \in \Re_+$ there exists $v \in \pi$ such that $\sum_{u \in pre(v)} \tau(u) \geq x$ (progress property). Nodes of a timed ω-tree become available as the time elapses, that is, at a given time only a finite portion of the tree is available. Each node of a timed ω-tree is labelled by a pair (symbol, real number): for the root the real number is the absolute time of occurrence, while for the other nodes is the time which has elapsed since their parent node was read. Positiveness implies that a positive delay occurs between any two consecutive nodes along a path. Progress property guarantees that infinitely many events (i.e. nodes appearing at input) cannot occur in a finite slice of time (nonzenoness). We denote with γ_v the absolute time at which a node v is available, that is $\gamma_v = \sum_{u \in pre(v)} \tau(u)$. In the rest of the paper, we will consistently use γ to denote absolute time, i.e. time elapsed from the beginning of a computation, and τ to denote delays between events. Moreover, we will use the term tree to refer to a Σ-valued ω-tree for some alphabet Σ and the term timed tree to refer to a timed Σ-valued ω-tree.

Now we recall the definition of timed Büchi tree automaton. It is possible to extend this paradigm by considering other acceptance conditions such as Muller,

Rabin, or Streett [Tho90]. Timed Muller tree automata as well as timed Büchi tree automata were introduced and studied in [LN97]. To define timed automata we introduce the notion of clock, timing constraint, and clock valuation. A finite set of *clock variables* (or simply *clocks*) is used to test timing constraints. Each clock can be seen as a chronograph which is synchronized to a unique system clock. Clocks can be read or set to zero (reset): after a reset, a clock automatically restarts. Timing constraints are expressed by clock constraints. Let C be a set of clocks, the set of clock constraints $\Xi(C)$ contains boolean combinations of simple clock constraints of type $x \leq y + c$, $x \geq y + c$, $x \leq c$, and $x \geq c$, where $x, y \in C$ and c is a rational number. A *clock valuation* is a mapping $\nu : C \longrightarrow \Re_+$. If ν is a clock valuation, λ is a set of clocks and d is a real number, we denote with $[\lambda \to 0](\nu + d)$ the clock valuation that gives 0 for each clock $x \in \lambda$ and $\nu(x) + d$ for each clock $x \notin \lambda$.

A *Büchi timed tree automaton* is a 6-tuple $A = (\Sigma, S, S_0, C, \Delta, F)$, where:

- Σ is an alphabet;
- S is a finite set of locations;
- $S_0 \subseteq S$ is the set of starting locations;
- C is a finite set of clocks;
- Δ is a finite subset of $\bigcup_{k \geq 0}(S \times \Sigma \times S^k \times (2^C)^k \times \Xi(C))$;
- $F \subseteq S$ is the set of accepting locations.

A timed Büchi tree automaton A is *deterministic* if $|S_0| = 1$ and for each pair of different tuples $(s, \sigma, s_1, \ldots, s_k, \lambda_1, \ldots, \lambda_k, \delta)$ and $(s, \sigma, s'_1, \ldots, s'_k, \lambda'_1, \ldots, \lambda'_k, \delta')$ in Δ, δ and δ' are inconsistent (i.e., $\delta \wedge \delta' =$false for all clock valuations).

A state system is completely determined by a location and a clock valuation, thus it is denoted by a pair (s, ν). A transition rule $(s, \sigma, s_1, \ldots, s_k, \lambda_1, \ldots, \lambda_k, \delta) \in \Delta$ can be described as follows. Suppose that the system is in the state (s, ν), and after a time τ the symbol σ is read. The system can take the transition $(s, \sigma, s_1, \ldots, s_k, \lambda_1, \ldots, \lambda_k, \delta)$ if the current clock valuation (i.e. $\nu + \tau$) satisfies the clock constraint δ. As a consequence of the transition, the system will enter the states $(s_1, \nu_1), \ldots, (s_k, \nu_k)$ where $\nu_1 = [\lambda_1 \to 0](\nu + \tau), \ldots, \nu_k = [\lambda_k \to 0](\nu + \tau)$. Each node of a timed tree has thus a location and a clock valuation assigned, according to the transition rules in Δ. Formally, this is captured by the concept of run. A *run* of A on a timed tree (t, τ) is a pair (r, ν), where:

- $r : dom(t) \longrightarrow S$ and $\nu : dom(t) \longrightarrow \Re_+^C$;
- $r(\varepsilon) \in S_0$ and $\nu(\varepsilon) = \nu_0$, where $\nu_0(x) = 0$ for any $x \in C$;
- for $v \in dom(t)$, $k = deg(v)$: $(r(v), t(v), r(v1), \ldots, r(vk), \lambda_1, \ldots, \lambda_k, \delta) \in \Delta$, $\nu(v) + \tau(v)$ fulfils δ and $\nu(vi) = [\lambda_i \to 0](\nu(v) + \tau(v)) \; \forall i \in \{1, \ldots, k\}$.

Clearly, deterministic timed automata have at most one run for each timed tree. A timed tree (t, τ) is accepted by A if and only if there is a run (r, ν) of A on (t, τ) and a path π such that $r(u) \in F$ for infinitely many u on π. The language accepted by A, denoted by $T(A)$, is the set of all timed trees accepted by A. In the following we refer to (timed) Büchi tree automata simply as (timed) tree automata.

For a timed tree automaton the set of states is infinite. However, they can be finitely partitioned according to a finite-index equivalence relation over the clock valuations. Each equivalence class, called *clock region*, is defined in such a way that all the clock valuations in an equivalence class satisfy the same set of clock constraints from a given timed automaton (see [AD94] for a precise definition). Given a clock valuation ν, $[\nu]$ denotes the clock region containing ν. A clock region α' is said to be a *time-successor* of a clock region α if and only if for any $\nu \in \alpha$ there is a $d \in \Re_+$ such that $\nu + d \in \alpha'$. The *region automaton* of a timed tree automaton A is a transition system defined by:

- the set of states $R(S) = \{\langle s, \alpha \rangle \mid s \in S$ and α is a clock region for $A\}$;

- the set of starting states $R(S_0) = \{\langle s_0, \alpha_0 \rangle \mid s_0 \in S_0$ and α_0 satisfies $x = 0$ for all $x \in C\}$;

- the transition rules $R(\Delta)$ such that: $(\langle s, \alpha \rangle, \sigma, \langle s_1, \alpha_1 \rangle, \ldots, \langle s_k, \alpha_k \rangle) \in R(\Delta)$ if and only if $(s, \sigma, s_1, \ldots, s_k, \lambda_1, \ldots, \lambda_k, \delta) \in \Delta$ and there is a time-successor α' of α such that α' satisfies δ and $\alpha_i = [\lambda_i \to 0]\alpha'$ for all $i \in \{1, \ldots, k\}$.

The region automaton is the key to reduce the emptiness problem of timed tree automata to the emptiness problem of tree automata. Given a timed tree language T, Untime$(T(A))$ is the tree language $\{t \mid (t, \tau) \in T\}$. We will denote by $R(A)$ the timed tree automaton accepting Untime$(T(A))$ and obtained by the region automaton (see [LN97] for more details).

Theorem 1. *[LN97] For timed Büchi tree automata:*

- *Emptiness problem is decidable in time exponential in the length of timing constraints and polynomial in the number of locations.*

- *Closure under union and intersection holds.*

We end this section by introducing for timed tree automata a concept which captures some of the properties that regular trees have in the context of tree languages. We will use this notion to relate timed tree automata to timed graphs. A timed tree automaton $A = (\Sigma, S, S_0, \Delta, C, F)$ is said to be *highly deterministic* if Untime$(T(A))$ contains a unique tree, and for $s \in S$, $e = (s, \sigma, s_1, \ldots, s_k, \lambda_1, \ldots, \lambda_k, \delta) \in \Delta$ and $e' = (s, \sigma', s'_1, \ldots, s'_h, \lambda'_1, \ldots, \lambda'_h, \delta') \in \Delta$ imply that $e = e'$. The second property of highly-deterministic timed tree automata simply states that there is at most one transition rule that can be executed in each location $s \in S$. A timed tree automaton $A' = (\Sigma, S', S'_0, \Delta', C, F')$ is *contained* in $A = (\Sigma, S, S_0, \Delta, C, F)$ if $S' \subseteq S$, $S'_0 \subseteq S_0$, $\Delta' \subseteq \Delta$, and $F' \subseteq F$. Clearly, $T(A') \subseteq T(A)$ holds. We recall that a regular tree contains a finite number of subtrees. Given a timed tree automaton $A = (\Sigma, S, S_0, \Delta, C, F)$, and a regular run r of $R(A)$ on a regular tree $t \in T(R(A))$, we define a *shrink* of r and t as the labelled directed finite graph $G = (V, E, lab)$ such that there is a mapping $\theta : dom(t) \longrightarrow V$ such that:

- for any $u, u' \in dom(t)$, $\theta(u) = \theta(u')$ implies that $deg(u) = deg(u')$, and for each $i = 1, \ldots, deg(u)$, $\theta(ui) = \theta(u'i)$;

- $E = \{(\theta(u), \theta(ui), i) \mid u \in dom(t)$ and $i \leq deg(u)\}$, and $(v, v', i) \in E$ is an edge from v to v' labelled by i;

- for $v \in V$, $lab(v) = (r(u), t(u))$ for any u such that $v = \theta(u)$.

From the definition of regular tree, such a graph G always exists. Thus, the following theorem holds.

Theorem 2. *Given a timed tree automaton A, $T(A)$ is not empty if and only if there exists a highly-deterministic timed tree automaton contained in A.*

Later in the paper we will use the following property. Given a highly-deterministic timed tree automaton A, there exists a highly-deterministic timed tree automaton A' such that $T(A) = T(A')$ and for each transition rule $(s, \sigma, s_1, \ldots, s_k, \lambda_1, \ldots, \lambda_k, \delta)$ of A' we have that $s_i \neq s_j$ for $i \neq j$. We call such an automaton a *graph-representable* timed tree automaton, since it corresponds to a labelled directed graph such that for any ordered pair of locations (s, s') there is exactly an edge connecting s to s' in the graph.

3   Timed Computation Tree Logic

In this section we recall the real-time branching-time temporal logics TCTL [ACD93] and STCTL [LN97].

Let AP be a set of atomic propositions, the syntax of TCTL-formulae is given by the following grammar:

$$\varphi := p \mid \neg\varphi \mid \varphi \wedge \varphi \mid \exists[\varphi U_{\approx c}\varphi] \mid \forall[\varphi U_{\approx c}\varphi]$$

where $p \in AP$, $\approx\, \in \{<, \leq, >, \geq\}$, and c is a rational number. Notice that the TCTL-syntax given in [ACD93] allows the use of equality in the timing constraints. Here we restrict the syntax to obtain our decidability result.

Before giving the semantics of TCTL, we introduce some common notation. The constant FALSE is equivalent to $\varphi \wedge \neg\varphi$, the constant TRUE is equivalent to $\neg$ FALSE, $\Diamond_{\approx c}\varphi$ and $\Box_{\approx c}\varphi$ are equivalent to TRUE$U_{\approx c}\varphi$ and $\neg\Diamond_{\approx c}\neg\varphi$, respectively. In the rest of the paper with AP we denote the set of atomic propositions of the considered TCTL-formulae. If it is not differently stated, with $\approx$ we refer to a relational operator in $\{<, \leq, >, \geq\}$, and with c to a rational number. We define a dense path through a set of nodes S as a function $\rho : \Re_+ \longrightarrow S$. With ρ_I we denote the restriction of ρ to an interval I and with $\rho_{[0,b)} \cdot \rho'$ the dense path defined as $(\rho_{[0,b)} \cdot \rho')(d) = \rho(d)$, if $d < b$, and $(\rho_{[0,b)} \cdot \rho')(d) = \rho'(d-b)$, otherwise. The semantics of TCTL is given with respect to a *dense tree*. A Σ-valued dense tree M is a triple (S, μ, f) where:

- S is a set of nodes;

- $\mu : S \longrightarrow \Sigma$ is a labelling function;

- f is a function assigning to each $s \in S$ a set of dense paths through S, starting at s, and satisfying the *tree constraint*: $\forall \rho \in f(s)$ and $\forall t \in \Re_+$, $\rho_{[0,t)} \cdot f(\rho(t)) \subseteq f(s)$.

Given a 2^{AP}-valued dense tree $M = (S, \mu, f)$, a state s, and a formula φ, φ is satisfied at s in M if and only if $M, s \models \varphi$, where the relation $\models$ is defined as follows:

- for $p \in AP$, $M, s \models p$ if and only if $p \in \mu(s)$;
- $M, s \models \neg \psi$ if and only if $not(M, s \models \psi)$;
- $M, s \models \psi_1 \wedge \psi_2$ if and only if $M, s \models \psi_1$ and $M, s \models \psi_2$;
- $M, s \models \exists[\psi_1 U_{\approx c}\psi_2]$ if and only if $\exists \rho \in f(s)$ and $\exists d \approx c$ such that $M, \rho(d) \models \psi_2$ and for each d' such that $0 \le d' < d$, $M, \rho(d') \models \psi_1$;
- $M, s \models \forall[\psi_1 U_{\approx c}\psi_2]$ if and only if $\forall \rho \in f(s)$, $\exists d \approx c$ such that $M, \rho(d) \models \psi_2$ and $M, \rho(d') \models \psi_1$ for each d' such that $0 \le d' < d$.

We say that M is a TCTL-*model* of φ if and only if $M, s \models \varphi$ for some $s \in S$. Moreover, a TCTL-formula φ is said to be *satisfiable* if and only if there exists a TCTL-model of φ.

We define the *closure* of a TCTL-formula φ, denoted by $cl(\varphi)$, as the set of all the subformulae of φ and the *extended closure*, denoted by $ecl(\varphi)$, as the set $cl(\varphi) \cup \{\neg\psi \mid \psi \in cl(\varphi)\}$. Moreover, we define $S_\varphi \subseteq 2^{ecl(\varphi)}$ as the collection of sets Ψ with the following properties:

- $\psi \in \Psi \implies \neg\psi \notin \Psi$;
- $\psi_1 \wedge \psi_2 \in \Psi \implies \psi_1 \in \Psi$ and $\psi_2 \in \Psi$;
- $\alpha[\psi_1 U_{\approx c}\psi_2] \in \Psi$, $\alpha \in \{\forall, \exists\} \implies \psi_1 \in \Psi$ or ($\psi_2 \in \Psi$ and $(0 \approx c)$);
- Ψ is maximal, that is for each $\psi \in ecl(\varphi)$: either $\psi \in \Psi$ or $\neg\psi \in \Psi$.

Note that S_φ contains the maximal sets of formulae in $ecl(\varphi)$ which are consistent, in the sense that given an $\Psi \in S_\varphi$ and a dense tree (S, μ, f), the fulfilment at a given $s \in S$ of a formula in Ψ does not prevent all the other formulae in Ψ from being satisfied at s. From now on we only consider TCTL-formulae, thus we will refer to them simply as formulae. In the rest of this section we recall two semantic restrictions to TCTL that have been considered in literature.

3.1   Finite Satisfiability

A first restriction of TCTL-semantics consists of considering only dense trees defined by runs of a timed graph [ACD93]. A *timed graph* is a tuple $G = (V, \mu, s_0, E, C, \Lambda, \xi)$, where:

- V is a finite set of vertices;
- $\mu : V \longrightarrow 2^{AP}$ is a labelling function;
- s_0 is the start vertex;
- $E \in V \times V$ is the set of edges;
- C is a finite set of clocks;
- $\Lambda : E \longrightarrow 2^C$ maps each edge to a set of clocks to reset;
- $\xi : E \longrightarrow \Xi(C)$ maps each edge to a clock constraint.

A timed graph is a timed transition system, where vertices correspond to locations and edges to transitions. A state is given by the current location and the array of all clock values. When a clock constraint is satisfied by the clock valuation of the current state, the corresponding transition can be taken. A transition e forces the system to move, instantaneously, to a new state which is described by the target location of e, and the clock values obtained by resetting the clocks in the reset set of e. Any computation of the system maps reals to states. This concept is captured by the notion of run. Given a state (s, ν) of a timed graph G, an (s, ν)-run of G is an infinite sequence of triples $(s_1, \nu_1, \tau_1), (s_2, \nu_2, \tau_2), \ldots$ where:

- $s_1 = s$, $\nu_1 = \nu$, and $\tau_1 = 0$;

- for $i > 1$ $s_i \in S$, $\tau_i \in \Re_+$, and ν_i is a clock valuation;

- $e_i = (s_i, s_{i+1}) \in E$, $\nu_{i+1} = [\Lambda(e_i) \rightarrow 0](\nu_i + \tau_{i+1})$, $(\nu_i + \tau_{i+1})$ satisfies the enabling condition $\xi(e_i)$, and the series of reals τ_i is divergent (*progress condition*).

An (s, ν)-run can be also seen as a real-valued mapping $\rho(d)$ defined as $\rho(d) = (s_i, \nu_i + d - \gamma_i)$ for $d \in \Re_+$ such that $\gamma_i \leq d < \gamma_{i+1}$ (ρ is also said to be a dense path of G). Notice that a dense path ρ gives for each time a truth assignment of the atomic propositions. Moreover, the truth values stay unchanged in intervals of type $[\gamma_i, \gamma_{i+1})$. The dense tree M defined by a timed graph G is a tuple $(S \times \Re^n, \mu', f)$ where $\mu'(s, \nu) = \mu(s)$ and $f(s, \nu)$ is the set of all the paths corresponding to (s, ν)-runs of G. For a formula φ, we say that $G \models \varphi$ if and only if $M, (s_0, \nu_0) \models \varphi$ where $\nu_0(x) = 0$ for any clock $x \in C$. Thus a formula φ is *finitely satisfiable* if and only if there exists a timed graph G such that $G \models \varphi$.

3.2   Restricting the Semantics to Timed Trees

In this section we recall the temporal logic STCTL which is obtained restricting the TCTL-semantics to dense trees obtained from $2^{AP} \times 2^{AP}$-valued ω-trees. An STCTL-*structure* is a timed $2^{AP} \times 2^{AP}$-valued ω-tree (t, τ) with $\tau(\varepsilon) = 0$. Given an STCTL-structure (t, τ) we denote by t_{open} and t_{sing} the functions defined as $(t_{open}(v), t_{sing}(v)) = t(v)$ for each $v \in dom(t)$. An open and a singular interval along the paths in (t, τ) correspond to each node $v \neq \varepsilon$: $t_{open}(v)$ and $t_{sing}(v)$ are the sets of the atomic propositions which are true in these two intervals. For $v = \varepsilon$, only $t_{sing}(\varepsilon)$ is meaningful. Given a path $\pi = v_0, v_1, v_2, \ldots$ in an STCTL-structure (t, τ), a dense path in (t, τ) corresponding to π and shifted by d is a function $\rho_d^\pi : \Re_+ \longrightarrow 2^{AP}$ such that for any natural number i:

$$\rho_d^\pi(d') = \begin{cases} t_{sing}(v_i) & \text{if } d + d' = \gamma_{v_i} \\ t_{open}(v_{i+1}) & \text{if } \gamma_{v_i} < d + d' < \gamma_{v_{i+1}}. \end{cases}$$

Thus any dense path in (t, τ) corresponds to a sequence of alternatively open and singular intervals where the truth values stay unchanged. Clearly, an STCTL-structure has a dense time semantics on paths and a discrete branching-time structure. In particular, an STCTL-structure (t, τ) defines the dense tree $M^{t, \tau} = (S, \mu, f)$ where (1) $S = \{(vi, d) \mid v \in dom(t) \text{ and } 0 < d \leq \tau(vi)\} \cup \{(\varepsilon, 0)\}$, (2) $\mu(\varepsilon, 0) = t_{sing}(\varepsilon)$, $\mu(vi, d) = t_{open}(vi)$ if $d < \tau(vi)$, and $\mu(vi, d) = t_{sing}(vi)$ otherwise, and (3) $f(\varepsilon, 0)$ is the set of all dense paths ρ_0^π of (t, τ) and $f(vi, d)$ is the set of all the dense paths $\rho_{\gamma_v + d}^\pi$ of (t, τ). For a formula φ, we say that $(t, \tau) \models \varphi$, i.e. (t, τ) is an STCTL-model of φ, if and only if $M^{t, \tau}, (\varepsilon, 0) \models \varphi$.

Thus a formula φ is STCTL-*satisfiable* if and only there exists an STCTL-model of φ.

In [LN97] the problem of STCTL-satisfiability is reduced to the emptiness problem of timed tree automata. In particular, given a formula φ it is possible to construct a timed tree automaton accepting a nonempty language if and only if φ is STCTL-satisfiable. Moreover, all the accepted trees are STCTL-models of φ. The corresponding construction leads to the following results.

Theorem 3. *[LN97] Given a formula φ, if φ is STCTL-satisfiable then there exists an STCTL-model (t, τ) of φ such that:*

- *for each $v \in dom(t)$, $deg(v) \leq 2 \max_{s \in S_\varphi} |\{\exists \psi \mid \exists \psi \in s\}| + 1$, and*

- *there exists a mapping $\eta : dom(t) \longrightarrow S_\varphi \times S_\varphi$ such that $M^{t,\tau}, (v, d) \models \psi$ for each $\psi \in \mu(v, d)$, where $M^{\eta,\tau} = (S, \mu, f)$.*

Moreover, there exists a timed ω-tree automaton A_φ with $O(2^{|\varphi|})$ states and timing constraints of total size $O(|\varphi|)$ such that (t, τ) is an STCTL-model of φ satisfying the above properties if and only if $(t, \tau) \in T(A_\varphi)$.

By the above Theorems 1 and 3 the satisfiability problem in STCTL is decidable in exponential time.

4   Decidability of Finite Satisfiability

In this section we prove the main result of this paper. We show that the finite satisfiability of formulae is decidable. This result is obtained by proving that a formula is finitely satisfiable if and only if is satisfiable over a particular class of STCTL-structures, the left-closed right-open STCTL-structures. Then we show that the satisfiability of formulae on these structures is decidable via a reduction to the emptiness problem of timed tree automata. Finally, we prove that the set of formulae which are STCTL-satisfiable strictly contains the set of finitely-satisfiable formulae.

Let L_{Fin} be the language of formulae that are finitely-satisfiable. We start providing a characterization of L_{Fin} based on a subclass of STCTL-structures. Let (t, τ) be an STCTL-structure, (t, τ) is said to be a *left-closed right-open* STCTL-structure if $t_{sing}(v) = t_{open}(vi)$ for any $v \in dom(t)$ and $1 \leq i \leq deg(v)$. Before to show that the set of formulae which are finitely-satisfiable is exactly the set of formulae which are satisfiable over left-closed right-open STCTL-structures, we prove that the existence of a left-closed right-open STCTL-model of a formula is decidable. The decision procedure we give is obtained, as for the STCTL-satisfiability, via a reduction to the emptiness problem of timed tree automata.

Lemma 1. *Given a formula φ, there exists a timed tree automaton A such that (1) $T(A)$ is not empty if and only if there is a left-closed right-open STCTL-model of φ, and (2) for each $(t, \tau) \in T(A)$ there exists a function $\eta : dom(t) \longrightarrow S_\varphi \times S_\varphi$ such that $M^{t,\tau}, (v, d) \models \psi$ for each $\psi \in \mu(v, d)$, where $M^{\eta,\tau} = (S, \mu, f)$. Moreover, the existence of a left-closed right-open STCTL-model of φ can be checked in exponential time.*

The next two lemmata show that the finitely-satisfiable formulae are exactly the formulae which are satisfiable over left-closed right-open STCTL-structures.

Lemma 2. *Given a formula φ, if φ is finitely satisfiable then φ is satisfiable on a left-closed right-open STCTL-structure.*

Proof. Let G be a timed graph such that $G \models \varphi$. For each subformula $\psi = \exists \psi'$ of φ such that $G \models \psi$, we denote by ρ_ψ a dense path in G such that ψ is satisfied on ρ_ψ. Let Π be the set of all these paths ρ_ψ. If Π is empty, then we add to Π an arbitrary dense path of G. Consider the dense tree obtained deleting all the paths from G but the paths in Π. Since there are only a finite number of such paths, this tree can be mapped into a left-closed right-open STCTL-structure (t, τ) such that $(t, \tau) \models \varphi$.

Lemma 3. *Given a formula φ, if φ has a left-closed right-open STCTL-model then φ is finitely satisfiable.*

Proof. From Lemma 1 we have that there exists a timed tree automaton A_φ accepting left-closed right-open STCTL-models of φ, if there are any. We can consider a new timed tree automaton A'_φ accepting 2^{AP}-valued ω-trees obtained from the timed trees $(t, \tau) \in T(A_\varphi)$ by disregarding $t_{open}(v)$ for each $v \in dom(t)$ (we recall that for left-closed right-open STCTL-structures $t_{sing}(v) = t_{open}(vi)$). Clearly, $T(A'_\varphi)$ is not empty, and hence by Theorem 2, there exists a highly-deterministic timed tree automaton contained in A'_φ and, as a consequence, there exists a graph-representable timed tree automaton $A' = (2^{AP}, S', s_0, \Delta', S')$ such that $T(A') \subseteq T(A'_\varphi)$. Let G be a timed graph $(S', \mu, s_0, E, C, \Lambda, \Delta)$ such that $\mu(s) = \sigma$, $\Delta(e) = \delta$ for any $e = (s', s) \in E$, $e_i = (s, s_i) \in E$ and $\Lambda(e_i) = \lambda_i$ for $i = 1, \ldots, k$ if and only if $(s, \sigma, s_1, \ldots, s_k, \lambda_1, \ldots, \lambda_k, \delta) \in \Delta'$. Notice that, due to the properties of A', G is well defined. Denoted as ν_0 the clock valuation mapping each clock to 0, by the above construction we have that each $\langle s_0, \nu_0 \rangle$-run ρ of G is a continuous path of a timed tree $(t, \tau) \in T(A')$, and on the other hand, for each $(t, \tau) \in T(A')$ any continuous path ρ' in (t, τ) is also an $\langle s_0, \nu_0 \rangle$-run of G. Moreover, by Lemma 1 since $T(A') \subseteq T(A'_\varphi)$, for $(t, \tau) \in T(A')$ there is a timed tree (η, τ) such that $\eta : dom(t) \longrightarrow S_\varphi \times S_\varphi$ and $M^{t,\tau}, (v, d) \models \psi$ for each $\psi \in \mu(v, d)$, where $M^{\eta,\tau} = (S_\eta, \mu, f)$ is the dense tree corresponding to (η, τ). Notice that η is independent by the choice of $(t, \tau) \in T(A')$, since A' is highly deterministic. Thus, since A' is graph-representable, η defines in an obvious way a labelling function η' of the G vertices such that $G \models \psi$ for each $\psi \in \eta'(s_0)$. Since $(t, \tau) \models \varphi$, it holds that $M^{\eta,\tau}, (v, d) \models \varphi$ and thus $\varphi \in \eta'(s_0)$. Hence $G \models \varphi$, and φ is finitely satisfiable.

Directly from the last two lemmata we have the following theorem.

Theorem 4. *A TCTL-formula φ is finitely satisfiable if and only if φ has a left-closed right-open STCTL-model.*

As a consequence of the above results, the membership problem in L_{Fin} is decidable in exponential time and can be reduced to the emptiness problem of timed tree automata.

Theorem 5. *The finite satisfiability of* TCTL-*formulae is decidable in exponential time.*

Proof. By Theorem 4, we have that φ is finitely satisfiable in TCTL if and only if φ has a left-closed right-open STCTL-model. Thus by Lemma 1, the finite satisfiability of TCTL-formulae is decidable in exponential time.

We end this section by proving that the set L_{Fin} is a proper subset of L_{STCTL}, where L_{STCTL} is the language of the STCTL-satisfiable formulae. By Theorem 4 we have that $L_{Fin} \subseteq L_{STCTL}$. The strict containment can be proved by showing that there exists a Formula φ such that φ is STCTL-satisfiable but is not finitely satisfiable.

Example 1. Consider the formula $\varphi = \forall\square_{\leq c}\, p \wedge \forall\square_{>c}\, \neg p$. Let (t, τ) be an STCTL-structure such that (1) for any $i \leq deg(\varepsilon)$, $t_{sing}(\varepsilon) = t_{open}(i) = t_{sing}(i) = p$ and $\tau(i) = c$, and (2) $t_{sing}(v) = t_{open}(v) = \neg p$ for any other $v \in dom(t)$. Clearly (t, τ) is an STCTL-model of φ, and thus we have that $\varphi \in L_{STCTL}$. Moreover $\varphi \notin L_{Fin}$ since truth assignments of a dense path in a timed graph vary on left-closed right-open intervals.

Thus we have the following lemma.

Lemma 4. L_{Fin} *is strictly contained in* L_{STCTL}.

5 Conclusions

In this paper we have proved the decidability of the finite satisfiability of the TCTL-formulae that do not contain the equality in the timing constraints. The result is obtained by reducing this problem to the emptiness problem for timed tree automata. The presented construction uses as intermediate step the decidability of formulae on left-closed right-open STCTL-structures. According to the previously known results there were two possible causes of the undecidability of TCTL-finite satisfiability: the denseness of the underlying structure and the equality in the timing constraints. Our results prove that the only source of undecidability when the structures are defined by timed graphs is the presence of the equality in the timing constraints. We have also compared TCTL to STCTL, via the language of the formulae which are satisfiable in each of them. The interesting result we obtained is that the satisfiability problem in TCTL is decidable on a set of structures more general than those obtained from timed graphs. As a consequence there exists a more general formulation of dense trees with dense branching time that matches the language of formulae which are satisfiable in STCTL. Finally, we prove our results by relating to the theory of timed tree automata, so strengthening the already well-founded connections between the field of logics and the field of finite automata.

Acknowledgements

We would like to thank Rajeev Alur for helpful discussions and suggestions.

References

[ACD93]   R. Alur, C. Courcoubetis, and D.L. Dill. Model-checking in dense real-time. *Information and Computation*, 104(1):2 – 34, 1993.

[AD94]   R. Alur and D.L. Dill. A theory of timed automata. *Theoretical Computer Science*, 126:183 – 235, 1994.

[AFH96]   R. Alur, T. Feder, and T.A. Henzinger. The benefits of relaxing punctuality. *Journal of the ACM*, 43(1):116 – 146, 1996.

[AH93]   R. Alur and T.A. Henzinger. Real-time logics: complexity and expressiveness. *Information and Computation*, 104(1):35 – 77, 1993.

[CE81]   E.M. Clarke and E.A. Emerson. Design and synthesis of synchronization skeletons using branching time temporal logic. In *Proceedings of Workshop on Logic of Programs*, LNCS 131, pages 52 – 71. Springer-Verlag, 1981.

[DW99]   M. Dickhofer and T. Wilke. Timed alternating tree automata: the automata-theoretic solution to the TCTL model checking problem. In *Proceedings of the 26th International Colloquium on Automata, Languages and Programming*, LNCS 1644, pages 281 – 290. Springer-Verlag, 1999.

[Eme90]   E.A. Emerson. Temporal and modal logic. In J. van Leeuwen, editor, *Handbook of Theoretical Computer Science*, volume B, pages 995 – 1072. Elsevier Science Publishers, 1990.

[EMSS90] E.A. Emerson, A.K. Mok, A.P. Sistla, and J. Srinivasan. Quantitative temporal reasoning. In *Proceedings of the 2nd International Conference on Computer Aided Verification*, LNCS 531, pages 136 – 145. Springer-Verlag, 1990.

[Hen98]   T.A. Henzinger. It's about time: Real-time logics reviewed. In *Proceedings of the 9th International Conference on Concurrency Theory, CONCUR'98*, LNCS 1466, pages 439 – 454. Springer-Verlag, 1998.

[HP85]   D. Harel and A. Pnueli. On the development of reactive systems. In *Logics and Models of Concurrent Systems*, volume F-13 of *NATO Advanced Summer Institutes*, pages 477 – 498. Springer-Verlag, 1985.

[JM86]   F. Jahanian and A.K. Mok. Safety analysis of timing properties in real-time systems. *IEEE Transactions on Software Engineering*, SE - 12(9):890 – 904, 1986.

[Koy90]   R. Koymans. Specifying real-time properties with metric temporal logic. *Journal of Real-Time Systems*, 2:255 – 299, 1990.

[LN97]   S. LaTorre and M. Napoli. Timed tree automata with an application to temporal logic. Technical report, Dipartimento di Informatica ed Applicazioni, Università degli Studi di Salerno, Italy, 1997. URL:"http://www.cis.upenn.edu/~latorre/Papers/stctl.ps.gz".

[PH88]   A. Pnueli and E. Harel. Applications of temporal logic to the specification of real-time systems. In *Formal Techniques in Real-time and Fault-tolerant Systems*, LNCS 331, pages 84 – 98. Springer-Verlag, 1988.

[Pnu77]   A. Pnueli. The temporal logic of programs. In *Proceedings of the 18th IEEE Symposium on Foundations of Computer Science*, pages 46 – 77, 1977.

[Rab72]   M.O. Rabin. Automata on infinite objects and Church's problem. *Trans. Amer. Math. Soc.*, 1972.

[Tho90]   W. Thomas. Automata on infinite objects. In J. van Leeuwen, editor, *Handbook of Theoretical Computer Science*, volume B, pages 133 – 191. Elsevier Science Publishers, 1990.

[URL]   URL: "http://www.cis.upenn.edu/~latorre/Papers/fsttcs.ps.gz".

Fair Equivalence Relations

Orna Kupferman[1], Nir Piterman[2], and Moshe Y. Vardi[3]*

[1] Hebrew University, School of Engineering and Computer Science, Jerusalem 91904, Israel
orna@cs.huji.ac.il,
http://www.cs.huji.ac.il/~orna
[2] Weizmann Institute of Science, Department of Computer Science, Rehovot 76100, Israel
nirp@wisdom.weizmann.ac.il,
http://www.wisdom.weizmann.ac.il/~nirp
[3] Rice University, Department of Computer Science, Houston, TX 77251-1892, U.S.A.
vardi@cs.rice.edu,
http://www.cs.rice.edu/~vardi

Abstract. Equivalence between designs is a fundamental notion in verification. The linear and branching approaches to verification induce different notions of equivalence. When the designs are modeled by fair state-transition systems, equivalence in the linear paradigm corresponds to fair trace equivalence, and in the branching paradigm corresponds to fair bisimulation.

In this work we study the expressive power of various types of fairness conditions. For the linear paradigm, it is known that the Büchi condition is sufficiently strong (that is, a fair system that uses Rabin or Streett fairness can be translated to an equivalent Büchi system). We show that in the branching paradigm the expressiveness hierarchy depends on the types of fair bisimulation one chooses to use. We consider three types of fair bisimulation studied in the literature: ∃-bisimulation, game-bisimulation, and ∀-bisimulation. We show that while game-bisimulation and ∀-bisimulation have the same expressiveness hierarchy as tree automata, ∃-bisimulation induces a different hierarchy. This hierarchy lies between the hierarchies of word and tree automata, and it collapses at Rabin conditions of index one, and Streett conditions of index two.

1 Introduction

In formal verification, we check that a system is correct with respect to a desired behavior by checking that a mathematical model of the system satisfies a formal specification of the behavior. In a concurrent setting, the system under consideration is a composition of many components, giving rise to state spaces of exceedingly large size. One of the ways to cope with this state-explosion problem is *abstraction* [BCG88, CFJ93, BG00]. By abstracting away parts of the system that are irrelevant for the specification being checked, we hope to end up with manageable state-spaces. Technically, abstraction may cause different states s and s' of the system to become equivalent. The abstract system then has as its state space the equivalence classes of the equivalence relation between the states. In particular, s and s' are merged into the same state.

* Supported in part by NSF grants CCR-9700061 and CCR-9988322, and by a grant from the Intel Corporation.

S. Kapoor and S. Prasad (Eds.): FST TCS 2000, LNCS 1974, pp. 151–163, 2000.
© Springer-Verlag Berlin Heidelberg 2000

We distinguish between two types of equivalence relations between states. In the *linear* approach, we require s and s' to agree on linear behaviors (i.e., properties satisfied by all the computations that start in s and s'). In the *branching* approach, we require s and s' to agree on branching behaviors (i.e., properties satisfied by the computation trees whose roots are s and s'). When we model systems by *state-transition systems*, two states are equivalent in the linear approach iff they are *trace equivalent*, and they are equivalent in the branching approach iff they are *bisimilar* [Mil71]. The branching approach is stronger, in the sense that bisimulation implies trace equivalence but not vice versa [Mil71, Pnu85].

Of independent interest are the one-way versions of trace equivalence and bisimulation, namely *trace containment* and *simulation*. There, we want to make sure that s does not have more behaviors than s'. This corresponds to the basic notion of verification, where an implementation cannot have more behaviors than its specification [AL91]. In the *hierarchical refinement* top-down methodology for design development, we start with a highly abstract specification, and we construct a sequence of "behavior descriptions". Each description refers to its predecessor as a specification, and the last description is sufficiently concrete to constitute the implementation (cf. [LT87, Kur94]).

The theory behind trace equivalence and bisimulation is well known. We know that two states are trace equivalent iff they agree on all LTL specifications, and the problem of deciding whether two states are trace equivalent is PSPACE-complete [MS72, KV98b]. In the branching approach, two states are bisimilar iff they agree on all CTL* formulas, which turned out to be equivalent to agreement on all CTL and μ-calculus formulas [BCG88, JW96]. The problem of deciding whether two states are bisimilar is PTIME-complete [Mil80, BGS92], and a witnessing relation for bisimulation can be computed using a symbolic fixpoint procedure [McM93, HHK95]. Similar results hold for trace containment and simulation. The computational advantage of simulation makes it a useful precondition to trace containment [CPS93].

State-transition systems describe only the *safe* behaviors of systems. In order to model *live* behaviors, we have to augment systems with *fairness conditions*, which partition the infinite computations of a system into fair and unfair computations [MP92, Fra86]. It is not hard to extend the linear approach to account for fairness: s and s' are equivalent if every sequence of observations that is generated along a fair computation that starts in s can also be generated along a fair computation that starts in s', and vice versa. Robustness with respect to LTL, and PSPACE-completeness extend to the fair case. It is less obvious how to generalize the branching approach to account for fairness. Several proposals for *fair bisimulation* can be found in the literature. We consider here three: ∃-*bisimulation* [GL94], *game-bisimulation* [HKR97, HR00], and ∀-*bisimulation* [LT87]. In a bisimulation relation between S and S' with no fairness, two related states s and s' agree on their observable variables, every successor of s is related to some successor of s', and every successor of s' is related to some successor of s. In all the definitions of fair bisimulation, we require related states to agree on their observable variables. In ∃-bisimulation, we also require every fair computation starting at s to have a related fair computation starting at s', and vice versa. In game-bisimulation, the related fair computations should be generated by strategies that depend on the states visited so far, and in ∀-bisimulation, the relation is a bisimulation in which related computations agree on their fairness (we review the formal definitions in Section 2).

The different definitions induce different relations: $\forall$-bisimulation implies game-bisimulation, which implies $\exists$-bisimulation, but the other direction does not hold [HKR97]. The difference in the distinguishing power of the definitions is also reflected in their logical characterization: while $\exists$-bisimulation corresponds to fair-CTL* (that is, two systems are $\exists$-bisimilar iff they agree on all fair-CTL* formulas, where path quantifiers range over fair computations only [CES86]), game-bisimulation corresponds to fair-alternation-free μ-calculus[1]. Thus, unlike the non-fair case, where almost all modal logics corresponds to bisimulation, here different relations correspond to different logics [ASB^{+}94] [2]. Finally, the different definitions induce different computational costs. The exact complexity depends on the fairness condition being used. For the case of the Büchi fairness condition, for example, the problem of checking whether two systems are bisimilar is PSPACE-complete for $\exists$-bisimulation [KV98b], NP-complete for $\forall$-bisimulation [Hoj96], and PTIME-complete for game-bisimulation [HKR97, HR00].

There are various types of fairness conditions with which we can augment labeled state-transition systems [MP92]. Our work here relates fair transition systems and automata on infinite objects, and we use the types and names of fairness conditions that are common in the latter framework [Tho90]. The simplest condition is *Büchi* (also known as *unconditional* or *impartial* fairness), which specifies a set of states that should be visited infinitely often along fair computations. In its dual condition, *co-Büchi*, the specified set should be visited only finitely often. More involved are *Streett* (also known as *strong* fairness or *compassion*), *Rabin* (Streett's dual), and *parity* conditions, which can restrict both the set of states visited infinitely often and the set of states visited finitely often. Rabin and parity conditions were introduced for automata and are less frequent in the context of state-transition systems. Rabin conditions were introduced by Rabin and were used to prove that the logic S2S is decidable [Rab69]. Parity conditions can be easily translated to both Rabin and Streett conditions. They have gained their popularity as they are suitable for modeling behaviors that are given by means of fixed-points [EJ91]. As we formally define in Section 2, Rabin, Streett, and parity conditions are characterized by their *index*, which is the number of pairs (in the case of Rabin and Streett) or sets (in the case of parity) they contain. When we talk about a *type* of a system, we refer to its fairness condition and, in the case of Rabin, Streett, and parity, also to its index. For example, a Rabin[1] system is a system whose fairness condition is a Rabin condition with a single pair.

The relations between the various types of fairness conditions are well known in the linear paradigm. There, we can regard fair transition systems as a notational variant of automata on infinite words, and adopt known results about translations among the various types and about the complexity of the trace-equivalence and the trace-containment problems [Tho90]. In particular, it is known that the Büchi fairness condition is sufficiently strong, in the sense that every system can be translated to an equivalent Büchi system, where equivalence here means that the systems are trace equivalent.

---

[1] A semantics of fair-alternation-free μ-calculus is given in [HR00].

[2] As shown in [ASB^{+}94], the logic CTL induces yet another definition, strictly weaker than $\exists$-bisimulation. Also, no logical characterization is known for $\forall$-bisimulation.

In the branching paradigm, tight complexity bounds are known for the fair-bisimulation problem with respect to the three definitions of fair bisimulation and the various types of fairness conditions [Hoj96, HKR97, KV98b], but nothing is known about their expressive power, and about the possibilities of translations among them. For example, it is not known whether every system can be translated to an equivalent Büchi system, where now equivalence means fair bisimulation. In particular, it is not clear whether one can directly apply results from the theory of *automata on infinite trees* in order to study fair-bisimulation, and whether the different definitions of fair bisimulation induce different expressiveness hierarchies.

In this paper, we study the expressive power of the various types of fairness conditions in the context of fair bisimulation. For each of the three definitions of fair bisimulation, we consider the following question: given types γ and γ' of fairness conditions, is it possible to translate every γ-system to a fair-bisimilar γ'-system? If this is indeed the case, we say that γ' is *at least as strong as* γ. Then, γ is *stronger than* γ' if γ is at least as strong as γ', but γ' is not at least as strong as γ. When γ is stronger than γ', we also say that γ' is *weaker than* γ. We show that the expressiveness hierarchy for game-bisimulation and $\forall$-bisimulation is strict, and it coincides with the expressiveness hierarchy of tree automata. Thus, Büchi and co-Büchi systems are incomparable and are the weakest, and for all $i \geq 1$, Rabin$[i+1]$, Streett$[i+1]$, and parity$[i+1]$, are stronger than Rabin$[i]$, Streett$[i]$, and parity$[i]$, respectively [Rab70, DJW97, Niw97, NW98]. In contrast, the expressiveness hierarchy for $\exists$-bisimulation is different, and it is not strict. We show that Büchi and co-Büchi systems are incomparable, and they are both weaker than Streett[1] systems. Streett[1] systems are in turn weaker than Streett[2] and Rabin[1] systems, which are both at least as strong as Rabin$[i]$ and Streett$[i]$, for all $i \geq 1$.

Our results imply that the different definitions of fair bisimulation induce different expressiveness relations between the various types of fairness conditions. These relations are different than those known for the linear paradigm, and, unlike the case there, they do not necessarily coincide with the relations that exist in the context of automata on infinite trees. A decision of which fairness condition and which type of fair-bisimulation relation to use in a modeling and verificatiuon process should take into an account all the characteristics of these types, and it cannot be assumed that what is well known for one type is true for another.

Due to space limitations, most of the proofs are omitted. A full version can be found in the homepages of the authors.

2   Definitions

A *fair state-transition system* (*system*, for short) $S = \langle \Sigma, W, R, W_0, L, \alpha \rangle$ consists of an alphabet Σ, a finite set W of states, a total transition relation $R \subseteq W \times W$ (i.e., for every $w \in W$ there exists $w' \in W$ such that $R(w, w'))$, a set W_0 of initial states, a labeling function $L : W \to \Sigma$, and a fairness condition α. We will define several types of fairness conditions shortly. A *computation* of S is a sequence $\pi = w_0, w_1, w_2, \ldots$ of states such that for every $i \geq 0$, we have $R(w_i, w_{i+1})$. Each computation $\pi = w_0, w_1, w_2, \ldots$ induces the word $L(\pi) = L(w_0) \cdot L(w_1) \cdot L(w_2) \cdots \in \Sigma^\omega$. In order to determine whether a computation is *fair*, we refer to the set $inf(\pi)$ of states that π visits infinitely often. Formally, $inf(\pi) = \{w \in W : \text{for infinitely many } i \geq 0, \text{ we have } w_i = w\}$. The way we refer to $inf(\pi)$ depends on the fairness condition of S. Several types of fairness conditions are studied in the literature:

- *Büchi* (*unconditional* or *impartial*), where $\alpha \subseteq W$, and π is fair iff $inf(\pi) \cap \alpha \neq \emptyset$.
- *co-Büchi*, where $\alpha \subseteq W$, and π is fair iff $inf(\pi) \cap \alpha = \emptyset$.
- *Parity*, where α is a partition of W, and π is fair in $\alpha = \{F_1, F_2, \ldots, F_k\}$ if the minimal index i for which $inf(r) \cap F_i \neq \emptyset$ exists and is even.
- *Rabin*, where $\alpha \subseteq 2^W \times 2^W$, and π is fair in $\alpha = \{\langle G_1, B_1 \rangle, \ldots, \langle G_k, B_k \rangle\}$ if there is a $1 \leq i \leq k$ such that $inf(\pi) \cap G_i \neq \emptyset$ and $inf(\pi) \cap B_i = \emptyset$.
- *Streett* (*compassion* or *strong fairness*), where $\alpha \subseteq 2^W \times 2^W$, and π is fair in $\alpha = \{\langle G_1, B_1 \rangle, \ldots, \langle G_k, B_k \rangle\}$ if for all $1 \leq i \leq k$, we have that $inf(\pi) \cap G_i \neq \emptyset$ implies $inf(\pi) \cap B_i \neq \emptyset$.

The number k of sets in a parity fairness condition or of pairs in a Rabin or Streett fairness condition is the *index* of α. When we talk about the *type* of a system, we refer to its fairness condition and, in the case of Rabin, Streett, and parity, also to its index. For example, a Rabin[1] system is a system whose fairness condition is a Rabin condition with a single pair. For a state w, a w-computation is a computation $w_0, w_1, w_2, \ldots$ with $w_0 = w$. We use $T(S^w)$ to denote the set of all traces $\sigma_0 \cdot \sigma_1 \cdots \in \Sigma^\omega$ for which there exists a fair w-computation $w_0, w_1, \ldots$ in S with $L(w_i) = \sigma_i$ for all $i \geq 0$. The *trace set* $T(S)$ of S is then defined as $\bigcup_{w \in W_0} T(S^w)$.

We now formalize what it means for two systems (or two states of the same system) to be equivalent. We give the definitions with respect to two systems $S = \langle \Sigma, W, R, W_0, L, \alpha \rangle$ and $S' = \langle \Sigma, W', R', W'_0, L', \alpha' \rangle$, with the same alphabet.[3] We consider two equivalence criteria: *trace equivalence* and *bisimulation*. While the first criterion is clear ($T(S) = T(S')$), several proposals are suggested in the literature for bisimulation in the case of systems with fairness. Before we define them, let us first recall the definition of bisimulation for the non-fair case.

Bisimulation [Mil71] A relation $H \subseteq W \times W'$ is a *bisimulation relation* between S and S' iff the following conditions hold for all $\langle w, w' \rangle \in H$.

1. $L(w) = L'(w')$.
2. For all $s \in W$ with $R(w, s)$, there is $s' \in W'$ such that $R'(w', s')$ and $H(s, s')$.
3. For all $s' \in W$ with $R'(w', s')$, there is $s \in W$ such that $R(w, s)$ and $H(s, s')$.

We now describe three extensions of bisimulation relations to the fair case. In all definitions, we extend a relation $H \subseteq W \times W'$, over the states of S and S', to a relation over infinite computations of S and S': for two computations $\pi = w_0, w_1, \ldots$ in S, and $\pi' = w'_0, w'_1, \ldots$ in S', we have $H(\pi, \pi')$ iff $H(w_i, w'_i)$, for all $i \geq 0$.

∃-bisimulation [GL94] A relation $H \subseteq W \times W'$ is an *∃-bisimulation relation* between S and S' iff the following conditions hold for all $\langle w, w' \rangle \in H$.

---

[3] In practice, S and S' are given as systems over alphabets 2^{AP} and $2^{AP'}$, when AP and AP' are the sets of atomic propositions used in S and S', and possibly $AP \neq AP'$. When we compare S with S', we refer only to the common atomic propositions, thus $\Sigma = 2^{AP \cap AP'}$.

1. $L(w) = L'(w')$.
2. Each fair w-computations π in S has a fair w'-computation π' in S' with $H(\pi, \pi')$.
3. Each fair w'-computations π' in S' has a fair w-computation π in S with $H(\pi, \pi')$.

Game bisimulation [HKR97, HR00] Game bisimulation is defined by means of a game between a protagonist against an adversary. The positions of the game are pairs in $W \times W'$. A *strategy* τ for the protagonist is a partial function from $(W \times W')^* \times (W \cup W')$ to $(W' \cup W)$, such that for all $\rho \in (W \times W')^*$, $w \in W$, and $w' \in W'$, we have that $\tau(\rho \cdot w) \in W'$ and $\tau(\rho \cdot w') \in W$. Thus, if the game so far has produced the sequence ρ of positions, and the adversary moves to w in S, then the strategy τ instructs the protagonist to move to $w' = \tau(\pi \cdot w)$, resulting in the new position $\langle w, w' \rangle$. If the adversary chooses to move to w' in S', then τ instructs the protagonist to move to $w = \tau(\pi \cdot w')$, resulting in the new position $\langle w, w' \rangle$. A sequence $\overline{w} = \langle w_0, w_0' \rangle \cdot \langle w_1, w_1' \rangle \cdots \in (W \times W')^\omega$ is an *outcome* of the strategy τ if for all $i \geq 0$, either $w_{i+1}' = \tau(\langle w_0, w_0' \rangle \cdots \langle w_i, w_i' \rangle \cdot w_{i+1})$, or $w_{i+1} = \tau(\langle w_0, w_0' \rangle \cdots \langle w_i, w_i' \rangle \cdot w_{i+1}')$.

A binary relation $H \subseteq W \times W'$ is a *game bisimulation relation* between S and S' if there exists a strategy τ such that the following conditions hold for all $\langle w, w' \rangle$ in H.

1. $L(w) = L(w')$.
2. Every outcome $\overline{w} = \langle w_0, w_0' \rangle \cdot \langle w_1, w_1' \rangle \cdots$ of τ with $w_0 = w$ and $w_0' = w'$ has the following two properties: (1) for all $i \geq 0$, we have $\langle w_i, w_i' \rangle \in H$, and (2) the projection $w_0 \cdot w_1 \cdots$ of $\overline{w}$ to W is a fair w_0-computation of S iff the projection $w_0' \cdot w_1' \cdots$ of $\overline{w}$ to W' is a fair w_0'-computation of S'.

∀-bisimulation [LT87, DHW91] A binary relation $H \subseteq W \times W'$ is a *∀-bisimulation relation* between S and S' if the following conditions hold:

1. H is a bisimulation relation between S and S'.
2. If $H(w, w')$, then for every fair w-computation π of S and for every w'-computation π' of S', if $H(\pi, \pi')$, then π' is fair.
3. If $H(w, w')$, then for every fair w'-computation π' of S' and for every w-computation π of S, if $H(\pi, \pi')$, then π is fair.

It is not hard to see that if H is a ∀-bisimulation relation, then H is also a game-bisimulation relation. Also, if H is a game-bisimulation relation, then H is also an ∃-bisimulation relation. As demonstrated in [HKR97], the other direction is not true.

For all types β of bisimulation relations (that is $\beta \in \{\exists, game, \forall\}$), a β-bisimulation relation H is a β-*bisimulation between S and S'* if for every $w \in W_0$ there exists $w' \in W_0'$ such that $H(w, w')$, and for every $w' \in W_0'$ there exists $w \in W_0$ such that $H(w, w')$. If there is a β-bisimulation between S and S', we say that S and S' are β-*bisimilar*. Intuitively, bisimulation implies that S and S' have the same behaviors. Formally, two bisimilar systems with no fairness agree on the satisfaction of all branching properties that can be specified in a conventional temporal logic (in particular, CTL* and μ-calculus) [BCG88, JW96]. When we add fairness, the logical characterization becomes less robust: ∃-simulation corresponds to fair-CTL*, and game-simulation corresponds to fair-alternation-free μ-calculus [ASB$^+$94, GL94, HKR97, HR00].

For ∃-bisimulation and ∀-bisimulation, a relation $H \subseteq W \times W'$ is a β-*simulation relation* from S to S' if conditions 1 and 2 for H being a β-bisimulation relation hold.

For game-bisimulation, a relation H is a *game-simulation relation* from S to S' if we restrict the moves of the adversary to choose only states from S. A β-simulation relation H is a *β-simulation from S to S'* iff for every $w \in W_0$ there exists $w' \in W_0'$ such that $H(w, w')$. If there is a β-simulation from S to S', we say that S' β-simulates S, and we write $S \leq_\beta S'$. Intuitively, while bisimulation implies that S and S' have the same behaviors, simulation implies that S has less behaviors than S'.

It is easy to see that bisimulation implies trace equivalence. The other direction, however, is not true [Mil71]. Hence, our equivalence criteria induce different equivalence relations. When attention is restricted to trace equivalence, it is known how to translate all fair systems to an equivalent Büchi system. In this paper we consider the problem of translations among systems that preserve bisimilarity.

3   Expressiveness with ∃-Bisimulation

In the linear case, it follows from automata theory that co-Büchi systems are weaker than Büchi systems, which are as strong as parity, Rabin, and Streett systems. In the branching case, nondeterministic Büchi and co-Büchi tree automata are both weaker than Rabin tree automata, and, for all $i \geq 1$, parity$[i]$, Rabin$[i]$, and Streett$[i]$ are weaker than parity$[i+1]$, Rabin$[i+1]$, and Streett$[i+1]$, respectively [Rab70, DJW97, Niw97, NW98]. In this section we show that the expressiveness hierarchy in the context of ∃-bisimulation is located between the hierarchies of word and tree automata.[4]

We first show that Büchi and co-Büchi systems are weak. The arguments we use are similar to these used by Rabin in the context of tree automata [Rab70]. Our proofs use the notion of maximal models [GL94, KV98c]. A system M_ψ is a *maximal model* for an $\forall$CTL* formula ψ if $M_\psi \models \psi$ and for every module M we have that $M \leq_\exists M_\psi$ iff $M \models \psi$. It can be shown that there is no Büchi system that is ∃-bisimilar to the maximal model of the formula $\forall \Diamond \Box p$ and that there is no co-Büchi system that is ∃-bisimilar to the maximal model of the formula $\forall \Box \Diamond p$. Hence, we have:

Theorem 1. *Büchi is not at least as ∃-strong as co-Büchi and co-Büchi is not at least as ∃-strong as Büchi.*

Note that Theorem 1 implies that the Büchi condition is too weak for defining maximal models for $\forall$CTL* formulas. On the other hand, the Büchi condition is sufficiently strong for defining maximal models for $\forall$CTL formulas [GL94, KV98a]. Since parity, Rabin, and Streett are at least as ∃-strong as Büchi and co-Büchi, it follows from Theorem 1 that parity, Rabin, and Streett are all ∃-stronger than Büchi and co-Büchi.

So far things seem to be very similar to tree automata, where Büchi and co-Büchi conditions are incomparable [Rab70]. In particular, the ability of the Büchi condition to define maximal models for $\forall$CTL and its inability to define maximal models for $\forall$CTL* seems related to the ability to translate CTL formulas to Büchi tree automata and the inability to translate CTL* formulas to Büchi tree automata (as follows from Rabin's result [Rab70]). In tree automata, the hierarchy of expressive power stays strict

---

[4] Here and in the sequel, we use terms like γ is ∃-stronger than γ' to indicate that γ is stronger than γ' in the context of ∃-bisimulation.

also when we proceed to parity (or Rabin or Streett) fairness condition with increasing indices [DJW97, Niw97, NW98]. We now show that, surprisingly, in the context of ∃-bisimulation, Rabin conditions of index one are at least as strong as parity, Rabin, and Streett conditions with an unbounded index. In particular, it follows that maximal models for ∀CTL* can be defined with Rabin[1] fairness. The idea behind the construction is similar to the conversion of Rabin and Streett automata on infinite words to Büchi automata on infinite words.

Lemma 1. *Every Rabin system with n states and index k has an ∃-bisimilar Rabin system with $O(nk)$ states and index 1.*

Proof: Let $S = \langle \Sigma, W, W_0, R, L, \alpha \rangle$ be a Rabin system with $\alpha = \{\langle G_1, B_1 \rangle, \ldots, \langle G_k, B_k \rangle\}$. We define $S' = \langle \Sigma, W', W_0', R', L', \alpha' \rangle$ as follows.

- For every $1 \leq i \leq k$, let $W_i = (W \setminus B_i) \times \{i\}$. Then, $W' = (W \times \{0\}) \cup \bigcup_{1 \leq i \leq k} W_i$, and $W_0' = W_0 \times \{0\}$.
- $R' = \bigcup_{0 \leq i \leq k} \{\langle (w, 0), (w', i) \rangle, \langle (w, i), (w', 0) \rangle, \langle (w, i), (w', i) \rangle : \langle w, w' \rangle \in R\} \cap (W' \times W')$. Note that R' is total.
- For all $w \in W$ and $0 \leq i \leq k$, we have $L'((w, i)) = L(w)$.
- $\alpha' = \{\langle \bigcup_{1 \leq i \leq k} G_i \times \{i\}, W \times \{0\} \rangle\}$.

Thus, S' consists of $k + 1$ copies of S. One copy ("the idle copy") contains all the states in W, marked with 0. Then, k copies are partial: every such copy is associated with a pair $\langle G_i, B_i \rangle$, its states are marked with i, and it contains all the states in $W \setminus B_i$. A computation of S' can return to the idle copy from all copies, where it can choose between staying in the idle copy or moving to one of the other k copies. The acceptance condition forces a fair computation to visit the idle copy only finitely often, forcing the computation to eventually get trapped in a copy associated with some pair $\langle G_i, B_i \rangle$. There, the computation cannot visit states from B_i (indeed, W_i does not contain such states), and it has to visit infinitely many states from G_i. It is not hard to see that the relation $H = \{\langle w, (w, i) \rangle : w \in W \text{ and } 0 \leq i \leq k\}$ is an ∃-bisimulation between S and S', thus S and S' are ∃-bisimilar. □

In the case of transforming Rabin[k] word automata to Rabin[1] (or Büchi) automata, runs of the automaton on different computations are independent of each other, so there is no need for the automaton to "change its mind" about the pair in α with respect to which the computation is fair. Accordingly, there is no need to return to an idle copy. In the case of tree automata, runs on different computations of the tree depend on each other, and the run of the automaton along a computation may need to postpone its choice of a suitable pair in α ad infinitum, which cannot be captured with a Rabin[1] condition. The crucial observation about ∃-bisimulation is that here, if π_1 and π_2 are different fair w-computations, then the fair computations π_1' and π_2' for which $H(\pi_1, \pi_1')$ and $H(\pi_2, \pi_2')$ are independent. Thus, each computation eventually reaches a state where it can stick to its suitable pair in α. Accordingly, a computation needs to change its mind only finitely often. A visit to the idle copy corresponds to the computation changing its mind, and the fairness condition guarantees that there are only finitely many visits to the idle copy.

We now describe a similar transformation for Streett systems. While in Rabin systems each copy of the original system corresponds to a guess of a pair $\langle G_i, B_i \rangle$ for which G_i is visited infinitely often and B_i is visited only finitely often, here each copy would correspond to a subset $I \subseteq \{1, \ldots, k\}$ of pairs, where the copy associated with I corresponds to a guess that B_i and G_i are visited infinitely often for all $i \in I$, and G_i is visited only finitely often for all $i \notin I$.

Lemma 2. *Every Streett system with n states and index k has an $\exists$-bisimilar Rabin system with $O(n \cdot 2^{O(k)})$ states and index 1.*

Note that while the blow up in the construction in Lemma 1 is linear in the index of the Rabin system, the blow up in the construction in Lemma 2 is exponential in the index of the Streett system. The above blow ups are tight for the linear paradigm [SV89][5]. Since $\exists$-bisimulation implies trace equivalence, it follows that these blow ups are tight also for the $\exists$-bisimulation case.

Since the parity condition is a special case of Rabin, Lemma 1 also implies a translation of parity systems to $\exists$-bisimilar Rabin[1] systems. Also, a Rabin[1] condition $\{\langle G, B \rangle\}$ can be viewed as a parity condition $\{B, G \setminus B, W \setminus (G \cup B)\}$. Hence, parity[3] is as $\exists$-strong as Rabin[1] [6]. A Rabin[1] condition $\{\langle G, B \rangle\}$ is equivalent to the Streett[2] condition $\{\langle W, G \rangle, \langle B, \emptyset \rangle\}$. So, Streett[2] is also as $\exists$-strong as Rabin[1]. It turns out that we can combine the arguments for Büchi and co-Büchi in Theorem 1 to prove that Streett[1] is $\exists$-weaker than Streett[2]. To sum up, we have the following.

Theorem 2. *For every fairness type γ, the types Rabin[1], Streett[2], and parity[3] are all at least as $\exists$-strong as γ.*

Note that the types described in Theorem 2 are tight, in the sense that, as discussed above, Büchi, co-Büchi, Streett[1], and parity[2] may be $\exists$-weaker than γ.

In the full version, we also show that a system with a *generalized Büchi condition* or with a *justice condition* [MP92] can be translated to an $\exists$-bisimilar Büchi system, implying that generalized Büchi and justice conditions are also too weak.

4 Expressiveness with Game-Bisimulation and $\forall$-Bisimulation

We now study the expressiveness hierarchy for game-bisimulation and $\forall$-bisimulation. We show that unlike $\exists$-simulation, here the hierarchy coincides with the hierarchy of tree automata. Thus, Rabin[i+1] is stronger than Rabin[i], and similarly for Streett and parity. In order to do so, we define game-bisimulation between tree automata, and define transformations preserving game-bisimulation between tree automata and fair systems. We show that game-bisimilar tree automata agree on their languages (of trees), which enables us to relate the expressiveness hierarchies in the two frameworks.

---

[5] [SV89] shows that the transition from Streett word automata to Büchi word automata is exponential in the index of the Streett automaton. Since the transition from Rabin[1] to Büchi is linear, a lower bound for the transition from Streett to Rabin[1] follows.

[6] Recall that a parity fairness condition is a partition of the state set. Hence, a parity[2] condition can be translated to an equivalent co-Büchi fairness condition and vice versa, implying that Rabin[1] is $\exists$-stronger than parity[2].

Due to lack of space we only give an outline of the proof. We define a special type of tree automata, called *loose tree automata*. Unlike conventional tree automata [Tho90], the transition function of loose tree automata does not distinguish between the successors of a node, it does not force states to be visited, and it only restricts the set of states that each of the successors may visit. When $\mathcal{A}$ runs on a labeled tree $\langle T, V \rangle$ and it visits a node x with label σ at state q, then $\delta(q, \sigma) = S$ (where S is a subset of the states of $\mathcal{A}$) means that $\mathcal{A}$ should send to all the successors of x copies in states in S. Loose tree automata can use all types of fairness. A run of a loose tree automaton is accepting if all the infinite paths of the run tree satisfy the fairness condition.

We can define game-bisimulation for loose tree automata. Given two loose tree automata, we define a game whose positions are pairs of states. A strategy for the game is similar to the strategy defined for systems, but this time the adversary gets to choose an alphabet letter and a successor corresponding to this letter. The protagonist has to follow with a successor corresponding to the same letter in the other automaton. A relation is a game-bisimulation relation if all the outcomes of such plays starting at related states have both projections fair or have both projections unfair. Two loose tree automata are game-bisimilar if there exists a game-bisimulation between them that relates the starting states of each one of the automata to starting states of the other.

Recall that game-bisimulation between systems implies trace equivalence. Game-bisimulation between loose tree automata implies not only agreement on traces that may label paths of accepted trees, but also agreement on the accepted trees! The idea is that given an accepting run tree of one automaton, we use the strategy to build an accepting run tree of its game-bisimilar counterpart. This property of game-bisimulation between loose tree automata enables us to relate the hierarchy of loose tree automata with that of game-bisimulation. Formally, we have the following.

Theorem 3. *Let γ and γ' be two types of fairness conditions. If γ is at least as strong as γ' in the context of game-bisimulation or $\forall$-bisimulation, then γ is at least as strong as γ' also in the context of loose tree automata.*

While loose tree automata are weaker than conventional tree automata [Tho90],the expressiveness hierarchy of loose tree automata coincides with that of tree automata (this is beacause the latter coincides with the hierarchy of deterministic word automata [Wag79, Kam85], and is proven in [KSV96, DJW97, Niw97, NW98] by means of languages that can be recognized by loose tree automata). It follows that the expressiveness hierarchy in the context of game-bisimulation and $\forall$-bisimulation coincides with that of tree automata.

5   Discussion

We considered two equivalence criteria — bisimulation and trace equivalence — between fair state-transition systems. We studied the expressive power of various fairness conditions in the context of fair bisimulation. We showed that while the hierarchy in the context of trace equivalence coincides with the one of nondeterministic word automata, the hierarchy in the context of bisimulation depends on the exact definition of fair bisimulation, and it does not necessarily coincide with the hierarchy of tree automata. In particular, we showed that Rabin[1] systems are sufficiently strong to model all systems up to $\exists$-bisimilarity.

There is an intermediate equivalence criterion: *two-way simulation* (that is $S \leq S'$ and $S' \leq S$) is implied by bisimulation, it implies trace equivalence, and it is equal to neither of the two [Mil71]. Two-way simulation is a useful criterion: S and S' are two-way similar iff for every system S'' we have $S'' \leq S$ iff $S'' \leq S'$ and $S \leq S''$ iff $S' \leq S''$. Hence, in hierarchical refinement, or when defining maximal models for universal formulas, we can replace S with S'. A careful reading through our proofs shows that all the results described in the paper for bisimulation hold also for two-way simulation.

Finally, the study of $\exists$-bisimulation in Section 3 has led to a simple definition of parallel compositions for Rabin and parity systems, required for modular verification of concurrent systems. In the linear paradigm, the composition $S = S_1 \| S_2$ of S_1 and S_2 is defined so that $\mathcal{T}(S) = \mathcal{T}(S_1) \cap \mathcal{T}(S_2)$ (cf. [Kur94]). In the branching paradigm [GL94], Grumberg and Long defined the parallel compositions of two Streett systems. As studied in [GL94, KV98a], in order to be used in modular verification, a definition of composition has to satisfy the following two conditions, for all systems S, S', and S''. First, if $S' \leq_\exists S''$, then $S \| S' \leq_\exists S \| S''$. Second, $S \leq_\exists S' \| S''$ iff $S \leq_\exists S'$ and $S \leq_\exists S''$. In particular, it follows that $S \| S' \leq_\exists S'$, thus every universal formula that is satisfied by a component of a parallel composition, is satisfied also by the composition. When S_1 and S_2 are Streett systems, the definition of $S_1 \| S_2$ is straightforward, and is similar to the product of two Streett word automata. When, however, S_1 and S_2 are Rabin systems, the definition of product of word automata cannot be applied, and a definition that follows the ideas behind a product of tree automata is very complicated and complex. In the full paper we show that the fact that $\exists$-bisimulation is located between word and tree automata enables a simple definition of parallel composition that obeys the two conditions above.

References

[AL91]   M. Abadi and L. Lamport. The existence of refinement mappings. *TCS*, 82(2):253–284, 1991.

[ASB+94] A. Aziz, V. Singhal, F. Balarin, R. Brayton, and A.L. Sangiovanni-Vincentelli. Equivalences for fair kripke structures. In *Proc. 21st ICALP*, Jerusalem, Israel, July 1994.

[BCG88]   M.C. Browne, E.M. Clarke, and O. Grumberg. Characterizing finite Kripke structures in propositional temporal logic. *TCS*, 59:115–131, 1988.

[BG00]   D. Bustan and O. Grumberg. Simulation based minimization. In *Proc. 17th ICAD*, Pittsburgh, PA, June 2000.

[BGS92]   J. Balcazar, J. Gabarro, and M. Santha. Deciding bisimilarity is P-complete. *Formal Aspects of Computing*, 4(6):638–648, 1992.

[CES86]   E.M. Clarke, E.A. Emerson, and A.P. Sistla. Automatic verification of finite-state concurrent systems using temporal logic specifications. *ACM Transactions on Programming Languages and Systems*, 8(2):244–263, January 1986.

[CFJ93]   E.M. Clarke, T. Filkorn, and S. Jha. Exploiting symmetry in temporal logic model checking. In *Proc. 5th CAV*, LNCS 697, 1993.

[CPS93]   R. Cleaveland, J. Parrow, and B. Steffen. The concurrency workbench: A semantics-based tool for the verification of concurrent systems. *ACM Trans. on Programming Languages and Systems*, 15:36–72, 1993.

[DHW91]   D.L. Dill, A.J. Hu, and H. Wong-Toi. Checking for language inclusion using simulation relations. In *Proc. 3rd CAV*, LNCS 575, pp. 255–265, 1991.

[DJW97]   S. Dziembowski, M. Jurdzinski, and I. Walukiewicz. How much memory is needed to win infinite games. In *Proc. 12th LICS*, pp. 99–110, 1997.

[EJ91]   E.A. Emerson and C. Jutla. Tree automata, μ-calculus and determinacy. In *Proc. 32nd FOCS*, pp. 368–377, 1991.

[Fra86]   N. Francez. *Fairness*. Texts and Monographs in Computer Science. Springer-Verlag, 1986.

[GL94]   O. Grumberg and D.E. Long. Model checking and modular verification. *ACM Trans. on Programming Languages and Systems*, 16(3):843–871, 1994.

[HHK95]   M.R. Henzinger, T.A. Henzinger, and P.W. Kopke. Computing simulations on finite and infinite graphs. In *Proc. 36th FOCS*, pp. 453–462, 1995.

[HKR97]   T.A. Henzinger, O. Kupferman, and S. Rajamani. Fair simulation. In *Proc. 8th Conference on Concurrency Theory*, LNCS 1243, pp. 273–287, 1997.

[Hoj96]   R. Hojati. *A BDD-based Environment for Formal Verification of Hardware Systems*. PhD thesis, University of California at Berkeley, 1996.

[HR00]   T. Henzinger and S. Rajamani. Fair bisimulation. In *Proc. 4th TACAS*, LNCS 1785, pp. 299–314, 2000.

[JW96]   D. Janin and I. Walukiewicz. On the expressive completeness of the propositional μ-calculus with respect to the monadic second order logic. In *Proc. 7th Conference on Concurrency Theory*, LNCS 1119, pp. 263–277, 1996.

[Kam85]   M. Kaminski. A classification of ω-regular languages. *TCS*, 36:217–229, 1985.

[KSV96]   O. Kupferman, S. Safra, and M.Y. Vardi. Relating word and tree automata. In *Proc. 11th LICS*, pp. 322–333, 1996.

[Kur94]   R.P. Kurshan. *Computer Aided Verification of Coordinating Processes*. Princeton Univ. Press, 1994.

[KV98a]   O. Kupferman and M.Y. Vardi. Modular model checking. In *Proc. Compositionality Workshop*, LNCS 1536, pp. 381–401, 1998.

[KV98b]   O. Kupferman and M.Y. Vardi. Verification of fair transition systems. *Chicago Journal of TCS*, 1998(2).

[KV98c]   O. Kupferman and M.Y. Vardi. Weak alternating automata and tree automata emptiness. In *Proc. 30th STOC*, pp. 224–233, 1998.

[LT87]   N. A. Lynch and M.R. Tuttle. Hierarchical correctness proofs for distributed algorithms. In *Proc. 6th PODC*, pp. 137–151, 1987.

[McM93]   K.L. McMillan. *Symbolic Model Checking*. Kluwer Academic Publishers, 1993.

[Mil71]   R. Milner. An algebraic definition of simulation between programs. In *Proc. 2nd International Joint Conference on Artificial Intelligence*, pp. 481–489, 1971.

[Mil80]   R. Milner. *A Calculus of Communicating Systems*, LNCS 92, 1980.

[MP92]   Z. Manna and A. Pnueli. *The Temporal Logic of Reactive and Concurrent Systems: Specification*. Springer-Verlag, Berlin, January 1992.

[MS72]   A.R. Meyer and L.J. Stockmeyer. The equivalence problem for regular expressions with squaring requires exponential time. In *Proc. 13th SWAT*, pp. 125–129, 1972.

[Niw97]   D. Niwiński. Fixed point characterization of infinite behavior of finite-state systems. *TCS*, 189(1–2):1–69, December 1997.

[NW98]   D. Niwinski and I. Walukiewicz. Relating hierarchies of word and tree automata. In *Symposium on Theoretical Aspects in Computer Science*, LNCS 1373, 1998.

[Pnu85]   A. Pnueli. Linear and branching structures in the semantics and logics of reactive systems. In *Proc. 12th ICALP*, LNCS 194 pp. 15–32, 1985.

[Rab69]   M.O. Rabin. Decidability of second order theories and automata on infinite trees. *Transaction of the AMS*, 141:1–35, 1969.

[Rab70]   M.O. Rabin. Weakly definable relations and special automata. In *Proc. Symp. Math. Logic and Foundations of Set Theory*, pp. 1–23. North Holland, 1970.

[SV89]   S. Safra and M.Y. Vardi. On ω-automata and temporal logic. In *Proc. 21st STOC*, pp. 127–137, 1989.

[Tho90]   W. Thomas. Automata on infinite objects. *Handbook of Theoretical Computer Science*, pp. 165–191, 1990.

[Wag79]   K. Wagner. On ω-regular sets. *Information and Control*, 43:123–177, 1979.

Arithmetic Circuits and Polynomial Replacement Systems

Pierre McKenzie[1]*, Heribert Vollmer[2], and Klaus W. Wagner[2]

[1] Informatique et recherche opérationnelle, Université de Montréal, C.P. 6128, Succ. Centre-Ville, Montréal (Québec), H3C 3J7 Canada.
[2] Theoretische Informatik, Universität Würzburg, Am Hubland, 97074 Würzburg, Germany.

Abstract. This paper addresses the problems of counting proof trees (as introduced by Venkateswaran and Tompa) and counting proof circuits, a related but seemingly more natural question. These problems lead to a common generalization of straight-line programs which we call polynomial replacement systems. We contribute a classification of these systems and we investigate their complexity. Diverse problems falling in the scope of this study include, for example, counting proof circuits, and evaluating $\{\cup, +\}$-circuits over the natural numbers. The former is shown #P-complete, the latter to be equivalent to a particular problem for replacement systems.

1 Introduction

1.1 Motivation

When $+$ and $\times$ replace $\vee$ and $\wedge$ in the adjacent figure, the gate g_1 on input $x_1 = x_2 = 1$ evaluates to 9. Equivalently, the tree-like Boolean circuit T obtained from the circuit drawn has 9 *proof trees* [VT89], i.e. 9 different minimal subcircuits witnessing that T outputs 1 (gates replicated to form T are independent). This relationship between proof tree counting and monotone arithmetic circuits was used by Venkateswaran [Ven92] to characterize nondeterministic time classes, including #P [Val79], and by Vinay [Vin91] to characterize the counting version of LOGCFL [Sud78]. The same relationship triggered the investigation of #NC^1 by Caussinus et al. [CMTV98], and that of #AC^0 by Allender et al. [AAD97]. See [All98] for recent results and for motivation to study such "small" arithmetic classes.

A recent goal has been to capture small arithmetic classes by counting objects other than proof trees, notably paths in graphs. Allender et al. [AAB+99]

---

* Research performed in part while on leave at the Universität Tübingen. Supported by the (German) DFG, the (Canadian) NSERC and the (Québec) FCAR.

S. Kapoor and S. Prasad (Eds.): FST TCS 2000, LNCS 1974, pp. 164–175, 2000.

succeeded in identifying appropriate graphs for $\#AC^0$. Given the growing importance of counting classes, our motivation for the present work was the desire to avoid unwinding circuits into trees before counting their "proofs". Define a proof *circuit* to be a minimal subcircuit witnessing that a circuit outputs 1. More precisely, for a Boolean circuit C and an input x, a proof circuit is an edge-induced connected subcircuit of C which evaluates to 1 on x. This subcircuit must contain the output gate of C, as well as exactly one C-edge into each $\vee$-gate and all C-edges into each $\wedge$-gate. The reader should convince herself that the circuit depicted above, which had 9 proof trees on input $x_1 = x_2 = 1$, has only 7 proof circuits on that input.

What counting classes arise from counting proof circuits instead of trees? This question held in stock two surprises, the first of which is the following algorithm:

1. replace $\vee$ by $+$ and $\wedge$ by $\times$ in a negation-free Boolean circuit C,
2. view C as a straight-line program prescribing in the usual way a formal polynomial in the input variables $x_1, \ldots, x_n$,
3. compute the polynomial top-down, with an important proviso: at each step, knock any nontrivial exponent down to 1 in the intermediate sum-of-monomials representation.

We get the number of proof circuits of C on an input x by evaluating the final polynomial at x! For example, the circuit depicted above had 7 proof circuits on input $x_1 = x_2 = 1$ because

$$g_1 \rightarrow g_2 g_3 \tag{1}$$
$$\rightarrow (x_1 + g_4) g_3 \tag{2}$$
$$\rightarrow (x_1 + g_4)(g_4 + x_2) \quad \rightarrow \quad x_1 g_4 + x_1 x_2 + g_4 + g_4 x_2 \tag{3}$$
$$\rightarrow x_1(x_1 + x_2) + x_1 x_2 + (x_1 + x_2) + (x_1 + x_2)x_2, \tag{4}$$

where g_4^2 became g_4 in the middle of step 3.

One's intuition might be that such a simple strategy could be massaged into an arithmetic circuit or at least into a sublinear parallel algorithm [VSBR83]. Our second surprise was that counting proof circuits, even for depth-4 semi-unbounded circuits, is $\#P$-complete. Hence, not only is our strategy hard to parallelize, it likely genuinely requires exponential time!

Our three-step algorithm above thus counts proof *trees* in the absence of the idempotent rules $y^2 \rightarrow y$, and it counts proof *circuits* in their presence. Moreover, whereas an arithmetic circuit computing the number of proof *trees* of a circuit is readily available, producing such a circuit to compute proof *circuits* seems intractable. What is special about the idempotent rules? What would the effect of multivariate rules be? Which nontrivial rules would nonetheless permit expressing the counting in the form of an arithmetic circuit? What is a general framework in which complexity questions such as these can be investigated?

1.2 Results

We view our results as forming three main contributions.

Our first contribution is to define and classify *polynomial replacement systems* (prs for short). Prs provide the answer to the framework question. A prs in its full generality is a start polynomial $q \in \mathbb{N}[x_1, \ldots, x_m]$ together with a set of replacement rules. A replacement rule is a pair of polynomials (p_1, p_2). Informally, (p_1, p_2) is applicable to a polynomial q if q can be written in a form in which p_1 appears. Applying (p_1, p_2) to q then consists of replacing p_1 by p_2 (see Sect. 3 for formal definitions).

A prs generally defines a *set* of polynomials, since the choice and sequencing of the rules, and the way in which the rules are applied, may generate different polynomials. Computational problems of interest include computing the polynomials themselves (POLY), evaluating the polynomials at specific points (EVAL), and testing membership in their ranges (RANGE). We identify four natural families of prs: *simple* if the rules only replace variables, *deterministic* if no two rules have the same left-hand side, *acyclic* if no nontrivial infinite sequence of rules is applicable, and *idempotent* if the rules (y^2, y) are present.

For general prs, we obtain canonical forms and we outline broad complexity issues. Our detailed complexity analysis involves *simple* prs. For instance, we exhibit simple and deterministic prs for which RANGE is NP-complete. When the prs is given as part of the input, POLY is P-hard and in coRP, while RANGE is NP-complete and EVAL is P-complete.

Our second contribution concerns the specific case of proof trees and proof circuits. We prove that, to any Boolean circuit C and input x, corresponds an easily computable idempotent, simple, deterministic and acyclic prs S having the property that the number of proof trees (resp. proof circuits) of C on x is the maximum (resp. minimum) value of the EVAL problem for S on x, and vice versa (see Lemma 20). This offers one viewpoint on the reason why our algorithm from Subsect. 1.1 counts proof circuits correctly. We also prove that computing the minimum of the EVAL problem for idempotent, simple, deterministic and acyclic prs is #P-complete, or equivalently, that counting proof circuits is #P-complete under Turing reductions (but not under many-one reductions unless P = NP). This provides a new characterization of #P which is to be contrasted with Venkateswaran's (poly-degree, poly-depth) characterization [Ven92] and with the retarded polynomials characterization of Babai and Fortnow [BF91]. We also prove that detecting whether a circuit has more proof trees than proof circuits is NP-complete.

Our third contribution concerns the specific case of simple and acyclic prs. We prove that the EVAL problem for such prs is the evaluation problem for $\{\cup, +, \times\}$-circuits. These circuits have been considered previously (under the name hierarchical descriptions) in [Wag84, Wag86]. They are obtained by generalizing, from trees to general circuits, the $\{\cup, +, \times\}$-expressions (a.k.a. integer expressions), whose evaluation problem was shown NP-complete 25 years ago by Stockmeyer and Meyer [SM73]. From a PSPACE upper bound given in [Wag84] we conclude that evaluation of simple acyclic prs has a polynomial space algorithm, and from a PSPACE-hardness result given in [Yan00] we then conclude PSPACE-completeness of our problem.

1.3   Paper Organization

The main result of Sect. 2, in which proof trees and proof circuits are defined formally, is that counting proof circuits is #P-complete. Section 3 introduces polynomial replacement systems and their canonical form and defines the relevant computational problems. Section 4 classifies prs and links them to arithmetic circuit problems. Section 5 contains the bulk of our complexity results, Section 6 concludes. For lack of space, formal proofs of all claims made in this abstract have to be omitted, but can be found in ftp://ftp-info4.informatik.uni-wuerzburg.de/pub/ftp/TRs/mc-vo-wa00.ps.gz.

2   Counting Circuits vs. Counting Trees

By a circuit C, in this paper, we will mean a circuit over the basis $\{\wedge, \vee\}$ in the usual sense, with $2n$ inputs labeled $x_1, \ldots, x_n, \neg x_1, \ldots, \neg x_n$.

Fix an input x to C. Unwind C into a tree C' by (repeatedly) duplicating gates with fan-out greater than 1. Define a *proof tree* as a subgraph H of C' whose gates evaluate to 1 and which additionally fulfills the following properties: H must contain the output gate of C. For every $\wedge$ gate v in H, all the input wires of v must be in H, and for every $\vee$ gate v in H, exactly one input wire of v must be in H. Only wires and nodes obtained in this way belong to H. By $\#C(x)$ we denote the number of proof trees of C. Define a *proof circuit* as a subcircuit H of C with the same properties as above. (I.e., the only difference is that now we do *not* start by unwinding C into a tree.) Given an input x, let $\#_c C(x)$ denote the number of proof circuits of C on x. We will consider the following problems:

> *Problem:* PT
> *Input:*   circuit C over $\{\wedge, \vee\}$, an input $x \in \{0,1\}^*$, a number k in unary
> *Output:* $\#C(x) \bmod 2^k$

> *Problem:* PC
> *Input:*   circuit C over $\{\wedge, \vee\}$, an input $x \in \{0,1\}^*$
> *Output:* $\#_c C(x)$

Observe that if we unwind a circuit into a tree there may be an exponential blowup in size, which has the consequence that the number of proof trees may be doubly-exponential in the size of the original circuit. This is not possible for the problem PC; the values of this function can be at most exponential in the input length. In order to achieve a fair comparison of the complexity of the problems, we therefore count proof trees only modulo an exponential number.

Theorem 1.   *1. PC is complete for #P under $\leq_{1\text{-T}}^{\log}$, but not under $\leq_m^P$ unless* P = NP.

2. PT *is complete for* FP *under* $\leq_m^{\log}$.
3. *The following problem is* NP-*complete under* $\leq_m^{\log}$: *Given a circuit* C, *is there an input* x *such that* $\#C(x) \neq \#_c C(x)$?
4. *The following problem is* P-*complete under* $\leq_m^{\log}$: *Given a circuit* C *and an input* $x \in \{0,1\}^*$, *is* $\#C(x) \neq \#_c C(x)$?

3   How to Generate Polynomials

A straight line program P over variables $x_1, \ldots, x_m$ is a set of instructions of one of the following types: $x_i \leftarrow x_j + x_k$, $x_i \leftarrow x_j \cdot x_k$, $x_i \leftarrow 0$, $x_i \leftarrow 1$, where $j, k < i$. Every variable appears at most once on the left hand side of the $\leftarrow$. Those variables that never appear on the left hand side of the $\leftarrow$ are the *input variables*. The variable x_m is the *output variable*. Given values for the input variables, the values of all other variables are computed in the obvious way. The value computed by P is the value of the output variable. Let p_P be the number-theoretic function computed in this way by P.

A straight line program hence is just another way of looking at an arithmetic circuit. The connection between counting proof trees and evaluating arithmetic circuits yields an obvious algorithm to compute the number of proof trees of a circuit C on input x: evaluate the straight line program obtained from C in the order of its variables, and plug in x. To compute the number of proof circuits instead, a mere variant of this algorithm was sketched in Sect. 1.1: do as for proof trees, but at each replacement step, express the intermediate polynomial as a sum of monomials and replace any occurrence of g^2 by g, for any variable g.

Theorem 2. *The algorithm sketched in Sect. 1.1 correctly computes the number of proof circuits of a circuit* C *on input* x.

Both the proof tree and the proof circuit counting algorithms prescribe a unique intermediate formal polynomial in the circuit input variables. These algorithms originate from special types of *polynomial replacement systems*, which we now define. Polynomial replacement systems will produce sets of polynomials from a given start polynomial, using rules replacing certain polynomials by other polynomials. This will be very similar to the way formal grammars produce sets of words from a start symbol, applying production rules.

In this paper we almost exclusively consider polynomials with nonnegative integer coefficients. This is motivated by the application to proof trees and proof circuits discussed above. We write $p(z_1, \ldots, z_s)$ to denote that p is such a polynomial in variables $z_1, \ldots, z_s$.

Below, the variable vector $\overline{x}$ will always be defined to consist of $\overline{x} = (x_1, \ldots, x_m)$. Let us say that the variable x_i is *fictive* (or, *inessential*) in the polynomial $p(\overline{x})$ if for all $a_1, \ldots, a_m, a_i' \in \mathbb{N}$ we have $p(a_1, \ldots, a_{i-1}, a_i, a_{i+1}, \ldots, a_m) = p(a_1, \ldots, a_{i-1}, a_i', a_{i+1}, \ldots, a_m)$. This means that x_i is fictive in p if and only if p can be written as a term in which x_i does not appear.

Definition 3. *A polynomial replacement system (for short: prs) is defined as a quadruple $S = \left(\{x_1, \ldots, x_n\}, \{x_{n+1}, \ldots, x_m\}, q, R\right)$ where*

- *$\{x_1, \ldots, x_n\}$ is the set of* terminal variables,
- *$\{x_{n+1}, \ldots, x_m\}$ is the set of* nonterminal variables,
- *q is a polynomial in the variables $x_1, \ldots, x_m$, the* start polynomial, *and*
- *R is a finite set of* replacement rules, *i. e., a finite set of pairs of polynomials in the variables $x_1, \ldots, x_m$.*

How does such a system generate polynomials?

Definition 4. *Let $S = \left(\{x_1, \ldots, x_n\}, \{x_{n+1}, \ldots, x_m\}, q, R\right)$ be a prs, let p_1, p_2 be polynomials in the variables $\overline{x}$.*

$$p_1 \underset{S}{\Longrightarrow} p_2 \iff_{\text{def}} \text{ there exist } (p_3, p_4) \in R \text{ and a polynomial } p_5(\overline{x}, y) \text{ such that}$$
$$p_1(\overline{x}) = p_5(\overline{x}, p_3(\overline{x})) \text{ and } p_2(\overline{x}) = p_5(\overline{x}, p_4(\overline{x})).$$

Let $\underset{S}{\overset{*}{\Longrightarrow}}$ be the reflexive and transitive closure of $\underset{S}{\Longrightarrow}$.

It turns out that the above form for derivations can be simplified:

Definition 5. *Let S, p_1, p_2 be as above.*

$$p_1 \underset{S}{\to} p_2 \iff_{\text{def}} \text{ there exist } (p_3, p_4) \in R \text{ and polynomials } p_5(\overline{x}), p_6(\overline{x}) \text{ such that}$$
$$p_1(\overline{x}) = p_5(\overline{x}) \cdot p_3(\overline{x}) + p_6(\overline{x}) \text{ and } p_2(\overline{x}) = p_5(\overline{x}) \cdot p_4(\overline{x}) + p_6(\overline{x}).$$

Let $\underset{S}{\overset{*}{\to}}$ be the reflexive and transitive closure of $\underset{S}{\to}$.

Lemma 6 (Normal Form of Replacement). *For any prs $S = \left(\{x_1, \ldots, x_n\}, \{x_{n+1}, \ldots, x_m\}, q, R\right)$ and any polynomials $p_1(\overline{x}), p_2(\overline{x})$, we have:*

$$p_1 \underset{S}{\overset{*}{\Longrightarrow}} p_2 \quad \textit{iff} \quad p_1 \underset{S}{\overset{*}{\to}} p_2.$$

A prs thus generates a set of polynomials; hence we define:

Definition 7. *For a prs $S = \left(\{x_1, \ldots, x_n\}, \{x_{n+1}, \ldots, x_m\}, q, R\right)$, let*

$$\text{POLY}(S) = \{p(x_1, \ldots, x_n) \mid \text{ there exists } p'(\overline{x}) \text{ such that } q \underset{S}{\overset{*}{\Longrightarrow}} p' \text{ and}$$
$$p(x_1, \ldots, x_n) = p'(x_1, \ldots, x_n, a_{n+1}, \ldots, a_m)$$
$$\text{for all } a_{n+1}, \ldots, a_m \in \mathbb{N}\}.$$

To determine the complexity of the sets $\text{POLY}(\cdot)$, we have to fix an encoding of polynomials. We choose to represent polynomials by straight-line programs (as, e.g., in [IM83, Kal88]), and state our result below for this particular representation. Other representations have been considered in the literature (e.g., formula representation, different sparse representations where a polynomial is given as a sequence of monomials, etc.). We remark that our results remain valid for most of these, as we will prove in the full version of this paper.

From the set $\text{POLY}(S)$ of polynomials we derive several sets of natural numbers, whose complexities we will determine in the upcoming sections.

Definition 8. *Let $S = (\{x_1, \ldots, x_n\}, \{x_{n+1}, \ldots, x_m\}, q, R)$ be a prs. Define*

- RANGE$(S) =_{\text{def}} \{ p(a) \mid p \in \text{POLY}(S) \wedge a \in \mathbb{N}^n \}$;
- EVAL$(S) =_{\text{def}} \{ (a, p(a)) \mid p \in \text{POLY}(S) \text{ and } a \in \mathbb{N}^n \}$.

Observe that if we also allow negative numbers as coefficients for our polynomials, then there are prs S such that RANGE(S) is not decidable. This is seen as follows. By the Robinson-Matiasjevič result (see [Mat93]), every recursively enumerable set can be represented in the form $\{ p(a) \mid a \in \mathbb{N}^n \}$ where p is a suitable n-ary polynomial with integer coefficients. Now let p be such an n-ary polynomial such that $\{ p(a) \mid a \in \mathbb{N}^n \}$ is not decidable. Defining the prs $S_p =_{\text{def}} (\{x_1, \ldots, x_n\}, \emptyset, p, \emptyset)$ we obtain POLY$(S_p) = \{p\}$ and RANGE$(S_p) = \{ p(a) \mid a \in \mathbb{N}^n \}$.

Besides the membership problems POLY(S), RANGE(S), and EVAL(S), we also consider the corresponding *variable membership problems*.

Definition 9.   – POLY$(\cdot) =_{\text{def}} \{ (S, p) \mid S \text{ prs and } p \in \text{POLY}(S) \}$;
- RANGE$(\cdot) =_{\text{def}} \{ (S, a) \mid S \text{ prs and } a \in \text{RANGE}(S) \}$;
- EVAL$(\cdot) =_{\text{def}} \{ (S, a, p(a)) \mid S \text{ prs}, p \in \text{POLY}(S), \text{ and } a \in \mathbb{N}^* \}$.

4   Different Types of Replacement Systems

Prs are very general. Here, we introduce a number of natural restrictions. Our approach is similar to the way different restrictions of grammar types were introduced, e.g., in the definition of the classes of the Chomsky hierarchy. We will later view the problems of counting proof trees and proof circuits as two instances of a problem about these restricted prs types.

4.1   Simple Polynomial Replacement Systems

Definition 10. *A prs $S = (\{x_1, \ldots, x_n\}, \{x_{n+1}, \ldots, x_m\}, q, R)$ is* simple *(or* context-free*), if the polynomials in the left-hand sides of the rules of R are variables from $\{x_{n+1}, \ldots, x_m\}$.*

All definitions made in the preceding section for general prs carry over to the special cases of simple systems. However, for simple prs we additionally define a particular type of replacement, where the application of a rule (z, q) results in the replacement of *all occurrences* of z with q. This latter form is denoted by $\underset{S}{\Longmapsto}$, in contrast to the notation $\underset{S}{\Longrightarrow}$ for the derivations defined so far. Formally:

Definition 11. *Let $S = (\{x_1, \ldots, x_n\}, \{x_{n+1}, \ldots, x_m\}, q, R)$ be a simple prs.*

$$p_1 \underset{S}{\Longmapsto} p_2 \Longleftrightarrow_{\text{def}} \text{there exist } (x_i, p_3) \in R \text{ such that}$$
$$p_2(\overline{x}) = p_1(x_1, \ldots, x_{i-1}, p_3(\overline{x}), x_{i+1}, \ldots, x_m).$$

Let $\underset{S}{\overset{*}{\Longmapsto}}$ be the reflexive and transitive closure of $\underset{S}{\Longmapsto}$.

For the sets of polynomials and numbers derived from simple systems using our new derivation type, we use the same names as before but now use square brackets $[\cdots]$ instead of parentheses $(\cdots)$; e.g., POLY$[S]$, POLY$[\cdot]$, etc.

4.2   Simple Deterministic or Acyclic Polynomial Replacement Systems

Definition 12. *A prs* $S = (\{x_1,\ldots,x_n\}, \{x_{n+1},\ldots,x_m\}, q, R)$ *is said to be* deterministic, *if no two different rules in* R *have the same left-hand side.*

Definition 13. *Let* $S = (\{x_1,\ldots,x_n\}, \{x_{n+1},\ldots,x_m\}, q, R)$ *be a prs. The* dependency graph G_S *of* S *is the directed graph* $G_S = (\{1,\ldots,m\}, E_S)$, *where* E_S *consists of all edges* (j,i) *for which there exists a rule* $(p_1, p_2) \in R$ *such that* x_i *is essential in* p_1 *and* x_j *is essential in* p_2. *The prs* S *is said to be* acyclic, *if its dependency graph* G_S *is acyclic.*

Lemma 14. *For every simple and deterministic prs* S, *there exists a simple, deterministic, and acyclic prs* S', *computable in polynomial time, such that* $\mathrm{POLY}(S) = \mathrm{POLY}(S')$ *and* $\mathrm{POLY}[S] = \mathrm{POLY}[S']$.

We also obtain the following easy properties:

Lemma 15.   *1. If* S *is a simple and deterministic prs then* $\mathrm{POLY}(S) = \mathrm{POLY}[S]$, *and this set consists of at most one polynomial.*
  2. If S *is a simple and acyclic prs then* $\mathrm{POLY}(S)$ *and* $\mathrm{POLY}[S]$ *are finite.*

Note that there are simple and acyclic prs S such that $\mathrm{POLY}[S] \subsetneq \mathrm{POLY}(S)$. For example take $S = (\{x\}, \{z\}, 2z, \{(z,x), (z,2x)\})$ where $\mathrm{POLY}[S] = \{2x, 4x\}$ and $\mathrm{POLY}(S) = \{2x, 3x, 4x\}$. Thus, the requirement that S is deterministic is necessary in Lemma 15.1.

In the remainder of this subsection, we relate simple deterministic and simple acyclic prs to different forms of circuits operating over the natural numbers.

First, it is intuitively clear that there is some connection between simple, deterministic, and acyclic systems and straight-line programs. This is made precise in the following lemma.

Lemma 16.   *1. If* S *is a simple, deterministic, and acyclic prs and* $\mathrm{POLY}(S) \neq \emptyset$, *then there exists a slp* P, *computable in logarithmic space, such that* $\mathrm{POLY}(S) = \{p_P\}$.
  2. If P *is a slp then there exists a simple, deterministic, and acyclic prs* S, *computable in logarithmic space, such that* $\{p_P\} = \mathrm{POLY}(S)$.

Next we show that acyclic systems are strongly related to a certain type of arithmetic circuit we now define. These circuits are immediate generalizations of *integer expressions*, introduced by Stockmeyer and Meyer [SM73]. Therefore we call our circuits *integer circuits* (not to be confused with ordinary arithmetic circuits), or, referring to the operations allowed, $(\cup, +, \times)$-circuits.

An integer circuit with n inputs is a circuit C where the inner nodes compute one of the operations $\cup, +, \times$. Such a circuit C has a specified output gate g_s. It computes a function $f_C \colon \mathbb{N}^n \to 2^{\mathbb{N}}$ as follows: We first define for every gate $g \in C$ the function f_g computed by g.

1. If g is an input gate x_i, then $f_g(a_1, \ldots, a_n) = \{a_i\}$ for all $a_1, \ldots, a_n \in \mathbb{N}$.
2. If g is $+$ gate with predecessors g_l, g_r, then $f_g(a_1, \ldots, a_n) = \{\, k + m \mid k \in f_{g_l}(a_1, \ldots, a_n), m \in f_{g_r}(a_1, \ldots, a_n) \,\}$. The function computed by a $\times$ gate is defined analogously.
3. If g is a $\cup$ gate with predecessors g_l, g_r, then $f_g(a_1, \ldots, a_n) = f_{g_l}(a_1, \ldots, a_n) \cup f_{g_r}(a_1, \ldots, a_n)$.

Finally, the function computed by C is $f_C = f_{g_s}$.

The following relation between simple, acyclic replacement systems and integer circuits is obtained by an easy induction:

Lemma 17. *1. For every simple, acyclic prs S, there is an integer circuit C with n inputs, computable in logarithmic space, such that $f_C(a) = \{\, b \mid (a, b) \in \mathrm{EVAL}(S) \,\}$ for all $a \in \mathbb{N}^n$.*
2. For every integer circuit C with n inputs, there is a simple, acyclic prs S, computable in logarithmic space, such that $\{\, b \mid (a, b) \in \mathrm{EVAL}(S) \,\} = f_C(a)$ for all $a \in \mathbb{N}^n$.

We consider the following problems:

$$\text{N-MEMBER}(\cup, +, \times) =_{\mathrm{def}} \{\, (C, a, b) \mid C \text{ is an integer circuit with } n \text{ inputs,} \atop a \in \mathbb{N}^n, b \in \mathbb{N} \text{ and } b \in f_C(a) \,\}$$

$$\text{N-RANGE}(\cup, +, \times) =_{\mathrm{def}} \{\, (C, b) \mid C \text{ is an integer circuit with } n \text{ inputs,} \atop b \in \mathbb{N} \text{ and } (\exists a \in \mathbb{N}^n) b \in f_C(a) \,\}$$

Analogous notations will be used when we restrict the gate types allowed.

The following lemma is immediate from Lemma 17:

Lemma 18. *1.* $\text{N-MEMBER}(\cup, +, \times) \equiv_m^{\log} \mathrm{EVAL}(\cdot)$.
2. $\text{N-RANGE}(\cup, +, \times) \equiv_m^{\log} \mathrm{RANGE}(\cdot)$.

4.3   Idempotent Polynomial Replacement Systems

Definition 19. *For a prs $S = (\{x_1, \ldots, x_n\}, \{x_{n+1}, \ldots, x_m\}, q, R)$, we define $S_{\mathrm{idem}} = (\{x_1, \ldots, x_n\}, \{x_{n+1}, \ldots, x_m\}, q, R \cup \{\, (x_i^2, x_i) \mid 1 \le i \le m \,\})$ to be the* idempotent prs *derived from S.*

In the case that S is simple (deterministic, acyclic, resp.), we will say that S_{idem} is an idempotent simple (deterministic, acyclic, resp.) prs.

For a prs $S = (\{x_1, \ldots, x_n\}, \{x_{n+1}, \ldots, x_m\}, q, R)$ and $a \in \mathbb{N}^n$, we write $\min \mathrm{EVAL}(S, a)$ as a shorthand for $\min\{\, p(a) \mid p \in \mathrm{POLY}(S) \,\}$ (analogously, we use $\max \mathrm{EVAL}(S, a)$).

Lemma 20. *1. For every Boolean circuit C, input x, and $k \in \mathbb{N}$, there exists a simple, deterministic and acyclic polynomial replacement system S, computable in logarithmic space, such that $\min \mathrm{EVAL}\big(S_{\mathrm{idem}}, (1, \ldots, 1)\big) = \#_c C(x)$, and $\max \mathrm{EVAL}\big(S_{\mathrm{idem}}, (1, \ldots, 1)\big) = \#C(x)$.*
2. For every simple, deterministic, and acyclic prs S_{idem} there exists a Boolean circuit C, computable in logarithmic space, such that $\min \mathrm{EVAL}\big(S_{\mathrm{idem}}, (1, \ldots, 1)\big) = \#_c C(x)$, and $\max \mathrm{EVAL}\big(S_{\mathrm{idem}}, (1, \ldots, 1)\big) = \#C(x)$.

5 Complexity Results for Simple Replacement Systems

5.1 Deterministic Systems

In this section, we consider the complexity of the above defined sets for simple replacement systems. Let us start with the complexity of fixed membership problems.

Theorem 21. *Let S be simple and deterministic.*

1. $\text{POLY}(S), \text{POLY}[S]$ *are P-complete.*
2. $\text{RANGE}(S), \text{RANGE}[S] \in \text{NP}$, *and there are systems S such that the problems* $\text{RANGE}(S)$ *and* $\text{RANGE}[S]$ *are NP-complete.*
3. $\text{EVAL}(S), \text{EVAL}[S] \in \text{TC}^0$.

Concerning variable membership problems of simple, deterministic systems, we obtain:

Theorem 22. *For simple and deterministic prs,*

1. $\text{POLY}(\cdot)$ *and* $\text{POLY}[\cdot]$ *are in* coRP *and P-hard under* $\leq_m^{\log}$,
2. $\text{RANGE}(\cdot)$ *and* $\text{RANGE}[\cdot]$ *are NP-complete under* $\leq_m^{\log}$,
3. $\text{EVAL}(\cdot)$ *and* $\text{EVAL}[\cdot]$ *are P-complete under* $\leq_m^{\log}$.

5.2 Acyclic Systems

For simple and acyclic systems S which are not necessarily deterministic, the sets $\text{POLY}(S)$ and $\text{POLY}[S]$ are finite (by Lemma 15), hence the statement of Theorem 21 also holds in this case. Again, interesting questions arise when we examine variable membership problems.

Theorem 23. *For simple and acyclic prs,*

1. $\text{POLY}[\cdot]$ *is contained in* MA *and is NP-hard under* $\leq_m^{\log}$,
2. $\text{RANGE}[\cdot]$ *and* $\text{EVAL}[\cdot]$ *are NP-complete under* $\leq_m^{\log}$.

Next, we turn to different variable membership problems for simple, acyclic systems under "$\Longrightarrow$"- derivations.

Stockmeyer and Meyer considered integer expressions (in our terminology, these are integer circuits with fan-out of non-input gates at most 1) where the only operations allowed are $\cup$ and $+$. They proved that the membership problem in that case is NP-complete. It is easy to see that their result carries over to the case that we also allow multiplication, i. e., the problems $\text{N-MEMBER}(\cup, +)$ and $\text{N-MEMBER}(\cup, +, \times)$ for expressions are NP-complete.

The corresponding problems for circuits were not considered in their paper, but in later papers by Wagner [Wag84, Wag86] (under the name *hierarchical descriptions*). Only PSPACE as upper bound for membership is known from there, but recently it was shown by Ke Yang that both circuit problems are PSPACE-hard [Yan00].

Since (by Lemma 18), the member and range problems for these circuits are equivalent to the $\text{EVAL}(\cdot)$ and $\text{RANGE}(\cdot)$ problems for simple acyclic prs, we conclude:

Theorem 24. *For simple and acyclic prs and for all representations,*

1. POLY$(\cdot) \in$ EXPTIME,
2. RANGE$(\cdot)$, EVAL$(\cdot)$ *are PSPACE-complete.*

5.3   Idempotent Systems

Again, since also here, POLY(S) and POLY$[S]$ are finite, we obtain results analogous to Theorem 21. For the variable membership problems the following can be said:

Theorem 25. *For idempotent, simple, deterministic, and acyclic systems, we obtain* POLY$(\cdot)$, RANGE$(\cdot)$, EVAL$(\cdot) \in$ EXPTIME.

Lemma 20 shows the importance of the minimization and maximization operations in the case of idempotent systems. We obtain from Theorem 1:

Theorem 26. *For idempotent, simple, deterministic, and acyclic replacement systems,*

1. *the functions* $\min$ EVAL$(\cdot)$ *and* $\min$ EVAL$[\cdot]$ *are #P-complete under* $\leq_{1\text{-}T}^{\log}$-*reductions,*
2. *the functions* $\max$ EVAL$(\cdot)$ *and* $\max$ EVAL$[\cdot]$ *are FP-complete under* $\leq_m^P$.

Remark 27. For simple, deterministic and for simple, acyclic prs, the functions $\min$ EVAL$(\cdot)$, $\min$ EVAL$[\cdot]$, $\max$ EVAL$(\cdot)$, $\max$ EVAL$[\cdot]$ are FP-complete.

6   Conclusion

Our original motivation was the PC problem for circuits of restricted depth. Our proof (omitted in this proceedings version) shows that the problem is #P-complete even for circuits of depth 4. For depth-2 circuits, the problem is easily seen to be in FP. The case of depth 3 remains open.

The complexity of the sets RANGE(S), RANGE$[S]$ for fixed S is equivalent to determining the complexity of the range of a multivariate polynomial with nonnegative integer coefficients. While this is always an NP-problem, the proof of our result (again, unfortunately, omitted here) shows that there is a 4-variable polynomial of degree 6 whose range is NP complete. Can this be improved?

A lot of interesting questions about prs remain open. To come back to some of the problems posed in Subsect. 1.1, we did not look at all at multivariate rules. Also, it seems worthwhile to examine if, besides idempotent systems, other prs families can be related to various types of arithmetic circuits and counting problems in Boolean circuits.

Acknowledgment. We are grateful to Sven Kosub (Würzburg) and Thomas Thierauf (Ulm) for helpful discussions.

References

[AAB+99] E. Allender, A. Ambainis, D. Mix Barrington, S. Datta, and H. LêThanh. Bounded depth arithmetic circuits: Counting and closure. In *Proceedings 26th International Colloquium on Automata, Languages and Programming*, Lecture Notes in Computer Science, Berlin Heidelberg, 1999. Springer Verlag. To appear.

[AAD97] M. Agrawal, E. Allender, and S. Datta. On TC^0, AC^0, and arithmetic circuits. In *Proceedings 12th Computational Complexity*, pages 134–148. IEEE Computer Society, 1997.

[All98] E. Allender. Making computation count: arithmetic circuits in the nineties. *SIGACT News*, 28(4):2–15, 1998.

[BF91] L. Babai and L. Fortnow. Arithmetization: a new method in structural complexity theory. *Computational Complexity*, 1:41–66, 1991.

[CMTV98] H. Caussinus, P. McKenzie, D. Thérien, and H. Vollmer. Nondeterministic NC^1 computation. *Journal of Computer and System Sciences*, 57:200–212, 1998.

[IM83] O. Ibarra and S. Moran. Probabilistic algorithms for deciding equivalence of straight-line programs. *Journal of the ACM*, 30:217–228, 1983.

[Kal88] E. Kaltofen. Greatest common divisors of polynomials given by straight-line programs. *Journal of the ACM*, 35:231–264, 1988.

[Mat93] Y. V. Matiasjevič. *Hilbert's Tenth Problem*. Foundations of Computing Series. MIT Press, Cambridge, MA, 1993.

[SM73] L. J. Stockmeyer and A. R. Meyer. Word problems requiring exponential time. In *Proceedings 5th ACM Symposium on the Theory of Computing*, pages 1–9. ACM Press, 1973.

[Sud78] I. H. Sudborough. On the tape complexity of deterministic context-free languages. *Journal of the Association for Computing Machinery*, 25:405–414, 1978.

[Val79] L. G. Valiant. The complexity of enumeration and reliability problems. *SIAM Journal of Computing*, 8(3):411–421, 1979.

[Ven92] H. Venkateswaran. Circuit definitions of non-deterministic complexity classes. *SIAM Journal on Computing*, 21:655–670, 1992.

[Vin91] V. Vinay. Counting auxiliary pushdown automata and semi-unbounded arithmetic circuits. In *Proceedings 6th Structure in Complexity Theory*, pages 270–284. IEEE Computer Society Press, 1991.

[VSBR83] L. Valiant, S. Skyum, S. Berkowitz, and C. Rackoff. Fast parallel computation of polynomials using few processors. *SIAM Journal on Computing*, 12:641–644, 1983.

[VT89] H. Venkateswaran and M. Tompa. A new pebble game that characterizes parallel complexity classes. *SIAM J. on Computing*, 18:533–549, 1989.

[Wag84] K. W. Wagner. The complexity of problems concerning graphs with regularities. Technical report, Friedrich-Schiller-Universität Jena, 1984. Extended abstract in *Proceedings 11th Mathematical Foundations of Computer Science*, Lecture Notes in Computer Science 176, pages 544–552, 1984.

[Wag86] K. W. Wagner. The complexity of combinatorial problems with succinct input representation. *Acta Informatica*, 23:325–356, 1986.

[Yan00] K. Yang. Integer circuit evaluation is PSPACE-complete. In *Proceedings 15th Computational Complexity Conference*, pages 204–211. IEEE Computer Society Press, 2000.

Depth-3 Arithmetic Circuits for $S_n^2(X)$ and Extensions of the Graham-Pollack Theorem

Jaikumar Radhakrishnan[1], Pranab Sen[1], and Sundar Vishwanathan[2]

[1] School of Technology and Computer Science, Tata Institute of Fundamental Research, Mumbai 400005, India
{jaikumar, pranab}@tcs.tifr.res.in
[2] Department of Computer Science and Engineering, Indian Institute of Technology, Mumbai 400076, India
sundar@cse.iitb.ernet.in

Abstract. We consider the problem of computing the second elementary symmetric polynomial $S_n^2(X) \stackrel{\triangle}{=} \sum_{1 \le i < j \le n} X_i X_j$ using depth-three arithmetic circuits of the form $\sum_{i=1}^{r} \prod_{j=1}^{s_i} L_{ij}(X)$, where each L_{ij} is a linear form. We consider this problem over several fields and determine *exactly* the number of multiplication gates required. The lower bounds are proved for inhomogeneous circuits where the L_{ij}'s are allowed to have constants; the upper bounds are proved in the homogeneous model. For reals and rationals the number of multiplication gates required is exactly $n - 1$; in most other cases, it is $\lceil \frac{n}{2} \rceil$.
This problem is related to the Graham-Pollack theorem in algebraic graph theory. In particular, our results answer the following question of Babai and Frankl: what is the minimum number of complete bipartite graphs required to cover each edge of a complete graph an odd number of times? We show that for infinitely many n, the answer is $\lceil \frac{n}{2} \rceil$.

1 Introduction

1.1 The Graham-Pollack Theorem

Let K_n denote the complete graph on n vertices. By a *decomposition* of K_n, we mean a set $\{G_1, G_2, \ldots, G_r\}$ of subgraphs of K_n such that

1. Each G_i is a complete bipartite graph (on some subset of the vertex set of K_n); and
2. Each edge of K_n appears in precisely one of the G_i's.

It is easy to see that there is such a decomposition of the complete graph with $n - 1$ bipartite graphs. Graham and Pollack [4] showed that this is tight.

Theorem. *If $\{G_1, G_2, \ldots, G_r\}$ is a decomposition of K_n, then $r \ge n-1$.*

The original proof of this theorem, and other proofs discovered since then [3, 10, 14], used algebraic reasoning in one form or another; no combinatorial proof of this fact is known.

S. Kapoor and S. Prasad (Eds.): FST TCS 2000, LNCS 1974, pp. 176–187, 2000.
© Springer-Verlag Berlin Heidelberg 2000

One of the goals of this paper is to obtain extensions of this theorem. To better motivate the problems we study, we first present a proof of this theorem. This will also help us explain how algebraic reasoning enters the picture. Consider polynomials in variables $X = X_1, X_2, \ldots, X_n$ with rational coefficients. Let

$$S_n^2(X) \triangleq \sum_{1 \leq i < j \leq n} X_i X_j;$$

$$T_n^2(X) \triangleq \sum_{i=1}^{n} X_i^2.$$

Then, we can reformulate the question as follows. What is the smallest r for which there exist sets $L_i, R_i \subseteq [n]$ ($L_i \cap R_i = \emptyset$) for $i = 1, 2, \ldots, r$, such that

$$S_n^2(X) = \sum_{i=1}^{r} (\sum_{j \in L_i} X_j) \times (\sum_{j \in R_i} X_j)? \tag{1}$$

Notice that the two sums in the product on the right correspond to homogeneous linear forms. One may generalise this question, and ask: What is the smallest r for which there exist homogeneous linear forms $L_i(X), R_i(X)$ for $i = 1, 2 \ldots, r$, such that

$$S_n^2(X) = \sum_{i=1}^{r} L_i(X) R_i(X)? \tag{2}$$

Graham and Pollack gave the following elegant argument to show that r must be at least $n - 1$. Observe that $S_n^2(X) = \frac{1}{2}[(\sum_{i=1}^{n} X_i)^2 - T_n^2(X)]$. Thus, (2) implies

$$T_n^2(X) = (\sum_{i=1}^{n} X_i)^2 - 2 \sum_{i=1}^{r} L_i(X) R_i(X). \tag{3}$$

Now, if r is less than $n-1$, then there exists a non-zero $\alpha = (\alpha_1, \alpha_2, \ldots, \alpha_n) \in \mathbb{Q}^n$ such that $L_i(\alpha) = 0$, for $i = 1, 2 \ldots, r$, and $\sum_{i=1}^{n} \alpha_i = 0$ (because at most $n - 1$ homogeneous equations in n variables always have a non-zero solution). Under this assignment to the variables, the right hand side of (3) is zero but the left hand side is not.

With this introduction to the Graham-Pollack theorem and its proof, we are now ready to state the questions we consider in this paper. Observe that the lower bound above depended crucially on the field being $\mathbb{Q}$, and there are two main difficulties in generalising it to other fields. First, over fields of characteristic two, the relationship between $S_n^2(X)$ and $T_n^2(X)$ does not hold, for we cannot divide by 2. Second, even if we are not working over fields of characteristic two, $T_n^2(X)$ can vanish at some non-zero points. Equations similar to the ones considered above have been studied in the past in at least two different contexts viz. covering a complete graph by complete bipartite graphs such that each edge is covered an odd number of times (the *odd cover problem*) and depth–3 arithmetic circuits for $S_n^2(X)$.

1.2   The Odd Cover Problem

Suppose in the problem on graphs above, we drop the condition that the bipartite graphs be edge-disjoint, but instead ask for each edge of the complete graph to be covered an odd number of times. How many bipartite graphs are required in such a cover? This question was stated by Babai and Frankl [2], who also observed a lower bound of $\lfloor \frac{n}{2} \rfloor$. Note that this problem is equivalent to considering (1) over the field GF(2).

1.3 $\Sigma\Pi\Sigma$ Arithmetic Circuits

By a $\Sigma\Pi\Sigma$ arithmetic circuit we mean an expression of the form

$$\sum_{i=1}^{r} \prod_{j=1}^{s_i} L_{ij}(X), \tag{4}$$

where each L_{ij} is a (possibly inhomogeneous) linear form in variables $X_1, \ldots, X_n$. Such 'depth-three' circuits play an important role in the study of arithmetic complexity [8, 9, 6, 13]. If each linear form $L_{ij}(X)$ is homogeneous (i.e. has constant term zero) then the circuit is said to be homogeneous, or else, it is said to be inhomogeneous. Although, depth-three circuits appear to be rather restrictive, these are the strongest model of circuits for which superpolynomial lower bounds are known; no such lower bounds are known at present for depth-four circuits.

The k-th elementary symmetric polynomial on n variables is defined by

$$S_n^k(X) \triangleq \sum_{T \in \binom{[n]}{k}} \prod_{i \in T} X_i.$$

Elementary symmetric polynomials are the most commonly studied candidates for showing lower bounds in arithmetic circuits. Nisan and Wigderson [9] showed that any homogeneous $\Sigma\Pi\Sigma$ circuit for computing $S_n^{2k}(X)$ has size $\Omega((n/4k)^k)$. In their paper, they explicitly stated the method of partial derivatives (but see also Alon [1]). Although, a superpolynomial lower-bound was obtained in [9], the lower bound applied only to homogeneous circuits. Indeed, Ben-Or (see [9]) showed that any elementary symmetric polynomial can be computed by an inhomogeneous $\Sigma\Pi\Sigma$ formula of size $O(n^2)$. Thus inhomogeneous circuits are significantly more powerful than homogeneous circuits. Shpilka and Wigderson [13] (and later, Shpilka [12]) addressed this shortcoming of the Nisan-Wigderson result and showed an $\Omega(n^2)$ lower bound on the size of inhomogeneous formulae computing certain elementary symmetric polynomials, thus showing that Ben-Or's construction is optimal. To obtain their results they augmented the method of partial derivatives by an analysis of (affine) subspaces where elementary symmetric polynomials vanish. Many of the lower bounds in this paper are inspired by the insights from these papers. All the results cited above work over fields of characteristic zero. At present no super-quadratic lower bounds are known for computing some explicitly defined polynomial in the inhomogeneous model over infinite fields. Over finite fields the situation is better. Karpinski and Grigoriev [5] showed an exponential lower bound for computing the determinant polynomial using (inhomogeneous) $\Sigma\Pi\Sigma$ circuits over any finite field. Grigoriev and Razborov [6] showed an exponential lower bound for any (inhomogeneous) $\Sigma\Pi\Sigma$ circuit computing a *generalised majority* function over any finite field. Thus, the elementary symmetric polynomials have been studied with considerable success in the past in this arithmetic model of computation.

Organisation of This Paper

In the next section, we give a summary of our results. In Sect. 3, we present formal proofs of our upper bound results for GF(2). Section 4 contains formal proofs for our lower bound results for GF(2). A summary of the proof methods for our upper and lower bound results, as well as the proofs of the lemmas and theorems which have been omitted here, can be found in the full version [11] at http://www.tcs.tifr.res.in/~pranab/papers/s2n.ps.

2   Our Results

We study the computation of the symmetric polynomial $S_n^2(X)$ using $\Sigma\Pi\Sigma$ arithmetic circuits over several fields, with the aim of obtaining tight bounds on the number of multiplication gates required. Many of the techniques developed earlier (in particular, the method of partial derivatives), in fact, give lower bounds on the number of multiplication gates. Unlike the previous results in arithmetic circuits, we will not be satisfied with obtaining bounds up to constant factors; instead, we shall try to get the *exact* answer, in the spirit of the Graham-Pollack theorem.

As described above, computations of elementary symmetric polynomials have been considered for several kinds of $\Sigma\Pi\Sigma$ circuits. For the polynomial $S_n^2(X)$, we have three different models.

1. *The graph model:* This is the most restrictive model. Here the linear forms L_i and R_i must correspond to bipartite graphs; that is, all coefficients must be 1 (or 0), no variable can appear in both L_i and R_i (with coefficient 1), and no constant term is allowed in these linear forms. This is the setting for the Graham-Pollack theorem and its generalisations.
2. *The homogeneous model:* Here the linear forms are required to be homogeneous, that is, no constant term is allowed in them. However, any element from the field is allowed as a coefficient in the linear forms. This model was studied by Nisan and Wigderson [9], using the method of partial derivatives.
3. *The inhomogeneous model:* This is the most general model; there is no restriction on the coefficients or the constant term.

We show our upper bounds in the graph and the homogeneous model; our lower bounds hold even in the stronger inhomogeneous model. We juxtapose our results against the previously known results, highlighting our contribution. Note that the previous lower bounds were only for the homogeneous circuit model which were proved using the method of partial derivatives [9] (see also the rank arguments of Babai and Frankl [2]). The notation $\exists^\infty n$ used below stands for 'for infinitely many n'.

2.1 Computing $S_n^2(X)$ over Finite Fields of Odd Characteristic

Field		Our Bounds		Previous Bounds	
		Upper Bnds. Hom.	Lower Bnds. Inhom.	Upper Bnds. Hom.	Lower Bnds. Hom.
$GF(p^r)$ r even $p>3$	n even	$\frac{n}{2}\forall n$	$\frac{n}{2}\forall n$	$\frac{n}{2}+1\forall n$	$\frac{n}{2}\forall n$
	n odd	$\lceil\frac{n}{2}\rceil\forall n$	$\lceil\frac{n}{2}\rceil\exists^\infty n$ $\lfloor\frac{n}{2}\rfloor\forall n$	$\lceil\frac{n}{2}\rceil\forall n$	$\lceil\frac{n}{2}\rceil\exists^\infty n$ $\lfloor\frac{n}{2}\rfloor\forall n$
$GF(3^r)$ r even	n even	$\frac{n}{2}\forall n$	$\frac{n}{2}\forall n$	$\frac{n}{2}+1\forall n$	$\frac{n}{2}\forall n$
	n odd	$\lceil\frac{n}{2}\rceil\forall n$	$\lfloor\frac{n}{2}\rfloor\forall n$	$\lceil\frac{n}{2}\rceil\forall n$	$\lceil\frac{n}{2}\rceil\exists^\infty n$ $\lfloor\frac{n}{2}\rfloor\forall n$
$GF(p^r)$ r odd $p\equiv 1 \bmod 4$	n even	$\frac{n}{2}\exists^\infty n$	$\frac{n}{2}\forall n$	$\frac{n}{2}+1\forall n$	$\frac{n}{2}\forall n$
	n odd	$\lceil\frac{n}{2}\rceil\forall n$	$\lceil\frac{n}{2}\rceil\exists^\infty n$ $\lfloor\frac{n}{2}\rfloor\forall n$	$\lceil\frac{n}{2}\rceil\forall n$	$\lceil\frac{n}{2}\rceil\exists^\infty n$ $\lfloor\frac{n}{2}\rfloor\forall n$
$GF(p^r)$ r odd $p\equiv 3 \bmod 4$	n even	$\frac{n}{2}\exists^\infty n$	$\frac{n}{2}\forall n$	$n-1\forall n$	$\frac{n}{2}\forall n$
	n odd	$\lceil\frac{n}{2}\rceil\exists^\infty n$	$\lfloor\frac{n}{2}\rfloor\forall n$	$n-1\forall n$	$\lfloor\frac{n}{2}\rfloor\forall n$

2.2 The Odd Cover Problem and Computing $S_n^2(X)$ over GF(2)

	Our Bounds			Previous Bounds	
	Upper Bounds Graph	Hom.	Lower Bounds Inhom.	Upper Bounds Graph	Lower Bounds Hom.
$n\equiv 0 \bmod 4$	$\frac{n}{2}\exists^\infty n$	$\frac{n}{2}\exists^\infty n$	$\frac{n}{2}\forall n$	$n-1\forall n$	$\frac{n}{2}\forall n$
$n\equiv 2 \bmod 4$	$\frac{n}{2}\exists^\infty n$	$\frac{n}{2}\exists^\infty n$	$\frac{n}{2}\forall n$	$n-1\forall n$	$\frac{n}{2}\forall n$
$n\equiv 3 \bmod 4$	$\lceil\frac{n}{2}\rceil\exists^\infty n$	$\lceil\frac{n}{2}\rceil\exists^\infty n$	$\lceil\frac{n}{2}\rceil\forall n$	$n-1\forall n$	$\lfloor\frac{n}{2}\rfloor\forall n$
$n\equiv 1 \bmod 4$	$\lceil\frac{n}{2}\rceil\exists^\infty n$	$\lfloor\frac{n}{2}\rfloor\exists^\infty n$	$\lfloor\frac{n}{2}\rfloor\forall n$	$n-1\forall n$	$\lfloor\frac{n}{2}\rfloor\forall n$

2.3 Computing $S_n^2(X)$ over $\mathbb{C}$

	Our Bounds		Previous Bounds	
	Upper Bounds	Lower Bounds	Upper Bounds	Lower Bounds
	Hom.	Inhom.	Hom.	Hom.
$\forall n$	$\frac{n}{2}$	$\frac{n}{2}$	$\frac{n+1}{2}$	$\frac{n}{2}$

2.4 1 mod p Cover Problem, p an Odd Prime

	Our Bounds	Previous Bounds	
	Upper Bounds	Upper Bounds	Lower Bounds
	Graph	Graph	Hom.
n odd	$\lceil \frac{n}{2} \rceil \, \exists^\infty n$	$n - 1 \, \forall n$	$\lfloor \frac{n}{2} \rfloor \, \forall n$
n even	$\frac{n}{2} \exists^\infty n$	$n - 1 \, \forall n$	$\frac{n}{2} \forall n$

2.5 Computing $S_n^2(X)$ over $\mathbb{R}$ and $\mathbb{Q}$

	Our Bounds		Previous Bounds	
	Upper Bounds	Lower Bounds	Upper Bounds	Lower Bounds
	Graph	Inhom.	Graph	Hom.
$\forall n$	$n - 1$	$n - 1$	$n - 1$	$n - 1$

3 Upper Bounds

3.1 The Odd Cover Problem and Computing $S_n^2(X)$ over $\mathrm{GF}(2)$

In this section, we will show that there is an odd cover of K_{2n} by n bipartite graphs whenever there exists a $n \times n$ matrix satisfying certain properties. For this, we describe a scheme for producing an odd cover of K_{2n}.

We want to cover the edges of K_{2n} with n bipartite graphs such that each edge is covered an odd number of times. Each complete bipartite graph is specified by specifying the two parts A and B. Partition the vertex set $[2n]$ (of K_{2n}) into ordered pairs $(1, 2), (3, 4), \ldots, (2n - 1, 2n)$. In our construction, if the vertex $2i - 1$ of the pair $(2i - 1, 2i)$ appears in part A of a bipartite graph, then $2i$ appears in part B; similarly, if $2i$ appears in part A, then $2i - 1$ appears in part B. In particular, if one element of the pair does not participate in the bipartite graph, then the other element does not participate in it either. We shall call such a construction a *pairs construction*.

Hence, to describe a bipartite graph, it suffices to specify for each pair $(2i - 1, 2i)$, whether the pair participates in the bipartite graph, and when it does, whether $2i$ appears in part A or part B. The n bipartite graphs are specified using an $n \times n$ matrix M with entries in $\{-1, 0, 1\}$. The rows of the matrix are indexed by pairs; the ith row is for the pair $(2i - 1, 2i)$. The columns are indexed by bipartite graphs. If $M_{ij} = 0$, then the pair $(2i - 1, 2i)$ does not participate in the jth bipartite graph; if $M_{ij} = 1$, then $2i$ appears in part B; if $M_{ij} = -1$, then $2i$ appears in part A. We now identify properties of the matrix M that ensure that the bipartite graphs arising from it form an odd cover of K_{2n}.

Definition 1. *A matrix with entries from* $\{-1, 0, 1\}$ *is* good *if it satisfies the following conditions:*

1. *In every row, the number of non-zero entries is odd.*
2. *For every pair of distinct rows, the number of columns where they both have non-zero entries is congruent to* 2 mod 4.
3. *Any two distinct rows are orthogonal over the integers.*

Lemma 1. *If a* $n \times n$ *matrix is good, then the* n *complete bipartite graphs that arise from it form an odd cover of* K_{2n}.

Proof. The proof appears in the full version of the paper [11].   □

Thus, to obtain odd covers, it is enough to construct good matrices. We now give two methods for constructing such matrices.

Construction 1: Skewsymmetric Conference Matrices

A *Hadamard matrix* H_n is an $n \times n$ matrix with entries in $\{-1, 1\}$ such that $H_n H_n^T = nI$. A *conference matrix* C_n is an $n \times n$ matrix, with 0's on the diagonal and $-1, +1$ elsewhere, such that $C_n C_n^T = (n-1)I$. The following fact can be verified easily.

Lemma 2. $n \times n$ *conference matrices, where* $n \equiv 0$ mod 4 *are good matrices.*

Skewsymmetric conference matrices can be obtained from *skew Hadamard matrices*. A skew Hadamard matrix is defined as a Hadamard matrix that one gets by adding the identity matrix to a skewsymmetric conference matrix. Several constructions of skew Hadamard matrices can be found in [7, p. 247]. In particular, the following theorem is proved there.

Theorem 1. *There is a skew Hadamard matrix of order* n *if* $n = 2^t k_1 \cdots k_s$, *where* $n \equiv 0$ mod 4, *each* $k_i \equiv 0$ mod 4 *and each* k_i *is of the form* $p^r + 1$, p *an odd prime.*

Corollary 1. *There is a good matrix of order* n *if* n *satisfies the conditions in the above theorem. Note that the conditions hold for infinitely many* n.

An easy construction of a skew Hadamard matrix for $n = 2^t, t > 1$ is given in the full version of the paper [11].

Construction 2: Symmetric Designs

The matrices we now construct are based on a well-known construction for symmetric designs. These matrices are not conference matrices; in fact, they have more than one zero in every row.

Let q be a prime power congruent to 3 mod 4. Let $\mathbb{F} = \mathrm{GF}(q)$ be the finite field of q elements. Index the rows with the lines and the columns with points of the projective 2-space over $\mathbb{F}$. That is, the projective points and lines are the one dimensional and two dimensional subspaces respectively, of $\mathbb{F}^3$. A projective point is represented by a vector in $\mathbb{F}^3$ (out of $q-1$ possible representatives) in the one dimensional subspace corresponding to it. A projective line is also represented by a vector in $\mathbb{F}^3$ (out of $q-1$ possible representatives). The representative for a projective line can be thought of as a 'normal vector' to the two dimensional subspace corresponding to it. We associate with each projective line L a linear form on the vector space $\mathbb{F}^3$, given by

$$L(w) = v^T w,$$

where $w \in \mathbb{F}^3$ and v is the representative for L. For a projective line L and a projective point Q, let $L(Q) \stackrel{\triangle}{=} L(w)$, where w is the representative for Q.

Now, the matrix M is defined as follows. If $L(Q) = 0$, then we set $M_{L,Q} = 0$; if $L(Q)$ is a (non-zero) square in $\mathbb{F}$, set $M_{L,Q} = 1$; otherwise, set $M_{L,Q} = -1$.

The proof that M is a good matrix appears in the full version of the paper [11]. We thus have proved the following lemma.

Lemma 3. *If* $q \equiv 3 \bmod 4$ *is a prime power then there is a good matrix of order* $q^2 + q + 1$. *Note that infinitely many such q exist.*

We can now easily prove the following theorem and its corollary.

Theorem 2. *For infinitely many* $n \equiv 0, 2 \bmod 4$ *we have an odd cover of* K_n *using* $\frac{n}{2}$ *complete bipartite graphs.*

Corollary 2. *For infinitely many* $n \equiv 1, 3 \bmod 4$ *we have an odd cover of* K_n *using* $\left\lceil \frac{n}{2} \right\rceil$ *complete bipartite graphs.*

We can also prove the following lemma.

Lemma 4. *If* $S_n^2(X), n \equiv 0 \bmod 4$, *can be computed over* $GF(2)$ *by a homogeneous* $\Sigma\Pi\Sigma$ *circuit using* $\frac{n}{2}$ *multiplication gates, then* $S_{n+1}^2(X)$ *can be computed over* $GF(2)$ *by a homogeneous* $\Sigma\Pi\Sigma$ *circuit using* $\frac{n}{2}$ *multiplication gates.*

Proof. The proof appears in the full version of the paper [11].   □

From the above, we can now prove the following theorem.

Theorem 3. *For infinitely many* $n \equiv 0, 2, 3 \bmod 4$ *we have homogeneous* $\Sigma\Pi\Sigma$ *circuits computing* $S_n^2(X)$ *over* $GF(2)$ *using* $\left\lceil \frac{n}{2} \right\rceil$ *multiplication gates. For infinitely many* $n \equiv 1 \bmod 4$ *we can compute* $S_n^2(X)$ *over* $GF(2)$ *using homogeneous* $\Sigma\Pi\Sigma$ *circuits having* $\left\lfloor \frac{n}{2} \right\rfloor$ *multiplication gates.*

4   Lower Bounds

4.1   Preliminaries

We first develop a framework for showing lower bounds for $S_n^2(X)$ based on the method of substitution [13, 12]. Suppose that over a field $\mathbb{F}$

$$S_n^2(X) = \sum_{i=1}^{r} \prod_{j=1}^{s_i} L_{ij}(X), \tag{5}$$

where each L_{ij} is a linear form, not necessarily homogeneous. We wish to show that r must be large. Following the proof of the Graham-Pollack theorem that was sketched in the introduction, we could try to force some of the L_{ij}'s to zero by setting the variables to appropriate field elements. There are two difficulties with this plan. First, since the L_{ij}'s are not necessarily homogeneous, we may not be able to set all of them to zero; we can do so if the linear forms have linearly independent homogeneous parts. The second difficulty arises from the nature of the underlying field: as observed earlier, $S_n^2(X)$ might vanish on non-trivial subspaces of $\mathbb{F}^n$.

In this section, our goal is to first show that if r is small, then $S_n^2(X)$ must be zero over a linear subspace of large dimension. A similar observation was used by Shpilka and Wigderson [13] and Shpilka [12]. Our second goal is to examine linear subspaces over which $S_n^2(X)$ is forced to be zero. We derive conditions on such subspaces, and relate them to the existence of a certain family of vectors. Later on, we will exploit these equations, based on the field in question, and derive our lower bounds for r.

Goal 1: Obtaining the Subspace.

Lemma 5. *If $S_n^2(X)$ can be written in the form of (5) over a field $\mathbb{F}$, then there exist homogeneous linear forms $\ell_1, \ell_2, \ldots, \ell_r$ in variables $X_1, X_2, \ldots, X_{n-r}$ such that*

$$S_n^2(X_1, X_2, \ldots, X_{n-r}, \ell_1, \ell_2, \ldots, \ell_r) = 0.$$

Proof. The proof appears in the full version of the paper [11].   □

Goal 2: The Nature of the Subspace. Our goal now is to understand the algebraic structure of the coefficients that appear in the linear forms $\ell_1, \ell_2, \ldots, \ell_r$ promised by Lemma 5. Let $\ell_i = \sum_{j=1}^{n-r} \ell_{ij} X_j$, $\ell_{ij} \in \mathbb{F}$, and let L be the $r \times (n-r)$ matrix (ℓ_{ij}). Let $Y_1, Y_2, \ldots, Y_{n-r} \in \mathbb{F}^r$ be the $n-r$ columns of L. We will obtain conditions on the columns by computing the coefficients of monomials X_j^2 for $1 \le j \le n-r$, and $X_i X_j$ for $1 \le i < j \le n-r$. For X_j^2 ($1 \le j \le n-r$), we obtain the following equation over $\mathbb{F}$,

$$\sum_{m=1}^{r} \ell_{mj} + \sum_{1 \le m < m' \le r} \ell_{mj} \ell_{m'j} = 0. \tag{6}$$

For monomials of the form $X_i X_j$ ($1 \le i < j \le n-r$), we obtain the following equation over $\mathbb{F}$,

$$1 + \sum_{m=1}^{r} \ell_{mi} + \sum_{m=1}^{r} \ell_{mj} + \sum_{1 \le m < m' \le r} (\ell_{mi} \ell_{m'j} + \ell_{m'i} \ell_{mj}) = 0. \tag{7}$$

For a positive integer m, let $\mathbb{1}_m$ be the all 1's column vector and $\mathbf{0}_m$ be the all 0's column vector of dimension m. Let U_m be the $m \times m$ matrix with 1's above the diagonal and zero elsewhere. Let J_m be the $m \times m$ matrix with all 1's, and let I_m be the $m \times m$ identity matrix. Using this notation, we can rewrite (6) and (7) as follows:

$$\mathbb{1}_r^t Y_j + Y_j^t U_r Y_j = 0, \quad \text{for } 1 \le j \le n - r; \tag{8}$$

$$1 + \mathbb{1}_r^t Y_i + \mathbb{1}_r^t Y_j + Y_i^t (J_r - I_r) Y_j = 0, \quad \text{for } 1 \le i < j \le n - r. \tag{9}$$

If the characteristic of the field is not two, we may rewrite (8) as

$$2\mathbb{1}_r^t Y_j + Y_j^t (J_r - I_r) Y_j = 0, \quad \text{for } 1 \le j \le n - r; \tag{10}$$

With this, we are now ready to prove lower bounds. We will exploit (8), (9) and (if the characteristic is not 2) (10) to derive lower bounds for various fields.

4.2   Lower Bounds for GF(2)

Let $\mathbb{Z}$ stand for the integers. For $Y \in \mathbb{Z}^r$, let $|Y|$ be the number of odd components in Y. For $Y, Y' \in \mathbb{Z}^r$, let $Y \cdot Y' \triangleq \sum_{m=1}^{r} Y_m Y_m'$ be the dot product of Y and Y' over $\mathbb{Z}$.

Lemma 6. *Suppose $\ell_1, \dots, \ell_r$ are homogeneous linear forms in the variables $X_1, \dots, X_{n-r}$ such that*

$$S_n^2(X_1, \dots, X_{n-r}, \ell_1, \dots, \ell_r) = 0$$

over $GF(2)$. Then $r \ge \lfloor \frac{n}{2} \rfloor$. If $n \equiv 3 \bmod 4$, then $r \ge \lceil \frac{n}{2} \rceil$.

Proof. We use the arguments of Sect. 4.1. If there exist homogeneous linear forms $\ell_1, \dots, \ell_r$ over variables $X_1, \dots, X_{n-r}$ so that $S_n^2(X_1, \dots, X_{n-r}, \ell_1, \dots, \ell_r) = 0$ over GF(2), we have (8) and (9). We treat the vectors Y_j as elements of $\mathbb{Z}^r$ and the equalities as equalities over integers (mod 2). By counting the number of odd components (i.e. 1's) on the left and right hand side of (8), we obtain, for $1 \le j \le n - r$,

$$|Y_j| + \binom{|Y_j|}{2} \equiv 0 \pmod 2,$$

from which it follows that

$$|Y_j| \equiv 0 \text{ or } 3 \pmod 4. \tag{11}$$

Since $Y_i^t (J_r - I_r) Y_j = |Y_i| \, |Y_j| - Y_i \cdot Y_j$ over $\mathbb{Z}$, we conclude from (9) that, for $1 \le i < j \le n - r$,

$$|Y_i| + |Y_j| + |Y_i| \, |Y_j| + Y_i \cdot Y_j \equiv 1 \pmod 2,$$

that is,

$$Y_i \cdot Y_j \equiv (1 + |Y_i|)(1 + |Y_j|) \pmod 2. \tag{12}$$

Let $W_1, \dots, W_s$ be the vectors among $Y_1, \dots, Y_{n-r}$ with $|Y_j|$ odd, and let $E_1, \dots, E_t$ be the remaining $t = n - r - s$ vectors, with $|Y_j|$ even.

Claim. If $Y_1, Y_2, \dots, Y_{n-r}$ are not linearly independent over GF(2), then t is odd and the only dependency over GF(2) among them is $\sum_{k=1}^{t} E_j = \mathbf{0}_r$.

Proof. The proof appears in the full version of the paper [11].   □

By the claim above, we see that there are at least $n-r-1$ linearly independent vectors over GF(2) among the Y_j's. Since $Y_j \in \mathbb{Z}^r$, we get $r \geq n - r - 1$ i.e. $r \geq \lfloor \frac{n}{2} \rfloor$.

To obtain a better bound for $n \equiv 3 \mod 4$, we make better use of our equations. So suppose $n = 2r + 1$ and $n \equiv 3 \mod 4$. We shall derive a contradiction.

If $n = 2r + 1$, then $n - r > r$, and since $Y_j \in \mathbb{Z}^r$, the vectors Y_j are not linearly independent over GF(2). Then by the claim above, t is odd, $\sum_{k=1}^{t} E_j \equiv 0_r \mod 2$, and $W_1, \ldots, W_s, E_1, \ldots, E_{t-1}$ are linearly independent over GF(2). Since, $s + t - 1 = n - r - 1 = r$, these vectors span (over GF(2)) the entire vector space $GF(2)^r$; in particular, $\mathbb{1}_r$ is in their span:

$$\sum_{i=1}^{s} \alpha_i W_i + \sum_{k=1}^{t-1} \beta_k E_k \equiv \mathbb{1}_r \quad (\text{mod } 2)$$

for some $\alpha_i, \beta_k \in \mathbb{Z}$. Taking dot products with W_i and E_k, we conclude (using (12)) that $\alpha_i \equiv 1 \mod 2$ for $1 \leq i \leq s$, and $(J_{t-1} - I_{t-1})\beta \equiv 0_{t-1} \mod 2$, where $\beta \in \mathbb{Z}^{t-1}$ and the kth component of β is β_k. Since t is odd, $J_{t-1} - I_{t-1}$ is full rank over GF(2) and $\beta \equiv 0_{t-1} \mod 2$. Thus,

$$\sum_{i=1}^{s} W_i \equiv \mathbb{1}_r \quad (\text{mod } 2).$$

It is easy to verify that for all integer vectors Y,

$$|Y| \equiv Y \cdot Y \quad (\text{mod } 4) \tag{13}$$

Thus, $(\sum_{i=1}^{s} W_i) \cdot (\sum_{i=1}^{s} W_i) \equiv |\sum_{i=1}^{s} W_i| \equiv r \mod 4$, that is

$$\sum_{i=1}^{s} W_i \cdot W_i + 2 \sum_{1 \leq i < j \leq s} W_i \cdot W_j \equiv r \quad (\text{mod } 4).$$

By (11) and (13), $W_i \cdot W_i \equiv 3 \mod 4$, and by (12), $W_i \cdot W_j \equiv 0 \mod 2$ for $i \neq j$. Thus,

$$3s \equiv r \quad (\text{mod } 4). \tag{14}$$

Similarly, starting with $\sum_{k=1}^{t} E_j \equiv 0_r \mod 2$, we obtain $t(t-1) \equiv 0 \mod 4$; since t is odd, $t \equiv 1 \mod 4$. But then, using (14),

$$n \equiv r + s + t \equiv 3s + s + 1 \equiv 1 \quad (\text{mod } 4)$$

which is a contradiction.

Since $r \geq \lfloor \frac{n}{2} \rfloor$ holds for all n, we have shown that if $n \equiv 3 \mod 4$, then $r \geq \lceil \frac{n}{2} \rceil$. $\qquad \square$

Using lemma 5 and the above lemma, we can now prove the following theorem.

Theorem 4. *Any (inhomogeneous) $\Sigma\Pi\Sigma$ circuit computing $S_n^2(X_1, \ldots, X_n)$ over GF(2) requires at least $\lceil \frac{n}{2} \rceil$ multiplication gates if $n \equiv 0, 2, 3 \mod 4$ and at least $\lfloor \frac{n}{2} \rfloor$ multiplication gates if $n \equiv 1 \mod 4$.*

Acknowledgements

We thank Amir Shpilka for sending us the preliminary version of [12] and generously sharing his insights with us.

References

[1] N. Alon. Decomposition of the complete r-graph into complete r-partite r-graphs. *Graphs and Combinatorics*, 2:95–100, 1986.

[2] L. Babai and P. Frankl. Linear algebra methods in combinatorics (with applications to geometry and computer science). Preliminary Version 2, Department of Computer Science, The University of Chicago, September 1992.

[3] D. de Caen and D.G. Hoffman. Impossibility of decomposing the complete graph on n points into $n - 1$ isomorphic complete bipartite graphs. *SIAM Journal of Discrete Mathematics*, 2:48–50, 1989.

[4] R.L. Graham and H.O. Pollack. On embedding graphs in squashed cubes. In *Graph Theory and Applications*, pages 99–110. Springer-Verlag, 1972. Lecture Notes in Mathematics, 303.

[5] D. Grigoriev and M. Karpinski. An exponential lower bound for depth-3 arithmetic circuits. In *Proceedings of the 30th Annual ACM Symposium on Theory of Computing*, pages 577–582, 1998.

[6] D. Grigoriev and A.A. Razborov. Exponential lower bounds for depth-3 arithmetic circuits in algebras of functions over finite fields. In *Proceedings of the 39th Annual IEEE Symposium on Foundations of Computer Science*, pages 269–278, 1998.

[7] M. Hall Jr. *Combinatorial Theory*. Wiley Interscience series in Discrete Mathematics, 1986.

[8] N. Nisan. Lower bounds for non-commutative computation. In *Proceedings of the 23rd Annual ACM Symposium on Theory of Computing*, pages 410–418, 1991.

[9] N. Nisan and A. Wigderson. Lower bounds on arithmetic circuits via partial derivatives. *Computational Complexity*, 6:217–234, 1996.

[10] G.W. Peck. A new proof of a theorem of Graham and Pollack. *Discrete Mathematics*, 49:327–328, 1984.

[11] J. Radhakrishnan, P. Sen, and S. Vishwanathan. Depth-3 arithmetic circuits for $S_n^2(X)$ and extensions of the Graham-Pollack theorem. Full version. Manuscript available at http://www.tcs.tifr.res.in/~pranab/papers/s2n.ps, September 2000.

[12] A. Shpilka. Symmetric computation. Manuscript. Personal Communication, January 2000.

[13] A. Shpilka and A. Wigderson. Depth-3 arithmetic formulae over fields of characteristic zero. In *Proceedings of the 14th Annual IEEE Conference on Computational Complexity*, pages 87–96, 1999. Also ECCC report no. 23, 1999, available at http://www.eccc.uni-trier.de/eccc.

[14] H. Tverberg. On the decomposition of K_n into complete bipartite graphs. *Journal of Graph Theory*, 6:493–494, 1982.

The Bounded Weak Monadic Quantifier Alternation Hierarchy of Equational Graphs Is Infinite

Olivier Ly

LaBRI, Université Bordeaux I
ly@labri.u-bordeaux.fr

Abstract. Here we deal with the question of definability of infinite graphs up to isomorphism by weak monadic second-order formulæ. In this respect, we prove that the quantifier alternation bounded hierarchy of equational graphs is infinite. Two proofs are given: the first one is based on the Ehrenfeucht-Fraissé games; the second one uses the arithmetical hierarchy. Next, we give a new proof of the Thomas's result according to which the bounded hierarchy of the weak monadic second-order logic of the complete binary tree is infinite.

Introduction

Logic is by now a classical mean in theoretical computer science to describe complexity issues; it has been used in many areas, for instance effective computability (cf. [20]), descriptive complexity (cf. [7,15]), or else formal language theory (cf. [16]).

This paper deals with the question of definability of infinite graphs up to isomorphism by logical formulæ. The graphs which are studied here are *equational graphs* which have been introduced in [4] as models of program schemes. Such a graph is the inductive limit of a sequence of finite graphs generated by a deterministic hyperedge replacement grammar (cf. [5,21]). They extend strictly *context-free graphs* which have been introduced in [14] as configuration graphs of pushdown automata. Note that these kinds of graphs generalize the concept of *regular trees*; such a tree is defined as the tree of all the runs of a finite state automaton (see [3]).

We deal with *monadic second-order formulæ* on graphs (MS-formulæ for short) which are the logical formulæ which deal with graphs as relational logical structures using individual and set variables ranging over vertices and edges. We consider the *weak monadic second-order logic* (WMS logic for short) which consists in interpreting the MS-formulæ by considering set variables as ranging over *finite* sets of vertices and edges. WMS logic is a classical extention of the first-order logic. It is a variation of the well-known *monadic second-order logic* (MS-logic for short) (cf. [21,8]). The monadic second-order logic is related to the concept of equational graphs because of the two following results: first, one can decide in an effective way whether a given equational graph satisfies a given

S. Kapoor and S. Prasad (Eds.): FST TCS 2000, LNCS 1974, pp. 188–200, 2000.

MS-formula according to MS-logic, which is also true for WMS-logic (cf. [4], for generalizations cf. [21,2]). Second, the equational graphs are exactly the graphs of bounded tree-width (cf. [17]) which are definable up to isomorphism by MS-formulæ (cf. [5]) according to MS-logic. Note that these results generalise the fundamental results of [1] and [18,19] where MS and WMS were studied in the contexts of infinite words and infinite trees.

The present work is motivated by the conjecture of [5] according to which any equational graph is definable up to isomorphism by a formula of WMS-logic. This is true if one considers MS-logic (cf. [5]); let us note that this implies that the isomorphism problem for equational graphs is decidable. Concerning WMS-logic, some steps have been ever raised: the conjecture has been firstly proved to be true for context-free graphs (cf. [22]), and then for the equational graphs which have covering trees of finite degree (cf. [23]).

Here, we investigate equational WMS-definable graphs through the study of the *quantifier alternation bounded hierarchy*. A graph is said to be in the n-th level of the bounded WMS-hierarchy if there exists a WMS-formula which defines it up to isomorphism and which has $n-1$ alternations of existential and universal unbounded quantifiers. One says that the hierarchy is infinite if for each integer n, the n-th level is *strictly* included in the $n+1$-th one. The main result of this paper is the following:

Theorem 1. *The bounded WMS-hierarchy of equational graphs is infinite.*

In order to see how this theorem fits in existing works, let us now mention some results about hierarchies relating to monadic second-order logical systems. Firstly, the bounded MS and WMS-hierachies of languages of infinite words, i.e. the one successor theory, are finite, which follows from [1] and [12]. Next, the bounded MS-hierarchy of languages of infinite binary trees, i.e. the two successors theory (MS2S for short) is also finite, which follows from Rabin's Theorem (cf. [18], see also e.g. [9] for further results), while the weak one (the bounded WMS2S-hierarchy) is infinite (cf. [24], see also [13]). Concerning definability of graphs up to isomorphism, it follows from the results of [5] that the bounded MS-hierarchy of equational graphs is finite. Theorem 1 shows that the situation is different when considering WMS-logic, even though it follows from [22] that the bounded WMS-hierarchy of context-free graphs is finite and stops at most at the fifth level.

As we mentioned above the best result concerning the conjecture of WMS-definability of equational graphs has been raised in [23]. The WMS-formulæ which are constructed in this work have unbounded numbers of quantifier alternations; Theorem 1 shows that one can not get away from this fact: equational graphs are hard to define up to isomorphism by using weak monadic second-order formulæ.

We give two proofs of Theorem 1. The first one is based on an extention to WMS-logic of the classical technique of Ehrenfeucht-Fraissé games [8,10]). The method is similar to the well-known construction used to show that $FO(LFP)$ is strictly more expressive than $FO(TC)$ (see e.g. [6]). The second one is based on infinity of arithmetical hierarchy (see [20]). It allows us to recover the fact that WMS2S-hierarchy is infinite, which is deserved in [24] where it was proved using arithmetical hierarchy as well together with Rabin's Theorem (cf. [18]). The first proof is in some sense stronger than the second one because the graphs which are constructed in it have covering trees of finite degree, which is not true for the second proof. But in other respects, this last one allows us to recover the result of [24] without using Rabin's Theorem, which does not seem to be possible with the game-based proof.

Acknowledgements. The author is greatly indebted to his supervisor G. Sénizergues for having spent much time in very helpful discussions. He also wants to thank the reviewers for their work which led to important improvements of this paper.

1   Preliminary

For an introduction to monadic second-order logical systems, the reader is refered to [6,8], cf. also [21, chap. 5].

We deal with *labelled directed multi-graphs* (graphs for short), i.e. tuple $\langle V, E, vert, A, lab \rangle$ where V is an at most countable set whose elements are the vertices; E is the set of edges; $vert : E \to V \times V$ is the map defining the target and the origin of each edge. Vertices and edges are labelled by elements of A according to the map $lab : V \cup E \to A$. Such a graph is represented by a *logical relational structure* $\langle D, (D_a)_{a \in A}, \text{inc} \rangle$ where $D = V \cup E$; for all $a \in A$, $D_a \subset D$ is a unary relation on D defining elements labelled by a and inc is a ternary relation defining incidence in the sense that $(x, y, z) \in$ inc if and only if $x \in E$ is an edge of origin $y \in V$ and target $z \in V$. In order to simplify notations, the relational structure associated to a graph G is still denoted by G. We consider *monadic second-order formulæ* on graphs. These formulæ are constructed using individual variables (usually denoted by latin letters x, y, z,...) and set variables (usually denoted by greek letters α, β, ...). Atomic formulæ are of the forms $x \in \alpha$ or $\alpha \subset \beta$ or $\text{inc}(x, y, z)$ or else $x \in D_a$ with $a \in A$. Syntax is not restricted: we allows existential and universal quantifiers over individual and set variables, conjunctions and negations. The *semantics* of such formulæ differs in the *monadic second-order logic* and in the *weak monadic second-order logic*: in the former one set variables range over all the subsets of the domain, while in the last one they range over finite subsets only.

Remark 1. What we call here MS-logic of graphs is usually denoted by MS_2-logic in order to distinguish it from MS_1-logic in which quantifications are done over vertex sets only (cf. e.g. [21]). MS_1-logic deals with simple graphs; such a graph is encoded by a relational structure whose domain is its set of vertices only, instead of the union of its set of vertices and its set of edges as we consider here. MS_2 is more expressive than MS_1. This is why we have chosen to prove Theorem 1 for MS_2. As we shall see, it is true for MS_1 as well (see Remark 3).

A formula is said to be in *prenex form* if it is of the form $Q_1 X_1 ... Q_n X_n \, \psi$ where for each i, $Q_i \in \{\exists, \forall\}$, the X_i's are variables and ψ does not contain any quantifier. Any formula can be put in prenex form in an effective way (cf. e.g. [8]).

Let α and β be some set variables; let $\varphi(\alpha, \beta)$ be a formula with α and β as free variables. The quantifier over α in the formula $\exists \alpha\, \varphi(\alpha, \beta)$ ($\forall \alpha\, \varphi(\alpha, \beta)$ respectively) is said to be *bounded* if this last formula is equivalent to $\exists \alpha\, (\alpha \subset \beta \wedge \varphi(\alpha, \beta))$ ($\forall \alpha\, (\alpha \subset \beta \Rightarrow \varphi(\alpha, \beta))$ respectively). In this case it is denoted by $\exists \alpha \subset \beta\, \varphi(\alpha)$ ($\forall \alpha \subset \beta\, \varphi(\alpha)$ respectively). A formula in which all the quantifiers are bounded is said to be bounded. As stated by next Lemma, bounded quantifications can always be put after the unbounded ones (cf. [13] for a proof).

Lemma 1. *Let $\psi(\alpha, \beta, \gamma)$ be a formula, then we have:*
- $\exists \alpha \subset \beta\, \forall \gamma\, \psi(\alpha, \beta, \gamma) \equiv \forall \gamma'\, \exists \alpha \subset \beta\, \forall \gamma \subset \gamma'\, \psi(\alpha, \beta, \gamma)$
- $\exists \alpha \subset \beta\, \exists \gamma\, \psi(\alpha, \beta, \gamma) \equiv \exists \gamma\, \exists \alpha \subset \beta\, \psi(\alpha, \beta, \gamma)$
- $\forall \alpha \subset \beta\, \forall \gamma\, \psi(\alpha, \beta, \gamma) \equiv \forall \gamma\, \forall \alpha \subset \beta\, \psi(\alpha, \beta, \gamma)$
- $\forall \alpha \subset \beta\, \exists \gamma\, \psi(\alpha, \beta, \gamma) \equiv \exists \gamma'\, \forall \alpha \subset \beta\, \exists \gamma \subset \gamma'\, \psi(\alpha, \beta, \gamma)$

A formula is called a Σ_n-formula (a Π_n-formula respectively) if it has the form $\exists X_1\, \forall X_2\, ... \, \substack{\exists \\ \forall} X_n\, \psi(X_1, X_2, ..., X_n)$ ($\forall X_1\, \exists X_2\, ... \, \substack{\exists \\ \forall} X_n\, \psi(X_1, X_2, ..., X_n)$ respectively) where ψ is bounded. The above lemma implies that any formula is equivalent to a Σ_n-formula (a Π_n-formula respectively) for a suitable n.

This classification of formulæ provides a classification of definable graph properties, i.e. the properties which can be expressed by logical formulæ. What we call here a graph property is formally a class of graphs, for instance the class of all the connected graphs. Let $\mathcal{G}^{\Sigma_n}$ ($\mathcal{G}^{\Pi_n}$ respectively) denote the set of families of graphs which are WMS-definable by some Σ_n-formulæ (by Π_n-formulæ respectively). Note that for all $n \geq 0 : \mathcal{G}^{\Sigma_n} \cup \mathcal{G}^{\Pi_n} \subset \mathcal{G}^{\Sigma_{n+1}} \cap \mathcal{G}^{\Pi_{n+1}}$. This classification of WMS-definable graph families, i.e. the sequence $\mathcal{G}^{\Sigma_n}$, is called the *bounded weak monadic quantifier alternation hierarchy*.

2   Ehrenfeucht-Fraïssé Games

2.1   Ehrenfeucht-Fraïssé Games for Weak Monadic Second-Order Logic

Let $G = \langle V_G, E_G, vert_G, A, lab_G \rangle$ and $G' = \langle V_{G'}, E_{G'}, vert_{G'}, A, lab_{G'} \rangle$ be two A-labelled graphs. Let $\bar{\alpha}$ and $\bar{\alpha}'$ be some finite parts of G and G' respectively, i.e. finite subsets of vertices and edges. We call *partial isomorphism* between $\bar{\alpha}$ and $\bar{\alpha}'$ a one-to-one mapping $\sigma : \bar{\alpha} \to \bar{\alpha}'$ which preserves the adjacency, i.e $\forall\, a, x, y \in \bar{\alpha} : vert_G(a) = (x, y)$ if and only if $vert_{G'}(\sigma(a)) = (\sigma(x), \sigma(y))$, and the labels, i.e. $\forall\, a \in \bar{\alpha} : lab(a) = lab(\sigma(a))$. We shall identify such a mapping with its graph, i.e. the subset of $(V_G \cup E_G) \times (V_{G'} \cup E_{G'})$ which encodes it. If σ and σ' are two partial isomorphisms, we say that σ *extends* σ' if $\sigma' \subset \sigma$ as subsets of $(V_G \cup E_G) \times (V_{G'} \cup E_{G'})$. The set of partial isomorphisms between two finite parts $\bar{\alpha}$ and $\bar{\alpha}'$ is denoted by PartIsom$(\bar{\alpha}, \bar{\alpha}')$.

Let A and B be two players, a session in the *Ehrenfeucht-Fraïssé game* associated to the pair of graphs (G, G') goes on as follows : at the first round, A chooses a finite part $\bar{\alpha}_1$ of G, and then B replies with a finite part $\bar{\alpha}'_1$ of G' together with a partial isomorphism $\sigma_1 \in$ PartIsom$(\bar{\alpha}_1, \bar{\alpha}'_1)$. At the second round,

A chooses a finite part $\bar{\alpha}_2'$ of G' and B replies with a finite part $\bar{\alpha}_2$ of G together with a partial isomorphism $\sigma_2 \in \text{PartIsom}(\bar{\alpha}_1 \cup \bar{\alpha}_2, \bar{\alpha}_1' \cup \bar{\alpha}_2')$ which extends σ_1; and so on, A chooses finite parts alternatively in G and G' and B extends the isomorphism as A goes along. The game stops when B can not find a good answer. We say that B has a strategy of order n if he can play the first n rounds whatever choices A makes.

The following results show the classical link between this game view point and logic.

Lemma 2. *Let G and G' be two graphs and let $\psi(\alpha_1, .., \alpha_n)$ be a bounded formula. Let $\bar{\alpha}_1, ..., \bar{\alpha}_n$ ($\bar{\alpha}_1', ..., \bar{\alpha}_n'$ respectively) be some finite parts of G (G' respectively) and let $\sigma \in \text{PartIsom}(\bigcup \bar{\alpha}_i, \bigcup \bar{\alpha}_i')$ be a partial isomorphism which exchanges $\bar{\alpha}_i$ and $\bar{\alpha}_i'$ for all i, then we have $(G, \bar{\alpha}_1, .., \bar{\alpha}_n) \models \psi(\alpha_1, ..., \alpha_n)$ if and only if $(G', \bar{\alpha}_1', ..., \bar{\alpha}_n') \models \psi(\alpha_1, ..., \alpha_n)$*

Lemma 3. *Let G and G' be two graphs such that B has a strategy of order n in the Ehrenfeucht-Fraïssé game associated to (G, G'); then for all Σ_n-formula φ, $G \models \varphi$ implies that $G' \models \varphi$.*

2.2   First Proof of Theorem 1

This section is devoted to the proof of Theorem 1.

We begin by defining two sequences of graphs $(G_n)_{n \geq 1}$ and $(G_n')_{n \geq 1}$ such that for all $n \geq 1$, G_n and G_n' are not isomorphic and B has a strategy of order n in the Ehrenfeucht-Fraïssé game associated to (G_n, G_n').

Let us set $V = (\mathbb{Z})^*$, i.e. the set of finite sequences of integers, and $E = (\mathbb{Z})^+ \times \{r, t\}$, where $(\mathbb{Z})^+$ denotes the set of finite non empty sequences of integers, r and t are two symbols (r like *r*adial and t like *t*ransversal); let us consider the mapping $vert : E \rightarrow V \times V$ defined as following:

- $\forall e = ((x_1, .., x_l), r) \in E,\ vert(z) = ((x_1, .., x_{l-1}), (x_1, .., x_{l-1}, x_l)),$
- $\forall e = ((x_1, .., x_l), t) \in E,\ vert(z) = ((x_1, .., x_l), (x_1, .., x_l + 1)).$

Let A be a set; let $\mathcal{Z}(A)$ denote the set of A-labelled graphs whose vertex set, edge set and edge mapping respectively are V, E and $vert$. Let $G \in \mathcal{Z}(A)$ and $z_0 \in V$, we denote by $G(z_0)$ the graph of $\mathcal{Z}(A)$ defined by : $\forall z \in V \cup E$: $lab_{G(z_0)}(z) = lab_G(z_0.z)$, where $z_0.z$ denotes the concatenation of z_0 and z if $z \in V$, and by abuse of notation $(z_0.u, r)$ or $(z_0.u, t)$ if $z = (u, r) \in E$ or $z = (u, t) \in E$ respectively.

G_n and G_n' are now defined as elements of $\mathcal{Z}(\{0, 1\})$. First, all edges are labelled by 0: for all $n \geq 1$ and for all $e \in E$, let $lab_{G_n}(e) = lab_{G_n'}(e) = 0$. The labels of vertices are defined inductively as following:

- For all $z \in V$, let $lab_{G_1}(z) = 0$
- For all $z \in V$, $lab_{G_1'}(z) = \begin{cases} 1 \text{ if } z = (0), \\ 0 \text{ otherwise.} \end{cases}$

- For $n \geq 2$, G_n is defined from G'_{n-1} as following: $lab_{G_n}(()) = 0$ and $\forall z \in V$ such that $|z| = 1 : G_n(z) = G'_{n-1}$.
- For $n \geq 2$, G'_n is defined from G_{n-1} and G'_{n-1} as following: $lab_{G_n}(()) = 0$, $\forall z \in V \setminus \{(0)\}$ such that $|z| = 1 : G_n(z) = G'_{n-1}$ and $G_n((0)) = G_{n-1}$.

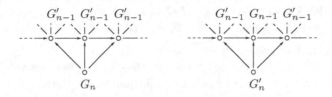

Lemma 4. *For all $n \geq 1$, B has a strategy of order n in the Ehrenfeucht-Fraïssé game associated to (G_n, G'_n).*

Sketch of proof : (induction on n)

• *Case $n = 1$:* Let $\bar{\alpha}_1 \subset V_{G_1} \cup E_{G_1}$ be the choice of A in the first round of the game. All the elements of $\bar{\alpha}_1$ are labelled by 0 and the unique element of $V_{G'_1} \cup E_{G'_1}$ labelled by 1 is $(0) \in V_{G'_1}$; so B performs a shifting on the left of $m = \min\{x_1 \in \mathbb{Z} \mid \exists x_2, .., x_l \text{ such that } (x_1, .., x_l) \in \bar{\alpha}\}$ first level vertices; more precisely he uses the one-to-one mapping $\lambda_m : V \rightarrow V$ defined by $\lambda_m((x_1, .., x_l)) = (x_1 - m + 1, .., x_l)$ which extends in a natural way to $V \cup E$ to define an automorphism of $(V, E, vert)$ which shall be still denoted by λ_m; B then chooses $\bar{\alpha}'_1 - \lambda(\bar{\alpha}_1)$ and $\sigma_1 = \lambda_m^{-1}|_{\bar{\alpha}_1}$.

• *Case $n > 1$:* Suppose now that B has a strategy ζ of order $n - 1$ relatively to (G_{n-1}, G'_{n-1}); we will then define a strategy of order n relatively to (G_n, G'_n). The first round proceeds exactly like in the case $n = 1$; if $\bar{\alpha}_1 \subset V_{G_n} \cup E_{G_n}$ is the first choice of A , we define m, λ_m, $\bar{\alpha}'_1$ and σ_1 as above.

Let us consider $\bar{\alpha}'_2 \subset V_{G'_n} \cup E_{G'_n}$ the second choice of A which we divide into two parts $\bar{\alpha}_2^{,1} = \{(x_1, .., x_l) \in \bar{\alpha}'_2 \mid x_1 \neq 0\}$ and $\bar{\alpha}_2^{,2} = \{(x_1, .., x_l) \in \bar{\alpha}'_2 \mid x_1 = 0\}$. For $\bar{\alpha}_2^{,1}$ B uses the above shifting λ_m : let $\bar{\alpha}_2^1 = \lambda_m^{-1}(\bar{\alpha}_2^{,1})$ and $\sigma_2^1 = \lambda_m^{-1}|_{\bar{\alpha}_2^{,1}} \cup \sigma_1$. For $\bar{\alpha}_2^{,2}$, let us remark that $G'_n((0)) = G_{n-1}$ and $G_n((m-1)) = G'_{n-1}$, so $\bar{\alpha}_2^{,2}$ induces a finite part of G_{n-1} which can be considered as the first choice of A in the game associated to (G_{n-1}, G'_{n-1}) then, ζ gives a answer, i.e. a finite part $\tilde{\alpha}_2^2 \subset G'_{n-1}$ and a partial isomorphism which induce a finite part $\bar{\alpha}_2^2$ of the subgraph of G_n of root $(m - 1)$ (the subset of G_n of words of which $(m - 1)$ is a prefix) and a partial isomorphism $\sigma_2^2 \in \text{PartIsom}(\bar{\alpha}_2^2, \bar{\alpha}_2^{,2})$.

Finally $\bar{\alpha}_2^1 \cup \bar{\alpha}_2^2$ and $\sigma_2^1 \cup \sigma_2^2$ is a correct answer of B .

For the succeeding rounds, B uses in the same way the strategy ζ in the game relative to the pair $(G'_n((0)), G_n((m-1)))$ and λ_m in the rest. Since ζ the strategy of B of order $n - 1$, intervenes in the second and later rounds, the above method gives a strategy of order n. $\square$

Since G_n and G'_n are not isomorphic, the preceding lemma together with Lemma 3 proves the next result:

Lemma 5. *For all $n \geq 1$, G_n is not definable up to isomorphism by a Σ_n-formula.*

We have now to see that G_n is indeed WMS-definable up to isomorphism, which is stated by next Lemma.

Lemma 6. *There exists a MS-formula Φ_n such that for any graph G, $G \models \Phi_n$ according to WMS-logic if and only if G is isomorphic to G_n; moreover, there exists $N \geq 1$ such that for all $n \geq N$, Φ_n can be constructed to be Σ_{n+1}.*

Because of the lack of space, Proof is omited (cf. [11]).

Hence, for all $n \geq N$, the isomorphism class of G_n belongs to $\mathcal{G}^{\Sigma_{n+1}} \backslash \mathcal{G}^{\Sigma_n}$ which proves that the bounded weak monadic quantifier alternation hierarchy is infinite. To conclude the proof of Theorem 1, it remains to show that the bounded hierarchy is still infinite when restricted to isomorphism classes of equational graphs. This is true seeing that the G_n's actually are equational. Equational graphs can be seen as canonical solutions of systems of graph equations (cf. [5]). The lack of space makes impossible to construct in details such systems for the G_n's. Nevertheless, we give the main ideas: first, we have to note that the graph which is made of one vertex of infinite degree which is connected to all the vertices of an infinite linear graph is equational; let us denote it by G. Then, the equations defining G_1 say that it is obtained by gluing one copy of itself on each vertex of G except the root, i.e. the vertex of G of infinite degree. On the other hand, G'_1 is obviously equational as it is obtained from G_1 by modifying the label of only one vertex. Then, the equations defining G_n and G'_n are constructed by induction from those defining G and those defining G_{n-1} and G'_{n-1} by following the definition scheme given above. For instance, G_n is obtained by gluing a copy of G'_{n-1} on each vertex of G except the root (cf. [11] for more details).

Let us note that instead of constructing directly such systems of graph equations, one can notice that the G_n's are of bounded tree width (cf. [17]) and WMS-definable, as we saw in Lemma 6. In view of the results of [5], this implies that they are equational.

Remark 2. Transversal edges of G_n are useless in the proof of Theorem 1. However, they give the existence of a covering tree of finite degree, which, to some extend, is meaningful because of the following: the equational graphs with covering trees of finite degree have been proved to be WMS-definable up to isomorphism (cf. [23]); but the numbers of quantifier alternations of the formulæ which are constructed in this proof are not bounded. We have thus shown that one can not get away from this fact.

Remark 3. Let us note that MS$_1$-logic (see Remark 1) gives rise to another bounded quantifier alternation hierarchy. In this respect, it turns out that our construction also shows that this hierarchy is also infinite. First, one can verify that a MS$_1$-formula can be translated into an MS$_2$-formula, adding at most one quantifier alternation. Therefore, in view of Lemma 5, G_n can not be defined up to isomorphism by a MS$_1$-formula with less than $n - 1$ quantifier alternation. Second, one shows that G_n is MS$_1$-definable.

Remark 4. The bounded weak monadic quantifier alternation hierarchy differs from the *weak monadic quantifier alternation hierarchy* (weak monadic hierarchy for short) which is defined in an analogous way by considering *first-order formulæ* instead of bounded formulæ. In this respect, one shows that there exists a fixed integer k such that for all n, the isomorphism class of G_n belongs to the k-th level of the weak monadic hierarchy. Indeed, the formula Φ_n given in Lemma 6 actually consists in the conjunction of a fixed weak monadic formula Φ which defines the family $\mathcal{Z}(\{0,1\})$, and a first-order formula φ_n which checks the labels of G_n. So, the level of Φ_n in the weak monadic hierarchy is the one of Φ, which is fixed. We hence obtain that the k-th level of the weak monadic hierarchy contains instances beyond any given level of the bounded weak monadic hierarchy.

3 Arithmetical Hierarchy and Graph Hierarchy

3.1 Arithmetical Hierarchy

For basics about effective computability, the reader is refered to [20].

Let $\tau^* : (\mathbb{N})^* \to \mathbb{N}$ be a Gödel numbering of finite integer sequences, i.e. a bi-recursive one-one mapping (cf. [20, p. 70] for such a construction); as usually, $\tau^*((x_1, ..., x_k))$ shall be sometimes denoted by $< x_1, ..., x_k >$.

Let $(M_i^X)_{i \geq 0}$ be a Gödel numbering of *the set oracle Turing machines*. For any $A \subset \mathbb{N}$ and $i \geq 0$, let $f_i^A : W_i^A \subset \mathbb{N} \to \{0,1\}$ be the partial function computed by M_i^X with A as oracle, where $W_i^A = \{x \in \mathbb{N} \mid M_i^X$ stops on the instance x using A as oracle $\}$ and $\forall x \in W_i^A$, $f_i^A(x) = 0$ if and only if M_i^X give 0 on the instance x using A as oracle. Classical *Turing machines* are identified with oracle machines with $\emptyset$ as oracle. For all k, $f_k^\emptyset$ and $W_k^\emptyset$ shall be denoted by f_k and W_k respectively; f_k is called the k-th *recursive partial function*. Let *Rec* denote the set of all the recursive functions.

Let Σ_n^{arith} be the set of subsets of $\mathbb{N}$ of the form $\{k \mid \exists k_1 \forall k_2 ... \overset{\exists}{\forall} k_n \, f_{i_0}(< k_1, k_2, ..., k_n, k >) = 1\}$ for any fixed i_0. We also consider Π_n^{arith} which denotes the set of subsets of $\mathbb{N}$ of the form $\{k \mid \forall k_1 \exists k_2 ... \overset{\forall}{\exists} k_n \, f_{i_0}(< k_1, k_2, ..., k_n, k >) = 1\}$ for any fixed i_0. $\bigcup_n \Sigma_n^{arith}$ is called the *arithmetical hierarchy*.

Let us consider the *jump operation* which associates to any subset $A \subset \mathbb{N}$ the set $A' = \{x \in \mathbb{N} \mid x \in W_x^A\}$. For all $n \geq 1$, let $A^n = (A^{n-1})'$ where $A^0 = A$. We then consider the sequence $(\emptyset^n)_{n \geq 0}$, which is called the sequence of *jumps*.

Let A and B be two subsets of $\mathbb{N}$, let us recall that B is said to be *recursive in A* if and only if its characteristic function is equal to f_k^A for some k. B is said to be *recursively enumerable in A* if and only if there exists $k \in \mathbb{N}$ such that $B = W_k^A$.

The proofs of the two following results can be found in [20].

Lemma 7. *For all $n \geq 1$, $B \in \Sigma_n^{arith}$ if and only if B is recursively enumerable in $\emptyset^n$.*

The next result implies that the arithmetical hierarchy is infinite, i.e. for each n, $\Sigma_n^{arith} \subsetneq \Sigma_{n+1}^{arith}$.

Lemma 8. *For all $n \geq 0$, $\emptyset^{n+1}$ is not recursive in $\emptyset^n$.*

3.2   Second Proof of Theorem 1

The first part of this second proof consists in constructing a sequence $(t_n)_{n \geq 1}$ of recursive trees such that $\emptyset^n$ is reducible to the problem of being isomorphic to t_n. The second argument is that the problem of determining whether a recursive tree satisfies a Σ_n-formula according to weak monadic second-order logic is recursive in $\emptyset^n$ (see Lemma 10 bellow). Seeing that t_n is constructed in order to be equational and WMS-definable up to isomorphism, Theorem 1 follows.

Let us make precise what we shall call a *recursive tree*. Let $\mathcal{T}^\infty$ be the set of trees of the form $T = \langle V_T, E_T, Edg_T, \{0,1\}, lab_T \rangle$ where

- $V_T \subset (\mathbb{N})^*$ is prefix closed, i.e. $x \in V_T$ and $y <_{pref} x$ implies $y \in V_T$;
- E_T is a copy of $V_T \backslash \{()\}$;
- For $x = (x_1, ..., x_n) \in E_T : Edg_T(x) = ((x_1, ..., x_{n-1}), (x_1, ..., x_n))$;
- $lab_T : V_T \cup E_T \to \{0,1\}$ and $lab_T|_{E_T} \equiv 0$.

For $T \in \mathcal{T}^\infty$ and $x_0 \in V_T$, let $T(x_0)$ be the *subtree* of T of root x_0: $T(x_0) \in \mathcal{T}^\infty$, $V_{T(x_0)} = \{x \in (\mathbb{N})^* \, | \, x_0.x \in V_T\}$ and $\forall x \in V_{T(x_0)}$, $lab_{T(x_0)}(x) = lab_T(x_0.x)$. Let us consider for any partial function $f : \text{Dom}(f) \subset \mathbb{N} \to \{0,1\}$ the element $\text{Tr}(f)$ of $\mathcal{T}^\infty$ defined as following: $V_{\text{Tr}(f)}$ is the greater subset of $\text{Dom}(f \circ \tau^*) = \tau^{*-1}(\text{Dom}(f))$ which is prefix-closed; and for all $x \in V_{\text{Tr}(f)} : lab_{\text{Tr}(f)}(x) = f \circ \tau^*(x)$.

Definition 1 (Recursive Trees). *A **recursive tree** is an element of $\mathcal{T}^\infty$ of the form $\text{Tr}(f_i)$ for some i where f_i denotes the i-th recursive partial function. Let $T_i = \text{Tr}(f_i)$ denote the i-th recursive tree.*

For each tree $t \in \mathcal{T}^\infty$ let us consider the set of integer $\text{RecIsom}(t) = \{i \in \mathbb{N} \, | \, T_i$ is isomorphic to $t\}$.

Let us turn to the construction of $(t_n)_{n \geq 1}$. We need for that an auxilliary sequence $(t'_n)_{n \geq 1}$ of trees. First, for all $n \in \mathbb{N}$, $V_{t_n} = V_{t'_n} = (\mathbb{N})^*$; then

- $\forall x \in V_{t_1}$, $lab_{t_1}(x) = \begin{cases} 1 \text{ if } z = (i) \text{ for any even integer i} \\ 0 \text{ otherwise} \end{cases}$
- $lab_{t'_1} \equiv 0$
- For $n \geq 1$, t_{n+1} is defined by: $lab_{t_{n+1}}(()) = 0$ and $\forall x \in V_{t_{n+1}}$ such that $|x| = 1$: $t_{n+1}(x) = t_n$ if $x = (i)$ with i even and $t_{n+1}(x) = t'_n$ otherwise.
- For $n \geq 1$, t'_{n+1} is defined by: $lab_{t'_{n+1}}(()) = 0$ and $\forall x \in V_{t'_{n+1}}$ such that $|x| = 1 : t'_{n+1}(x) = t_n$.

Lemma 9. *For all $n \geq 1$, $\emptyset^n$ is recursive in $\text{RecIsom}(t_n)$.*

Proof. Let $B(n,i) \subset \mathbb{N}$ be the i-th set of the n-th level of the arithmetical hierarchy, i.e. $\{k \, | \, \exists k_1 \forall k_2 ... \overset{\forall}{\exists} k_n \; : \; f_i(< k_1, k_2, ..., k_n, k >) = 1\}$. Let us consider the following induction hypothesis :
HR_n: *there exists a computable recursive function $\tilde{\rho}_n$ such that $\forall i, k \in \mathbb{N} : \tilde{\rho}_n(< i, k >) \in \text{RecIsom}(t_n) \cup \text{RecIsom}(t'_n)$ and $\tilde{\rho}_n(< i, k >) \in \text{RecIsom}(t_n)$ iff $k \in B(n,i)$. In other words, for all n there is an algorithm which associates to any*

pair $i, k \in \mathbb{N}$ an algorithm which computes a tree which is isomorphic to t_n or t'_n and is isomorphic to t_n if and only if $k \in B(n, i)$.

It follows from Lemma 7 that there exists an index i_n such that $\emptyset^n = B(n, i_n)$. Therefore, HR_n implies that $\emptyset^n$ is Turing reducible to $\text{RecIsom}(t_n)$ by the function $\rho_n : k \mapsto \tilde{\rho}(< i_n, k >)$. This implies the lemma.

• Proof of HR_1: Here we describe the algorithm which computes $T_{\tilde{\rho}_1(<i,k>)}$:

Instance : $x \in (\mathbb{N})^*$
if $|x| \neq 1$ then $lab_T(x) = 0$
   else let $x_1, x_2 \in \mathbb{N}$ be such that $x = (< x_1, x_2 >)$
    if $x_1 = 0$ then $lab_T(x) = 0$

$$\text{else } lab_T(x) = \begin{cases} 1 \text{ if } M_i \text{ stops and not give } 0 \\ \quad \text{before the } x_2\text{th calculus steps} \\ \quad \text{on } < x_1 - 1, k > \\ 0 \text{ otherwise} \end{cases}$$

$\tilde{\rho}_1(< i, k >)$ is then defined to be the index of the tree which is described by this algorithm. Because of the lack of space, we shall omit the proof that $\tilde{\rho}_1$ indeed satisfies HR_1.

• Let us suppose that HR_n is true.
We begin the proof of the induction step with some preliminaries: Let $i, k \in \mathbb{N}$; we have $B(n + 1, i) = \{k \mid \exists k_1 \forall k_2 ... \overset{\vee}{\exists} k_{n+1} : f_i(< k_1, k_2, ..., k_{n+1}, k >) = 1\}$. Let us consider the integer set $\tilde{B}(n + 1, i) = \{< k, k_1 > \mid \forall k_2 ... \overset{\vee}{\exists} k_{n+1} : f_i(< k_1, k_2, ..., k_{n+1}, k >) = 1\}$. Note that $B(n + 1, i) = \{k \mid \exists k_1 :< k, k_1 >\in \tilde{B}(n + 1, i)\}$. This is the complement of an integer set $B(n, \delta(i, n))$ which belongs to the n-th level of the hierarchy; note that $\delta(i, n)$ is computable. Therefore, it follows from HR_n that $< k, k_1 >\in \tilde{B}(n + 1, i)$ if and only if $\tilde{\rho}_n(< \delta(i, n), < k, k_1 >>) \in \text{RecIsom}(t'_n)$.
Let us now describe the algorithm computing $T_{\tilde{\rho}_{n+1}(<i,k>)}$:

Instance : $x = (x_1, ..., x_l) \in (\mathbb{N})^*$
if $x = ()$ then $lab_{T_{\tilde{\rho}_{n+1}(<i,k>)}}(x) = 0$
Let $x_{11}, x_{12} \in \mathbb{N}$ be such that $x_1 =< x_{11}, x_{12} >$
if $x_{11} = 0$ then $lab_{T_{\tilde{\rho}_{n+1}(<i,k>)}}(x) = lab_{t_n}((x_2, .., x_l))$ (i)
   else $lab_{T_{\tilde{\rho}_{n+1}(<i,k>)}}(x) = lab_{T_{\tilde{\rho}_n(<\delta(i,n),<k,x_{11}-1>>)}}((x_2, .., x_l))$ (ii)

First, (i) guarantees that there are infinitely many sub-trees of level 1 $T_{\tilde{\rho}_{n+1}(<i,k>)}$ which are isomorphic to t_n.
Second, (ii) guarantees by induction that each sub-tree of level 1 is isomorphic to t_n or to t'_n. One verifies that if there is at least one sub-tree of level 1 which is isomorphic to t'_n then there are infinitely many one.

Now $T_{\tilde{\rho}_{n+1}(<i,k>)}$ is isomorphic to t_{n+1} if and only if at least one of its sub-trees of level 1 is isomorphic to t'_n. This is true if and only if there exists $x_{11} \neq 0$ such that $\tilde{\rho}_n(< \delta(i, n), < k, x_{11} - 1 >>) \in Isom(t'_n)$, which is equivalent, by using preliminaries, to $k \in B(n + 1, i)$.

Because of the lack of space, we state the next lemma without proof.

Lemma 10. *The problem of determining whether a recursive tree satisfies a Σ_n-formula according to weak monadic second-order logic is recursive in $\emptyset^n$.*

It follows from the previous lemma that whatever $n \geq 1$ is, there is no Σ_n-formula which defines t_{n+1} up to isomorphism. Indeed, the existence of such a formula would be a contradiction with Lemma 8. On the other hand, like in section 2.2, t_n is equational and it is WMS-definable. We shall omit to prove that it is WMS-definable. The construction of some systems of graph equations which define the t_n's can be performed by using the ideas of section 2.2. First, transversal edges are no longer considered here. And even if the definition schemes of the t_n's and the t'_n's are slightly different from the ones of the G_n's and the G'_n's, they mainly follow the same idea. And one can verify that the idea for the construction of the systems which define the G_n's applies here as well to construct some systems defining the t_n's. Theorem 1 then follows.

3.3   Thomas Theorem

Here we deal with *labelled complete binary trees*, i.e. mapping $t : \{l, r\}^* \to \{0, 1\}$; their set is denoted by $T_{inf}(\{0, 1\})$. In the context of the weak monadic second-order logic of the binary tree (WMS2S for short), the usual concept of bounded quantifier is different: following [24] and [13], bounded quantifiers are indeed those of the form $\exists \alpha \leq_{pref} \beta...$ or $\forall \alpha \leq_{pref} \beta...$ where $\leq_{pref}$ denote the prefix ordering of $\{l, r\}^*$. This defines an other concept of bounded hierarchy. However, one verifies that levels are the same.

In [24], it is proved that the WMS2S bounded hierarchy is infinite, i.e. Theorem 2 below; the proof involves infiniteness of arithmetical hierarchy together with Rabin's theorem (cf. [18]). By using the tools introduced in the preceding section, we give an alternative proof of this result which does not use Rabin's theorem.

Theorem 2 (Thomas 82). *The bounded WMS2S hierarchy is infinite.*

Proof. Let $\Lambda : \{l, r\}^* \to \mathbb{N}^*$ be the mapping defined as follows: for any $x_1, .., x_{l+1}$, $\Lambda(l^{x_1} r l^{x_2} r.. l^{x_l} r l^{x_{l+1}}) = (x_1, .., x_l)$. Λ allows us to consider a mapping from $T_{inf}(\{0, 1\})$ to $\mathcal{T}^\infty$, which shall be still denoted by Λ, defined as following: for $t \in T_{inf}(\{0, 1\})$, $V_{\Lambda(t)} = (\mathbb{N})^*$ and for any $x_1, .., x_l$, $lab_{\Lambda(t)}(x_1, .., x_l) = t(l^{x_1} r l^{x_2}.. l^{x_l} r)$. Λ contracts all the left edges of t. We also consider the partial converse Λ^{-1} defined on the set of trees of $\mathcal{T}^\infty$ whose domains are equal to $(\mathbb{N})^*$.

Let us set $\mathcal{T}_n = \{t \in T_{inf}(\{0, 1\}) \mid \Lambda(t) \text{ is isomorphic to } t_n\}$. We will see that the tree family $\mathcal{T}_n$ is definable, but not at a lower level than the n-th of the hierarchy. Let ρ_n be the Turing reduction of $\emptyset^n$ to $\text{RecIsom}(t_n)$ which has been constructed in the proof of Lemma 9; let us note that for each integer k, $V_{T_{\rho_n(k)}} = (\mathbb{N})^*$ and thus $\Lambda^{-1}(T_{\rho_n(k)})$ is defined. Now, for each integer k, we have:

$$k \in \emptyset^n \text{ if and only if } \Lambda^{-1}(T_{\rho_n(k)}) \in \mathcal{T}_n.$$

On the other hand, one can verify that the family $\mathcal{T}_n$ can be WMS2S-defined by a Σ_ℓ-formula φ_n for a suitable ℓ. Then $k \in \emptyset^n$ if and only if $\Lambda^{-1}(T_{\rho_n(k)}) \models \varphi_n$. By a result similar to Lemma 10, one verifies that this last predicate is recursive in $\emptyset^\ell$, which implies that $\ell \geq n$. Theorem 2 is proved.

References

1. J.R. Büchi, On a decision method in restricted second order arithmetic. Proc. Internat. Congr. on Logic, Methodology and Philosophy of Science, E. Nagel and al. eds. p1-11 (1960).
2. D. Caucal, On Infinite Transition Graphs having Decidable Monadic Theory, ICALP'96 - LNCS 1099:194-205 (1996).
3. B. Courcelle, Fundamental Properties of Infinite Trees, TCS 25:95-169 (1983).
4. B. Courcelle, The monadic second-order logic of graphs II : Infinite Graphs of Bounded Width, Math. System Theory 21:187-221 (1989).
5. B. Courcelle, The monadic second-order logic of graphs IV : Definability properties of Equational Graphs, Annals of Pure and Applied Logic 49:193-255 (1990).
6. Ebbinghaus H.-D and Flum J., Finite Model Theory, second edition, Springer Verlag (1999).
7. R. Fagin, Generalized first-order spectra and polynomial-time recognizable sets. In "Complexity of Computation", SIAM-AMS Proceedings 43-73 (1974).
8. Y. Gurevich, Monadic Second-Order Theories, in Model Theoretic Logic, Barwise and Ferferman eds. Springer (1985).
9. D. Janin and G. Lenzi, On the structure of the monadic logic of the binary tree. MFCS'99, LNCS 1672:310 320 (1999).
10. R. Lassaigne and M. de Rougemond, Logique et Complexité, Hermes - collection informatique (1996).
11. O. Ly, On Hierarchy of Graphs, internal report no 1178-97 LaBRI - 1997.
12. R. McNaughton, Testing and Generating Infinite Sequences by a Finite Automaton, Inf. Contr. 9:521-530 (1966)
13. A.W. Mostowski Hierarchies of Weak Monadic Formulas for Two Successors Arithmetic, J. Inf. Process. Cybern. EIK 23 10/11, 509-515 (1987).
14. D.E. Muller and P.E. Schupp, The Theory of Ends, Pushdown Automata, and Second-Order Logic, TCS 37: 51-75 (1985).
15. C.H. Papadimitriou, Computational Complexity, Addison-Wexley Pub. Comp. (1994).
16. D. Perrin and J. E. Pin, First-Order Logic and Star-Free Sets, J. Comp. Syst. Sci. 32:393-406 (1986).
17. N. Robertson and P. Seymour, Some new results on Well-Quasi Ordering of Graphs. Annals of Discrete Math., 23:343-354 (1984).
18. M.O. Rabin, Decidability of second order theories and automata on infinite trees, Trans. Amer. Math. Soc. 141 (1969).
19. M.O. Rabin, Weakly Definable Relations and Special Automata
20. H. Rogers, Theory of Recursive Functions and Effective Computability, McGraw-Hill, Series in Higher Mathematics (1967).
21. G. Rozenberg ed., Handbook of Graph Grammars and Computing by Graph Transformation, vol. 1, World Scientific (1997).
22. G. Sénizergues, Définissabilité des graphes context-free, unpublished work.

23. G. Sénizergues, Definability in weak monadic second order logic of some infinite Graphs, Dagstuhl seminar on Automata theory : Infinite Computations 28:16-16 (1992).

24. W. Thomas, A Hierarchie of Sets of Infinite Trees, LNCS 145:335-342 (1982).

Combining Semantics with Non-standard Interpreter Hierarchies

Sergei Abramov[1] and Robert Glück[2]*

[1] Program Systems Institute, Russian Academy of Sciences
RU-152140 Pereslavl-Zalessky, Russia,
abram@botik.ru
[2] PRESTO, JST, Institute for Software Production Technology
Waseda University, Tokyo 169-8555, Japan,
glueck@acm.org

Abstract. This paper reports on results concerning the combination of non-standard semantics via interpreters. We define what a semantics combination means and identify under which conditions a combination can be realized by computer programs (robustness, safely combinable). We develop the underlying mathematical theory and examine the meaning of several non-standard interpreter towers. Our results suggest a technique for the implementation of a certain class of programming language dialects by composing a hierarchy of non-standard interpreters.

1 Introduction

The definition of programming language semantics from simpler, more elementary parts is an intriguing question [6,11,17,18]. This paper reports on new results concerning the combination of semantics via non-standard interpreters. Instead of using the familiar tower of interpreters [13] for implementing the standard semantics of a programming language, we generalize this idea to implement the non-standard semantics of a programming language by combining one or more non-standard interpreters.

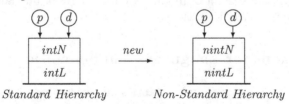

Standard Hierarchy   Non-Standard Hierarchy

The essence of the *interpreter tower* is to evaluate an N-interpreter $intN$ written in L by an L-interpreter $intL$ written in some ground language. This means, we give standard semantics to N-programs via L's standard semantics. But what does it mean to build a tower involving one or more non-standard interpreters? For example, what does it mean for the semantics of an N-program p if we replace interpreter $intL$ by an *inverse*-interpreter $nintL$?

---

* On leave from DIKU, Department of Computer Science, University of Copenhagen.

S. Kapoor and S. Prasad (Eds.): FST TCS 2000, LNCS 1974, pp. 201–213, 2000.

A formal answer to this question and related *non-standard towers* will be given in this paper. Using the mathematical foundations developed here some well-known results about the combination of standard interpreters are shown, as well as new results about the combination of non-standard semantics. This extends our previous work on semantics modifiers [1] and inverse computation [2] where we have observed that some non-standard semantics can be ported via standard interpreters. We can now formalize a class of non-standard semantics that can serve as semantics modifiers, and reason about new non-standard combinations, some of which look potentially useful. We focus on deterministic programming languages as an important case for practice. Since this includes universal programming languages, there is no loss of generality. Extending our results to other computation models can be considered for future work.

In practice, the implementation of a non-standard tower will be inefficient because each level of interpretation adds extra computational overhead. To improve efficiency we assume program specialization techniques. Program specialization, or *partial evaluation* [9,5], was shown to be powerful enough to collapse towers of standard interpreters and to drastically reduce their interpretive overhead. We believe powerful program transformation tools will enable us in the future to combine non-standard interpreters with less concern about efficiency, which may make this approach more practical for the construction of software.

Finally, note that we use the term 'programming language' in a broad sense, that is, not only for *universal* programming languages, such as Fortran, C or ML, but also for *domain-specific* languages (*e.g.*, networks, graphics), and for languages which are *computationally incomplete* (*e.g.*, regular grammars). This means, potentially our results apply to a broad spectrum of application areas.

The main contributions of this paper are: (i) *a mathematical foundation for a theory about semantics combination*: we define what semantics combinations mean and we identify several theoretical combinations; (ii) *an approach to implementing programming language dialects* by building towers of non-standard interpreters and their correctness; (iii) *explaining the essence of known results* such as interpreter towers and the Futamura projections [7], and giving *novel insights* regarding the semantics modification of programming languages. Proofs are omitted due to space limitations.

2   Foundations for Languages and Semantics

Before introducing languages and semantics we give two preliminary definitions. We define a *projection* $(A.b)$ to form new sets given a set of tuples A and an element b, and a *preserving set definedness* relation $(A \nsubseteq B)$ which requires $A \subseteq B$ and A to be non-empty when B is non-empty.

Definition 1 (projection). *Let A, B, C be sets, let $A \subseteq B \times C$, and let $b \in B$, then we define* projection $A.b \stackrel{\text{def}}{=} \{ c' \mid (b', c') \in A, b' = b \}$.

Example 1. Let $A = \{(2, 3, 1), (5, 6, 7), (2, 4, 1)\}$ then $A.2 = \{(3, 1), (4, 1)\}$.

Definition 2 (preserving set definedness). *Let A, B be sets, then we define relation* preserving set definedness $(A \nsubseteq B) \overset{\text{def}}{\Longleftrightarrow} ((A \subseteq B) \wedge (B \neq \emptyset \Rightarrow A \neq \emptyset))$.

We define languages, semantics and functional equivalence using a relational approach. When we speak of languages we mean formal languages. As is customary, we use the same universal data domain (D) for all languages and for representing all programs. Mappings between different data domains and different program representations are straightforward to define and not essential for our discussion.

The reader should be aware of the difference between the abstract language definitions given in this section, which may be non-constructive, and the definitions for programming languages in Sect. 3 which are constructive. The formalization is geared towards the definition of deterministic programming languages.

Definition 3 (language). *A language L is a triple $L = (P_L, D_L, [\![\ ]\!]_L)$, where $P_L \subseteq D$ is the set of L-programs, $D_L = D$ is the data domain for L, and $[\![\ ]\!]_L$ is the semantics of L: $[\![\ ]\!]_L \subseteq P_L \times D \times D$. We denote by $\mathcal{L}$ the set of all languages.*

Definition 4 (program semantics, application). *Let $L = (P_L, D, [\![\ ]\!]_L)$ be a language, let $p \in P_L$ be an L-program, let $d \in D$ be data, then the semantics of p is defined by $[\![\ ]\!]_L.p \subseteq D \times D$, and the application of p to d by $[\![\ ]\!]_L.p.d \subseteq D$.*

Definition 5 (functional equivalence). *Let $L_1 = (P_{L_1}, D, [\![\ ]\!]_{L_1})$ and $L_2 = (P_{L_2}, D, [\![\ ]\!]_{L_2})$ be languages, let $p_1 \in P_{L_1}$ and $p_2 \in P_{L_2}$ be programs, then p_1 and p_2 are functionally equivalent iff $[\![\ ]\!]_{L_1}.p_1 = [\![\ ]\!]_{L_2}.p_2$.*

Note that we defined the semantics $[\![\ ]\!]_L$ of a language as a relation $(P_L \times D \times D)$. For convenience, we will sometimes use notation $[\![p]\!]_L\, d$ for *application* $[\![\ ]\!]_L.p.d$, and notation $[\![p]\!]_L$ for *program semantics* $[\![\ ]\!]_L.p$. As Def. 4 shows, the result of an application is always a set of data, and we can distinguish three cases:

$[\![p]\!]_L\, d = \emptyset$ — application *undefined*,

$[\![p]\!]_L\, d = \{a\}$ — application *defined* (deterministic case),

$[\![p]\!]_L\, d = \{a_1, a_2, \ldots\}$ — application *defined* (non-deterministic case).

Definition 6 (deterministic language). *A language $L = (P_L, D, [\![\ ]\!]_L)$ is deterministic iff $\forall (p_1, d_1, a_1), (p_2, d_2, a_2) \in [\![\ ]\!]_L : (p_1 = p_2 \wedge d_1 = d_2) \Rightarrow (a_1 = a_2)$. We denote by $\mathcal{D}$ the set of all deterministic languages $(\mathcal{D} \subseteq \mathcal{L})$.*

Relations $\subseteq$ and $\nsubseteq$ have a clear meaning for application: $[\![p]\!]_{L_1}\, d \subseteq [\![p]\!]_{L_2}\, d$ tells us that the left application may be undefined even when the right application is defined (*definedness is not preserved*); $[\![p]\!]_{L_1}\, d \nsubseteq [\![p]\!]_{L_2}\, d$ tells us that both applications are either defined or undefined (*definedness is preserved*). In Def. 7 we use $\nsubseteq$ to define a definedness preserving relation between semantics ($\underline{\underline{\subseteq}}$).

Definition 7 (preserving semantics definedness). *Let $L_1 = (P_{L_1}, D, [\![\ ]\!]_{L_1})$ and $L_2 = (P_{L_2}, D, [\![\ ]\!]_{L_2})$ be languages such that $P_{L_1} = P_{L_2}$, then we define relation* preserving semantics definedness ($\underline{\underline{\subseteq}}$) *as follows:*

$$([\![\ ]\!]_{L_1} \underline{\underline{\subseteq}} [\![\ ]\!]_{L_2}) \overset{\text{def}}{\Longleftrightarrow} (\forall p \in P_{L_1}\ \forall d \in D : [\![\ ]\!]_{L_1}.p.d \nsubseteq [\![\ ]\!]_{L_2}.p.d).$$

2.1   Semantics Properties and Language Dialects

A *property* S is a central concept for the foundations of non-standard semantics. It specifies a modification of the standard semantics of a language. When we speak of an S-*dialect* L' of a language L, then the relation of input/output of all L-programs applied under L' must satisfy property S. For example, we require that the output of applying an L-program under an *inverse*-dialect L' of L [2] is a possible input of that program applied under L's standard semantics. Given a *request* r for S-computation, there may be infinitely many *answers* a that satisfy property S.[1] We consider each of them as a correct *wrt* S.

A property describes a semantics modification for a set of languages. The specification can be non-constructive and non-deterministic. We specify a property S for a set of languages as a set of tuples (L, p, r, a). We say a language L' is an S-dialect of L if both languages have the same syntax, and the semantics of L' is a subset of $S.L$. We define three types of dialects that can be derived from S. Later in Sect. 3 we consider only those dialects that are constructive.

Definition 8 (property). *Let* $\mathcal{N} \subseteq \mathcal{L}$, *then set* S *is a property for* $\mathcal{N}$ *iff*

$$S \subseteq \bigcup_{L \in \mathcal{N}} \{L\} \times P_L \times D \times D .$$

Example 2 (properties). Let $\mathcal{N} = \mathcal{L}$ and $R \in \mathcal{L}$, then Id, Inv, $Trans_R$ and $Copy$ are properties for $\mathcal{L}$, namely identity, inversion, translation, and copying of programs. Other, more sophisticated properties may be defined that way.

$$Id \stackrel{\text{def}}{=} \{ (L, p, r, a) \mid L \in \mathcal{L}, p \in P_L, r \in D, a \in [\![p]\!]_L\, r \}$$
$$Inv \stackrel{\text{def}}{=} \{ (L, p, r, a) \mid L \in \mathcal{L}, p \in P_L, a \in D, r \in [\![p]\!]_L\, a \}$$
$$Trans_R \stackrel{\text{def}}{=} \{ (L, p, r, p') \mid L \in \mathcal{L}, p \in P_L, r \in D, p' \in P_R : [\![p]\!]_L = [\![p']\!]_R \}$$
$$Copy \stackrel{\text{def}}{=} \{ (L, p, r, p) \mid L \in \mathcal{L}, p \in P_L, r \in D \}$$

Definition 9 (dialects). *Let* S *be a property for* $\mathcal{N}$, *let* $L \in \mathcal{N}$, *then* $S.L \subseteq P_L \times D \times D$ *is the* most general S-semantics *for* L. *Let* $L = (P_L, D, [\![\ ]\!]_L)$, *then a language* $L' = (P_L, D, [\![\ ]\!]_{L'}) \in \mathcal{L}$ *is (i) the* most general S-dialect *of* L *iff* $[\![\ ]\!]_{L'} = S.L$, *(ii) an* S-dialect *of* L *iff* $[\![\ ]\!]_{L'} \subsetneq S.L$, *and (iii) an* S-semi-dialect *of* L *iff* $[\![\ ]\!]_{L'} \subseteq S.L$. *We denote by* $S|L$ *the* most general S-dialect *of* L *and by* $\mathcal{D}_{S|L}$ *the set of all deterministic* S-dialects *of* L.

The most general S-semantics $S.L$ specifies all correct answers for an application $S.L.p.r$ given S, L, p, r. In general, the most general dialect $S|L$ of a language L will be non-deterministic. This allows the definition of different S-dialects for L.

---

[1] When we talk about non-standard semantics, we use the terms *request* and *answer* to distinguish them from input and output of standard computation.

Example 3 (dialects). There are usually infinitely many deterministic and non-deterministic *Inv*-dialects of L (they differ in which and how many inverse answers they return). For property *Copy*, the most general dialect *Copy*$|L$ is always deterministic and there exists only one *Copy*-dialect for each L. Another example is property *Id*. If L is non-deterministic, then there are usually infinitely many deterministic and non-deterministic *Id*-dialects. But if L is deterministic, then there exists only *one* deterministic *Id*-dialect L' and $L' = Id|L = L$.

Definition 10 (robust property). *Let $\mathcal{N}$ be a set of languages, let S be a property for $\mathcal{N}$, then S is robust iff all functionally equivalent programs are also functionally equivalent under the most general S-dialect:*

$$\forall L_1, L_2 \in \mathcal{N} \ \forall p_1 \in P_{L_1} \ \forall p_2 \in P_{L_2} : (\llbracket p_1 \rrbracket_{L_1} = \llbracket p_2 \rrbracket_{L_2}) \Rightarrow (\llbracket p_1 \rrbracket_{S|L_1} = \llbracket p_2 \rrbracket_{S|L_2}) \ .$$

Example 4 (robustness). All properties in Ex. 2 are robust $(Id, Inv, Trans_R)$, except *Copy*, which returns different results for fct. equivalent programs $p \neq p'$.

The motivation for defining *robustness* is that we are mainly interested in a class of properties that can be combined by interpreters. When we use a robust property S we cannot distinguish by the semantics of the most general dialect $S|L$ two programs which are functionally equivalent under L's standard semantics. A robust property specifies an extensional modification of a language semantics which is independent of the particular operational features of a program.

2.2 Combining Properties

Two properties S' and S'' can be combined into a new property $S' \circ S''$. Intuitively speaking, one gets an $(S' \circ S'')$-dialect of a language L by taking an S'-dialect of an S''-dialect of L. This combination is captured by projection $S'.L''.p.r$ in the following definition. The reason for choosing language L'' from the set of deterministic S''-dialects $\mathcal{D}_{S''|L}$ of L is that later we will use deterministic programming languages for implementing property combinations.

Definition 11 (combination). *Let S', S'' be properties for $\mathcal{D}$, then we define*

$$S' \circ S'' \stackrel{\text{def}}{=} \{(L, p, r, a) \mid L \in \mathcal{D}, p \in P_L, r \in D, a \in D, L'' \in \mathcal{D}_{S''|L}, a \in S'.L''.p.r\} \ .$$

Example 5 (combination). Let S be a property for $\mathcal{D}$, then some of the combinations of the properties in Example 2 are as follows:

$\boxed{S \circ Id = S}$: Right combination with identity does not change property S.

$\boxed{Id \circ S = S}$: Left combination with identity does not change property S.

$\boxed{Trans_R \circ S = S\_Trans_R}$: S-translation to R (will be explained in Sec. 4.3).

$\boxed{Inv \circ S = S^{-1}}$: Inversion of property S (will be explained in Sec. 4.4).

$\boxed{Copy \circ S = Copy}$: "Left zero" for property S.

In addition, we are interested in combinations $(S' \circ S'')$ that guarantee that all applications $S'.L''.p.r$ are defined for the same set of program-request pairs (p, r) regardless which deterministic S''-dialect L'' we select for L. This requires that S' and S'' satisfy the condition given in the following definition. In this case we say, S' and S'' are *safely combinable* $(S' \bowtie S'')$.

Definition 12 (safely combinable). *Let S', S'' be properties for $\mathcal{D}$, then S' is safely combinable with S'' $(S' \bowtie S'')$ iff*

$$\forall L \in \mathcal{D}, \ \forall L''_1, L''_2 \in \mathcal{D}_{S''|L}, \ \forall p \in P_L, \ \forall d \in D :$$
$$(S'.L''_1.p.d \neq \emptyset) \Leftrightarrow (S'.L''_2.p.d \neq \emptyset) .$$

Example 6 (safely combinable). Let S', S'' be properties, and let S' be robust, then the following combinations are always safely combinable:
$\boxed{Id \bowtie S''}$, $\boxed{S' \bowtie Id}$, $\boxed{S' \bowtie Copy}$.

3 Programming Languages

We now turn to *programming languages*, and focus on *deterministic* programming languages as an important case for practice. Since this includes universal programming languages, there is no loss of generality. All computable functions can be expressed. First, we give definitions for programming languages and interpreters, then we introduce *non-standard interpreters* which we define as programs that implement non-standard dialects.

As before we assume a universal data domain D for programming languages, but require D to be constructive (recursively enumerable) and to be closed under tupling: $d_1, \dots, d_k \in D \Rightarrow [\, d_1, \dots, d_k \,] \in D$. For instance, a suitable choice for D is the set of S-expressions familiar from Lisp [13]. Since we consider only deterministic programming languages, the result of an application is either a singleton set or the empty set.

Definition 13 (programming language). *A programming language L is a deterministic language $L = (P_L, D_L, [\![\ ]\!]_L)$ where $P_L \subseteq D$ is the recursively enumerable set of L-programs, $D_L = D$ is the recursively enumerable data domain for L, and $[\![\ ]\!]_L$ is the recursively enumerable semantics of L: $[\![\ ]\!]_L \subseteq P_L \times D \times D$. We denote by $\mathcal{P}$ the set of all programming languages.*

Definition 14 (interpreter). *Let $L = (P_L, D, [\![\ ]\!]_L)$, $M = (P_M, D, [\![\ ]\!]_M)$ be programming languages, then an M-program $intL$ is an interpreter for L in M iff*

$$\forall p \in P_L, \ \forall d \in D : \ [\![intL]\!]_M [\, p, d \,] = [\![p]\!]_L \, d .$$

Definition 15 (partially fixed argument). *Let $L = (P_L, D, [\![\ ]\!]_L)$ be a programming language, let $p, p' \in P_L$, and let $d_1 \in D$ such that*

$$\forall d_2 \in D : \ [\![p']\!]_L \, d_2 = [\![p]\!]_L [\, d_1, d_2 \,] .$$

If program p' exists we denote it by "$[\,p,[\,d_1,\bullet\,]\,]$", and we have

$$\forall d_2 \in D : \;\; [\![\,[\,p,[\,d_1,\bullet\,]\,]\,]\!]_L\, d_2 = [\![p]\!]_L[\,d_1,d_2\,]\;.$$

In a universal programming language we can always write program $[\,p,[\,d_1,\bullet\,]\,]$ given $p \in P_L$ and $d_1 \in D$ (this is similar to Kleene's S-m-n theorem). In a programming language that supports abstraction and application as in the lambda-calculus we can define: $[\,p,[\,d_1,\bullet\,]\,] \stackrel{\text{def}}{=} \lambda d_2.p\,[\,d_1,d_2\,]$.

Definition 16 (prog. lang. dialects). *Let $\mathcal{P}' \subseteq \mathcal{P}$, let S be a property for $\mathcal{P}'$, and let $L = (P_L, D, [\![\ ]\!]_L) \in \mathcal{P}'$, then a prog. language $L' = (P_L, D, [\![\ ]\!]_{L'}) \in \mathcal{P}$ is an S-dialect of L iff $[\![\ ]\!]_{L'} \subseteq\!\!\!\!\!\cdot\; [\![\ ]\!]_{S|L}$, and an S-semi-dialect of L iff $[\![\ ]\!]_{L'} \subseteq [\![\ ]\!]_{S|L}$.*

Definition 17 ($S|L/M$-interpreter). *Let L, M be programming languages, let $\mathcal{P}' \subseteq \mathcal{P}$, let S be a property for $\mathcal{P}'$, and let $L \in \mathcal{P}'$, then an M-program $nintL$ is an S-interpreter for L in M ($S|L/M$-interpreter) if there exists an S-dialect L' of L such that $nintL$ is an interpreter for L' in M.*

An interpreter for a language L is an implementation of the standard semantics of L, while an S-interpreter is an implementation, if it exists, of an S-dialect L' of L. Since a property S may specify infinitely many S-dialects for L (see Sect. 2.1), we say that *any* program that implements one of these dialects is an S-interpreter.[2]

In general, not every non-standard S-dialect is computable. Some dialects may be undecidable, others (semi-)decidable. A non-standard interpreter $nintL$ realizes an S-dialect for a given language L, and having $nintL$ we can say that S can be realized constructively for L. If this is the case for two properties S' and S'', then $(S' \circ S'')$ can be implemented by a tower of non-standard interpreters.

4   Towers of Non-standard Interpreters

Definition 18 (non-standard tower). *Let $\mathcal{P}' \subseteq \mathcal{P}$, let $M \in \mathcal{P}$, let $N, L \in \mathcal{P}'$, let S', S'' be properties for $\mathcal{P}'$, let S' be robust, let M-program $nintL'$ be an S'-interpreter for L in M, let L-program $nintN''$ be an S''-interpreter for N in L, and let $p \in P_N$, $d \in D$, then a non-standard tower is defined by application*

$$[\![nintL']\!]_M[\,[\,nintN'', [\,p, \bullet\,]\,], d\,]\;.$$

Theorem 1 (correctness of non-standard tower). *Let $\mathcal{P}' \subseteq \mathcal{P}$, let $M \in \mathcal{P}$, let $N, L \in \mathcal{P}'$, let S', S'' be properties for $\mathcal{P}'$, let M-program $nintL'$ be an S'-interpreter for L in M, let L-program $nintN''$ be an S''-interpreter for N in L:*

- *If S' is robust then the following non-standard tower implements an $(S' \circ S'')$-semi-dialect of N in M (cf. Fig. 1):*

$$\forall p \in P_N,\; \forall d \in D : [\![nintL']\!]_M[\,[\,nintN'', [\,p, \bullet\,]\,], d\,] \subseteq (S' \circ S'').N.p.d\;.$$

---

[2] In general, deterministic programs cannot implement all S-dialects since some dialects may be non-deterministic (*e.g.*, *Inv*-dialects).

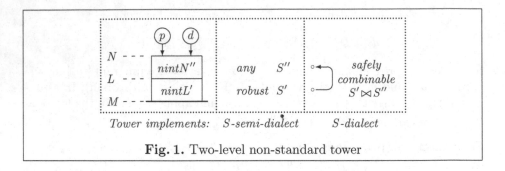

Tower implements: *S-semi-dialect*   *S-dialect*

Fig. 1. Two-level non-standard tower

- *If S' is robust and safely combinable with S'' ($S' \bowtie S''$) then the following non-standard tower implements an ($S' \circ S''$)-dialect of N in M (cf. Fig. 1):*
$$\forall p \in P_N, \ \forall d \in D : [\![nintL']\!]_M [\, [\, nintN'', [\, p, \bullet \,]\,], \, d\,] \not\subseteq (S' \circ S'').N.p.d \ .$$

The theorem guarantees that a non-standard tower consisting of an S'-interpreter $nintL'$ and an S''-interpreter $nintN''$ returns a result (if defined) that is correct *wrt* $S' \circ S''$, provided property S' is robust. Regardless of how the two interpreters are implemented, we obtain an implementation of (at least) an ($S' \circ S''$)-*semi-dialect*. If in addition S' and S'' are safely combinable ($S' \bowtie S''$), we obtain an implementation of an ($S' \circ S''$)-*dialect*. In contrast to the mathematical combination of two properties (Sect. 2.2), a combination of two non-standard interpreters requires that the source language of $nintL'$ and the implementation language of $nintN''$ match (*i.e.* language L). This is illustrated in Fig. 1. We showed which properties are robust (Sect. 2.1) and which are safely combinable (Sect. 2.2).

Figure 2 summarizes relation safely combinable for combinations of properties defined in Ex. 2. For $Trans_R$ we assume R is a universal language. Property Inv is not always safely combinable. While some properties S' and S'' are not safely combinable for all languages, they may be safely combinable for some languages. Two cases when properties are safely combinable for a subset of $\mathcal{D}$:

1. *Only one S''-dialect exists for N.* For instance, for property Inv this condition is satisfied for programming languages in which all programs are injective (this is not true for all programming languages).
2. *Property S' is total for N''.* For example, if R is a universal programming language in property $Trans_R$, then every source program can be translated to R. Thus, $Trans_R$ is totally defined. More formally, S' is a total property for N'' if we have: $\forall p \in P_N, \ \forall d \in D : S'.N''.p.d \neq \emptyset$.

We now examine several semantics combinations and their non-standard towers. The results are summarized in Fig. 3. (Multi-level towers can be constructed by repeating the construction of a two-level tower.)

4.1   Classical Interpreter Tower

Two classical results about standard interpreters can be obtained in our framework using two facts: (i) property Id is robust (Sect. 2.1), and (ii) Id is safely

$S' \bowtie S''$	Id	Inv	$Trans_Q$	$Copy$
Id	Yes	Yes	Yes	Yes
Inv	Yes	No	No	Yes
$Trans_R^*$	Yes	Yes	Yes	Yes

$^{(*)}$ R is a universal programming language

Fig. 2. S' and S'' are safely combinable

combinable with any property S for $\mathcal{D}$: $Id \bowtie S$ (Sect. 2.2). We also observe that an interpreter $intL$ is an Id-interpreter because $Id|L = L$ is an Id-dialect of L, and accord. to Def. 17 $intL$ is an interpreter for this Id-dialect. Thus we have:

Corollary 1 (Id-interpreter). *Let* $L, M \in \mathcal{P}$, *and let* M-*program* $intL$ *be an interpreter for* L *in* M, *then* $intL$ *is an* Id-*interpreter for* L *in* M.

$\boxed{Id \circ Id = Id}$ Since we consider only deterministic programming languages, there exists only one deterministic Id-dialect, and since $Id \bowtie Id$ is safely combinable, we can build the following non-standard tower consisting of an L/M-interpreter $intL$ and an N/L-interpreters $intN$:

$$\forall p \in P_N, \ \forall d \in D : \ [\![intL]\!]_M \, [\, [\, intN, [\, p, \bullet \,]\,], d\,] = [\![p]\!]_N \, d \ .$$

It is easy to see that this combination is the classical *interpreter tower*. The key point is that the semantics of N is preserved by combination $Id \circ Id$. Property Id can be regarded as *identity operation* in the algebra of semantics combination. $\boxed{Id \circ S = S}$ More generally, any S-interpreter $nintN$ for N in L can be evaluated in M given an Id-interpreter $intL$ for L in M. The non-standard tower is a faithful implementation of an S-dialect in M. Not surprisingly, an S-interpreter can be ported from L to M using an Id-interpreter $intL$.

$$\forall p \in P_N, \ \forall d \in D : \ [\![intL]\!]_M \, [\, [\, nintN, [\, p, \bullet \,]\,], d\,] \nsqsubseteq S.N.p.d = [\![p]\!]_{S|N} \, d \ .$$

4.2 Semantics Modifiers

A novel application of Id-interpreters can be obtained from combination $S \circ Id$. $\boxed{S \circ Id = S}$ If property S for $\mathcal{D}$ is robust then $S \bowtie Id$ is safely combinable, and we can write the following non-standard tower consisting of an Id-interpreter $intN$ for N in L and an S-interpreter $nintL$ for L in M:

$$\forall p \in P_N, \ \forall d \in D : \ [\![nintL]\!]_M \, [\, [\, intN, [\, p, \bullet \,]\,], d\,] \nsqsubseteq S.N.p.d = [\![p]\!]_{S|N} \, d \ .$$

The equation asserts that an S-interpretation of N-programs can be performed by combining an Id-interpreter for N in L and an S-interpreter for L. The non-standard tower implements an S-interpreter for N. Every S-interpreter captures

$S' \circ S''$	Id	Inv	$Trans_Q$	$Copy$
Id	Id int-tower	Inv porting	$Trans_Q$ porting	$Copy$ porting
Inv	Inv semmod	Id identity	$Cert_Q$ certifier	$Recog$ recognizer
$Trans_R$	$Trans_R$ semmod	$InvTrans_R$ inverter	$G_{Q/R}$	$Arch_R$ archiver

Fig. 3. Examples of property combinations

the essence of S-computation regardless of its source language. This is radically different from other forms of program reuse because all interpreters implementing robust properties can be ported to new programming languages by means of Id-interpreters. In other words, the entire class of robust properties is suited as *semantics modifiers* [1]. This idea was demonstrated for the following examples.

$\boxed{Inv \circ Id = Inv}$ Since Inv is a robust property (Sect. 2.1), we can reduce the problem of writing an Inv-interpreter for N to the simpler problem of writing an Id-interpreter for N in L, provided an inverse interpreter for L exists. For experimental results see [16,1,2].

$\boxed{Trans_R \circ Id = Trans_R}$ A translator is a classical example of an equivalence transformer. Since property $Trans_R$ is robust for all universal programming languages R, this equations asserts that translation from N to R can be performed by combining a standard interpreter for N in L and a translator from L to R. A realization of this idea are the *Futamura projections* [7]: it was shown [9] that partial evaluation can implement this equation efficiently (for details see also [1]).

4.3   Non-standard Translation

$\boxed{Trans_R \circ S = S\_Trans_R}$ where S is a property for $\mathcal{D}$ and

$$S\_Trans_R \stackrel{\text{def}}{=}$$
$$\{ (L, p, r, p') \mid L \in \mathcal{D}, p \in P_L, r \in D, L' \in \mathcal{D}_{S|L}, p' \in P_R, [\![p']\!]_R = [\![p]\!]_{L'} \}$$

This combination describes the semantics of translating an L-program p into a standard R-program p' which is functionally equivalent to p evaluated under a deterministic S-dialect of N. In other words, non-standard computation of p is performed by standard computation of p' in R. We say S_Trans_R is the semantics of S-*compilation into* R. This holds regardless of S. We have already met the case of Id-translation (Sect. 4.2). Let us examine two examples:

$\boxed{Trans_R \circ Inv = InvTrans_R}$: semantics of an *program inverter* which produces an inverse R-program p^{-1} given an L-program p.

$\boxed{Trans_R \circ Copy = Arch_R}$: semantics of an *archival program* which converts an L-program p into a "self-extracting archive" written in R.

4.4   Semantics Inversion

$\boxed{Inv \circ S = S^{-1}}$ where S is a property for $\mathcal{D}$ and

$$S = \{ (L, p, a, r) \mid L \in \mathcal{D}, p \in P_L, a \in D, r \in S.L.p.a \}$$
$$S^{-1} = \{ (L, p, r, a) \mid L \in \mathcal{D}, p \in P_L, a \in D, r \in S.L.p.a \}$$

The combination describes the inversion of a property S. Three examples:

$\boxed{Inv \circ Trans_Q = Cert_Q}$: semantics of a *program certifier* which, given Q-program p' and L-programs p, verifies whether p' is a translated version of p.

$\boxed{Inv \circ Copy = Recog}$: semantics of a *recognizer*, a program checking whether two L-programs are textually identical – a rather simple-minded semantics.

$\boxed{Inv \circ Inv = Id}$: the inverse semantics of an inverse semantics is the Id-semantics (in general they are not safely combinable and a tower of two Inv-interpreters ensures only a semi-dialect).

5   Related Work

Interpreters are a convenient way for designing and implementing programming languages [13,6,14,19,10]. Early operational semantics [12] and definitional interpreters [15] concerned the definition of one programming language using another which, in our terms, relies on the robustness of Id-semantics.

Monadic interpreters have been studied recently to support features of a programming language, such as profiling, tracing, and error messages (*e.g.*, [11,18]). These works are mostly concerned with modifying operational aspects of a particular language, rather than modifying extensional semantics properties of a class of languages. We studied language-independent conditions for analyzing semantics changes and provided a solid mathematical basis for their correctness.

Meta-interpreters have been used in logic programming for instrumenting programs and for changing ways of formal reasoning [20,3]. These modifications usually change the inference rules of the underlying logic system, and in general do not attempt the deep semantics changes covered by our framework.

Reflective languages have been advocated to enable programs to semantically extend the source language itself, by permitting them to run at the level of the language implementation with access to their own context [4,8]. The reflective tower [17] is the principle architecture of such languages. More should be known to what extent reflective changes can be captured by robust semantics properties.

Experimental evidence for porting S-semantics via Id-interpreters ($S \circ Id$) has been given for inverse semantics (Inv) [16,1,2], and for translation semantics ($Trans_R$) in the area of partial evaluation [9,5]. We are not aware of other work developing mathematical foundations for a theory about semantics combinations, but should mention related work [1] studying the subclass of semantics modifiers.

6   Conclusion and Future Work

The semantics conditions we identified, allow us to reason about the combination of semantics on an abstract level without referring to a particular implementation, and to examine a large class of non-standard semantics instead of particular instances (*e.g.*, a specializer and a translator both implement a translation semantics $Trans_R$). Our results suggest a technique for the implementation of a certain class of programming language dialects by composing a hierarchy of non-standard interpreters (*e.g.*, inverse compilation by $Trans_R \circ Inv$).

Among others, we can now answer the question raised in the introduction, namely what it means for the semantics of a language N if the implementation language L of its standard interpreter $intN$ is interpreted in a non-standard way ($S \circ Id = ?$). As an example we showed that an inverse interpretation of L implements an inverse interpreter for N (even though we have *never* written an inverse interpreter for N, only a standard interpreter $intN$). This is possible because Inv is a *robust property* that can be *safely combined* with Id.

For some of the properties presented in this paper, practical demonstrations of their combination exist (*e.g.*, $Id \circ Id$, $Trans_R \circ Id$, $Inv \circ Id$). In fact, for the first two combinations it was shown that partial evaluation is strong enough to achieve efficient implementations. It is clear that more experimental work will be needed to examine to what extent these and other transformation techniques can optimize non-standard towers, and to what extent stronger techniques are required. We presented a dozen property combinations. Which of these combinations will be useful for which application is another practical question for future work.

Acknowledgements Special thanks to Neil D. Jones for suggesting a relational semantics in an earlier version of this paper. The second author was partly supported by an Invitation Fellowship of the Japan Society for the Promotion of Science (JSPS). Thanks to David Sands and the four anonymous reviewers for constructive feedback.

References

1. S. M. Abramov, R. Glück. From standard to non-standard semantics by semantics modifiers. *International Journal of Foundations of Computer Science*. to appear.
2. S. M. Abramov, R. Glück. The universal resolving algorithm: inverse computation in a functional language. In R. Backhouse, J. N. Oliveira (eds.), *Mathematics of Program Construction. Proceedings, LNCS* 1837, 187–212. Springer-Verlag, 2000.
3. K. Apt, F. Turini. *Meta-Logics and Logic Programming*. MIT Press, 1995.
4. O. Danvy. Across the bridge between reflection and partial evaluation. In D. Bjørner, A. P. Ershov, N. D. Jones (eds.), *PEMC*, 83–116. North-Holland, 1988.
5. O. Danvy, R. Glück, P. Thiemann (eds.). *Partial Evaluation. Proceedings, LNCS* 1110. Springer-Verlag, 1996.
6. J. Earley, H. Sturgis. A formalism for translator interactions. *CACM*, 13(10):607–617, 1970.
7. Y. Futamura. Partial evaluation of computing process – an approach to a compiler-compiler. *Systems, Computers, Controls*, 2(5):45–50, 1971.

8. S. Jefferson, D. P. Friedman. A simple reflective interpreter. *Lisp and Symbolic Computation*, 9(2/3):181–202, 1996.
9. N. D. Jones, C. K. Gomard, P. Sestoft. *Partial Evaluation and Automatic Program Generation*. Prentice-Hall, 1993.
10. S. N. Kamin. *Programming Languages: An Interpreter-Based Approach*. Addison-Wesley, 1990.
11. A. Kishon, P. Hudak. Semantics directed program execution monitoring. *Journal of Functional Programming*, 5(4):501–547, 1995.
12. P. Lucas, P. Lauer, H. Stigleitner. Method and notation for the formal definition of programming languages. Technical report, IBM Lab Vienna, 1968.
13. J. McCarthy. Recursive functions of symb. expressions. *CACM*, 3(4):184–195, 1960.
14. F. G. Pagan. On interpreter-oriented definitions of programming languages. *Computer Journal*, 19(2):151–155, 1976.
15. J. C. Reynolds. Definitional interpreters for higher-order programming languages. In *ACM Annual Conference*, 717–740. ACM, 1972.
16. B. J. Ross. Running programs backwards: the logical inversion of imperative computation. *Formal Aspects of Computing*, 9:331–348, 1997.
17. B. C. Smith. Reflection and semantics in Lisp. In *POPL*, 23–35. ACM Press, 1984.
18. G. L. Steele. Building interpreters by composing monads. In *POPL*, 472–492. ACM Press, 1994.
19. G. L. Steele, G. J. Sussman. The art of the interpreter or, the modularity complex (parts zero, one, two). MIT AI Memo 453, MIT AI Laboratory, 1978.
20. L. Sterling, E. Shapiro. *The Art of Prolog*. MIT Press, 1986.

Using Modes to Ensure Subject Reduction for Typed Logic Programs with Subtyping[*]

Jan–Georg Smaus[1], François Fages[2], and Pierre Deransart[2]

[1] CWI, Kruislaan 413, 1098 SJ Amsterdam, The Netherlands,
jan.smaus@cwi.nl
[2] INRIA-Rocquencourt, BP105, 78153 Le Chesnay Cedex, France,
{francois.fages, pierre.deransart}@inria.fr

Abstract. We consider a general prescriptive type system with parametric polymorphism and subtyping for logic programs. The property of
subject reduction expresses the consistency of the type system w.r.t. the
execution model: if a program is "well-typed", then all derivations starting in a "well-typed" goal are again "well-typed". It is well-established
that without subtyping, this property is readily obtained for logic programs w.r.t. their standard (untyped) execution model. Here we give
syntactic conditions that ensure subject reduction also in the presence
of general subtyping relations between type constructors. The idea is to
consider logic programs with a fixed dataflow, given by modes.

1 Introduction

Prescriptive types are used in logic and functional programming to restrict the
underlying syntax so that only "meaningful" expressions are allowed. This allows
for many programming errors to be detected by the compiler. Gödel [7] and
Mercury [15] are two implemented typed logic programming languages.

A natural stability property one desires for a type system is that it is consistent with the execution model: once a program has passed the compiler, it
is guaranteed that "well-typed" configurations will only generate "well-typed"
configurations at runtime. Adopting the terminology from the theory of the λ-
calculus [17], this property of a typed program is called *subject reduction*. For
the simply typed λ-calculus, subject reduction states that the type of a λ-term
is invariant under reduction. This translates in a well-defined sense to functional
and logic programming.

Semantically, a type represents a set of terms/expressions [8, 9]. Now subtyping makes type systems more expressive and flexible in that it allows to express
inclusions among these sets. For example, if we have types *int* and *real*, we might
want to declare $int \leq real$, i.e., the set of integers is a subset of the set of reals.
More generally, subtype relations such as $list(u) < term$ make it possible to type
Prolog meta-programming predicates [5], as shown in Ex. 1.4 below and Sec. 6.

---

[*] A long version of this paper, containing all proofs, is available in [14].

S. Kapoor and S. Prasad (Eds.): FST TCS 2000, LNCS 1974, pp. 214–226, 2000.
© Springer-Verlag Berlin Heidelberg 2000

In functional programming, a type system that includes subtyping would then state that wherever an expression of type σ is expected as an argument, any expression having a type $\sigma' \leq \sigma$ may occur. The following example explains this informally, using an ad hoc syntax.

Example 1.1. Assume two functions *sqrt* : *real* $\rightarrow$ *real* and *fact* : *int* $\rightarrow$ *int* which compute the square root and factorial, respectively. Then *sqrt* (*fact* 3) is a legal expression, since *fact* 3 is of type *int* and may therefore be used as an argument to *sqrt*, because *sqrt* expects an argument of type *real*, and *int* < *real*.

Subject reduction in functional programming crucially relies on the fact that there is a clear notion of dataflow. It is always the *arguments* (the "input") of a function that may be smaller than expected, whereas the result (the "output") may be greater than expected. This is best illustrated by a counterexample, which is obtained by introducing *reference* types.

Example 1.2. Suppose we have a function f : *real ref* $\rightarrow$ *real* defined by *let* $f(x) = x := 3.14$; *return* x. So f takes a *reference* (pointer) to a real as argument, assigns the value 3.14 to this real, and also return 3.14. Even though *int* < *real*, this function cannot be applied to an *int ref*, since the value 3.14 cannot be assigned to an integer.

In the example, the variable x is used both for input and output, and hence there is no clear direction of dataflow. While this problem is marginal in functional programming, it is the main problem for subject reduction in logic programming with subtypes.

Subject reduction for logic programming means that resolving a "well-typed" goal with a "well-typed" clause will always result in a "well-typed" goal. It holds for parametric polymorphic type systems without subtyping [9, 10].[1]

Example 1.3. In analogy to Ex. 1.1, suppose `Sqrt/2` and `Fact/2` are predicates of declared type (`Real, Real`) and (`Int, Int`), respectively. Consider the program
    `Fact(3,6).`
    `Sqrt(6,2.45).`
and the derivations
    `Fact(3,x), Sqrt(x,y)` $\rightsquigarrow$ `Sqrt(6,y)` $\rightsquigarrow$ $\square$
    `Sqrt(6,x), Fact(x,y)` $\rightsquigarrow$ `Fact(2.45,y)`
In the first derivation, all arguments have a type that is less than or equal to the declared type, and so we have subject reduction. In the second derivation, the argument 2.45 to `Fact` has type `Real`, which is greater than the declared type. The atom `Fact(2.45, y)` is illegal, and so we do not have subject reduction.

Here we address this problem by giving a fixed direction of dataflow to logic programs, i.e., by introducing modes [1] and replacing unification with double matching [2], so that the dataflow is always from the input to the output positions in an atom. We impose a condition on the terms in the output positions, or more

---

[1] Note however that the first formulation of subject reduction [10] was incorrect [8].

precisely, on the types of the *variables* occurring in these terms: each variable must have *exactly* the declared (expected) type of the position where it occurs.

In Ex. 1.3, let the first argument of each predicate be input and the second be output. In both derivations, x has type Int. For Fact(3, x), this is exactly the declared type, and so the condition is fulfilled for the first derivation. For Sqrt(6, x), the declared type is Real, and so the condition is violated.

The contribution of this paper is a statement that programs that are typed according to a type system with subtyping, and respect certain conditions concerning the modes, enjoy the subject reduction property, i.e., the type system is consistent w.r.t. the (untyped) execution model. This means that effectively the types can be ignored at runtime, which has traditionally been considered as desirable, although there are also reasons for keeping the types during execution [11]. In Sec. 6, we discuss the conditions on programs.

There are few works on prescriptive type systems for logic programs with subtyping [3, 4, 5, 6, 8]. Hill and Topor [8] give a result on subject reduction for systems without subtyping, and study general type systems with subtyping. However their results on the existence of principal typings turned out to be wrong [3]. Beierle [3] shows the existence of principal typings for systems with subtype relations between constant types, and provides type inference algorithms. Beierle and also Hanus [6] do not claim subject reduction for their systems. Fages and Paltrinieri [5] have shown a weak form of subject reduction for constraint logic programs with subtyping, where equality constraints replace substitutions in the execution model.

The idea of introducing modes to ensure subject reduction for logic programs was proposed previously by Dietrich and Hagl [4]. However they do not study the decidability of the conditions they impose on the subtyping relation. Furthermore since each result type must be transparent (a condition we will define later), subtype relations between type constructors of different arities are forbidden.

Example 1.4. Assume types Int, String and List(u) defined as usual, and a type Term that contains all terms (so all types are subtypes of Term). Moreover, assume Append as usual with declared type (List(u), List(u), List(u)), and a predicate Functor with declared type (Term, String), which gives the top functor of a term. In our formalism, we could show subject reduction for the query Append([1], [], x), Functor(x, y), whereas this is not possible in [4] because the subtype relation between List(Int) and Term cannot be expressed.

The plan of the paper is as follows. Section 2 mainly introduces the type system. In Sec. 3, we show how expressions can be typed assigning different types to the variables, and we introduce *ordered substitutions*, which are substitutions preserving types. In Sec. 4, we show under which conditions substitutions obtained by unification are indeed ordered. In Sec. 5, we show how these conditions on unified terms can be translated into conditions on programs and derivations.

Table 1. The subtyping order on types

$(Par) \quad u \leq u \qquad\qquad\qquad\qquad u$ is a parameter

$$(Constr) \quad \frac{\tau_{\iota(1)} \leq \tau'_1 \ \cdots \ \tau_{\iota(m')} \leq \tau'_{m'}}{K(\tau_1,...,\tau_m) \leq K'(\tau'_1,...,\tau'_{m'})} \qquad K \leq K', \iota = \iota_{K,K'}.$$

2   The Type System

We use the type system of [5]. First we recall some basic concepts [1]. When we refer to a *clause in a program*, we mean a copy of this clause whose variables are renamed apart from any other variables in the context. A query is a sequence of atoms. A query Q' is a **resolvent of** a query Q and a clause $h \leftarrow B$ if $Q = a_1,\ldots,a_m$, $Q' = (a_1,\ldots,a_{k-1},B,a_{k+1},\ldots,a_m)\theta$, and h and a_k are unifiable with MGU θ. **Resolution steps** and **derivations** are defined in the usual way.

2.1   Type Expressions

The set of types $\mathcal{T}$ is given by the term structure based on a finite set of **constructors** $\mathcal{K}$, where with each $K \in \mathcal{K}$ an arity $m \geq 0$ is associated (by writing K/m), and a denumerable set $\mathcal{U}$ of **parameters**. A **flat type** is a type of the form $K(u_1,\ldots,u_m)$, where $K \in \mathcal{K}$ and the u_i are distinct parameters. We write $\tau[\sigma]$ to denote that the type τ strictly contains the type σ as a subexpression.

A **type substitution** Θ is a mapping from parameters to types. The **domain** of Θ is denoted by $dom(\Theta)$, the parameters in its range by $ran(\Theta)$. The set of parameters in a syntactic object o is denoted by $pars(o)$.

We assume an order $\leq$ on type constructors such that: $K/m \leq K'/m'$ implies $m \geq m'$; and, for each $K \in \mathcal{K}$, the set $\{K' \mid K \leq K'\}$ has a maximum. Moreover, we associate with each pair $K/m \leq K'/m'$ an injection $\iota_{K,K'} : \{1,\ldots,m'\} \rightarrow \{1,\ldots,m\}$ such that $\iota_{K,K''} = \iota_{K,K'} \circ \iota_{K',K''}$ whenever $K \leq K' \leq K''$. This order is extended to the **subtyping order** on types, denoted by $\leq$, as the least relation satisfying the rules in Table 1.

Proposition 2.1. If $\sigma \leq \tau$ then $\sigma\Theta \leq \tau\Theta$ for any type substitution Θ.

Proposition 2.2. For each type σ, the set $\{\tau \mid \sigma \leq \tau\}$ has a maximum, which is denoted by $Max(\sigma)$.

For Prop. 2.2, it is crucial that $K/m \leq K'/m'$ implies $m \geq m'$. For example, if we allowed for `Emptylist`$/0 \leq$ `List`$/1$, then we would have `Emptylist` $\leq$ `List`(τ) for all τ, and so Prop. 2.2 would not hold. Note that the possibility of "forgetting" type parameters, as in `List`$/1 \leq$ `Anylist`$/0$, may provide solutions to inequalities of the form `List`$(u) \leq u$, e.g. $u = $ `Anylist`. However, we have:

Proposition 2.3. An inequality of the form $u \leq \tau[u]$ has no solution. An inequality of the form $\tau[u] \leq u$ has no solution if $u \in vars(Max(\tau))$.

Table 2. The type system.

(Var) $\{x : \tau, \ldots\} \vdash x : \tau$

$(Func)$ $\dfrac{U \vdash t_i : \sigma_i \quad \sigma_i \leq \tau_i \Theta \quad (i \in \{1,\ldots,n\})}{U \vdash f_{\tau_1 \ldots \tau_n \to \tau}(t_1,\ldots,t_n) : \tau\Theta}$ Θ is a type substitution

$(Atom)$ $\dfrac{U \vdash t_i : \sigma_i \quad \sigma_i \leq \tau_i \Theta \quad (i \in \{1,\ldots,n\})}{U \vdash p_{\tau_1 \ldots \tau_n}(t_1,\ldots,t_n) Atom}$ Θ is a type substitution

$(Headatom)$ $\dfrac{U \vdash t_i : \sigma_i \quad \sigma_i \leq \tau_i \quad (i \in \{1,\ldots,n\})}{U \vdash p_{\tau_1 \ldots \tau_n}(t_1,\ldots,t_n) Headatom}$

$(Query)$ $\dfrac{U \vdash A_1 \; Atom \; \ldots \; U \vdash A_n \; Atom}{U \vdash A_1,\ldots,A_n \; Query}$

$(Clause)$ $\dfrac{U \vdash Q \; Query \quad U \vdash A \; Headatom}{U \vdash A \leftarrow Q \; Clause}$

2.2   Typed Programs

We assume a denumerable set $\mathcal{V}$ of **variables**. The set of variables in a syntactic object o is denoted by $vars(o)$. We assume a finite set $\mathcal{F}$ (resp. $\mathcal{P}$) of **function** (resp. **predicate**) symbols, each with an arity and a **declared type** associated with it, such that: for each $f \in \mathcal{F}$, the declared type has the form $(\tau_1, \ldots, \tau_n, \tau)$, where n is the arity of f, $(\tau_1, \ldots, \tau_n) \in \mathcal{T}^n$, τ is a flat type and satisfies the *transparency condition* [8]: $pars(\tau_1, \ldots, \tau_n) \subseteq pars(\tau)$; for each $p \in \mathcal{P}$, the declared type has the form $(\tau_1, \ldots, \tau_n)$, where n is the arity of p and $(\tau_1, \ldots, \tau_n) \in \mathcal{T}^n$. The declared types are indicated by writing $f_{\tau_1 \ldots \tau_n \to \tau}$ and $p_{\tau_1 \ldots \tau_n}$. We assume that there is a special predicate symbol $=_{u,u}$ where $u \in \mathcal{U}$.

We assume that $\mathcal{K}$, $\mathcal{F}$, and $\mathcal{P}$ are fixed by declarations in a **typed program**, where the syntactical details are insignificant for our results. In examples we loosely follow Gödel syntax [7].

A **variable typing** is a mapping from a finite subset of $\mathcal{V}$ to $\mathcal{T}$, written as $\{x_1 : \tau_1, \ldots, x_n : \tau_n\}$. The restriction of a variable typing U to the variables in o is denoted as $U{\upharpoonright}_o$. The type system, which defines terms, atoms etc. relative to a variable typing U, consists of the rules shown in Table 2.

3   The Subtype and Instantiation Hierarchies

3.1   Modifying Variable Typings

We now show that if we can derive that some object is in the typed language using a variable typing U, then we can always modify U in three ways: extending its domain, instantiating the types, and making the types smaller.

Definition 3.1. Let U, U' be variable typings. We say that U is **smaller or equal** U', denoted $U \leq U'$, if $U = \{x_1 : \tau_1, \ldots, x_n : \tau_n\}$, $U' = \{x_1 : \tau_1', \ldots, x_n :$

$\tau'_n\}$, and for all $i \in \{1, \ldots, n\}$, we have $\tau_i \leq \tau'_i$. We write $U' \supseteq\leq U$ if there exists a variable typing U'' such that $U' \supseteq U''$ and $U'' \leq U$.

Lemma 3.1. Let U, U' be variable typings and Θ a type substitution such that $U' \supseteq\leq U\Theta$. If $U \vdash t : \sigma$, then $U' \vdash t : \sigma'$ where $\sigma' \leq \sigma\Theta$. Moreover, if $U \vdash A$ *Atom* then $U' \vdash A$ *Atom*, and if $U \vdash Q$ *Query* then $U' \vdash Q$ *Query*.

3.2   Typed Substitutions

Typed substitutions are a fundamental concept for typed logic programs.

Definition 3.2. If $U \vdash x_1 = t_1, \ldots, x_n = t_n$ *Query* where $x_1, \ldots, x_n$ are distinct variables and for each $i \in \{1, \ldots, n\}$, t_i is a term distinct from x_i, then $(\{x_1/t_1, \ldots, x_n/t_n\}, U)$ is a **typed (term) substitution**.

To show that applying a typed substitution preserves "well-typedness" for systems with subtyping, we need a further condition. Given a typed substitution (θ, U), the type assigned to a variable x by U must be sufficiently big, so that it is compatible with the type of the term replaced for x by θ.

Example 3.1. Consider again Ex. 1.3. Taking $U = \{x : \text{Int}, y : \text{Int}\}$, we have $U \vdash x : \text{Int}$, $U \vdash 2.45 : \text{Real}$, and hence $U \vdash x = 2.45$ *Atom*. So $(\{x/2.45\}, U)$ is a typed substitution. Now $U \vdash \text{Fact}(x, y)$ *Atom*, but $U \not\vdash \text{Fact}(2.45, y)$ *Atom*. The type of x is too small to accommodate for instantiation to 2.45.

Definition 3.3. A typed (term) substitution $(\{x_1/r_1, \ldots, x_n/r_n\}, U)$ is an **ordered substitution** if, for each $i \in \{1, \ldots, n\}$, where $x_i : \tau_i \in U$, there exists σ_i such that $U \vdash r_i : \sigma_i$ and $\sigma_i \leq \tau_i$.

We now show that expressions stay "well-typed" when ordered substitutions are applied [8, Lemma 1.4.2].

Lemma 3.2. Let (θ, U) be an ordered substitution. If $U \vdash t : \sigma$ then $U \vdash t\theta : \sigma'$ for some $\sigma' \leq \sigma$. Moreover, if $U \vdash A$ *Atom* then $U \vdash A\theta$ *Atom*, and likewise for queries and clauses.

4   Conditions for Ensuring Ordered Substitutions

In this section, we show under which conditions it can be guaranteed that the substitutions applied in resolution steps are ordered substitutions.

4.1   Type Inequality Systems

The substitution of a resolution step is obtained by unifying two terms, say t_1 and t_2. In order for the substitution to be typed, it is necessary that we can derive $U \vdash t_1 = t_2 \; Atom$ for some U. We will show that if U is, in a certain sense, maximal, then it is guaranteed that the typed substitution is ordered.

We first formalise paths leading to subterms of a term.

Definition 4.1. A term t has the subterm t in **position** ϵ. If $t = f(t_1, \ldots, t_n)$ and t_i has subterm s in position ζ, then t has subterm s in position $i.\zeta$.

Example 4.1. The term $\mathtt{F(G(C), H(C))}$ has subterm $\mathtt{C}$ in position 1.1, but also in position 2.1. The position 2.1.1 is undefined for this term.

Let us write $\_ \vdash t :\leq \sigma$ if there exist U and σ' such that $U \vdash t : \sigma'$ and $\sigma' \leq \sigma$. To derive $U \vdash t_1 = t_2 \; Atom$, clearly the last step has the form

$$\frac{U \vdash t_1 : \tau_1 \quad U \vdash t_2 : \tau_2 \quad \tau_1 \leq \mathsf{u}\Theta \quad \tau_2 \leq \mathsf{u}\Theta}{U \vdash t_1 =_{\mathsf{u},\mathsf{u}} t_2 \; Atom}$$

So we use an *instance* $(\mathsf{u}, \mathsf{u})\Theta$ of the declared type of the equality predicate, and the types of t_1 and t_2 are both less then or equal to $\mathsf{u}\Theta$. This motivates the following question: Given a term t such that $\_ \vdash t :\leq \sigma$, what are the maximal types of subterm positions of t with respect to σ?

Example 4.2. Let $\mathtt{List}/1, \mathtt{Anylist}/0 \in \mathcal{K}$ where $\mathtt{List}(\tau) \leq \mathtt{Anylist}$ for all τ, and $\mathtt{Nil}_{\rightarrow\mathtt{List}(\mathsf{u})}, \mathtt{Cons}_{\mathsf{u},\mathtt{List}(\mathsf{u})\rightarrow\mathtt{List}(\mathsf{u})} \in \mathcal{F}$. Consider the term $[\mathtt{x}, [\mathtt{y}]]$ (in usual list notation) depicted in Fig. 1, and let $\sigma = \mathtt{Anylist}$. Each functor in $[\mathtt{x}, [\mathtt{y}]]$ is introduced using Rule *(Func)*. E.g., any type of $\mathtt{Nil}$ in position 2.1.2 is necessarily an instance of $\mathtt{List}(\mathsf{u}^{2.1.2})$, its declared type.[2] To derive that $\mathtt{Cons(y, Nil)}$ is a typed term, this instance must be smaller than some instance of the second declared argument type of $\mathtt{Cons}$ in position 2.1, i.e., $\mathtt{List}(\mathsf{u}^{2.1})$.

So in order to derive that $[\mathtt{x}, [\mathtt{y}]]$ is a term of a type smaller than $\mathtt{Anylist}$, we need an instantiation of the parameters such that for each box (position), the type in the *lower* subbox is smaller than the type of the *upper* subbox.

We see that in order for $\_ \vdash t :\leq \sigma$ to hold, a solution to a certain *type inequality system* (set of inequalities between types) must exist.

Definition 4.2. Let t be a term and σ a type such that $\_ \vdash t :\leq \sigma$. For each position ζ where t has a non-variable subterm, we denote the function in this position by $f^{\zeta}_{\tau^{\zeta}_1, \ldots, \tau^{\zeta}_{n_\zeta} \rightarrow \tau^{\zeta}}$ (assuming that the parameters in $\tau^{\zeta}_1, \ldots, \tau^{\zeta}_{n_\zeta}, \tau^{\zeta}$ are fresh, say by indexing them with ζ). For each variable $x \in vars(t)$, we introduce a parameter u^x (so $\mathsf{u}^x \notin pars(\sigma)$). The **type inequality system** of t and σ is

$$\mathcal{I}(t, \sigma) = \{\tau^{\epsilon} \leq \sigma\} \cup \{\tau^{\zeta.i} \leq \tau^{\zeta}_i \mid \text{Position } \zeta.i \text{ in } t \text{ is non-variable}\} \cup$$
$$\{\mathsf{u}^x \leq \tau^{\zeta}_i \mid \text{Position } \zeta.i \text{ in } t \text{ is variable } x\}.$$

---

[2] We use the positions as superscripts to parameters in order to obtain fresh copies.

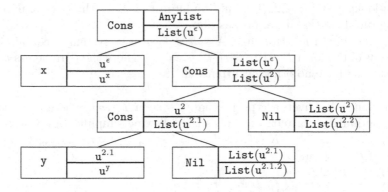

Fig. 1. The term $[x, [y]]$ and associated inequalities

A **solution** of $\mathcal{I}(t, \sigma)$ is a type substitution Θ such that $dom(\Theta) \cap pars(\sigma) = \emptyset$ and for each $\tau \leq \tau' \in \mathcal{I}(t, \sigma)$, the inequality $\tau\Theta \leq \tau'\Theta$ holds. A solution Θ to $\mathcal{I}(t, \sigma)$ is **principal** if for every solution $\tilde{\Theta}$ for $\mathcal{I}(t, \sigma)$, there exists a Θ' such that for each $\tau \leq \tau' \in \mathcal{I}(t, \sigma)$, we have $\tau\tilde{\Theta} \leq \tau\Theta\Theta'$ and $\tau'\tilde{\Theta} \leq \tau'\Theta\Theta'$.

Proposition 4.1. Let t be a term and σ a type. If $U \vdash t :\leq \sigma$ for some variable typing U, then there exists a solution Θ for $\mathcal{I}(t, \sigma)$ (called the **solution for** $\mathcal{I}(t, \sigma)$ **corresponding to** U) such that for each subterm t' in position ζ in t, we have $U \vdash t' : \tau^\zeta\Theta$ if $t' \notin \mathcal{V}$, and $U \vdash t' : u^{t'}\Theta$ if $t' \in \mathcal{V}$.

In the next subsection, we present an algorithm, based on [5], which computes a principal solution to a type inequality system, provided t is linear. In Subsec. 4.3, our interest in principal solutions will become clear.

4.2 Computing a Principal Solution

The algorithm transforms the inequality system, thereby computing bindings to parameters which constitute the solution. It is convenient to consider system of both *inequalities*, and *equations* of the form $u = \tau$. The inequalities represent the current type inequality system, and the equalities represent the substitution accumulated so far. We use $\tilde{\leq}$ for $\leq$ or $=$.

Definition 4.3. A system is **left-linear** if each parameter occurs at most once on the left hand side of an equation/inequality. A system is **acyclic** if it does not have a subset $\{\tau_1 \tilde{\leq} \sigma_1, ..., \tau_n \tilde{\leq} \sigma_n\}$ with $pars(\sigma_i) \cap pars(\tau_{i+1}) \neq \emptyset$ for all $1 \leq i \leq n - 1$, and $pars(\sigma_n) \cap pars(\tau_1) \neq \emptyset$.

Proposition 4.2. If t is a linear term, then any inequality system $\mathcal{I}(t, \sigma)$ is acyclic and left-linear.

By looking at Ex. 4.2, it should be intuitively clear that assuming linearity of t is crucial for the above proposition.

We now give the algorithm. A *solved form* is a system I containing only equations of the form $I = \{u_1 = \tau_1, ..., u_n = \tau_n\}$ where the parameters u_i are all different and have no other occurrence in I.

Definition 4.4. Given a type inequality system $\mathcal{I}(t, \sigma)$, where t is linear, the **type inequality algorithm** applies the following simplification rules:

(1) $\{K(\tau_1, ..., \tau_m) \leq K'(\tau_1', ..., \tau_{m'}')\} \cup I \longrightarrow \{\tau_{\iota(i)} \leq \tau_i'\}_{i=1,..,m'} \cup I$
  if $K \leq K'$ and $\iota = \iota_{K,K'}$

(2) $\{u \leq u\} \cup I \longrightarrow I$

(3) $\{u \leq \tau\} \cup I \longrightarrow \{u = \tau\} \cup I[u/\tau]$
  if $\tau \neq u$, $u \notin vars(\tau)$.

(4) $\{\tau \leq u\} \cup I \longrightarrow \{u = Max(\tau)\} \cup I[u/Max(\tau)]$
  if $\tau \notin V$, $u \notin vars(Max(\tau))$ and $u \notin vars(l)$ for any $l \leq r \in \Sigma$.

Intuitively, left-linearity of $\mathcal{I}(t, \sigma)$ is crucial because it renders the binding of a parameter (point (3)) unique.

Proposition 4.3. Given a type inequality system $\mathcal{I}(t, \sigma)$, where t is linear, the type inequality algorithm terminates with either a solved form, in which case the associated substitution is a principal solution, or a non-solved form, in which case the system has no solution.

4.3   Principal Variable Typings

The existence of a principal solution Θ of a type inequality system $\mathcal{I}(t, \sigma)$ and Prop. 4.1 motivate defining the variable typing U such that Θ is exactly the solution of $\mathcal{I}(t, \sigma)$ corresponding to U.

Definition 4.5. Let $\_ \vdash t :\leq \sigma$, and Θ be a principal solution of $\mathcal{I}(t, \sigma)$. A variable typing U is **principal** for t and σ if $U \supseteq \{x : u^x \Theta \mid x \in vars(t)\}$.

By the definition of a principal solution of $\mathcal{I}(t, \sigma)$ and Prop. 4.1, if U is a principal variable typing for t and σ, then for any U' such that $U'(x) > U(x)$ for some $x \in vars(t)$, we have $U' \nvdash t :\leq \sigma$.

The following key lemma states conditions under which a substitution obtained by unifying two terms is indeed ordered.

Lemma 4.4. Let s and t be terms, s linear, such that $U \vdash s :\leq \rho$, $U \vdash t :\leq \rho$, and there exists a substitution θ such that $s\theta = t$. Suppose U is principal for s and ρ. Then there exists a type substitution Θ such that for $U' = U\Theta\lceil_{vars(s)} \cup U\lceil_{\mathcal{V}\backslash vars(s)}$, we have that (θ, U') is an ordered substitution.

Example 4.3. Consider the term vectors (since Lemma 4.4 generalises in the obvious way to term vectors) $s = (3, \mathbf{x})$ and $t = (3, 6)$, let $\rho = (\mathtt{Int}, \mathtt{Int})$ and $U_s = \{\mathbf{x} : \mathtt{Int}\}$, $U_t = \emptyset$ (see Ex. 1.3). Note that U_s is principal for s and ρ, and so $(\{\mathbf{x}/6\}, U_s \cup U_t)$ is an ordered substitution (Θ is empty).

In contrast, let $s = (6, \mathbf{x})$ and $t = (6, 2.45)$, let $\rho = (\mathtt{Real}, \mathtt{Real})$ and $U_s = \{\mathbf{x} : \mathtt{Int}\}$, $U_t = \emptyset$. Then U_s is not principal for s and ρ (the principal variable typing would be $\{\mathbf{x}/\mathtt{Real}\}$), and indeed, there exists no Θ such that $(\{\mathbf{x}/2.45\}, U_s \Theta \cup U_t)$ is an ordered substitution.

5   Nicely Typed Programs

So far we have seen that *matching*, *linearity*, and *principal variable typings* are crucial to ensure that unification yields ordered substitutions. Note that those results generalise in the obvious way from terms to term vectors. We now define three corresponding conditions on programs and the execution model.

First, we define modes [1]. For $p/n \in \mathcal{P}$, a **mode** is an atom $p(m_1, \ldots, m_n)$, where $m_i \in \{I, O\}$ for $i \in \{1, \ldots, n\}$. Positions with I (resp. O) are called **input** (resp. **output**) **positions** of p. We assume that a mode is associated with each $p \in \mathcal{P}$. The notation $p(\bar{s}, \bar{t})$ means that $\bar{s}$ (resp. $\bar{t}$) is the vector of terms filling the input (resp. output) positions of $p(\bar{s}, \bar{t})$. Moded unification is a special case of *double matching* [2].

Definition 5.1. Consider a resolution step where $p(\bar{s}, \bar{t})$ is the selected atom and $p(\bar{w}, \bar{v})$ is the renamed apart clause head. The equation $p(\bar{s}, \bar{t}) = p(\bar{w}, \bar{v})$ is **solvable by moded unification** if there exist substitutions θ_1, θ_2 such that $\bar{w}\theta_1 = \bar{s}$ and $vars(\bar{t}\theta_1) \cap vars(\bar{v}\theta_1) = \emptyset$ and $\bar{t}\theta_1\theta_2 = \bar{v}\theta_1$. A derivation where all unifications are solvable by moded unification is a **moded derivation**.

Definition 5.2. A query $Q = p_1(\bar{s}_1, \bar{t}_1), \ldots, p_n(\bar{s}_n, \bar{t}_n)$ is **nicely moded** if $\bar{t}_1, \ldots, \bar{t}_n$ is a linear vector of terms and for all $i \in \{1, \ldots, n\}$

$$vars(\bar{s}_i) \cap \bigcup_{j=i}^{n} vars(\bar{t}_j) = \emptyset. \tag{1}$$

The clause $C = p(\bar{t}_0, \bar{s}_{n+1}) \leftarrow Q$ is **nicely moded** if Q is nicely moded and

$$vars(\bar{t}_0) \cap \bigcup_{j=1}^{n} vars(\bar{t}_j) = \emptyset. \tag{2}$$

An atom $p(\bar{s}, \bar{t})$ is **input-linear** if $\bar{s}$ is linear, **output-linear** if $\bar{t}$ is linear.

Definition 5.3. Let $C = p_{\bar{\tau}_0, \bar{\sigma}_{n+1}}(\bar{t}_0, \bar{s}_{n+1}) \leftarrow p^1_{\bar{\sigma}_1, \bar{\tau}_1}(\bar{s}_1, \bar{t}_1), \ldots, p^n_{\bar{\sigma}_n, \bar{\tau}_n}(\bar{s}_n, \bar{t}_n)$ be a clause. If C is nicely moded, $\bar{t}_0$ is input-linear, and there exists a variable typing U such that $U \vdash C$ *Clause*, and for each $i \in \{0, \ldots, n\}$, U is principal for $\bar{t}_i$ and $\bar{\tau}'_i$, where $\bar{\tau}'_i$ is the instance of $\bar{\tau}_i$ used for deriving $U \vdash C$ *Clause*, then we say that C is **nicely typed**. A query $U_Q : Q$ is **nicely typed** if the clause $\mathtt{Go} \leftarrow Q$ is nicely typed.

We can now state the main result.

Theorem 5.1. Let C and Q be a nicely typed clause and query. If Q' is a resolvent of C and Q where the unification of the selected atom and the clause head is solvable by moded unification, then Q' is nicely typed.

Example 5.1. Consider again Ex. 1.3. The program is nicely typed, where the declared types are given in that example, and the first position of each predicate is input, and the second output. Both queries are nicely moded. The first query is also nicely typed, whereas the second is not (see also Ex. 4.3). For the first query, we have subject reduction, for the second we do not have subject reduction.

6   Discussion

In this paper, we have proposed criteria for ensuring subject reduction for typed logic programs with subtyping under the untyped execution model. Our starting point was a comparison between functional and logic programming: In functional programs, there is a clear notion of dataflow, whereas in logic programming, there is no such notion a priori, and arguments can serve as input arguments and output arguments. This difference is the source of the difficulty of ensuring subject reduction for logic programs.

It is instructive to divide the numerous conditions we impose into four classes: (1) "basic" type conditions on the program (Sec. 2), (2) conditions on the execution model (Def. 5.1), (3) mode conditions on the program (Def. 5.2), (4) "additional" type conditions on the program (Def. 5.3).

Concerning (1), our notion of subtyping deserves discussion. Approaches differ with respect to conditions on the *arities* of type constructors for which there is a subtype relation. Beierle [3] assumes that the (constructor) order is only defined for type constants, i.e. constructors of arity 0. Thus we could have $Int \leq Real$, and so by extension $List(Int) \leq List(Real)$, but not $List(Int) \leq Tree(Real)$. Many authors assume that only constructors of the same arity are comparable. Thus we could have $List(Int) \leq Tree(Real)$, but not $List(Int) \leq Anylist$. We assume, as [5], that if $K/m \leq K'/m'$, then $m \geq m'$. We think that this choice is crucial for the existence of principal types.

Stroetmann and Glaß [16] argue that comparisons between constructors of arbitrary arity should be allowed in principle. Their formalism is such that the subtype relation does not automatically correspond to a subset relation. Nevertheless, the formalism heavily relies on such a correspondence, although it is not said how it can be decided. We refer to [14] for more details.

Technically, what is crucial for subject reduction is that substitutions are *ordered*: each variable is replaced with a term of a smaller type. In Section 4, we gave conditions under which unification of two terms yields an ordered substitution: the unification is a matching, the term that is being instantiated is linear and is typed using a *principal* variable typing. The linearity requirement ensures that a principle variable typing exists and can be computed (Subsec. 4.2).

In Sec. 5, we showed how those conditions translate to conditions on the program and the execution model. We introduce modes and assume that programs are executed using moded unification (2). This might be explicitly enforced by the compiler, or it might be verified statically [2]. Moded unification can actually be very beneficial for efficiency, as witnessed by the language Mercury [15]. Apart from that, (3) nicely-modedness states the linearity of the terms being instantiated in a unification. Finally, (4) nicely-typedness states that the instantiated terms must be typed using a principal variable typing.

Nicely-modedness has been widely used for verification purposes (e.g. [2]). In particular, the linearity condition on the output arguments is natural: it states that every piece of data has at most one producer. Input-linearity of clause heads however can sometimes be a demanding condition [13, Section 10.2].

Note that introducing modes into logic programming does not mean that logic programs become functional. The aspect of non-determinacy (possibility of computing several solutions for a query) remains.

Even though our result on subject reduction means that it is possible to execute programs without maintaining the types at runtime, there are circumstances where keeping the types at runtime is desirable, for example for memory management, printing, or in higher-order logic programming where the existence and shape of unifiers depends on the types [11].

There is a relationship between our notion of subtyping and *transparency* (see Subsec. 2.2). Transparency ensures that two terms of the same type have identical types in all corresponding subterms, e.g. if [1] and [x] are both of type List(Int), we are sure that x is of type Int. Now in a certain way, allowing for a subtyping relation that "forgets" parameters undermines transparency. For example, we can derive $\{x : \text{String}\} \vdash [x] = [1]$ *Atom*, since List(String) $\leq$ Anylist and List(Int) $\leq$ Anylist, even though Int and String are incomparable. We compensate for this by requiring principal variable typings. A principal variable typing for [x] and Anylist contains $\{x : u^x\}$, and so u^x can be instantiated to Int. Our intuition is that whenever this phenomenon ("forgetting" parameters) occurs, requiring principal variable typings is very demanding; but otherwise, subject reduction is likely to be violated. As a topic for future work, we want to substantiate this intuition by studying examples.

Acknowledgements. We thank Erik Poll and François Pottier for interesting discussions. Jan-Georg Smaus was supported by an ERCIM fellowship.

References

[1] K. R. Apt. *From Logic Programming to Prolog.* Prentice Hall, 1997.

[2] K. R. Apt and S. Etalle. On the unification free Prolog programs. In A. Borzyszkowski and S. Sokolowski, editors, *Proceedings of MFCS*, LNCS, pages 1–19. Springer-Verlag, 1993.

[3] C. Beierle. Type inferencing for polymorphic order-sorted logic programs. In L. Sterling, editor, *Proceedings of ICLP*, pages 765–779. MIT Press, 1995.

[4] R. Dietrich and F. Hagl. A polymorphic type system with subtypes for Prolog. In H. Ganzinger, editor, *Proceedings of ESOP*, LNCS, pages 79–93. Springer-Verlag, 1988.

[5] F. Fages and M. Paltrinieri. A generic type system for CLP($\mathcal{X}$). Technical report, Ecole Normale Supérieure LIENS 97-16, December 1997.

[6] M. Hanus. *Logic Programming with Type Specifications*, chapter 3, pages 91–140. MIT Press, 1992. In [12].

[7] P. M. Hill and J. W. Lloyd. *The Gödel Programming Language*. MIT Press, 1994.

[8] P. M. Hill and R. W. Topor. *A Semantics for Typed Logic Programs*, chapter 1, pages 1–61. MIT Press, 1992. In [12].

[9] T.K. Lakshman and U.S. Reddy. Typed Prolog: A semantic reconstruction of the Mycroft-O'Keefe type system. In V. Saraswat and K. Ueda, editors, *Proceedings of ILPS*, pages 202–217. MIT Press, 1991.

[10] A. Mycroft and R. O'Keefe. A polymorphic type system for Prolog. *Artificial Intelligence*, 23:295–307, 1984.

[11] G. Nadathur and F. Pfenning. *Types in Higher-Order Logic Programming*, chapter 9, pages 245–283. MIT Press, 1992. In [12].

[12] F. Pfenning, editor. *Types in Logic Programming*. MIT Press, 1992.

[13] J.-G. Smaus. *Modes and Types in Logic Programming*. PhD thesis, University of Kent at Canterbury, 1999.

[14] J.-G. Smaus, F. Fages, and P. Deransart. Using modes to ensure subject reduction for typed logic programs with subtyping. Technical report, INRIA, 2000. Available via CoRR: http://arXiv.org/archive/cs/intro.html.

[15] Z. Somogyi, F. Henderson, and T. Conway. The execution algorithm of Mercury, an efficient purely declarative logic programming language. *Journal of Logic Programming*, 29(1–3):17–64, 1996.

[16] K. Stroetmann and T. Glaß. A semantics for types in Prolog: The type system of PAN version 2.0. Technical report, Siemens AG, ZFE T SE 1, 81730 München, Germany, 1995.

[17] Simon Thompson. *Type Theory and Functional Programming*. Addison-Wesley, 1991.

Dynamically Ordered Probabilistic Choice Logic Programming

Marina De Vos* and Dirk Vermeir

Dept. of Computer Science
Free University of Brussels, VUB
Pleinlaan 2, Brussels 1050, Belgium
{marinadv,dvermeir}@.vub.ac.be
http://tinf2.vub.ac.be

Abstract. We present a framework for decision making under uncertainty where the priorities of the alternatives can depend on the situation at hand. We design a logic-programming language, DOP-CLP, that allows the user to specify the static priority of each rule and to declare, dynamically, all the alternatives for the decisions that have to be made. In this paper we focus on a semantics that reflects all possible situations in which the decision maker takes the most rational, possibly probabilistic, decisions given the circumstances. Our model theory, which is a generalization of classical logic-programming model theory, captures uncertainty at the level of total Herbrand interpretations. We also demonstrate that DOP-CLPs can be used to formulate game theoretic concepts.

1 Introduction

Reasoning with priorities and reasoning under uncertainty play an important role in human behavior and knowledge representation. Recent research has been focused on either priorities, [14, 8, 6][1], or uncertainty, [10, 9, 12, 1] and many others.

We present a framework for decision making under uncertainty where the priorities of the alternatives depend on the different (**probabilistic**) situations. This way we obtain a semantics that reflects all possible situations in which the most rational (probabilistic) decisions are made, given the circumstances. The basic idea for the framework, a logic programming language called "Dynamically Ordered Probabilistic Choice Logic Programming" or DOP-CLP for short, incorporates the intuition behind both ordered logic programs ([8]) and choice logic programs ([4, 5]). The former models the ability of humans to reason with defaults[2] in a logic programming context, using a static ordering of the rules in the program. This works well, as long as probabilities stay out of the picture, but once they are present something extra is needed to express order. Take the famous "Tweety example" for instance: if you are sure that Tweety is indeed a penguin, you should derive that she cannot fly. But suppose you believe for only 30%

---

* Wishes to thank the FWO for its support.

[1] [8] uses the word order instead of priority.

[2] Intuitively, something is true by default unless there is evidence to the contrary.

S. Kapoor and S. Prasad (Eds.): FST TCS 2000, LNCS 1974, pp. 227–239, 2000.

that the bird you are holding is indeed a penguin. Is it then sensible to derive that she is a non-flying bird?

By also taking into account the probabilities of the antecedents of the rules, in addition to their static order, we can overcome this problem. This leads to a **dynamic ordering** of rules, where the priority of a rule depends on the actual situation.

We aim for a decision making framework that allows decisions to have possibly more than two alternatives, as in the case of ordered logic[3]. To accomplish this, we turn to a variant of **Choice Logic Programs**[4, 5], in which the possible alternatives for the decisions are described by choice rules. This approach has two nice side effects. First of all, there is not necessarily a partition of the Herbrand base: atoms can belong to more than one decision or to no decision at all. In the former case, there is a probability distribution over the various alternatives. In the latter case, an atom is either true or false, as in classical logic programming. The second advantage of our approach is that we allow a "lazy" evaluation of the alternatives which become active only when they are present in the head of an applicable choice rule.

An interesting application of DOP-CLP is Game Theory. We provide a transformation from strategic games to DOP-CLPs such that a one-to-one mapping is established between the mixed strategy Nash Equilibria of the game and the stable models of its corresponding DOP-CLP.

2   Dynamically Ordered Probabilistic Choice Logic Programs

In this paper, we identify a program with its grounded version, i.e. the set of all ground instances of its clauses. In addition we do not allow function symbols (i.e. we stick to datalog) so the number of literals is finite.

Definition 1. *A **Dynamically Ordered Probabilistic Choice Logic Program**, or **DOP-CLP** for short, is a finite set of rules of the form $A \leftarrow^p B$, where A and B are (possibly empty) sets of atoms and $p \in \mathbb{R}^+$. For a rule $r \in P$, the set A is called the **head**, denoted H_r, while the set B is called the **body** of the rule r, denoted B_r. The $p \in \mathbb{R}^+$ denotes the **priority** this rule. A rule without a priority number has an infinite priority. We will denote the priority of rule r as $\rho(r)$. The **Herbrand base** of P, denoted $\mathcal{B}_P$, is the set of all atoms appearing in P.*

A rule $H \leftarrow^p B$ can be read as:

> "The occurrence of the events in B forces a probabilistic decision between the elements $h \in H$ and supports each h with a priority p."

This means that rules with more than one head atom express no preference among the different alternatives they create.

The priority of a rule r, $\rho(r)$, indicates the maximal impact of the situation, described by the events of the body, on the preference of the head atom over the other alternatives. The dynamic priority of a rule, which we will define later, adjusts this impact according to the probability of the situation described in the body. By combining

---

[3] In ordered logic the two alternatives are represented using negation.

all the dynamic priorities of rules sharing a common head atom, we obtain an evaluation of the total impact on that atom which can then be used for comparison with other alternatives.

Example 1 (Jakuza). A young member of the Jakuza, the Japanese Mafia, is faced with his first serious job. It will be his duty to control a new victim. Since it is his first job, he goes to his oyabun, head of his clan and mentor, for advice. He tells him that the Jakuza has three methods for controlling its victims: blackmail, intimidation and bribing. The victim can either give in immediately or can put up a stand. In the latter case, she does this by just ignoring the threats of the organization or she can threaten to go to the police. The oyabun is only able to give some information about previous encounters, which still needs to be interpreted in the current situation. So he starts telling about his previous successes. "Every time when I knew that the victim was willing to give in, I resorted to intimidation as this is the easiest technique and each time it worked perfectly. In case we would know that the victim was planning to stand up to us, we looked in to the possibility of bribing. Nine out of ten times, we were successful when we just offered enough money. If you are sure that the victim will run to the police from the moment that you approach him, you have to try to bribe her. Unfortunately this technique worked only 4 times out of 10. When your victim tries to ignore you, you should find something to blackmail her with. However, finding something interesting is not that easy as reflected by a success rate of 3 out of 10 times.
So now it is up to you to make a good estimation of the victim's reaction in order to succeed with your assignment."

All this information can easily be represented as the next DOP-CLP:

$$jakuza \leftarrow$$
$$blackmail \oplus intimidate \oplus bribe \leftarrow^0 jakuza$$
$$stand - up \oplus give - in \leftarrow^0$$
$$ignore \oplus police \leftarrow^0 stand - up$$
$$intimidate \leftarrow^{10} give - in$$
$$bribe \leftarrow^4 police$$
$$blackmail \leftarrow^3 ignore$$
$$enough \oplus more \leftarrow^0 stand - up$$
$$bribe \leftarrow^9 enough$$

An interpretation assigns a probability distribution over every state of affairs[4].

Definition 2. *Let P be a DOP-CLP. A (probabilistic) **interpretation** is a probability distribution $\mathbf{I} : 2^{\mathcal{B}_P} \to [0..1]$.*

In our examples, we will mention only the probabilities of those states that have a positive probability in the interpretation.

---

[4] Each state corresponds to a total interpretation of the choice logic program obtained from P by omitting the priorities. Because we are working with total interpretations we only have to mention the positive part of the interpretation.

Example 2. Recall the Jakuza program of Example 1. The following functions **I**, **J** and **K** are interpretations for this program[5]:

$$\mathbf{I}(\{j, br, s, ig, e\}) = \tfrac{1}{4} \quad \mathbf{J}(\{j, in, p, e, s\}) = \tfrac{3}{20} \quad \mathbf{K}(\{j, bl, s, ig, e\}) = \tfrac{30}{441}$$
$$\mathbf{I}(\{j, br, s, ig, m\}) = \tfrac{5}{12} \quad \mathbf{J}(\{j, in, p, m, g\}) = \tfrac{7}{10} \quad \mathbf{K}(\{j, bl, s, ig, m\}) = \tfrac{240}{441}$$
$$\mathbf{I}(\{j, br, s, ig\}) = \tfrac{1}{6} \quad \mathbf{J}(\{j, in, p, m, s\}) = \tfrac{2}{20} \quad \mathbf{K}(\{j, bl, s, p, e\}) = \tfrac{12}{441}$$
$$\mathbf{I}(\{j, br, s, p\}) = \tfrac{1}{24} \quad \mathbf{J}(\{j, br, p, m, s\}) = \tfrac{1}{20} \quad \mathbf{K}(\{j, bl, s, p, m\}) = \tfrac{96}{441}$$
$$\mathbf{I}(\{j, br, s, m\}) = \tfrac{1}{8} \quad\quad \mathbf{K}(\{j, bl, g, ig, e\}) = \tfrac{5}{441}$$
$$\mathbf{K}(\{j, bl, g, ig, m\}) = \tfrac{40}{441}$$
$$\mathbf{K}(\{j, bl, g, p, e\}) = \tfrac{2}{441}$$
$$\mathbf{K}(\{j, bl, g, p, m\}) = \tfrac{16}{441}$$

Given an interpretation, we can compute the probability of a set of atoms, as the sum of the probabilities assigned to those situations which contain this set of atoms.

Definition 3. *Let* **I** *be an interpretation for a DOP-CLP P. The **probability of set** $A \subseteq \mathcal{B}_P$, denoted $\vartheta_{\mathbf{I}}(A)^6$, is $\vartheta_{\mathbf{I}}(A) = \sum_{A \subseteq Y \subseteq \mathcal{B}_P} \mathbf{I}(Y)$.*

In choice logic programs, the basis of DOP-CLP, a rule is applicable when the body is true, and is applied when both the body and a single head atom are true. This situation becomes more tricky when probabilities come into play. Applicability is achieved when the body has a non-zero probability. In order for a rule to be applied it must be applicable. In addition, we demand that at least one head element has a chance of happening and that no two of them can happen simultaneously.

Definition 4. *Let* **I** *be an interpretation for a DOP-CLP P.*

1. *A rule $r \in P$ is called **applicable** iff $\vartheta_{\mathbf{I}}(B_r) > 0$.*
2. *An applicable rule $r \in P$ is **applied** iff $\exists a \in H_r \cdot \vartheta_{\mathbf{I}}(a) > 0$ and $\forall S \in 2^{H_r}$ with $|S| > 1 : \vartheta_{\mathbf{I}}(S) = 0$.*

We have been referring to alternatives of decisions without actually defining them. Two atoms are alternatives if they appear together in the head of an applicable choice rule. Alternatives are thus dynamic, since the applicability of the rules depends on the interpretation.

Definition 5. *Let* **I** *be an interpretation for a DOP-CLP P.*

- *Two atoms $a, b \in \mathcal{B}_P$ are **alternatives** wrt* **I** *iff $\exists$ applicable $r \in P \cdot \{a, b\} \subseteq H_r$.*
- *The set of all alternatives of an atom $a \in \mathcal{B}_P$ wrt* **I** *is denoted $\Omega_{\mathbf{I}}(a)^7$.*
- *A set $D \subseteq \mathcal{B}_P$ is a **maximal alternative set** wrt* **I** *iff $\forall a, b \in D \cdot a$ and b are alternatives and $\forall c \notin D \cdot \exists a \in D \cdot a$ and c are no alternatives.*
- *$\Delta_{\mathbf{I}}$ is the set of all maximal alternative sets wrt* **I** *.*
- *An atom $a \in \mathcal{B}_P$ is called **single** iff $\Omega_{\mathbf{I}}(a) = a$.*

---

[5] For brevity, the names of the atoms are abbreviated.
[6] When the set A contains just one element a we omit the brackets and write $\vartheta_{\mathbf{I}}(a)$.
[7] Notice that $a \in \Omega_{\mathbf{I}}(a)$. The set $\Omega_{\mathbf{I}}(a) \setminus \{a\}$, denoted $\Omega_{\mathbf{I}}^-(a)$, is the set of all alternatives of a excluding itself.

A naive approach to defining a probability distribution is to insist that the sum of probabilities of the multiple elements in the head of a choice rule must be one. This approach fails in situations of the following kind:

$$a \oplus b \oplus c \leftarrow^0 \ldots$$
$$a \oplus b \leftarrow^4 \ldots$$
$$\ldots$$

In this situation, the atom c would not stand a chance of obtaining a positive probability, although this might be the most favorable alternative.

To overcome this problem, we introduced maximum alternative sets. They group all the atoms that have an alternative relation with each other. It is those sets that will be used for the probability distribution. In the next definition we call an interpretation total if it defines a probability distribution in which the probabilities of the elements of any maximal alternative set add up to one. Furthermore, for all decisions that need to be made, an alternative is selected for every possible outcome.

Definition 6. *A interpretation* $\mathbf{I}$ *for a DOP-CLP* P *is called* **total** *iff* $\forall D \in \Delta_{\mathbf{I}}$,

- $\sum_{a \in D} \vartheta_{\mathbf{I}}(a) = 1$, *and*

- $\forall A \subseteq \mathcal{B}_P$ *such that* $\vartheta_{\mathbf{I}}(A) > 0 \cdot |A \cap D| = 1$.

Example 3. Reconsider the Jakuza program of Example 1 and the interpretations of Example 2. The interpretation $\mathbf{I}$ is not total. Indeed, consider the maximal alternative set $\{ig, p\}$. We have $\{ig, p\} \cap \{br, j, s, m\} = \emptyset$, while $\mathbf{I}(\{br, j, s, m\}) > 0$, and $\vartheta_{\mathbf{I}}(ig) + \vartheta_{\mathbf{I}}(p) = 5/6 + 1/12 \neq 1$. The interpretations $\mathbf{J}$ and $\mathbf{K}$ are total.

As we mentioned earlier, the dynamic priority of a rule adjusts the (static) preference of the rule to the probability that this situation might actually occur. It does this by giving the maximal contribution of the body atoms to the general preference of the head atoms. The dynamic priority of an atom is obtained by taking into account every real contribution of any situation that provides a choice for this atom.

Definition 7. *Let* $\mathbf{I}$ *be an interpretation for a DOP-CLP* P. *The* **dynamic priority** *of a rule* $r \in P$, *denoted* $\varrho_{\mathbf{I}}(r)$, *equals* $\varrho_{\mathbf{I}}(r) = \rho(r) * \vartheta_{\mathbf{I}}(B_r)$.
The **dynamic priority** *of an atom* $a \in \mathcal{B}_P$, *denoted* $\varrho_{\mathbf{I}}(a)$, *is* $\varrho_{\mathbf{I}}(a) = \sum_{r \in P : a \in H_r} \varrho_{\mathbf{I}}(r)$.

The dynamic priority will be used to determine which alternatives of a decision are eligible candidates and which ones are not. An atom is said to be blocked if there exists an alternative which has higher dynamic priority. Preferred atoms are those that block every other alternative. The competitors of an atom are those alternatives which are not blocked by this atom. Their dynamic priority is thus at least as high as that of the atom.

Definition 8. *Let* $\mathbf{I}$ *be an interpretation for a DOP-CLP* P. *An atom* $a \in \mathcal{B}_P$ *is* **blocks** *by* $b \in \Omega_{\mathbf{I}}^-(a)$ *w.r.t.* $\mathbf{I}$ *iff* $\varrho_{\mathbf{I}}(b) > \varrho_{\mathbf{I}}(a)$.
An atom $a \in \mathcal{B}_P$ *is called* **preferred** *in* $\mathbf{I}$ *iff* $\forall b \in \Omega_{\mathbf{I}}^-(a) \cdot \varrho_{\mathbf{I}}(a) > \varrho_{\mathbf{I}}(b)$.
The atom a *is a* **competitor** *of the atom* $b \in \Omega_{\mathbf{I}}^-(a)$ *w.r.t.* $\mathbf{I}$ *if* b *does not block* a *w.r.t* $\mathbf{I}$.

In standard logic programming an interpretation is a model if every rule is either not applicable or applied. When priorities are involved, in order for an interpretation to become a model, it must be possible to assign a zero-probability to atoms which have a more favorable alternative with non-zero probability.

Definition 9. *Let* **I** *be an interpretation for the DOP-CLP P.* **I** *is a **model** for P iff* $\forall r \in P$:

- $\vartheta_{\mathbf{I}}(B_r) = 0$, *i.e. r is not applicable, or*
- *r is applied, or*
- $\forall a \in H_r \cdot \exists b$ *competitor of a w.r.t.* **I** $\cdot \vartheta_{\mathbf{I}}(b) > 0$.

Example 4. Consider again the Jakuza program of Example 1 and its interpretations of Example 2. The interpretation **I** is not a model, since the rule *blackmail* $\leftarrow^{10}$ *ignore* does not satisfy any of the above conditions. This rule is applicable, since $\vartheta_{\mathbf{I}}(ignore) = 5/6$; it is not applied, since $\vartheta_{\mathbf{I}}(blackmail) = 0$; it does not have any competitors since $\varrho_{\mathbf{I}}(blackmail) = 5/2$ while $\varrho_{\mathbf{I}}(intimidate) = 0$ and $\varrho_{\mathbf{I}}(bribe) = 29/12$. The interpretations **J** and **K** are both models.

Proposition 1. *Let P be a DOP-CLP and let* **I** *be a model for it. If* $a \in \mathcal{B}_P$ *is a preferred atom then* $\exists r \in P : a \in H_r \cdot r$ *is applied* .

In some cases, atoms receive a probability which they actually do not deserve. This happens when there is some better qualified alternative (i.e., an alternative that has a higher dynamic priority) that should obtain this probability. Such atoms are called assumptions, since they were just "assumed" to have a chance of happening.

Definition 10. *Let* **I** *be an interpretation for a DOP-CLP P. An atom* $a \in \mathcal{B}_P$ *is called an **assumption** w.r.t.* **I** *iff* $\vartheta_{\mathbf{I}}(a) > 0$ *when either a is blocked or a is single and* $\varrho_{\mathbf{I}}(a) = 0$. **I** *is **assumption-free** iff it contains no assumptions.*

Example 5. Consider once more the Jakuza program of Example 1 and its interpretations in Example 2. The interpretation **J** is not assumption-free, as $\vartheta_{\mathbf{J}}(intimidate) > 0$ and the alternative *bribe* blocks *intimidate*, since $\varrho_{\mathbf{J}}(bribe) = 4 + 27/20 > 3.5 = \varrho_{\mathbf{J}}(intimidate)$. Intuitively, because bribing is more successful than intimidation, intimidation should not be considered at all. The interpretation **K** is assumption-free.

Proposition 2. *Let P be a DOP-CLP and let* **I** *be a total assumption-free interpretation for P. If* $a \in \mathcal{B}_P$ *is a preferred atom then* $\exists r \in P : a \in H_r \cdot r$ *is applied and* $\vartheta_{\mathbf{I}}(a) = 1$.

Interpretations evaluate the likelihood of every possible outcome, by assigning a probability distribution to every situation. These probabilities are influenced by the atoms which are present in each situation. In order to quantify this influence one must know whether the events which occur in any such interpretation are independent of each other. An interpretation which assumes that there is no inter-dependence between atoms, is said to be "independent", as follows:

Definition 11. *Let* **I** *be an interpretation for a DOP-CLP P. We say that* **I** *is **independent** iff* $\forall A \subseteq \mathcal{B}_P$:

- $\vartheta_{\mathbf{I}}(A) = 0$, *or*
- $\forall D \in \Delta_{\mathbf{I}} \ s.t \ |D \cap A| \leq 1 \cdot \vartheta_{\mathbf{I}}(A) = \prod_{a \in A} \vartheta_{\mathbf{I}}(a)$.

Example 6. Consider the interpretations **J** and **K** of Example 2. The interpretation **J** is not independent as $\vartheta_{\mathbf{J}}(\{y, in, p, e, s\}) = 3/20 \neq \vartheta_{\mathbf{J}}(y) * \vartheta_{\mathbf{J}}(in) * \vartheta_{\mathbf{J}}(p) * \vartheta_{\mathbf{J}}(e) * \vartheta_{\mathbf{J}}(s) = 1 * 19/20 * 1 * 3/20 * 8/20$. The interpretation **K** is independent.

Definition 12. *Let* P *be a DOP-CLP. A total independent assumption-free model is said to be **stable**. A stable model is **crisp** if it assigns probability one to a single subset of the Herbrand base.*

Example 7. For the last time we return to the Jakuza example and its three interpretations **I**,**J** and **K** from Example 2. Combining the results from Examples 3, 4, 5 and 6, we can conclude that **K** is the only stable model of the three.

A stable model for the Jakuza example represents a rational choice where the probability of the action is consistent with the estimates on the victim's reactions. In general, stable models reveal all possible situations in which the decisions are made rationally, considering the likelihood of the events that would force such decisions.

3   An Application of DOP-CLPs: Equilibria of Strategic Games

3.1   Strategic Games

A strategic game models a situation where several agents (called players) independently choose which action they should take, out of a limited set of possibilities. The result of the actions is determined by the combined effect of the choices made by each of the players. Players have a preference for certain outcomes over others. Often, preferences are modeled indirectly using the concept of *payoff* where players are assumed to prefer outcomes where they receive a higher payoff.

Example 8 (Bach or Stravinsky). Two people wish to go out together to a music concert. They have a choice between a Bach or Stravinsky concert. Their main concern is to be together, but one person prefers Bach and the other prefers Stravinsky. If they both choose Bach then the person who preferred Bach gets a payoff of 2 and the other a payoff of 1. If both go for Stravinsky, it is the other way around. If they pick different concerts, they both get a payoff of zero.

The game is represented in Fig. 1. One player's actions are identified with the rows and the other player's with the columns. The two numbers in the box formed by row r and column c are the players' payoffs when the row player chooses r and the column player chooses c. The first of the two numbers is the payoff of the row player.

	Bach	Stravinsky
Bach	2, 1	0, 0
Stravinsky	0, 0	1, 2

Fig. 1. Bach or Stravinsky (BoS)

Definition 13 ([11]). *A **strategic game** is a tuple* $\langle N, (A_i), (u_i) \rangle$ *where*

- N *is a finite set of **players**;*
- *for each player* $i \in N$, A_i *is a nonempty set of **actions** that are available to her* [8] *and,*
- *for each player* $i \in N$, $u_i : A = \times_{j \in N} A_j \rightarrow \mathbb{R}$ *is a **utility function** which describes the players' preferences.*

An element $\mathbf{a} \in A$ *is called a **profile**. For a profile* $\mathbf{a}$ *we use* $\mathbf{a}_i$ *to denote the component of* $\mathbf{a}$ *in* A_i. *For any player* $i \in N$, *we define* $A_{-i} = \times_{j \in N \setminus \{i\}} A_j$. *Similarly, an element of* A_{-i} *will often be denoted as* $\mathbf{a}_{-i}$. *For any* $\mathbf{a}_{-i} \in A_{-i}$ *and* $a_i \in A_i$, $(\mathbf{a}_{-i}, a_i)$ *is the profile* $\mathbf{a}' \in A$ *in which* $\mathbf{a}'_i = a_i$ *and* $\mathbf{a}'_j = \mathbf{a}_j$ *for all* $j \neq i$.

A game $\langle N, (A_i), (u_i) \rangle$ is played when each player $i \in N$ selects a single action from the set A_i of actions available to her. Since players are thought to be rational, it is assumed that a player will select an action that, to the best of her knowledge, leads to a "preferred" profile. Of course, this is limited by the fact that she must decide without knowing what the other players will choose.

The notion of Nash equilibrium shows that, in many cases, it is possible to limit the possible outcomes (profiles) of the game.

Definition 14 ([11]). *A **Nash equilibrium** of a strategic game* $\langle N, (A_i), (u_i) \rangle$ *is a profile* $\mathbf{a}^*$ *satisfying* $\forall a_i \in A_i \cdot (\mathbf{a}^*_{-i}, \mathbf{a}^*_i) \geq_i (\mathbf{a}^*_{-i}, a_i)$.

Intuitively, a profile $\mathbf{a}^*$ is a Nash equilibrium if no player can unilaterally improve upon his choice. This means that, given the other players' actions $\mathbf{a}^*_{-i}$, $\mathbf{a}^*_i$ is the best player i can do[9].

Although the notion of Nash equilibrium is intuitive, it does not provide a solution to every game. Take for example the Matching Pennies game.

Example 9 (Matching Pennies). Two people toss a coin. Each of them has to choose head or tail. If the choices differ, person 1 pays person 2 a Euro; if they are the same, person 2 pays person 1 a Euro. Each person cares only about the amount of money that she receives. The game modeling this situation is depicted in Fig. 2. This game does not have a Nash equilibrium.

The intuitive strategy to choose head or tail with 50% frequency each (yielding a profit in 25% of the cases) corresponds with a mixed strategy Nash equilibrium where agents assign a probability distribution over their actions.

---

[8] We assume that $A_i \cap A_j = \emptyset$ whenever $i \neq j$.
[9] Note that the actions of the other players are not known to i.

	Head	Tail
Head	1, 0	0, 1
Tail	0, 1	1, 0

Fig. 2. Matching Pennies (Example 9).

Definition 15 ([11]). *The **mixed extension** of the strategic game $\langle N, (A_i), (u_i) \rangle$ is the strategic game $\langle N, (\Delta(A_i)), (U_i) \rangle$ in which $\Delta(A_i)$ is the set of probability distributions over A_i, and $U_i : \times_{j \in N} \Delta(A)_j \rightarrow \mathbb{R}$ assigns to each $\alpha \in \times_{j \in N} \Delta(A)_j$ the expected value under u_i of the lottery over A that is induced by α (so that $U_i(\alpha) = \sum_{\mathbf{a} \in A} (\prod_{j \in N} \alpha_j(\mathbf{a})) u_i(\mathbf{a})$ if A is finite).*

Note that $U_i(\alpha) = \sum_{a_i \in \alpha_i} \alpha_i(a_i) U_i(\alpha_{-i}, e(a_i))$, for any mixed strategy profile α, where $e(a_i)$ is the degenerate mixed strategy of player i that attaches probability one to $a_i \in A_i$. This because we are working with finite sets of actions (e.g. A_i).

Definition 16 (Mixed Strategy Nash Equilibrium). *A **mixed strategy Nash equilibrium** of a strategic game is a Nash equilibrium of its mixed extension.*

Example 10. Although the matching pennies game (Example 9) does not have a Nash equilibrium, it has the single mixed strategy Nash equilibrium $\{\{Head : 1/2, Tail : 1/2\}, \{Head : 1/2, Tail : 1/2\}\}$, which corresponds to how humans would reason. Apart from its two Nash equilibria, the Bach and Stravinsky game (Example 8) also has the extra mixed strategy Nash equilibrium $\{\{Bach : 2/3, Stravinsky : 1/3\}, \{Bach : 1/3, Stravinsky : 2/3\}\}$.

Each strategic game has at least one mixed strategy Nash equilibrium. Furthermore, each Nash equilibrium is also a mixed strategy Nash equilibrium and every crisp mixed strategy Nash equilibrium (where all the probabilities are either 0 or 1) responds to a Nash equilibrium.

3.2   Transforming Strategic Games to DOP-CLPs

In this subsection, we combine propose an intuitive transformation from strategic games to DOP-CLPs such that the stable models of the former correspond with the mixed strategy Nash equilibria of the latter.

Definition 17. *Let $\langle N, (A_i), (u_i) \rangle$ be a strategic game. The corresponding DOP-CLP P equals $P = \{A_i \leftarrow^0 | \ \forall i \in N\} \cup \{\mathbf{a_i} \leftarrow^{u_i(\mathbf{a})} \mathbf{a}_{-i} \mid \mathbf{a} \in A, \forall i \in N\}$.*

The corresponding DOP-CLP contains two types of rules. First, there are the real choice rules which represent, for each player, the actions she can choose from. The zero priority assures that the choice itself does not contribute to the decision making process. Rules of the second type represent all the decisions a player can make (the heads) according to the situations that the other players can create (the bodies). A rule's priority corresponds with the payoff that the deciding player would receive for the pure strategy profile corresponding to the head and body of the rule.

Example 11. The Bach and Stravinsky game (Example 8) can be mapped to the DOP-CLP P:

$$
\begin{array}{ll}
b_1 \oplus s_1 \leftarrow^0 & b_2 \oplus s_2 \leftarrow^0 \\
b_1 \leftarrow^2 b_2 & b_2 \leftarrow^1 b_1 \\
b_1 \leftarrow^0 s_2 & b_2 \leftarrow^0 s_1 \\
s_1 \leftarrow^0 b_2 & s_2 \leftarrow^0 b_1 \\
s_1 \leftarrow^1 s_2 & s_2 \leftarrow^2 s_1
\end{array}
$$

This program has three stable models:

$$
\begin{array}{lll}
\mathbf{I_1}(b_1, b_2) = 2/9 & \mathbf{I_2}(b_1, b_2) = 1 & \mathbf{I_3}(b_1, b_2) = 0 \\
\mathbf{I_1}(b_1, s_2) = 4/9 & \mathbf{I_2}(b_1, s_2) = 0 & \mathbf{I_3}(b_1, s_2) = 0 \\
\mathbf{I_1}(s_1, b_2) = 1/9 & \mathbf{I_2}(s_1, b_2) = 0 & \mathbf{I_3}(s_1, b_2) = 0 \\
\mathbf{I_1}(s_1, s_2) = 2/9 & \mathbf{I_2}(s_1, s_2) = 0 & \mathbf{I_3}(s_1, s_2) = 1
\end{array}
$$

In this example, the probabilities of the actions correspond with the one given for mixed strategy Nash equilibria. The following theorem demonstrates that this is generally true.

Theorem 1. *Let $\langle N, (A_i), (u_i) \rangle$ be a strategic game and let P be its corresponding DOP-CLP and let $\mathbf{I}$ and α^* be respectively an interpretation for P and a mixed strategy profile for $\langle N, (A_i), (u_i) \rangle$ such that $\forall \mathbf{a} \in A$, $\forall i \in N$, $\alpha_i(\mathbf{a}) = \vartheta_{\mathbf{I}}(\mathbf{a}_i)$. Then, $\mathbf{I}$ is a stable model iff α^* is a mixed strategy Nash equilibrium.*

4   Relationships to Other Approaches

4.1   Logic Programming

It is easy to see that positive logic programs are a subclass of the dynamically ordered choice logic programs, and that the stable models for both systems coincide. All necessary properties follow immediately from the way we handle single atoms. With the current semantics it is impossible to have a mapping between the stable models of a choice logic program ([4]) and the crisp stable models of the corresponding DOP-CLP. Indeed, our system is more credulous, since it allows a pure choice (probability 1) when two alternatives are equally preferred. However, we have that every stable model of a CLP is also a crisp stable model of the corresponding DOP-CLP.

4.2   Priorities

The logic programming language using priorities that corresponds best to our approach is dynamically ordered choice logic programming (OCLP) introduced in [6]. Although OCLP does not work with probabilities, these two systems have a common approach to and a similar notion of alternatives, in the sense that alternatives appear in the head of an applicable choice rule. OCLP also requires that this choice rule has a higher priority than the rule for which one computes the head atoms' alternatives. So, the main difference with our approach is the way that OCLP uses priority to create alternatives. Ordered logic programs ([8]) can easily be transformed to DOP-CLPs in such a way that the credulous stable models of the former correspond with the crisp stable models of the latter. For the same reason that we mentioned for CLPs, it is not yet possible to represent the skeptical stable model semantics for ordered logic programs. In [3], preference in extensive disjunctive logic programming is considered. As far as overriding is concerned the technique corresponds rather well with skeptical defeating of [6], but alternatives are fixed as an atom and its (classical) negation. Dynamic preference in extended logic programs is introduced in [2] in order to obtain a better suited well-founded semantics. Preferences/priorities are incorporated here as rules in the program. While alternatives make our system dynamic, [2] introduces the dynamics via a stability criterion that overrules preference information but the alternatives remain static. A totally different approach is proposed in [14]. Here the preferences are defined between atoms without references to the program. After defining models in the usual way, one then uses preferences to filter out the less preferred models.

4.3 Uncertainty

A lot of researchers [1, 10, 9, 12, 13] have tackled the problem of bringing probabilities into logic programming. The probabilities used can be divided into two categories depending on the type of knowledge symbolized: statistical or belief. [1] concentrates on the first type while [10, 12, 13] are more interested in the latter. [9] is one of the few that is able to handle both types. Our formalism focuses mainly on knowledge of belief although it is possible to use statistical knowledge for defining the static priorities. An other difference between the various systems is the way they introduce probabilities and handle conjunctions. For example, [9] works with probability intervals and then uses the rules of probability to compute the probability of formulae. In this respect, we adopt the possible world/model theory of [10, 12]. However, we introduce probabilities at the level of interpretations, while they hard-code the alternatives by means of disjoint declarations together with probabilities, and the other atoms are computed by means of the minimal models of the logic program.

4.4 Games and Logic Programming

The logical foundations of game theory have been studied for a long time in epistemic logic. Only recently, researchers have become interested in the relationships between game theory and logic programming. The first to do so was [7]. It was shown that n-person games or coalition games can be transformed into an argumentation framework such that the NM-solutions of the game correspond with the stable extensions of the corresponding argumentation framework. [7] illustrated also that every argumentation framework can be transformed into a logic program such that the stable extensions of the former coincide with the stable models of the latter. In [4] it was demonstrated that each strategic game could be transformed into a CLP such that the Nash equilibria of the former correspond with the stable models of the latter. [6] shows that OCLPs can be used for an elegant representation of extensive games with perfect information such that, depending on the transformation, either the Nash or the subgame perfect equilibria of the game correspond with the stable models of the program. Concerning mixed strategy Nash equilibria of strategic games, the approach which is the most related to ours is the Independent Choice Logic of [13]. [13] uses (acyclic) logic programs to deterministically model the *consequences* of choices made by agents. Since choices are external to the logic program, [13] restricts the programs further, not only to be deterministic (i.e. each choice leads to a unique stable model) but also to be independent, in the sense that literals representing alternatives may not influence each other, e.g. they may not appear in the head of rules. ICL is further extended to reconstruct much of classical game theory and other related fields. The main difference with our approach is that we do not go outside of the realm of logic programming to recover the notion of equilibrium. The basis of his formalism does not contain probabilities but works with selector functions over the hypotheses and then works with the (unique) stable model that comes from the program itself. This way one creates a possible world semantics. Our transformation makes sure that every atom is an alternative of a choice/decision for which a probability can be computed.

References

[1] Fahiem Bacchus. **LP**, a Logic for Representing and Reasoning with Statistical Knowledge. *Computational Intelligence*, 6:209–231, 1990.

[2] Gerhard Brewka. Well-Founded Semantics for Extended Logic Programs with Dynamic Preferences. *Journal of Articficial Intelligence Research*, 4:19–36, 1996.

[3] Francesco Buccafurri, Wolfgang Faber, and Nicola Leone. Disjunctive Logic Programs with Inheritance. In Danny De Schreye, editor, *International Conference on Logic Programming (ICLP)*, pages 79–93, Las Cruces, New Mexico, USA, 1999. The MIT Press.

[4] Marina De Vos and Dirk Vermeir. Choice Logic Programs and Nash Equilibria in Strategic Games. In Jörg Flum and Mario Rodríguez-Artalejo, editors, *Computer Science Logic (CSL'99)*, volume 1683 of *Lecture Notes in Computer Science*, pages 266–276, Madrid, Spain, 1999. Springer Verslag.

[5] Marina De Vos and Dirk Vermeir. On the Role of Negation in Choice Logic Programs. In Michael Gelfond, Nicola Leone, and Gerald Pfeifer, editors, *Logic Programming and Non-Monotonic Reasoning Conference (LPNMR'99)*, volume 1730 of *Lecture Notes in Artificial Intelligence*, pages 236–246, El Paso, Texas, USA, 1999. Springer Verslag.

[6] Marina De Vos and Dirk Vermeir. A Logic for Modelling Decision Making with Dynamic Preferences. Accepted at Jelia 2000. Lecture Notes in Artificial Intelligence. Springer Verslag.

[7] Phan Minh Dung. On the acceptability of arguments and its fundamental role in nonmonotonic reasoning, logic programming and n-person games. *Artificial Intelligence*, 77(2):321–358, 1995.

[8] D. Gabbay, E. Laenens, and D. Vermeir. Credulous vs. Sceptical Semantics for Ordered Logic Programs. In J. Allen, R. Fikes, and E. Sandewall, editors, *Proceedings of the 2nd International Conference on Principles of Knowledge Representation and Reasoning*, pages 208–217, Cambridge, Mass, 1991. Morgan Kaufmann.

[9] Raymond Ng and V.S. Subrahmanian. A semantical framework for supporting subjective and conditional probabilities in deductive databases. In Koichi Furukawa, editor, *Proceedings of the 8th International Conference on Logic Programming*, pages 565–580. MIT, June 1991.

[10] Liem Ngo. Probabilistic Disjunctive Logic Programming. In Eric Horvitz and Finn Jensen, editors, *Proceedings of the 12th Conference on Uncertainty in Artificial Intelligence (AI-96)*, pages 387–404, San Francisco, aug 1996. Morgan Kaufmann Publishers.

[11] Martin J. Osborne and Ariel Rubinstein. *A Course in Game Theory*. The MIT Press, Cambridge, Massachusets, London, Engeland, third edition, 1996.

[12] David Poole. Logic programming, abduction and probability. In Institute for New Generation Computer Technology (ICOT), editor, *Proceedings for the International Conference on Fifth Generation Computer Systems*, pages 530–538. IOS Press, 1992.

[13] David Poole. The independent choice logic for modeling multiple agents under uncertainty. *Artificial Intelligence*, 94(1–2):7–56, 1997.

[14] Chiaki Sakama and Katsumi Inoue. Representing Priorities in Logic Programs. In Michael Maher, editor, *Proceedings of the 1996 Joint International Conference and Symposium on Logic Programming*, pages 82–96, Cambridge, September2–6 1996. MIT Press.

Coordinatized Kernels and Catalytic Reductions: An Improved FPT Algorithm for Max Leaf Spanning Tree and Other Problems

Michael R. Fellows[1], Catherine McCartin[1], Frances A. Rosamond[1], and
Ulrike Stege[2]*

[1] School of Mathematical and Computing Sciences, Victoria University
Wellington, New Zealand
{Mike.Fellows,Catherine.Mccartin,Fran.Rosamond}@mcs.vuw.ac.nz
[2] Department of Computer Science, University of Victoria
Victoria, B.C., Canada
stege@csr.uvic.ca

Abstract. We describe some new, simple and apparently general methods for designing *FPT* algorithms, and illustrate how these can be used to obtain a significantly improved *FPT* algorithm for the MAXIMUM LEAF SPANNING TREE problem. Furthermore, we sketch how the methods can be applied to a number of other well-known problems, including the parametric dual of DOMINATING SET (also known as NONBLOCKER), MATRIX DOMINATION, EDGE DOMINATING SET, and FEEDBACK VERTEX SET FOR UNDIRECTED GRAPHS. The main payoffs of these new methods are in improved functions $f(k)$ in the *FPT* running times, and in general systematic approaches that seem to apply to a wide variety of problems.

1 Introduction

The investigations on which we report here are carried out in the framework of parameterized complexity, so we will begin by making a few general remarks about this context of our research. The subject is concretely motivated by an abundance of natural examples of two different kinds of complexity behaviour. These include the well-known problems MIN CUT LINEAR ARRANGEMENT, BANDWIDTH, VERTEX COVER, and MINIMUM DOMINATING SET (for definitions the reader may refer to [GJ79]).

All four of these problems are *NP*-complete, an outcome that is now so routine that we are almost never surprised. In the classical complexity framework that pits polynomial-time solvability against the ubiquitous phenomena of *NP*-hardness, they are therefore indistinguishable. All four of these decision problems take as input a pair consisting of a graph G and a positive integer k. The positive integer k is the *natural parameter* for all four problems, although one might also wish to consider eventually other problem parameterizations, such as treewidth. We have the following contrasting facts:

* Ulrike Stege is supported by the Pacific Institute for the Mathematical Sciences (PIMS), where she is a postdoctoral fellow.

S. Kapoor and S. Prasad (Eds.): FST TCS 2000, LNCS 1974, pp. 240–251, 2000.

1. MIN CUT LINEAR ARRANGEMENT and VERTEX COVER are solvable in linear time for any fixed k.
2. The best known algorithms for BANDWIDTH and MINIMUM DOMINATING SET are respectively $O(n^k)$ and $O(n^{k+1})$.

In fact, we now have very strong evidence, in the framework of parameterized complexity, that probably BANDWIDTH and MINIMUM DOMINATING SET do not admit algorithms of the qualitative type that MIN CUT LINEAR ARRANGEMENT and VERTEX COVER have.

1.1   What Is the Nature of This Evidence?

We offer a new view of the study of computability and of its sequel concerns with efficient computability. We motivate how this naturally divides into three main zones of discussion, anchored by variations on the HALTING PROBLEM.

In the first zone, we have the unsolvability of the HALTING PROBLEM, the unsolvability of other problems following by recursive reductions, and Gödel's Theorem as a corollary.

In the second zone, we have "classical complexity" where the reference problem is the HALTING PROBLEM FOR NONDETERMINISTIC P-TIME TURING MACHINES. This problem is trivially NP-complete and essentially defines the class NP. Another way to look at this problem is as a generic embodiment of computation that is potentially exponential. If at each nondeterministic step there were two possible choices of transition, then it is a reasonable conjecture that, in general, we will not be able to analyze Turing machines of size n for the possibility of halting in n steps in much less than the $O(2^n)$ time.[1]

The third zone of negotiation with intractability is anchored by the k-STEP HALTING PROBLEM FOR NONDETERMINISTIC TURING MACHINES (k-NDTM), where in this case we mean Turing machines with an unrestricted alphabet size, and with unrestricted nondeterminism at each step. This is a generic embodiment of n^k computational complexity. For the same reasons as in the second zone of negotiation, we would not expect any method of solving this problem that greatly improves on exhaustively exploring the n-branching depth-k tree of possible computation paths (for a Turing machine of size n).

This leads to the following three basic definitions of the parameterized complexity framework.

Definition 1. *A parameterized language is a subset $L \subseteq \Sigma^* \times \Sigma^*$. For notational convenience, and without any loss of generality, we can also consider that $L \subseteq \Sigma^* \times I\!N$.*

---

[1] It would take $O(2^n)$ time to exhaustively explore the possible computation paths — because nondeterministic Turing machines are so unstructured and opaque, i.e., such a generic embodiment of exponential possibility.

Definition 2. *A parameterized language L is fixed-parameter tractable (FPT) if there is an algorithm to determine if $(x, k) \in L$ in time $f(k) + n^c$ where $|x| = n$, c is a constant, and f is a function (unrestricted).*

Definition 3. *A parameterized language L is many:1 parametrically reducible to a parameterized language L' if there is an FPT algorithm that transforms (x, k) into (x', k') so that:*

1. $(x, k) \in L$ *if and only if* $(x', k') \in L'$, *and*
2. $k' \leq g(k)$ *(where g is an unrestricted function; k' is purely a function of k)*

The analog of *NP* in the third zone of dicussion is the parameterized complexity class $W[1]$ [DF95b]. That the k-NDTM problem is complete for $W[1]$ was proven by Cai, Chen, Downey and Fellows in [CCDF97]. Since BANDWIDTH and DOMINATING SET are hard for $W[1]$, we thus have strong natural evidence that they are not fixed-parameter tractable, as VERTEX COVER and MIN CUT LINEAR ARRANGEMENT are.[2]

1.2   What Is the Current Status of This "Third Zone" of Discussion?

As an example of how useful *FPT* algorithms can be, we know that VERTEX COVER can be solved (after several rounds of improvement) in time $O((1.27)^k + n)$ for graphs of size n [CKJ99]. The problem is thus well-solved for $k \leq 100$ analytically. In practice, this worst-case analytical bound appears to be pessimistic. The current best *FPT* algorithms for this problem (which deliver an optimal solution if they terminate) appear to have completely solved the problem for input graphs *in toto*, so long as the parameter value is $k \leq 200$. These new effective algorithms for small ranges of k have applications in computational biology [Ste99].

Another promising result is the fixed-parameter-tractable algorithm for 3-HITTING-SET presented in [NR00b]. The running time is $O(2.270^k + n)$ where k is the size of the hitting set to determine and n denotes the length of the encoding of the input. As VERTEX COVER 3-HITTING SET has applications in computational biology.

The Biologist, knowing a bit about algorithms, asks for an algorithm that is **"like sorting"**, i.e., a polynomial-time algorithm for her problem. Working with an *NP*-hard problem, if the fixed-parameter for the problem to solve is rather small (e.g., PROTEIN FOLDING involves as the (natural) parameter the number

---

[2] Further background on parameterized complexity can be found in [DF98]. We remark in passing that if the fundamental mission of theoretical computer science is conceived of as *empirical and explanatory*, like theoretical physics, then a two- (or more-) dimensional theoretical framework might well be more suitable than the one-dimensional framework inherited from recursion theory, to the task of explaining the crucial differences in intrinsic problem complexity encountered in natural computational practice, even if the explanatory framework involves phenomenologically "unequal" dimensions — a situation frequently encountered in physics.

of adjacencies between hydrophobic constituents of the protein sequence, that is the parameter is less than 100 for interesting applications) the knowledgeable biologist will henceforth ask, "Can I get an algorithm **like Vertex Cover**?" There is no way to answer this question without taking the discussion into the third natural zone.

We remark, that the parameterize complexity of PROTEIN FOLDING as well as TOPLOGICAL CONTAINMENT FOR GRAPHS and Directed Feedback Vertex Set is still open. All three problems are conjectured to be in *FPT*. For *NP*-completeness of PROTEIN FOLDING and TOPLOGICAL CONTAINMENT FOR GRAPHS we refer to [CGPPY98, BL98] and [DF98], respectively. *NP*-completeness of Directed Feedback Vertex Set is shown in [GJ79].

1.3   The Substantial Open Question About Parameterized Complexity

Despite the fact that *logically, in some sense,* *NP*-completeness can now reasonably be considered a rather unimportant issue for problems that are, when naturally parameterized, fixed-parameter tractable, and *for which the main applications are covered by small or moderate parameter ranges.* From a practical point of view there is still an important unresolved question that motivates our work in this paper:

What are typical functions $f(k)$ for problems in *FPT*?

We make two contributions.

- We substantially improve the best known *FPT* algorithm for the MAX LEAF SPANNING TREE problem. The best previous algorithm due to Downey and Fellows in [DF95a, DF98] has a running time of $O(n + (2k)^{4k})$. Our algorithm runs in time $O(n + (k+1)(14.23)^k)$. In the concluding section we discuss the fine-grained significance of this improvement.
- We introduce new methods that appear to be widely useful in designing improved *FPT* algorithms. The first new method is that of *coordinatized kernelization arguments* for establishing problem kernelizations. The second new method, *catalytic reduction*, employs a small amount of partial information about potential solutions to guide the efficient development of a search tree.

2   Prototype: An Improved FPT Algorithm for the Max Leaf Spanning Tree Problem

The history of the problem and some recent complexity developments can be found in [D74, GJ79, GMM94, GMM97, LR98]. An interesting application of the problem is described in [KKRUW95]. The flagship problem for the new techniques we introduce here is defined as follows.

MAX LEAF SPANNING TREE

Input: A graph G and a positive integer k.
Parameter: k
Question: Does G have a spanning tree with at least k leaves?

One of the remarkably nice properties of *FPT* is that the following is an equivalent definition of the tractable class of parameterized problems [DFS99].

Definition 4. *A parameterized language L is in FPT if and only if there is:*

1. *A function $g(k)$.*
2. *A 2-variable polynomial $q(n, k)$.*
3. *A many:1 parametric reduction Φ of L to itself, requiring time at most $q(n, k)$, that transforms an instance (x, k), where $|x| = n$, to an instance (x', k') with $|x'| \leq g(k)$ and $k' \leq k$, so that $(x, k) \in L$ if and only if $(x', k') \in L$.*

In other words, a problem with parameter k is in *FPT* if and only if an input to the problem can be reduced *in ordinary polynomial time* to an equivalent input whose size is bounded by a function (only) of the parameter. For most problems in *FPT*, moreover, a *natural* set of reduction rules are known that accomplish the transformation Φ by a series of "local simplifications". This process is termed *kernelization* in the terminology of [DF95a][3] and, currently, the main practical methods of *FPT*-algorithm design are based on *kernelization* and *the method of bounded search trees*.

The idea of kernelization is relatively simple and can be quickly illustrated for the VERTEX COVER problem. If the instance is (G, k) and G has a pendant vertex v of degree 1 connected to the vertex u, then it would be silly to include v in any solution (it would be better, and equally necessary, to include u), so (G, k) can be reduced to $(G', k - 1)$, where G' is obtained from G by deleting u and v. Some more complicated and much less obvious reduction rules for the VERTEX COVER problem can be found in the current state-of-the-art *FPT* algorithms (see [BFR98, DFS99, CKJ99, NR99b, Ste99]). The basic schema of this method of *FPT* algorithm design is that reduction rules are applied until an *irreducible* instance (G', k') is obtained. At this point in the *FPT* algorithm, a *Kernelization Lemma* is invoked to decide all those instances where the reduced instance G' is larger than $g(k')$ for some function g. For example, in the cases of VERTEX COVER and PLANAR DOMINATING SET, if a reduced graph is *large* then (G', k') is a no-instance for a suitable linear function g. In the case of MAX LEAF SPANNING TREE and NONBLOCKER, large reduced instances are automatically yes-instances.

---

[3] These natural kernelization algorithms have significant applications in the design of heuristics for hard problems, since they are a reasonable preprocessing step for *any* algorithmic attack on an intractable problem [DFS99].

In first phase of our algorithm a set of reduction rules transforms an instance (G, k) of MAX LEAF SPANNING TREE to another instance (G', k') where $k' \leq k$ and $|G'| \leq 5.75k$. By exploring all k-subsets of the problem kernel G', this immediately implies an *FPT* algorithm with running time $O(n + k2^{5.75k}) = O(n + k(33.1)^k)$. But we will do substantially better than that, namely we present an algorithm running in time $O(n + (k + 1)(14.23)^k)$.

Our algorithm has three phases:

Phase 1: Reduction to a problem kernel of size $5.75k$.
Phase 2: The introduction of catalytic vertices.
Phase 3: A search tree based on catalytic branching (section 2.2) and coordinatized reduction (section 2.1).

Our algorithm is actually based on a slight variation on MAX LEAF SPANNING TREE defined as follows.

CATALYTIC MAX LEAF SPANNING TREE

Input: A graph $G = (V, E)$ with a distinguished *catalytic* vertex $t \in V$, $k \in Z^+$.
Parameter: k
Question: Does G have a spanning tree T having at least k leaves, such that t is an internal vertex of T?

In the following subsection we prove that this variant of the originl problem has a kernel of size $5.75k$. Because the reduction rules used are almost identical to the reduction rules used in the proof that MAX LEAF SPANNING TREE has a kernel of size $5.75k$, we will concentrate on the kernelization of CATALYTIC MAX LEAF SPANNING TREE only.

2.1 The Kernelization Lemma and the Method of Coordinatized Kernels

How does one proceed to discover an adequate set of reduction rules, or elucidate (and prove) a bounding function $g(k)$ that insures for instances larger than this bound, that the question can be answered simply?

The technique of *coordinatized kernels* is aimed at these difficulties, and we will illustrate it by example with the MAX LEAF SPANNING TREE problem. We seek a Lemma of the following form:

Lemma 1. *If $(G = (V, E), k)$ is a reduced instance of* CATALYTIC MAX LEAF SPANNING TREE *with catalytic vertex $t \in V$, and G has more than $g(k)$ vertices, then (G, k) with catalytic vertex t is a yes-instance.*

Proof. Suppose that:

(1) G has more than $g(k)$ vertices. (We will eventually determine $g(k)$, cf. page 247.)

(2) G is connected and reduced. (As we make the argument, we will see how to define the reduction rules.)

(3) G is a yes-instance for k, witnessed by a subtree T (with t internal; not necessarily spanning) having k leaves.

(4) G is a no-instance for $k + 1$.

(5) Among all such G satisfying (1-4), the witnessing tree T has a minimum possible number of vertices.

(6) Among all such G and T satisfying (1-5), the quantity $\sum_{l \in L} d(t, l)$ is minimized, where L is the set of leaves of T and $d(t, l)$ is the distance in T to the root vertex t.

Then we argue for a contradiction.

Comment. The point of all this is to set up a framework for argument that will allow us to see what reduction rules are needed, and what $g(k)$ can be achieved. In essence we are setting up a (possibly elaborate, in the spirit of extremal graph theory) argument by minimum counterexample — and using this as a discovery process for the *FPT* algorithm design. Condition (3) gives us a way of "coordinatizing" the situation by giving us the structure of a solution to refer to (how this is used will become clear as we proceed).

Since G is connected, any tree subgraph T of G with k leaves extends to a spanning tree with k leaves. This witnessing subgraph given by condition (3) is minimized by condition (5). Refer to the vertices of $V - T$ as *outsiders*. The following claims are easily established. The first five claims are enforced by condition (4).

Claim 1: No outsider is adjacent to an internal vertex of T.
Claim 2: No leaf of T can be adjacent to two outsiders.
Claim 3: No outsider has three or more outsider neighbors.
Claim 4: No outsider with 2 outsider neighbors is connected to a leaf of T.
Claim 5: The graph induced by the outsider vertices has no cycles.

It follows from Claims (1-5) that the subgraph induced by the outsiders consists of a collection of paths, where the internal vertices of the paths have degree 2 in G. Since we are ultimately attempting to bound the size of G, this suggests (as a discovery process) the following reduction rule for kernelization.

Kernelization Rule 1: If (G, k) has two adjacent vertices u and v of degree 2, neither of which is the catalyst t, then:
(Rule 1.1) If uv is a bridge, then contract uv to obtain G' and let $k' = k$.
(Rule 1.2) If uv is not a bridge, then delete the edge uv to obtain G' and let $k' = k$.

The soundness of this reduction rule is not completely obvious, although not difficult. Having now partly clarified condition (2), we can continue the argument. The components of the subgraph induced by the outsiders must consist of paths having either 1,2 or 3 vertices.

The first possibility leads to another reduction rule which eliminates pendant vertices. This leads to a situation where the only possibilities for a component C of the outsider graph are:

1. The component C consists of a single vertex and C has at least 2 leaf neighbors in T.
2. The component C consists of two vertices, and C has at least 3 leaf neighbors in T.
3. The component C has three vertices, and has at least four leaf neighbors in T.

The weakest of the population ratios for our purposes in bounding the kernel size is given by case (3). We can conclude, using Claim 2, that the number of outsiders is bounded by $3k/4$.

The next step is to study the tree T. Since it has k leaves it has at most $k-2$ branch vertices. Using conditions (5) and (6), it is not hard to see that:

1. Any path in T between a leaf and its parental branch vertex has no subdivisions.
2. Any other path in T between branch vertices has at most 3 subdivisions (with respect to T).

Consequently T has at most $5k$ vertices, unless there is a contradiction. This yields our $g(k)$ of $5.75k$. We believe that this bound can be improved by a more detailed structural analysis in this same framework.

2.2   Catalytic Reduction in Search Tree Branching

The catalytic branching technique is described as follows. Let $c = 5.75$ for convenience. Assume that any instance G for parameter k can be reduced in linear time to an instance G' of size at most ck. Suppose we are considering an instance (G, k) with catalytic vertex t. We can assume that G is connected. Consider a neighbor u of t in G.

Catalytic Branching. We have the following basic branching procedure: (G, k) with catalytic vertex t is a *yes*-instance if and only if one of the following two branch instances is a *yes*-instance. The first branch is developed on the assumption that u is also an internal vertex of a k-leaf spanning tree T (for which t is internal). The second branch is developed on the assumption that u is a leaf for such a tree T.

First Branch: Here we have (G', k), where G' is obtained from G by contracting the edge between t and u. The resulting combined vertex is the catalytic vertex for G'.

Second Branch: Here we begin with $(G', k-1)$, where G' is obtained by deleting u. But now, since the parameter has been decreased, we may *re-kernelize* so that the resulting graph has size at most $c(k-1)$. Depending on the size of G', the size of the instance that we reduce to on this branch is somewhere between $n-1$ and $n-c-1$, when G has size n, in the worst case.

The key to the efficiency of this technique is in the re-kernelization on the second branch. Because the amount of re-kernelization varies, this leads to a somewhat complicated recurrence. Our bound on the running time is based on a simpler recurrence that provides an upper bound that is probably not particularly tight.

We have thus described Phase 3 of our algorithm. We must still describe Phase 2.

Introducing Catalytic Vertices. A simple way to accomplish this task is to simply choose a set of $k+1$ vertices in G. If (G, k) is a yes-instance for MAX LEAF SPANNING TREE then one of these vertices can be assumed to be an internal vertex of a solution k-leaf spanning tree. The $k+1$ branches must all be explored.

2.3   Analysis of the Running Time

We define an abstract *value* $v(n, k)$ for a node (G, k) in the search tree, where G is a graph on n vertices. Choosing an appropriate *abstract weighting* w (by computational experiment) for the parameter k, in order to capture some of the information about the efficiency of catalytic branching, we define $v(n, k) = 8k+n$ (that is, $w = 8$ seems to work best for our current kernelization bound). Our kernelization bound of $n \leq 5.75k$ means that we require an upper bound on the size of a search tree with root value $v(n, k)$ of at most $13.75k$. The catalytic branching gives the recurrence

$$f(v) \leq f(v - 1) + f(v - 9)$$

which yields a positive real root of $\alpha = 1.2132$. Evaluating α^v at $v = 13.75k$, and noting that the nodes of the search tree require $O(k)$ time to process, we immediately obtain a parameter function of $k(14.23)^k$ (for each of the $k + 1$ search trees initiated in Phase 2 by the introduction of a catalytic vertex). By the speedup technique of Niedermeier and Rossmanith [NR00a], we get a running time of $O(n + (k + 1)(14.23)^k)$ for our algorithm.

3   Catalytic Branching as General *FPT* Technique

The catalytic branching strategy can easily be adapted to a number of other *FPT* problems. The following are some sketches of further applications. (Note, however, that to make use of catalytic branching, it is first necessary to prove a kernelization procedure that respects the presence of a catalytic vertex.)

*Example 1 (*FEEDBACK VERTEX SET FOR UNDIRECTED GRAPHS*).* The catalytic vertex t is required *not* to be in the feedback vertex set. If the neighbor u is also *not* in the feedback vertex set, then the edge tu can be contracted. If u is in the fvs, then (G', k') is obtained by deleting u, setting $k' = k - 1$ and re-kernelizing.

*Example 2 (*PLANAR DOMINATING SET*).* The catalytic vertex t is required to belong to the dominating set. If the neighbor u is not in the dominating set then it can be deleted (first branch). On the second branch, the edge tu can be contracted, and the resulting graph can be re-kernelized for $k - 1$.

*Example 3 (*EDGE DOMINATING SET). (The *FPT* algorithms for MATRIX DOMINATION are currently based on a reduction to this problem.) Very similar to Example 2.

*Example 4 (*NONBLOCKER *(Also called enclaveless sets [HHS98]).* The parametric dual of MINIMUM DOMINATING SET.)

> *Input:* A graph $G = (V, E)$ where $|V| = n$ and a positive integer k.
> *Parameter:* k
> *Question:* Does G admit a dominating set of size at most $n - k$? Equivalently, is there a set $N \subseteq V$ of size k with the property that for every element $x \in N$, there is a neighbor y of x in $V - N$?

For this problem, a kernelization respecting a catalytic vertex is known. We require that the catalytic vertex t be a member of $V - N$. On the first branch of the search tree, we can contract tu if the neighbor u of t is also in $V - N$. On the second branch, we delete u and re-kernelize for $k - 1$. The detailed algorithm is going to be published elsewehere [FMRS00].

Catalytic branching is a general idea that might be applied in other settings besides graph problems — what it really amounts to is a search tree branching method based on retaining a small amount of partial information about potential solutions.

4   Concluding Remarks

How does one evaluate the goodness of an *FPT* algorithm? Since every problem in *FPT* can be solved in time $f(k) + n^c$ where c is a fixed constant (usually $c \leq 3$), and there are no hidden constants, we can measure the success of an *FPT* algorithm by its *klam value*, defined to be the maximum k such that the parameter function $f(k)$ for the algorithm (where $c \leq 3$) is bounded by some universal limit U on the number of basic operations any computation in our practical universe can perform. We will (perhaps too optimistically) take $U = 10^{20}$. This might appear a bit strange at first, but parameterized complexity in many ways represents a welding of engineering sensibilities (with the attendant sensitivity to particular finite ranges of magnitudes), and mathematical complexity analysis. Engineers have never been too keen on asymptotic analysis for practical situations.

MAX LEAF SPANNING TREE was first observed to be in *FPT* nonconstructively via the Robertson-Seymour graph minors machinery by Fellows and Langston [FL88]. This approach had a klam value of zero! Bodlaender subsequently gave a constructive *FPT* algorithm based on depth-first search methods with a parameter function of around $17k^4!$ which has a klam value of 1 [Bod89]. This was improved by Downey and Fellows [DF95a, DF98] to $f(k) = (2k)^{4k}$ which has a klam value of 5. Our algorithm here has a klam value of 16, according to our current analysis, which is probably not very tight.[4]

At this point in time, there are many examples of trajectories of this sort in the design of *FPT* algorithms. VERTEX COVER is another classic example of such a trajectory of (eventually) striking improvements (see [DF98]). What these algorithm design trajectories really show is that we are still discovering the basic elements, tricks and habits of mind required to devise efficient *FPT* algorithms. It is a *new game* and it is a rich game. After many rounds of improvements the best known algorithm for VERTEX COVER runs in time $O((1.27)^k + n)$ [CKJ99] and has a klam value of 192. Will MAX LEAF SPANNING TREE admit klam values of more than 50? How much more improvement is possible? Can any plausible mathematical limits to such improvements be established?

References

[Bod89] H. L. Bodlaender. "On linear time minor tests and depth-first search." In F. Dehne et al. (eds.), *Proc. First Workshop on Algorithms and Data Structures*, LNCS 382, pp. 577–590, 1989.

[BFR98] R. Balasubramanian, M. R. Fellows, and V. Raman. "An Improved Fixed-Parameter Algorithm for Vertex Cover." *Information Processing Letters* 65:3, pp. 163–168, 1998.

[BL98] B. Berger, T. Leighton. "Protein Folding in the Hydrophobic-Hydrophilic (HP) Model is NP-Complete." In S. Istrail, P. Pevzner, and M. Waterman (eds.), *Proceedings of the Second Annual International Conference on Computational Molecular Biology (RECOMB98)*, pp. 30–39, 1998.

[CCDF97] L. Cai, J. Chen, R. Downey, and M. Fellows. "The parameterized complexity of short computation and factorization." *Archive for Mathematical Logic* 36, pp. 321–338, 1997.

[CGPPY98] P. Crescenzi, D. Goldman, C. Papadimitriou, A. Piccolboni and M. Yannakakis. "On the complexity of protein folding." In S. Istrail, P. Pevzner, and M. Waterman (eds.), *Proceedings of the Second Annual International Conference on Computational Molecular Biology (RECOMB98)*, 1998.

[CKJ99] J. Chen, I. Kanj, and W. Jia. "Vertex cover: Further Observations and Further Improvements." *25th International Workshop on Graph-Theoretic Concepts in Computer Science (WG'99)* Ascona, Switzerland, June 1999.

[DF95a] R. Downey and M. Fellows. "Parameterized Computational Feasibility." *P. Clote, J. Remmel (eds.): Feasible Mathematics II* Boston: Birkhauser, pp. 219–244, 1995.

[DF95b] R. Downey and M. Fellows. "Fixed-parameter tractability and completeness II: completeness for $W[1]$." *Theoretical Computer Science A* 141, pp. 109–131, 1995.

[DF98] R. Downey and M. Fellows. *Parameterized Complexity*. Springer-Verlag, 1998.

[DFS99] R. Downey, M. Fellows, and U. Stege. "Parameterized complexity: a framework for systematically confronting computational intractability." In: *Contemporary Trends in Discrete Mathematics* (R. Graham, J. Kratochvil, J. Nesetril and

---

[4] Keep in mind that this is still worst-case analysis, via theoretical estimates on the sizes of search trees. In practice, some *FPT* algorithms seem to have much larger empirical klam values [Ste99].

F. Roberts, eds.), *AMS-DIMACS Series in Discrete Mathematics and Theoretical Computer Science*, 49, pp. 49–99, 1999.

[D74] E. Dijkstra. "Self-Stabilizing Systems in Spite of Distributed Control." *Communications of the ACM* 17, pp. 643–644, 1974.

[FL88] M. Fellows and M. Langston. "On Well-Partial-Order Theory and Its Applications to Combinatorial Problems of VLSI Design." *Technical Report CS-88-188*, Department of Computer Science, Washington State University, 1988.

[FMRS00] M. Fellows, C. McCartin, F. Rosamond, and U. Stege. "The parametric dual of DOMINATING SET is fixed-parameter tractable," 2000.

[GMM94] G. Galbiati, F. Maffioli, and A. Morzenti. "A Short Note on the Approximability of the Maximum Leaves Spanning Tree Problem." *Information Processing Letters* 52, pp. 45–49, 1994.

[GMM97] G. Galbiati, A. Morzenti and F. Maffioli. "On the Approximability of some Maximum Spanning Tree Problems." *Theoretical Computer Science* 181, pp. 107–118, 1997.

[GJ79] M. Garey and D. Johnson. *Computers and Intractability: A Guide to the Theory of NP-Completeness*. W.H. Freeman, San Francisco, 1979.

[HHS98] T. Haynes, S. Hedetniemi, and P. Slater. *Fundamentals of Domination in Graphs*. Marcel Dekker, Inc, 1998.

[KKRUW95] E. Kranakis, D. Krizanc, B. Ruf, J. Urrutia, G. Woeginger. "VC-dimensions for graphs." In M. Nagl, editor, *Graph-theoretic concepts in computer science*, LNCS 1017, pp. 1–13, 1995.

[LR98] H.-I. Lu and R. Ravi. "Approximating Maximum Leaf Spanning Trees in Almost Linear Time." *Journal of Algorithms* 29, pp. 132–141, 1998.

[NR99b] R. Niedermeier and P. Rossmanith. "Upper Bounds for Vertex Cover Further Improved." In C. Meinel and S. Tison, editors, *Proceedings of the 16th Symposium on Theoretical Aspects of Computer Science*, LNCS 1563, pp. 561–570, 1999.

[NR00a] R. Niedermeier and P. Rossmanith. "A General Method to Speed Up Fixed-Parameter-Tractable Algorithms." *Information Processing Letters*, 73, pp. 125–129, 2000.

[NR00b] R. Niedermeier and P. Rossmanith. "An efficient fixed parameter algorithm for 3-Hitting Set." accepted for publication in *Journal of Discrete Algorithms*, August 2000.

[Ste99] U. Stege. *Resolving Conflicts from Computational Biology.* Ph.D. thesis, Department of Computer Science, ETH Zürich, Switzerland, 1999.

Planar Graph Blocking for External Searching

Surender Baswana[*] and Sandeep Sen[**]

Department of Computer Science and Engineering,
I.I.T. Delhi, Hauz Khas, New Delhi-110016, India.
{sbaswana, ssen}@cse.iitd.ernet.in

Abstract. We present a new scheme for storing a planar graph in external memory so that any online path can be traversed in an I-O efficient way. Our storage scheme significantly improves the previous results for planar graphs with bounded face size. We also prove an upper bound on I-O efficiency of any storage scheme for well-shaped triangulated meshes. For these meshes, our storage scheme achieves optimal performance.

1   Introduction

There are many search problems in computer science which require efficient ways of online traversal in an undirected graph e.g. robot motion planning, searching in constraint networks. There are some important problems in computational geometry which are also reducible to the efficient online traversal in a graph. For example, ray shooting problem in a simple polygon [4] and reporting intersection of a line segment with a triangulated mesh. Since most of the applications of these problems are of very large scales, it is important to ensure I-O efficient traversal in undirected graphs. Graph blocking corresponds to storing of a graph in external memory so that the number of I-O operations i.e. block-transfers required to perform any arbitrary online walk, is minimized. Efficiency of a blocking scheme is measured by *speed-up* σ which is the worst case average number of steps traversed between two I-O operations. The efficiency of a blocking scheme is proportional to the value of σ. We address the problem of blocking of planar undirected graphs.

We assume that the graph is of bounded degree and a vertex is allowed to be present in more than one block. These assumptions are valid for most of the applications of the graph-blocking problem. Let us assume that a block can hold B vertices and the internal memory can hold M vertices. The parameters B and M are related to the block size and the internal-memory size(by a constant factor). At any stage of the online walk, the next node to be visited can be any neighbor of the most recently visited node. Therefore, it is natural that with every node we store its associated adjacency list. In case a neighbor w of a node v is not present in the same block as that of v, we must also store the block address of w in adjacency list of v. A simple observation is as follows. At any

---

[*] Work was supported in part by an Infosys PhD fellowship.
[**] Work was supported in part by an AICTE career award.

S. Kapoor and S. Prasad (Eds.): FST TCS 2000, LNCS 1974, pp. 252–263, 2000.

time while traversing a graph, let v be the most recently visited node. If the nearest node which is not present in internal memory lies at distance k from v, then a block transfer can be forced in next k steps. Based on this observation, here is a naive blocking scheme: *Let B_v be the set of nodes lying in a breadth-first-search tree (BFS tree) of size B rooted at node v. For every node v of the graph, store B_v in a block on disk, and whenever walk extends to v and v is not present in internal memory, bring in the block storing B_v.* This blocking scheme ensures speed-up of $r^-(B)$, which denotes the minimum depth of a BFS tree of size B in the given graph. However, this speed-up is achieved at the expense of B-fold blow up in storage requirement. For a blocking scheme to be practical, such a large increase in storage requirement is not acceptable. A blocking scheme is said to be *space optimal* if the number of blocks it requires to store a graph is at most a constant multiple of the minimum number of blocks required to store the graph. Like other previous approaches[1,3] to graph blocking, we do not take into account the cost of preprocessing.

Goodrich et al. [3] gave space optimal blocking scheme for grid graphs and complete d−trees. They also gave a nontrivial upper bound of $r^+(B)$ on the speed-up that can be achieved in a graph where $r^+(B)$ denotes the maximum depth of B size BFS tree in the given graph. They also gave a space optimal blocking scheme for the family of graphs with bounded $\frac{r^+(B)}{r^-(B)}$. They did not give blocking scheme for general graphs although they conjectured a scheme which ensures a speed-up of $r^-(B)$ with optimal storage. Agarwal et al. [1] gave a space optimal blocking scheme for planar graphs that achieves a speed-up of $r^-(\sqrt{B})$. For the family of planar graphs $B \geq r^-(B) \geq \log_d B$, where d is the maximum degree of a node in the graph. For the smaller extreme(i.e. $r^-(B) \approx \log_d B$) the speed-up achieved by blocking scheme of Agarwal et al.[1] is close to $r^-(B)$ - more precisely $\sigma = \frac{r^-(B)}{2}$ in this case. But the speed-up deteriorates steadily from $r^-(B)$ as we move away from the smaller extreme of $r^-(B)$. As a case in point, note that for a planar graph with $r^-(k) = k^\alpha$ (where α is some positive fraction), the speed-up achieved is just $B^{\frac{\alpha}{2}}$th fraction of $r^-(B)$. Therefore, in the case of planar graphs with $r^-(B) = \sqrt{B}$, the speed-up achieved is $B^{\frac{1}{4}}$ and the gap between $r^-(B)$ and the speed-up achieved widens even further for the planar graphs with larger values of $r^-(B)$.

We present an efficient blocking scheme for general planar graphs with improved speed-up over that of Agarwal et al.[1]. Our blocking scheme guarantees a speed-up of $r^-(s)$ where $s = \min(r^-(\frac{s}{c})\sqrt{B}, B)$ and c is the maximum face size in the graph. It can be seen that for the family of planar graphs having small face size and $r^-(B) \geq \sqrt{B}$, the speed-up achieved by our blocking scheme is $\Omega(r^-(B))$. Whereas the speed-up achieved by blocking scheme of Agarwal et al. [1] deviates further from $r^-(B)$ as $r^-(B)$ increases, the speed-up achieved by our blocking scheme approaches $r^-(B)$ as $r^-(B)$ increases and the speed-up matches $r^-(B)$ from the point $r^-(B) = \sqrt{B}$ onwards. There are a large number of applications that employ planar graphs with small face size like trees and geometric graphs(grids and meshes). Thus for such planar graphs our blocking scheme outperforms earlier blocking scheme. We also make an observation that the blocking scheme for undirected planar graphs can be used to achieve good speed-up even in case of directed graphs.

We also prove a bound on the best speed-up achievable in planar mesh in terms of degree of local uniformity of mesh and block size. Most of the meshes in practical applications possess good degree of local uniformity. Intuitively speaking, these meshes are well-shaped. We prove that the best worst-case speed-up achievable in a planar mesh is $O(\sqrt{B})$, where the constant of proportionality depends upon the degree of well-shapedness of the mesh. We use our blocking scheme of planar graphs to achieve speed-up in planar mesh that matches this bound.

2   Efficient Blocking of Planar Graphs

In this section, we shall devise efficient blocking scheme for planar graphs that achieves improved speed-up over earlier blocking scheme given by Agarwal et al. [1]. First we present a terminology given by them: Set of nodes lying in a BFS tree of size k rooted at a node v is called *k-neighborhood* for the node v. We extend the following idea of partitioning planar graph employed in their blocking scheme. Consider a planar graph of size N partitioned into $O(N/B)$ regions with each region containing at most B nodes and surrounded by boundary nodes such that every path going from a node in one region to a node in another region will pass through one or more of these boundary nodes. By storing *B-neighborhood* around every boundary node in a block and storing nodes of a region together in a block, the following block-transfer strategy(referred as τ henceforth) will ensure speed-up of $\sigma = \Omega(r^-(B))$ for any online traversal.

Strategy τ : *Whenever walk extends to a node, say v not present in internal memory, if v is a boundary node read the block storing B-neighborhood of v from the disk otherwise read the block corresponding to the region in which v lies.*

It is apparently clear that for every two blocks that we load from disk, we traverse at least $r^-(B)$ nodes of the path and thus speed-up achieved is $\Omega(r^-(B))$. For meeting the space optimality constraint, we have to make sure that the space used for storing *B-neighborhoods* around the boundary nodes is $O(\frac{N}{B})$ blocks.

Agarwal et al.[1] gave an efficient blocking scheme along above lines. They used a technique developed by Fredrickson [2] for partitioning planar graphs. Based on the separator theorem of Lipton and Tarjan [5], Fredrickson gave an algorithm for partitioning a planar graph into $O(\frac{N}{B})$ regions, with each region having at most B nodes and the total number of boundary nodes being $O(\frac{N}{\sqrt{B}})$. Storing each of $O(\frac{N}{B})$ regions in blocks, and storing $\sqrt{B}$-*neighborhood* around every boundary node, it can be seen that using block-transfer strategy τ, the speed-up achieved is $\Omega(r^-(\sqrt{B}))$ and storage space required is optimal.

For the class of planar graphs with bounded degree, $r^-(B)$ can be as small as $\log_d B$ on one extreme(where d is the maximum degree of a node in a graph) and as large as B on the other extreme. For planar graphs with $r^-(B) \approx \log_d B$, speed-up achieved using above blocking scheme is $\frac{1}{2}\log_d B$, which is indeed close to $r^-(B)$. Though the speed-up achieved is close to $r^-(B)$ for small values of

$r^-(B)$, it degrades drastically as $r^-(B)$ increases. To appreciate this point, note that for planar graphs having $r^-(k) = k^\alpha$ (where α is some constant ≤ 1), the speed-up is just $B^{\alpha/2}$ th fraction of $r^-(B)$.

We devise a refinement of the above mentioned blocking scheme for achieving better speed-up. Note that using the above blocking scheme speed-up achieved is $r^-(\sqrt{B})$, because we stored $\sqrt{B}$-*neighborhood* around every boundary node in the partition. To improve the speed-up, we should store neighborhood of size greater than $\sqrt{B}$ around every boundary node. But the number of boundary nodes being $\Omega(\frac{N}{\sqrt{B}})$, any attempt to increase the size of neighborhood around a boundary node beyond $\sqrt{B}$ will lead to nonlinear space(an undesirable situation). Here we make a useful observation: We need not store separate $\sqrt{B}$-*neighborhoods* for boundary nodes which are *closely* placed. For example, let v be a boundary node, and $v_1, v_2, \cdots, v_j$ be other boundary nodes which lie within distance of $\frac{r^-(\sqrt{B})}{2}$ from v. Starting from any of these boundary nodes, we must traverse at least $\frac{r^-(\sqrt{B})}{2}$ steps to cross $\sqrt{B}$-*neighborhood* of v (it follows from triangle inequality). Therefore, instead of storing $\sqrt{B}$-*neighborhood* around every node $v_i \in v_1, \cdots, v_j$, we can just store $\sqrt{B}$-*neighborhood* around v only (as a common neighborhood for $v_1, \cdots, v_j$). In doing so, the speed-up is reduced at most by half; but we will be storing less than $\frac{N}{\sqrt{B}}$ neighborhoods. This reduction in total number of neighborhoods allows us to increase corresponding size of neighborhood(while still maintaining the linear space constraint). For this idea to be useful, the partitioning scheme must ensure that the separator nodes be contiguous. The separator computed using Lipton Tarjan separator theorem does not guarantee a separator with sufficiently *clustered* nodes. The planar-separator theorem given by Miller [6] shows existence of a node or a cycle as a balanced separator for a planar graph. For the case when the separator is a cycle we can form clusters of the separator-nodes by breaking the cycle into appropriate number of equally long chains. The set of nodes belonging to a chain define a cluster. We finally store just one neighborhood per cluster. Having given this basic idea we shall now describe the new blocking scheme and calculate the speed-up that can be achieved in various planar graphs based on it. First we state the planar-separator theorem given by Miller [6]

Theorem 1 (Miller). *If G is embedded planar graph consisting of N nodes, then there exists a balanced separator which is a vertex or a simple cycle of size at most $2\sqrt{2.\lfloor\frac{c}{2}\rfloor N}$, where c is maximum face size. Such a separator is constructible in linear sequential time.*

Based on the above separator theorem, we present a new blocking scheme which gives improved speed-up for planar graphs with bounded face size.

Let $G(V, E)$ be the given planar graph with maximum face size equal to c and N be the number of nodes of the graph. If $N < B$ we store G in a block on disk; otherwise we proceed as follows: we compute separator using Miller's theorem given above. If the separator is a vertex v, we store B-*neighborhood* around v; otherwise (separator is a cycle $\mathcal{C}$ of size $\leq 2\sqrt{cN}$) let s be a number in the range $(\sqrt{B}, B)$ (depending on the underlying graph) that will be specified later.

Pick every $\frac{r^-(s)}{2}$th node of C to form a set S. For every node $v \in S$, store the *s-neighborhood* of v in a block. Associate every node w of separator C with the block which contains the *s-neighborhood* of $v \in S$ nearest to w (let us denote the block associated with a boundary node w by B_w). Whenever path extends to a boundary node w, and w is not present in internal memory we shall bring the block containing B_w into internal memory. Now let the separator C partition V into two subsets P_1 and P_2, each of size at most $\frac{2}{3}N$. We recursively carry out blocking of subgraphs induced by P_1 and P_2.

It can be verified that using block-transfer strategy τ, the speed-up achieved using above blocking scheme is $\Omega(r^-(s))$. So larger value of s will result in larger speed-up. For a given s, total space used for blocking according to above described scheme can be expressed by the following recurrence:

$$S(N) = S(N_1) + S(N_2) + \frac{4\sqrt{cN}}{r^-(s)}s \qquad \text{where} \qquad N_1, N_2 < \frac{2N}{3}$$

the solution of which is

$$S(N) = c_1 N + c_2 \frac{\sqrt{c}N}{\sqrt{B}r^-(s)}s$$

where c_1, c_2 are constants independent of s.

To maximize s and keeping linear space constraint($S(N) = O(N)$), we choose $s = \min\left(r^-(s)\sqrt{\frac{B}{c}}, B\right)$; i.e. s is the largest number $k \leq B$ such that $k \leq r^-(k)\sqrt{\frac{B}{c}}$.

Theorem 2. *A planar graph of size N and maximum face size c can be stored in $O(\frac{N}{B})$ blocks so that any online path of length t can be traversed using $O\left(\frac{t}{r^-(s)}\right)$ I-O operations where $s = \min\left(r^-(s)\sqrt{\frac{B}{c}}, B\right)$.*

The new blocking scheme gives improvement in speed-up for planar graphs with bounded face size. The important point is that the improvement achieved is most significant in case of graphs with $r^-(k) = k^\alpha$(the graphs for which the previous blocking scheme fails).

Remark 1. For planar graphs with $r^-(B) = \Omega(\sqrt{B})$ and maximum face size $= c$, value of s is equal to $\frac{B}{c}$; and we get speed-up equal to $r^-\left(\frac{B}{c}\right)$, which is a significant improvement for planar graphs with *small c*, over previous speed-up of $r^-(\sqrt{B})$, achieved by blocking scheme of Agarwal et al. [1].

Remark 2. For a tree we get speed-up which is equal to $r^-(B)$.

Remark 3. For planar graphs with $r^-(k) = k^\alpha$, for some constant $\alpha \leq \frac{1}{2}$, value of s is $\left(\frac{B}{c}\right)^{\frac{1}{2(1-\alpha)}}$. Thus speed up achieved by the new blocking scheme is $\left(\frac{B}{c}\right)^{\frac{\alpha}{2(1-\alpha)}}$ which is a significant improvement for planar graphs with *small c* (maximum face size) over the previous speed-up of $B^{\frac{\alpha}{2}}$ achieved by blocking scheme of Agarwal et al. [1].

A Blocking scheme for undirected graphs can be employed for directed graphs in the following way. Let G_d be given directed planar graph and G_u be the undirected graph constructed by ignoring the direction of edges in G_d. Let $Adj_d(x)$ and $Adj_u(x)$ be adjacency lists of a node x in the graphs G_d and G_u respectively. Based on the given blocking scheme for undirected graphs, let ρ be the storage description of G_u in blocks of external memory where each block stores adjacency lists of some B nodes of graph G_u. For every node x lying in a block b, replace $Adj_u(x)$ by $Adj_d(x)$, and carry out this step for all the blocks storing G_u. This gives a storage description ρ' for G_d. Since $Adj_d(x) \subset Adj_u(x)$, we shall be able to keep all the nodes earlier belonging to a block, still in a block(thus preserving the locality defined by ρ). Also note that a path in G_d exists in G_u as well. Therefore, if ρ ensures that k is the worst-case average-number of steps of walk performed on G_u between two block transfers, ρ' will ensure that the worst-case average-number of steps of walk in G_d between two block transfers is at least k. Thus the speed-up achieved in G_d by the new(adapted) blocking scheme is at least as much as the speed-up achieved in G_u by the given blocking scheme for undirected graphs. We can combine this observation with our blocking scheme for planar undirected graph to state the following theorem.

Theorem 3. *A planar directed graph G of size N and maximum face size c can be stored in $O(\frac{N}{B})$ blocks so that any online path of length t can be traversed using $O(\frac{t}{r^-(s)})$ I-O operations where $s = \min\left(r^-(s)\sqrt{\frac{B}{c}}, B\right)$ and $r^-(s)$ is the minimum depth of a BFS-tree of size s in the undirected graph formed by ignoring the direction of edges in G.*

The above blocking scheme is useful for I-O efficient traversal in a planar directed graph especially when the underlying undirected graph has significantly large value of $r^-(B)$.

3   Blocking of Planar Mesh

In various problems of scientific computing, graphs are often defined geometrically; for example, grid graphs and graphs in VLSI technology. In addition to combinatorial structure, these graphs also have geometric structure associated with them. One such family of graphs is *mesh*. A mesh in d-dimensional space is a subdivision of a d-dimensional domain into simplices which meet only at shared faces e.g. a mesh in 2-dimension is a triangulation of a planar region where triangles intersect only at shared edges and vertices.

Unlike a grid graph, where edges are of same length throughout and positioning of vertices has high degree of symmetry, a mesh need not be uniform and symmetric. We define two parameters α, γ to be associated with a planar mesh which (intuitively speaking) measure its *well-shapedness*. We address blocking of planar mesh. We prove two results: First, we show that the maximum worst-case speed-up achievable in a planar mesh is $O(\sqrt{B})$, where the constant of proportionality depends upon parameters α, γ. Next, we use the blocking scheme of planar graphs described in previous section to achieve a speed-up of $\Omega(\sqrt{B})$, where the constant of proportionality that depends upon α, γ becomes smaller as well-shapedness of mesh reduces. Thus for meshes having good degree of well-shapedness, the speed-up achieved by the blocking scheme matches the best possible.

3.1   Well-Shaped Planar Meshes

A planar mesh is a triangulation of a region in 2-dimensions where the triangles meet only at shared edges and vertices. For simplicity, we assume that a planar mesh extends infinitely in all directions (e.g. a mesh embedded on a torus or a sphere). A planar mesh need not possess perfect uniformity and symmetry like a grid graph. There may be variation in edge-lengths and density of vertices as we move from one region to another region in the mesh. But as observed in most of practical applications, there is certain degree of local uniformity present in mesh i.e. in a neighborhood around a vertex there is not *too much* variation in edge-lengths and vertex-density though the variation may be unbounded for the whole mesh. Visualizing mesh as a triangulation, this local uniformity can be viewed in the following way: the triangles constituting the mesh are fat and the variation of size(area) of these triangles is bounded in a finite neighborhood. This local uniformity captures formally the notion of well-shapedness of a planar mesh. We now define parameters to measure the local uniformity of a planar mesh. We parameterize fatness of triangles by the smallest angle α of a triangle in planar mesh. We parameterize the variation in sizes of triangles within a B-neighborhood in the following way : Let u be a node and B_u be the set of nodes of a BFS tree of size B rooted at u. Let $\triangle_B^u$ be the set of triangles with at least one vertex belonging to B_u. γ is defined as the ratio of area of the largest-area triangle to area of the smallest-area triangle belonging to the set $\triangle_B^u$. The parameters (α, γ) thus defined measure local uniformity of a planar mesh.

The area of any triangle in the set $\triangle_B^u$ will lie in the range $[A, \gamma A]$ for some A. Using elementary geometry it can be shown that length, l of any edge in the subgraph induced by B_u has the following bounds:

$$e_{min} = 2\sqrt{A \tan \frac{\alpha}{2}} \leq l \leq 2\sqrt{\gamma A \cot \alpha} = e_{max}$$

Lemma 4. $r^-(B)$ *of a planar mesh with parameters* α, γ *is* $\Omega(\frac{\sqrt{\tan \alpha}}{\sqrt{\gamma}}\sqrt{B})$.

Proof. Consider an arbitrary node u of planar mesh and let l be the depth of BFS tree of size B rooted at u. Let w be a node of B_u at maximum Euclidean distance d_{max} from u. Consider a circle C centered at u with radius d_{max}. The number of nodes lying in the circle is at least B since B_u lies inside it. Thus the number of triangles lying inside the circle is at least $\frac{B}{3}$. Note that the maximum number of triangles lying inside a circle of radius d_{max} is $\leq \frac{\pi d_{max}^2}{A}$. Hence the following inequality must hold: $\frac{\pi d_{max}^2}{A} \geq \frac{B}{3}$, i.e., $d_{max} \geq \sqrt{\frac{A}{\pi}}\sqrt{\frac{B}{3}}$. Also note that

d_{max} is bounded by le_{max}. Thus $l \geq \frac{\sqrt{A}}{\sqrt{\pi}e_{max}}\sqrt{\frac{B}{3}}$. Using the bound on e_{max}, and the definition of $r^-(B)$, it follows that $r^-(B) = \Omega(\frac{\sqrt{\tan\alpha}}{\sqrt{\gamma}}\sqrt{B})$.

For a planar mesh, the following theorem gives a lower bound on Euclidean distance between two nodes in terms of the length of the shortest path separating them.

Theorem 5. *Let v be a node belonging to B_u in a planar mesh. If p_{uv} is the shortest path-length from u to v, the Euclidean distance, d_{uv} between u and v is $\Omega(p_{uv}\sqrt{A})$, where the constant of proportionality depends upon the well-shapedness parameters (α, γ) of the mesh.*

Proof. Let w be a boundary node of the neighborhood B_u which is at the closest Euclidean distance from u and $\mathcal{N}$ be the set of nodes lying within distance d_{uw} from u. It can be seen that $\mathcal{N} \subset B_u$. To prove the theorem it would suffice if we can show that for a node $v \in \mathcal{N}$ separated by shortest path of length l from u, the Euclidean distance d_{uv} is $\Omega(l\sqrt{A})$. We proceed as follows: Let v be a node belonging to $\mathcal{N}$ and S be the line segment joining u and v. We build a path z_{uv}

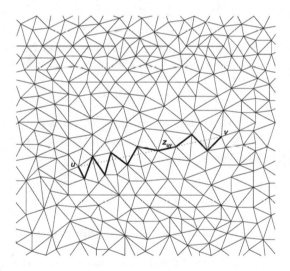

Fig. 1. Zig-zag path z_{uv} between nodes u and v of neighborhood B_u in a planar mesh

as we move along S from u. Let p denotes the vertex most recently added to the path we are building (initially $p = u$). We add edges to our path maintaining the following invariant :
$I : p$ *lies on the edge most recently intersected by S; and every edge forming the path has some point in common with S.*

While moving along S building the path, let e be an edge intersected by S. If e contains p, we keep p unchanged. Otherwise (e does not contain p), there is one end point say q of edge e adjacent to p such that pq is the valid edge to be added to our path maintaining the invariant I. So we extend our path by adding the edge pq to it and p gets updated to q now. We continue this process until we reach v. In the special case when S passes through a node, say x, we update p to x. We shall now bound length of this zig-zag shaped path z_{uv}. The segment S intersects triangles of $\triangle_B^u$ only and so the ratio of areas of intersected triangles is bounded by γ. Path z_{uv} divides the segment S into subsegments whose total number is equal to number of nodes lying on path z_{uv} excluding u and v. Consider any three consecutive edges of z_{uv}. There will be one (or two adjacent) subsegment(s) of S intercepted between these three edges. Because of the constraints imposed by bounded α and γ, the length of the intercepted subsegment(or sum of the lengths of two intercepted subsegments) is at least $e_{min} \sin \alpha$. Hence the number of subsegments into which the segment S is divided by the path z_{uv} (and so length of the path z_{uv}) is at most $\frac{d_{uv}}{e_{min} \sin \alpha}$. Using the bound on e_{min} (given before), it follows that length of the path z_{uv} is at most $\frac{d_{uv}}{c_\alpha \sqrt{A}}$, where $c_\alpha = 2\sqrt{\tan \frac{\alpha}{2}} \sin \alpha$.

Since p_{uv}, length of the shortest path between u, v is less than or equal to the length of path z_{uv}. Thus

$$p_{uv} \leq \left(\frac{d_{uv}}{c_\alpha \sqrt{A}} \right) \tag{1}$$

In other words, if v is a node belonging to B_u separated from u by Euclidean distance d_{uv} and p_{uv} be the length of the shortest path between u and v, then

$$d_{uv} \geq c_\alpha p_{uv} \sqrt{A} \tag{2}$$

We showed in Lemma 4 that shortest path-length from u to a boundary node of B_u is $r^-(B) = \Omega(\frac{\sqrt{\tan \alpha}}{\sqrt{\gamma}} \sqrt{B})$. By substituting the value of $r^-(B)$ for p_{uv} in equation 2, we get the following Corollary:

Corollary 6. *The minimum Euclidean distance between u and boundary node of B_u in a planar mesh is $\rho \geq c_\alpha \sqrt{\frac{\tan \alpha}{\gamma}} \sqrt{B} \sqrt{A}$.*

We now state the following Lemma which gives an upper bound on $r^+(B)$:

Lemma 7. $r^+(B)$ *for a planar mesh with parameters α, γ is $O\left(\frac{\sqrt{\gamma}}{c_\alpha} \sqrt{B} \right)$*

Above Lemma is based on equation 2. The arguments used to prove the Lemma are similar to those used in Lemma 4 and thus the details are omitted.

3.2   Upper Bound on Speed-Up in a Planar Mesh

Goodrich et al. [3] proved an upper bound of $r^+(B)$ on the best worst-case speed-up achievable in a graph. We showed in previous subsection(Lemma 7)that $r^+(B)$ for a planar mesh is $O(\frac{\sqrt{\gamma}}{c_\alpha} \sqrt{B})$. We can now state the following theorem:

Theorem 8. *For a planar mesh, the best worst-case speed-up that can be a-chieved is $\sigma = O(\mathcal{C}_{\alpha\gamma}\sqrt{B})$, where $\mathcal{C}_{\alpha\gamma}$ is a constant depending upon the parameters (α, γ) which capture the well-shapedness of the mesh.*

For sake of completeness, we give an alternate proof for upper bound on the speed-up in a planar mesh. Consider a planar mesh in x-y plane. We present a traversal strategy which will ensure one block transfer on an average for every $O(\sqrt{B})$ steps traversed, irrespective of the underlying blocking scheme. We introduce a terminology here : a node is said to be covered if it happens to be in internal memory at least once. Initially, before starting traversal, no node is present in internal memory, and so all the nodes are uncovered. Let u be the most recently visited node(path-front). It is obvious that if there is an uncovered node separated by a path of length $\leq \sqrt{B}$ from u, we can extend our path to that uncovered node(and thus force a block transfer in $\sqrt{B}$ steps). But what if all the nodes separated by paths of length $\leq \sqrt{B}$ from u are covered? Note that at least one block-transfer is required to cover a set of B uncovered nodes. So in case there is no uncovered node separated by path of length $\sqrt{B}$ from the path-front, we move the next $\sqrt{B}$ steps in such a way that we can associate distinct $\Omega(B)$ covered nodes to these steps. This would still imply that there is a block transfer after every $O(\sqrt{B})$ steps on average. This is the basic idea underlying the traversal strategy.

For a node u of mesh, $Cell_u$ denotes a square with base parallel to x-axis and with u lying on its left vertical side. Length of each of its four sides is chosen to be $\rho/2$, where ρ is the minimum Euclidean distance between u and a boundary node of B_u. It follows from Corollary 6 that the number of nodes lying in $Cell_u$ is more than $c_o B$ and every node of $Cell_u$ is reachable from u by path of length less than $c\sqrt{B}$ for some constants c, c_o depending upon well-shapedness of the mesh. Here is the traversal strategy :

We start from any vertex and always move rightward within the mesh. The path can be visualized as a sequence of sub-paths of type p' and p^x. At a point, let v be the most recently visited vertex. If there is any uncovered node inside $Cell_v$, we extend our path to that uncovered node(we call it sub-path of type p') and thus force a block-read from disk; otherwise we extend our path to a covered node lying closest to the right edge of $Cell_v$ (we call it sub-path of type p^x).

Let b be the number of block transfers encountered in traversing t steps in the mesh according to above strategy. Every sub-path of type p' causes a block transfer. So the number of sub-paths of type p' is at most b. Also note that for every sub-path of type p^x, the number of covered nodes lying to the left of path-front in the mesh increases by $c_o B$. Thus we can associate a set of unique $c_o B$ covered nodes to a sub-path of type p^x (uniqueness follows from the unidirectionality of motion). Since a block transfer can cover at most B nodes, it follows that the number of sub-paths of type p^x is at most $\frac{b}{c_o}$. So the total number of sub-paths(of type p' and p^x) is bounded by $\frac{2b}{c_o}$. Also note that the length of each sub-path is no more than $c\sqrt{B}$(from definition of $Cell$ above). Hence $t \leq \frac{2c\sqrt{B}b}{c_o}$ or in other words, the number of block-transfers, b required

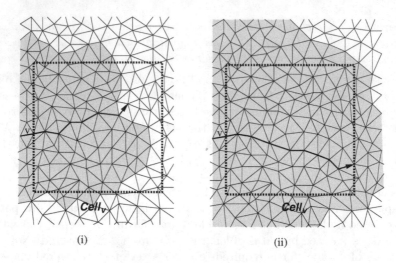

(i)   (ii)

(i) Subpath of type **p'** if there is any uncovered node in **$Cell_v$**

(ii) Subpath of type **p^x** if all the nodes of **$Cell_v$** are covered

Fig. 2. Two types of sub-paths from a node v in the mesh(the nodes lying in the shaded region are covered nodes)

to traverse t steps is $\Omega(\frac{t}{\sqrt{B}})$. Hence we can conclude that the traversal strategy described above will ensure a block transfer after every $O(\sqrt{B})$ steps on an average, irrespective of the underlying blocking scheme of the mesh.

3.3   Efficient Blocking of Planar Meshes

We can block planar meshes efficiently using our blocking scheme for planar graphs described in Section 2. From Lemma 4 it follows that $r^-(k) = \Omega(\sqrt{\frac{\tan\alpha}{\gamma}}\sqrt{k})$ for $k \leq B$. Also note that face size in a planar mesh is 3. So It follows easily from Theorem 2 that our blocking scheme guarantees speed-up of $\Omega(\frac{\tan\alpha}{\gamma}\sqrt{B})$ in a planar mesh with parameters α, γ. In previous subsection we established an upper bound of $O(\sqrt{B})$ on the speed-up in a planar mesh. Thus our blocking scheme achieves optimal speed-up in a planar mesh having good degree of well-shapedness.

Theorem 9. *There is a space optimal blocking scheme which ensures a speed-up of $\Omega(\frac{\tan\alpha}{\gamma}\sqrt{B})$ in a planar mesh, where the parameters (α, γ) measure the well-shapedness of the mesh.*

4   Conclusions

We addressed the problem of planar graph blocking in this paper. We described a blocking scheme which guarantees improved speed-up in planar graphs of bounded face size over previous blocking schemes. We also established a bound on the best worst-case speed-up that can be achieved in a planar mesh. For planar meshes with good degree of well-shapedness(local uniformity) our blocking scheme achieves optimal speed-up.

There is still no space optimal blocking scheme which can ensure I-O efficient traversal in general graphs(not necessarily planar). Such a scheme will help solve a large number of problems which require I-O efficient traversal in general graphs.

References

1. Pankaj K. Agarwal, Lars Arge, T.M. Murli, Kasturi R. Varadarajan, J.S. Vitter. I/O-efficient algorithms for contour-line extraction and planar graph blocking. 9th *ACM-SIAM Symposium on Discrete Algorithms*, 1998.
2. G.N. Fredrickson. Fast algorithms for shortest paths in planar graphs, with applications. *SIAM Journal of Computing*, 16, pp. 1004–1022, 1987.
3. M.T. Goodrich, M.H. Nodine and J.S. Vitter. Blocking for external graph searching. *Algorithmica* , 16, pp. 181–214, August 1996.
4. John Hershberger and Subhash Suri. A pedestrian approach to ray shooting : shoot a ray, take a walk. *Journal of Algorithms*, 18, pp. 403–432, 1995.
5. R.J. Lipton and R.E. Tarjan. A separator theorem for planar graphs. *SIAM journal of Applied Math.*, 36, pp. 177–189, 1979.
6. G. Miller. Balanced cyclic separator for 2-connected planar graphs. *Journal of Computer and System Sciences*, 32(3), pp. 265–279, 1986.

A Complete Fragment of Higher-Order Duration μ-Calculus

Dimitar P. Guelev

International Institute for Software Technology
of the United Nations University
(UNU/IIST), Macau, P.O.Box 3058.
Institute of Mathematics and Informatics
Bulgarian Academy of Sciences Bl. 8, Akad G. Bonchev blvd., Sofia.
gelevdp@bgnet.bg, dg@iist.unu.edu

Abstract. The paper presents an extension μHDC of Higher-order Duration Calculus (HDC,[ZGZ99]) by a polyadic least fixed point (μ) operator and a class of non-logical symbols with a *finite variability* restriction on their interpretations, which classifies these symbols as intermediate between *rigid* symbols and *flexible* symbols as known in DC. The μ operator and the new kind of symbols enable straightforward specification of recursion and data manipulation by HDC. The paper contains a completeness theorem about an extension of the proof system for HDC by axioms about μ and symbols of finite variability for a class of *simple* μHDC *formulas*. The completeness theorem is proved by the method of local elimination of the extending operator μ, which was earlier used for a similar purpose in [Gue98].

Introduction

Duration calculus(DC, [ZHR91]) has been proved to be a suitable formal system for the specification of the semantics of concurrent real-time programming languages[SX98, ZH00]. The introduction of a least fixed point operator to DC was motivated by the need to specify recursive programming constructs simply and straightforwardly. Recursive control structures as available in procedural programming languages are typically approximated through translation into iterative ones with explicit special storage (stacks). This blurs intuition and can add a significant overhead to the complexity of deductive verification. It is also an abandonment of the principle of abbreviating away routine elements of proof in specialised notations. That is why it is worth having an immediate way not only to specify but also to be able to reason about this style of recursion as it appears in high level programming languages.

Recently, an extension of DC by quantifiers which bind state variables (boolean valued functions of time) was introduced[ZGZ99]. Systematic studies regarding the application of this sort of quantification in DC had gained speed earlier, cf. [Pan95]; HDC allowed the integration of some advanced features of DC, such as *super-dense chop* [ZH96, HX99], into a single general system, called

S. Kapoor and S. Prasad (Eds.): FST TCS 2000, LNCS 1974, pp. 264–276, 2000.

Higher-order Duration Calculus (HDC), and enabled the specification of the semantics of temporal specification and programming languages such as Verilog and Timed RAISE[ZH00, LH99] by DC. The kind of completeness of the proof system of HDC addressed in [ZGZ99], which is ω-completeness, allowed to conclude the study of the expressive power of some axioms about the state quantifier.

In this paper we present some axioms about the least fixed point operator in HDC and show that adding them to a proof system for HDC yields a complete proof system for a fragment of the extension of HDC with this operator, μHDC.

The axioms we study are obtained by paraphrasing of the inference rules known about the propositional modal μ-calculus(cf. [Koz83, Wal93]), which were first introduced to DC in [PR95]. The novelty in our approach is the way we use the expressive power of the axioms about the μ-operator in our completeness argument, because, unlike the propositional μ-calculus, μHDC is a first-order logic with a binary modal operator.

Our method was first developed and applied in [Gue98] to so-called simple DC^* formulas which were introduced in [DW94] as a DC counterpart of a class of finite timed automata. That class was later significantly extended in [DG99, Gue00]. In this paper we show the completeness of an extension of a proof system for HDC for a corresponding class of simple μHDC formulas.

Our method of proof significantly relies on the exact form of the completeness of the proof system for HDC, which underlies the extension in focus. The completeness theorem about the original proof system for DC[HZ92] applies to the derivability of individual formulas only, and we need to have equivalence between the satisfiability of the infinite sets of instances of our new axioms and the consistency of these sets together with some other formulas, i.e. we need an ω-complete proof system for HDC. That is why we use a modification of the system from [ZGZ99], which is ω-complete with respect to a semantics for HDC, shaped after the abstract semantics of ITL, as presented in [Dut95]. Material to suggest an ω-completeness proof for this modification can be found starting from completion of Peano arithmetics by an ω-rule (cf. e.g. [Men64]) to [ZNJ99]. The completeness result presumed in this paper applies to the class of abstract HDC frames with their duration domains satisfying the principle of Archimedes. Informally, this principle states that there are no infinitely small positive durations and it holds for the real-time based frame.

The purpose of the modification of HDC here is to make a form of finite variability which is preserved under logical operations explicitly appear in this system. The choice to work with Archimedean duration domains is just to provide the convenience to axiomatise this kind of finite variability (axiom $HDC5$ below).

The fragment of μHDC language that our completeness result applies to is sufficient to provide convenience of the targetted kind for the design and use of HDC semantics of practically significant timed languages which admit recursive procedure invocations.

1   Preliminaries on HDC with Abstract Semantics

In this section we briefly introduce a version of HDC with abstract semantics[ZGZ99], which closely follows the abstract semantics for ITL given in [Dut95]. It slightly differs from the one presented in [ZGZ99]. Along with quantification over state, we allow quantifiers to bind so-called temporal variables and temporal propositional letters with the finite variability property.

1.1   Languages

A language for HDC is built starting from some given sets of *constant symbols* a, b, c, ..., *function symbols* f, g, ..., *relation symbols* R, S, ..., *individual variables* x, y, ... and *state variables* P, Q, ... Function symbols and relation symbols have *arity* to indicate the number of arguments they take in terms and formulas. Relation symbols and function symbols of arity 0 are also called *temporal propositional letters* and *temporal variables* respectively. Constant symbols, function symbols and relation symbols can be either *rigid* or *flexible*. Flexible symbols can be either symbols of *finite variability* (*fv* symbols) or not. Rigid symbols, fv symbols and (general) flexible and symbols are subjected to different restrictions on their interpretations. Every HDC language contains countable sets of individual variables, fv temporal propositional letters and fv temporal variables, the rigid constant symbol 0, the flexible constant symbol ℓ, the rigid binary function symbol $+$ and the rigid binary relation symbol $=$. Given the sets of symbols, *state expressions S, terms t and formulas φ* in a HDC language are defined by the BNFs:

$$S ::= \mathbf{0}\,|\,P\,|\,S \Rightarrow S$$
$$t ::= c\,|\int S\,|\,f(t,\ldots,t)\,|\,\overleftarrow{t}\,|\,\overrightarrow{t}$$
$$\varphi ::= \bot\,|\,R(t,\ldots,t)\,|\,\varphi \Rightarrow \varphi\,|\,(\varphi;\varphi)\,|\,\exists x\varphi\,|\,\exists v\varphi\,|\,\exists P\varphi$$

In BNFs for formulas here and below v stands for a fv temporal variable or a fv temporal propositional letter.

Terms and formulas which contain no flexible symbols are called *rigid*. Terms and formulas which contain only fv flexible symbols, rigid symbols and subformulas of the kind $\int S = \ell$ are called *fv terms* and *fv formulas* respectively. Terms of the kinds $\overleftarrow{t}$ and $\overrightarrow{t}$ are well-formed only if t is a fv term. We call individual variables, temporal variables, temporal propositional letters and state variables just variables, in case the exact kind of the symbol is not significant.

1.2   Frames, Models, and Satisfaction

Definition 1. *A* time domain *is a linearly ordered set with no end points. Given a time domain $\langle T, \leq \rangle$, we denote the set $\{[\tau_1, \tau_2] : \tau_1, \tau_2 \in T, \tau_1 \leq \tau_2\}$ of intervals in T by $\mathbf{I}(T)$. Given $\sigma_1, \sigma_2 \in \mathbf{I}(T)$, where $\langle T, \leq \rangle$ is a time domain, we denote $\sigma_1 \cup \sigma_2$ by $\sigma_1; \sigma_2$, in case $\max \sigma_1 = \min \sigma_2$. A* duration domain *is a system of the type $\langle D, 0^{(0)}, +^{(2)}, \leq^{(2)} \rangle$ which satisfies the following axioms*

$(D1)$ $x + (y + z) = (x + y) + z$ $(D6)$ $x \leq x$
$(D2)$ $x + 0 = x$ $(D7)$ $x \leq y \wedge y \leq x \Rightarrow x = y$
$(D3)$ $x + y = x + z \Rightarrow y = z$ $(D8)$ $x \leq y \wedge y \leq z \Rightarrow x \leq z$
$(D4)$ $\exists z(x + z = y)$ $(D9)$ $x \leq y \Leftrightarrow \exists z(x + z = y \wedge 0 \leq z)$
$(D5)$ $x + y = y + x$ $(D10)$ $x \leq y \vee y \leq x$

Given a time domain $\langle T, \leq \rangle$, and a duration domain $\langle D, 0, +, \leq \rangle$, $m : \mathbf{I}(T) \rightarrow D$ is a measure if

$(M0)$ $x \geq 0 \Leftrightarrow \exists \sigma(m(\sigma) = x)$
$(M1)$ $\min \sigma = \min \sigma' \wedge m(\sigma) = m(\sigma') \Rightarrow \max \sigma = \max \sigma'$
$(M2)$ $\max \sigma = \min \sigma' \Rightarrow m(\sigma) + m(\sigma') = m(\sigma \cup \sigma')$
$(M3)$ $0 \leq x \wedge 0 \leq y \wedge m(\sigma) = x + y \Rightarrow \exists \tau \in \sigma \; m([\min \sigma, \tau]) = x.$

Definition 2. *A HDC frame is a tuple of the kind $\langle \langle T, \leq \rangle, \langle D, 0, +, \leq \rangle, m \rangle$, where $\langle T, \leq \rangle$ is a time domain, $\langle D, 0, +, \leq \rangle$ is a duration domain, and $m : \mathbf{I}(T) \rightarrow D$ is a measure.*

Definition 3. *Given a HDC frame $F = \langle \langle T, \leq \rangle, \langle D, 0, +, \leq \rangle, m \rangle$ and a HDC language $\mathbf{L}$, a function I which is defined on the set of the non-logical symbols of $\mathbf{L}$ is called* interpretation *of $\mathbf{L}$ into F, if*

  ∘ $I(c), I(x) \in D$ *for constant symbols c and individual variables x*
  ∘ $I(f) : D^n \rightarrow D$ *for rigid n-place function symbols f*
  ∘ $I(f) : \mathbf{I}(T) \times D^n \rightarrow D$ *for flexible n-place function symbols f*
  ∘ $I(R) : D^n \rightarrow \{0, 1\}$ *for rigid n-place relation symbols R*
  ∘ $I(R) : \mathbf{I}(T) \times D^n \rightarrow \{0, 1\}$ *for flexible n-place relation symbols R*
  ∘ $I(P) : T \rightarrow \{0, 1\}$ *for state variables P*
  ∘ $I(0) = 0$, $I(\ell) = m$, $I(+) = +$ *and $I(=)$ is $=$.*

The following finite variability *condition is imposed on interpretations of state variables P:*

  Every $\sigma \in \mathbf{I}(T)$ can be represented in the form $\sigma_1; \ldots; \sigma_m$ so that $I(P)$ is constant on $[\min \sigma_i, \max \sigma_i)$, $i = 1, \ldots, m$.

A similar condition is imposed on the interpretations of fv symbols s. Given a frame F and an interpretation I as above, and $\sigma \in \mathbf{I}(T)$, a function (predicate) A on $\mathbf{I}(T) \times D^n$ is called fv *in F, I with respect to $\sigma_1, \ldots, \sigma_m \in \mathbf{I}(T)$ iff $\sigma = \sigma_1; \ldots; \sigma_m$ for some interval σ and for all $d_1, \ldots, d_n \in D$, $i, j \leq m$, $i \leq j$, $\sigma' \in \mathbf{I}(T)$:*

  ∘ *if $\min \sigma' \in (\min \sigma_i, \max \sigma_i)$ and $\max \sigma' \in (\min \sigma_j, \max \sigma_j)$, $A(\sigma', d_1, \ldots, d_n)$ is determined by $d_1, \ldots, d_n$ i and j only;*
  ∘ *if $\min \sigma' = \min \sigma_i$ and $\max \sigma' \in (\min \sigma_j, \max \sigma_j)$, $A(\sigma', d_1, \ldots, d_n)$ is determined by $d_1, \ldots, d_n$ i and j only, possibly in a different way;*
  ∘ *if $\min \sigma' \in (\min \sigma_i, \max \sigma_i)$ and $\max \sigma' = \min \sigma_j$, $A(\sigma', d_1, \ldots, d_n)$ is determined by $d_1, \ldots, d_n$ i and j only, possibly in a different way;*
  ∘ *if $\min \sigma' = \min \sigma_i$ and $\max \sigma' = \min \sigma_j$, $A(\sigma', d_1, \ldots, d_n)$ is determined by $d_1, \ldots, d_n$ i and j only, possibly in a different way.*

A symbol s is fv with respect to $\sigma_1, \ldots, \sigma_m$ in some F, I as above, if $I(s)$ has the corresponding property. Given a fv symbol s, for every $\sigma \in \mathbf{I}(T)$ there should be $\sigma_1, \ldots, \sigma_m \in \mathbf{I}(T)$ such that s is fv with respect to $\sigma_1, \ldots, \sigma_m$ in F, I.

Given a language $\mathbf{L}$, a pair $\langle F, I \rangle$ is a model for $\mathbf{L}$ if F is a frame and I is an interpretation of $\mathbf{L}$ into F.

Interpretations I and J of language $\mathbf{L}$ into frame F are said to s-agree, if they assign the same values to all non-logical symbols from $\mathbf{L}$, but possibly s.

Given a frame F (model M) we denote its components by $\langle T_F, \leq_F \rangle$, $\langle D_F, 0_F, +_F, \leq_F \rangle$ and m_F ($\langle T_M, \leq_M \rangle$, $\langle D_M, 0_M, +_M, \leq_M \rangle$ and m_M) respectively. We denote the frame and the interpretation of a given model M by I_M and F_M respectively.

Definition 4. *Given a model $M = \langle F, I \rangle$ for the language $\mathbf{L}$, $\tau \in T_M$ and $\sigma \in \mathbf{I}(T_M)$ the values $I_\tau(S)$ and $I_\sigma(t)$ of state expressions S and terms t and the satisfaction of formulas φ are defined by induction on their construction as follows:*

$$I_\tau(\mathbf{0}) = 0$$
$$I_\tau(P) = I(P)(\tau)$$
$$I_\tau(S_1 \Rightarrow S_2) = \max\{1 - I_\tau(S_1), I_\tau(S_2)\}.$$

$$I_\sigma(c) = I(c) \text{ for rigid } c$$
$$I_\sigma(c) = I(c)(\sigma) \text{ for flexible } c$$
$$I_\sigma(\textstyle\int S) = \int_{\min\sigma}^{\max\sigma} I_\tau(S) d\tau$$
$$I_\sigma(f(t_1, \ldots, t_n)) = I(f)(I_\sigma(t_1), \ldots, I_\sigma(t_n)) \text{ for rigid } f$$
$$I_\sigma(f(t_1, \ldots, t_n)) = I(f)(\sigma, I_\sigma(t_1), \ldots, I_\sigma(t_n)) \text{ for flexible } f$$
$$I_\sigma(\overleftarrow{t}) = d \text{ if } I_{\sigma'}(t) = d \text{ for some } \tau < \min\sigma \text{ and all } \sigma' \subset (\tau, \min\sigma)$$
$$I_\sigma(\overrightarrow{t}) = d \text{ if } I_{\sigma'}(t) = d \text{ for some } \tau > \max\sigma \text{ and all } \sigma' \subset (\max\sigma, \tau)$$

$$M, \sigma \not\models \bot$$
$$M, \sigma \models R(t_1, \ldots, t_n) \text{ iff } I(R)(I_\sigma(t_1), \ldots, I_\sigma(t_n)) = 1 \text{ for rigid } R$$
$$M, \sigma \models R(t_1, \ldots, t_n) \text{ iff } I(R)(\sigma, I_\sigma(t_1), \ldots, I_\sigma(t_n)) = 1 \text{ for flexible } R$$
$$M, \sigma \models \varphi \Rightarrow \psi \quad \text{iff either } M, \sigma \models \psi \text{ or } M, \sigma \not\models \varphi$$
$$M, \sigma \models (\varphi; \psi) \quad \text{iff there exist } \sigma_1, \sigma_2 \in \mathbf{I}(T_F) \text{ such that } \sigma = \sigma_1; \sigma_2,$$
$$M, \sigma_1 \models \varphi \text{ and } M, \sigma_2 \models \psi$$
$$M, \sigma \models \exists x \varphi \quad \text{iff } \langle F, J \rangle, \sigma \models \varphi \text{ for some } J \text{ which } x\text{-agrees with } I$$

Note that discrete time domains, which make the above definitions of $\overleftarrow{t}$ and $\overrightarrow{t}$ incorrect, also render any "corrected" definition for these operators grossly non-introspective, and therefore these operators should be disregarded in the case of discrete domains. In the clause about $\exists x$ above x stands for variable of an arbitrary kind, temporal variables and propositional temporal letters included. The integral used to define values of terms of the kind $\int S$ above is defined as follows. Given σ and S, there exist $\sigma_1, \ldots, \sigma_n \in \mathbf{I}(T_F)$ such that $\sigma = \sigma_1; \ldots; \sigma_n$ and $I_\tau(S)$ is constant in $[\min\sigma_i; \max\sigma_i)$, $i = 1, \ldots, n$. Given such a partitition $\sigma_1, \ldots, \sigma_n$ of σ, we put:

$$\int_{\min \sigma}^{\max \sigma} I_\tau(S)d\tau = \sum_{i=1,\ldots,n,\ I_{\min \sigma_i}(S)=1} m_F(\sigma_i)$$

Clearly, the value thus defined does not depend on the choice of $\sigma_1, \ldots, \sigma_n$.

1.3 Abbreviations

Infix notation and propositional constant $\top$, connectives $\neg$, $\wedge$, $\vee$ and $\Leftrightarrow$ and quantifier $\forall$ are introduced as abbreviations in the usual way. $\mathbf{1}$ stands for $\mathbf{0} \Rightarrow \mathbf{0}$ in state expressions. The relation symbol $\leq$ is defined by the axiom $x \leq y \Leftrightarrow \exists z(x+z=y)$. The related symbols $\geq$, $<$ and $>$ are introduced in the usual way. We use the following DC-specific abbreviations:

$\lceil S \rceil \rightleftharpoons \int S = \ell \wedge \ell \neq 0$, $\Diamond\varphi \rightleftharpoons ((\top; \varphi); \top)$, $\Box\varphi \rightleftharpoons \neg\Diamond\neg\varphi$, $n.t \rightleftharpoons \underbrace{t + \ldots + t}_{n \text{ times}}.$

$\Diamond_i\varphi \rightleftharpoons ((\ell \neq 0; \varphi); \ell \neq 0)$, $\Box_i\varphi \rightleftharpoons \neg\Diamond_i\neg\varphi$ $\xi_{t_1,t_2}(\varphi) \rightleftharpoons ((\ell = t_1; \varphi) \wedge \ell = t_2; \top).$

1.4 Proof System

Results in the rest of this paper hold for the class of DC models which satisfy the principle of Archimedes. It states that given positive durations d_1 and d_2, there exists a natural number n such that $n.d_1 \geq d_2$.

Here follows a proof system for HDC which is ω-complete with respect to the class of HDC models which satisfy the principle of Archimedes:

$(A1_l)$ $(\varphi; \psi) \wedge \neg(\chi; \psi) \Rightarrow (\varphi \wedge \neg\chi; \psi)$ $\qquad \dfrac{\varphi \quad \varphi \Rightarrow \psi}{\qquad} \qquad \dfrac{\varphi}{\qquad}$

$(A1_r)$ $(\varphi; \psi) \wedge \neg(\varphi; \chi) \Rightarrow (\varphi; \psi \wedge \neg\chi)$ $\quad (MP) \quad \dfrac{}{\psi} \qquad (G) \ \dfrac{}{\forall x \varphi}$

$(A2)$ $((\varphi; \psi); \chi) \Leftrightarrow (\varphi; (\psi; \chi))$ $\qquad\qquad\qquad\qquad\quad \dfrac{\varphi}{\varphi}$

(R_l) $(\varphi; \psi) \Rightarrow \varphi$ if φ is rigid $\qquad (N_l) \ \dfrac{}{\neg(\neg\varphi; \psi)} \quad (N_r) \ \dfrac{}{\neg(\psi; \neg\varphi)}$

(R_r) $(\varphi; \psi) \Rightarrow \psi$ if ψ is rigid $\qquad\qquad\qquad \dfrac{\varphi \Rightarrow \psi}{\qquad}$

(B_l) $(\exists x\varphi; \psi) \Rightarrow \exists x(\varphi; \psi)$ if $x \notin FV(\varphi)$ $\ (Mono_l) \ \dfrac{}{(\varphi; \chi) \Rightarrow (\psi; \chi)}$

(B_r) $(\varphi; \exists x\psi) \Rightarrow \exists x(\varphi; \psi)$ if $x \notin FV(\psi)$ $\qquad\qquad \dfrac{\varphi \Rightarrow \psi}{\qquad}$

$(L1_l)$ $(\ell = x; \varphi) \Rightarrow \neg(\ell = x; \neg\varphi)$ $\qquad (Mono_r) \ \dfrac{}{(\chi; \varphi) \Rightarrow (\chi; \psi)}$

$(L1_r)$ $(\varphi; \ell = x) \Rightarrow \neg(\neg\varphi; \ell = x)$ $\qquad\qquad \dfrac{\forall k < \omega \ [(\ell = 0 \vee \lceil S \rceil \vee \lceil \neg S \rceil)^k / R]\varphi}{\qquad}$

$(L2)$ $\ell = x + y \Leftrightarrow (\ell = x; \ell = y)$ $\qquad (\omega) \qquad\qquad \dfrac{}{[\top / R]\varphi}$

$(L3_l)$ $\varphi \Rightarrow (\ell = 0; \varphi)$ $\qquad\qquad\qquad\qquad\qquad \dfrac{\forall n < \omega \ \varphi \Rightarrow n.x \leq y}{\qquad}$

$(L3_r)$ $\varphi \Rightarrow (\varphi; \ell = 0)$ $\qquad\qquad (Arch) \qquad \dfrac{}{\varphi \Rightarrow x \leq 0}$

$(DC0)$ $\ell = 0 \Rightarrow \int S = 0$ $\qquad\qquad (DC4)$ $(\int S = x; \lceil \neg S \rceil) \Rightarrow \int S = x$

$(DC1)$ $\int \mathbf{0} = 0$ $\qquad\qquad\qquad\qquad (DC5)$ $\lceil S_1 \rceil \wedge \lceil S_2 \rceil \Leftrightarrow \lceil S_1 \wedge S_2 \rceil$

$(DC2)$ $\lceil \mathbf{1} \rceil \vee \ell = 0$ $\qquad\qquad\qquad (DC6)$ $\lceil S_1 \rceil \Leftrightarrow \lceil S_2 \rceil$, if $\vdash_{PC} S_1 \Leftrightarrow S_2$.

$(DC3)$ $(\int S = x; \lceil S \rceil \wedge \ell = y) \Rightarrow \int S = x + y$ $\ (DC7)$ $\lceil S \rceil \Rightarrow \Box(\lceil S \rceil \vee \ell = 0)$

$(PV1)$ $(\ell \neq 0; \overleftarrow{t} = x \wedge \ell = y) \Leftrightarrow (\top; (\Box_i(t = x) \wedge \ell \neq 0; \ell = y))$

$(PV2)$ $(\overrightarrow{t} = x \wedge \ell = y; \ell \neq 0) \Leftrightarrow ((\ell = y; \Box_i(t = x) \wedge \ell \neq 0); \top)$

$\qquad\qquad \dfrac{((\ell = a; \varphi); \ell = b) \Rightarrow ((\ell = a; \psi); \ell = b)}{\qquad}$

$(NL) \qquad\qquad\qquad\qquad \dfrac{}{\varphi \Rightarrow \psi}$

$(\exists_v)$ $\quad [t/v]\varphi \Rightarrow \exists v\varphi$ for fv-terms t and temporal variables v;

$(\exists_p)$ $\quad [\psi/p]\varphi \Rightarrow \exists p\varphi$ for fv-formulas ψ and temporal propositional letters p;

$(HDC1)$ $\exists v(\overleftarrow{v} = x)$

$(HDC2)$ $\exists v(\overrightarrow{v} = x)$

$(HDC3)$ $(\exists S\varphi; \exists S\psi) \Leftrightarrow \exists S(\varphi; \psi)$

$(HDC3_{v,l})$ $x \leq \ell \Rightarrow \exists v \forall y_1 \forall y_2 (\overleftarrow{v} = \overleftarrow{t_1} \wedge \overrightarrow{v} = \overrightarrow{t_2} \wedge$
$\qquad \wedge (y_1 \leq x \wedge y_2 \leq x \wedge y_1 \leq y_2 \Rightarrow \xi_{y_1,y_2}(v = t_1)) \wedge$
$\qquad \wedge (y_1 > x \wedge y_2 > x \wedge y_1 \leq y_2 \wedge y_2 \leq \ell \Rightarrow \xi_{y_1,y_2}(v = t_2)) \wedge$
$\qquad \wedge (y_1 \leq x \wedge y_2 > x \wedge y_2 \leq \ell \Rightarrow \xi_{y_1,y_2}(v = t_3)))$

$(HDC3_{v,r})$ $x \leq \ell \Rightarrow \exists v \forall y_1 \forall y_2 (\overleftarrow{v} = \overleftarrow{t_1} \wedge \overrightarrow{v} = \overrightarrow{t_2} \wedge$
$\qquad \wedge (y_1 < x \wedge y_2 < x \wedge y_1 \leq y_2 \Rightarrow \xi_{y_1,y_2}(v = t_1)) \wedge$
$\qquad \wedge (y_1 \geq x \wedge y_2 \geq x \wedge y_1 \leq y_2 \wedge y_2 \leq \ell \Rightarrow \xi_{y_1,y_2}(v = t_2)) \wedge$
$\qquad \wedge (y_1 < x \wedge y_2 \geq x \wedge y_2 \leq \ell \Rightarrow \xi_{y_1,y_2}(v = t_3)))$

$(HDC3_{p,l})$ $x \leq \ell \Rightarrow \exists p \forall y_1 \forall y_2 ($
$\qquad (y_1 \leq x \wedge y_2 \leq x \wedge y_1 \leq y_2 \Rightarrow \xi_{y_1,y_2}(p \Leftrightarrow \psi_1) \wedge$
$\qquad \wedge (y_1 > x \wedge y_2 > x \wedge y_1 \leq y_2 \wedge y_2 \leq \ell \Rightarrow \xi_{y_1,y_2}(p \Leftrightarrow \psi_2)) \wedge$
$\qquad \wedge (y_1 \leq x \wedge y_2 > x \wedge y_2 \leq \ell \Rightarrow \xi_{y_1,y_2}(p \Leftrightarrow \psi_3)))$

$(HDC3_{p,r})$ $x \leq \ell \Rightarrow \exists p \forall y_1 \forall y_2 ($
$\qquad (y_1 < x \wedge y_2 < x \wedge y_1 \leq y_2 \Rightarrow \xi_{y_1,y_2}(p \Leftrightarrow \psi_1) \wedge$
$\qquad \wedge (y_1 \geq x \wedge y_2 \geq x \wedge y_1 \leq y_2 \wedge y_2 \leq \ell \Rightarrow \xi_{y_1,y_2}(p \Leftrightarrow \psi_2)) \wedge$
$\qquad \wedge (y_1 < x \wedge y_2 \geq x \wedge y_2 \leq \ell \Rightarrow \xi_{y_1,y_2}(p \Leftrightarrow \psi_3)))$

$(HDC4)$ $\forall x \forall y ((\varphi \wedge \ell = x; \psi) \wedge \neg(\varphi \wedge \ell = y; \psi) \Rightarrow x < y) \Rightarrow$
$\qquad \Rightarrow \exists x (\forall y ((\varphi \wedge \ell = y; \psi) \Leftrightarrow y < x) \vee \exists x (\forall y ((\varphi \wedge \ell = y; \psi) \Leftrightarrow y \leq x)$

$(HDC5)$ $\ell \neq 0 \Rightarrow \exists x (x \neq 0 \wedge$
$\qquad \forall y \Box (\varphi \wedge \Diamond(\neg\varphi \wedge \Diamond(\varphi \wedge \Diamond(\neg\varphi \wedge \Diamond(\varphi \wedge \ell = y)))) \Rightarrow \ell \leq x + y))$

The symbol x denotes a variable of an arbitrary kind in the rule G and the axioms B_l and B_r. Instances of $HDC3_*$, $HDC4$ and $HDC5$ are valid only if $v, p, x, y, y_1, y_2 \notin FV(t_1), FV(t_2), FV(t_3), FV(\psi_1), FV(\psi_2), FV(\psi_3), FV(\varphi)$, $FV(\psi)$ and $t_1, t_2, t_3, \psi_1, \psi_2, \psi_3, \varphi$ and ψ are fv terms and formulas respectively.

The proof system also includes the axioms $D1$-$D10$ for duration domains, first order axioms and equality axioms. Substitution $[t/x]\varphi$ of variable x by term t in formula φ is allowed in proofs only if either t is rigid, or x is not in the scope of a modal operator.

Note that this proof system is slightly different from the original HDC one, as fv symbols are not considered in HDC as in [ZGZ99]. Nevertheless, its ω-completeness can be shown in way that is similar to the one taken in [ZNJ99].

The meaning of the new axioms $HDC1$, $HDC2$ and $HDC3_*$ is to enable the construction of fv functions and predicates on the set of intervals of the given model (from simpler ones). Given that a language $\mathbf{L}$ has rigid constants to name all the durations in a model M for it, as in the case of canonical models which are used in the completeness argument for this system, the existence of every fv function and predicate on $\mathbf{I}(T_M)$ can be shown using these axioms. The axioms $HDC4$ and $HDC5$ express the restrictions on the interpretations of fv formulas, and hence - the fv symbols occurring in them. The following ω-completeness theorem holds about this proof system:

Theorem 1. *Let Γ be a consistent set of formulas from the language $\mathbf{L}$ of HDC. Then there exists a model M for $\mathbf{L}$ and an interval $\sigma \in \mathbf{I}(T_M)$ such that $M, \sigma \models \varphi$ for all $\varphi \in \Gamma$.*

2 μHDC

In this section we briefly introduce the extension of HDC by a least fixed point operator.

2.1   Languages of μHDC

A language of μHDC is built using the same sets of symbols as for HDC languages and a distinguished countable set of propositional variables X, Y, $\ldots$. Terms are defined as in HDC. The BNF for formulas is extended to allow fixed point operator formulas as follows:

$$\varphi ::= \bot |X| R(t, \ldots, t) | \varphi \Rightarrow \varphi | (\varphi; \varphi) | \mu_i X \ldots X.\varphi, \ldots, \varphi | \exists x \varphi | \exists v \varphi | \exists P \varphi$$

Formulas of the kind $\mu_i X_1 \ldots X_m.\varphi_1, \ldots, \varphi_n$ are well-formed only if $m = n$, all the occurrences of the variables $X_1, \ldots, X_n$ in $\varphi_1, \ldots, \varphi_n$ are *positive*, i.e. each of these occurrences is in the scope of an even number of negations, $X_1, \ldots, X_n$ are distinct variables and $i \in \{1, \ldots, n\}$. Formulas which contain μ are not regarded as fv. Note that we work with a vector form of the least fixed point operator. This has some technical advantages, because it enables elimination of nested occurrences of μ under some additional conditions.

2.2   Frames, Models, and Satisfaction

Frames and models for μHDC languages are as for HDC languages. The only relative novelty is the extension of the satisfaction relation $\models$, which captures μ-formulas too.

Let $M = \langle \Gamma, I \rangle$ be a model for the (μHDC) language $\mathbf{L}$. Let $\tilde{I}(\varphi)$ denote the set $\{\sigma \in \mathbf{I}(T_F) : M, \sigma \models \varphi\}$ for an arbitrary formula φ from $\mathbf{L}$. Let s be a non-logical symbol in $\mathbf{L}$ and a be a constant, function or predicate of the type of s. We denote the interpretation of $\mathbf{L}$ into F which s-agrees with I and assigns a to s by I_s^a. Given a set $A \subseteq \mathbf{I}(T_M)$, we define the function $\chi_A : \mathbf{I}(T_M) \to \{0, 1\}$ by putting $\chi_A(\sigma) = 1$ iff $\sigma \in A$.

Now assume that the propositional variables $X_1, \ldots, X_n$ occur in φ. We define the function $f_\varphi : \left(2^{\mathbf{I}(T_F)}\right)^n \to 2^{\mathbf{I}(T_F)}$ by the equality $f_\varphi(A_1, \ldots, A_n) = (I_{X_1, \ldots, X_n}^{\widetilde{\chi_{A_1}, \ldots, \chi_{A_n}}})(\varphi)$. Assume that the variables $X_1, \ldots, X_n$ have only positive occurrences in φ. Then f_φ is monotone on each of its arguments, i.e. $A_i \subseteq A_i'$ implies $f_\varphi(A_1, \ldots, A_i, \ldots, A_n) \subseteq f_\varphi(A_1, \ldots, A_i', \ldots, A_n)$.

Now consider a sequence of n formulas, $\varphi_1, \ldots, \varphi_n$, which have only positive occurrences of the variables $X_1, \ldots, X_n$ in them. Then the system of inclusions

$$f_{\varphi_i}(A_1, \ldots, A_n) \subseteq A_i, \; i = 1, \ldots, n$$

has a least solution, which is also a least fixed point of the operator

$$\lambda A_1 \ldots A_n.\langle f_{\varphi_1}(A_1, \ldots, A_n), \ldots, f_{\varphi_n}(A_1, \ldots, A_n)\rangle.$$

Let this solution be $\langle B_1, \ldots, B_n \rangle$, $B_i \subseteq \mathbf{I}(T_F)$. We define the satisfaction relation for
$\mu_i X_1 \ldots X_n.\varphi_1, \ldots, \varphi_n$ by putting:
$$M, \sigma \models \mu_i X_1 \ldots X_n.\varphi_1, \ldots, \varphi_n \text{ iff } \sigma \in B_i.$$

3   Simple μHDC Formulas

The class of formulas which we call *simple* in this paper is a straightforward extension to the class of simple DC^* formulas considered in [Gue98]. We extend that class by allowing μ instead of iteration, positive formulas built up of fv symbols and existential quantification over the variables which occur in these formulas.

3.1   Super-Dense Chop

The *super-dense chop* operator $(.\circ .)$ was introduced in [ZH96] to enable the expression of sequential computation steps which consume negligible time, yet occur in some specified causal order, by DC. Given that $v_1, \ldots, v_n$ are all the free temporal variables of formulas φ and ψ, $(\varphi \circ \psi)$ is equivalent to

$$\exists v_1' \ldots \exists v_n' \exists v_1'' \ldots \exists v_n'' \exists x_1 \ldots \exists x_n \left(\begin{array}{l} [v_1'/v_1, \ldots, v_n'/v_n]\varphi \wedge \bigwedge_{i=1}^{n} \left(\begin{array}{l} \overleftarrow{v_i'} = \overleftarrow{v_i} \wedge \\ \overrightarrow{v_i'} = x_i \wedge \\ \Box v_i' = v_i \end{array} \right) ; \\ ; ([v_1''/v_1, \ldots, v_n''/v_n]\psi \wedge \bigwedge_{i=1}^{n} \left(\begin{array}{l} \overrightarrow{v_i''} = \overrightarrow{v_i} \wedge \\ \overleftarrow{v_i''} = x_i \wedge \\ \Box v_i'' = v_i \end{array} \right) \end{array} \right)$$

3.2   Simple Formulas

Definition 1. *Let* **L** *be a language for* μHDC *as above. We call* μHDC *formulas* γ *which can be defined by the BNF*
$$\gamma ::= \bot | R(t, \ldots, t) | X | (\gamma \wedge \gamma) | \gamma \vee \gamma | \neg \gamma | (\gamma; \gamma) | (\gamma \circ \gamma) | \mu_i X \ldots X.\gamma, \ldots, \gamma$$
where R and t stand for either rigid or fv relation symbols and terms respectively, open fv formulas*. We call an* open fv formula *strictly positive if it has no occurrences of propositional variables in the scope of $\neg$. An open fv formula is* propositionally closed *if it has no free occurrences of propositional variables. Simple μHDC formulas are defined by the BNF*
$$\varphi ::= \ell = 0 | X | \lceil S \rceil | \lceil S \rceil \wedge \ell \prec a | \lceil S \rceil \wedge \ell \succ a | \lceil S \rceil \wedge \ell \prec a \wedge \ell \succ b |$$
$$\varphi \vee \varphi | (\varphi; \varphi) | (\varphi \circ \varphi) | \varphi \wedge \gamma | \mu_i X \ldots X.\varphi, \ldots, \varphi | \exists x \varphi | \exists v \varphi$$
where a and b denote rigid constants, γ denotes a a propositionally closed strictly positive open fv formula, x denotes a variable of arbitrary kind, $\prec \in \{\leq, <\}$ and $\succ \in \{\geq, >\}$. Additionally, a simple formula should not have subformulas of the kind $\exists x \varphi$ where x has a free occurrence in the scope of a μ-operator in φ.

4   A Complete Proof System for the Simple Fragment of μHDC

In this section we show the completeness of a proof system for the fragment of μHDC where the application of μ is limited to simple formulas. We add the following axioms and rule to the proof system for HDC with abstract semantics:

(μ_1) $\Box(\mu_i X_1 \ldots X_n.\varphi_1, \ldots, \varphi_n \Leftrightarrow$
$\qquad [\mu_1 X_1 \ldots X_n \varphi_1, \ldots, \varphi_n/X_1, \ldots, \mu_n X_1 \ldots X_n \varphi_1, \ldots, \varphi_n/X_n]\varphi_i)$

(μ_2) $\bigwedge\limits_{i=1}^{n} \Box([\psi_1/X_1, \ldots \psi_n/X_n]\varphi_i \Rightarrow \psi_i) \Rightarrow \Box(\mu_i X_1 \ldots X_n.\varphi_1, \ldots, \varphi_n \Rightarrow \psi_i)$ The

(μ_3) $\mu_i X_1 \ldots X_m.\varphi_1, \ldots, [\mu Z_1 \ldots Z_n.\psi_1, \ldots, \psi_n/Y]\varphi_k, \ldots, \varphi_m \Leftrightarrow$
$\qquad \Leftrightarrow \mu_i X_1 \ldots X_n Y Z_1 \ldots Z_n \varphi_1, \ldots, \varphi_m, \psi_1, \ldots, \psi_n$

variable Y should not have negative free occurrences in φ_k in the instances of μ_3.

4.1   The Completeness Theorem

Lemma 1. *Let φ, α and β be HDC formulas and X be a propositional temporal letter. Let Y not occur in φ in the scope of quantifiers which bind any of the variables from $FV(\alpha) \cup FV(\beta)$. Then $\vdash_{\mu HDC} \Box(\alpha \Leftrightarrow \beta) \Rightarrow ([\alpha/Y]\varphi \Leftrightarrow [\beta/Y]\varphi)$.*

The following two propositions have a key role in our completeness argument. Detailed proofs are given in [Gue00b].

Proposition 1. *Let γ be a propositionally closed strictly positive open fv formula. Let M be a model for the language $\mathbf{L}$ of γ and $\sigma \in \mathbf{I}(T_M)$. Then there exists a μ-free propositionally closed strictly positive open fv formula γ' such that $M, \sigma \models \Box(\gamma \Leftrightarrow \gamma')$.*

This proposition justifies regarding μ formulas with fv subformulas as fv formulas.

Proposition 2 (local elimination of μ from simple formulas). *Let φ be a propositionally closed simple μHDC formula. Let M be a model for the language of φ and $\sigma \in \mathbf{I}(T_M)$. Then there exists a μ-free formula ψ such that $M, \sigma \models \Box(\varphi \Leftrightarrow \psi)$.*

Theorem 1 (completeness). *Let Γ be a set of formulas in a μHDC language $\mathbf{L}$. Let every μ-subformula of a formula $\varphi \in \Gamma$ be simple, and moreover occur in φ as a subformula of some propositionally closed μ-subformula of φ. Let Γ be consistent with respect to $\vdash_{\mu HDC}$. Then there exists a model M for $\mathbf{L}$ and an interval $\sigma \in \mathbf{I}(M)$ such that $M, \sigma \models \Gamma$.*

Proof. Proposition 1 entails that every fv μ-subformula of a formula from Γ is locally equivalent to a μ free fv formula. Hence occurrences of μ in fv subformulas can be eliminated using Lemma 1 and we may assume that there are no such subformulas. Since nested occurrences of μ in μ-subformulas from Γ can be eliminated by appropriate use of μ_3, we may assume that there are no such occurrences.

Let $S = \{s_{\mu_i X_1 \ldots X_n.\varphi_1, \ldots, \varphi_n} : 1 \leq i \leq n < \omega, \mu_i X_1 \ldots X_n.\varphi_1, \ldots, \varphi_n$ is a formula from $\mathbf{L}\}$ be a set of fresh 0-place flexible relation symbols. Let $\mathbf{L}(S)$ be the HDC language built using the non-logical symbols of $\mathbf{L}$ and the symbols from S. Every formula φ from $\mathbf{L}$ can be represented in the form $[\psi_1/X_1, \ldots, \psi_n/X_n]\psi$

where ψ does not contain μ and contains $X_1, \ldots, X_n$, and ψ_i, $i = 1, \ldots, n$ are distinct μ-formulas. This representation is unique. Given this representation of φ, we denote the formula $[s_{\psi_1}/X_1, \ldots, s_{\psi_n}/X_n]\psi$ from $\mathbf{L}(S)$ by $\mathsf{t}(\varphi)$. Note that the translation t is invertible and its converse of is defined on the whole $\mathbf{L}(S)$.

Let $\Delta = \{\Box(\alpha) : \alpha \text{ is an instance of } \mu_1, \mu_2 \text{ in } \mathbf{L}\}$. Then the set $\Gamma' = \{\mathsf{t}(\varphi) : \varphi \in \Gamma \cup \Delta\}$ is consistent with respect to $\vdash_{HDC}$. Assume the contrary. Then there exists a proof of $\bot$ with its premisses in Γ' in $\vdash_{HDC}$. Replacing each formula ψ in this proof by $\mathsf{t}^{-1}(\psi)$ gives a proof of $\bot$ from Γ in $\vdash_{\mu HDC}$.

Hence there exists a model M for $\mathbf{L}(S)$ and an interval $\sigma \in \mathbf{I}(T_M)$ such that $M, \sigma \models \Gamma'$.

Now let us prove that $M, \sigma \models \Box(\varphi \Leftrightarrow s_\varphi)$ for every closed simple formula φ from $\mathbf{L}$. Let φ be $\mu_i X_1 \ldots X_n.\psi_1, \ldots, \psi_n$. Let $\varphi_k \rightleftharpoons \mu_k X_1 \ldots X_n.\psi_1, \ldots, \psi_n$, $k = 1, \ldots, n$, for short. Then M satisfies the t-translations

$$\Box(s_{\varphi_k} \Leftrightarrow [s_{\varphi_1}/X_1, \ldots, s_{\varphi_n}/X_n]\psi_k)$$
$$\bigwedge_{j=1}^{n} \Box(\mathsf{t}(\theta_j) \Leftrightarrow [\mathsf{t}(\theta_1)/X_1, \ldots \mathsf{t}(\theta_n)/X_n]\psi_j) \Rightarrow \Box(s_{\varphi_k} \Rightarrow \mathsf{t}(\theta_k))$$

of the instances of μ_1 and μ_2 for all n-tuples of formulas $\theta_1, \ldots, \theta_n$ from $\mathbf{L}$. The first of these instances implies that $\langle s_{\varphi_1}, \ldots, s_{\varphi_n} \rangle$ evaluates to a fixed point of the operator represented by $\langle \psi_1, \ldots, \psi_n \rangle$. Consider the instance of μ_2. Let θ_k be a μ-free formula from $\mathbf{L}$ such that $M, \sigma \models \Box(\theta_k \Leftrightarrow \varphi_k)$ for $k = 1, \ldots, n$. Such formulas exist by Proposition 2. Then $\mathsf{t}(\theta_k)$ is θ_k and the above instance of μ_2 is actually

$$\bigwedge_{j=1}^{n} \Box(\theta_j \Leftrightarrow [\theta_1/X_1, \ldots, \theta_n/X_n]\psi_j) \Rightarrow \Box(s_{\varphi_k} \Rightarrow \theta_k)$$

Besides $M, \sigma \models \Box(\theta_j \Leftrightarrow [\theta_1/X_1, \ldots, \theta_n/X_n]\psi_j)$, $j = 1, \ldots, n$, by the choice of θ_k. Hence $M, \sigma \models \Box(s_{\varphi_k} \Rightarrow \theta_k)$. This means that $\langle s_{\varphi_1}, \ldots, s_{\varphi_n} \rangle$ evaluates to the least fixed point of the operator represented by $\langle \psi_1, \ldots, \psi_n \rangle$. Hence $M, \sigma \models \Box(s_\varphi \Leftrightarrow \varphi)$ for every μ-formula φ with no nested occurrences of μ. This entails that $M, \sigma \models \Box(\varphi \Leftrightarrow \mathsf{t}(\varphi))$ for every $\varphi \in \Gamma$. Hence, $M, \sigma \models \Gamma$.

Acknowledgements

Guidance towards the topic addressed here, and a sequel of invigorating and pitfall marking discussions are thanks to He Jifeng. Some mistakes were detected in an early version of the paper thanks to Dang Van Hung and indirectly by Dimiter Skordev. Among other flaws, an undeliberate overclaim, which was also inconsistent with the announced purpose of the article, was avoided due to the efforts of anonymous referees.

References

[DG99]   DANG VAN HUNG AND D. P. GUELEV. Completeness and Decidability of a Fragment of Duration Calculus with Iteration. In: P.S. THIAGARAJAN AND R. YAP (EDS), *Advances in Computing Science*, LNCS 1742, Springer-Verlag, 1999, pp. 139-150.

[DW94]   DANG VAN HUNG AND WANG JI. On The Design of Hybrid Control Systems Using Automata Models. In: CHANDRU, V. AND V. VINAY (EDS.) LCNS 1180, *Foundations of Software Technology and Theoretical Computer Science*, 16th Conference, Hyderabad, India, December 1996, Springer, 1996.

[Dut95]   DUTERTRE, B. On First Order Interval Temporal Logic. Report no. CSD-TR-94-3 Department of Computer Science, Royal Holloway, University of London, Egham, Surrey TW20 0EX, England, 1995.

[Gue98]   GUELEV, D. P. Iteration of Simple Formulas in Duration Calculus. Tech. report 141, UNU/IIST, June 1998.

[Gue00]   GUELEV, D. P. *Probabilistic and Temporal Modal Logics*, Ph.D. thesis, submitted, January 2000.

[Gue00b] GUELEV, D. P. *A Complete Fragment of Higher-Order Duration μ-Calculus*. Tech. Report 195, UNU/IIST, April 2000.

[HX99]   HE JIFENG AND XU QIWEN. *Advanced Features of DC and Their Applications. Proceedings of the Symposium in Celebration of the Work of C.A.R. Hoare*, Oxford, 13-15 September, 1999.

[HZ92]   M. R. HANSEN AND ZHOU CHAOCHEN. Semantics and Completeness of Duration Calculus. *Real-Time: Theory and Practice*, LNCS 600, Springer-Verlag, 1992, pp. 209-225.

[Koz83]   KOZEN, D. Results on the propositional μ-calculus. TCS 27:333-354, 1983.

[LH99]   LI LI AND HE JIFENG. *A Denotational Semantics of Timed RSL using Duration Calculus. Proceedings of RTCSA'99*, pp. 492-503, IEEE Computer Society Press, 1999.

[Men64]   MENDELSON, E. *Introduction to Mathematical Logic*. Van Nostrand, Princeton, 1964.

[Pan95]   PANDYA, P. K. Some Extensions to Propositional Mean-Value Calculus. Expressiveness and Decidability. *Proceedings of CSL'95*, Springer-Verlag, 1995.

[PR95]   PANDYA, P. K. AND Y RAMAKRISHNA. *A Recursive Duration Calculus*. Technical Report TCS-95/3, TIFR, Bombay, 1995.

[PWX98]   PANDYA, P. K, WANG HANPING AND XU QIWEN. Towards a Theory of Sequential Hybrid Programs. *Proc. IFIP Working Conference PROCOMET'98* D. GRIES AND W.-P. DE ROEVER (EDS.), Chapman & Hall, 1998.

[SX98]   SCHNEIDER, G. AND XU QIWEN. Towards a Formal Semantics of Verilog Using Duration Calculus. *Proceedings of FTRTFT'98*, ANDERS P. RAVN AND HANS RISCHEL (EDS.), LNCS 1486, pp. 282-293, Springer-Verlag, 1998.

[Wal93]   WALURKIEWICZ, I. *A Complete Deductive System for the μ-Calculus.*, Ph.D. Thesis, Warsaw University, 1993.

[ZGZ99]   ZHOU CHAOCHEN, D. P. GUELEV AND ZHAN NAIJUN. *A Higher-Order Duration Calculus. Proceedings of the Symposium in Celebration of the Work of C.A.R. Hoare*, Oxford, 1999.

[ZH96]   ZHOU CHAOCHEN AND M. HANSEN Chopping a Point. *Proceedings of BCS FACS 7th Refinement Workshop, Electronic Workshop in Computer Sciences*, Springer-Verlag, 1996.

[ZH00]   ZHU HUIBIAO AND HE JIFENG. *A DC-based Semantics for Verilog* Tech. Report 183, UNU/IIST, January 2000.

[ZHR91]   ZHOU CHAOCHEN, C. A. R. HOARE AND A. P. RAVN. A Calculus of
            Durations. *Information Processing Letters*, 40(5), pp. 269-276, 1991.
[ZNJ99]   ZHAN NAIJUN. *Completeness of Higher-Order Duration Calculus*. Research
            Report 175, UNU/IIST, August 1999.

Cited UNU/IIST reports can be found at `http://www.iist.unu.edu`.

A Complete Axiomatisation for Timed Automata

Huimin Lin[1]* and Wang Yi[2]

[1] Laboratory for Computer Science
Institute of Software, Chinese Academy of Sciences
lhm@ios.ac.cn
[2] Department of Computer Systems
Uppsala University
yi@docs.uu.se

Abstract. In this paper we present a complete proof system for timed automata. It extends our previous axiomatisation of timed bisimulation for the class of loop-free timed automata with *unique fixpoint induction*. To our knowledge, this is the first algebraic theory for the whole class of timed automata with a completeness result, thus fills a gap in the theory of timed automata. The proof of the completeness result relies on the notion of *symbolic timed bisimulation*, adapted from the work on value–passing processes.

1 Introduction

The last decade has seen a growing interest in extending various concurrency theories with timing constructs so that real-time aspects of concurrent systems can be modeled and analysed. Among them timed automata [AD94] has stood out as a fundamental model for real-timed systems.

A timed automaton is a finite automaton extended with a finite set of real-valued *clock variables*. A node of a timed automata is associated with an *invariant* constraint on the clock variables, while an edge is decorated with a clock constraint, an action label, and a subset of clocks to be reset after the transition. At each node a timed automaton may perform two kinds of transitions: it may let time pass for any amount (a delay transition), as long as the invariant is satisfied, or choose an edge whose constraint is met, make the move, reset the relevant clocks to zero, and arrive at the target node (an action transition). Two timed automata are *timed bisimilar* if they can match each other's action transitions as well as delay transitions, and their residuals remain timed bisimilar. By now most theoretical aspects of timed automata have been well studied, but they still lack a satisfactory algebraic theory.

In this paper we shall develop a complete axiomatisation for timed automata, in the form of an inference system, in which the equalities between pairs of timed automata that are timed bisimilar can be derived. To this end we first propose

---

* Supported by a grant from National Science Foundation of China.

S. Kapoor and S. Prasad (Eds.): FST TCS 2000, LNCS 1974, pp. 277–289, 2000.

a language, in CCS style, equipping it with a symbolic transitional semantics in such a way that each term in the language denotes a timed automaton. The language has a conditional construct $\phi \rightarrow t$, read "if ϕ then t", and an action prefixing $a(\mathbf{x}).t$, meaning "perform the action a, reset the clocks in $\mathbf{x}$ to zero, then behave like t". The proof system consists of a set of inference rules and the standard monoid laws for bisimulation. Roughly speaking the monoid laws characterise bisimulation, while the inference rules deal with specific constructs in the language. The judgments of the inference system are of the form

$$\phi \triangleright t = u$$

where ϕ is a time constraint and t, u are terms. Intuitively it means: t and u are timed bisimilar over clock evaluations satisfying ϕ. A proof system of this nature already appeared in our previous work on axiomatising timed automata, [LW00.1], with a serious limitation: it is complete only over the recursion-free subset of the language, i.e. the subset of timed automata without loops. A standard way of extending such an axiomatisation to deal with recursion, is to add the following *unique fixpoint induction* rule [Mil84]:

$$\text{UFI} \quad \frac{t = u[t/X]}{t = \mathbf{fix}Xu} \quad X \text{ guarded in } u$$

However, this rule is incompatible with the inference system presented in [LW00.1]: the key inference rule handling action prefixing and clock resetting takes the form

$$\text{ACTION-DJ} \quad \frac{\phi\downarrow_{\mathbf{xy}}\Uparrow \triangleright t = u}{\phi \triangleright a(\mathbf{x}).t = a(\mathbf{y}).u} \quad \mathbf{x} \cap \mathcal{C}(u) = \mathbf{y} \cap \mathcal{C}(t) = \emptyset$$

where $\phi\downarrow_{\mathbf{xy}}\Uparrow$ is a clock constraint obtained from ϕ by first setting the clocks in $\mathbf{xy}$ to zero (operator $\downarrow_{\mathbf{xy}}$), then removing upper bounds on all clocks of ϕ (operator $\Uparrow$); $\mathcal{C}(t)$ and $\mathcal{C}(u)$ are the sets of clocks appearing in t and u, respectively. The side condition is needed to ensure that the clocks of one process do not get reset by the other. Because of this the inference system of [LW00.1] is mainly used for reasoning between terms with disjoint sets of clocks. It is difficult to fit the UFI rule into such contexts: it does not make sense to perform the substitution $t[u/X]$ when the clock set of u is different from that of t. To overcome this difficulty, we replace ACTION-DJ with two rules:

$$\text{ACTION} \quad \frac{\phi\downarrow_{\mathbf{x}}\Uparrow \triangleright t = u}{\phi \triangleright a(\mathbf{x}).t = a(\mathbf{x}).u}$$

and

$$\text{THINNING} \quad \frac{}{a(\mathbf{xy}).t = a(\mathbf{x}).t} \quad \mathbf{y} \cap \mathcal{C}(t) = \emptyset$$

The first rule does not require a side condition. However, the subset of clocks reset by the action prefix on both sides of the equation must be the same. This guarantees soundness. On the other hand, "redundant" clocks can be removed by THINNING. It is not difficult to see that ACTION-DJ is derivable from the above two rules.

The completeness proof relies on the introduction of the notion of *symbolic timed bisimulation*, $t \sim^\phi u$, which captures timed bisimulation in the following sense: $t \sim^\phi u$ if and only if $t\rho$ and $u\rho$ are timed bisimilar for any clock evaluation ρ satisfying ϕ. Following [Mil84], to show that the inference system is complete, that is $t \sim^\phi u$ implies $\vdash \phi \rhd t = u$, we first transform t and u into *standard equation sets* which are the syntactical representations of timed automata. We then construct a new equation set out of the two and prove that t and u both satisfy the product equation set, by exploiting the assumption that t and u are symbolically timed bisimilar. Finally we show that, with the help of UFI, if two terms satisfy the same set of standard equations then they are provably equal.

The result of this paper fills a gap in the theory of timed automata. It demonstrates that bisimulation equivalences of timed automata are as mathematically tractable as those of standard process algebras.

The rest of the paper is organised as follows: In the next section we first recall the definition of timed automata, then present a language to describe them. Section 3 introduces symbolic timed bisimulation. The inference system is put forward in Section 4. Section 5 is devoted to proving the completeness of the proof system. The paper concludes with Section 6 where related work is also briefly discussed.

Due to space limitation, many details and proofs have been omitted. They can be found in the full version of the paper [LW00.2].

2   A Language for Timed Automata

We assume a finite set $\mathcal{A}$ for synchronization actions and a finite set $\mathcal{C}$ for real-valued clock variables. We use a, b etc. to range over $\mathcal{A}$ and x, y etc. to range over $\mathcal{C}$. We use $\mathcal{B}(C)$, ranged over by ϕ, ψ etc., to denote the set of conjunctive formulas of atomic constraints in the form: $x_i \bowtie m$ or $x_i - x_j \bowtie n$, where $x_i, x_j \in \mathcal{C}$, $\bowtie \in \{\leq, <, \geq, >\}$ and m, n are natural numbers. The elements of $\mathcal{B}(C)$ are called *clock constraints*.

Definition 2.1. *A timed automaton over actions $\mathcal{A}$ and clocks $\mathcal{C}$ is a tuple $\langle N, l_0, E \rangle$ where*

- *N is a finite set of nodes,*
- *$l_0 \in N$ is the initial node,*
- *$E \subseteq N \times \mathcal{B}(C) \times \mathcal{A} \times 2^C \times N$ is the set of edges.*

When $\langle l, g, a, r, l' \rangle \in E$, we write $l \xrightarrow{g,a,r} l'$.

We shall present the operational semantics for timed automata in terms of a process algebraic language in which each term denotes an automaton.

$$\text{DELAY} \frac{}{t\rho \xrightarrow{d} t(\rho + d)} \quad \rho + d \models Inv(t) \quad \text{CHOICE} \frac{t\rho \xrightarrow{a} t'\rho'}{(t+u)\rho \xrightarrow{a} t'\rho'}$$

$$\text{ACTION} \frac{}{(a(\mathbf{x}).t)\rho \xrightarrow{a} t\rho\{\mathbf{x} := 0\}} \quad \text{GUARD} \frac{t\rho \xrightarrow{a} t'\rho'}{(\phi \to t)\rho \xrightarrow{a} t'\rho'} \quad \rho \models \phi$$

$$\text{REC} \frac{(t[\mathbf{fix}Xt/X])\rho \xrightarrow{a} t'\rho'}{(\mathbf{fix}Xt)\rho \xrightarrow{a} t'\rho'} \quad \text{INV} \frac{t\rho \xrightarrow{a} t'\rho'}{(\{\phi\}t)\rho \xrightarrow{a} t'\rho'} \quad \rho \models \phi$$

Fig. 1. Standard Transitional Semantics

We preassume a set of process variables, ranged over by X, Y, Z, ... The language for timed automata over $\mathcal{C}$ can be given by the following BNF grammar:

$$s ::= \{\phi\}t$$
$$t ::= \mathbf{0} \quad | \quad \phi \to t \quad | \quad a(\mathbf{x}).s \quad | \quad t + t \quad | \quad X \quad | \quad \mathbf{fix}Xt$$

$\mathbf{0}$ is the inactive process which can do nothing, except for allowing time to pass. $\phi \to t$, read "if ϕ then t", is the usual (one-armed) conditional construct. $a(\mathbf{x}).s$ is action prefixing. $+$ is nondeterministic choice. The $\{\phi\}t$ construct introduces an invariant.

A recursion $\mathbf{fix}Xt$ binds X in t. This is the only binding operator in this language. It induces the notions of bound and free process variables as usual. Terms not containing free process variables are *closed*. A recursion $\mathbf{fix}Xt$ is *guarded* if every occurrence of X in t is within the scope of an action prefixing.

The set of clock variables used in a term t is denoted $\mathcal{C}(t)$.

A *clock valuation* is a function from $\mathcal{C}$ to $\mathbf{R}^{\geq 0}$ (non-negative real numbers), and we use ρ to range over clock valuations. The notations $\rho\{\mathbf{x} := 0\}$ and $\rho + d$ are defined thus

$$\rho\{\mathbf{x} := 0\}(y) = \begin{cases} 0 & \text{if } y \in \mathbf{x} \\ \rho(y) & \text{otherwise} \end{cases}$$
$$(\rho + d)(x) = \rho(x) + d \quad \text{for all } x$$

To give a transitional semantics to our language, we first assign each term t an invariant constraint $Inv(t)$ by letting

$$Inv(t) = \begin{cases} \phi & \text{if } t \text{ has the form } \{\phi\}s \\ \text{tt} & \text{otherwise} \end{cases}$$

We shall require all invariants to be *downward-closed*:

$$\text{For all } d \in \mathbf{R}^{\geq 0}, \ \rho + d \models \phi \text{ implies } \rho \models \phi$$

Given a clock valuation $\rho : \mathcal{C} \to \mathbf{R}^{\geq 0}$, a term can be interpreted according to the rules in Figure 1, where the symmetric rule for $+$ has been omitted. The transitional semantics uses two types of transition relations: action transition $\xrightarrow{a}$ and delay transition $\xrightarrow{d}$. We call $t\rho$ a *process*, where t is a term and ρ a valuation; we use p, q, ... to range over the set of processes. We also write μ for either an action or a delay (a real number).

$$\text{ACTION} \xrightarrow{\hspace{1.5em}} a(\mathbf{x}).t \xrightarrow{tt,a,\mathbf{x}} t \qquad \text{CHOICE} \; \frac{t \xrightarrow{b,a,\mathbf{x}} t'}{t + u \xrightarrow{b,a,\mathbf{x}} t'} \qquad \text{INV} \; \frac{t \xrightarrow{\psi,a,\mathbf{x}} t'}{\{\phi\}t \xrightarrow{\psi,a,\mathbf{x}} t'}$$

$$\text{GUARD} \; \frac{t \xrightarrow{\psi,a,\mathbf{x}} t'}{\phi \rightarrow t \xrightarrow{\phi \wedge \psi,a,\mathbf{x}} t'} \qquad \text{REC} \; \frac{t[\mathbf{fix}Xt/X] \xrightarrow{b,a,\mathbf{x}} t'}{\mathbf{fix}Xt \xrightarrow{b,a,\mathbf{x}} t'}$$

Fig. 2. Symbolic Transitional Semantics

Definition 2.2. *A symmetric relation R over processes is a timed bisimulation if $(p,q) \in R$ implies*

whenever $p \xrightarrow{\mu} p'$ then $q \xrightarrow{\mu} q'$ for some q' with $(p',q') \in R$.

We write $p \sim q$ if $(p,q) \in R$ for some timed bisimulation R.

The symbolic transitional semantics of this language is listed in Figure 2. Again the symmetric rule for $+$ has been omitted. Note that invariants are simply forgotten in the symbolic transitional semantics. This reflects our intention that symbolic transitions correspond to edges in timed automata, while invariants reside in nodes.

According to the symbolic semantics, each guarded closed term of the language gives rise to a timed automaton; on the other hand, it is not difficult to see that every timed automaton can be generated from a guarded closed term in the language. In the sequel we will use the phrases "timed automata" and "terms" interchangeably.

3   Symbolic Timed Bisimulation

In this section we shall define a symbolic version of timed bisimulation. To simplify the presentation we fix two timed automata. To avoid clock variables of one automaton being reset by the other, we assume the sets of clocks of the two timed automata under consideration are disjoint, and write C for the union of the two clock sets. [1] Let N be the largest natural number occurring in the constraints of the two automata. An atomic constraint over C with ceiling N has one of the three forms: $x > N$, $x \bowtie m$ or $x - y \bowtie n$ where $x, y \in C, \bowtie \in \{\leq, <, \geq, >\}$ and $m, n \leq N$ are natural numbers.

In the following, "atomic constraint" always means "atomic constraint over C with ceiling N". Note that given two timed automata there are only finite number of such atomic constraints. We shall use c to range over atomic constraints.

A constraint, or *zone*, is a boolean combination of atomic constraints. A constraint ϕ is consistent if there is some ρ such that $\rho \models \phi$. Let ϕ and ψ be two constraints. We write $\phi \models \psi$ to mean $\rho \models \phi$ implies $\rho \models \psi$ for any ρ. Note that the relation $\models$ is decidable.

---

[1] This does not put any restriction on our results, because we can always rename clock variables of an automaton without affecting its behaviour.

A *region constraint*, or *region* for short, over n clock variables $x_1, \ldots, x_n$ is a consistent constraint containing the following atomic conjuncts:

- For each $i \in \{1, \ldots, n\}$ either $x_i = m_i$ or $m_i < x_i < m_i + 1$ or $x_i > N$;
- For each pair of $i, j \in \{1, \ldots, n\}$, $i \neq j$, such that both x_i and x_j are not greater than N, either $x_i - m_i = x_j - m_j$ or $x_i - m_i < x_j - m_j$ or $x_j - m_j < x_i - m_i$.

where the m_i in $x_i - m_i$ of the second clause refers to the m_i related to x_i in the first clause. In words, m_i is the integral part of x_i and $x_i - m_i$ its fractional part.

Given a finite set of clock variables C and a ceiling N, the set of region constraints over C is finite and is denoted $\mathcal{RC}_N^C$. In the sequel we will omit the sub- and super-scripts when they can be supplied by the context.

Fact 1 *Suppose that ϕ is a region constraint and ψ a zone. Then either $\phi \Rightarrow \psi$ or $\phi \Rightarrow \neg\psi$.*

So a region is either entirely contained in a zone, or is completely outside a zone.

A *canonical* constraint is a disjunction of regions. Given a constraint we can first transform it into disjunctive normal form, then decompose each disjunct into a disjoint set of regions. Both steps can be effectively implemented. As a corollary to Fact 1, if we write $\mathcal{RC}(\phi)$ for the set of regions contained in the zone ϕ, then $\bigvee \mathcal{RC}(\phi) = \phi$, i.e. $\bigvee \mathcal{RC}(\phi)$ is the canonical form of ϕ.

We will need two (postfixing) operators $\downarrow_{\mathbf{x}}$ and $\Uparrow$ to deal with resetting and time passage. Here we only define them semantically. The syntactical and effective definitions can be found in the full version of the paper.

For any ϕ, let $\phi\downarrow_{\mathbf{x}} = \{ \rho\{\mathbf{x} := 0\} \mid \rho \models \phi \}$ and $\phi\Uparrow = \{ \rho + d \mid \rho \models \phi, d \in \mathbf{R}^{\geq 0} \}$. Call a constraint ϕ $\Uparrow$-closed if $\phi\Uparrow = \phi$. It is easy to see that $\phi\Uparrow$ is $\Uparrow$-closed, and if ϕ is a region constraint then so is $\phi\downarrow_{\mathbf{x}}$.

Symbolic bisimulation will be defined as a family of binary relations indexed by clock constraints. Following [Cer92] we use constraints over the union of the (disjoint) clock sets of two timed automata as indices. Given a constraint ϕ, a finite set of constraints Φ is called a ϕ-*partition* if $\bigvee \Phi = \phi$. A ϕ-partition Φ is called *finer* than another such partition Ψ if Φ can be obtained from Ψ by decomposing some of its elements. By the corollary to Fact 1, $\mathcal{RC}(\phi)$ is a ϕ-partition, and is the finest such partition. In particular, if ϕ is a region constraint then $\{\phi\}$ is the only partition of ϕ.

Definition 3.1. *A constraint indexed family of symmetric relations over terms $\mathbf{S} = \{ S^\phi \mid \phi \Uparrow\text{-}closed \}$ is a symbolic timed bisimulation if $(t, u) \in S^\phi$ implies*

1. $\phi \models Inv(t) \Leftrightarrow Inv(u)$ *and*
2. *whenever $t \xrightarrow{\psi, a, \mathbf{x}} t'$ then there is an $(Inv(t) \wedge \phi \wedge \psi)$-partition Φ such that for each $\phi' \in \Phi$ there is $u \xrightarrow{\psi', a, \mathbf{y}} u'$ for some ψ', $\mathbf{y}$ and u' such that $\phi' \Rightarrow \psi'$ and $(t', u') \in S^{\phi'\downarrow_{\mathbf{xy}}\Uparrow}$.*

We write $t \sim^\phi u$ if $(t, u) \in S^\phi$ and $S^\phi \in \mathbf{S}$ for some symbolic bisimulation $\mathbf{S}$.

$$\text{S1 } X + \mathbf{0} = X \qquad \text{S2 } X + X = X$$
$$\text{S3 } X + Y = Y + X \qquad \text{S4 } (X + Y) + Z = X + (Y + Z)$$

Fig. 3. The Equational Axioms

Symbolic timed bisimulation captures $\sim$ in the following sense:

Theorem 3.2. *For $\Uparrow-closed$ ϕ, $t \sim^\phi u$ iff $t\rho \sim u\rho$ for any $\rho \models \phi \land Inv(t) \land Inv(u)$.*

4   The Proof System

The proposed proof system consists of a set of equational axioms in Figure 3 and a set of inference rules in Figure 4 where the standard rules for equational reasoning have been omitted. The judgments of the inference system are *conditional equations* of the form

$$\phi \rhd t = u$$

where ϕ is a constraint and t, u terms. Its intended meaning is "$t \sim^\phi u$", or "$t\rho \sim u\rho$ for any $\rho \models \phi \land Inv(t) \land Inv(u)$". $\mathrm{tt} \rhd t = u$ will be abbreviated as $t = u$.

The axioms are the standard monoid laws for bisimulation in process algebras. More interesting are the inference rules. For each construct in the language there is a corresponding introduction rule. CHOICE expresses the fact that timed bisimulation is preserved by $+$. The rule GUARD permits a case analysis on conditional. The rule INV deals with invariants. It also does a case analysis and appears very similar to GUARD. However, there is a crucial difference: When the guard ψ is false $\psi{\rightarrow}t$ behaves like $\mathbf{0}$, the process which is inactive but can allow time to pass; On the other hand, when the invariant ψ is false $\{\psi\}t$ behaves like $\{\mathrm{ff}\}\mathbf{0}$, the process usually referred to as *time-stop*, which is not only inactive but also "still", can not even let time elapse. ACTION is the introduction rule for action prefixing (with clock resetting). The THINNING rule allows to introduce/remove redundant clocks. REC is the usual rule for folding/unfolding recursions, while UFI says if X is guarded in u then $\mathbf{fix}Xu$ is the unique solution of the equation $X = u$. UNG can be used to remove unguarded recursion. Finally the two rules PARTITION and ABSURD do not handle any specific constructs in the language. They are so-called "structural rules" used to "glue" pieces of derivation together.

Let us write $\vdash \phi \rhd t = u$ to mean $\phi \rhd t = u$ can be derived from this proof system.

Some useful properties of the proof system are summarised in the following proposition:

Proposition 4.1.   *1.* $\vdash \phi{\rightarrow}(\psi{\rightarrow}t) = \phi \land \psi{\rightarrow}t$
2. $\vdash t = t + \phi{\rightarrow}t$
3. *If $\phi \models \psi$ then $\vdash \phi \rhd t = \psi{\rightarrow}t$*

$$\text{GUARD } \frac{\phi \wedge \psi \rhd t = u \quad \phi \wedge \neg\psi \rhd \mathbf{0} = u}{\phi \rhd \psi{\to}t = u} \qquad \text{CHOICE } \frac{\phi \rhd t = t'}{\phi \rhd t + u = t' + u}$$

$$\text{INV } \frac{\phi \wedge \psi \rhd t = u \quad \phi \wedge \neg\psi \rhd \{\text{ff}\}\mathbf{0} = u}{\phi \rhd \{\psi\}t = u} \qquad \text{ACTION } \frac{\phi{\downarrow}_{\mathbf{x}}{\Uparrow} \rhd t = u}{\phi \rhd a(\mathbf{x}).t = a(\mathbf{x}).u}$$

$$\text{THINNING } \frac{}{a(\mathbf{xy}).t = a(\mathbf{x}).t} \; \mathbf{y} \cap \mathcal{C}(t) = \emptyset \qquad \text{REC } \frac{}{\mathbf{fix}Xt = t[\mathbf{fix}Xt/X]}$$

$$\text{UFI } \frac{t = u[t/X]}{t = \mathbf{fix}Xu} \; X \text{ guarded in } u \qquad \text{UNG } \frac{}{\mathbf{fix}X(X + t) = \mathbf{fix}Xt}$$

$$\text{PARTITION } \frac{\phi_1 \rhd t = u \quad \phi_2 \rhd t = u}{\phi \rhd t = u} \; \phi{\models}\phi_1{\vee}\phi_2 \quad \text{ABSURD } \frac{}{\text{ff} \rhd t = u}$$

Fig. 4. The Inference Rules

4. $\vdash \phi \wedge \psi \rhd t = u$ *implies* $\vdash \phi \rhd \psi{\to}t = \psi{\to}u$
5. $\vdash \phi{\to}(t + u) = \phi{\to}t + \phi{\to}u$
6. $\vdash \phi{\to}t + \psi{\to}t = \phi \vee \psi{\to}t$

The following lemma shows how to "push" a condition through an action prefix. It can be proved using ACTION, INV and Proposition 4.1.3.

Lemma 4.2. $\vdash \phi \rhd a(\mathbf{x}).\{\psi\}t = a(\mathbf{x}).\{\psi\}\phi{\downarrow}_{\mathbf{x}}{\Uparrow}{\to}t.$

The UFI rule, as presented in Figure 4, is unconditional. However, a conditional version can be easily derived from Proposition 4.1.4, REC and UFI:

Proposition 4.3. *Suppose X is guarded in u. Then* $\vdash \phi{\rhd}t = u[\phi{\to}t/X]$ *implies* $\vdash \phi \rhd t = \mathbf{fix}X(\phi{\to}u).$

The rule PARTITION has a more general form:

Proposition 4.4. *Suppose Ψ is a ϕ-partition and $\vdash \psi \rhd t = u$ for each $\psi \in \Psi$, then $\vdash \phi \rhd t = u.$*

Soundness of the proof system is stated below:

Theorem 4.5. *If $\vdash \phi \rhd t = u$ and ϕ is $\Uparrow$-closed then $t\rho \sim u\rho$ for any $\rho \models \phi \wedge Inv(t) \wedge Inv(u).$*

The standard approach to the soundness proof is by induction on the length of derivations, and perform a case analysis on the last rule/axiom used. However, this does not quite work here. The reason is that the definition of timed bisimulation requires two processes to simulate each other after any time delays. To reflect this in the proof system, we apply the $\Uparrow$ operator, after $\downarrow$ for clock resetting, in the premise of the ACTION rule. But not all the inference rules preserve the $\Uparrow$-closeness property. An example is GUARD. In order to derive $\phi \rhd \psi{\to}t = u$, we need to establish $\phi \wedge \psi \rhd t = u$ and $\phi \wedge \neg\psi \rhd \mathbf{0} = u$. Even if ϕ is $\Uparrow$-closed, $\phi \wedge \psi$ may not be so.

To overcome this difficulty, we need a notion of "timed bisimulation up to a time bound", formulated as follows:

Definition 4.6. *Two processes p and q are timed bisimular up to $d_0 \in \mathbf{R}^{\geq 0}$, written $p \sim^{d_0} q$, if for any d such that $0 \leq d \leq d_0$*

— whenever $p \xrightarrow{d} p'$ then $q \xrightarrow{d} q'$ for some q' and $p' \dot{\sim} q'$

(and symmetrically for q), where $p \dot{\sim} q$ is defined thus

— whenever $p \xrightarrow{a} p'$ then $q \xrightarrow{a} q'$ for some q' and $p' \sim q'$

(and symmetrically for q).

Note that $\dot{\sim}$ is the same as $\sim^0$, and $\sim^{d_0} \subseteq \dot{\sim}$ in general.

Now the following proposition, of which Theorem 4.5 is a special case when ϕ is ⇑-closed, can be proved by standard induction on the length of derivations :

Proposition 4.7. *If $\vdash \phi \triangleright t = u$ then $t\rho \sim^{d_0} u\rho$ for any ρ and d_0 such that $\rho + d \models \phi \wedge Inv(t) \wedge Inv(u)$ for all $0 \leq d \leq d_0$.*

5 Completeness

This section is devoted to proving the completeness of the proof system which is stated thus: if $t \sim^\phi u$ then $\vdash \phi \triangleright t = u$. The structure of the proof follows from that of [Mil84]. The intuition behind the proof is as follows: A timed automaton is presented as a set of standard equations in which the left hand-side of each equation is a formal process variable corresponding to a node of the automaton, while the right hand-side encodes the outgoing edges from the node. We first transform, within the proof system, both t and u into such equation sets (Proposition 5.1). We then construct a "product" of the two equation sets, representing the product of the two underlying timed automata. Because t and u are timed bisimilar over ϕ, each should also bisimilar to the product over ϕ. Using this as a guide we show that such bisimilarity is derivable within the proof system, i.e. both t and u provably satisfy the product equation set (Proposition 5.2). Finally we demonstrate that a standard set of equations has only one solution, therefore the required equality between t and u can be derived. The unique fixpoint induction is only employed in the last step of the proof, namely Proposition 5.3.

Let $\mathbf{X} = \{ X_i \mid i \in I \}$ and $\mathbf{W}$ be two disjoint sets of process variables and $\mathbf{x}$ a set of clock variables. Let also u_i, $i \in I$, be terms with free process variables in $\mathbf{X} \cup \mathbf{W}$ and clock variables in $\mathbf{x}$. Then

$$E : \quad \{ X_i = u_i \mid i \in I \}$$

is an equation set with formal process variables $\mathbf{X}$ and free process variables in $\mathbf{W}$. E is closed if $\mathbf{W} = \emptyset$. E is a *standard* equation set if each u_i has the form

$$\{\psi_i\}(\sum_{k \in K_i} \phi_{ik} \rightarrow a_{ik}(\mathbf{x}_{ik}).X_{f(i,k)} + \sum_{k' \in K_i'} \psi_{ik'} \rightarrow W_{f'(i,k')})$$

A term t provably ϕ-satisfies an equation set E if there exist a vector of terms $\{ t_i \mid i \in I \}$, each t_i being of the form $\{\psi_i'\}t_i'$, and a vector of conditions $\{ \phi_i \mid i \in I \}$ such that $\phi_1 = \phi$, $\vdash \phi \triangleright t_1 = t$, $\phi_i \models \psi_i \Leftrightarrow \psi_i'$, and

$$\vdash \phi_i \triangleright t_i = u_i[\{\psi_i\}(\phi_i \to t_i')/X_i \mid i \in I]$$

for each $i \in I$. We will simply say "t provably satisfies E" when $\phi_i = \mathrm{tt}$ for all $i \in I$.

Proposition 5.1. *For any term t with free process variables $\mathbf{W}$ there exists a standard equation set E, with free process variables in $\mathbf{W}$, which is provably satisfied by t. In particular, if t is closed then E is also closed.*

Proof. We first show that, by using UNG, for any term t there is a guarded term t' such that $FV(t) = FV(t')$ and $\vdash t = t'$. The proposition can then be proved by structural induction on guarded terms.

Proposition 5.2. *For guarded, closed terms t and u, if $t \sim^{\phi} u$ then there exists a standard, closed equation set E which is provably ϕ-satisfied by both t and u.*

Proof. Let the sets of clock variables of t, u be $\mathbf{x}$, $\mathbf{y}$, respectively, with $\mathbf{x} \cap \mathbf{y} = \emptyset$. Let also E_1 and E_2 be the standard equation sets for t and u, respectively:

$$E_1 : \{ X_i = \{\phi_i\} \sum_{k \in K_i} \phi_{ik} \to a_{ik}(\mathbf{x}_{ik}).X_{f(i,k)} \mid i \in I \}$$

$$E_2 : \{ Y_j = \{\psi_j\} \sum_{l \in L_j} \psi_{jl} \to b_{jl}(\mathbf{y}_{jl}).Y_{g(j,l)} \mid j \in J \}$$

So there are $t_i \equiv \{\phi_i'\}t_i'$, $u_j \equiv \{\psi_j'\}u_j'$ with $\vdash t_1 = t$, $\vdash u_1 = u$ such that $\models \phi_i \Leftrightarrow \phi_i'$, $\models \psi_i \Leftrightarrow \psi_i'$, and

$$\vdash t_i = \{\phi_i\} \sum_{k \in K_i} \phi_{ik} \to a_{ik}(\mathbf{x}_{ik}).t_{f(i,k)} \qquad \vdash u_j = \{\psi_j\} \sum_{l \in L_j} \psi_{jl} \to b_{jl}(\mathbf{y}_{jl}).u_{g(j,l)}$$

Without loss of generality, we may assume $a_{ik} = b_{jl} = a$ for all i, k, j, l.

For each pair of i, j, let

$$\Phi_{ij} = \{ \Delta \in \mathcal{RC}(\mathbf{xy}) \mid t_i \sim^{\Delta \Uparrow} u_j \}$$

Set $\phi_{ij} = \bigvee \Phi_{ij}$. By the definition of Φ_{ij}, ϕ_{ij} is the weakest condition over which t_i and u_j are symbolically bisimilar, that is, $\psi \Rightarrow \phi_{ij}$ for any ψ such that $t_i \sim^{\psi} u_j$. Also for each $\Delta \in \Phi_{ij}$, $\Delta \models Inv(t_i) \Leftrightarrow Inv(u_j)$, i.e., $\Delta \models \phi_i' \Leftrightarrow \psi_j'$, hence $\Delta \models \phi_i \Leftrightarrow \psi_j$.

For each $\Delta \in \Phi_{ij}$ let $I_{ij}^{\Delta} = \{ (k,l) \mid t_{f(i,k)} \sim^{\Delta \downarrow \mathbf{x}_{ik} \mathbf{y}_{jl} \Uparrow} u_{g(j,l)} \}$. Define

$$E : \quad Z_{ij} = \{\phi_i\}(\sum_{\Delta \in \Phi_{ij}} \Delta \to \sum_{(k,l) \in I_{ij}^{\Delta}} a(\mathbf{x}_{ik}\mathbf{y}_{jl}).Z_{f(i,k)g(j,l)})$$

We claim that E is provably ϕ-satisfied by t when each Z_{ij} is instantiated with t_i over ϕ_{ij}. We need to show

$$\vdash \phi_{ij} \rhd t_i = \{\phi_i\} \sum_{\Delta \in \Phi_{ij}} \Delta \rightarrow \sum_{(k,l) \in I_{ij}^\Delta} a(\mathbf{x}_{ik}\mathbf{y}_{jl}).\{\phi_{f(i,k)}\}(\phi_{f(i,k)g(j,l)} \downarrow_{\mathbf{x}_{ik}\mathbf{y}_{jl}} \Uparrow \rightarrow t'_{f(i,k)})$$

Since the elements of Φ_{ij} are mutually disjoint, by Propositions 4.4 and 4.1, it is sufficient to show that, for each $\Delta \in \Phi_{ij}$,

$$\vdash \Delta \rhd t_i = \{\phi_i\} \sum_{(k,l) \in I_{ij}^\Delta} a(\mathbf{x}_{ik}\mathbf{y}_{jl}).\{\phi_{f(i,k)}\}(\phi_{f(i,k)g(j,l)} \downarrow_{\mathbf{x}_{ik}\mathbf{y}_{jl}} \Uparrow \rightarrow t'_{f(i,k)})$$

By the definition of I_{ij}^Δ, we have $t_{f(i,k)} \sim^{\Delta \downarrow_{\mathbf{x}_{ik}\mathbf{y}_{jl}} \Uparrow} u_{g(j,l)}$. Hence, from the definition of Φ_{ij}, $\Delta \downarrow_{\mathbf{x}_{ik}\mathbf{y}_{jl}} \Uparrow \Rightarrow \phi_{f(i,k)g(j,l)} \downarrow_{\mathbf{x}_{ik}\mathbf{y}_{jl}} \Uparrow$. Therefore

$$\vdash \Delta \rhd \quad \{\phi_i\} \sum_{(k,l) \in I_{ij}^\Delta} a(\mathbf{x}_{ik}\mathbf{y}_{jl}).\{\phi_{f(i,k)}\}(\phi_{f(i,k)g(j,l)} \downarrow_{\mathbf{x}_{ik}\mathbf{y}_{jl}} \Uparrow \rightarrow t'_{f(i,k)})$$

$$\overset{Lemma\ 4.2}{=} \{\phi_i\} \sum_{(k,l) \in I_{ij}^\Delta} a(\mathbf{x}_{ik}\mathbf{y}_{jl}).\{\phi_{f(i,k)}\}(\Delta \downarrow_{\mathbf{x}_{ik}\mathbf{y}_{jl}} \Uparrow \rightarrow \phi_{f(i,k)g(j,l)} \downarrow_{\mathbf{x}_{ik}\mathbf{y}_{jl}} \Uparrow \rightarrow t'_{f(i,k)})$$

$$\overset{Prop.\ 4.1}{=} \{\phi_i\} \sum_{(k,l) \in I_{ij}^\Delta} a(\mathbf{x}_{ik}\mathbf{y}_{jl}).\{\phi_{f(i,k)}\}(\Delta \downarrow_{\mathbf{x}_{ik}\mathbf{y}_{jl}} \Uparrow \rightarrow t'_{f(i,k)})$$

$$\overset{Lemma\ 4.2}{=} \{\phi_i\} \sum_{(k,l) \in I_{ij}^\Delta} a(\mathbf{x}_{ik}\mathbf{y}_{jl}).\{\phi_{f(i,k)}\}t'_{f(i,k)}$$

$$\overset{THINNING}{=} \{\phi_i\} \sum_{(k,l) \in I_{ij}^\Delta} a(\mathbf{x}_{ik}).t_{f(i,k)} \overset{S1\text{-}S4}{=} \{\phi_i\} \sum_{k \in K_i} a(\mathbf{x}_{ik}).t_{f(i,k)} = t_i$$

Symmetrically we can show E is provably ϕ-satisfied by u when Z_{ij} is instantiated with u_j over ϕ_{ij}.

Proposition 5.3. *If both t and u provably ϕ-satisfy standard equation set E then $\vdash \phi \rhd t = u$.*

Proof. By induction on the size of E.

Combining Propositions 5.1, 5.2 and 5.3 we obtain the main theorem:

Theorem 5.4. *If $t \sim^\phi u$ then $\vdash \phi \rhd t = u$.*

6   Conclusion and Related Work

We have presented an axiomatisation, in the form of an inference system, of timed bisimulation for timed automata, and proved its completeness. To the best of our knowledge, this is the first complete axiomatisation for the full set of timed automata. As already mentioned in the introduction, the precursor to this work is [LW00.1], in which an inference system complete over the loop-free subset of timed automata is formulated. The key ingredient of the current extension is unique fixpoint induction. Although the form of this rule is syntactically the same as that used for parameterless processes [Mil84], here it is implicitly parameterised on clock variables, in the sense that the rule deals with terms involving clock variables which do not appear explicitely.

The most interesting development so far in algebraic characterizations for timed automata are presented in [ACM97, BP99]. As the main result, they established that each timed automaton is equivalent to an algebraic expression built out of the standard operators in formal languages, such as union, intersection, concatenation and variants of Kleene's star operator, in the sense that the automaton recognizes the same timed language as denoted by the expression. However, the issue of axiomatisation was not considered there. In [DAB96] a set of equational axioms was proposed for timed automata, but no completeness result was reported. [HS98] presents an algebraic framework for real-time systems which is similar to timed automata where "invariants" are replaced by "deadlines" (to express "urgency"), together with some equational laws. Apart from these, we are not aware of any other published work on axiomatising timed automata. On the other hand, most timed extensions of process algebras came with equational axiomatisations. Of particular relevance are [Bor96] and [AJ94]. The former developed a symbolic theory for a timed process algebra, while the later used the unique fixpoint induction to achieve a complete axiomatisation for the regular subset of the timed-CCS proposed in [Wan91].

References

[ACM97] E. Asarin, P. Caspi and O. Maler. A Kleene theorem for timed automata. In proceedings of LICS'97, 1997.

[AD94]   R. Alur and D.L. Dill. A theory of timed automata. *Theoretical Computer Science*, 126:183–235. 1994.

[AJ94]   L. Aceto and A. Jeffrey. A complete axiomatization of timed bisimulation for a class of timed regular behaviours. *Theoretical Computer Science*, 152(2):251–268. 1995.

[Bor96] M. Boreale. Symbolic Bisimulation for Timed Processes. In *AMAST'96*, LNCS 1101 pp.321-335. Springer–Verlag. 1996.

[BP99]   P. Bouyer and A. Petit. Decomposition and Composition of Timed Automata. In *ICALP'99*, LNCS 1644, pp. 210-219. Springer–Verlag. 1999.

[HS98]   S. Bornot and J. Sifakis. An Algebraic Framework for Urgency. In *Calculational System Design*, NATO Science Series, Computer and Systems Science 173, Marktoberdorf, July 1998.

[Cer92] K. Čerāns. Decidability of Bisimulation Equivalences for Parallel Timer Processes. In *CAV'92*, LNCS 663, pp.302-315. Springer–Verlag. 1992.

[DAB96] P.R. D'Argenio and Ed Brinksma. A Calculus for Timed Automata (Extended Abstract). In *FTRTFTS'96*, LNCS 1135, pp.110-129. 1996.

[HL95]   M. Hennessy and H. Lin. Symbolic bisimulations. *Theoretical Computer Science*, 138:353–389, 1995.

[HL96]   M. Hennessy and H. Lin. Proof systems for message-passing process algebras. *Formal Aspects of Computing*, 8:408–427, 1996.

[LW00.1] H. Lin and Y. Wang. A proof system for timed automata. Fossacs'2000, LNCS 1784. March 2000.

[LW00.2] H. Lin and Y. Wang. A complete proof system for timed automata (Full version). Available at: http://www.it.uu.se/research/reports/.

[Mil84] R. Milner. A complete inference system for a class of regular behaviours. *J. Computer and System Science*, 28:439–466, 1984.

[Mil89] R. Milner. *Communication and Concurrency*. Prentice-Hall, 1989.

[Wan91] Wang Yi. *A Calculus of Real Time Systems*. Ph.D. thesis, Chalmers University, 1991.

[WPD94] Wang Yi, Paul Pettersson, and Mats Daniels. Automatic Verification of Real-Time Communicating Systems By Constraint-Solving. In *Proc. of the 7th International Conference on Formal Description Techniques*, 1994.

Text Sparsification via Local Maxima[*]

Extended Abstract

Pilu Crescenzi[1], Alberto Del Lungo[2], Roberto Grossi[3], Elena Lodi[2],
Linda Pagli[3], and Gianluca Rossi[1]

[1] Dipartimento di Sistemi e Informatica, Università degli Studi di Firenze
Via C. Lombroso 6/17, 50134 Firenze, Italy
`piluc@dsi.unifi.it, rossig@dsi.unifi.it`
[2] Dipartimento di Matematica, Università degli Studi di Siena
Via del Capitano 15, 53100 Siena, Italy
`dellungo@unisi.it, lodi@unisi.it`
[3] Dipartimento di Informatica, Università degli Studi di Pisa
Corso Italia 40, 56125 Pisa, Italy
`grossi@di.unipi.it, pagli@di.unipi.it`

Abstract. In this paper, we investigate a text sparsification technique based on the identification of local maxima. In particular, we first show that looking for an order of the alphabet symbols that minimizes the number of local maxima in a given string is an NP-hard problem. Successively, we describe how the local maxima sparsification technique can be used to filter the access to unstructured texts. Finally, we experimentally show that this approach can be successfully used in order to create a space efficient index for searching a DNA sequence as quickly as a full index.

1 Introduction

The Run Length Encoding (in short, RLE) is a well-known lossless compression technique [7] based on the following simple idea: a sequence of k equal characters (also called *run*) can be encoded by a pair whose first component is the character and whose second component is k. RLE turns out to be extremely efficient in some special cases: for example, it can reach a 8-to-1 compression factor in the case of scanned text. Moreover, it is also used in the JPEG image compression standard [10]. We can view RLE as a (lossless) *text sparsification* technique. Indeed, let us imagine that the position of each character of a text is an access point to the text itself from which it is then possible to navigate either on the left or on the right. The RLE technique basically sparsifies these access points by selecting only the first position of each run.

Another well known form of text sparsification applies to structured texts, that is, texts in which the notion of *word* is precisely identifiable (for example,

---

[*] Research partially supported by ITALIAN MURST project "Algoritmi per Grandi Insiemi di Dati: Scienza ed Ingegneria".

S. Kapoor and S. Prasad (Eds.): FST TCS 2000, LNCS 1974, pp. 290–301, 2000.

by means of delimiter symbols). In this case, a natural way of defining the access points consists of selecting the position of the first character of each word (for example, each character following a delimiter symbol). Differently from RLE, this technique does not allow us to obtain a lossless compression of the original text but it can be very useful in order to create a space efficient index for searching the text itself. For example, if we are looking for a given word X, this technique allows us (by means of additional data structures such as suffix arrays) to analyze only the words starting with the first character of X. The main drawback of this approach is that it does not generalize to unstructured texts, that is, texts for which there is no clear notion of word available (for example, DNA sequences).

The main goal of this paper is to study an alternative text sparsification technique, which might be used to create space efficient indices for searching unstructured texts. In particular, we will consider the following technique. Given a string X over a finite alphabet Σ, let us assume that a total order of the symbols in Σ is specified. We then say that an occurrence of a symbol x in X is an access point to X if x is a *local maximum*; that is, both symbols adjacent to x are smaller than x. As in the case of the previous technique based on words, this sparsification technique is *lossy*. Indeed, assume that x and y are two local maxima: from this information, we can deduce that the symbol immediately after x is smaller than x and the symbol immediately before y is smaller than y. Of course, in general this information is not sufficient to uniquely identify the sequence of symbols between x and y.

Nevertheless, the notion of local maxima has been proven very useful in string matching [1, 4, 8, 9] and dynamic data structures [6, 9] as an extension of the deterministic coin tossing technique [3]. It is well understood in terms of local similarities, by which independent strings that share equal portions have equal local maxima in those portions. In this paper we will consider the following two questions.

- *How much can a given text be sparsified by applying the local maxima technique?* Note that this question is different from the ones previously studied in the literature on local maxima, as we would like to minimize the number of local maxima while previous results aimed at minimizing the distance between consecutive maxima.

In order to answer this question, we will introduce the following combinatorial problem: given a text of length n over an alphabet Σ, find an order of Σ which minimizes the number of local maxima (that is, the number of access points). We will then prove that this problem is NP-hard (see Sect. 2) for non-constant sized alphabets (clearly, the problem can be solved in time $O(|\Sigma|!n)$ that, for constant sized alphabets, is $O(n)$).

- *Can the local maxima sparsification technique be used to filter the access to unstructured texts in practice?*

In order to answer this question we first describe how the technique can be used to create an index for searching a given text (see Sect. 3). We will then give a positive answer to the above question by experiments, in the case of texts that are DNA sequences (see Sect. 4). In particular, we show that each run of the sparsification algorithm reduces the number of maxima by a factor of three. We exploit this to create a space efficient index for searching the sequence as quickly as a full index by means of additional data structures such as suffix arrays.

2   The NP-Hardness Result

In this section, we first describe the combinatorial problem associated with the local maxima sparsification technique and we then show that this problem is Np-hard.

2.1   The Combinatorial Problem

Let $X = x_1 \cdots x_n$ be a string over a finite alphabet Σ and assume that an order π of the symbols in Σ (that is, a one-to-one function π from Σ to $\{1, \ldots, |\Sigma|\}$) has been fixed. The *local maxima measure* $\mathcal{M}(X, \pi)$ of X with respect to π is then defined as the number of local maxima that appear in X, that is,

$$\mathcal{M}(X, \pi) = |\{i : (1 < i < n) \wedge (\pi(x_{i-1}) \leq \pi(x_i)) \wedge (\pi(x_i) > \pi(x_{i+1}))\}|.$$

The MINIMUM LOCAL MAXIMA NUMBER decision problem is then defined as follows:

Instance A string X over a finite alphabet Σ and an integer value K.
Question Does there exist an order π of Σ such that $\mathcal{M}(X, \pi) < K$?

Clearly, MINIMUM LOCAL MAXIMA NUMBER belongs to Np (since we just have to non-deterministically try all possible orders of the alphabet symbols).

2.2   The Reduction

We now define a polynomial-time reduction from the MAXIMUM EXACTLY-TWO SATISFIABILITY decision problem to MINIMUM LOCAL MAXIMA NUMBER. Recall that MAXIMUM EXACTLY-TWO SATISFIABILITY is defined as follows: given a set of clauses with exactly two literals per clause and given an integer H, does there exist a truth-assignment that satisfies at least H clauses? It is well-known that MAXIMUM EXACTLY-TWO SATISFIABILITY is Np-complete [5].

The basic idea of the reduction is to associate two symbols with each variable and one symbol with each clause and to force each pair of variable-symbols to be either smaller or greater than all clause-symbols. The variables whose both corresponding symbols are greater (respectively, smaller) than the clause-symbols will be assigned the true (respectively, false) value. The implementation of this basic idea will require several additional technicalities which are described in the next two sections.

The instance mapping. Let $C = \{c_1, \ldots, c_m\}$ be a set of clauses over the set of variables $U = \{u_1, \ldots, u_n\}$ such that each clause contains exactly two literals, and let H be a positive integer. The alphabet $\Sigma(C)$ of the corresponding instance of MINIMUM LOCAL MAXIMA NUMBER contains the following symbols:

- Two special symbols σ_m and σ_M (which, intuitively, will be the extremal symbols in any "reasonable" order of the alphabet).
- For each variable u_i with $1 \leq i \leq n$, two symbols σ_i^u and $\sigma_i^{\bar{u}}$.
- For each clause c_j with $1 \leq j \leq m$, a symbol σ_j^c.

The string $X(C)$ of the instance of MINIMUM LOCAL MAXIMA NUMBER is formed by several substrings with different goals. In order to define it, we first introduce the following gadget: given three symbols a, b, and c, let $g(a, b, c) = abcbcba$. The next result states the basic property of the previously defined gadget.

Lemma 1. *Let Σ be an alphabet and let a, b, and c three symbols in Σ. For any order π of Σ and for any integer $r > 0$, the following hold:*

1. *If $\pi(c) < \pi(b) < \pi(a)$, then $\mathcal{M}(g(a, b, c)^r, \pi) = 2r - 1$.*
2. *If $\pi(a) < \pi(b) < \pi(c)$, then $\mathcal{M}(g(a, b, c)^r, \pi) = 2r$.*
3. *If none of the previous two cases applies, then $\mathcal{M}(g(a, b, c)^r, \pi) \geq 3r - 1$.*

Proof. The proof of the lemma is done by examining all possible cases. Indeed, in Table 1 the occurrences of maxima produced by one gadget in correspondence of the six possible orders of the three symbols a, b, and c are shown.

Minimum	Medium	Maximum	a	b	c	b	c	b	a
a	b	c	+	+	+	-	+	-	-
a	c	b	+	+	-	+	-	+	-
b	a	c	+	-	+	-	+	-	+
b	c	a	+	-	+	-	+	-	+
c	a	b	+	+	-	+	-	+	-
c	b	a	+	-	-	+	-	+	+

Table 1. The possible orders of the gadget symbols

By looking at the table, it is easy to verify the correctness of the three statements of the lemma.   □

The first $m + 2n$ substrings of the instance X will force σ_m and σ_M to be the extremal symbols of any efficient order of Σ. They are then defined as follows:

- For $j = 1, \ldots, m$, $X_1^j = g(\sigma_m, \sigma_j^c, \sigma_M)^{m^3}$.
- For $i = 1, \ldots, n$, $X_2^i = g(\sigma_m, \sigma_i^u, \sigma_M)^{m^3}$.
- For $i = 1, \ldots, n$, $X_3^i = g(\sigma_m, \sigma_i^{\bar{u}}, \sigma_M)^{m^3}$.

The next nm substrings will force each pair of variable symbols to be either both on the left or both on the right of all clause symbols. In particular, for $i = 1, \ldots, n$ and for $j = 1, \ldots, m$, we define

$$Y_i^j = (\sigma_{\mathtt{m}} g(\sigma_i^u, \sigma_i^{\bar{u}}, \sigma_j^c) \sigma_{\mathtt{M}})^{m^2}.$$

Finally, for each clause c_j with $1 \leq j \leq m$, we have one substring Z_j^c whose definition depends on the type of the clause and whose goal is to decide the truth-value of each variable depending on its symbol's position relatively to the clause symbols. In particular:

- If $c_j = u_i \vee u_k$, then $Z_j^c = \sigma_{\mathtt{m}} \sigma_j^c \sigma_i^u \sigma_j^c \sigma_k^u \sigma_j^c \sigma_{\mathtt{m}}$.
- If $c_j = \neg u_i \vee u_k$, then $Z_j^c = \sigma_{\mathtt{m}} \sigma_i^u \sigma_j^c \sigma_i^u \sigma_k^u \sigma_j^c \sigma_{\mathtt{m}}$.
- If $c_j = u_i \vee \neg u_k$, then $Z_j^c = \sigma_{\mathtt{m}} \sigma_j^c \sigma_i^u \sigma_k^u \sigma_j^c \sigma_k^u \sigma_{\mathtt{m}}$.
- If $c_j = \neg u_i \vee \neg u_k$, then $Z_j^c = \sigma_{\mathtt{m}} \sigma_i^u \sigma_j^c \sigma_k^u \sigma_j^c \sigma_{\mathtt{M}} \sigma_{\mathtt{m}}$.

In conclusion, the instance X is defined as:

$$X = X_1^1 \cdots X_1^m X_2^1 \cdots X_2^n X_3^1 \cdots X_3^n Y_1^1 \cdots Y_1^m \cdots Y_n^1 \cdots Y_n^m \sigma_{\mathtt{m}} Z_1 \cdots Z_m.$$

The proof of correctness We now prove the following result.

Lemma 2. *For any set C of m clauses with exactly two literals per clause and for any integer H, there exists a truth-assignment that satisfies H clauses in C if and only if there exists an order π of $\Sigma(C)$ such that*

$$\mathcal{M}(X(C), \pi) = (2m + 7n)m^3 + 3m - H.$$

Proof. Assume that there exists a truth-assignment τ to the variables $u_1, \ldots, u_n$ that satisfies H clauses in C. We then define the corresponding order π of the symbols in $\Sigma(C)$ as follows.

1. For any symbol a different from $\sigma_{\mathtt{m}}$ and $\sigma_{\mathtt{M}}$, $\pi(\sigma_{\mathtt{m}}) < \pi(a) < \pi(\sigma_{\mathtt{M}})$.
2. For any i with $1 \leq i \leq n$ and for any j with $1 \leq j \leq m$, if $\tau(u_i) = $ FALSE, then $\pi(\sigma_i^u) < \pi(\sigma_i^{\bar{u}}) < \sigma_j^c$, otherwise $\sigma_j^c < \pi(\sigma_i^{\bar{u}}) < \pi(\sigma_i^u)$.

From Lemma 1 it follows that

$$\mathcal{M}(X_1^1 \cdots X_1^m X_2^1 \cdots X_2^n X_3^1 \cdots X_3^n, \pi) = (2m + 4n)m^3.$$

Moreover, because of the same lemma and since Y_n^m ends with the maximal symbol, we have that

$$\mathcal{M}(X_1^1 \cdots X_1^m X_2^1 \cdots X_2^n X_3^1 \cdots X_3^n Y_1^1 \cdots Y_n^m, \pi) = (2m + 4n)m^3 + 3nm^3 - 1$$
$$= (2m + 7n)m^3 - 1.$$

The concatenation of Y_n^m with $\sigma_{\mathtt{m}}$ produces one more maximum. Successively, the number of maxima will depend on whether one clause is satisfied: indeed, it is possible to prove that if the jth clause is satisfied, then Z_j produces two

u_i	u_k	σ_m	σ_j^c	σ_i^u	σ_j^c	σ_k^u	σ_j^c	σ_m
FALSE	FALSE	+	+	-	+	-	+	-
FALSE	TRUE	+	+	-	+	+	-	-
TRUE	FALSE	+	+	+	-	-	+	-
TRUE	TRUE	+	+	+	-	+	-	-

Table 2. The occurrences of maxima corresponding to $u_i \vee u_k$

maxima, otherwise it produces three maxima. For example, assume that c_j is the disjunction of two positive literals $u_i \vee u_k$. The occurrence of maxima according to the truth-values of the two variables is then shown in Table 2 (recall that by definition of π, for any $h = 1, \ldots, n$ and for any $l = 1, \ldots, m$, $\pi(\sigma_h^u) < \pi(\sigma_l^c)$ if and only if $\tau(\sigma_h^u) = $ FALSE, and that, in this case, $Z_j = \sigma_m \sigma_j^c \sigma_i^u \sigma_j^c \sigma_k^u \sigma_j^c \sigma_m$). As it can be seen from the table, we have two maxima in correspondence of the three truth-assignments that satisfy the clause and three maxima in the case of the non-satisfying assignment. The other types of clauses can be dealt in a similar way.

In summary, we have that the number of maxima generated by π on the string $X(C)$ is equal to $(2m + 7n)m^3 + 2H + 3(m - H) = (2m + 7n)m^3 + 3m - H$.

Conversely, assume that an order π of the symbols in $\Sigma(C)$ is given such that the number of maxima generated on $X(C)$ is equal to $(2m + 7n)m^3 + 3m - H$. Because of Lemma 1, the substring $X_1^1 \cdots X_1^m X_2^1 \cdots X_2^n X_3^1 \cdots X_3^n$ produces at least $(2m + 4n)m^3 - 1$ maxima and it ensures that either σ_m is the minimal symbol and σ_M is the maximal one or σ_M is the minimal symbol and σ_m is the maximal one. Assume that the former case holds so that $(2m + 4n)m^3$ are generated (the latter case can be dealt in a similar way). The next substring $Y_1^1 \cdots Y_1^m \cdots Y_n^1 \cdots Y_n^m \sigma_m$ instead produces at least $3nm^3$ maxima and it ensures that, for any i with $1 \le i \le n$ and for any j with $1 \le j \le m$, either $\pi(\sigma_i^u) < \pi(\sigma_i^{\bar u}) < \pi(\sigma_j^c)$ or $\pi(\sigma_j^c) < \pi(\sigma_i^{\bar u}) < \pi(\sigma_i^u)$. We then assign the value TRUE to variable u_i if the former case holds, otherwise we assign to it the value FALSE. It is then easy to verify that the remaining $3m - H$ maxima are produced by H clauses that are satisfied and $m - H$ clauses that are not satisfied. For example, assume that $c_j = \neg u_i \vee \neg u_j$ so that $Z_j = \sigma_m \sigma_i^u \sigma_j^c \sigma_k^u \sigma_j^c \sigma_M \sigma_m$. The truth-values of u_i and u_j corresponding to the six possible order of σ_i^u, σ_k^u, and σ_j^c and the resulting truth-value of c_j along with the number of generated maxima are shown in Table 3. We have thus shown that H clauses of C can be satisfied if and only if $\mathcal{M}(X(C), \pi) = (2m + 7n)m^3 + 3m - H$ and the lemma is proved. $\square$

From the above lemma and from the fact that MINIMUM LOCAL MAXIMA NUMBER belongs to NP it follows the following theorem.

Theorem 1. MINIMUM LOCAL MAXIMA NUMBER *is* NP-*complete.*

Minimum	Medium	Maximum	$\tau(u_i)$	$\tau(u_k)$	$\tau(c_j)$	Maxima
σ_i^u	σ_k^u	σ_j^c	FALSE	FALSE	TRUE	2
σ_i^u	σ_j^c	σ_k^u	FALSE	TRUE	TRUE	2
σ_k^u	σ_i^u	σ_j^c	FALSE	FALSE	TRUE	2
σ_k^u	σ_j^c	σ_i^u	TRUE	FALSE	TRUE	2
σ_j^c	σ_i^u	σ_k^u	TRUE	TRUE	FALSE	3
σ_j^c	σ_k^u	σ_i^u	TRUE	TRUE	FALSE	3

Table 3. The possible orders corresponding to $c_j = \neg u_i \vee \neg u_k$

3   The Sparsification Algorithm

We have seen in the previous section that the problem of assigning an ordering to the characters of a sequence which minimizes the number of local maxima is a hard problem. Clearly, for any fixed string, the number of local maxima produced by any ordering is at most half of the length of the string. The following lemma, instead, guarantees that, for any fixed ordering π, the number of local maxima produced by π, on a randomly chosen string, is at most one third of the length of the string.

Lemma 3. *Let π be an order over an alphabet Σ. If X is a randomly chosen string over Σ of length n, then the expected value of $\mathcal{M}(X,\pi)$ is at most $n/3$.*

Proof. Let $X = x_1 \cdots x_n$ be the randomly chosen string over Σ and let $T(x_k)$ be the random variable that equals to 1 if x_k is a maximum and 0 otherwise, for any k with $1 \leq k \leq n$. Clearly, for any k with $2 \leq k \leq n-1$,

$$\Pr[T(x_k) = 1] = \Pr[\pi(x_{k-1}) \leq \pi(x_k)] \Pr[\pi(x_{k+1}) < \pi(x_k)].$$

Hence, the probability that x_k is a maximum, assuming that $\pi(x_k) = i$, is

$$\Pr[T(x_k) = 1 | \pi(x_k) = i] = \sum_{j=1}^{i} \Pr[\pi(x_{k-1}) = j] \sum_{j=1}^{i-1} \Pr[\pi(x_{k+1}) = j]$$

$$= \frac{i(i-1)}{|\Sigma|^2}.$$

Finally, the probability that x_k is a maximum is

$$\Pr[T(x_k) = 1] = \sum_{i=1}^{|\Sigma|} \Pr[T(x_k) = 1 | \pi(x_k) = i] \Pr[\pi(x_k) = i]$$

$$= \frac{1}{|\Sigma|^3} \sum_{i=1}^{|\Sigma|} i(i-1) = \frac{1}{3} - \frac{1}{3|\Sigma|^2}.$$

By linearity of expectation, the expected number of local maxima is

$$\frac{n-2}{3}\left(1-\frac{1}{|\Sigma|^2}\right)$$

and the lemma follows.   □

The above lemma suggests that random strings (that is, strings which are not compressible) can be sparsified by means of the local maxima technique so that the number of resulting access points is at most one third of the length of the original string. We wish to exploit this property in order to design a *sparsification procedure* that replaces a given string with a shorter one made up of only the local maxima (the new string will not clearly contain the whole original information). We repeat this simple procedure by computing the local maxima of the new string to obtain an even shorter string. We iterate this shortening several times until the required sparsification is obtained. That is, the compressed string is short enough to be efficiently processed, but it still contains enough information to solve a given problem, as we will see shortly. For example, let us consider the very basic problem of searching a pattern string in a text string. We can compress the two strings by means of our sparsification procedure. Then, we search for the pattern by matching the local maxima only. Whenever a match is detected, we check that it is not a false occurrence by a full comparison of the pattern and the text substring at hand. It is worth pointing that the number of times we apply the sparsification on the text must be related to the length of the patterns we are going to search for. Indeed, performing too many iterations of the sparsification could drop too many characters between two consecutive local maxima selected in the last iteration. As a result, we could not find the pattern because it is too short (see Lemma 4).

Another care must be taken with alphabets of small size, such as binary strings or DNA sequences. At each iteration of the algorithm, at least one character of the alphabet disappears from the new string, since the smallest character in the alphabet is not selected as local maximum. This fact can be a limitation, for instance in DNA sequences, where $|\Sigma|$ is only 4. Indeed, we can apparently apply the sparsification less than $|\Sigma|$ times. We can circumvent this problem by storing each local maximum along with its offset to (i.e., the number of characters before) the next maximum. Each local maximum in Σ is replaced by a new character given by the pair (local maximum, offset) in the new alphabet $\Sigma \times N$ undergoing the lexicographic order.

In order to explain the sparsification algorithm, let us consider the following string over the alphabet Σ of four characters A, C, G, and T:

$$T_0 = \text{TGACACGTGACGAGCACACACGTCGCAGATGCATA}.$$

Assuming that the characters are ordered according to the lexicographical order, the number of local maxima contained in the above string is 11 (which is approximately one third of the total length of the string, i.e., 35). The new string obtained after the first iteration of the sparsification algorithm is then the following one:

$$T_1 = (\texttt{C},4)\,(\texttt{T},4)\,(\texttt{G},2)\,(\texttt{G},3)\,(\texttt{C},2)\,(\texttt{C},4)\,(\texttt{T},2)\,(\texttt{G},3)\,(\texttt{G},2)\,(\texttt{T},4)\,(\texttt{T},2).$$

Observe that the new alphabet consists of six characters, each one composed by a character of $\Sigma - \{\texttt{A}\}$ and a natural number in $\{2,\ldots,4\}$. Assuming that these characters are ordered according to the lexicographical order, we have that the number of local maxima contained in the above string is 4 (which is approximately one third of the total length of the string, i.e., 11). The new string obtained after the second iteration of the sparsification algorithm is then the following one:

$$T_2 = (\texttt{T},6)\,(\texttt{G},9)\,(\texttt{T},7)\,(\texttt{T},6).$$

Assume we are looking for the pattern $\texttt{ACACGTGACGAGCA}$ which occurs in T_0 starting at position 3. By applying the first iteration of the sparsification algorithm to the pattern, we obtain the string $(\texttt{C},4)\,(\texttt{T},4)\,(\texttt{G},2)\,(\texttt{G},3)$ which occurs in T_1 starting at position 1. However, if we apply the second iteration of the sparsification algorithm to the pattern, we obtain the new pattern $(\texttt{T},9)$ which does not occur in T_2. Indeed, as we have already observed, the size of the pattern to be searched bounds the number of the iterations of the algorithm that can be performed. Formally, let T_i, for $i \geq 1$, be the text after the ith iteration and let m_i be the maximum value of the offset of a local maxima in T_i. It can be easily verified the following:

Lemma 4. *A pattern P of size m is successfully found in a text T_i, as long as $m \geq 2m_i$.*

In the previous example, we have that $m = 14$, $m_1 = 4$, and $m_2 = 9$. According to the above lemma, the pattern is successfully found in T_1 but it is not successfully found in T_2.

4   Experimental Results

In our experiments, we consider DNA sequences, where $\Sigma = \{\texttt{A}, \texttt{T}, \texttt{C}, \texttt{G}\}$. In Table 4, we report the number of local maxima obtained for the three DNA sequences: *Saccharomyces Cervisiae* (file $\texttt{IV.fna}$), *Archeoglobus Fulgidus* (file $\texttt{aful.fna}$) and *Escherichia Coli* (file $\texttt{ecoli.fna}$), for three consecutive iterations of the algorithm.

In the ith iteration, $i = 1,\ldots,3$, we have n_i local maxima with maximum distance m_i among two consecutive of them. The values of m_i is not exactly the maximum among all possible values. There are very few values that are very much larger than the majority. The additive term in the figures for n_i accounts for those local maxima that are at distance greater than m_i. For example, after the first iteration on *Saccharomyces Cervisiae* (file $\texttt{IV.fna}$), there are $n_i = 459027$ local maxima having offset at most $m_1 = 18$, and only 616 local maxima with offset larger than 18 (actually, much larger). It goes without saying that it is better to treat these 616 maxima independently from the rest

	n_0	n_1	m_1	n_2	m_2	n_3	m_3
IV.fna	1532027	459027+616	18	146408+1814	37	47466+6618	92
aful.fna	2178460	658396+101	13	213812+952	35	69266+3781	94
ecoli.fna	4639283	1418905+61	14	458498+851	37	148134+4572	98

Table 4. Sample values for three DNA sequences

of the maxima. Finally, we observe a reduction of about 1/3 at each iteration on the values of n_i.

In Fig. 1, we report the distribution of the distances between consecutive maxima in the sequence after each of three iterations of the sparsification algorithm. After the first iteration almost all the values are concentrated in a small range of values (see Fig. 1a); the distribution curve is maintained and flattened after the next two iterations (see Fig. 1b-c).

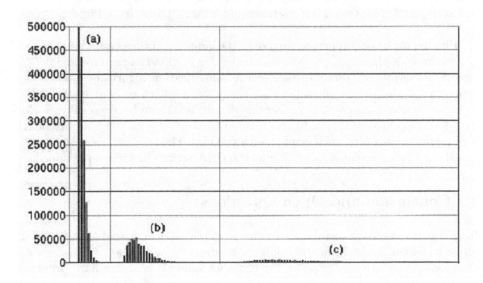

Fig. 1. Distribution of distances among the local maxima for file `ecoli.fna`. Data for iteration $i = 1$ is reported in (a) where distances range from 1 to 16, for $i = 2$ in (b) where distances range from 3 to 46, and for $i = 3$ in (c) where distances range from 8 to 144.

As a result of the application of our sparsification technique to construct a text index on the suffixes starting at the local maxima (for this purpose, we use a *suffix array* in our experiments), the occupied space is small compared to the text size itself, and this seems to be a rather interesting feature. Here, we are considering *the exact string matching problem*. The application of our method to other important string problems, such as multiple sequence alignment and matching with errors, seems promising but it is still object of study. As the search of patterns in DNA applications has to be performed considering the possibility of errors, one should use the approximate, rather than the exact string matching. However, in several algorithms used in practice, findings the exact occurrences of the pattern in the text [2] is a basic filtering step towards solving the approximate problem due to the large size of the text involved.

Three final considerations are in order. First, the threshold of $2m_i$ on the minimum pattern length in Lemma 4 is overly pessimistic. In our experiments, we successfully found all patterns of length at least $m_i/2$. For example, in file `ecoli.fna`, we had $m_3 = 98$. We searched for patterns of length ranging from 50 to 198, and all searches were successful.

Second, it may seem that a number of false matches are caused by discarding the first characters in the pattern in each iteration of the sparsification algorithm. Instead, the great majority of these searches did not give raise to false matches due to the local maxima, except for a minority. Specifically, on about 150,000 searches, we counted only about 300 searches giving false matches, and the average ratio between good matches and total matches (including false ones) was 0.28.

Finally, the most important feature of the index is that it saves a lot of space. For example, a plain suffix array for indexing file `ecoli.fna` requires about 17.7 megabytes. Applying one iteration of the sparsification algorithm reduces the space to 5.4 megabytes, provided that the pattern length m is at least 14; the next two iterations give 1.8 megabytes (for $m \geq 37$) and 0.6 megabytes (for $m \geq 98$), respectively. These figures compare favorably with the text size of 1.1 megabytes by encoding each symbol with two bits. The tradeoff between pattern length and index space is inevitable as the DNA strings are incompressible.

5   Conclusion and Open Questions

In this paper, we have investigated some properties of a text sparsification technique based on the identification of local maxima. In particular, we have shown that looking for the best order of the alphabet symbols is an NP-hard problem. Successively, we have described how the local maxima sparsification technique can be used to filter the access to unstructured texts. Finally, we have experimentally shown that this approach can be successfully used in order to create a space efficient index for searching a DNA sequence as quickly as a full index.

Regarding the combinatorial optimization problem, the main question left open by this paper is whether the optimization version of MINIMUM LOCAL MAXIMA NUMBER admits a polynomial-time approximation algorithm. It would also be interesting to accompany the experimental results obtained with DNA sequences by some theoretical results, such as the evaluation of the expected maximal distance between two local maxima or the expected number of false matches.

References

[1] A. Alstrup, G. S. Brodal, and T. Rauhe. Pattern matching in dynamic texts. In *Proceedings of the 11th ACM-SIAM Annual Symposium on Discrete Algorithms*, pages 819–828, San Francisco, CA, 2000.

[2] S. Burkhardt, A. Crauser, H.P. Lenhof, P. Ferragina, E. Rivals, and M. Vingron. Q-gram based database searching using a suffix array (QUASAR). In *Proceedings of the Annual International Conference on Computational Biology (RECOMB)*, 1999.

[3] R. Cole and U. Vishkin. Deterministic coin tossing with applications to optimal parallel list ranking. *Information and Control*, 70(1):32–53, July 1986.

[4] G. Cormode, M. Paterson, S.C. Sahinalp, and U. Vishkin. Communication complexity of document exchange. In *Proceedings of the 11th ACM-SIAM Annual Symposium on Discrete Algorithms*, 2000.

[5] M.R. Garey and D.S. Johnson. *Computers and Intractability: A Guide to the Theory of NP-completeness*. Freeman, San Francisco, 1979.

[6] K. Mehlhorn, R. Sundar, and C. Uhrig. Maintaining dynamic sequences under equality tests in polylogarithmic time. *Algorithmica*, 17(2):183–198, February 1997.

[7] M. Nelson and J.-L. Gailly. *The Data Compression Book*. M&T Books, 1996.

[8] S.C. Şahinalp and U. Vishkin. Symmetry breaking for suffix tree construction (extended abstract). In *Proceedings of the Twenty-Sixth Annual ACM Symposium on the Theory of Computing*, pages 300–309, Montréal, Québec, Canada, 23–25 May 1994.

[9] S.C. Şahinalp and U. Vishkin. Efficient approximate and dynamic matching of patterns using a labeling paradigm (extended abstract). In *37th Annual Symposium on Foundations of Computer Science*, pages 320–328. IEEE, 14–16 October 1996.

[10] G.K. Wallace. The JPEG still picture compression standard. *Communications of the ACM*, 34(1):30–44, April 1991.

Approximate Swapped Matching

Amihood Amir[1], Moshe Lewenstein[2], and Ely Porat[3]

[1] Department of Mathematics and Computer Science,
Bar-Ilan University, 52900 Ramat-Gan, Israel
and Georgia Tech
Tel. (972-3)531-8770;
amir@cs.biu.ac.il

[2] Department of Mathematics and Computer Science,
Bar-Ilan University, 52900 Ramat-Gan, Israel
Tel. (972-3)531-8407;
moshe@cs.biu.ac.il

[3] Department of Mathematics and Computer Science,
Bar-Ilan University, 52900 Ramat-Gan, Israel
and Weizmann Institute.
Tel. (972-3)531-8407;
porately@cs.biu.ac.il

Abstract. Let a text string T of n symbols and a pattern string P of m symbols from alphabet Σ be given. A *swapped version* P' of P is a length m string derived from P by a series of *local swaps*, (i.e. $p'_\ell \leftarrow p_{\ell+1}$ and $p'_{\ell+1} \leftarrow p_\ell$) where each element can participate in *no more than one swap*. The *Pattern Matching with Swaps* problem is that of finding all locations i of T for which there exists a swapped version P' of P with an exact matching of P' in location i of T.

Recently, some efficient algorithms were developed for this problem. Their time complexity is better than the best known algorithms for pattern matching with mismatches. However, the *Approximate Pattern Matching with Swaps* problem was not known to be solved faster than the pattern matching with mismatches problem.

In the *Approximate Pattern Matching with Swaps* problem the output is, for every text location i where there is a swapped match of P, the *number* of swaps necessary to create the swapped version that matches location i. The fastest known method to-date is that of counting mismatches and dividing by two. The time complexity of this method is $O(n\sqrt{m \log m})$ for a general alphabet Σ.

In this paper we show an algorithm that counts the number of swaps at every location where there is a swapped matching in time $O(n \log m \log \sigma)$, where $\sigma = min(m, |\Sigma|)$. Consequently, the total time for solving the approximate pattern matching with swaps problem is $O(f(n, m) + n \log m \log \sigma)$, where $f(n, m)$ is the time necessary for solving the pattern matching with swaps problem.

Key Words: Design and analysis of algorithms, combinatorial algorithms on words, pattern matching, pattern matching with swaps, non-standard pattern matching, approximate pattern matching.

S. Kapoor and S. Prasad (Eds.): FST TCS 2000, LNCS 1974, pp. 302–311, 2000.
© Springer-Verlag Berlin Heidelberg 2000

1   Introduction

The *Pattern Matching with Swaps* problem (the *Swap Matching* problem, for short) requires finding all occurrences of a pattern of length m in a text of length n. The pattern is said to match the text at a given location i if adjacent pattern characters can be swapped, if necessary, so as to make the pattern identical to the substring of the text starting at location i. All the swaps are constrained to be disjoint, i.e., each character is involved in at most one swap.

The importance of the swap matching problem lies in recent efforts to understand the complexity of various *generalized pattern matching* problems. The textbook problem of *exact string matching* that was first shown to be solvable in linear time by Knuth, Morris and Pratt [10] does not answer the growing requirements stemming from advances in Multimedia, Digital libraries and Computational Biology. To this end, pattern matching has to adapt itself to increasingly broader definitions of "matching" [18, 17]. In computational biology one may be interested in finding a "close" mutation, in communications one may want to adjust for transmission noise, in texts it may be desirable to allow common typing errors. In multimedia one may want to adjust for lossy compressions, occlusions, scaling, affine transformations or dimension loss.

The above applications motivated research of two new types – *Generalized Pattern Matching*, and *Approximate Pattern Matching*. In generalized matching the input is still a text and pattern but the "matching" relation is defined differently. The output is all locations in the text where the pattern "matches" under the new definition of match. The different applications define the matching relation. An early generalized matching was the *string matching with don't cares* problem defined by Fischer and Paterson [8]. Another example of a generalized matching problem is the *less-than matching* [4] problem defined by Amir and Farach. In this problem both text and pattern are numbers. One seeks all text locations where every pattern number is less than its corresponding text number. Amir and Farach showed that the less-than-matching problem can be solved in time $O(n\sqrt{m \log m})$.

Muthukrishnan and Ramesh [15] prove that practically all general matching relations, where the generalization is in the definition of single symbol matches, are equivalent to the boolean convolutions, i.e. it is unlikely that they could be solved in time faster than $O(n \log m)$, where n is the text length and m is the pattern length. As we have seen, some examples have significantly worse upper bound than this.

The swap matching problem is also a generalized matching problem. It arises from one of the edit operations considered by Lowrance and Wagner [14, 19] to define a distance metric between strings.

Amir et al [3] obtained the first non-trivial results for this problem. They showed how to solve the problem in time $O(nm^{1/3} \log m \log \sigma)$, where $\sigma = \min(|\Sigma|, m)$. Amir et al. [5] also give certain special cases for which $O(m\text{polylog}(m))$ time can be obtained. However, these cases are rather restrictive. Cole and Hariharan [6] give a randomized algorithm that solves the swap matching problem over a binary alphabet in time $O(n \log n)$.

The second important pattern matching paradigm is that of *approximate matching*. Even under the appropriate matching relation there is still a distinction between *exact matching* and *approximate matching*. In the latter case, a distance function is defined on the text. A text location is considered a match if the distance between it and the pattern, under the given distance function, is within the tolerated bounds.

The fundamental question is what type of approximations are inherently hard computationally, and what types are faster to compute. This question motivated much of the pattern matching research in the last couple of decades.

The earliest and best known distance function is Levenshtein's *edit distance* [13]. The edit distance between two strings is the smallest number of edit operations, in this case insertions, deletions, and mismatches, whereby one string can be converted to the other. Let n be the text length and m the pattern length. A straightforward $O(nm)$ dynamic programming algorithm computes the edit distance between the text and pattern. Lowrance and Wagner [14, 19] proposed an $O(nm)$ dynamic programming algorithm for the extended edit distance problem, where the swap edit operation is added. In [9, 11, 12] $O(kn)$ algorithms are given for the edit distance with only k allowed edit operations. Recently, Cole and Hariharan [7] presented an $O(nk^4/m + n + m)$ algorithm for this problem.

Since the upper bound for the edit distance seems very tough to break, attempts were made to consider the edit operations separately. If only mismatches are counted for the distance metric, we get the *Hamming distance*, which defines the *string matching with mismatches* problem. A great amount of work was done on finding efficient algorithms for string matching with mismatches. By methods similar to those of Fischer and Paterson [8] it can be shown that the string matching with mismatches problem can be solved in time $O(\min(|\Sigma|, m)n \log m)$. For given finite alphabets, this is $O(n \log m)$. Abrahamson [1] developed an algorithm that solves this problem for general alphabets in time $O(n\sqrt{m \log m})$.

The *approximate pattern matching with swaps* problem considers the swaps as the only edit operation and seeks to compute, for each text location i, the number of swaps necessary to convert the pattern to the substring of length m starting at text location i (provided there is a swap match at i). In [2] it was shown that the approximate pattern matching with swaps problem can be reduced to the string matching with mismatches problem. For every location where there is a swap match, the number of swaps is precisely equal to half the number of mismatches (since a swap is two mismatches). Although swap matching as a generalized matching proved to be more efficient than counting mismatches, it remained open whether swap matching as an approximation problem can be done faster than mismatches.

In this paper we answer this question in the affirmative. We show that if all locations where there is a swap match are known, the approximate swap matching problem can be solved in time $O(n \log m \log \sigma)$, where $\sigma = \min(|\Sigma|, m)$. Therefore, assuming swap matching can be done in time $f(n, m)$, approximate swap matching can be done in time $O(f(n, m) + n \log m \log \sigma)$.

Paper organization. This paper is organized in the following way. In section 2 we give basic definitions. In sections 3, we outline the key idea and intuition behind our algorithm. In section 4 we give a randomized algorithm, which easily highlights the idea of our solution. It turns out that rather than using a generic derandomization strategy, a simple, problem specific, method can be used to obtain the deterministic counterpart. Section 5 presents an easy and efficient code that solves our problem deterministically.

2   Problem Definition

Definition: Let $S = s_1 \ldots s_n$ be a string over alphabet Σ. A *swap permutation* for S is a permutation $\pi : \{1, \ldots, n\} \to \{1, \ldots, n\}$ such that

1. if $\pi(i) = j$ then $\pi(j) = i$ (characters are swapped).
2. for all i, $\pi(i) \in \{i - 1, i, i + 1\}$ (only adjacent characters are swapped).
3. if $\pi(i) \neq i$ then $s_{\pi(i)} \neq s_i$ (identical characters are not swapped).

For a given string $S = s_1 \ldots s_n$ and swap permutation π for S we denote $\pi(S) = s_{\pi(1)} s_{\pi(2)} \cdots s_{\pi(n)}$. We call $\pi(S)$ a *swapped version* of S.

The *number of swaps* in swapped version $\pi(S)$ of S is the number of pairs $(i, i+1)$ where $\pi(i) = i + 1$ and $\pi(i + 1) = i$.

For pattern $P = p_1 \ldots p_m$ and text $T = t_1 \ldots t_n$, we say that P *swap matches at location* i if there exists a swapped version P' of P that matches T starting at location i, i.e. $p'_j = t_{i+j-1}$ for $j = 1, \ldots, m$. It is not difficult to see that if P swap matches at location i there is a *unique* swap permutation for that location.

The *Swap Matching Problem* is the following:
INPUT: Pattern $P = p_1 \ldots p_m$ and text $T = t_1 \ldots t_n$ over alphabet Σ.
OUTPUT: All locations i where P swap matches T.

We note that the definition in [3] and the papers that followed is slightly different, allowing the swaps in the text rather than the pattern. However, it follows from Lemma 1 in [3] that both versions are of the same time complexity.

The *Approximate Swap Matching Problem* is the following:
INPUT: Pattern $P = p_1 \ldots p_m$ and text $T = t_1 \ldots t_n$ over alphabet Σ.
OUTPUT: For every location i where P swap matches T, write the number of swaps in the swapped version of P that matches the text substring of length m

starting at location i. If there is no swap matching of P at i, write $m + 1$ at location i.

Observation 1 *Assume there is a swap match at location i. Then the number of swaps is equal to half the number of mismatches at location i.*

3   Intuition and Key Idea

It would seem from Observation 1 that finding the number of swaps is of the same difficulty as finding the number of mismatches. However, this is not the case. Note that if it is known that there is a swap match at location i, this puts tremendous constraints on the mismatches. It means that if there is a mismatch between pattern location j and text location $i + j - 1$ we also know that either $t_{i+j-1} = p_{j-1}$ or $t_{i+j-1} = p_{j+1}$. There is no such constraint in a general mismatch situation!

Since we can "anticipate" for every pattern symbol what would be the mismatch, it gives us some flexibility to change the alphabet to reflect the expected mismatches. Thus we are able to reduce the alphabet to one with a small constant size. For such alphabets, the Fischer and Paterson algorithm [8] allows counting mismatches in time $O(n \log m)$.

In order to be able to anticipate the mismatching symbol, we need to isolate every pattern symbol from its right and left neighbors. This can be done by splitting a pattern P to three patterns, P_1, P_2 and P_3, where each subpattern counts mismatches only in the central element of each triple. P_1, P_2 and P_3 represent the three different offsets of triples in the pattern. For a schema of these three patterns, see Fig. 1.

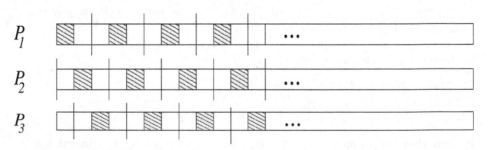

Fig. 1. The three patterns resulting from different triple offsets.

For each one of P_1, P_2 and P_3, the central symbol in every triple (the one shaded in Fig. 1) has the same value as the respective element of P. All other symbols are "don't care"s (ϕ). The sum of the mismatches of P_i in T, $i = 1, 2, 3$ is precisely the mismatches of P in T. Therefore, half of this sum is the desired number of swaps.

Throughout the remainder of this paper we will concentrate on counting the mismatches of P_2. The cases of P_1 and P_3 are similar. In the next section we will show a randomized algorithm that allows efficient counting of mismatches of P_2 by reducing the alphabet. Section 5 will show a deterministic alphabet reduction.

4   Randomized Alphabet Reduction

Let $h : \Sigma \to \{1, 2, ..., 4\}$ be chosen randomly. For string $S = s_1, ..., s_m$ define $h(S) = h(s_1), ..., h(s_m)$. Consider $h(P_2)$. Let (x, y, z) be a triple such that $x \neq y$ or $y \neq z$ (i.e. a swap *could* happen). Call such a triple a *potential swap triple*. We say that h *separates* the triple (x, y, z) if $h(x) \neq h(y)$ when $x \neq y$ and $h(y) \neq h(z)$ when $y \neq z$.

If h happens to separate every potential swap triple in the pattern, then the number of mismatches of P_2 in T equals the number of mismatches of $h(P_2)$ in $h(T)$. However, the alphabet of $h(P_2)$ and $h(T)$ is of size 4, hence the mismatches can be counted in time $O(n \log m)$.

We need to be quite lucky to achieve the situation where all potential swap triples get separated by h. However, we really do not need such a drastic event. Every potential swap triple that gets separated, counts all its mismatches. From now on it can be replaced by "don't care"s and never add mismatches. Conversely, every potential swap triple that does not get separated can be masked by "don't care"s and not contribute anything.

Our algorithm, then, is the following.

Algorithm

Let $P_t \leftarrow P_2$
Replace all non potential swap triples of P_t with "don't care"s
while not all triples have been masked do:
    choose a random $h : \Sigma \to \{1, 2, ..., 4\}$
    Let $P_q \leftarrow h(P_t)$
    Replace all non-separated triples of P_q with "don't care"s
    Count all mismatches of P_q in $h(T)$
    Replace all triples of P_t that were separated by h with "don't care"s

end Algorithm

Since counting all mismatches of P_q in $h(T)$ can be done in time $O(n \log m)$, it is sufficient to know the expected number of times we run through the while loop to calculate the expected running time of the algorithm.

Claim. The expected number of times the while loop executed in the above algorithm is $O(\log \sigma)$, where $\sigma = \min(|\Sigma|, m)$.

Proof: The probability that a given potential swap triple gets separated is

$$\left(\frac{3}{4}\right)^2 = \frac{9}{16} > \frac{1}{2}.$$

Therefore, the expectation is that at least half of the triples will be separated in the first execution of the while loop, with every subsequent execution of the while loop separating half of the remaining triples. Since there are no more than $\min(\frac{m}{3}, |\Sigma|^3)$ triples, then the expectation is that in $O(\log(\min(\frac{m}{3}, |\Sigma|^3))) = O(\log \sigma)$ executions of the while loop all triples will be separated.   □

Conclude: The expected running time of the algorithm is $O(n \log m \log \sigma)$.

5   Deterministic Alphabet Reduction

Recall that our task is really to separate all triples. There exists in the literature a powerful code that does this separation. Subsequently we show a simple code that solves our problem.

Definition: A $(\Sigma, 3)$-*universal set* is a set $S = \{\chi_1, \ldots, \chi_k\}$ of characteristic functions, $\chi_j : \Sigma \to \{0, 1\}$ such that for every $a, b, c \in \Sigma$, and for each of the eight possible combinations of $0 - 1$s, there exists χ_j such that $\chi_j(a), \chi_j(b), \chi_j(c)$ equals this combination.

We extend the definition of the functions χ_j to strings in the usual manner, i.e. for $S = s_1 \ldots s_n$, $\chi_j(S) = \chi_j(s_1)\chi_j(s_2) \ldots \chi_j(s_n)$.

Let $S = \{\chi_1, \ldots, \chi_k\}$ be a be a $(\Sigma, 3)$-universal set such that for every potential swap triple (a, b, c) there exist a j for which $\chi_j(a) = 0, \chi_j(b) = 1, \chi_j(c) = 0$. We run the following algorithm, which is very similar to the randomized algorithm in Section 4.

Algorithm

Let $P_t \leftarrow P_2$
Replace all non potential swap triples of P_t with "don't care"s
for $j = 1$ to k do:
    Let $P_q \leftarrow \chi_j(P_t)$
    Replace all non-separated triples of P_q with "don't care"s
    Count all mismatches of P_q in $\chi_j(T)$
    Replace all triples of P_t that were separated by χ_j with "don't care"s

end Algorithm

In [16] it was shown how to construct $(\Sigma, 3)$-universal set of cardinality $k = O(\log \sigma)$ yielding the following.

Corollary 1. *The deterministic algorithm's running time is* $O(n \log m \log \sigma)$.

The Naor and Naor construction of [16] is quite heavy. We conclude with an extremely simple coding of the alphabet that separates triples sufficiently well for our purposes.

First note the following.

Claim. It is sufficient to provide a set $S = \{\chi_1, \ldots, \chi_k\}$ of characteristic functions, $\chi_j : \Sigma \to \{0, 1\}$ such that for every potential swap triple (a, b, c) there either exists a χ_j such that $\chi_j(a) = x, \chi_j(b) = 1 - x$ and $\chi_j(c) = x$, where $x \in \{0, 1\}$, or there exist χ_{j_1}, χ_{j_2} such that $\chi_{j_1}(a) = x, \chi_{j_1}(b) = 1 - x$ and $\chi_{j_1}(c) = 1 - x$, and $\chi_{j_2}(a) = y, \chi_{j_2}(b) = y$ and $\chi_{j_2}(c) = 1 - y$, where $x, y \in \{0, 1\}$.

Call such a set a *swap separating set*.

Proof: Let $S = \{\chi_1, \ldots, \chi_k\}$ be a swap separating set. Every potential swap triple for which there exists a χ_j such that $\chi_j(a) = x, \chi_j(b) = 1 - x$ and $\chi_j(c) = x$, where $x \in \{0, 1\}$, will be separated by χ_j and masked with "don't care"s for all other characteristic functions. In other words, we initially decide, for each function, what are the triples it separates, and mark those triples. If several functions separate the same triple we will, of course, only use one of them.

For every other potential swap triple, there are χ_{j_1}, χ_{j_2} such that $\chi_{j_1}(a) = x, \chi_{j_1}(b) = 1 - x$ and $\chi_{j_1}(c) = 1 - x$, and $\chi_{j_2}(a) = y, \chi_{j_2}(b) = y$ and $\chi_{j_2}(c) = 1 - y$, where $x, y \in \{0, 1\}$. Every such triple will participate in the separation of χ_{j_1} and χ_{j_2}. For all other characteristic functions it will be masked with "don't care"s.

Note that if there is a match of such a triple with a text location without a swap, then neither χ_{j_1} nor χ_{j_2} will contribute a mismatch. However, if the triple's match requires a swap, then **exactly** one of χ_{j_1} or χ_{j_2} will contribute a mismatch. $\square$

Our remaining task is to provide a simple construction for swap separating set of size $O(\log \sigma)$.

Swap Separating Set Construction

Consider a $\sigma \times \log \sigma$ bit matrix B where the rows are a binary representation of the alphabet elements $(P \cap \Sigma)$. Take $\chi_j(a) = B[a, j]$. In words, the characteristic functions are the columns of B.

For every potential swap triple (a, b, c), if there is a column where the bits of a, b, c are $x, 1 - x, x$ then this column provides the function in which the triple participates. If no such column exists, then there clearly are two columns j_1, j_2 such that $B[a, j_1] \neq B[b, j_1]$ and $B[c, j_2] \neq B[b, j_2]$. It is clear that $B[c, j_1] = B[b, j_1]$ and $B[a, j_2] = B[b, j_2]$ (otherwise the first condition holds). The columns j_1 and j_2 provide the functions where triple (a, b, c) participates.

6 Conclusion and Open Problems

We have shown a faster algorithm for the approximate swap matching problem than that of the pattern matching with mismatches problem. This is quite a surprising result considering that it was thought that swap matching may be even harder than pattern matching with mismatches. However, this leads us to conjecture that the current upper bound on the mismatches problem ($O(n\sqrt{m}\log m)$) is not the final word.

The swap operation and the mismatch operation have proven to be relatively "easy" to solve. However, insertion and deletion are still not known to be solvable in time faster than the dynamic programming $O(nm)$ in the worst case. A lower bound or a better upper bound on the complexity of edit distance would be of great interest.

7 Acknowledgments

The first author was partially supported by NSF grant CCR-96-10170, BSF grant 96-00509, and a BIU internal research grant. The second author was partially supported by the Israel Ministry of Science Eshkol Fellowship 061-1-97, and was visiting the Courant Institute at NYU while working on some of the work in this paper.

References

[1] K. Abrahamson. Generalized string matching. *SIAM J. Comp.*, 16(6):1039–1051, 1987.

[2] A. Amir, Y. Aumann, G. Landau, M. Lewenstein, and N. Lewenstein. Pattern matching with swaps. Submitted for publication.

[3] A. Amir, Y. Aumann, G. Landau, M. Lewenstein, and N. Lewenstein. Pattern matching with swaps. *Proc. 38th IEEE FOCS*, 144–153, 1997.

[4] A. Amir and M. Farach. Efficient 2-dimensional approximate matching of half-rectangular figures. *Information and Computation*, 118(1):1–11, April 1995.

[5] A. Amir, G.M. Landau, M. Lewenstein, and N. Lewenstein. Efficient special cases of pattern matching with swaps. *Information Processing Letters*, 68(3):125–132, 1998.

[6] R. Cole and R. Harihan. Randomized swap matching in $o(m\log m\log|\sigma|)$ time. Technical Report TR1999-789, New York University, Courant Institute, September 1999.

[7] R. Cole and R. Hariharan. Approximate string matching: A faster simpler algorithm. In *Proc. 9th ACM-SIAM Symposium on Discrete Algorithms (SODA)*, 463–472, 1998.

[8] M.J. Fischer and M.S. Paterson. String matching and other products. *Complexity of Computation, R.M. Karp (editor), SIAM-AMS Proceedings*, 7:113–125, 1974.

[9] Z. Galil and K. Park. An improved algorithm for approximate string matching. *SIAM J. Comp.*, 19(6):989–999, 1990.

[10] D.E. Knuth, J.H. Morris, and V.R. Pratt. Fast pattern matching in strings. *SIAM J. Comp.*, 6:323–350, 1977.

[11] G. M. Landau and U. Vishkin. Fast parallel and serial approximate string matching. *Journal of Algorithms*, 10(2):157–169, 1989.

[12] G.M. Landau, E. W. Myers, and J. P. Schmidt. Incremental string comparison. *SIAM J. Comp.*, 27(2):557–582, 1998.

[13] V. I. Levenshtein. Binary codes capable of correcting, deletions, insertions and reversals. *Soviet Phys. Dokl.*, 10:707–710, 1966.

[14] R. Lowrance and R. A. Wagner. An extension of the string-to-string correction problem. *J. of the ACM*, 177–183, 1975.

[15] S. Muthukrishnan and H. Ramesh. String matching under a general matching relation. *Information and Computation*, 122(1):140–148, 1995.

[16] J. Naor and M. Naor. Small-bias probability spaces: Efficient constructions and applications. *SIAM J. Comp.*, 838–856, 1993.

[17] M. V. Olson. A time to sequence. *Science*, 270:394–396, 1995.

[18] A. Pentland. Invited talk. NSF Institutional Infrastructure Workshop, 1992.

[19] R. A. Wagner. On the complexity of the extended string-to-string correction problem. In *Proc. 7th ACM STOC*, 218–223, 1975.

A Semantic Theory for Heterogeneous System Design⋆

Rance Cleaveland[1] and Gerald Lüttgen[2]

[1] Department of Computer Science, State University of New York at Stony Brook,
Stony Brook, New York 11794–4400, USA,
rance@cs.sunysb.edu
[2] Department of Computer Science, Sheffield University, 211 Portobello Street,
Sheffield S1 4DP, England,
g.luettgen@dcs.shef.ac.uk

Abstract. This paper extends DeNicola and Hennessy's testing theory
from labeled transition system to Büchi processes and establishes a tight
connection between the resulting Büchi must–preorder and satisfaction of
linear–time temporal logic (LTL) formulas. An example dealing with the
design of a communications protocol testifies to the utility of the theory
for heterogeneous system design, in which some components are specified
as labeled transition systems and others are given as LTL formulas.

1 Introduction

Approaches to formally verifying reactive systems typically follow one of two
paradigms. The first paradigm is founded on notions of *refinement* and is employed in process algebra [2]. In such approaches one formulates specifications
and implementations in the same notation and then proves that the latter refine
the former. The underling semantics is usually given operationally, and refinement relations are formalized as preorders. *Testing/failure preorders* [4, 8] have
attracted particular attention because of their intuitive formulations in terms of
responses a system exhibits to tests. Their strength is their support for *compositional reasoning*, i.e., one may refine part of a system design independently of
others, and their *full abstractness* with respect to trace inclusion [18].

The other paradigm relies on the use of *temporal logics* [22] to formulate
specifications, with implementations being given in an operational notation. One
then verifies a system by establishing that it is a model of its specification; *model
checkers* [5] automate this task for finite–state systems. Temporal logics support
the definition of properties that constrain single aspects of expected system
behavior and, thus, allow a "property–at–a–time" approach. Such logics also
have connections with automata over infinite words. For example, *linear–time*

⋆ Research support was provided under NASA Contract No. NAS1–97046 and by
NSF grant CCR–9988489. The first author was also supported by AFOSR Grant
F49620–95–1–0508, ARO Grant P–38682–MA, and NSF Grants CCR–9505562,
CCR–9996086, and INT–9996095.

S. Kapoor and S. Prasad (Eds.): FST TCS 2000, LNCS 1974, pp. 312–324, 2000.

temporal logic (LTL) specifications may be translated into *Büchi automata* [27] which allow semantic constraints on infinite behavior to be expressed.

The objective of this paper is to develop a semantic framework that seamlessly unifies testing–based refinement and LTL, thereby enabling the development of design formalisms that provide support for both styles of verification. Using Büchi automata and the testing framework of DeNicola and Hennessy [8] as starting points, we approach this task by developing *Büchi may–* and *must–preorders* that relate *Büchi processes* on the basis of their responses to *Büchi tests*. Alternative characterizations are provided and employed for proving conservative–extension results regarding DeNicola and Hennessy's testing theory. We then apply this framework to defining a semantics for heterogeneous design notations, where systems are specified using a mixture of labeled transition systems and LTL formulas. This is done in two steps: first, we show that our Büchi must–preorder is compositional for *parallel composition* and *scoping* operators that are inspired by CCS [19]. Second, we establish that the Büchi must–preorder reduces to a variant of reverse trace inclusion when its first argument is purely nondeterministic. Consequently, the Büchi must–preorder permits a uniform treatment of traditional notions of process refinement and LTL satisfaction. The utility of our new theory is illustrated by means of a small example featuring the heterogeneous design of a generic communications protocol.

2   Büchi Testing

We extend the *testing theory* of DeNicola and Hennessy [8], which was developed for labeled transition systems in a process–algebraic setting, to *Büchi automata*. Traditional testing relates labeled transition systems via two preorders, the *may–* and *must–preorders*, which distinguish systems on the basis of the tests they might be able to, or are necessarily able to, pass. Büchi automata generalize labeled transition systems by means of an acceptance condition for infinite traces. However, the classical Büchi semantics, which identifies automata having the same infinite languages, is in general not compositional with respect to parallel composition operators, since it is insensitive to the potential for deadlock. Our testing semantics is intended to overcome this problem. In the sequel, we refer to Büchi automata as *Büchi processes* to emphasize that we are equipping Büchi automata with a different semantics than the traditional one.

Basic Definitions. Our semantic framework is defined relative to some *alphabet* $\mathcal{A}$, i.e., a countable set of *actions* which does not include the distinguished *unobservable, internal action* τ. In the remainder, we let $a, b, \ldots$ range over $\mathcal{A}$ and $\alpha, \beta, \ldots$ over $\mathcal{A} \cup \{\tau\}$. Büchi processes are distinguished from labeled transition systems in their treatment of infinite traces. Whereas in labeled transition systems all infinite traces are typically deemed possible, in Büchi processes only those infinite traces that go through designated *Büchi states* infinitely often are considered actual executions.

Definition 1 (Büchi process). *A Büchi process is a tuple $\langle P, \longrightarrow, \sqrt{}, p \rangle$, where P is a countable set of states, $\longrightarrow \subseteq P \times (\mathcal{A} \cup \{\tau\}) \times P$ is the transition relation,*

$\sqrt{} \subseteq S$ *is the* Büchi *set, and* $p \in P$ *is the* start state. *If* $\sqrt{} = P$ *we refer to the* Büchi process *as a* labeled transition system.

For convenience, we often write (i) $p' \xrightarrow{\alpha} p''$ instead of $\langle p', \alpha, p'' \rangle \in \longrightarrow$, (ii) $p' \xrightarrow{\alpha}$ for $\exists p'' \in P.\, p' \xrightarrow{\alpha} p''$, (iii) $p' \longrightarrow$ for $\exists \alpha \in \mathcal{A} \cup \{\tau\}$, $p'' \in P$. $p' \xrightarrow{\alpha} p''$, and (iv) $p' \sqrt{}$ for $p' \in \sqrt{}$. If no confusion arises, we abbreviate the Büchi process $\langle P, \longrightarrow, \sqrt{}, p \rangle$ by its start state p and refer to its transition relation and Büchi set as $\longrightarrow_p$ and $\sqrt{}_p$, respectively. Moreover, we denote the set of all Büchi processes by $\mathcal{P}$. Note that we do not require Büchi processes to be finite–state.

Definition 2 (Path & trace). *Let* $\langle P, \longrightarrow, \sqrt{}, p \rangle$ *be a* Büchi process. *A* path π *starting from state* $p' \in P$ *is a potentially infinite sequence* $(\langle p_{i-1}, \alpha_i, p_i \rangle)_{0 < i \leq k}$, *where* $k \in \mathbb{N} \cup \{\infty\}$, *such that* $k = 0$, *or* $p_0 = p'$ *and* $p_{i-1} \xrightarrow{\alpha_i} p_i$, *for all* $0 < i \leq k$. *We use* $|\pi|$ *to refer to* k, *the* length *of* π. *If* $|\pi| = \infty$, *we say that* π *is* infinite; *otherwise,* π *is* finite. *If* $|\pi| \in \mathbb{N}$ *and* $p_{|\pi|} \not\longrightarrow$, *i.e.,* $p_{|\pi|}$ *is a deadlock state, path* π *is called* maximal. *Path* π *is referred to as a* Büchi path *if* $|\pi| = \infty$ *and* $|\{i \in \mathbb{N} \,|\, p_i \sqrt{}\}| = \infty$. *The (visible)* trace $\mathit{trace}(\pi)$ *of* π *is defined as the sequence* $(\alpha_i)_{i \in I_\pi} \in \mathcal{A}^* \cup \mathcal{A}^\infty$, *where* $I_\pi =_{df} \{0 < i \leq |\pi| \,|\, \alpha_i \neq \tau\}$.

We denote the sets of all finite paths, all maximal paths, and all Büchi paths starting from state $p' \in P$ by $\Pi_{\mathsf{fin}}(p')$, $\Pi_{\mathsf{max}}(p')$, and $\Pi_{\mathsf{B}}(p')$, respectively. The empty path π with $|\pi| = 0$ is symbolized by $()$ and its trace by ϵ. We sometimes write α for the empty or single-element sequence trace (α) and use the notation $p' \xRightarrow{w}_p p''$ to indicate that state p' of Büchi process p may evolve to state p'' when observing trace w for some path $\pi \in \Pi_{\mathsf{fin}}(p')$. Formally, $p' \xRightarrow{w}_p p''$ if $\exists \pi = (\langle p_{i-1}, \alpha_i, p_i \rangle)_{0 < i \leq k} \in \Pi_{\mathsf{fin}}(p)$. $p_0 = p'$, $p_k = p''$, and $\mathit{trace}(\pi) = w$. Moreover, $\mathcal{I}_p(p') =_{df} \{a \in \mathcal{A} \,|\, \exists p''.\, p' \xRightarrow{a}_p p''\}$ is the set of initial actions of p in state $p' \in P$. We may also introduce different languages for Büchi process p.

$$\begin{aligned} \mathcal{L}_{\mathsf{fin}}(p) &=_{df} \{\mathit{trace}(\pi) \,|\, \pi \in \Pi_{\mathsf{fin}}(p)\} \subseteq \mathcal{A}^* && \textit{finite–trace language of } p \\ \mathcal{L}_{\mathsf{max}}(p) &=_{df} \{\mathit{trace}(\pi) \,|\, \pi \in \Pi_{\mathsf{max}}(p)\} \subseteq \mathcal{A}^* && \textit{maximal–trace language of } p \\ \mathcal{L}_{\mathsf{B}}(p) &=_{df} \{\mathit{trace}(\pi) \,|\, \pi \in \Pi_{\mathsf{B}}(p)\} \subseteq \mathcal{A}^* \cup \mathcal{A}^\infty && \textit{Büchi–trace language of } p \end{aligned}$$

A key notion in testing–based semantics is *divergence*, i.e., a system's ability to engage in an infinite internal computation. In this paper, we use adaptations of the traditional notions of DeNicola and Hennessy [8]; more sophisticated definitions may be found elsewhere in the literature [3, 21, 23] but are not considered here. We say that state p' of Büchi process p is *(Büchi) divergent*, in symbols $p' \Uparrow_p$, if $\exists \pi \in \Pi_{\mathsf{B}}(p')$. $\mathit{trace}(\pi) = \epsilon$. State p' is called w–*divergent* for some $w = (a_i)_{0 < i \leq k} \in \mathcal{A}^* \cup \mathcal{A}^\infty$, in symbols $p' \Uparrow_p w$, if one can reach a divergent state starting from p' when executing a finite prefix of w, i.e., if $\exists l \leq k$, $p'' \in P$, $w' \in \mathcal{A}^*$. $w' = (a_i)_{0 < i \leq l}$, $p' \xRightarrow{w'} p''$, and $p'' \Uparrow_p$. For convenience, we write $\mathcal{L}_{\mathsf{div}}(p')$ for the *divergent–trace language* of p', i.e., $\mathcal{L}_{\mathsf{div}}(p') =_{df} \{w \in \mathcal{A}^* \cup \mathcal{A}^\infty \,|\, p' \Uparrow_p w\}$. State p' is convergent or w–convergent, in symbols $p' \Downarrow_p$ and $p' \Downarrow_p w$, if not $p' \Uparrow_p$ and not $p' \Uparrow_p w$, respectively. Note that a finite trace $w \in \mathcal{L}_{\mathsf{B}}(p)$ indicates that p is divergent exactly after executing w. In the following, we often omit the indices of the divergence and convergence predicates, as well as of the transition relations, whenever these are obvious from the context. Finally, we write $w \cdot w'$ for the *concatenation* of finite trace $w \in \mathcal{A}^*$ with the finite or infinite trace $w' \in \mathcal{A}^* \cup \mathcal{A}^\infty$.

Testing Theory. The testing framework of DeNicola and Hennessy defines behavioral preorders that relate labeled transition systems with respect to their responses to *tests* [8]. Tests are employed to witness the interactions a system may have with its environment. In our setting, a test is a Büchi process in which certain states are designated as *successful*. In order to determine whether a system passes a test, one has to examine the finite and infinite *computations* that result when the test runs in lock–step with the system under consideration.

Definition 3 (Test, computation, success). *A Büchi test* $\langle T, \longrightarrow, \sqrt{}, t, Suc \rangle$ *is a Büchi process* $\langle T, \longrightarrow, \sqrt{}, t \rangle$ *together with a set* $Suc \subseteq T$ *of success states. If* $\sqrt{} = \emptyset$, *we call the test* classical. *The set of all Büchi tests is denoted by* $\mathcal{T}$.

A potential computation *c with respect to a Büchi process p and a Büchi test t is a potentially infinite sequence* $(\langle p_{i-1}, t_{i-1} \rangle \xmapsto{\alpha_i}_{r_i} \langle p_i, t_i \rangle)_{0 < i \leq k}$, *where* $k \in \mathbb{N} \cup \{\infty\}$, *such that (1) $p_i \in P$ and $t_i \in T$, for all $0 \leq i \leq k$, and (2) $\alpha_i \in \mathcal{A} \cup \{\tau\}$ and $r_i \in \{\blacktriangleleft, \blacktriangleright, \blacklozenge\}$, for all $0 < i \leq k$. The relation $\longmapsto$ is defined by:*

- $\langle p_{i-1}, t_{i-1} \rangle \xmapsto{\alpha_i}_{\blacktriangleleft} \langle p_i, t_i \rangle$ *if* $\alpha_i = \tau$, $t_{i-1} = t_i$, $p_{i-1} \xrightarrow{\tau}_p p_i$, *& $t_{i-1} \notin Suc$.*
- $\langle p_{i-1}, t_{i-1} \rangle \xmapsto{\alpha_i}_{\blacktriangleright} \langle p_i, t_i \rangle$ *if* $\alpha_i = \tau$, $p_{i-1} = p_i$, $t_{i-1} \xrightarrow{\tau}_t t_i$, *& $t_{i-1} \notin Suc$.*
- $\langle p_{i-1}, t_{i-1} \rangle \xmapsto{\alpha_i}_{\blacklozenge} \langle p_i, t_i \rangle$ *if* $\alpha_i \in \mathcal{A}$, $p_{i-1} \xrightarrow{\alpha_i}_p p_i$, $t_{i-1} \xrightarrow{\alpha_i}_t t_i$, *& $t_{i-1} \notin Suc$.*

The potential computation c is finite *if $|c| < \infty$ and* infinite *if $|c| = \infty$. The projection* $\mathsf{proj}_p(c)$ *of c on p is defined as* $(\langle p_{i-1}, \alpha_i, p_i \rangle)_{i \in I_p^c} \in \Pi(p)$, *where* $I_p^c =_{df} \{0 < i \leq k \mid r_i \in \{\blacktriangleleft, \blacklozenge\}\}$, *and the projection* $\mathsf{proj}_t(c)$ *of c on t as* $(\langle t_{i-1}, \alpha_i, t_i \rangle)_{i \in I_t^c} \in \Pi(p)$, *where* $I_t^c =_{df} \{0 < i \leq k \mid r_i \in \{\blacktriangleright, \blacklozenge\}\}$. *A potential computation c is called a* computation *if it satisfies the following properties: (1) c is maximal, i.e. if $|c| < \infty$ then $\langle p_{|c|}, t_{|c|} \rangle \not\xmapsto{\alpha}_r$ for any α and r; and (2) if $|c| = \infty$ then $\mathsf{proj}_p(c) \in \Pi_B(p)$. The set of all computations of p and t is denoted by $\mathcal{C}(p, t)$.*

Computation c is called successful *if $t_{|c|} \in Suc$, in case $|c| < \infty$, or if $\mathsf{proj}_t(c) \in \Pi_B(t)$, in case $|c| = \infty$. We say that p may pass t, in symbols $p\ \mathsf{may}_{\mathrm{CL}}\ t$, if there exists a successful computation $c \in \mathcal{C}(p, t)$. Analogously, p must pass t, in symbols $p\ \mathsf{must}_{\mathrm{CL}}\ t$, if every computation $c \in \mathcal{C}(p, t)$ is successful.*

Intuitively, an infinite computation of process p and test t differs from an infinite potential computation in that in the former the process is required to enter a Büchi state infinitely often. An infinite computation is then successful if the test also passes through a Büchi state infinitely often. Hence, in contrast with the original theory of DeNicola and Hennessy, some infinite computations can be successful in our setting. Also, since Büchi processes and Büchi tests potentially exhibit nondeterministic behavior, one may distinguish between the *possibility* and *inevitability* of success. This is captured in the following definitions of the Büchi *may–* and *must–*preorders.

Definition 4 (Büchi Preorders). *For Büchi processes p and q we define:*

- $p \sqsubseteq_{CL}^{may} q$ *if* $\forall t \in \mathcal{T}$. p $may_{CL} t$ *implies* q $may_{CL} t$.
- $p \sqsubseteq_{CL}^{must} q$ *if* $\forall t \in \mathcal{T}$. p $must_{CL} t$ *implies* q $must_{CL} t$.

It is easy to check that $\sqsubseteq_{CL}^{may}$ and $\sqsubseteq_{CL}^{must}$ are preorders. The classical may– and must–preorders of DeNicola and Hennessy are defined analogously, but with respect to transition systems and classical tests [8]. Note that in this paper we consider the Büchi may–preorder only for the sake of completing the Büchi testing theory; it is not used in our semantic framework for heterogeneous system specification.

3   Alternative Characterizations and Conservativity

We now present characterizations of our Büchi preorders and use these characterizations as a basis for comparing DeNicola-Hennessy testing theory [8] ours.

Theorem 1. *Let p and q be Büchi processes. Then*

1. $p \sqsubseteq_{CL}^{may} q$ *if and only if* $\mathcal{L}_{fin}(p) \subseteq \mathcal{L}_{fin}(q)$ *and* $\mathcal{L}_B(p) \subseteq \mathcal{L}_B(q)$.
2. $p \sqsubseteq_{CL}^{must} q$ *if and only if for all* $w \in \mathcal{A}^* \cup \mathcal{A}^\infty$ *such that* $p \Downarrow w$:
   - (a) $q \Downarrow w$
   - (b) $|w| < \infty$: $\forall q'. q \stackrel{w}{\Longrightarrow} q'$ *implies* $\exists p'. p \stackrel{w}{\Longrightarrow} p'$ *and* $\mathcal{I}_p(p') \subseteq \mathcal{I}_q(q')$.

     $\boxed{|w| = \infty}$: $w \in \mathcal{L}_B(q)$ *implies* $w \in \mathcal{L}_B(p)$.

With respect to finite traces, the characterizations are virtually the same as the ones of DeNicola and Hennessy's preorders [8]. However, we need to refine the classical characterizations in order to capture the sensitivity of Büchi may– and must–testing to infinite behavior. The proof of this theorem relies on the properties of several specific Büchi tests. Some of them are standard [8]; the other ones are depicted in Fig. 1, where (i) $w = (a_i)_{0 < i \leq k} \in \mathcal{A}^*$ for tests $t_w^{may,div}$ and $t_w^{must,max}$ and (ii) $w = (a_i)_{i \in \mathbb{N}} \in \mathcal{A}^\infty$ for tests $t_w^{may,\infty}$, $t_w^{\Downarrow}$, and $t_w^{must,\infty}$. In Fig. 1, Büchi states are marked by the symbol $\sqrt{}$, and success states are distinguished by thick borders.

Intuitively, while Büchi test $t_w^{may,\infty}$ tests for the presence of Büchi trace w, Büchi tests $t_w^{may,div}$ and $t_w^{\Downarrow}$ are capable of detecting divergent behavior when executing trace w. Moreover, Büchi tests $t_w^{must,max}$ and $t_w^{must,\infty}$ are concerned with the absence of maximal trace and Büchi trace w, respectively. These intuitions are made precise by the following properties, which hold for any Büchi process p.

1. Let $w \in \mathcal{A}^\infty$. Then $w \in \mathcal{L}_B(p)$ if and only if p $may_{CL} t_w^{may,\infty}$.
2. Let $w \in \mathcal{A}^*$. Then $w \in \mathcal{L}_B(p)$ if and only if p $may_{CL} t_w^{may,div}$.
3. Let $w \in \mathcal{A}^\infty$. Then $p \Downarrow w$ if and only if p $must_{CL} t_w^{\Downarrow}$.
4. Let $w \in \mathcal{A}^*$ s.t. $p \Downarrow w$. Then $w \notin \mathcal{L}_{max}(p)$ if and only if p $must_{CL} t_w^{must,max}$.
5. Let $w \in \mathcal{A}^\infty$ s.t. $p \Downarrow w$. Then $w \notin \mathcal{L}_B(p)$ if and only if p $must_{CL} t_w^{must,\infty}$.

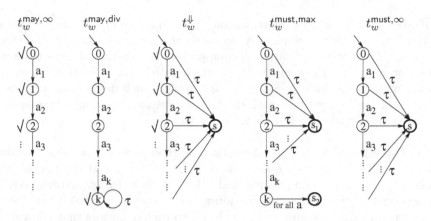

Fig. 1. Büchi tests used for characterizing the Büchi may– and must–preorders.

The proof of Thm. 1, which can be found in [7], relies on these properties of Büchi tests. Specifically, it uses the *infinite–state* tests $t_w^{\mathsf{may},\infty}$, $t_w^{\Downarrow}$, and $t_w^{\mathsf{must},\infty}$. The employment of infinite–state tests — even when relating finite–state Büchi processes — is justified by our view that Büchi tests represent the arbitrary, potentially irregular behavior of the unknown system environment.

Using the above characterizations, we investigate the relation of our Büchi preorders to the corresponding classical preorders, $\sqsubseteq_{\mathsf{DH}}^{\mathsf{may}}$ and $\sqsubseteq_{\mathsf{DH}}^{\mathsf{must}}$, respectively, as defined by DeNicola and Hennessy [8]. It should be noted that their framework is restricted to *image–finite* labeled transition systems and classical, image–finite tests; a labeled transition system or Büchi process is called image–finite if every state has only a finite number of outgoing transitions for any action.

Theorem 2. *Let p and q be image–finite labeled transition systems.*

1. *If p and q are convergent, then $p \sqsubseteq_{\mathsf{CL}}^{\mathsf{may}} q$ if and only if $p \sqsubseteq_{\mathsf{DH}}^{\mathsf{may}} q$.*
2. *$p \sqsubseteq_{\mathsf{CL}}^{\mathsf{must}} q$ if and only if $p \sqsubseteq_{\mathsf{DH}}^{\mathsf{must}} q$.*

We refer the reader to [7] for the proof of this theorem. In a nutshell, the second part follows by inspection of the alternative characterizations of $\sqsubseteq_{\mathsf{CL}}^{\mathsf{must}}$ and $\sqsubseteq_{\mathsf{DH}}^{\mathsf{must}}$. Thm. 2(1) is invalid if one allows *divergent* labeled transition systems. As a counterexample consider the transition systems $\langle\{p\}, \{\langle p, \tau, p\rangle\}, \{p\}, p\rangle$ and $\langle\{q\}, \emptyset, \{q\}, q\rangle$, as well as the Büchi test $\langle\{t\}, \{\langle t, \tau, t\rangle\}, \{t\}, t, \emptyset\rangle$. Then, $p \sqsubseteq_{\mathsf{DH}}^{\mathsf{may}} q$ since $\mathcal{L}_{\mathsf{fin}}(p) = \mathcal{L}_{\mathsf{fin}}(q) = \{\epsilon\}$, but $p \not\sqsubseteq_{\mathsf{CL}}^{\mathsf{may}} q$ since $p\,\mathsf{may}_{\mathsf{CL}}\,t$ and $q\,\mathsf{m\acute{a}y}_{\mathsf{CL}}\,t$.

4   Büchi Testing and Heterogeneous System Design

In this section we investigate the utility of our theory as a semantic framework for heterogeneous design notations that mix labeled transition systems and formulas in LTL. The design methodology which we wish to support is *component–based*, where a system designer starts off with a system architecture, with components given either as automata or, more abstractly, as LTL formulas. Then the system is refined by successively implementing each component as a labeled transition system satisfying its specification. To support such a methodology mathematically, one needs a *refinement preorder* which satisfies at least two properties. First, it must be compositional for key operators of such design languages. Second, it must be "compatible" with the LTL satisfaction relation. We show that our Büchi must–preorder obeys both properties.

Büchi Testing and Compositionality. In the component–based design framework we wish to study, two operators are central: (i) *parallel composition*, for connecting concurrent components and allowing them to interact via system channels, and (ii) *restriction*, for restricting access to channels to certain system components. In the following, we introduce two such operators that allow us to give the reader hints about the application of the semantic theory developed so far. While other operators are of course possible, the ones considered here suffice for the purposes of the example in the next section.

Our parallel composition operator "$|$" and the restriction operator $\backslash A$, where $A \subseteq \mathcal{A}$, are inspired by the ones in the process algebra CCS [19]. We assume that alphabet $\mathcal{A}$ is composed of two sets $\mathcal{A}!$ and $\mathcal{A}?$, representing *sending* and *receiving actions*, such that for every $a! \in \mathcal{A}!$ there exists a corresponding $a? \in \mathcal{A}?$, and vice versa. Here, a should be interpreted as a *channel name*. The intuition for parallel composition in CCS is that a process willing to send a message on channel a and another one able to receive a message on a can do so by performing the actions $a!$ and $a?$ in synchrony with each other. This *handshake* is invisible to an external observer, i.e., it results in the distinguished, unobservable action τ. When adapting the CCS parallel operator to our framework of Büchi processes, the question that naturally arises concerns the interpretation of Büchi traces. We adopt the following point of view: intuitively, "fair merges" of Büchi traces of p and q should also be Büchi traces of $p|q$. Moreover, a Büchi trace of one process, when merged with a finite trace of the other process, should also result in a Büchi trace of $p|q$.

Formally, our parallel composition of Büchi processes $\langle P, \longrightarrow_p, \sqrt{}_p, p \rangle$ and $\langle Q, \longrightarrow_q, \sqrt{}_q, q \rangle$ is defined as the Büchi process $\langle P|Q, \longrightarrow_{p|q}, \sqrt{}_{p|q}, p|q \rangle$, where $P|Q =_{df} \{p'|q' \mid p' \in P, q' \in Q\} \cup \{q'|p' \mid p' \in P, q' \in Q\}$ and where $\longrightarrow_{p|q}$ is the least relation such that:

(1) $p' \xrightarrow{\alpha}_p p''$ implies $p'|q' \xrightarrow{\alpha}_{p|q} q'|p''$ if $p'\sqrt{}_p$

(2) $p' \xrightarrow{\alpha}_p p''$ implies $p'|q' \xrightarrow{\alpha}_{p|q} p''|q'$ if not $p'\sqrt{}_p$

(3) $q' \xrightarrow{\alpha}_q q''$ implies $p'|q' \xrightarrow{\alpha}_{p|q} q''|p'$

(4) $p' \xrightarrow{a!}_p p''$ and $q' \xrightarrow{a?}_q q''$ implies $p'|q' \xrightarrow{\tau}_{p|q} q''|p''$ if $p'\sqrt{}_p$

(5) $p' \xrightarrow{a!}_p p''$ and $q' \xrightarrow{a?}_q q''$ implies $p'|q' \xrightarrow{\tau}_{p|q} p''|q''$ if not $p'\sqrt{}_p$

(6) $p' \xrightarrow{a?}_p p''$ and $q' \xrightarrow{a!}_q q''$ implies $p'|q' \xrightarrow{\tau}_{p|q} q''|p''$ if $p'\sqrt{}_p$

(7) $p' \xrightarrow{a?}_p p''$ and $q' \xrightarrow{a!}_q q''$ implies $p'|q' \xrightarrow{\tau}_{p|q} p''|q''$ if not $p'\sqrt{}_p$

These rules are in accordance with our above–mentioned intuition of system behavior. The "switching" of the states of p and q in Rules (1), (3), (4), and (6) allows us to fairly merge "Büchi traces with Büchi traces" and "Büchi traces with finite traces" of the argument Büchi processes. Finally, the Büchi predicate $\sqrt{}_{p|q}$ is defined by $p'|q'\sqrt{}_{p|q}$ if $p'\sqrt{}_p$, for any $p' \in P$ and $q' \in Q$. The unary *restriction* operator $\backslash A$, for $A \subseteq \mathcal{A}$, essentially is a *scoping mechanism* on channel names. Intuitively, $p \setminus A$ is defined as the Büchi process p, except that all transitions labeled by actions $a!$ and $a?$, where $a \in A$, are eliminated.

By referring to the characterizations of the Büchi may– and must–preorders one can establish the desired compositionality results: the Büchi may– and must–preorders are substitutive under parallel composition and restriction.

Büchi Must–testing and LTL Satisfaction. We now show that the Büchi must–preorder is compatible with the LTL satisfaction relation $\models$, which relates labeled transition systems and LTL formulas [22]. By "compatible" we mean that, for every LTL formula ϕ, there exists a Büchi process B_ϕ such that the following holds for any labeled transition system p: $p \models \phi$ if and only if $B_\phi \sqsubseteq_{\mathrm{CL}}^{must} p$, i.e., the 'implementation' p *refines* the 'specification' ϕ.

To achieve this goal, we characterize the Büchi must–preorder for a certain class of Büchi processes by means of trace inclusion. We call a Büchi process p *purely nondeterministic*, if for all $p' \in P$: (i) $p' \xrightarrow{\tau}_p$ implies $p' \xrightarrow{a}\!\!\!\!/\,_p$, for all $a \in \mathcal{A}$, and (ii) $|\{\langle a, p''\rangle \in \mathcal{A} \times P \,|\, p' \xrightarrow{a}_p p''\}| = 1$. Note that a Büchi process p can be transformed to a purely nondeterministic Büchi process p', such that $\mathcal{L}_{\mathsf{div}}(p) = \mathcal{L}_{\mathsf{div}}(p')$, $\mathcal{L}_{\mathsf{fin}}(p) = \mathcal{L}_{\mathsf{fin}}(p')$, $\mathcal{L}_{\mathsf{max}}(p) = \mathcal{L}_{\mathsf{max}}(p')$, and $\mathcal{L}_{\mathsf{B}}(p) = \mathcal{L}_{\mathsf{B}}(p')$, by splitting every transition $p' \xrightarrow{a}_p p''$ into two transitions $p' \xrightarrow{\tau}_p p_{\langle p',a,p''\rangle} \xrightarrow{a}_p p''$, where $p_{\langle p',a,p''\rangle} \notin P$ is a new, distinguished state.

Theorem 3. *Let p and q be Büchi processes and p be purely nondeterministic. Then, $p \sqsubseteq_{\mathrm{CL}}^{must} q$ if and only if (i) $\mathcal{L}_{\mathsf{div}}(q) \subseteq \mathcal{L}_{\mathsf{div}}(p)$, (ii) $\mathcal{L}_{\mathsf{fin}}(q) \setminus \mathcal{L}_{\mathsf{div}}(p) \subseteq \mathcal{L}_{\mathsf{fin}}(p)$, (iii) $\mathcal{L}_{\mathsf{max}}(q) \setminus \mathcal{L}_{\mathsf{div}}(p) \subseteq \mathcal{L}_{\mathsf{max}}(p)$, and (iv) $\mathcal{L}_{\mathsf{B}}(q) \setminus \mathcal{L}_{\mathsf{div}}(p) \subseteq \mathcal{L}_{\mathsf{B}}(p)$.*

The necessity of the premise of this theorem, whose proof is in [7], may be demonstrated by Büchi processes $p =_{\mathrm{df}} \langle \{p_1, p_2\}, \{\langle p_1, a, p_1\rangle, \langle p_1, b, p_2\rangle\}, \emptyset, p_1\rangle$ and $q =_{\mathrm{df}} \langle \{q_1, q_2\}, \{\langle q_1, b, q_2\rangle\}, \emptyset, q_1\rangle$. Then p is *not* purely nondeterministic and Inclusions (i)–(iv) obviously hold, but $p \not\sqsubseteq_{\mathrm{CL}}^{must} q$ since $p \operatorname{must_{CL}} t$ and $q \operatorname{m\!/\!ust_{CL}} t$, for the Büchi test $t =_{\mathrm{df}} \langle \{t_1, t_2\}, \{\langle t_1, a, t_2\rangle\}, \emptyset, t_1, \{t_2\}\rangle$.

The above theorem is the key for establishing the desired connection between the Büchi must–preorder $\sqsubseteq_{\mathrm{CL}}^{must}$ and the satisfaction relation $\models$ for LTL. In particular, well–known constructions — starting with the seminal work of Vardi and Wolper [27] — exist for converting LTL formulas into Büchi automata whose languages consist precisely of the models of the corresponding formulas. These constructions may be adapted to yield purely nondeterministic Büchi processes. However, there are a few subtleties of our setting compared to the traditional one on which we need to comment. First of all, our framework is concerned with labeled transition systems, so we must be able to interpret LTL formulas with respect to sequences of actions rather than states. Also, our framework is not only concerned with Büchi traces but also with finite traces (i.e., deadlocks) and divergent traces. The syntax and semantics of LTL may be modified to cope with these new phenomena; the details are not difficult and are omitted. The classical constructions of Büchi automata from LTL formulas may then be adapted to cope with the modifications to the logic. Whereas the adaptation for deadlock is well–known [17], the handling of divergence requires some attention. Intuitively, in a Büchi process a divergent state may engage in arbitrary behavior; this is reflected in its divergence language, which is $\mathcal{A}^* \cup \mathcal{A}^\infty$ (cf. Sec. 2). The only LTL formulas satisfied by arbitrary behavior are tautologies. Hence, in the Büchi process construction for LTL formulas, every state which corresponds to a tautology needs to be made divergent. Having these twists in mind, one may obtain the following variant of the key theorem for automata–based LTL model checking (cf. [27]), where $\hat{B}_\phi$ denotes the Büchi process constructed for LTL formula ϕ.

Theorem 4. *Let p be a labeled transition system and ϕ be an* **LTL** *formula. Then, $p \models \phi$ if and only if Inclusions (i)–(iv) in Thm. 3 hold for $\hat{B}_\phi$ and p (i.e., replace p in Inclusions (i)–(iv) by $\hat{B}_\phi$ and q by p).*

Note that the "$\Longrightarrow$"–direction of Thm. 4 is invalid if p is allowed to be an arbitrary Büchi process rather than a labeled transition system. As a counterexample consider $p =_{df} \langle \{p_1, p_2, p_3\}, \{\langle p_1, a, p_2 \rangle, \langle p_1, b, p_3 \rangle, \langle p_3, b, p_3 \rangle\}, \emptyset, p_1 \rangle$ and $\phi =_{df} a$. Then $p \models a$ as $b^\infty \notin \mathcal{L}_\mathsf{B}(p)$ and $b \in \mathcal{L}_{\mathsf{fin}}(p) \setminus \mathcal{L}_{\mathsf{div}}(\hat{B}_\phi)$. But obviously $b \notin \mathcal{L}_{\mathsf{fin}}(\hat{B}_\phi)$. When transforming $\hat{B}_\phi$ to a purely nondeterministic Büchi process B_ϕ as outlined above, we may combine Thms. 3 and 4 to obtain our desired result.

Corollary 1 (Büchi must–testing and LTL). *Let p be a labeled transition system and ϕ be an* **LTL** *formula. Then, $p \models \phi$ if and only if $B_\phi \sqsubseteq_{\mathsf{CL}}^{must} p$.*

Hence, our notion of Büchi must–testing not only extends DeNicola and Hennessy's [8] and Narayan Kumar et al.'s [20] must–preorders to (arbitrary) Büchi processes, but is also compatible with the LTL satisfaction relation.

5   Example

As an example for the utility of our theory for heterogeneous system design, consider the design of a very simple communications protocol given in Fig. 2.

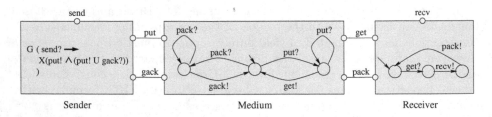

Fig. 2. A simple communications protocol.

The architecture of the protocol has already been fixed by the system designer and consists of a sender `Sender`, a medium `Medium`, and a receiver `Receiver`. The components communicate with the protocol's environment and among themselves via *channels*. In case of component `Sender`, these are the channels `send`, `put`, and `gack` (*get acknowledgment*). Each component in turn has its own specification. `Receiver` and `Medium` are given as labeled transition systems, reflecting the fact that their designs are relatively advanced. `Sender`, in contrast, is specified as an LTL formula stating that whenever a `send?` action occurs during an execution sequence of the sender, the remainder of the execution must begin with a sequence of `put!` actions followed by a `gack?` action. Finally, the overall specification of the protocol's required behavior may be given by the LTL formula `Spec` $=_{df}$ `G (send?` $\rightarrow$ `(F recv!))`. This formula dictates that in any sequence of actions which the system performs, whenever a `send?` action occurs, a `recv!` action eventually follows. An obvious question that a designer would be interested in is whether the specification of the sender is "strong enough" to ensure that the protocol satisfies `Spec`. The theory developed in this paper provides the semantic framework for answering this question. To do so, we first construct the purely nondeterministic Büchi process B_{Spec} for LTL formula `Spec`, as well as Büchi process B_{Sender} for LTL formula ϕ_{sender}. Next we assemble the overall system by employing our parallel composition and restriction operators.

$$\text{System} =_{df} (B_{\text{Sender}} \,|\, \text{Medium} \,|\, \text{Receiver}) \setminus \{\text{put}, \text{get}, \text{pack}, \text{gack}\}$$

Finally, we determine whether or not $B_{\text{Spec}} \sqsubseteq_{\text{CL}}^{\text{must}}$ `System`; this indeed holds.

The development of an efficient algorithm for automatically determining whether two Büchi processes are related by $\sqsubseteq_{\text{CL}}^{\text{must}}$ is future work. However, the alternative characterization of $\sqsubseteq_{\text{CL}}^{\text{must}}$ (cf. Thm. 1) already provides some hints about how this can be done. Due to the compositionality of the Büchi must–preorder, our positive answer is preserved when replacing B_{Sender} with any Büchi process p such that $B_{\text{Sender}} \sqsubseteq_{\text{CL}}^{\text{must}} p$.

If p is a labeled transition system and B_{Sender} is made purely nondeterministic, then $B_{\text{Sender}} \sqsubseteq_{\text{CL}}^{\text{must}} p$ holds exactly when $p \models \phi_{\text{sender}}$, according to Cor. 1. One such p is depicted to the right.

6   Related Work

Starting with the same motivation we did, Abadi and Lamport have developed ideas for heterogeneous specification for shared–memory systems [1]. Their technical setting is the logical framework of TLA [15], in which processes and temporal formulas are indistinguishable and logical implication serves as the refinement relation. TLA refinement coincides in some sense with trace inclusion in our testing scenario and is therefore insensitive to deadlock and divergence. Such issues are not of concern in the shared–memory world but must be dealt with in our setting, which is targeted towards specifying *distributed* systems in which components can interact directly, rather than indirectly via the shared memory.

Of direct relevance to this paper is the work of Kurshan [14], who developed a theory of ω–*word automata* that includes notions of synchronous and asynchronous composition. However, his underlying semantic model maps processes to their *maximal (infinite) traces*, and the associated notion of refinement is (reverse) trace inclusion. In theories of concurrency such as CCS [19] and CSP [4], in which deadlock is possible, maximal trace inclusion is not compositional [18]. In contrast, our must–preorder is compositional, at least for the operators presented here. Other work [6, 10, 13] investigated *modular* and *compositional model–checking* in similar non–deadlock environments.

Relatively more work has been devoted to analyzing relationships *between* refinement and logical approaches. One line of study relates temporal–logic specifications to refinement–based ones by establishing that one system refines another if and only if both satisfy the same properties. Results along these lines were pioneered by Hennessy and Milner [11] for *bisimulation* equivalence and a modal logic of their devising [19]. Similar ideas were also adapted regarding other behavioral equivalences and preorders and other temporal logics [9, 19, 25]. Congruences preserving "next-time–less" LTL have been studied by Kaivola and Valmari in [12]; the results have subsequently been extended to handle deadlock [26] and livelock [23]. Our work differs from theirs in that we want to have LTL formulas *embedded* in specifications.

Another line of research involves the encoding of labeled transition systems as logical formulas, and vice versa. Steffen and Ingólfsdóttir [24] defined an algorithm for converting finite–state labeled transition systems into formulas in the *mu–calculus*, while Larsen [16] demonstrated that certain mu–calculus formulas can be encoded as bisimulation–based *implicit specifications*. Finally, traditional testing has also been enriched with notions of *fairness* [3, 21] in order to constrain infinite computations in labeled transition systems.

7   Conclusions and Future Work

We conservatively extended the testing theories of DeNicola and Hennessy [8] and Narayan Kumar et al. [20] to Büchi processes. We then studied the derived Büchi may– and must–preorders, developed alternative characterizations for them, argued that the preorders are substitutive for several operators necessary for component–based system design, and showed that the Büchi must–preorder degrades to a variant of reverse trace inclusion when its first argument is purely nondeterministic. Using the latter result, we illustrated that Büchi must–testing provides a uniform basis for analyzing heterogeneous system designs given as a mixture of labeled transition systems and LTL formulas.

Regarding future work, we plan to develop specification languages mixing process algebras and LTL, which are given a semantics in terms of Büchi testing. We also intend to explore algorithms for computing our Büchi must–preorder.

References

[1] M. Abadi and L. Lamport. Composing specifications. *TOPLAS*, 15(1):73–132, 1993. See also: Conjoining Specifications, *TOPLAS*, 17(3):507–534, 1995.

[2] J.A. Bergstra, A. Ponse, and S.A. Smolka. *Handbook of Process Algebra*. Elsevier Science, 2000.

[3] E. Brinksma, A. Rensink, and W. Vogler. Fair testing. In *CONCUR '95*, volume 962 of *LNCS*, pages 313–328. Springer-Verlag, 1995.

[4] S.D. Brookes, C.A.R. Hoare, and A.W. Roscoe. A theory of communicating sequential processes. *J. of the ACM*, 31(3):560–599, 1984.

[5] E.M. Clarke, O. Grumberg, and D. Peled. *Model Checking*. MIT Press, 1999.

[6] E.M. Clarke, D.E. Long, and K.L. McMillan. Compositional model checking. In *LICS '89*, pages 353–362. IEEE Computer Society Press, 1989.

[7] R. Cleaveland and G. Lüttgen. Model checking is refinement: Relating Büchi testing and linear-time temporal logic. Technical Report 2000-14, Institute for Computer Applications in Science and Engineering, March 2000.

[8] R. DeNicola and M.C.B. Hennessy. Testing equivalences for processes. *TCS*, 34:83–133, 1983.

[9] R. DeNicola and F. Vaandrager. Three logics for branching bisimulation. *J. of the ACM*, 42(2):458–487, 1995.

[10] O. Grumberg and D.E. Long. Model checking and modular verification. *TOPLAS*, 16(3):843–871, 1994.

[11] M.C.B. Hennessy and R. Milner. Algebraic laws for nondeterminism and concurrency. *J. of the ACM*, 32(1):137–161, 1985.

[12] R. Kaivola and A. Valmari. The weakest compositional semantic equivalence preserving nexttime-less linear temporal logic. In *CONCUR '92*, volume 630 of *LNCS*, pages 207–221. Springer-Verlag, 1992.

[13] O. Kupferman and M.Y. Vardi. Modular model checking. In *Compositionality: The Significant Difference*, volume 1536 of *LNCS*. Springer-Verlag, 1997.

[14] R.P. Kurshan. *Computer-Aided Verification of Coordinating Processes: The Automata-Theoretic Approach*. Princeton University Press, 1994.

[15] L. Lamport. The temporal logic of actions. *TOPLAS*, 16(3):872–923, 1994.

[16] K.G. Larsen. The expressive power of implicit specifications. *TCS*, 114(1):119–147, 1993.

[17] O. Lichtenstein, A. Pnueli, and L. Zuck. The glory of the past. In *Workshop on Logics of Programs*, volume 193 of *LNCS*, pages 196–218. Springer-Verlag, 1985.

[18] M.G. Main. Trace, failure and testing equivalences for communicating processes. *J. of Par. Comp.*, 16(5):383–400, 1987.

[19] R. Milner. *Communication and Concurrency*. Prentice Hall, 1989.

[20] K. Narayan Kumar, R. Cleaveland, and S.A. Smolka. Infinite probabilistic and nonprobabilistic testing. In *FSTTCS '98*, volume 1530 of *LNCS*, pages 209–220. Springer-Verlag, 1998.

[21] V. Natarajan and R. Cleaveland. Divergence and fair testing. In *ICALP '95*, volume 944 of *LNCS*, pages 684–695. Springer-Verlag, 1995.

[22] A. Pnueli. The temporal logic of programs. In *FOCS '77*, pages 46–57. IEEE Computer Society Press, 1977.

[23] A. Puhakka and A. Valmari. Weakest-congruence results for livelock-preserving equivalences. In *CONCUR '99*, volume 1664 of *LNCS*, pages 510–524. Springer-Verlag, 1999.

[24] B. Steffen and A. Ingólfsdóttir. Characteristic formulae for CCS with divergence. *Inform. & Comp.*, 110(1):149–163, 1994.

[25] C. Stirling. Modal logics for communicating systems. *TCS*, 49:311–347, 1987.

[26] A. Valmari and M. Tiernari. Compositional failure-based semantics models for basic LOTOS. *FAC*, 7(4):440–468, 1995.

[27] M. Vardi and P. Wolper. An automata-theoretic approach to automatic program verification. In *LICS '86*, pages 332–344. IEEE Computer Society Press, 1986.

Formal Verification of the Ricart-Agrawala Algorithm*

Ekaterina Sedletsky[1], Amir Pnueli[1], and Mordechai Ben-Ari[2]

[1] Department of Computer Science and Applied Mathematics,
The Weizmann Institute of Science, Rehovot 76100, Israel.
{kate|amir}@wisdom.weizmann.ac.il
[2] Department of Science Teaching,
The Weizmann Institute of Science, Rehovot 76100, Israel.
moti.ben-ari@weizmann.ac.il

Abstract. This paper presents the first formal verification of the Ricart-Agrawala algorithm [RA81] for distributed mutual exclusion of an arbitrary number of nodes. It uses the Temporal Methodology of [MP95a]. We establish both the safety property of *mutual exclusion* and the liveness property of *accessibility*. To establish these properties for an arbitrary number of nodes, parameterized proof rules are used as presented in [MP95a] (for safety) and [MP94] (for liveness). A new and efficient notation is introduced to facilitate the presentation of liveness proofs by verification diagrams.

The proofs were carried out using the Stanford Temporal Prover (STeP) [BBC+95], a software package that supports formal verification of temporal specifications of concurrent and reactive systems.

1 Introduction

The Ricart-Agrawala algorithm (RA) [RA81] for achieving mutual exclusion in a network is one of the venerable and well-known algorithms in distributed computing. Nevertheless, the correctness of the algorithm has not been formally verified.

The only previous attempt to formally prove the RA algorithm is the unpublished work [Kam95], but it is restricted to the safety property of mutual exclusion and uses a simplified model. On the other hand, already [Lamp82] presented a non-mechanized proof of a similar algorithm.

The main motivation for this work was to attempt a fully mechanized formal deductive proof of the RA algorithm, establishing both its safety and liveness properties, and using the deductive methods of [MP91].

These methods have been mechanized in a software package called the Stanford Temporal Prover (STeP) [BBC+95]. A further motivation of this work was to push STeP to its limits, and see whether it could be used to prove an algorithm whose correctness proofs are quite complex.

---

* This research was supported in part by the Minerva Center for Verification of Reactive Systems, and a grant from the U.S.-Israel bi-national science foundation.

We successfully generated formal proofs of both mutual exclusion and accessibility using STEP. This research points the way for further improvements both in proof techniques and in software support for deductive verification methods.

2   Implementation of the Ricart-Agrawala Algorithm

To verify the RA algorithm, we have written it in a formal programming notation, the language SPL which is used in [MP91] as the programming language (Figure 1).

$$
\begin{array}{lll}
\textbf{in} & N & : \textbf{integer where } N \geq 2 \\
\textbf{local } chq, chp & : \textbf{array } [1..N, 1..N] \textbf{ of channel } [1..] \textbf{ of integer} \\
& & \textbf{where } chq = \Lambda, chp = \Lambda \\
y, z & : [1..N] \textbf{ where } y{=}1 \\
\textbf{type } Nar & = \textbf{array } [1..N] \textbf{ of integer} \\
Bar & = \textbf{array } [1..N] \textbf{ of boolean} \\
\textbf{value } mini & : Nar \times Bar \rightarrow [1..N]
\end{array}
$$

$$
\mathop{\Big\|}_{s=1}^{N} Node[s] ::
\left[
\begin{array}{l}
\textbf{local } osn, hsn, p, c : \textbf{integer where } osn = 0, hsn = 0, c = 0, p = 1 \\
rcs \qquad\qquad : \textbf{boolean where } rcs = F \\
rd \qquad\qquad\ \ : \textbf{array } [1..N] \textbf{ of boolean where } rd = F \\[4pt]
M :: \left[
\begin{array}{l}
\textbf{loop forever do} \\
\left[
\begin{array}{l}
m_1 : \textbf{noncritical} \\
m_2 : \langle rcs := T; osn := hsn + 1; c := N\text{-}1; p := 1; \\
\qquad\qquad\qquad\qquad\qquad\qquad y := mini(osn, rcs)\rangle \\
m_{31} : \textbf{while } p \leq N \textbf{ do} \\
m_{32} : \quad \langle \textbf{if } p \neq s \textbf{ then } chq[s,p] \Leftarrow osn; p := p+1 \rangle \\
m_4 : \textbf{await } c{=}0 \\
m_5 : \textbf{critical} \\
m_6 : \langle rcs := F; p := 1; y := mini(osn, rcs) \rangle \\
m_{71} : \textbf{while } p \leq N \textbf{ do} \\
m_{72} : \quad \langle \textbf{if } rd[p] \textbf{ then } [rd[p] := F; chp[s,p] \Leftarrow 1]; p := p+1 \rangle
\end{array}
\right]
\end{array}
\right] \\[4pt]
\| \\[4pt]
Q :: \mathop{\|}_{t=1}^{N} \left[
\begin{array}{l}
\textbf{local } rq : \textbf{integer} \\
\textbf{loop forever do} \\
q_1 : \left\langle
\begin{array}{l}
chq[t,s] \Rightarrow rq \\
\textbf{if } hsn < rq \textbf{ then } hsn := rq \\
\textbf{if } (rq,t) \prec (osn,s) \vee \neg rcs \\
\quad \textbf{then } chp[s,t] \Leftarrow 1 \textbf{ else } rd[t] := T
\end{array}
\right\rangle
\end{array}
\right] \\[4pt]
\| \\[4pt]
P :: \mathop{\|}_{u=1}^{N} \left[
\begin{array}{l}
\textbf{local } rp : \textbf{integer} \\
\textbf{loop forever do} \\
r_1 : \langle chp[u,s] \Rightarrow rp; c := c - 1 \rangle
\end{array}
\right]
\end{array}
\right]
$$

Fig. 1. Implementation of the RA algorithm.

The structure of the program is as follows: we assume that there are N nodes, where N is a parameter of the program which stays fixed during execution. Each node is a concurrent process: in the notation $Node[s] :: [\ldots]$, the ellipsis indicates the program text for the s'th node and s is the index of the node which may be referenced within the program text. The entire program is $\overset{N}{\underset{s=1}{\|}} Node[s]$, implying a concurrent execution of all the nodes.

The nodes are connected to each other in a complete graph: there is a pair of uni-directional asynchronous channels connecting each node to every other node, where chq is the outgoing channel for the REQUEST messages and chp is the incoming channel for the REPLY messages. The notation for output is $chq[a, b] \Leftarrow e$, meaning that node a sends the value of expression e to node b along channel chq, and similarly, $chq[a, b] \Rightarrow x$, means that node b removes the value coming from a and assigns it to x.

The additional global declarations are discussed below.

The program for process $Node$ is composed of three concurrent processes:

- M is the main process containing the critical section and the protocols to be executed upon entry and exit.
- P is the process which receives and counts replies.
- Q is the process which receives requests and decides if to reply or to defer the reply.

Note that P and Q are themselves composed of concurrent processes, one for each channel. Within $Node[s]$, process $Q[t]$ (which can also be identified as $Q[t, s]$) is responsible for reading messages from channel $chq[t, s]$. Similarly, process $P[u]$ ($P[u, s]$) is responsible for reading messages from channel $chp[u, s]$. The synchronization among the processes within the same node is based on shared variables, and we use the notation $< \ldots >$, to imply that the statements are to be executed atomically. This can be easily implemented using semaphores. We include the assignments c := N-1 and p := 1 within the atomic statement of line m_2, and p := 1 within line m_6, to reduce the number of verification conditions in the proof of accessibility.

Within each node there are global variables which are shared among the processes of that node:

- osn - the sequence number chosen by the node.
- hsn - the highest sequence number seen in any request message received by the node.
- rcs - a flag that is true if the node is requesting to enter the critical section.
- c - a counter of the number of outstanding reply messages.
- rd - an array that lists deferred requests. $rd[j]$ is true when the node is deferring a reply to the request from node j.
- p, rp, rq are auxiliary variables and could have been declared as local to the processes of the node.

The following variables are not needed by the algorithm; they were added to facilitate the proof.

- z is the index of a generic node, which is used to specify and verify accessibility.
- y is the index of the node with the minimal value of the rank $(osn[i], i)$, where the minimum is taken over all nodes i such that $rcs[i]$ is true. If $rcs[i]$ are all false, $y = 1$.
- $mini(osn, rcs)$ is a function that computes y, the index of the node with the minimal rank.

3   Proof of the Mutual Exclusion Property

Invariance properties of the form $\Box\, p$, where p is an *assertion* (a *state formula*) can be verified by the *invariance rule* B-INV, given by

$$
\boxed{
\begin{array}{ll}
\text{Rule B-INV} \qquad \text{I1.} \quad \Theta \;\;\rightarrow \varphi & \\[4pt]
\qquad\qquad\qquad \text{I2.} \; \dfrac{\varphi \wedge \rho_\tau \rightarrow \varphi'}{\Box\, p} & \text{For every transition } \tau \in \mathcal{T}
\end{array}
}
$$

where Θ is the initial condition and $\mathcal{T}$ is the set of transitions of the verified system. An assertion satisfying premises I1 and I2 of rule B-INV is called *inductive*.

In our case, the main invariance property is that of *Mutual exclusion*, which can be specified as

PROPERTY **excl**: $\Box\, \forall i, j : [1..N] : m_5[i] \wedge m_5[j] \rightarrow i = j$

Here and below, we use $m_5[i]$ to denote that process $M[i]$ is currently executing at location m_5.

3.1   Bottom Up Assertions

At first we use a *bottom-up* approach to deduce some simple properties of the program.

Locations at which $rcs = 1$ A first observed property is

PROPERTY **rcs_range**: $\Box(m_{31,32,4,5,6}[i] \leftrightarrow rcs[i])$.

Note, that whenever there is a free index, such as i in the above property, there is an implicit universal quantification, implying that the property holds for every $i \in [1..N]$.

Range of $p[i]$ The variable p serves as a loop counter for the loops at statements m_{31} and m_{71}. The upper limit of these two loops is N.

PROPERTY **p_range**: $\Box(1 \leq p[i] \leq N + 1 - m_{32,72}[i])$

This inductive assertion claims that $p[i] \leq N + 1$ at all locations, except for locations m_{32} and m_{72}, where the stronger inequality $p[i] \leq N$ holds.

The Message Chain Linkage

PROPERTY **message_chain**:

$$\Box((m_{31,32}[i] \land p[i] > j) + m_4[i] \geq |chq[i,j]| + rd[j,i] + |chp[j,i]|)$$

Here, $|chq[i,j]|$ and $|chp[j,i]|$ denote the sizes of the buffers of these asynchronous channels. This property states that the sum $|chq[i,j]| + rd[j,i] + |chp[j,i]|$ never exceeds 1, and can be positive only if process $M[i]$ is at location m_4 or at locations $m_{31,32}$ with $p[i] > j$.

The Reply Counter   The role of the counter $c[i]$ is to count the number of positive replies $Node[i]$ received since it last sent out requests for entering the critical section. We would expect that, at any point, the value of $c[i]$ will equal the number of pending replies. This is stated by

PROPERTY **c_range**: $\Box(c[i] =$

$$\sum_{k=1}^{N} |chq[i,k]| + rd[k,i] + |chp[k,i]|) + m_{31,32}[i] \cdot (N - p[i] + (p[i] > i))$$

Neither of properties **message_chain** or **c_range** is inductive by itself. However, their conjunction, to which we refer as **msg_range_coun- ter**, together with the previously established property **p_range** form an inductive assertion.

The Value of a Request Message   As the last bottom-up invariant, we formulate the following property:

PROPERTY **request_in_channel**:
$$\Box(|chq[i,j]| > 0 \quad \rightarrow \quad head(chq[i,j]) = osn[i])$$

This property states that if channel $chq[i,j]$ is not empty, then the value it contains is the *current* value of $osn[i]$.

3.2   Top Down Assertions

We now move to a set of assertions which are derived based on the goal we wish to prove, namely the property of mutual exclusion.

We start by introducing some definitions:

$requested(i,j)$: $i \neq j \land (m_{4,5}[i] \lor (m_{31,32}[i] \land p[i] > j))$
$request\_received(i,j)$: $requested(i,j) \land |chq[i,j]| = 0$
$granted(i,j)$: $request\_received(i,j) \land \neg rd[j,i]$

Variable *hsn* Retains the Highest Message Number Seen So Far
The following property states that, after having read the recent message from
$Node[i]$, the variable $hsn[j]$ ("highest seen") has a value which is not lower than
$osn[i]$.

PROPERTY hsn_highest:
$\square(request\_received(i,j) \;\rightarrow\; osn[i] \leq hsn[j])$

The Implication of *Node[j]* Granting Permission to *Node[i]*   The
following property describes the implications of a situation in which $Node[j]$ has
granted an entry permission to $Node[i]$ before $Node[i]$ exited its critical section:

PROPERTY permitted:
$\square(granted(i,j) \;\rightarrow\; \neg rcs[j] \;\vee\; (osn[i],i) \prec (osn[j],j))$

Finally, Mutual Exclusion Finally, we establish the property of mutual exclusion, specified by

PROPERTY excl: $\square(m_5[i] \;\wedge\; m_5[j] \;\rightarrow\; i=j)$

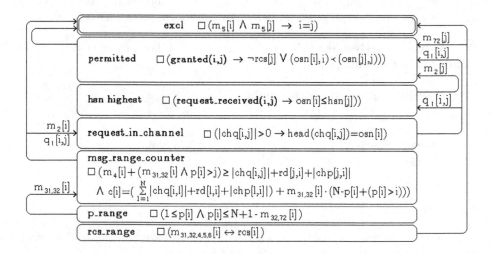

Fig. 2. Set of inductive properties leading to the proof of Mutual Exclusion. The
labels on the dependence edges identify the transitions for which the verification
of the higher placed property depends on the lower property.

4   A Proof Rule for Accessibility

Rule P-WELL   For assertions p and $q = \varphi_0, \varphi_1[k], \ldots, \varphi_m[k]$,
transitions $\tau_1[k], \ldots, \tau_m[k] \in \mathcal{J}$
a well-founded domain $(\mathcal{A}, \succ)$, and
ranking functions $\delta_0, \delta_1[k], \ldots, \delta_m[k] : \Sigma \mapsto \mathcal{A}$

$W1.$ $\qquad\qquad p \;\rightarrow\; \bigvee_{j=0}^{m} \exists k : \varphi_j[k]$

For $i = 1..m$

$W2.$ $\rho_\tau[k] \wedge \varphi_i[k] \;\rightarrow\; \bigvee_{j=0}^{m} \exists u : (\varphi'_j[u] \wedge \delta_i[k] \succ \delta'_j[u]) \vee (\varphi'_i[k] \wedge \delta_i[k] = \delta'_i[k])$

$\qquad\qquad\qquad\qquad\qquad\qquad\qquad\qquad\qquad\qquad\qquad\qquad$ for every $\tau \in \mathcal{T}$

$W3.$ $\rho_{\tau_i}[k] \wedge \varphi_i[k] \;\rightarrow\; \bigvee_{j=0}^{m} \exists u : (\varphi'_j[u] \wedge \delta_i[k] \succ \delta'_j[u])$

$W4.$ $\qquad\qquad \varphi_i[k] \;\rightarrow\; En(\tau_i[k])$

$\qquad\qquad\qquad\qquad\qquad p \Longrightarrow \Diamond q$

To verify liveness properties of parameterized programs, we can use a fixed number of intermediate formulas and helpful transitions but they may refer to an additional parameter k which is a process index. A parameterized rule for proving accessibility properties of parameterized systems has been presented in [MP95b]. However, to verify a complicated system such as the RA algorithm, it was necessary to introduce a new version of this rule, which we present here.

To improve readability of formulas, we write $\rho_{\tau_i[k]}$ as $\rho_{\tau_i}[k]$. Rule P-WELL uses parameterized intermediate assertion, parameterized helpful transitions, and parameterized ranking functions. For each $i = 1, \ldots m$, the parameter k in $\varphi_i[k]$, $\tau_i[k]$, and $\delta_i[k]$ ranges over some nonempty set, such as $[1..N]$.

The rule traces the progress of computations from an arbitrary p-state to an unavoidable q-state. With each non-terminal assertion φ_i, $i > 0$, the rule associates a *helpful transition* τ_i, such that the system is just (weakly fair) with respect to τ_i. Premise W2 requires that the application of any transition τ to a state satisfying a non-terminal assertion φ_i will never cause the rank of the state to increase. Premise W3 requires that if the applied transition is helpful for φ_i then the rank must decrease. Due to the well-foundedness of the ranking functions, we cannot have an infinite chain of helpful transitions, since this will cause the rank to decrease infinitely often. Premise W4 stipulates that the helpful transition τ_i is always enabled on every φ_i-state. Thus, we cannot have an infinite computation (which must be fair) which avoids reaching a $q = \varphi_0$ state.

4.1   Representation by Diagrams

The paper [MP94], introduced the graphical notation of *verification diagrams*. For our application here, verification diagrams can be viewed as a concise and optimized presentation of the components appearing in rule P-WELL. We refer the reader to Figure 3 for explanation of some of the main elements which typically appear in such diagrams.

The diagram contains a node for each assertion φ_i that appears in the rule. The helpful transition tau_i associated with φ_i is identified by the label of one or more directed edges departing from the node (labeled by) φ_i. Thus, in the diagram of Figure 3, $m_2[z]$ is identified as the transition helpful for assertion φ_{13}, and the helpful transition for φ_{10} is $q_1[z, i]$ even though it labels two edges departing from φ_{13}.

The interconnection topology in the diagram provides a more specialized (and efficient) version of the P-WELL rule. For a node φ_i, let $succ(i)$ be the set of indices of the nodes which are the targets of edges departing for φ_i. Then the diagram suggests that, instead of proving premises W2 and W3 as they appear in the rule, it is sufficient to prove their following stronger versions:

U2. $\rho_\tau[k] \wedge \varphi_i[k] \to \varphi_i'[k] \wedge \delta_i[k] = \delta_i'[k]$ for every $\tau \in \mathcal{T}$

U3. $\rho_{\tau_i}[k] \wedge \varphi_i[k] \to \bigvee_{j \in succ(i)} \exists u : (\varphi_j'[u] \wedge \delta_i[k] \succ \delta_j'[u])$

It is not difficult to see that U2 and U3 imply W2 and W3. For example, premise U3 for assertion φ_6 as implied by the diagram is

$$\varphi_6[i] \wedge \rho_{m_{72}}[i] \;\to\; (\varphi_7'[i] \wedge \delta_6[i] \succ \delta_7'[i]) \vee (\varphi_4'[i] \wedge \delta_6[i] \succ \delta_4'[i])$$

The more general notion of verification diagrams as presented in [MP94] admits two types of edges, one corresponding to the helpful transitions, which are the edges present in our diagram. The other type corresponds to unhelpful transitions. It is suggested there to use a double line for helpful edges. In our case, we only need to represent helpful transitions, so we draw them as single lines.

The rule also requires to associate with each non-terminal assertion φ_i a ranking function δ_i. By convention, whenever a ranking function is not explicitly defined within a node φ_i, the default value is the index of the node, i.e. $\delta_i = i$. For example, in the diagram, $\delta_{13} = 13$. However, as we will see below, this is not the end of the story.

4.2   Encapsulation Conventions

The diagram of Figure 3 contains, in addition to *basic nodes* such as those labeled by assertions, also *compound nodes* which are also called *blocks*. We may refer to compound nodes by the set of basic nodes they contain. For example, the successor of node φ_{13} is the compound node $\varphi_{31,32}$. Compound nodes may be annotated by λ-declarations, such as the compound node $\varphi_{4..7}$, by additional assertions, such as $m_4[y]$ for block $\varphi_{3..7}$, or ranking components, such as $(6, -p[i])$ for block $\varphi_{6,7}$. There are several encapsulation conventions associated with compound nodes.

- An edge stopping at the boundary of a block, is equivalent to individual edges which reach the basic nodes contained in the block. Thus, both φ_{11} and φ_{12} are immediate successors of node φ_{13}.

- For each basic node φ_i, the full assertion associated with this node is the conjunction of all the assertions labeling the blocks in which the basic node is contained. We denote this full assertion by $\widehat{\varphi}_i$. For example,

$$\widehat{\varphi}_7 \quad = \quad m_4[z] \; \wedge \; (\forall j : |chq[z,j]| = 0) \; \wedge \; m_4[y] \; \wedge \; rd[i,y] \; \wedge \; m_{71}[i]$$

- For each basic node φ_i, the full ranking function associated with this node is the left-to-right concatenation of all the ranking components labeling the blocks in which the basic node is contained. As the rightmost component, we add i. For example,

$$\delta_7 \quad = \quad (1, -osn[y], -y, 4, c[y], 6, -p[i], 7)$$

In Figure 3 we present the full ranking functions for each of the nodes alongside the diagram.

Note that whenever we have to compare ranking functions which are lexicographic tuples of different lengths, we add zeroes to the right of the shorter one. For example, to see that $\delta_{13} \succ \delta_{12}$, we confirm that $(13, 0, 0) \succ (11, -p[z], 12)$.

Note also that several components of the ranking functions are negative. When STEP is presented with any ranking function, one of the proof obligations which are generated require proving that all components are bounded from below. This has been done for all the components present in the diagram.

5   Proof of Accessibility Property

The property of accessibility may be written in the form

PROPERTY **m2m5**: $m_2[z] \Rightarrow \Diamond\, m_5[z]$

where $z \in [1..N]$.

5.1   Auxiliary Assertions Needed for the Proof

A crucial part in the proof is the computation of the index of the process y with minimal signature. We define

$$ismin(osn, rcs, y) : \left(\begin{array}{l} y = 1 \; \wedge \; \forall j : \neg rcs[j] \\ \vee \; rcs[y] \; \wedge \; \forall j : (rcs[j] \rightarrow (osn[y], y) \prec (osn[j], j)) \end{array} \right)$$

Thus, y has a minimal signature, if either there is no process j with $rcs[j]$ and then $y = 1$, or there exists some j with $rcs[j] = 1$ and y is such a j with the minimal signature. In fact, rather then defining the function $mini$ explicitly, we inform STEP of the following axiom:

AXIOM **mini:** $ismin(osn, rcs, mini(osn, rcs))$.

There were several auxiliary assertions whose invariance was necessary in order to establish the proof obligations generated by STEP, when being presented by the verification diagram of Figure 3. We list them below:

$$\text{rd\_osn} : \quad rd[i,j] \;\rightarrow\; (osn[i], i) \prec (osn[j], j)$$

$$\text{not\_rd\_range} : \quad m_{1,2}[i] \vee (m_{71,72}[i] \wedge p[i] > j) \;\rightarrow\; \neg rd[i,j]$$

$$\text{y\_eq\_mini} : \quad y = mini(osn, rcs)$$

$$\text{y\_is\_min} : \quad ismin(osn, rcs, y)$$

$$\text{rd\_to\_y} : \quad rd[i,y] \;\rightarrow\; m_{71,72}[i] \wedge p[i] \leq y$$

$$\text{y\_not\_change} : \quad (\forall j : |chq[z,j]| = 0) \wedge m_4[z] \wedge m_2[s] \;\rightarrow\;$$
$$(osn[y], y) \prec (hsn[s] + 1, s)$$

The last property **y_not_change** is very crucial in order to establish that y can stop being minimal only by retiring on exit from $m_6[y]$. In particular, no newcomer s can execute transition $m_2[s]$ and become minimal.

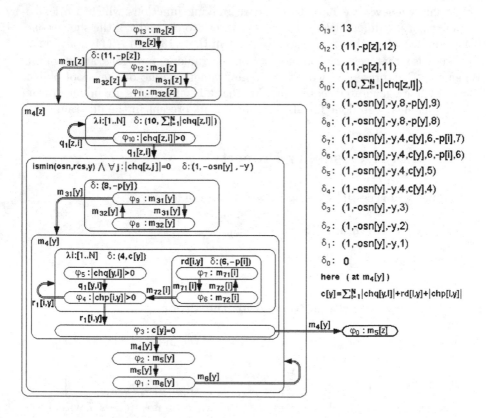

Fig. 3. Verification Diagram for the property $m_2[z] \Rightarrow \Diamond m_5[z]$.

5.2   Usage of STEP in the Proof

We used STEP version 1.4 from 2/XI/1999 in our proof. Some modifications of the source program were necessary in order for STEP to accept our SPL program. This version of STEP also fails to support *lambda*-blocks in the way there were used in Figure 3. To overcome this difficulty, we had to feed STEP with some processed fragments of this diagram and then modify manually some of the resulting verification conditions. We hope that future versions of STEP will provide direct support of *lambda*-blocks.

In spite of these minor inconveniences, we found STEP to be a very powerful and useful verification system, specially geared to the temporal verification of complex algorithms such as the Ricart Agrawala algorithm we considered here.

Acknowledgment

We gratefully acknowledge the help of Nikolaj Bjørner of the STeP research team for his advice and for his fast response to our requests for modifications in STeP.

References

[BA90]   M. Ben-Ari. *Principles of Concurrent and Distributed Programming.* Prentice-Hall International, Hemel Hempstead, 1990.

[BBC+95]   N. Bjørner, I.A. Browne, E. Chang, M. Colón, A. Kapur, Z. Manna, H.B. Sipma, and T.E. Uribe. STeP: The Stanford Temporal Prover, User's Manual. Technical Report STAN-CS-TR-95-1562, Computer Science Department, Stanford University, November 1995.

[Kam95]   J. Kamerer. Ricart and Agrawala's algorithm. Unpublished, http://rodin.stanford.edu/case-studies, 9 August 1995.

[Lamp82]   L. Lamport *An Assertional Correctness Proof of Distributed Program.* Science of Computer Programming, 2, 3, December 1982, pages 175–206.

[MP91]   Z. Manna and A. Pnueli. *The Temporal Logic of Reactive and Concurrent Systems: Specification.* Springer-Verlag, New York, 1991.

[MP94]   Z. Manna and A. Pnueli. Temporal verification diagrams. In T. Ito and A. R. Meyer, editors, *Theoretical Aspects of Computer Software*, volume 789 of *Lect. Notes in Comp. Sci.*, pages 726–765. Springer-Verlag, 1994.

[MP95a]   Z. Manna and A. Pnueli. *Temporal Verification of Reactive Systems: Safety.* Springer-Verlag, New York, 1995.

[MP95b]   Z. Manna and A. Pnueli. Verification of parameterized programs. In E. Börger, editor, *Specification and Validation Methods*, pages 167–230. Oxford University Press, Oxford, 1995.

[RA81]   G. Ricart and A.K. Agrawala. An optimal algorithm for mutual exclusion in computer networks. *Comm. ACM*, 24(1):9–17, 1981. Corr. ibid. 1981, p.581.

On Distribution-Specific Learning with Membership Queries versus Pseudorandom Generation

Johannes Köbler and Wolfgang Lindner*

Humboldt-Universität zu Berlin
Institut für Informatik
D-10099 Berlin, Germany

Abstract. We consider a weak version of pseudorandom function generators and show that their existence is equivalent to the non-learnability of Boolean circuits in Valiant's pac-learning model with membership queries on the uniform distribution. Furthermore, we show that this equivalence holds still for the case of non-adaptive membership queries and for any (non-trivial) p-samplable distribution.

1 Introduction

In computational learning theory, many non-learnability results for (the representation independent version of) Valiant's pac-learning model are based on cryptographic tools and assumptions. Already in [19], Valiant pointed out how the results of [8] can be used to show that (arbitrary) Boolean circuits are not efficiently pac-learnable if cryptographically secure one-way functions exist. Related results can be found in, e.g., [11, 3, 1, 12, 13].

On the other hand, it is known [4] that the non-learnability (in polynomial-time) of Boolean circuits provides a sufficient condition under which $\mathcal{RP}$ and, hence, $\mathcal{P}$ is different from $\mathcal{NP}$. Recently, Impagliazzo and Wigderson [9] showed that every problem in $\mathcal{BPP}$ can be solved deterministically in sub-exponential time on almost every input (for infinitely many input lengths), provided that $\mathcal{EXP} \neq \mathcal{BPP}$. Since $\mathcal{EXP} = \mathcal{BPP}$ implies $\mathcal{NP} = \mathcal{RP}$, this means that the non-learnability of Boolean circuits can further serve as a hypothesis to achieve derandomization results for probabilistic polynomial-time computations.

Interestingly, both implications, namely that Boolean circuits are not learnable (if one-way functions exist) as well as the derandomization of $\mathcal{BPP}$ (if Boolean circuits are not learnable) are based on the concept of *pseudorandom generation*, i.e., the possibility to expand a small number of random bits (also known as the *seed*) into a large amount of *pseudorandom* bits which cannot be significantly distinguished from truely random bits by any polynomial-time computation. An important difference between these two implications is, however, that in the first case the pseudorandom generator has to run in polynomial

---

* Supported by DFG grant Ko1053/1-1.

S. Kapoor and S. Prasad (Eds.): FST TCS 2000, LNCS 1974, pp. 336–347, 2000.
© Springer-Verlag Berlin Heidelberg 2000

time, while for the derandomization of $\mathcal{BPP}$ it suffices that the pseudorandom generator runs in exponential time.

A close connection between pseudorandom generation and non-learnability has been demonstrated already in [8], where the existence of a pseudorandom *function* generator (see Section 3) is shown to be in fact equivalent to the non-learnability of Boolean circuits. This, however, only holds if we consider learning agents that are successful with high probability for a randomly chosen target (instead of being successful for *all* targets as in the standard model of pac-learning). Furthermore, in the model of [8], the learning task has to be accomplished with the help of membership queries and with respect to the uniform distribution. Recall that the original model of [19] is *distribution-free* and *passive*, i.e., the learning algorithm has to be successful for any (unknown) distribution $\mathcal{D}$, and access to the target is given via random *examples* chosen according to $\mathcal{D}$ together with their classification by the target. Further connections between the average-case model of pac-learning and cryptographic primitives have been found in [2].

Recently [14], also the worst-case model of (passive) pac-learning under the uniform distribution has been closely related to pseudorandom function generation. In contrast to [8] the pseudorandomness condition in [14] is expressed in terms of the *worst-case* advantage of the distinguishing algorithms (see section 3). Moreover, the distinguishing algorithms for the function generator in [14] are required to be passive, i.e., the distinguishing algorithm may access its oracle only via random classified examples. This leads to an apparently weaker notion of pseudorandom function generation as compared to the standard definition [8].

Here we take a similar approach as in [14] for pac-learning *with* membership queries under the uniform distribution [1, 12, 10]. As our main result we show that the non-learnability of Boolean circuits is equivalent to the existence of a weak pseudorandom function generator, where as in [14] the pseudorandomness condition is expressed in terms of worst-case advantage of distinguishing algorithms. Further we show that this kind of weak pseudorandom generation is still strong enough to yield a derandomization result for $\mathcal{RP}$. In contrast to the above mentioned derandomization for $\mathcal{BPP}$, the function generator that we use is polynomial-time computable (instead of exponential time). Hence, we even get that every problem in $\mathcal{RP}$ can be solved nondeterministically in polynomial-time on almost every input (for infinitely many input lengths), where for every given $\epsilon > 0$, only n^ϵ many nondeterministic bits are used. Of course, this implies that every problem in $\mathcal{RP}$ can also be solved deterministically in sub-exponential time on almost every input.

As an application, we get that every learning algorithm for Boolean circuits which uses membership queries and which is successful for some arbitrary p-samplable distribution can be transformed into a learning algorithm which uses only *non-adaptive* membership queries and is successful for the uniform distribution. Thus, if Boolean circuits are not learnable with non-adaptive membership queries under the uniform distribution, then Boolean circuits are not learnable with adaptive membership queries under any non-trivial [12] p-samplable distribution.

2   Preliminaries

We use $\mathcal{B}_n$ to denote the set of all Boolean functions $f : \{0,1\}^n \to \{0,1\}$.

A *probability ensemble* $\mathcal{D} = \{\mathcal{D}_n : n \geq 0\}$ is a sequence of probability distributions $\mathcal{D}_n$ on $\{0,1\}^n$. The ensemble $\mathcal{D}$ is *p-samplable* if there is a polynomial time computable function f and a polynomial p such that for all n, $f(X)$ is distributed according to $\mathcal{D}_n$ when X is uniformly distributed over $\{0,1\}^{p(n)}$.

For a polynomially bounded function $f : \mathbb{N} \to \mathbb{R}$, let $\mathcal{NP}[f]$ [7] denote the class of all sets $A \in \mathcal{NP}$ for which there exists a set $B \in \mathcal{P}$ such that for all strings x, $x \in A \Leftrightarrow \exists y \in \{0,1\}^{\lceil f(|x|) \rceil} : (x,y) \in B$. For a function $\epsilon : \mathbb{N} \to [0,1]$ and any class $\mathcal{C}$ of sets, $\mathcal{H}eur_{\epsilon(n)}\mathcal{C}$ denotes the class of all sets A such that for any p-samplable distribution ensemble $\mathcal{D}$ there is a set $A' \in \mathcal{C}$ such that for all n and for $X \in_{\mathcal{D}_n} \{0,1\}^n$, $\Pr(A(X) \neq A'(X)) < \epsilon(n)$. Here, $A(x)$ is used to denote the characteristic function of a set A.

Let X and Y be independent and identically distributed random variables on $\{0,1\}^n$. The *Renyi entropy* of X is $\mathrm{ent}_{Ren}(X) = -\log(\Pr(X = Y))$, and the *minimum entropy* of X is $\mathrm{ent}_{min}(X) = \min\{-\log(\Pr(X = x)) : x \in \{0,1\}^n\}$. Note that for any random variable X defined on $\{0,1\}^n$, $\mathrm{ent}_{Ren}(X)/2 \leq \mathrm{ent}_{min}(X) \leq \mathrm{ent}_{Ren}(X)$. The Renyi entropy of a distribution $\mathcal{D}$ is the Renyi entropy of a random variable X chosen according to $\mathcal{D}$.

2.1   Predictability

We recall the learning model of efficient prediction with membership queries (cf. [1, 12]). A *representation of concepts* C is any subset of $\{0,1\}^* \times \{0,1\}^*$. A pair (u,x) of $\{0,1\}^* \times \{0,1\}^*$ is interpreted as consisting of a *concept name* u and an *example* x. The *concept* represented by u is $\kappa_C(u) = \{x : (u,x) \in C\}$.

A *prediction with membership queries algorithm*, or *pwm-algorithm*, is a possibly randomized algorithm A that takes as input two positive integers s and n, where s is the length of the target concept name and n is the length of examples n as well as a rational *accuracy bound* ϵ. It may make three different kinds of oracle calls, the responses to which are determined by the unknown target concept $c = \kappa_C(u)$ with $|u| = s$, and an unknown distribution $\mathcal{D}_n$ on $\{0,1\}^n$ as follows.

1. A *membership query* takes a string $x \in \{0,1\}^*$ as input and returns 1 if $x \in c$ and 0 otherwise.
2. A *request for a random classified example* takes no input and returns a pair (x,b), where x is a string chosen according to $\mathcal{D}_n$, and $b = 1$ if $x \in c$ and $b = 0$ otherwise.
3. A *request for an element to predict* takes no input and returns a string x chosen according to $\mathcal{D}_n$.

The algorithm A may make any number of membership queries or requests for random classified examples. However, A must eventually make one and only one request for an element to predict and then eventually halt with an output of 1 or 0 without making any further oracle calls.

A pwm-algorithm A is said to ϵ-*predict* $\kappa_C(u)$ on $\mathcal{D}_n$ if on input $s = |u|$, n and ϵ, when A is run with respect to the target concept $\kappa_C(u)$ and distribution $\mathcal{D}_n$, the probability is at most ϵ that the output of A is not equal to the correct classification of X by $\kappa_C(u)$, where $X \in_{\mathcal{D}_n} \{0,1\}^n$ is the request for an element to predict. Furthermore, a pwm-algorithm A is said to run in *polynomial time* if its running time is bounded by a polynomial in s, n, and $1/\epsilon$.

Definition 1. *Let C be a representation of concepts, and let $\mathcal{D} = \{\mathcal{D}_n : n \geq 0\}$ be a probability ensemble. Then C is* predictable on $\mathcal{D}$ with membership queries *in polynomial time if there exists a polynomial-time pwm-algorithm A such that for all positive integers s and n, for all concept names u of length s, and for all positive rationals ϵ, A ϵ-predicts $\kappa_C(u)$ on $\mathcal{D}_n$.*

Recall that in the *weak* model of learning [11] the learning algorithm has to predict the target only slightly better than a completely random guess.

Definition 2. *Let C be a representation of concepts, and let $\mathcal{D}$ be a probability ensemble. Then C is* weakly predictable on $\mathcal{D}$ with membership queries in polynomial time *if there exists a constant $c > 0$ and a polynomial-time pwm-algorithm A such that for all positive integers s and n, and for all concept names u of length s, A $(\frac{1}{2} - \frac{1}{(sn)^c})$-predicts $\kappa_C(u)$ on $\mathcal{D}_n$.*

In the distribution-free model of learning, weak and strong learning are shown to be equivalent in [18]. Based on Yao's XOR lemma [20], a similar result is shown in [5] also for learning on the uniform distribution.

Theorem 1. *[5] For every representation of concepts $C \in \mathcal{P}$ there exists a representation of concepts $C' \leq_{tt}^p C$ such that if C' is weakly predictable on the uniform distribution (with membership queries) in polynomial time, then also C is predictable on the uniform distribution (with membership queries) in polynomial time.*

We also consider predictability with *non-adaptive* membership queries, where by a non-adaptive membership query we mean a query that does not depend on the responses to previous queries. It may, however, still depend on the random coin tosses used by the prediction algorithm. Note that Theorem 1 holds also for non-adaptive membership queries.

Recall that in the pac-learning model the learning algorithm is required to output a concept name which approximates the target well rather than to guess the correct classification of the target by itself. For boolean circuits, however, polynomial-time pac-learnability and predictability coincide. Furthermore, if there exists some representation of concepts $C \in \mathcal{P}$ that is not predictable in polynomial time then boolean circuits are not predictable in polynomial time.

3   Weak Pseudorandom Generators

In this section we introduce our weak version of a pseudorandom function generator. But first let us recall the standard definition. Let $f : \{0,1\}^{l(n)} \times \{0,1\}^n \rightarrow$

$\{0,1\}$ be a polynomial-time computable function ensemble. Note that the polynomial-time computability of f implies that l is polynomially bounded. For a fixed string x of length $l(n)$ we can view $f(x,\cdot)$ as a function $f_x \in \mathcal{B}_n$ that is generated by f on the *seed* x, and therefore we refer to f also as a *function generator*. Now f is a *pseudorandom function generator* if the function f_X produced by f for a random seed $X \in_{\mathcal{U}} \{0,1\}^{l(n)}$ cannot be significantly distinguished from a truely random function $F \in_{\mathcal{U}} \mathcal{B}_n$ by any polynomial-time computation, i.e., for all probabilistic polynomial-time oracle algorithms T, for all positive integers c, and for infinitely many integers n, the *success*

$$\delta(n) = \left| \Pr\left(T^{f_X}(0^n) = 1\right) - \Pr\left(T^F(0^n) = 1\right) \right|$$

of T for f is less than $\frac{1}{n^c}$. The algorithm T is also called a *distinguishing algorithm* or *test* for f. Our definition of a weak pseudorandom function generator is based on the *advantage* of a test T for f (cf. [14]) which is defined with respect to a *fixed* seed $x \in \{0,1\}^{l(n)}$ as

$$\epsilon(x) = \Pr\left(T^{f_x}(0^n) = 1\right) - \Pr\left(T^F(0^n) = 1\right).$$

Note that the success of T for f can be expressed by the average advantage of T for f as $\delta(n) = |\mathsf{E}\left(\epsilon(X)\right)|$ for a random seed $X \in_{\mathcal{U}} \{0,1\}^{l(n)}$. So we refer to $\epsilon(n) = \min\{\epsilon(x) : x \in \{0,1\}^{l(n)}\}$ as the *worst-case advantage* of T for f.

Definition 3. *A* weak pseudorandom function generator *is a polynomial-time computable function ensemble* $f : \{0,1\}^{l(n)} \times \{0,1\}^n \to \{0,1\}$, *such that for all probabilistic polynomial-time oracle algorithms T, for all positive integers c, and for infinitely many integers n, the worst-case advantage of T for f is less than* $\frac{1}{n^c}$.

As a first property let us mention that a weak pseudorandom function generator is still useful for derandomizing probabilistic polynomial-time computations. We omit the proof.

Proposition 1. *If there exists a weak pseudorandom function generator, then for any constant $c \geq 1$, $\mathcal{RP} \subseteq$ io-$\mathcal{H}eur_{n^{-c}}\mathcal{NP}[n^{1/c}]$.*

As opposed to a pseudorandom generator, which has to expand its seed only by at least one bit, we can consider a pseudorandom function generator as a function which expands a seed of polynomial length into a pseudorandom bit-sequence of length 2^n. In the standard setting it is well known that a pseudorandom generator can be used to construct a pseudorandom function generator [8]. It is thus an immediate question whether a similar fact can also be shown with respect to worst-case advantage. Unfortunately, we are not able to answer this question completely. We can show, however, that there exists a weak pseudorandom function generator if there exist weak pseudorandom generators with an arbitrary polynomial expansion.

The proof is based on universal hashing, an ubiquitous tool in cryptography. A *linear hash function* h from $\{0,1\}^n$ to $\{0,1\}^m$ is given by a Boolean $n \times m$

matrix (a_{ij}), (or, equivalently, by a mn-bit string $a_{1,1} \ldots a_{1,m} \cdots a_{n,1} \ldots a_{n,m}$) and maps any string $x = x_1 \ldots x_n$ to a string $y = y_1 \ldots y_m$, where y_i is the inner product $a_i \cdot x = \sum_{j=1}^{n} a_{ij} x_j \pmod 2$ of the i-th row a_i and x. In [6] it is shown that the set of all linear hash functions from $\{0,1\}^n$ to $\{0,1\}^m$ is *universal*, i.e., for all n-bit strings x and y with $x \neq y$, $\Pr(H(x) = H(y)) = \frac{1}{2^m}$, when H is uniformly at random chosen from the set of all linear hash functions from $\{0,1\}^n$ to $\{0,1\}^m$. Now let us say that a hash function h from $\{0,1\}^n$ to $\{0,1\}^m$ *causes a collision* on a set $Q \subseteq \{0,1\}^n$ if there exist two different strings x and y in $\{0,1\}^m$ such that $h(x) = h(y)$. Otherwise, h is said to be *collision-free on Q*. Then a random linear hash function causes a collision on Q with probability at most $\frac{|Q|(|Q|-1)}{m}$.

Lemma 1. *Suppose that there exists a family $\{g_k : k \geq 1\}$ of function ensembles $g_k : \{0,1\}^n \to \{0,1\}^{n^k}$ with the following properties:*

1. *The i-th bit of $g_k(x)$ is computable in time polynomial in $|x|$ and k.*
2. *For all positive integers k and c, for all probabilistic polynomial-time algorithms T, and for infinitely many integers n, there exists some $x \in \{0,1\}^n$ such that for $Y \in_{\mathcal{U}} \{0,1\}^{n^k}$,*

$$\Pr(T(g(x)) = 1) - \Pr(T(Y) = 1) < \frac{1}{n^c}.$$

Then there exists a weak pseudorandom function generator.

Proof. Based on the family $\{g_k : k \geq 1\}$ we define a function generator f as follows. Given binary strings x and z of length n, and further two binary strings k and h of length $\log n$ and $n^2 \log n$, respectively, we think of k as encoding an integer $k \in \{1, \ldots, n\}$, and of h as a linear hash function mapping strings of length n to strings of length $m = \lfloor \log n^k \rfloor$ (by using only the first nm bits of the string h). Then we can interpret the m-bit string $h(z)$ as a positive integer $h(z) \in \{1, \ldots, n^k\}$, and define

$$f(k \circ h \circ x, z) = g_k(x)_{\{h(z)\}},$$

where $g_k(x)_{\{h(z)\}}$ denotes the $h(z)$-th bit of the n^k-bit string $g_k(x)$.

By the computability condition on the family $\{g_k : k \geq 1\}$ it follows that f is polynomial-time computable. To see that f is a weak pseudorandom function generator, assume to the contrary that there exist some $c \geq 1$ and a probabilistic oracle test T whose running-time on input 0^n and with any oracle in $\mathcal{B}_n$ is bounded by n^c, and such that for all (sufficiently large) integers n, the worst-case advantage of T for f is at least $\frac{1}{n^c}$. Note that this implies that for all $k \in \{0,1\}^{\log n}$ and $x \in \{0,1\}^n$, and for random $H \in_{\mathcal{U}} \{0,1\}^{n^2 \log n}$ and $F \in_{\mathcal{U}} \mathcal{B}_n$,

$$\Pr\left(T^{f_{k \circ H \circ x}}(0^n) = 1\right) - \Pr\left(T^F(0^n) = 1\right) \geq \frac{1}{n^c}.$$

Now fix $k = 4c$ and consider the test T' for g_k which on input $y \in \{0,1\}^{n^k}$ works as follows. First choose $h \in_{\mathcal{U}} \{0,1\}^{n^2 \log n}$. Then simulate T on input 0^n, where each query $z \in \{0,1\}^n$ is answered by the $h(z)$-th bit of y. Finally accept if and only if T accepts.

Under the condition that the hash function h chosen by T' is collision-free on the set of queries Q asked by T, the test T' accepts a random $Y \in_\mathcal{U} \{0,1\}^{n^k}$ with the same probability as T accepts with a random oracle $F \in_\mathcal{U} \mathcal{B}_n$. Even though the queries in Q might depend on h for some specific $y \in \{0,1\}^{n^k}$, it is not hard to see that for the random Y and H, the probability that H is collision-free on Q coincides with the probability that H is collision-free on the set of queries asked by T when T is run with the random oracle F. It follows that H causes a collision on Q with probability at most

$$\frac{|Q|^2}{2^m} \leq \frac{n^{2c}}{2^m} < \frac{1}{2n^c},$$

where the last inequality follows by the choice of $k = 4c$ which implies that $2^m \geq n^k/2 = n^{4c}/2 > 2n^{3c}$, and hence we have

$$\left| \Pr\left(T'(Y) = 1\right) - \Pr\left(T^F(0^n) = 1\right) \right| \leq \frac{1}{2n^c}.$$

On the other hand, for all $x \in \{0,1\}^n$, the probability that T' accepts $g_k(x)$ is just the probability that T accepts with oracle $f_{k \circ H \circ x}$ for a random hash function $H \in \{0,1\}^{n^2 \log n}$. It follows that for all $x \in \{0,1\}^n$,

$$\Pr\left(T'(g_k(x)) = 1\right) - \Pr\left(T'(Y) = 1\right) \geq \frac{1}{2n^c}.$$

But this contradicts the pseudorandomness condition of the lemma.   □

4   Weak Pseudorandomness versus Predictability

In this section we show that there exists some non-predictable representation of concepts $C \in \mathcal{P}$ if and only if there exists a weak pseudorandom function generator. The implication from non-predictability to the weak function generator is based on the (well-known) construction of a pseudorandom generator due to Nisan and Wigderson [17].

Definition 4. *A* (r,l,n,k)-*design is a collection* $\mathcal{S} = (S_1,\ldots,S_r)$ *of sets* $S_i \subseteq \{1,\ldots,l\}$, *each of which has cardinality* n, *such that for all* $i \neq j$, $|S_i \cap S_j| \leq k$. *Given a function* $f : \{0,1\}^n \to \{0,1\}$, *the* Nisan-Wigderson *generator (based on* f *and* $\mathcal{S}$), $f^\mathcal{S} : \{0,1\}^l \to \{0,1\}^r$, *is for every seed* $x = x_1 \ldots x_l$ *of length* l *defined as*

$$f^\mathcal{S}(x) = f(x_{S_1}) \ldots f(x_{S_r}),$$

where x_{S_i}, *for* $1 \leq i \leq r$, *denotes the restriction of* x *to* $S_i = \{i_1 < \ldots < i_n\}$ *defined as* $x_{S_i} = x_{i_1} \ldots x_{i_n}$.

It is shown in [17] that for all positive integers r and n we can use polynomials of degree at most $\log r$ over a suitably chosen field to construct a $(r, 4n^2, n, \log r)$-design $\mathcal{S} = (S_1, \ldots, S_r)$ such that each S_i can be computed in time polynomial in n and $\log r$ [17]. In the following, we refer to this design as the *low-degree polynomial design*.

Furthermore, as shown in [17], $f^{\mathcal{S}}$ is a pseudorandom generator with respect to non-uniformly computable distinguishing algorithms, provided that the function f is hard to approximate by polynomial-size circuits. This means that if there is a polynomial-size circuit T with sufficiently large distinguishing success for $f^{\mathcal{S}}$, then there is a polynomial-size circuit T' that approximates f. It is known that in fact T' can be uniformly obtained from T, though at the expense of polynomially many membership queries to f (cf. [2, 9]). By inspecting the proof in, e.g., [9] it is not hard to see that the required membership queries do not depend on the answers to previously asked queries, i.e., T' can be obtained from T by using only non-adaptive queries to f.

Lemma 2. *[17, 2, 9] There exists a probabilistic oracle algorithm A which given as input an integer n, a circuit T with input length r, a rational $\epsilon > 0$, and further access to a function $f : \{0,1\}^n \rightarrow \{0,1\}$ computes with probability at least $1 - \epsilon$ a circuit T' such that for $Z \in_{\mathcal{U}} \{0,1\}^n$,*

$$\mathsf{Pr}\left(T'(Z) = f(Z)\right) \geq \frac{1}{2} + \frac{\delta}{r} - \epsilon,$$

provided that for $X \in_{\mathcal{U}} \{0,1\}^{4n^2}$, $Y \in_{\mathcal{U}} \{0,1\}^r$, and the low-degree polynomial $(r, 4n^2, n, \log r)$-design $\mathcal{S}$,

$$\left|\, \mathsf{Pr}\left(T(f^{\mathcal{S}}(X)) = 1\right) - \mathsf{Pr}\left(T(Y) = 1\right)\, \right| \geq \delta.$$

Moreover, A runs in time polynomial in n, $|T|$ and $1/\epsilon$, and A asks only non-adaptive oracle queries to f.

Now, based on Lemmas 1 and 2 we can construct a weak pseudorandom function generator under non-predictability assumption.

Theorem 2. *If there exists a representation of concepts $C \in \mathcal{P}$ that is not weakly predictable with non-adaptive membership queries on the uniform distribution in polynomial time, then there exists a polynomial-time computable weak pseudorandom function generator.*

Proof. Let $l(n) = 4n^2$. W.l.o.g. we assume that also the $l(n)$-size bounded restriction of C defined as $C_{l(n)} = \{(u, x) \in C : |u| = l(|x|)\}$ is not weakly predictable with non-adaptive membership queries on the uniform distribution in polynomial time. (Otherwise this can be achieved by a simple padding argument.) Based on $C_{l(n)}$ we define a family $\{g_k : k \geq 1\}$ of function ensembles $g_k : \{0,1\}^{2l(n)} \rightarrow \{0,1\}^{n^k}$ satisfying the conditions of Lemma 1. The theorem will follow.

For all positive integers n and k with $k \leq n$, and for all binary strings u and x of length $l(n)$ define
$$g_k(u \circ x) = u_n^{\mathcal{S}}(x),$$
where u_n denotes the characteristic function of the set $\kappa_C(u) \cap \{0,1\}^n$, and $\mathcal{S} = (S_1, \ldots, S_{n^k})$ is the low-degree polynomial $(n^k, 4n^2, n, \log n^k)$-design. For $k > n$, put $g_k(u \circ x) = 0^{n^k}$.

Since $C \in \mathcal{P}$, and since each S_i for $i = 1, \ldots, n^k$ can be computed in time polynomial in n and $\log n^k$, it follows that the family $\{g_k : k \geq 1\}$ satisfies the computability condition of Lemma 1.

To see that the family $\{g_k : k \geq 1\}$ also satisfies the pseudorandomness condition of Lemma 1, assume to the contrary that there exist positive integers k and c and a polynomial-time computable test T, such that for all sufficiently large n, for all binary strings u and x of length $l(n)$, and for $Y \in_{\mathcal{U}} \{0,1\}^{n^k}$ it holds that
$$\Pr\left(T(g_k(u \circ x)) = 1\right) - \Pr\left(T(Y) = 1\right) \geq \frac{1}{n^c}.$$

Note that this implies that the test T distinguishes the random variable $g_k(U \circ X)$ with $U \in_{\mathcal{U}} \{0,1\}^{l(n)}$ and $X \in_{\mathcal{U}} \{0,1\}^{l(n)}$ from a uniformly distributed $Y \in_{\mathcal{U}} \{0,1\}^{n^k}$ by at least $\frac{1}{n^c}$.

Now consider the pwm-algorithm A for $C_{l(n)}$, which on input $s = l(n)$ and n and with respect to some concept $\kappa_C(u)$ with $|u| = l(n)$ works as follows. First obtain a probabilistic circuit T_{n^k} that computes T on input length n^k by using a (finite) description of the Turing machine computing T. Then run the algorithm of Lemma 2 with the circuit T_{n^k}, oracle u_n and parameter $\epsilon = \frac{1}{4n^{k+c}}$ to obtain a circuit T'. Finally request an element z to predict, and answer with $T'(z)$.

By the assumption on the test T, the algorithm of Lemma 2 produces, for all n and concept names u of length $l(n)$, and with probability at least $1 - \epsilon$, a circuit T' satisfying

$$\Pr\left(T'(Z) = u_n(Z)\right) \geq \frac{1}{2} + \frac{1}{n^{c+k}} - \epsilon$$

for $Z \in_{\mathcal{U}} \{0,1\}^n$. This implies that A's guess for the classification of Z by $\kappa_C(u)$ is correct with probability at least

$$\frac{1}{2} + \frac{1}{n^{c+k}} - 2\epsilon = \frac{1}{2} + \frac{1}{2n^{c+k}}.$$

Further note that A makes no additional membership queries except the non-adaptive queries to $\kappa_C(u)$ required to produce the circuit T'. Thus, $C_{l(n)}$ is weakly predictable with non-adaptive membership queries on the uniform distribution in polynomial time. This, however, contradicts our assumption on C, and hence, g_k is a weak pseudorandom generator for all $k \geq 1$. The theorem now follows from Lemma 1.   □

We now proceed to show the converse of Theorem 2 for any p-samplable distribution, provided that the distribution in question does not immediately imply a trivial prediction algorithm. Kharitonov [12] showed that any representation of concepts C is polynomially weakly predictable on any distribution (ensemble) $\mathcal{D} = \{\mathcal{D}_n\}$ with Renyi entropy $O(\log n)$. This motivates the following definition.

Definition 5. *[12] A distribution ensemble $\mathcal{D} = \{\mathcal{D}_n\}$ is* trivial *if for all $n \geq 1$, $\mathcal{D}_n$ has Renyi entropy $O(\log n)$.*

A function generator $f : \{0,1\}^{l(n)} \times \{0,1\}^n \to \{0,1\}$ is associated in a natural way with the representation of concepts $C^f = \{(x,z) : |x| = l(|z|), f(x,z) = 1\}$. Note that for a string x of length $l(n)$, f_x is just the characteristic function of the set $\kappa_C(x) \cap \{0,1\}^n$.

Theorem 3. *If $f : \{0,1\}^{l(n)} \times \{0,1\}^n \to \{0,1\}$ is a weak pseudorandom function generator, then C^f is not weakly predictable with membership queries on any non-trivial p-samplable distribution in polynomial time.*

Proof. Assume to the contrary that there exists a non-trivial p-samplable distribution $\mathcal{D}$ such that C^f is weakly predictable with membership queries on $\mathcal{D}$ in polynomial time, and let A be a pwm-algorithm and c be a positive integer such that for all $n \geq 1$, the running-time of A on inputs $s = l(n)$ and n is bounded by n^c, and such that for all strings x of length $l(n)$, A ϵ-predicts $\kappa_{C^f}(x)$ on $\mathcal{D}_n$ with $\epsilon = \frac{1}{2} - \frac{1}{n^c}$.

We now use A to construct a test T for f. Given input 0^n and an oracle $h : \{0,1\}^n \to \{0,1\}$, the test T simulates A with inputs n and $s = l(n)$. When A makes a membership query z, then T answers this query with $h(z)$. When A requests a random classified example or an element to predict, then T chooses a string $z \in_{\mathcal{D}_n} \{0,1\}^n$ and returns either the example $(z, h(z))$ or the prediction challenge z to A. Finally, T accepts if and only if the guess of A on the prediction challenge z coincides with $h(z)$.

Obviously, for all positive integers n and for all strings x of length $l(n)$, the test T accepts the oracle f_x with probability at least $\frac{1}{2} + \frac{1}{n^c}$. If on the other hand, the oracle for T is a random function $F \in_{\mathcal{U}} \mathcal{B}_n$, then there are two possibilities. If the prediction challenge z coincides with some previously seen labeled sample, then we have to assume that A's guess for z is correct. Since the running time of A, and hence the number of labeled samples it obtains (via example or membership queries) is bounded by n^c, and since $\mathcal{D}$ is a non-trivial distribution, this can happen for a random prediction challenge $Z \in_{\mathcal{D}_n} \{0,1\}^n$ only with probability at most $n^c \cdot 2^{-\text{ent}_{min}(\mathcal{D}_n)} \leq n^c \cdot 2^{-\text{ent}_{Ren}(\mathcal{D}_n)/2} < \frac{1}{2n^c}$. Otherwise, i.e. in the case where all previously seen examples are different from the prediction challenge z, the probability that A classifies z correctly is exactly $\frac{1}{2}$. It follows that T accepts the random oracle F with probability at most $\frac{1}{2} + \frac{1}{2n^c}$. Thus, for all strings x of length $l(n)$,

$$\Pr\left(T^{f_x}(0^n) = 1\right) - \Pr\left(T^F(0^n) = 1\right) \geq \frac{1}{2n^c},$$

contradicting the assumption that f is a weak pseudorandom function generator.
$\square$

Since Boolean circuits are not pac-learnable if and only if there exists a non-predictable representation of concepts $C \in \mathcal{P}$, we can combine Theorems 1, 2 and 3 to get the following corollary.

Corollary 1. *The following are equivalent.*

1. *There exists a weak pseudorandom function generator.*
2. *Boolean circuits are not pac-learnable with non-adaptive membership queries on the uniform distribution in polynomial time.*
3. *Boolean circuits are not pac-learnable with membership queries on any non-trivial p-samplable distribution in polynomial time.*

As an application of Corollary 1 to the theory of resource-bounded measure [16], let us mention that it is shown in [15] that if $\mathcal{P}/poly$ does not have p_2-measure zero , and furthermore $\mathcal{EXP} \neq \mathcal{MA}$, then Boolean circuits are not pac-learnable with non-adaptive queries on the uniform distribution in polynomial-time. By Corollary 1, this can be extended to pac-learnability with (adaptive) membership queries on arbitrary non-trivial p-samplable distributions.

Furthermore, by Proposition 1, the non-learnability of Boolean circuits with membership queries on any non-trivial p-samplable distribution implies that for any constant $c \geq 1$, $\mathcal{RP} \subseteq io\text{-}\mathcal{H}eur_{n-c}\mathcal{NP}[n^{1/c}]$.

Finally let us remark that the existence of a weak pseudorandom function generator can be expressed also in terms of (standard) success instead of worst-case advantage. But then we need to consider non-uniformly computable function generators and, moreover, the function generator might depend on the (still) uniformly computable distinguishing algorithm. More formally, we say that a test T *breaks* a function generator f if for some positive integer c and for all sufficiently large n, the success of T for f is at least $1/n^c$. Now a slight modification of the proof of Theorem 2 together with Theorem 3 yields the following equivalence.

Corollary 2. *The following are equivalent.*

1. *There exists a weak pseudorandom function generator.*
2. *There is some polynomial size-bound $s(n)$ such that no probabilistic polynomial-time computable test can break every function generator computable by circuits of size $s(n)$.*

References

[1] D. Angluin and M. Kharitonov. When won't membership queries help? In *Proc. 23rd ACM Symposium on Theory of Computing*, pages 444–454. ACM Press, 1991.

[2] A. Blum, M. Furst, M. Kearns, and R. J. Lipton. Cryptographic primitives based on hard learning problems. In *Proc. CRYPTO 93*, volume 773 of *Lecture Notes in Computer Science*, pages 278–291. Springer-Verlag, 1994.

[3] A. L. Blum. Separating distribution-free and mistake-bound learning models over the boolean domain. In *Proc. 31st IEEE Symposium on the Foundations of Computer Science*, pages 211–218. IEEE Computer Society Press, 1990.

[4] A. Blumer, A. Ehrenfeucht, D. Haussler, and M. K. Warmuth. Occam's razor. *Information Processing Letters*, 24(6):377–380, 1987.

[5] D. Boneh and R. J. Lipton. Amplification of weak learning under the uniform distribution. In *Proc. 6th ACM Conference on Computational Learning Theory*, pages 347–351. ACM Press, 1993.

[6] J. L. Carter and M. N. Wegman. Universal classes of hash functions. *Journal of Computer and System Sciences*, 18:143–154, 1979.

[7] J. Díaz and J. Torán. Classes of bounded nondeterminism. *Mathematical Systems Theory*, 23(1):21–32, 1990.

[8] O. Goldreich, S. Goldwasser, and S. Micali. How to construct random functions. *Journal of the ACM*, 33(4):792–807, 1986.

[9] R. Impagliazzo and A. Wigderson. Randomness vs. time: De-randomization under a uniform assumption. In *Proc. 39th IEEE Symposium on the Foundations of Computer Science*, pages 734–743. IEEE Computer Society Press, 1998.

[10] J. C. Jackson. An efficient membership-query algorithm for learning DNF with respect to the uniform distribution. *Journal of Computer and System Sciences*, 55(3):414–440, 1997.

[11] M. J. Kearns and L. G. Valiant. Cryptographic limitations on learning boolean formulae and finite automata. *Journal of the ACM*, 41:67–95, 1994.

[12] M. Kharitonov. Cryptographic hardness of distribution-specific learning. In *Proc. 25th ACM Symposium on Theory of Computing*, pages 372–381, 1993.

[13] M. Kharitonov. Cryptographic lower bounds for learnability of boolean functions on the uniform distribution. *Journal of Computer and System Sciences*, 50:600–610, 1995.

[14] M. Krause and S. Lucks. On learning versus distinguishing and the minimal hardware complexity of pseudorandom function generators. Technical Report TR00-014, Electronic Colloquium on Computational Complexity, 2000.

[15] W. Lindner, R. Schuler, and O. Watanabe. Resource-bounded measure and learnability. In *Proc. 13th Annual IEEE Conference on Computational Complexity*, pages 261–270, 1998.

[16] J. H. Lutz. Almost everywhere high nonuniform complexity. *Journal of Computer and System Sciences*, 44:220–258, 1992.

[17] N. Nisan and A. Wigderson. Hardness vs randomness. *Journal of Computer and System Sciences*, 49:149–167, 1994.

[18] R. E. Schapire. The strength of weak learnability. *Machine Learning*, 5(2):197–226, 1990.

[19] L. G. Valiant. A theory of the learnable. *Communications of the ACM*, 27(11):1134–1142, 1984.

[20] A. C. Yao. Theory and applications of trapdoor functions. In *Proc. 23rd IEEE Symposium on the Foundations of Computer Science*, pages 80–91. IEEE Computer Society Press, 1982.

Θ_2^p-Completeness: A Classical Approach for New Results*

Holger Spakowski[1] and Jörg Vogel[2]

[1] Ernst Moritz Arndt University, Dept. of Math. and Computer Science
D-17487 GREIFSWALD, Germany
spakow@mail.uni-greifswald.de
[2] Friedrich Schiller University, Computer Science Institute
D-07740 JENA, Germany
vogel@minet.uni-jena.de

Abstract. In this paper we present an approach for proving Θ_2^p-completeness. There are several papers in which different problems of logic, of combinatorics, and of approximation are stated to be complete for parallel access to NP, i.e. Θ_2^p-complete.

There is a special acceptance concept for nondeterministic Turing machines which allows a characterization of Θ_2^p as a polynomial-time bounded class.

This characterization is the starting point of this paper. It makes a master reduction from that type of Turing machines to suitable boolean formula problems possible. From the reductions we deduce a couple of conditions that are sufficient for proving Θ_2^p-hardness. These new conditions are applicable in a canonical way. Thus we are able to do the following: (i) we can prove the Θ_2^p-completeness for different combinatorial problems (e.g. max-card-clique compare) as well as for optimization problems (e.g. the Kemeny voting scheme), (ii) we can simplify known proofs for Θ_2^p-completeness (e.g. for the Dodgson voting scheme), and (iii) we can transfer this technique for proving Δ_2^p-completeness (e.g. TSPcompare).

1 Introduction

The complexity class $\Theta_2^p = \mathrm{P}^{\mathrm{NP}[\log]}$ was established as a constitutional level of the polynomial time hierarchy, e.g. Wagner in [Wag90] proved that $\Theta_2^p = \mathrm{L}^{\mathrm{NP}}$. Further characterizations of this complexity class are given by several authors: $\Theta_2^p = \mathrm{P}^{\mathrm{NP}[\log]} = \mathrm{L}^{\mathrm{NP}} = \mathcal{R}_{tt}^p(\mathrm{NP}) = \mathrm{P}_{||}^{\mathrm{NP}}$.

Krentel [Kre88] has stated a characterization of the complexity class $\Delta_2^p = \mathrm{P}^{\mathrm{NP}}$ as a polynomial-time bounded class by the so called MAX-acceptance concept: Given a nondeterministic polynomial-time bounded Turing machine with output device M and an input x, then M accepts x in the sense of MAX iff any computation path with (quasilexicographically) maximum output accepts x.

* Supported in part by grant NSF-INT-9815095/DAAD-315-PPP-gü-ab.

S. Kapoor and S. Prasad (Eds.): FST TCS 2000, LNCS 1974, pp. 348–360, 2000.

This paper starts with a characterization of Θ_2^p by the following acceptance concept, introduced in Spakowski/Vogel [SV99] :

Let M be a nondeterministic polynomially bounded Turing machine with output device and let x be an input. M accepts x in the sense of MAX-CH iff x is accepted on any computation path β of M on x with maximal number of mind-changes in the output. For $w \in \{0,1\}^*$, ch(w) denotes the number of mind-changes in w. It holds ch$(0) =$ ch$(1) =$ ch$(00) =$ ch$(11) = 0$, ch$(10) =$ ch$(01) = 1$ and e.g. ch$(10010) = 3$, ch$(10101) = 4$.

This concept means that the internal structure of the output is essential. It allows a characterization of Θ_2^p as a polynomial-time bounded complexity class: $\Theta_2^p =$ MAX-CH-P — which is in some sense analogous to $\Delta_2^p =$ MAX-P.

The theory of NP-completeness was initiated by Cook's master reduction [Cook71], continued by Karp's basic NP-complete problems [Karp72] and established by Garey/Johnson's guide to the the theory of NP-completeness [GJ79]. The aim of this paper is to give evidence that the theory of the class NP can be transcribed to the class Θ_2^p, and thus the theory of Θ_2^p-completeness becomes very canonical. We start our approach to that theory with the description of Θ_2^p as a polynomial-time bounded class. That allows the construction of "master reductions" even to two different basic satisfiability problems of boolean formulas:

MAX-TRUE-3SAT-COMPARE:

Given two 3-CNF formulas F_1 and F_2; is the maximum number of 1's in satisfying truth assignments for F_1 less than or equal to that for F_2 ?

and

ODD-MAX-TRUE-3SAT:

Given a 3-CNF formula F; is the maximum number of 1's in satisfying truth assignments for F odd ?

Of course, Wagner [Wag87] has provided a useful tool for proving Θ_2^p-hardness, and we state his result below as lemma 7. However, the "master reductions" stated in sect. 2.1 provide two conditions for proving Θ_2^p-hardness (lemma 3 and lemma 5) such that (first) Wagner's condition is a consequence of our condition and (second and more important) our condition is relatively simple to apply because we can make use of the classical constructions (see section 3).

Section 3.1 summarizes the results for some basic combinatorial problems like min-card-vertex cover compare and max-card-clique compare (given two graphs we compare the sizes of the smallest vertex covers and largest cliques, respectively). The completeness of these problems in Θ_2^p is a further argument for establishing this class.

In section 3.2 we re-translate our method for the class Δ_2^p, and we are able to prove that TSPcompare (given two instances of traveling-salesperson we ask if the optimal tour-length for the first TSP instance is shorter than that for the second instance) is complete for Δ_2^p— supplementing the list of Δ_2^p-complete problems given by Krentel.

Of special interest are the applications mentioned in section 3.3. Voting schemes are very well studied in the social choice literature. Bartholdi/Tovey/Trick [BTT89] investigated the computational complexity of such problems. They proved that the Dodgson voting scheme as well as the Kemeny voting scheme both are NP-hard. Hemaspaandra/Hemaspaandra/Rothe [HHR97] proved that Dodgson voting is Θ_2^p-complete using Wagner's lemma. The exact analysis of the Kemeny voting system was still an open problem.

We are able, following our method, to give a simplification of the involved proof of [HHR97] as well as to prove that Kemeny voting is Θ_2^p-complete. The result for Kemeny voting is also stated in a survey paper of Hemaspaandra/Hemaspaandra presented at MFCS 2000 very recently [HH00].

For the definitions and basic concepts we refer to the textbook [Pap94].

2   The Machine Based Technique

2.1   Basic Problems Being Complete for Θ_2^p

We gave in [SV99] a characterization of the complexity class Θ_2^p by the so called MAX-CH acceptance concept:

Given a nondeterministic polynomial-time bounded Turing machine M with output device and an input x, then M accepts x in the sense of MAX-CH iff on any computation path β of M on x with maximum number of mind-changes of the output x is accepted. It turns out that $\Theta_2^p = $ MAX-CH-P, where MAX-CH-P is the class of all sets decidable in the sense of MAX-CH by polynomial-time bounded machines.

A slight modification of these machines yields the following lemma:

Lemma 1. *For every $A \in \Theta_2^p$ there are a NPTM M with output device and polynomials p and q such that the following is true:*

1. *For any input x the output on every path β is of equal length $q(|x|)$.*
2. *For any input x every computation path β is of equal length $p(|x|)$.*
3. *For any input x, two paths β_1 and β_2 have the same acception behaviour whenever they have the same number of 1's in the output.*
4. *$x \in A$ if and only if M accepts x on β_{max}, where β_{max} is a computation path having the maximum number of 1's in the output.*

We call this acceptance concept MAX-1-acceptance in difference to Krentel [Kre88], who defined the so called MAX-acceptance.

We define the following three satisfiability problems for boolean formulas:

Decision Problem: MAX-TRUE-3SAT-COMPARE
Instance: Two 3-CNF formulas F_1 and F_2 having the same number of clauses and variables
Question: Is the maximum number of 1's in satisfying truth assignments for F_1 less than or equal to that for F_2 ?

Decision Problem: MAX-TRUE-3SAT-EQUALITY
Instance: Two 3-CNF formulas F_1 and F_2 having the same number of clauses and variables

Question: Is the maximum number of 1's in satisfying truth assignments for F_1 equal to that for F_2 ?

Decision Problem: ODD-MAX-TRUE-3SAT
Instance: A 3-CNF formula F
Question: Is the max number of 1's in satisf. truth assignments for F odd ?

In all cases it is straightforward to prove that the problem is in Θ_2^p using binary search. We concentrate on proving the hardness.

Theorem 2. MAX-TRUE-3SAT-COMPARE *and* MAX-TRUE-3SAT-EQUALITY *are complete in Θ_2^p under polynomial time many-one reduction.*

To show the hardness we need the following lemma.

Lemma 3. *For every $A \in \Theta_2^p$ there are $B_1, B_2 \in P$ having the following properties:*

1. $(x \in A \longrightarrow m_1(x) = m_2(x))$ and $(x \notin A \longrightarrow m_1(x) > m_2(x))$, where
   $m_1(x) =_{df} \max \{[w]_1 : \langle x, w \rangle \in B_1\}^1$ and
   $m_2(x) =_{df} \max \{[w]_1 : \langle x, w \rangle \in B_2\}$.
2. $x \in A \longrightarrow M_1(x) = M_2(x)$ where
   $M_1(x) =_{df} \{w : \langle x, w \rangle \in B_1 \wedge [w]_1 = m_1(x)\}$ and
   $M_2(x) =_{df} \{w : \langle x, w \rangle \in B_2 \wedge [w]_1 = m_2(x)\}$
3. *There is a pol. $\tilde{p}$ such that $\bigwedge_{x,w \in \Sigma^*} (\langle x, w \rangle \in B_i \longrightarrow |w| = \tilde{p}(|x|))$ $(i \in \{1, 2\})$*

Proof. We start from the characterization of Θ_2^p-sets given in lemma 1. Let $A \in \Theta_2^p$, M a NPTM deciding A in the sense of MAX-1, q and p' the polynomials determining the length of the output and the length of the computation paths of M, respectively. For each n let $p(n) =_{df} p'(n) + 1$.
We define B_1 and B_2:

$B_1 =_{df} \{\langle x, w \rangle : x, w \in \Sigma^* \text{ and}$
1. w has the form $\text{out}_M(x, \beta)^{p(|x|)}\beta$ and
2. The NPTM M has on input x on path β the output $\text{out}_M(x, \beta)$ $\}$

$B_2 =_{df} \{\langle x, w \rangle : x, w \in \Sigma^* \text{ and}$
1. w has the form $\text{out}_M(x, \beta)^{p(|x|)}\beta$ and
2. The NPTM M has on input x on path β the output $\text{out}_M(x, \beta)$ and
3. M accepts x on β $\}$

For any x let the set $B_{max}(x)$ contain all computation paths β of M on input x with the maximum number of 1's in the output, and let $\hat{B}_{max}(x)$ contain all paths from B_{max} having itself the maximum number of 1's.
$B_{max-acc}(x)$ and $\hat{B}_{max-acc}$ differ from $B_{max}(x)$ and $\hat{B}_{max}$ in that only accepting computation paths are considered.

---
[1] $[w]_1$ denotes the number of 1's in w.

It's not hard to see that

$$M_1(x) = \left\{ \text{out}_M(x, \beta')^{p(|x|)} \beta' : \beta' \in \hat{B}_{max}(x) \right\}, and \tag{1}$$

$$M_2(x) = \left\{ \text{out}_M(x, \beta')^{p(|x|)} \beta' : \beta' \in \hat{B}_{max-acc}(x) \right\} . \tag{2}$$

Now we can sketch the proofs of statements 1 and 2 of the lemma.

Let $x \in A$. Then M accepts x on all paths β from $B_{max}(x)$. Hence $B_{max-acc}(x) = B_{max}(x)$ and $\hat{B}_{max-acc}(x) = \hat{B}_{max}(x)$. Due to (1) and (2) follow $M_1(x) = M_2(x)$ and $m_1(x) = m_2(x)$.

Let $x \notin A$. Then M rejects x on all paths β from $B_{max}(x)$. Let $\hat{\beta}_{max}(x) \in \hat{B}_{max}(x)$ and $\hat{\beta}_{max-acc}(x) \in \hat{B}_{max-acc}(x)$.

Then

$$\left[\text{out}_M(x, \hat{\beta}_{max}(x)) \right]_1 > \left[\text{out}_M(x, \hat{\beta}_{max-acc}(x)) \right]_1 .$$

Therefore

$$m_1(x) \overset{eq.(1)}{=} \left[\text{out}_M \left(x, \hat{\beta}_{max}(x) \right)^{p(|x|)} \hat{\beta}_{max}(x) \right]_1$$

$$> \left[\text{out}_M \left(x, \hat{\beta}_{max-acc}(x) \right)^{p(|x|)} \hat{\beta}_{max-acc}(x) \right]_1 \overset{eq.(2)}{=} m_2(x) .$$

Proof of theorem 2. Let $A \in \Theta_2^p$ and B_1, B_2 the P-sets and $\tilde{p}$ a polynomial belonging to A having the properties of lemma 3. For a given $x \in \Sigma^*$ we devise two 3CNF-formulas $\hat{F}_1$ and $\hat{F}_2$ such that

$$M_1(x) = M_2(x) \longrightarrow \max\left\{ [\overline{y}]_1 : \hat{F}_1(\overline{y}) \right\} = \max\left\{ [\overline{y}]_1 : \hat{F}_2(\overline{y}) \right\}^2, and$$

$$m_1(x) > m_2(x) \longrightarrow \max\left\{ [\overline{y}]_1 : \hat{F}_1(\overline{y}) \right\} > \max\left\{ [\overline{y}]_1 : \hat{F}_2(\overline{y}) \right\} .$$

Our first step is to construct a boolean circuit C with two output gates C^1 and C^2 of polynomial size having the property

$$C^i(w) = 1 \longleftrightarrow \langle x, w \rangle \in B_i \qquad (i \in \{1, 2\}) :$$

C^1 yields 1 iff $\langle x, y_1, \ldots, y_{\tilde{p}(|x|)} \rangle \in B_1$

C^2 yields 1 iff $\langle x, y_1, \ldots, y_{\tilde{p}(|x|)} \rangle \in B_2$

---

[2] $\overline{y}$ denotes a sequence $(y_1, y_2, \ldots)$.

Using standard techniques (see e.g. [Pap94]) we build 3CNF-formulas $F_1(y_1,$ $\ldots, y_{\tilde{p}(|x|)}, h_1, \ldots, h_m)$ and $F_2(y_1, \ldots, y_{\tilde{p}(|x|)}, h_1, \ldots, h_m)$ such that

$$C^i(y_1, \ldots, y_{\tilde{p}(|x|)}) = 1 \longleftrightarrow \bigvee_{h_1, \ldots, h_m} F_i(y_1, \ldots, y_{\tilde{p}(|x|)}, h_1, \ldots, h_m)$$

$$\longleftrightarrow \bigvee_{h_1!, \ldots, h_m!} F_i(y_1, \ldots, y_{\tilde{p}(|x|)}, h_1, \ldots, h_m) \quad (i \in \{1, 2\})$$

and

$$C^1(y_1, \ldots, y_{\tilde{p}(|x|)}) = 1 \quad \wedge \quad C^2(y_1, \ldots, y_{\tilde{p}(|x|)}) = 1$$

$$\longleftrightarrow \tag{3}$$

$$\bigvee_{h_1!, \ldots, h_m!} \left(F_1(y_1, \ldots, y_{\tilde{p}(|x|)}, h_1, \ldots, h_m) \quad \wedge \quad F_2(y_1, \ldots, y_{\tilde{p}(|x|)}, h_1, \ldots, h_m) \right)$$

We define $\hat{F}$ to be equivalent to F, but with each variable y_i replicated $3m$ times (by adding clauses of the form $y_{i,1} \leftrightarrow y_{i,j}$ for each $j = 2$ to $3m$). This expansion is needed to pad out the number of 1's to maintain the required inequality. Note that

$$\max\left\{ [\overline{y}]_1 : \hat{F}_i(\overline{y}) \right\} = 3m \max\{ [\overline{y}]_1 : C_i(\overline{y}) = 1\} + \delta_i \quad (i \in \{1, 2\}) \tag{4}$$

for a $\delta_i \in \{0, \ldots, m\}$. Hence

$$m_1(x) > m_2(x) \longrightarrow \max\left\{ [\overline{y}]_1 : \hat{F}_1(\overline{y}) = 1 \right\} > \max\left\{ [\overline{y}]_1 : \hat{F}_2(\overline{y}) \right\},$$

and due to (3)

$$M_1(x) = M_2(x) \longrightarrow \max\left\{ [\overline{y}]_1 : \hat{F}_1(\overline{y}) \right\} = \max\left\{ [\overline{y}]_1 : \hat{F}_2(\overline{y}) \right\} . \tag{5}$$

$\square$

Theorem 4. ODD-MAX-TRUE-3SAT *is complete in Θ_2^p under polynomial time many-one reduction.*

For proving the hardness we need the following lemma.

Lemma 5. *For every $A \in \Theta_2^p$ there is $B \in P$ having the following properties:*

1. $x \in A \longleftrightarrow \max\{ [w]_1 : \langle x, w \rangle \in B \} \equiv 1(2)$
2. *There is a polynomial $\tilde{p}$ such that $\bigwedge_{x, w \in \Sigma^*} (\langle x, w \rangle \in B \longrightarrow |w| = \tilde{p}(|x|))$.*

Proof. In place of B_1 and B_2 defined in the proof of lemma 3 we define a single set B here:

$B =_{df} \{\langle x, w \rangle : x, w \in \Sigma^*$ and

1. w has the form $\text{out}_M(x, \beta)^{2p(|x|)} \beta^2 \text{acc}_M(x, \beta)$ and
2. The NPTM M has on input x on path β the output $\text{out}_M(x, \beta)$ }
3. $\text{acc}_M(x, \beta) = \begin{cases} 0 \text{ if } M \text{ rejects } x \text{ on } \beta \\ 1 \text{ if } M \text{ accepts } x \text{ on } \beta \end{cases}$

It's easy to see that

$$\max\{ [y]_1 : \langle x, y \rangle \in B \} = \left[\text{out}_M\left(x, \hat{\beta}_{max}(x)\right)^{2p(|x|)} \hat{\beta}_{max}^2(x) \text{acc}_M(x, \hat{\beta}_{max}(x)) \right]_1 .$$

We conclude:

$$\tag{6}$$

$$x \in A \Longleftrightarrow \text{acc}_M\left(x, \hat{\beta}_{max}(x)\right) = 1 \overset{eq.(6)}{\Longleftrightarrow} \max\{ [y]_1 : \langle x, y \rangle \in B \} \equiv 1(2) . \quad \square$$

The proof of theorem 4 follows the same ideas as the proof of theorem 2, but it makes use of lemma 5 instead of lemma 3.

2.2   Sufficient Conditions for Θ_2^p-Hardness

Wagner [Wag87] stated a sufficient condition for a set to be hard for Θ_2^p. It was applied subsequently in a number of papers to prove the Θ_2^p-hardness of various problems. Lemma 5 from section 2.1 implies immediately another sufficient condition for Θ_2^p-hardness which is given below in lemma 6. We show that Wagner's statement follows easily from our's. Thus we have evidence that lemma 6 is at least as strong as Wagner's condition.

Lemma 6. *A set $A \subseteq \Sigma^*$ is Θ_2^p-hard if the following property holds:*

$$(**) \quad \bigwedge_{B \in P} \bigwedge_{p \in Pol} \bigvee_{g \in FP} \bigwedge_{x \in \Sigma^*} (\max\{[y]_1 : |y| = p(|x|) \wedge \langle x, y \rangle \in B\} \equiv 1(2) \longleftrightarrow g(x) \in A)$$

$\square$

Lemma 7. [Wagner 1987][3] *A set $A \subseteq \Sigma^*$ is Θ_2^p-hard if the following property holds:*

(*) *There exists a polynomial-time computable function f and an NP-complete set D such that*

$$\|\{i : x_i \in D\}\| \equiv 1(2) \longleftrightarrow f(x_1, \dots, x_{2k}) \in A$$

for all $k \geq 1$ and all strings $x_1, \dots, x_{2k} \in \Sigma^*$ satisfying
$\chi_D(x_1) \geq \chi_D(x_2) \geq \dots \geq \chi_D(x_{2k})$.[4]

$\square$

Let A be an arbitrary set satisfying condition (*). We show that A satisfies (**) as well.
Let $B \in P$ and $p \in Pol$.
We define

$$E =_{df} \left\{ \langle x, y \rangle : \bigvee_{y', |y'|=p(|x|)} ([y']_1 \geq y \wedge \langle x, y' \rangle \in B) \right\} .$$

For a given $x \in \Sigma^*$ we have to distinguish two cases.
We prove here case 1: $p(|x|)$ is odd.
We set

$$x_1 = \langle x, 0 \rangle, x_2 = \langle x, 1 \rangle, x_3 = \langle x, 2 \rangle, \dots, x_{p(|x|)+1} = \langle x, p(|x|) \rangle .$$

Thus

$$\chi_E(x_1) \geq \chi_E(x_2) \geq \dots \geq \chi_E(x_{p(|x|)+1})$$

and $E \in NP$.
Let h be a polynomial-time reduction from E to the NP-complete set D. Then

$$\|\{i : x_i \in E\}\| \equiv 1(2) \longleftrightarrow \|\{i : h(x_i) \in D\}\| \equiv 1(2)$$
$$\longleftrightarrow f(h(x_1), \dots, h(x_{p(|x|)+1})) \in A$$
$$\longleftrightarrow g'(x_1, \dots, x_{p(|x|)+1}) \in A$$

---

[3] Wagner states hardness for P_{bf}^{NP}, a class which is now known to be equal to Θ_2^p.
[4] χ_D denotes the characteristic function of D.

for g' being the composition of h and f. Hence there is a $g \in$ FP such that

$$\|\{i : x_i \in E\}\| \equiv 1(2) \longleftrightarrow g(x) \in A \ .$$

To complete the proof note that

$$\|\{i : x_i \in E\}\| = \max\{[y]_1 : |y| = p(|x|) \wedge \langle x, y \rangle \in B\} \ . \qquad \square$$

If we apply the same reasoning steps to lemma 3 instead of lemma 5, we will get the following two lemmas.

Lemma 8. *A set $A \subseteq \Sigma^*$ is Θ_2^p-hard if the following property holds:*

$$\bigwedge_{B_1, B_2 \in P} \bigwedge_{p \in Pol} \bigvee_{g \in FP} \bigwedge_{x \in \Sigma^*} (\max\{[y]_1 : |y| = p(|x|) \wedge \langle x, y \rangle \in B_1\}$$

$$\leq \max\{[y]_1 : |y| = p(|x|) \wedge \langle x, y \rangle \in B_2\} \longleftrightarrow g(x) \in A) \qquad \square$$

Lemma 9. *A set $A \subseteq \Sigma^*$ is Θ_2^p-hard if the following property holds: There exist a polynomial-time computable function f and two NP-complete sets D_1 and D_2 such that*

$$\|\{i : x_i \in D_1\}\| \leq \|\{i : x_i \in D_2\}\| \longleftrightarrow f(x_1, \dots, x_{2k}) \in A$$

for all $k \geq 1$ and all strings $x_1, \dots, x_{2k} \in \Sigma^*$ satisfying
$\chi_{D_1}(x_1) \geq \chi_{D_1}(x_2) \geq \dots \geq \chi_{D_1}(x_{2k})$ *and*
$\chi_{D_2}(x_1) \geq \chi_{D_2}(x_2) \geq \dots \geq \chi_{D_2}(x_{2k})$ $\qquad \square$

This lemma is in style similar to Wagner's tool. We have a comparison in place of oddness.

2.3   Remark: We Can Use 2SAT Instead of 3SAT

What about the computational complexity of the "2SAT versions" for our comparison and oddness problems ?
It is well known that the satisfiability problem for 2CNF formulas is solvable in polynomial time. Using binary search it is possible to find the lexicographically largest satisfying assignment of a 2CNF formula in polynomial time. Hence MAX-LEX-2SAT-COMPARE and MAX-LEX-2SAT-EQUALITY[5] are in P.
In contrast to this we state here without proof:

Theorem 10. MAX-TRUE-2SAT-COMPARE,   MAX-TRUE-2SAT-EQUALITY,   *and* ODD-MAX-TRUE-2SAT *are complete in* Θ_2^p *under polynomial time many-one reduction.*

Nevertheless, the reductions presented in the remaining part of the paper are from the 3SAT versions since the 2SAT versions don't make the proofs easier.

---

[5] The definitions of the 3SAT versions are given in section 3.2.

3   Applications of the Method

3.1 Θ_2^p-Complete Combinatorial Problems

We define the following problem:

Decision Problem: min-PolynomiallyWeighted-vertex cover compare
Instance: Two graphs $G_1 = (V_1, E_1, w_1)$ and $G_2 = (V_2, E_2, w_2)$. Each $v \in V_i$ is assigned a weight $w_i(v) \in [0, \|V_i\|]$ $(i \in \{0, 1\})$.
Question: We define for each subset $V' \subseteq V_i$ the weight $w_i(V') = \Sigma_{v \in V'} w_i(v)$. Let $mwvc(G_i)$ be the weight of the vertex cover of G_i having minimum weight. Holds $mwvc(G_1) \leq mwvc(G_2)$?

Theorem 11. *There is a polynomial-time many-one reduction from* MAX-TRUE-3SAT-COMPARE *to* min-PolynomiallyWeighted-vertex cover compare. *Hence* min-PolynomiallyWeighted-vertex cover compare *is* Θ_2^p-*complete.*   □

Proof. Our proof is based on the reduction from 3SAT to vertex cover given in [GJ79]. We will say how the construction there is modified to obtain our result. Assume that we are given C_1 and C_2 both having n variables and m clauses. Our construction is accomplished in two steps:

1. Construct $G_1 = (V_1, E_1, w_1)$ from C_1 and $G_2 = (V_2, E_2, w_2)$ from C_2 as in the reduction from 3SAT to vertex cover. Note that $\|V_1\| = \|V_2\|$.
2. Assign weights to the vertices of G_k $(k \in \{1, 2\})$:

$$w_k(a_1[j]) = w_k(a_2[j]) = w_k(a_3[j]) = 2m + n + 1 \quad (0 \leq j \leq m)$$
$$w_k(u_i) = 2m + n + 1$$
$$w_k(\overline{u}_i) = 2m + n + 2 \quad (0 \leq i \leq n)$$

The weights of the vertices are chosen such that for each vertex cover $V_i' \subseteq V_i$ of G_i holds:

- If $\|V_i'\| = 2m + n$ then V_i' has weight of no more than $(2m + n)(2m + n + 2)$.
- If $\|V_i'\| > 2m + n$ then V_i' has weight of at least $(2m + n + 1)(2m + n + 1) = (2m + n)(2m + n + 2) + 1$.

Hence the vertex covers of G_i with minimum weight are among the vertex covers of G_i with cardinality $2m + n$. As discussed in the classical proof of [GJ79], each vertex cover of cardinality 2m+n defines a satisfying truth assignment. It remains to verify the following two assertions:

1. To each satisfying truth assignment t for U with $\|\{i : t(u_i) = 0\}\| = s$ belongs a vertex cover with weight $(2m+n+1)2m+(2m+n+1)(n-s)+(2m+n+2)s$.
2. To each vertex cover with weight $(2m + n + 1)2m + (2m + n + 1)(n - s) + (2m + n + 2)s$ and $0 \leq s \leq n$ belongs a satisfying truth assignment for U with $\|\{i : t(u_i) = 0\}\| = s$.   □

Decision Problem: `min-card-vertex cover compare`
Instance: Two graphs $G_1 = (V_1, E_1)$ and $G_2 = (V_2, E_2)$.
Question: Let $\kappa_i =_{df} \min\{\|V'\| : V' \subseteq V_i$ and V' is a vertex cover of $G_i\}$
$(i \in \{1, 2\})$. Holds $\kappa_1 \leq \kappa_2$?

Theorem 12. *There is a polynomial-time many-one reduction from* `min-PolynomiallyWeighted-vertex cover compare` *to* `min-card-vertex cover compare`. *Hence* `min-card-vertex cover compare` *is* Θ_2^p-*complete.*

Proof. Let $G = (V, E, w)$ be an arbitrary polynomially vertex-weighted graph as occurring in instances of `min-PolynomiallyWeighted-vertex cover compare`. We obtain a graph $G' = (V', E')$ such that

$$mwvc(G) = \min\{\|V''\| : V'' \subseteq V' \text{ and } V'' \text{ is a cover of } G'\}$$

by defining

$$V' =_{df} \{(u, 1), \ldots, (u, w(u)) : u \in V\}$$

and

$$E' =_{df} \{\{(u, i), (v, j)\} : \{u, v\} \in E \land 1 \leq i \leq w(u) \land 1 \leq j \leq w(v)\} .$$

The transformation from an instance $\langle G_1 = (V_1, E_1, w_1), G_2 = (V_2, E_2, w_2) \rangle$ of `min-PolynomiallyWeighted-vertex cover compare` to `min-card-vertex cover compare` is accomplished by applying this construction to G_1 and G_2. $\square$

Note that `min-card-vertex cover compare` remains Θ_2^p-complete for instances satisfying $\|V_1\| = \|V_2\|$.
Of course, theorem 12 can be stated in terms of `INDEPENDENT SET` and `CLIQUE` as well.

Theorem 13. `max-card-independent set compare` *and* `max-card-clique compare` *are* Θ_2^p-*complete.*

3.2   Transcription to Δ_2^p: TSPcompare and TSPequality

We transcribe the technique used in subsection 2.1 to $\Delta_2^p = \mathrm{P}^{\mathrm{NP}}$.
Decision Problem: `MAX-LEX-3SAT-COMPARE`
Instance: Two 3-CNF formulas F_1 and F_2 having the same number of clauses and variables
Question: Is the lexicographic maximum satisfying truth assignment for F_1 less than or equal to that for F_2 ?

In appropriate manner we define `MAX-LEX-3SAT-EQUALITY`.

Theorem 14. `MAX-LEX-3SAT-COMPARE` *and* `MAX-LEX-3SAT-EQUALITY` *are complete in* Δ_2^p *under polynomial time many-one reduction.*

For proving the hardness we need the following lemma.

Lemma 15. *For every $A \in \Delta_2^p$ there are $B_1, B_2 \in \mathrm{P}$ having the following properties:*

1. $x \in A \longrightarrow \max \{w : \langle x, w \rangle \in B_1\} = \max \{w : \langle x, w \rangle \in B_2\}$
   $x \notin A \longrightarrow \max \{w : \langle x, w \rangle \in B_1\} > \max \{w : \langle x, w \rangle \in B_2\}$

2. There is a pol. $\tilde{p}$ such that $\bigwedge_{x,w \in \Sigma^*} (\langle x, w \rangle \in B_i \longrightarrow |w| = \tilde{p}(|x|))$ $(i \in \{1, 2\})$

Proof. We start from the characterization of Δ_2^p-sets given by Krentel [Kre88] and define B_1 and B_2 as in the proof of lemma 3 where the strings "$\text{out}_M(x, \beta)^{p(|x|)}\beta$" are substituted by "$\text{out}_M(x, \beta)\beta$". $\qquad \square$

In order to complete the proof of theorem 14 the desired 3CNF formulas F_1 and F_2 can be constr. from B_1 and B_2 following the ideas of the proof of theorem 2. We are now able to prove that given two instances of traveling-salesperson, it is Δ_2^p-complete to decide if the optimal tour length for the first TSP instance is not longer than that for the second.

We assume the reader to be familiar with the NP-completeness proof for the Hamilton path problem given by Machtey/Young [MJ78], p. 244ff.

We define the following problems TSPcompare and TSPequality.

Decision Problem: TSPcompare

Instance: Two matrices (M_{ij}^1) and (M_{ij}^2) each consisting of nonnegative integer "distances" between s cities

Question: Let t_k be the length of the optimal tour for M_{ij}^k ($k \in \{1, 2\}$). Holds $t_1 \leq t_2$?

Analogously TSPequality is defined.

Theorem 16. TSPcompare *and* TSPequality *are* Δ_2^p-complete.

Proof. For intermediate steps we need the problems weighted directed Hamilton circuit compare/equality and weighted undirected Hamilton circuit compare/equality, which are defined in an obvious way.

Our reduction chain looks as follows:
MAX-LEX-3SAT-COMPARE/EQUALITY $\leq_m^p$ weighted directed Hamilton circuit compare/equality $\leq_m^p$ weighted undirected Hamilton circuit compare/equality $\leq_m^p$ TSPcompare/equality.

For the reduction from MAX-LEX-3SAT-COMPARE/EQUALITY to weighted directed Hamilton circuit compare/equality consider the reduction from 3SAT to the Hamilton path problem given in [MJ78]. Identifying the vertices v_{n+1} and v_1 we get a reduction from 3SAT to directed Hamilton circuit. Given two 3CNF formulas F_1 and F_2 let G_1/G_2 be the directed graphs belonging to F_2/F_1 according to the construction in the proof there. For each variable x_i there is a corresponding vertex v_i (in both graphs). Each such vertex v_i has two outgoing arcs: the "upper" arc corresponding to occurrences of x_i, the "lower" arc corresponding to occurences of $\neg x_i$. We set the weight of the lower arc of each vertex v_i to 2^{n-i}. All other arcs get weight 0.

The reduction from weighted directed Hamilton circuit compare/equality to weighted undirected Hamilton circuit compare/equality is accomplished by expanding each node into a trio of nodes: for incoming edges, outgoing edges, and a middle node to force us to go from in to out [MJ78].

For the reduction from `weighted undirected Hamilton circuit compare/e-`
`quality` to `TSPcompare/equality`
we set
$$M_{ij}^k =_{df} \begin{cases} w_k(\{i,j\}) & \text{if } \{i,j\} \in E_k \\ 2^n & \text{otherwise} \end{cases}$$

for undirected weighted graphs $G_1 = (V_1, E_1, w_1)$ and $G_2 = (V_2, E_2, w_2)$.   □

3.3   Voting Schemes

A lot of different voting systems are extensively treated in the social choice
literature. For an overview of that field consult e.g. [Fish77].

Bartholdi, Tovey, and Trick [BTT89] initiated the investigation of the computational complexity of voting systems:

For the Dodgson as well as the Kemeny voting systems they proved that it is
NP-hard to determine if a given candidate is a winner of an election under that
voting scheme.

Hemaspaandra/Hemaspaandra/Rothe [HHR97] improve this lower bound for the
Dodgson voting scheme by proving its Θ_2^p-hardness. Together with the almost
trivial Θ_2^p upper bound (holding for both problems) this allows an exact classification of the complexity of Dodgson voting.

A similar result for the Kemeny voting scheme has been unknown for a long
time. Applying theorem 2 we are not only able to settle this question showing
the completeness of the Kemeny voting scheme in Θ_2^p, but also to simplify the
proof given in [HHR97] considerably.

Hemaspaandra/Hemaspaandra [HH00] stated in an invited talk of the MFCS
2000 conference the result for Kemeny's voting scheme presented in a paper by
E. Hemaspaandra [Hem00].

An analysis of Dodgson's and Kemeny's voting systems reveals that the winner-
problems entail in fact a comparison between to instances of optimization problems:

Dodgson's voting scheme: Compare the Dodgson score of a candidate c_1
with the one of a second candidate c_2.

 The Dodgson score of a candidate is the minimum number of switches in
 the voters' preference orders such that this candidate becomes a Condorcet
 winner. See [Con85].

Kemeny's voting scheme: Compare the Kemeny score of a candidate c_1 with
the one of a second candidate c_2.

 The Kemeny score of a candidate c is the sum of the distances of a preference
 order P to the preferences of the voters, where P is a preference order with c
 in first place minimizing this sum. For the definition of the distance measure
 see [BTT89].

Thus both problems are similar in nature to `MAX-TRUE-3SAT-COMPARE` and
`min-card-vertex cover compare`.

A complete version of this topic is given in [SV00].

Acknowledgement   We want to thank the anonymous referees and J. Rothe
for their valuable hints and suggestions.

References

[BTT89] J. Bartholdi III, C.A. Tovey, M.A. Trick. *Voting schemes for which it can be difficult to tell who won the election.* Social Choice and Welfare 6 (1989), 157-165

[Con85] M. Condorcet. *Essai sur l'application de l'analyse à la probabilité des décisions rendues à la pluralité des voix.* Paris, 1785

[Cook71] S.A. Cook. *The complexity of theorem-proving procedures.* Proceedings of the 3rd IEEE Symp. on the Foundations of Computer Science 1971, 524-535

[Fish77] P.C. Fishburn. *Condorcet social choice functions.* SIAM J Appl Math, 33 (1977), 469-489

[GJ79] M.R. Garey, D.S.Johnson. *Computers and Intractability : A Guide to the Theory of NP-Completeness.* W.H. Freeman & Company,1979

[HHR97] E. Hemaspaandra, L. Hemaspaandra, J. Rothe. *Exact Analysis of Dodgson Elections: Lewis Carroll's 1876 Voting System is Complete for Parallel Access to NP.* JACM, 44(6) (1997), 806-825

[Hem00] E. Hemaspaandra. *The complexity of Kemeny elections.* In preparation.

[HH00] E. Hemaspaandra, L. Hemaspaandra. *Computational Politics: Electoral Systems.* Proceedings of MFCS 2000, LNCS 1893, 64-83

[Karp72] R.M. Karp. *Reducibility among combinatorial problems,* in R.E. Miller and J.W. Thatcher (eds.), *Complexity of Computer Computations,* Plenum Press, New York, 85-103

[Kre88] M. Krentel. *The complexity of optimization problems.* Journal of Computer and System Sciences, 36 (1988), 490-509

[MJ78] M. Machtey, P. Young. *An Introduction to General Theory of Algorithms.* North-Holland, New York, 1978

[Pap94] C. H. Papadimitriou. *Computational Complexity.* Addison-Wesley, 1994

[SV99] H. Spakowski, J.Vogel. *The Operators minCh and maxCh on the Polynomial Hierarchy.* Proceedings of FCT 99, LNCS 1684, 524-535

[SV00] H. Spakowski, J.Vogel. *The complexity of voting schemes — a method for proving completeness for parallel access to NP.* Friedrich-Schiller-Universität Jena, Jena, Germany, TR Math/Inf/00/16.

[Wag87] K.W. Wagner. *More complicated questions about maxima an minima, and some closures of NP.* Theoretical Computer Science, 51(1987), 53-80

[Wag90] K.W. Wagner. *Bounded query classes.* SIAM Journal on Computing, 19(1990), 833-846

Is the Standard Proof System for SAT P-Optimal?

(Extended Abstract)

Johannes Köbler[1] and Jochen Messner[2]

[1] Institut für Informatik, Humboldt-Universität zu Berlin, 10099 Berlin, Germany.
koebler@informatik.hu-berlin.de
[2] Abteilung Theoretische Informatik, Universität Ulm, 89069 Ulm, Germany.
messner@informatik.uni-ulm.de

Abstract. We investigate the question whether there is a (p-)optimal proof system for SAT or for TAUT and its relation to completeness and collapse results for nondeterministic function classes. A p-optimal proof system for SAT is shown to imply (1) that there exists a complete function for the class of all total nondeterministic multi-valued functions and (2) that any set with an optimal proof system has a p-optimal proof system. By replacing the assumption of the mere existence of a (p-)optimal proof system by the assumption that certain proof systems are (p-)optimal we obtain stronger consequences, namely collapse results for various function classes. Especially we investigate the question whether the standard proof system for SAT is p-optimal. We show that this assumption is equivalent to a variety of complexity theoretical assertions studied before, and to the assumption that every optimal proof system is p-optimal. Finally, we investigate whether there is an optimal proof system for TAUT that admits an effective interpolation, and show some relations between various completeness assumptions.

1 Introduction and Overview

Following Cook and Reckhow [3] we define the notion of an abstract proof system for a set $L \subseteq \{0,1\}^*$ as follows. A (possibly partial) polynomial-time computable function $h : \{0,1\}^* \to \{0,1\}^*$ with range $L = \{h(x) \mid x \in \{0,1\}^*\}$ is called a *proof system for L*. In this setting, an *h-proof* for the membership of φ to L is given by a string w with $h(w) = \varphi$. In order to compare the relative strength of different proof systems for the set TAUT of all propositional tautologies, Cook and Reckhow introduced the notion of p-simulation. A proof system h *p-simulates* a proof system g if g-proofs can be translated into h-proofs in polynomial time, i.e., there is a polynomial-time computable function f such that for each v in the domain of g, $h(f(v)) = g(v)$. Similarly, h is said to *simulate* g if for each g-proof v there is an h-proof w of length polynomial in the length of v with $h(w) = g(v)$. A proof system for a set L is called *(p-)optimal* if it (p-)simulates every proof system for L (cf. [12]). It's a natural question whether a set L has a p-optimal (or at least an optimal) proof system. Note that a p-optimal proof

S. Kapoor and S. Prasad (Eds.): FST TCS 2000, LNCS 1974, pp. 361–372, 2000.

system has the advantage that from any proof in an other proof system one can efficiently obtain a proof for the same instance in the p-optimal proof system. Hence, any method that is used to compute proofs in some proof system can be reformulated to yield proofs in the p-optimal proof system with little overhead.

It is observed in [19,16,11] that (p-)optimal proof systems for certain languages can be used to define complete sets for certain promise classes. For example, if TAUT has an optimal proof system then $\mathcal{NP} \cap Sparse$ has a many-one complete set, and if TAUT as well as SAT have a p-optimal proof system then $\mathcal{NP} \cap co\text{-}\mathcal{NP}$ has a complete set. We complete this picture here by showing that already a p-optimal proof system for SAT can be used to derive completeness consequences.

These results are however unsatisfactory in so far as they provide only necessary conditions for the existence of (p-)optimal proof systems. It appears that a much stronger assumption like $\mathcal{NP} = \mathcal{P}$ is needed to derive a p-optimal proof system for TAUT (actually, a somewhat weaker collapse condition suffices, namely that all tally sets in nondeterministic double exponential time are contained in deterministic double exponential time; see [11]). If, however, we consider proof systems with certain additional properties then we can indeed derive collapse consequences from the assumption that these proof systems are (p-)optimal. We consider two examples:

The best known proof system is probably the standard proof system for SAT where proofs are given by a satisfying assignment for the formula in question. Adapting this proof method to the current setting one obtains the following natural proof system for SAT:

$$ sat(x) = \begin{cases} \varphi & \text{if } x = \langle \alpha, \varphi \rangle \text{ and } \alpha \text{ is a satisfying assignment for } \varphi \\ \text{undef. otherwise.} \end{cases} $$

We consider the question whether sat is p-optimal[1] in Sect. 3. It turns out that the assumption of sat being p-optimal is equivalent to a variety of well studied complexity theoretic assumptions (which have unlikely collapse consequences as, e.g., that $\mathcal{NP} \cap co\text{-}\mathcal{NP} = \mathcal{P}$). Most of these assumptions were listed in [5] under "Proposition Q" (see also [6]). Proposition Q states for example that any function in the class $\mathcal{NPMV}_t$ of total multi-valued functions computable in nondeterministic polynomial time has a refinement in $\mathcal{FP}$. We further add to this list the statement that every optimal proof system is p-optimal.

As a second example, we consider in Sect. 4 proof systems that admit an effective interpolation. Due to Craig's Interpolation Theorem for Propositional Logic, for any tautology $\varphi \rightarrow \psi$ there is a formula ϕ that uses only common variables of φ and ψ such that $\varphi \rightarrow \phi$ and $\phi \rightarrow \psi$. A circuit C that computes the same function as ϕ is called an *interpolant* of $\varphi \rightarrow \psi$. Following [13] we say that a proof system h for TAUT *admits an effective interpolation* if there is a polynomial p such that for any h-proof w of a formula $h(w) = \varphi \rightarrow \psi$, the formula $\varphi \rightarrow \psi$ has an interpolant of size at most $p(|w|)$. We show that if TAUT

---

[1] Pavel Pudlák posed this question during the discussion after Zenon Sadowski's talk at CSL'98 [20].

has an optimal proof system with this property then any function in $\mathcal{NPSV}$ (the class of single valued functions computable in nondeterministic polynomial time) has a total extension in $\mathcal{FP}/poly$. The latter is equivalent to the statement that every disjoint pair of $\mathcal{NP}$-sets is $\mathcal{P}/poly$-separable which in turn implies that $\mathcal{NP} \cap co\text{-}\mathcal{NP} \subseteq \mathcal{P}/poly$ and that $\mathcal{UP} \subseteq \mathcal{P}/poly$.

The (likely) assumption that there are no p-optimal proof systems for SAT (as well as for TAUT) also has some practical implications due to its connection to the existence of optimal algorithms (see [12,20,15]). Note that usually a decision algorithm for SAT also provides a satisfying assignment for any positive instance. However, if sat is not p-optimal then there is a set $S \subseteq$ SAT of easy instances (i.e. $S \in \mathcal{P}$) for some of which it is hard to produce a satisfying assignment (i.e., there is no polynomial time algorithm that produces a satisfying assignment on all inputs from S, cf. Theorem 1). In fact, a stronger consequence can be derived: if sat is not p-optimal then there is a non-sparse set of easy instances from SAT for which it is hard to produce a satisfying assignment (see Theorem 5).

The observations from [19,16,11] that a p-optimal proof system for a set L implies the existence of a complete set for a certain promise class in fact shows a relationship between different completeness assumptions. Since the definition of p-simulation is equivalent to the definition of many-one reducibility between functions (in the sense of [11]), a proof system for L is p-optimal if and only if it is many-one complete for the (promise) function class $\mathcal{PS}_L$ that consists of all proof systems for L. Depending on the complexity of L, this completeness assumption can be used to derive the existence of complete sets for various other promise classes. This observation motivates us to further investigate whether there are relations between various completeness assumptions. Along this line of investigation we show in Sect. 5 that

- $\mathcal{NPSV}$ has a many-one complete function if and only if there is a strongly many-one complete disjoint $\mathcal{NP}$-pair,
- a complete function for $\mathcal{NPMV}_t$ implies a many-one complete pair for the class of disjoint $co\text{-}\mathcal{NP}$ pairs, and
- $\mathcal{NPSV}_t = \mathcal{NPMV}_t \cap \mathcal{NPSV}$ has a many-one complete function if and only if $\mathcal{NP} \cap co\text{-}\mathcal{NP}$ has a complete set.

The collapse consequences for the nondeterministic function classes $\mathcal{NPMV}_t$ and $\mathcal{NPSV}$ that are obtained in Sects 3, 4 from the assumption that sat is p-optimal, respectively that there is an optimal proof system for TAUT that admits an effective interpolation, are complemented by the following completeness consequences (presented in Sect. 6):

- If SAT has a p-optimal proof system then $\mathcal{NPMV}_t$ has a complete function.
- If TAUT has an optimal proof system then $\mathcal{NPSV}$ has a complete function.

Further we show that

- SAT has a p-optimal proof system if and only if any language with an optimal proof system also has a p-optimal proof system.

This result again complements the observation from Sect. 3 that *sat* is p-optimal if and only if every optimal proof system is p-optimal. As an application we can weaken the assumption used in [11] to show that $\mathcal{NP} \cap co\text{-}\mathcal{NP}$ has a complete set: it suffices to assume that SAT has a p-optimal proof system and TAUT has an optimal proof system.

Due to the limited space, several results are presented here without proof.

2   Preliminaries

Let $\Sigma = \{0, 1\}$. We denote the cardinality of a set A by $\|A\|$ and the length of a string $x \in \Sigma^*$ by $|x|$. A set S is called *sparse* if the cardinality of $S \cap \Sigma^n$ is bounded above by a polynomial in n. $\mathcal{FP}$ denotes the class of (partial) functions that can be computed in polynomial time. We use $\langle \cdot, \cdots, \cdot \rangle$ to denote a standard polynomial-time computable tupling function. The definitions of standard complexity classes like $\mathcal{P}$, $\mathcal{NP}$, etc. can be found in books like [1,17]. For a class $\mathcal{C}$ of sets we call a pair (A, B) of disjoint sets $A, B \in \mathcal{C}$ a *disjoint $\mathcal{C}$-pair*. If for a class $\mathcal{D}$, and some $D \in \mathcal{D}$ it holds $A \subseteq D$, and $B \cap D = \emptyset$ we call the pair (A, B) *$\mathcal{D}$-separable*.

A *nondeterministic polynomial time Turing machine* (*NPTM*, for short) is a Turing machine N such that for some polynomial p, every accepting path of N on any input of length n is at most of length $p(n)$. A *nondeterministic transducer* is a nondeterministic Turing machine T with a write-only output tape. On input x, T outputs $y \in \Sigma^*$ (in symbols: $T(x) \mapsto y$) if there is an accepting path on input x along which y is written on the output tape. Hence, the function f computed by T on Σ^* could be multi-valued and partial. Using the notation of [2,23] we denote the set $\{y \mid f(x) \mapsto y\}$ of all output values of T on input x by set-$f(x)$. $\mathcal{NPMV}$ denotes the class of all multi-valued, partial functions computable by some nondeterministic polynomial-time transducer. $\mathcal{NPSV}$ is the class of functions f in $\mathcal{NPMV}$ that are *single-valued*, i.e. $\|\text{set-}f(x)\| \leq 1$. (thus, a single-valued multi-valued function is a function in the usual sense, and we use $f(x)$ to denote the unique string in set-$f(x)$). The *domain* of a multi-valued function is the set of those inputs x where set-$f(x) \neq \emptyset$. A function is called *total* if its domain is Σ^*. For a function class $\mathcal{F}$ we denote by $\mathcal{F}_t$ the class of total functions in $\mathcal{F}$. We use $\mathcal{NPMV}_t \subseteq_c \mathcal{FP}$ to indicate that for any $g \in \mathcal{NPMV}_t$ there is a total function $f \in \mathcal{FP}$ that is a *refinement* of g, i.e. $f(x) \in \text{set-}g(x)$ for all $x \in \Sigma^*$. We say that a multi-valued function h *many-one reduces* to a multi-valued function g if there is a function $f \in \mathcal{FP}$ such that for every $x \in \Sigma^*$ set-$g(f(x)) = \text{set-}h(x)$.

For a function class $\mathcal{F}$ a function h is called *$\mathcal{F}$-invertible* if there is a function $f \in \mathcal{F}$ that *inverts* h, i.e. $h(f(y)) = y$ for each y in the range of h. A function h is *honest* if for some polynomial p, $p(|h(x)|) \geq |x|$ holds for all x in the domain of h. We call a function g an *extension* of a function f if $f(x) = g(x)$ for any x in the domain of f. A function $r : \mathbb{N} \to \mathbb{N}$ is called *super-polynomial* if for each polynomial p, $r(n) > p(n)$ for almost every $n \geq 0$. A set $B \in \mathcal{P}$ with $B \subseteq L$ is called a *$\mathcal{P}$-subset of L*.

3 Q and the P-Optimality of *sat*

In [5] the following statements were all shown to be equivalent. There, Q is defined to be the proposition that one (and consequently each) of these statements is true. In this section we show that Q is also equivalent to the p-optimality of *sat*.

Theorem 1 ([5], cf. [6]). *The following statements are equivalent*

1. *For each NPTM N that accepts SAT there is a function $f \in \mathcal{FP}$ such that for each α encoding an accepting path of N on input φ, $f(\alpha)$ is a satisfying assignment of φ.*
2. *Each honest function $f \in \mathcal{FP}$ with range Σ^* is $\mathcal{FP}$-invertible.*
3. *$\mathcal{NPMV}_t \subseteq_c \mathcal{FP}$.*
4. *For all $\mathcal{P}$-subsets S of SAT there exists a function $g \in \mathcal{FP}$ such that for all $\varphi \in S$, $g(\varphi)$ is a satisfying assignment of φ.*

Clearly, each nondeterministic Turing machine N corresponds to a proof system h with $h(\alpha) = \varphi$ if α encodes an accepting path of N on input φ. Now h is honest if, and only if, N is a NPTM. This leads to the observation that Statement 1 in Theorem 1 is equivalent to the condition that *sat* p-simulates every honest proof system for SAT. Hence, we just need to delete the term 'polynomial-time' in the Statement 1 of Theorem 1 to obtain the desired result that Q is equivalent to the p-optimality of *sat*. That this is possible without changing the truth of the theorem can be shown by a padding argument.

Theorem 2. *The following statements are equivalent.*

1. *For each NPTM N that accepts SAT there is a function $f \in \mathcal{FP}$ such that for each α encoding an accepting path of N on input φ, $f(\alpha)$ is a satisfying assignment of φ.*
2. *For each nondeterministic Turing machine N that accepts SAT there is a function $f \in \mathcal{FP}$ such that for each α encoding an accepting path of N on input φ, $f(\alpha)$ is a satisfying assignment of φ.*
3. *sat is a p-optimal proof system for SAT.*

It is known that the assumption $\mathcal{NP} = \mathcal{P}$ implies $\mathcal{NPMV}_t \subseteq_c \mathcal{FP}$ which in turn implies $\mathcal{NP} \cap \text{co-}\mathcal{NP} = \mathcal{P}$ (cf. [24]). Also, in [9] it has been shown that the converse of these implications is not true in suitable relativized worlds. The consequence $\mathcal{NP} \cap \text{co-}\mathcal{NP} = \mathcal{P}$ also shows that the assumption that *sat* is p-optimal is presumably stronger than the assumption that SAT has a p-optimal proof system. Namely the p-optimality of *sat* implies that $\mathcal{NP} \cap \text{co-}\mathcal{NP} = \mathcal{P}$, whereas the existence of a p-optimal proof system follows already (see [11]) if any super-tally set in Σ_2^P belongs to $\mathcal{P}$ (here, any set $L \subseteq \{0^{2^{2^n}} \mid n \geq 0\}$ is called super-tally).

The assumption that *sat* is a p-optimal proof system also has an effect on various reducibility degrees, as has been mentioned in [5] for Karp and Levin reducibility. Also in [14] it is shown that $\mathcal{NPMV}_t \subseteq_c \mathcal{FP}$ if and only if γ-reducibility equals polynomial time many-one reducibility. Furthermore it is shown in [4] that Statement 4 of Theorem 1 is equivalent to the assumption that the approximation class APX is closed under L-reducibility (see [4] for definitions).

The equivalence between the p-optimality of *sat* and $\mathcal{NPMV}_t \subseteq_c \mathcal{FP}$ directly leads to a proof of the following theorem.

Theorem 3. *The following statements are equivalent.*

1. *sat is p-optimal.*
2. *For any language L and all proof systems h and g for L*
   $\quad$ *h p-simulates g if and only if h simulates g*
   (i.e., the corresponding quasi-orders coincide).
3. *Every optimal proof system is p-optimal.*

In [15] it is observed that given a p-optimal proof system h for a language L the problem to find an h-proof for $y \in L$ is not much harder than deciding L, i.e. we can transform each deterministic Turing machine M with $L(M) = L$ to a deterministic Turing machine M' that on input $y \in L$ yields an h-proof of y in $\text{time}_{M'}(y) \leq p(|y| + \text{time}_M(y))$, for some polynomial p determined by M. Using this observation and the equivalences in Theorem 1 and 2 we obtain the following result: *sat* is p-optimal if and only if any deterministic Turing machine M that accepts SAT can be converted to a deterministic Turing machine that computes a satisfying assignment for any formula $\varphi \in$ SAT and runs not much longer than M on input φ.

Theorem 4. *The following statements are equivalent.*

1. *sat is p-optimal.*
2. *For any deterministic Turing machine M that accepts SAT, there is a deterministic Turing machine M' and a polynomial p such that for every $\varphi \in SAT$, M' produces a satisfying assignment of φ in $\text{time}_{M'}(\varphi) \leq p(|\varphi| + \text{time}_M(\varphi))$ steps.*

Under the assumption that *sat* is not p-optimal it follows from Theorem 4 that there is a Turing machine M that decides SAT such that any machine M' that on input $\varphi \in$ SAT has to produce a satisfying assignment for φ is much slower on some SAT instances. In some sense this appears counterintuitive as probably all SAT algorithms used in praxis produce a satisfying assignment in case the input belongs to SAT. Of course it follows from Theorem 4 that M is superior to any such M' on an infinite set of instances. As shown in the following theorem there is a deterministic Turing machine M accepting SAT that is more than polynomially faster than any deterministic transducer M' that produces satisfying assignments on a fixed non-sparse set of SAT instances. The result is due to the paddability of SAT, its proof uses ideas from the theory of complexity cores (see [21]).

Theorem 5. *The following statements are equivalent.*

1. *sat is not p-optimal.*
2. *There is a $\mathcal{P}$-subset S of SAT (i.e. there is a Turing machine M accepting SAT that has a polynomial time-bound on instances from S), a non-sparse subset L of S, and a super-polynomial function f such that for any deterministic Turing machine M' that on input of any $\varphi \in L$ produces a satisfying assignment of φ it holds $time_{M'}(\varphi) > f(|\varphi|)$ for almost every $\varphi \in L$.*

4   Collapse of $\mathcal{NPSV}$ and Effective Interpolation

In [22] the following hypothesis (called H2) has been examined.

> For every polynomial time uniform family of formulas $\{\varphi_n, \psi_n\}$ such that for every n, φ_n and ψ_n have n common variables and $\varphi_n \to \psi_n$ is a tautology, there is a polynomial p and a circuit family $\{C_n\}$ where for each n, C_n is of size at most $p(n)$ has n inputs and is an interpolant of $\varphi_n \to \psi_n$.

Theorem 6 ([22]). *The following statements are equivalent.*

1. *H2.*
2. *Every disjoint pair of $\mathcal{NP}$-sets is $\mathcal{P}/poly$-separable.*
3. *Every function in $\mathcal{NPSV}$ has a total extension in $\mathcal{FP}/poly$.*

As mentioned in [22] this in turn implies that H2 implies $\mathcal{NP} \cap co\text{-}\mathcal{NP} \subseteq \mathcal{P}/poly$ and $\mathcal{UP} \subseteq \mathcal{P}/poly$.

In [5] and [6] also a statement called Q' that is implied by Q is examined. Q' is equivalent to the statement that all disjoint $co\text{-}\mathcal{NP}$-pairs are $\mathcal{P}$-separable. Thus as proposed in [22] H2 is a nonuniform version of the dual condition to Q', namely that every disjoint pair of $\mathcal{NP}$-sets is $\mathcal{P}$-separable.

It is observed in [13] that extended Frege proof systems do not admit an effective interpolation if the RSA cryptosystem is secure. Partly generalizing this observation, one can state that the existence of an honest injective function in $\mathcal{FP}$ that is not $\mathcal{FP}/poly$-invertible (i.e., a one-way function that is secure against $\mathcal{FP}/poly$) implies the existence of a proof system for TAUT that does not admit an effective interpolation. Notice that each injective function in $\mathcal{FP}$ is invertible by a $\mathcal{NPSV}$-function. Thus the assumption that each $\mathcal{NPSV}$ function has a total extension in $\mathcal{FP}/poly$ implies that every injective function is $\mathcal{FP}/poly$-invertible. As the former assumption (that is equivalent to H2 by Theorem 6) implies $\mathcal{NP} \cap co\text{-}\mathcal{NP} \subseteq \mathcal{P}/poly$ and the latter is equivalent to $\mathcal{UP} \subseteq \mathcal{P}/poly$ (cf. [10,7]) it is presumably stronger. In the following Theorem we observe that H2, respectively the statement that every function in $\mathcal{NPSV}$ has a total extension in $\mathcal{FP}/poly$ is true if, and only if, every proof system for TAUT admits an effective interpolation.

Theorem 7. *The following statements are equivalent.*

1. *Every function in $\mathcal{NPSV}$ has a total extension in $\mathcal{FP}/poly$.*
2. *Every proof system for TAUT admits an effective interpolation.*
3. *For any set $S \subseteq TAUT$, $S \in \mathcal{NP}$, there is a polynomial p, such that any formula $\varphi \to \psi \in S$ has an interpolant of size at most $p(|\varphi \to \psi|)$.*

Proof. The implication *2* $\Longrightarrow$ H2 is easy to see, as for every polynomial time uniform family of tautologies $\varphi_n \to \psi_n$ one may define a proof system h for TAUT that has a short proof for any tautology of this family. Thus *2* $\Longrightarrow$ *1* using Theorem 6.

The proof of the implication *1* $\Longrightarrow$ *3* is obtained by extending an idea from [22] that was used to prove the implication *3* $\Longrightarrow$ *1* of Theorem 6. Let $S \subseteq$ TAUT, $S \in \mathcal{NP}$. Let f be a function such that for any formula $\varphi \in S$, $\varphi = \varphi_0(x, y) \to \varphi_1(x, z)$ (where x, y, z denote vectors of variables), it holds

$$f(\langle \alpha, \varphi \rangle) = \begin{cases} 1 & \text{if for some } \beta, \quad \varphi_0(\alpha, \beta) \text{ holds} \\ 0 & \text{if for some } \gamma, \quad \neg\varphi_1(\alpha, \gamma) \text{ holds.} \end{cases}$$

Otherwise, and for any other input let f be undefined. First observe that f is well defined, i.e. that f is single valued. This is due to the fact that $\varphi = \varphi_0(x, y) \to \varphi_1(x, z) \in$ TAUT. Further, f can be computed by a nondeterministic machine N that first (in deterministic polynomial time) validates that the input is of the appropriate form $\langle \alpha, \varphi \rangle$, $\varphi = \varphi_0(x, y) \to \varphi_1(x, z)$. Then N guesses a certificate for $\varphi \in S$ and, if successful, guesses some string w. Now if w is of an appropriate length and if $\varphi_0(\alpha, w)$ holds then N outputs 1, if $\varphi_1(\alpha, w)$ holds, N outputs 0. Hence $f \in \mathcal{NPSV}$. Assuming *1*, f has a total extension in $\mathcal{FP}/poly$. Thus there is a polynomial p and for any $n \geq 0$ a circuit C_n of size at most $p(n)$ such that for any tuple $v = \langle \alpha, \varphi \rangle$ of length n in the domain of f, $C_n(v) = f(v)$. Fixing the input bits of C_n that belong to the formula φ we obtain a circuit C_φ with $C_\varphi(\alpha) = C_n(\langle \alpha, \varphi \rangle) = f(\langle \alpha, \varphi \rangle)$ and thus C_φ is of size polynomial in $|\varphi|$. Now observe that C_φ is an interpolant for the formulas $\varphi_0(x, y)$ and $\varphi_1(x, z)$. If $\varphi_0(\alpha, y)$ is satisfiable then $C_\varphi(\alpha) = 1$, and if $C_\varphi(\alpha) = 1$ then for no γ it holds $\neg\varphi_1(\alpha, \gamma)$ and therefore $\varphi_1(\alpha, z)$ is a tautology.

Finally the proof of the implication *3* $\Longrightarrow$ *2* is obtained using padding techniques. We omit the details due to the limited space. □

It is easy to see that a proof system g admits an effective interpolation if there is a proof system h that simulates g, and h admits an effective interpolation. Hence, any proof system for TAUT admits an effective interpolation if there is an optimal proof system for TAUT that admits an effective interpolation. As a corollary we obtain

Corollary 8. *If there is an optimal proof system for TAUT that admits an effective interpolation then H2 holds.*

5   Relations between Completeness Assumptions

Using the characterization $\mathcal{NPSV}_t = \mathcal{FP}_t^{\mathcal{NP}\cap co\text{-}\mathcal{NP}}$ [23,8] we obtain the following result

Theorem 9. $\mathcal{NP}\cap co\text{-}\mathcal{NP}$ *has a many-one complete set iff* $\mathcal{NPSV}_t$ *has a many-one complete function.*

Now let us consider the function class $\mathcal{NPSV}$. In the same way as $\mathcal{NPSV}_t$ corresponds to the language class $\mathcal{NP} \cap co\text{-}\mathcal{NP}$, the function class $\mathcal{NPSV}$ corresponds to the class of all disjoint $\mathcal{NP}$-pairs. In fact, if we denote the class of all 0,1-valued functions in $\mathcal{NPSV}$ by $\mathcal{NPSV}_{\{0,1\}}$ then any function $h \in \mathcal{NPSV}_{\{0,1\}}$ can be identified with the $\mathcal{NP}$-pair (A_0, A_1) where $A_b = \{x \in \Sigma^* \mid h(x) \mapsto b\}$.

Razborov [18] introduced a notion of many-one reducibility between disjoint $\mathcal{NP}$-pairs a stronger version of which was studied in [11]. Let A_0, A_1, B_0, B_1 be $\mathcal{NP}$-sets with $A_0 \cap A_1 = B_0 \cap B_1 = \emptyset$. The pair (A_0, A_1) (strongly) many-one reduces to (B_0, B_1) if there is a function $f \in \mathcal{FP}$ such that $f(A_b) \subseteq B_b$ ($f^{-1}(B_b) = A_b$, respectively) for $b \in \{0, 1\}$. Actually, it is easy to see that f is a many-one reduction between two functions in $\mathcal{NPSV}_{\{0,1\}}$ if and only if f is a strong many-one reduction between the corresponding disjoint $\mathcal{NP}$-pairs. Thus, the class of disjoint $\mathcal{NP}$-pairs has a strongly many-one complete pair if and only if $\mathcal{NPSV}_{\{0,1\}}$ has a many-one complete function. As shown in the next theorem, this is even equivalent to the assumption that $\mathcal{NPSV}$ has a many-one complete function.

Theorem 10. *The following statements are equivalent.*

1. *$\mathcal{NPSV}$ has a many-one complete function.*
2. *$\mathcal{NPSV}_{\{0,1\}}$ has a many-one complete function.*
3. *There is a strongly many-one complete disjoint $\mathcal{NP}$-pair.*

Proof. (Sketch). Implication *1 $\Longrightarrow$ 2* is easy to prove, and the equivalence of *2* and *3* is clear by the preceding discussion. To see that *2* implies *1* we observe that $\mathcal{NPSV}$ can be characterized as $\mathcal{FP}^{\mathcal{NPSV}_{\{0,1\}}}$ where the value $M^f(x)$ computed by the deterministic oracle transducer M on input x is only defined if all oracle queries belong to the domain of the functional oracle f. $\square$

We conclude this section by observing that the class of disjoint $co\text{-}\mathcal{NP}$ pairs corresponds to the class $\mathcal{NPbV}_t$ of all 0,1-valued functions in $\mathcal{NPMV}_t$ studied in [5] (with the disjoint $co\text{-}\mathcal{NP}$-pair (A_0, A_1) associate the function $h \in \mathcal{NPbV}_t$ defined by set-$h(x) = \{b \mid x \notin A_b\}$). Similar to the implication $1 \Longrightarrow 3$ in Theorem 10 the following theorem can be proved.

Theorem 11. *If* $\mathcal{NPMV}_t$ *has a many-one complete function then there exists a strongly many-one complete disjoint $co\text{-}\mathcal{NP}$-pair.*

We leave it open whether the reverse implication also holds.

6   *Existence* of (P-)Optimal Proof Systems

In Theorem 3 it is observed that *sat* is p-optimal iff every optimal proof system is p-optimal. Although the assumption of the mere existence of a p-optimal proof system for SAT is presumably weaker than the assumption that *sat* is p-optimal, it is still equivalent to a quite similar statement, namely that any set with an optimal proof system has a p-optimal proof system. For the proof of this result we use the following observation from [11].

Lemma 12 ([11]). *If L has a (p-)optimal proof system, and $T \leq_m^p L$ then T has a (p-)optimal proof system (respectively).*

Theorem 13. *The following statements are equivalent.*

1. *SAT has a p-optimal proof system.*
2. *Any language L that has an optimal proof system also has a p-optimal proof system.*

Proof. Clearly *2 ⟹ 1*, as SAT has an optimal proof system. To see the inverse implication assume that SAT has a p-optimal proof system. Let T_L (cf. [11,15]) be the following language consisting of tuples $\langle M, x, 0^s \rangle$ where M is a deterministic Turing transducer, $s \geq 0$ and $x \in \Sigma^*$.

$$T_L = \{\langle M, x, 0^s \rangle \mid \text{if } \text{time}_M(x) \leq s \text{ then } M(x) \in L\}.$$

Notice that T_L is many-one reducible to L (without restriction assume $L \neq \emptyset$). Hence, the assumption that there is an optimal proof system for L implies that T_L has an optimal proof system, say h. Let

$$S = \{\langle \langle M, x, 0^s \rangle, 0^l \rangle \mid \exists w, |w| \leq l, h(w) = \langle M, x, 0^s \rangle\}.$$

Clearly $S \in \mathcal{NP}$. Therefore by assumption there is a p-optimal proof system system g for S. Let now f be the following proof system.

$$f(w) = \begin{cases} y & \text{if } g(w) = \langle \langle M, x, 0^s \rangle, 0^l \rangle, \text{ time}_M(x) \leq s, \text{ and } M(x) = y, \\ \text{undef.} & \text{otherwise.} \end{cases}$$

First notice that $y \in L$ if $f(w) = y$. This is due to the fact that $g(w) = \langle \langle M, x, 0^s \rangle, 0^l \rangle$ implies $\langle \langle M, x, 0^s \rangle, 0^l \rangle \in S$ which in turn implies $\langle M, x, 0^s \rangle \in T_L$. We now show that f p-simulates every proof system f' for L. Assume that f' is computed by the transducer $M_{f'}$ in polynomial time $p(n)$. Observe that $\langle M_{f'}, x, 0^{p(|x|)} \rangle \in T_L$ for any $x \in \Sigma^*$. Hence, one may define a proof system for T_L such that for any x the tuple $\langle M_{f'}, x, 0^{p(|x|)} \rangle$ has the short proof $1x$. Consequently, due to the optimality of h, there is a polynomial q such that $\langle M_{f'}, x, 0^{p(|x|)} \rangle$ has an h-proof of size $\leq q(|x|)$. Hence $\langle \langle M_{f'}, x, 0^{p(|x|)} \rangle, 0^{q(|x|)} \rangle \in S$ for any x, and one may define a proof system g' for S with $g'(1x) = \langle \langle M_{f'}, x, 0^{p(|x|)} \rangle, 0^{q(|x|)} \rangle$ for any x. As g is p-optimal, g p-simulates g', i.e. there is a function $t \in \mathcal{FP}$ such that $g(t(1x)) = g'(1x) = \langle \langle M_{f'}, x, 0^{p(|x|)} \rangle, 0^{q(|x|)} \rangle$. Observe now that $f(t(1x)) = f'(x)$ for any x. Hence f p-simulates f'.   □

As shown in [11], the assumption that SAT and TAUT both have p-optimal proof systems implies that $\mathcal{NP} \cap co\text{-}\mathcal{NP}$ has a many-one complete set. In fact, due to Theorem 13 it suffices to assume that SAT has a p-optimal proof system and TAUT only has an optimal proof system. Together with Theorem 9 we obtain

Corollary 14. *If SAT has a p-optimal and TAUT has an optimal proof system then $\mathcal{NP} \cap co\text{-}\mathcal{NP}$ has a many-one complete set, and $\mathcal{NPSV}_t$ has a many-one complete function.*

It is observed in [18] that the existence of an optimal proof system for TAUT implies the existence of a many-one complete pair for the class of disjoint $\mathcal{NP}$-pairs. In [11] it is shown that the same assumption allows one to infer the existence of a strongly many-one complete disjoint $\mathcal{NP}$-pair which by Theorem 10 implies that $\mathcal{NPSV}$ has a many-one complete function.

Corollary 15. *If TAUT has an optimal proof system then $\mathcal{NPSV}$ has a many-one complete function.*

Next we observe that a p-optimal proof system for SAT implies a complete function for the class $\mathcal{NPMV}_t$. The proof uses ideas from [11] (in fact, the only extension is that we are dealing with multi-valued functions here).

Theorem 16. *If SAT has a p-optimal proof system then $\mathcal{NPMV}_t$ has a many-one complete function.*

By Theorem 11 we obtain

Corollary 17. *If SAT has a p-optimal proof system then there exists a strongly many-one complete disjoint co-$\mathcal{NP}$-pair.*

7   Conclusion

We showed that the assumption that certain proof systems are (p-)optimal can be used to derive collapse results. Also we presented some relations between completeness assumptions for different classes. It would be interesting to know whether these observations could be extended to further proof systems and promise classes.

References

1. José Luis Balcázar, Josep Díaz, and Joaquim Gabarró. *Structural Complexity I.* Springer-Verlag, 2nd edition, 1995.
2. Ronald V. Book, Timothy J. Long, and Alan L. Selman. Quantitative relativizations of complexity classes. *SIAM Journal on Computing*, 13(3):461–487, 1984.
3. Stephen A. Cook and Robert A. Reckhow. The relative efficiency of propositional proof systems. *The Journal of Symbolic Logic*, 44(1):36–50, 1979.
4. Pierluigi Crescenzi, Viggo Kann, Riccardo Silvestri, and Luca Trevisan. Structure in approximation classes. *SIAM Journal on Computing*, 28(5):1759–1782, 1999.

5. Stephen A. Fenner, Lance Fortnow, Ashish V. Naik, and John D. Rogers. Inverting onto functions. In *Proceedings of the 11th Conference on Computational Complexity*, pages 213–222. IEEE, 1996.
6. Lance Fortnow and John D. Rogers. Separability and one-way functions. In *Proceedings of the 5th International Symposium on Algorithms and Computation*, *LNCS #834*, pages 396–404. Springer-Verlag, 1994.
7. Joachim Grollmann and Alan L. Selman. Complexity measures for public-key cryptosystems. *SIAM Journal on Computing*, 17(2):309–335, 1988.
8. Lane A. Hemaspaandra, Ashish V. Naik, Mitsunori Ogihara, and Alan L. Selman. Computing solutions uniquely collapses the polynomial hierarchy. In *Proceedings of the 5th International Symposium on Algorithms and Computation*, *LNCS #834*, pages 56–64. Springer-Verlag, 1994.
9. Russel Impagliazzo and Moni Naor. Decision trees and downward closures (extended abstract). In *Proceedings of the Third Conference on Structure in Complexity Theory*, pages 29–38. IEEE, 1988.
10. Ker-I Ko. On some natural complete operators. *Theoretical Computer Science*, 37:1–30, 1985.
11. Johannes Köbler and Jochen Messner. Complete problems for promise classes by optimal proof systems for test sets. In *Proceedings of the 13th Conference on Computational Complexity*, pages 132–140. IEEE, 1998.
12. Jan Krajíček and Pavel Pudlák. Propositional proof systems, the consistency of first order theories and the complexity of computations. *The Journal of Symbolic Logic*, 54(3):1063–1079, 1989.
13. Jan Krajíček and Pavel Pudlák. Some consequences of cryptographical conjectures for S_2^1 and EF. In D. Leivant, editor, *Logic and Computational Complexity*, *LNCS #960*, pages 210–220. Springer-Verlag, 1995.
14. Timothy J. Long. On γ-reducibility versus polynomial time many-one reducibility. *Theoretical Computer Science*, 14:91–101, 1981.
15. Jochen Messner. On optimal algorithms and optimal proof system. In *Proceedings of the 16th Symposium on Theoretical Aspects of Computer Science*, *LNCS #1563*. Springer-Verlag, 1999.
16. Jochen Messner and Jacobo Torán. Optimal proof systems for propositional logic and complete sets. In *Proceedings of the 15th Symposium on Theoretical Aspects of Computer Science*, *LNCS #1373*, pages 477–487. Springer-Verlag, 1998.
17. Christos H. Papadimitriou. *Computatational Complexity*. Addison-Wesley, 1994.
18. Alexander A. Razborov. On provably disjoint $\mathcal{NP}$-pairs. Technical Report RS-94-36, Basic Research in Computer Science Center, Aarhus, 1994.
19. Zenon Sadowski. On an optimal quantified propositional proof system and a complete language for $\mathcal{NP} \cap co\text{-}\mathcal{NP}$. In *Proceedings of the 11th International Symposium on Fundamentals of Computing Theory*, *LNCS #1279*, pages 423–428. Springer-Verlag, 1997.
20. Zenon Sadowski. On an optimal deterministic algorithm for SAT. In *Proceedings of the 12th Annual Conference of the European Association for Computer Science Logic*, *CSL '98*, *LNCS #1584*, pages 179–187. Springer-Verlag, 1999.
21. Uwe Schöning. *Complexity and Structure*, *LNCS #211*. Springer-Verlag, 1985.
22. Uwe Schöning and Jacobo Torán. A note on the size of craig interpolants, 1996. Unpublished manuscript.
23. Alan L. Selman. A taxonomy of complexity classes of functions. *Journal of Computer and System Sciences*, 48(2):357–381, 1994.
24. Leslie G. Valiant. Relative complexity of checking and evaluating. *Information Processing Letters*, 5(1):20–23, 1976.

A General Framework for Types in Graph Rewriting[*]

Barbara König

Fakultät für Informatik, Technische Universität München
(koenigb@in.tum.de)

Abstract. A general framework for typing graph rewriting systems is presented: the idea is to statically derive a type graph from a given graph. In contrast to the original graph, the type graph is invariant under reduction, but still contains meaningful behaviour information. We present conditions, a type system for graph rewriting should satisfy, and a methodology for proving these conditions. In two case studies it is shown how to incorporate existing type systems (for the polyadic π-calculus and for a concurrent object-oriented calculus) into the general framework.

1 Introduction

In the past, many formalisms for the specification of concurrent and distributed systems have emerged. Some of them are aimed at providing an encompassing theory: a very general framework in which to describe and reason about interconnected processes. Examples are action calculi [18], rewriting logic [16] and graph rewriting [3] (for a comparison see [4]). They all contain a method of building terms (or graphs) from basic elements and a method of deriving reduction rules describing the dynamic behaviour of these terms in an operational way.

A general theory is useful, if concepts appearing in instances of a theory can be generalised, yielding guidelines and relieving us of the burden to prove universal concepts for every single special case. An example for such a generalisation is the work presented for action calculi in [15] where a method for deriving a labelled transition semantics from a set of reaction rules is presented. We concentrate on graph rewriting (more specifically hypergraph rewriting) and attempt to generalise the concept of type systems, where, in this context, a type may be a rather complex structure.

Compared to action calculi[1] and rewriting logic, graph rewriting differs in a significant way in that connections between components are described explicitly (by connecting them by edges) rather than implicitly (by referring to the same channel name). We claim that this feature—together with the fact that it is easy

---

[*] Research supported by SFB 342 (subproject A3) of the DFG.
[1] Here we mean action calculi in their standard string notation. There is also a graph notation for action calculi, see e.g. [7].

S. Kapoor and S. Prasad (Eds.): FST TCS 2000, LNCS 1974, pp. 373–384, 2000.
© Springer-Verlag Berlin Heidelberg 2000

to add an additional layer containing annotations and constraints to a graph—
can simplify the design of a type system and therefore the static analysis of a
graph rewriting system.

After introducing our model of graph rewriting and a method for annotating
graphs, we will present a general framework for type systems where both—
the expression to be typed and the type itself—are hypergraphs and will show
how to reduce the proof obligations for instantiations of the framework. We
are interested in the following properties: correctness of a type system (if an
expression has a certain type, then we can conclude that this expression has
certain properties), the subject reduction property (types are invariant under
reduction) and compositionality (the type of an expression can always be derived
from the types of its subexpressions). Parts of the proofs of these properties can
already be conducted for the general case.

We will then show that our framework is realistic by instantiating it to two
well-known type systems: a type system avoiding run-time errors in the polyadic
π-calculus [17] and a type system avoiding "message not understood"-errors in
a concurrent object-oriented setting. A third example enforcing a security policy
for untrustworthy applets is included in the full version [11].

2   Hypergraph Rewriting and Hypergraph Annotation

We first define some basic notions concerning hypergraphs (see also [6]) and a
method for inductively constructing hypergraphs.

Definition 1. (Hypergraph) *Let L be a fixed set of labels. A hypergraph $H =$
$(V_H, E_H, s_H, l_H, \chi_H)$ consists of a set of nodes V_H, a set of edges E_H, a connection mapping $s_H : E_H \to V_H^*$, an edge labelling $l_H : E_H \to L$ and a string
$\chi_H \in V_H^*$ of external nodes. A hypergraph morphism $\phi : H \to H'$ (consisting
of $\phi_V : V_H \to V_{H'}$ and $\phi_E : E_H \to E_{H'}$) satisfies[2] $\phi_V(s_H(e)) = s_{H'}(\phi_E(e))$
and $l_H(e) = l_{H'}(\phi_E(e))$. A strong morphism (denoted by the arrow $\twoheadrightarrow$) additionally preserves the external nodes, i.e. $\phi_V(\chi_H) = \chi_{H'}$. We write $H \cong H'$ (H is
isomorphic to H') if there is a bijective strong morphism from H to H'.*

The arity of a hypergraph H is defined as $ar(H) = |\chi_H|$ while the arity of an
edge e of H is $ar(e) = |s_H(e)|$. External nodes are the interface of a hypergraph
towards its environment and are used to attach hypergraphs.

Notation: We call a hypergraph *discrete*, if its edge set is
empty. By **m** we denote a discrete graph of arity $m \in \mathbb{N}$ (a)
with m nodes where every node is external (see Figure (a)
to the right, external nodes are labelled (1), (2), ... in
their respective order).

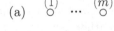

The hypergraph $H = [l]_n$ contains exactly one edge e with
label l where $s_H(e) = \chi_H$, $ar(e) = n$ and[3] $V_H = Set(\chi_H)$
(see (b), nodes are ordered from left to right).

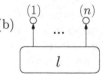

---

[2] The application of ϕ_V to a string of nodes is defined pointwise.
[3] $Set(\tilde{s})$ is the set of all elements of a string $\tilde{s}$

The next step is to define a method (first introduced in [10]) for the annotation of hypergraphs with lattice elements and to describe how these annotations change under morphisms. We use annotated hypergraphs as types where the annotations can be considered as extra typing information, therefore we use the terms *annotated hypergraph* and *type graph* as synonyms.

Definition 2. (Annotated Hypergraphs) *Let $\mathcal{A}$ be a mapping assigning a lattice $\mathcal{A}(H) = (I, \leq)$ to every hypergraph and a function $\mathcal{A}_\phi : \mathcal{A}(H) \to \mathcal{A}(H')$ to every morphism $\phi : H \to H'$. We assume that $\mathcal{A}$ satisfies:*

$$\mathcal{A}_\phi \circ \mathcal{A}_\psi = \mathcal{A}_{\phi \circ \psi} \quad \mathcal{A}_{id_H} = id_{\mathcal{A}(H)} \quad \mathcal{A}_\phi(a \vee b) = \mathcal{A}_\phi(a) \vee \mathcal{A}_\phi(b) \quad \mathcal{A}_\phi(\bot) = \bot$$

where $\vee$ is the join-operation, a and b are two elements of the lattice $\mathcal{A}(H)$ and $\bot$ is its bottom element.

If $a \in \mathcal{A}(H)$, then $H[a]$ is called an annotated hypergraph. And $\phi : H[a] \to_{\mathcal{A}} H'[a']$ is called an $\mathcal{A}$-morphism if $\phi : H \to H'$ is a hypergraph morphism and $\mathcal{A}_\phi(a) \leq a'$. Furthermore $H[a]$ and $H'[a']$ are called isomorphic if there is a strong bijective $\mathcal{A}$-morphism ϕ with $\mathcal{A}_\phi(a) = a'$ between them.

Example: We consider the following annotation mapping $\mathcal{A}$: let $(\{false, true\}, \leq)$ be the boolean lattice where $false < true$. We define $\mathcal{A}(H)$ to be the set of all mappings from V_H into $\{false, true\}$ (which yields a lattice with pointwise order). By choosing an element of $\mathcal{A}(H)$ we fix a subset of the nodes. So let $a : V_H \to \{false, true\}$ be an element of $\mathcal{A}(H)$ and let $\phi : H \to H'$, $v' \in V_H$. We define: $\mathcal{A}_\phi(a) = a'$ where $a'(v') = \bigvee_{\phi(v)=v'} a(v)$. That is, if a node v with annotation *true* is mapped to a node v' by ϕ, the annotation of v' will also be *true*.

From the point of view of category theory, $\mathcal{A}$ is a functor from the category of hypergraphs and hypergraph morphisms into the category of lattices and join-morphisms (i.e. functions preserving the join operation of the lattice).

We now introduce a method for attaching (annotated) hypergraphs with a construction plan consisting of discrete graph morphisms.

Definition 3. (Hypergraph Construction) *Let $H_1[a_1], \ldots, H_n[a_n]$ be annotated hypergraphs and let $\zeta_i : \mathbf{m_i} \to D$, $1 \leq i \leq n$ be hypergraph morphisms where $ar(H_i) = m_i$ and D is discrete. Furthermore let $\phi_i : \mathbf{m_i} \twoheadrightarrow H_i$ be the unique strong morphisms.*

For this construction we assume that the node and edge sets of $H_1, \ldots, H_n$ and D are pairwise disjoint. Furthermore let $\approx$ be the smallest equivalence on their nodes satisfying $\zeta_i(v) \approx \phi_i(v)$ if $1 \leq i \leq n$, $v \in V_{\mathbf{m_i}}$. The nodes of the constructed graph are the equivalence classes of $\approx$. We define

$$\textcircled{D}_{i=1}^{n}(H_i, \zeta_i) = ((V_D \cup \bigcup_{i=1}^{n} V_{H_i})/\approx, \bigcup_{i=1}^{n} E_{H_i}, s_H, l_H, \chi_H)$$

where $s_H(e) = [v_1]_\approx \ldots [v_k]_\approx$ if $e \in E_{H_i}$ and $s_{H_i}(e) = v_1 \ldots v_k$. Furthermore $l_H(e) = l_{H_i}(e)$ if $e \in E_{H_i}$. And we define $\chi_H = [v_1]_\approx \ldots [v_k]_\approx$ if $\chi_D = v_1 \ldots v_k$.
If $n = 0$, the result of the construction is D itself.

We construct embeddings $\phi : D \twoheadrightarrow H$ *and* $\eta_i : H_i \to H$ *by mapping every node to its equivalence class and every edge to itself. Then the construction of annotated graphs can be defined as follows:*

$$\textcircled{D}_{i=1}^{n}(H_i[a_i], \zeta_i) = \left(\textcircled{D}_{i=1}^{n}(H_i, \zeta_i)\right) \left[\bigvee_{i=1}^{n} \mathcal{A}_{\eta_i}(a_i)\right]$$

In other words: we join all graphs $D, H_1, \ldots, H_n$ and fuse exactly the nodes which are the image of one and the same node in the $\mathbf{m_i}$, χ_D becomes the new sequence of external nodes. Lattice annotations are joined if the annotated nodes are merged. In terms of category theory, $\textcircled{D}_{i=1}^{n}(H_i[a_i], \zeta_i)$ is the colimit of the ζ_i and the ϕ_i regarded as $\mathcal{A}$-morphisms (D and the $\mathbf{m_i}$ are annotated with the bottom element $\perp$). We do not mention this fact in the rest of the paper, but it is used extensively in the proofs (for the proofs and several examples see the full version [11]).

We also use another, more intuitive notation for graph construction. Let $\zeta_i : \mathbf{m_i} \to D$, $1 \leq i \leq n$. Then we depict $\textcircled{D}_{i=1}^{n}(H_i, \zeta_i)$ by drawing the hypergraph $(V_D, \{e_1, \ldots, e_n\}, s_H, l_H, \chi_D)$ where $s_H(e_i) = \zeta_i(\chi_{\mathbf{m_i}})$ and $l_H(e_i) = H_i$.

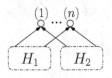

Example: we can draw $\textcircled{n}_{i=1}^{2}(H_i, \zeta_i)$ where $\zeta_1, \zeta_2 : \mathbf{n} \twoheadrightarrow \mathbf{n}$ as in the picture above (note that the edges have dashed lines). Here we fuse the external nodes of H_1 and H_2 in their respective order and denote the resulting graph by $H_1 \square H_2$. If there is an edge with a dashed line labelled with an edge $[l]_n$ we rather draw it with a solid line and label it with l (see e.g. the second figure in section 4.1).

Definition 4. (Hypergraph Rewriting) *Let $\mathcal{R}$ be a set of pairs (L, R) (called rewriting rules), where the left-hand side L and the right-hand side R are both hypergraphs of the same arity. Then $\to_{\mathcal{R}}$ is the smallest relation generated by the pairs of $\mathcal{R}$ and closed under hypergraph construction.*

In our approach we generate the same transition system as in the double-pushout approach to graph rewriting described in [2] (for details see [13]).

We need one more concept: a linear mapping which is an inductively defined transformation, mapping hypergraphs to hypergraphs and adding annotation.

Definition 5. (Linear Mapping) *A function from hypergraphs to hypergraphs is called arity-preserving if it preserves arity and isomorphism classes of hypergraphs.*

Let t be an arity-preserving function that maps hypergraphs of the form $[l]_n$ to annotated hypergraphs. Then t can be extended to arbitrary hypergraphs by defining $t(\textcircled{D}_{i=1}^{n}([l_i]_{n_i}, \zeta_i)) = \textcircled{D}_{i=1}^{n}(t([l_i]_{n_i}), \zeta_i)$ and is then called a linear mapping.

3   Static Analysis and Type Systems for Graph Rewriting

Having introduced all underlying notions we now specify the requirements for type systems. We assume that there is a fixed set $\mathcal{R}$ of rewrite rules, an annotation mapping $\mathcal{A}$, a predicate X on hypergraphs (representing the property we want to check) and a relation $\triangleright$ with the following meaning: if $H \triangleright T$ where H is a hypergraph and T a type graph (annotated wrt. to $\mathcal{A}$), then H has type T. It is required that H and T have the same arity.

We demand that $\triangleright$ satisfies the following conditions: first, a type should contain information concerning the properties of a hypergraph, i.e. if a hypergraph has a type, then we can be sure that the property X holds.

$$H \triangleright T \;\Rightarrow\; X(H) \quad \textbf{(correctness)} \tag{1}$$

During reduction, the type stays invariant.

$$H \triangleright T \wedge H \rightarrow_{\mathcal{R}} H' \;\Rightarrow\; H' \triangleright T \quad \textbf{(subject reduction property)} \tag{2}$$

From (1) and (2) we can conclude that $H \triangleright T$ and $H \rightarrow^{*}_{\mathcal{R}} H'$ imply $X(H')$, that is X holds during the entire reduction.

The strong $\mathcal{A}$-morphisms introduced in Definition 2 impose a preorder on type graphs. It should always be possible to weaken the type with respect to that preorder.

$$H \triangleright T \wedge T \twoheadrightarrow_{\mathcal{A}} T' \;\Rightarrow\; H \triangleright T' \quad \textbf{(weakening)} \tag{3}$$

We also demand that the type system is compositional, i.e a graph has a type if and only if this type can be obtained by typing its subgraphs and combining these types. We can not sensibly demand that the type of an expression is obtained by combining the types of the subgraphs in exactly the same way the expression is constructed, so we introduce a partial arity-preserving mapping f doing some post-processing.

$$\forall i \colon H_i \triangleright T_i \;\Rightarrow\; \textcircled{D}_{i=1}^{n}(H_i, \zeta_i) \triangleright f(\textcircled{D}_{i=1}^{n}(T_i, \zeta_i)) \quad \textbf{(compositionality)} \tag{4}$$

$$\textcircled{D}_{i=1}^{n}(H_i, \zeta_i) \triangleright T \;\Rightarrow\; \exists T_i \colon (H_i \triangleright T_i \text{ and } f(\textcircled{D}_{i=1}^{n}(T_i, \zeta_i)) \twoheadrightarrow_{\mathcal{A}} T)$$

A last condition—the existence of minimal types—may not be strictly needed for type systems, but type systems satisfying this condition are much easier to handle.

$$H \text{ typable} \;\Rightarrow\; \exists T \colon (H \triangleright T \wedge (H \triangleright T' \iff T \twoheadrightarrow_{\mathcal{A}} T')) \quad \textbf{(minimal types)} \tag{5}$$

Let us now assume that types are computed from graphs in the following way: there is a linear mapping t, such that $H \triangleright f(t(H))$, if $f(t(H))$ is defined, and all other types of H are derived by the weakening rule, i.e. $f(t(H))$ is the minimal type of H.

The meaning of the mappings t and f can be explained as follows: t is a transformation *local* to edges, abstracting from irrelevant details and adding annotation information to a graph. The mapping f on the other hand, is a *global* operation, merging or removing parts of a graph in order to anticipate future reductions and thus ensure the subject reduction property. In the example in section 4.1 f "folds" a graph into itself, hence the letter f. In order to obtain compositionality, it is required that f can be applied arbitrarily often at any stage of type inference, without losing information (see condition (6) of Theorem 1).

In this setting it is sufficient to prove some simpler conditions, especially the proof of (2) can be conducted locally.

Theorem 1. *Let $\mathcal{A}$ be a fixed annotation mapping, let f be an arity-preserving mapping as above, let t be a linear mapping, let X be a predicate on hypergraphs and let $H \triangleright T$ if and only if $f(t(H)) \twoheadrightarrow_{\mathcal{A}} T$. Let us further assume that f satisfies*[4]

$$f(\textstyle\bigodot_{i=1}^{n}(T_i, \zeta_i)) \cong f(\textstyle\bigodot_{i=1}^{n}(f(T_i), \zeta_i)) \quad (6) \qquad T \twoheadrightarrow_{\mathcal{A}} T' \Rightarrow f(T) \twoheadrightarrow_{\mathcal{A}} f(T') \quad (7)$$

Then the relation $\triangleright$ satisfies conditions (1)–(5) if and only if it satisfies

$$f(t(H)) \text{ defined} \Rightarrow X(H) \quad (8) \qquad (L, R) \in \mathcal{R} \Rightarrow f(t(R)) \twoheadrightarrow_{\mathcal{A}} f(t(L)) \quad (9)$$

The operation f can often be characterised by a universal property with the intuitive notion that $f(T)$ is the "smallest" type graph (wrt. the preorder $\twoheadrightarrow_{\mathcal{A}}$) for which $T \twoheadrightarrow_{\mathcal{A}} f(T)$ and a property C hold.

Proposition 1. *Let C be a property on type graphs such that $f(T)$ can be characterised in the following way: $f(T)$ satisfies C, there is a morphism $\phi : T \twoheadrightarrow_{\mathcal{A}} f(T)$ and for every other morphism $\phi' : T \twoheadrightarrow_{\mathcal{A}} T'$ where $C(T')$ holds, there is a unique morphism $\psi : f(T) \twoheadrightarrow_{\mathcal{A}} T'$ such that $\psi \circ \phi = \phi'$. Furthermore we demand that if there exists a morphism $\phi : T \twoheadrightarrow_{\mathcal{A}} T'$ such that $C(T')$ holds, then $f(T)$ is defined.*

Then if $f(T)$ is defined, it is unique up to isomorphism. Furthermore f satisfies conditions (6) and (7).

4   Case Studies

4.1   A Type System for the Polyadic π-Calculus

We present a graph rewriting semantics for the asynchronous polyadic π-calculus [17] without choice and matching, already introduced in [12]. Different ways of encoding the π-calculus into graph rewriting can be found in [21,5,4].

We apply the theory presented in section 3, introduce a type system avoiding runtime errors produced by mismatching arities and show that it satisfies the conditions of Theorem 1. Afterwards we show that a graph has a type if and only if the corresponding π-calculus process has a type in a standard type system with infinite regular trees.

Definition 6. (Process Graphs) *A process graph P is inductively defined as follows: P is a hypergraph with a duplicate-free string of external nodes. Furthermore each edge e is either labelled with $(k, n)Q$ where Q is again a process*

---

[4] In an equation $T \cong T'$ we assume that T is defined if and only if T' is defined. And in a condition of the form $T \twoheadrightarrow_{\mathcal{A}} T'$ we assume that T is defined if T' is defined.

graph, $1 \leq n \leq ar(Q)$ and $1 \leq k \leq ar(e) = ar(Q) - n$ (e is a process waiting for a message with n ports arriving at its k-th node), with $!Q$ where $ar(Q) = ar(e)$ (e is a process which can replicate itself) or with the constant M (e is a message sent to its last node).

The reduction relation is generated by the rules in (A) (replication) and by rule (B) (reception of a message by a process) and is closed under isomorphism and graph construction.

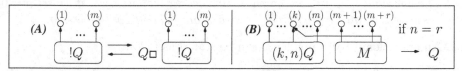

A process graph may contain a bad redex, if it contains a subgraph corresponding to the left-hand side of rule (B) with $n \neq r$, so we define the predicate X as follows: $X(P)$ if and only if P does not contain a bad redex.

We now propose a type system for process graphs by defining the mappings t and f. (Note that in this case, the type graphs are trivially annotated by $\bot$, and so we omit the annotation mapping.)

The linear t mapping is defined on the hyperedges as follows: $t([M]_n) = [\Diamond]_n$ ($\Diamond$ is a new edge label), $t([!Q]_m) = t(Q)$ and $t([[(k,n)Q]_m)$ is defined as in the image to the right (in the notation explained after Definition 3). It is only defined if $n + m = ar(Q)$.

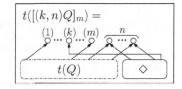

The mapping f is defined as in Proposition 1 where C is defined as follows[5]

$$C(T) \iff \forall e_1, e_2 \in E_T \colon (\lfloor s_T(e_1) \rfloor ar(e_1) = \lfloor s_T(e_2) \rfloor ar(e_2) \Rightarrow e_1 = e_2)$$

The linear mapping t extracts the communication structure from a process graph, i.e. an edge of the form $[\Diamond]_n$ indicates that its nodes (except the last) might be sent or received via its last node. Then f makes sure that the arity of the arriving message matches the expected arity and that nodes that might get fused during reduction are already fused in $f(t(II))$.

Proposition 2. *The trivial annotation mapping $\mathcal{A}$ (where every lattice consists of a single element $\bot$), the mappings f and t and the predicate X defined above satisfy conditions (6)–(9) of Theorem 1. Thus if $P \triangleright T$, then P will never produce a bad redex during reduction.*

We now compare our type system to a standard type system of the π-calculus. An encoding of process graphs into the asynchronous π-calculus can be defined as follows.

Definition 7. (Encoding) *Let P be a process graph, let $\mathcal{N}$ be the name set of the π-calculus and let $\tilde{t} \in \mathcal{N}^*$ such that $|\tilde{t}| = ar(P)$. We define $\Theta_{\tilde{t}}(P)$ inductively as follows:*

---

[5] $\lfloor s \rfloor_i$ extracts the i-th element of a string s.

$$\Theta_{a_1\dots a_{n+1}}([M]_{n+1}) = \overline{a_{n+1}}\langle a_1,\dots,a_n\rangle \qquad \Theta_{\tilde{t}}([!Q]_m) =! \Theta_{\tilde{t}}(Q)$$

$$\Theta_{a_1\dots a_m}([(k,n)Q]_m) = a_k(x_1,\dots,x_n).\Theta_{a_1\dots a_m x_1\dots x_n}(Q)$$

$$\Theta_{\tilde{t}}(\textcircled{D}_{i=1}^n (P_i,\zeta_i)) = (\nu\,\mu(V_D\backslash Set(\chi_D)))(\Theta_{\mu(\zeta_1(\chi_{\mathbf{m_1}}))}(P_1) \mid \dots \mid \Theta_{\mu(\zeta_n(\chi_{\mathbf{m_n}}))}(P_n))$$

where $\zeta_i : \mathbf{m_i} \to D$, $1 \leq i \leq n$ and $\mu : V_D \to \mathcal{N}$ is a mapping such that
μ restricted to $V_D\backslash Set(\chi_D)$ is injective, $\mu(V_D\backslash Set(\chi_D)) \cap \mu(Set(\chi_D)) = \emptyset$ and
$\mu(\chi_D) = \tilde{t}$. Furthermore the $x_1,\dots,x_n \in \mathcal{N}$ are fresh names.

The encoding of a discrete graph is included in the last case, if we set $n = 0$
and assume that the empty parallel composition yields the nil process $\mathbf{0}$.

An operational correspondence can be stated as follows:

Proposition 3. *Let p be an arbitrary expression in the asynchronous polyadic
π-calculus without summation. Then there exists a process graph P and a du-
plicate-free string $\tilde{t} \in \mathcal{N}^*$ such that $\Theta_{\tilde{t}}(P) \equiv p$. Furthermore for process graphs
P, P' and for every duplicate-free string $\tilde{t} \in \mathcal{N}^*$ with $|\tilde{t}| = ar(P) = ar(P')$ it is
true that:*

- *$P \cong P'$ implies $\Theta_{\tilde{t}}(P) \equiv \Theta_{\tilde{t}}(P')$*   *$P \to^* P'$ implies $\Theta_{\tilde{t}}(P) \to^* \Theta_{\tilde{t}}(P)$*
- *$\Theta_{\tilde{t}}(P) \to^* p \neq wrong$ implies that $P \to^* Q$ and $\Theta_{\tilde{t}}(Q) \equiv p$ for some process
  graph Q.*
- *$\Theta_{\tilde{t}}(P) \to^* wrong$ if and only if $P \to^* P'$ for some process graph P' containing
  a bad redex*

We now compare our type system with a standard type system of the π-
calculus: a *type tree* is a potentially infinite ordered tree with only finitely many
non-isomorphic subtrees. A type tree is represented by the tuple $[t_1,\dots,t_n]$ where
$t_1,\dots,t_n$ are again type trees, the children of the root. A type assignment $\Gamma =
x_1 : t_1,\dots,x_n : t_n$ assigns names to type trees where $\Gamma(x_i) = t_i$. The rules of the
type system are simplified versions of the ones from [19], obtained by removing
the subtyping annotations.

$$\Gamma \vdash \mathbf{0} \qquad \frac{\Gamma \vdash p \quad \Gamma \vdash q}{\Gamma \vdash p \mid q} \qquad \frac{\Gamma \vdash p}{\Gamma \vdash !p} \qquad \frac{\Gamma, a : t \vdash p}{\Gamma \vdash (\nu a)p}$$

$$\frac{\Gamma(a) = [t_1,\dots,t_m] \quad \Gamma, x_1 : t_1,\dots,x_m : t_m \vdash p}{\Gamma \vdash a(x_1,\dots,x_m).p} \qquad \frac{\Gamma(a) = [\Gamma(a_1)\dots,\Gamma(a_m)]}{\Gamma \vdash \overline{a}\langle a_1,\dots,a_m\rangle}$$

We will now show that if a process graph has a type, then its encoding has
a type in the π-calculus type system and vice versa. In order to express this we
first describe the unfolding of a type graph into type trees.

Proposition 4. *Let T be a type graph and let σ be a mapping from V_T into the
set of type trees. The mapping σ is called consistent, if it satisfies for every edge
$e \in E_T$: $s_T(e) = v_1\dots v_n v \Rightarrow \sigma(v) = [\sigma(v_1),\dots,\sigma(v_n)]$. Every type graph of
the form $f(t(P))$ has such a consistent mapping.*

*Let $P \triangleright T$ with $n = ar(T)$ and let σ be a consistent mapping for T. Then it
holds for every duplicate-free string $\tilde{t}$ of length n that $\lfloor\tilde{t}\rfloor_1 : \sigma(\lfloor\chi_T\rfloor_1),\dots,\lfloor\tilde{t}\rfloor_n :
\sigma(\lfloor\chi_T\rfloor_n) \vdash \Theta_{\tilde{t}}(P)$.*

Now let $\Gamma \vdash \Theta_{\tilde{i}}(P)$. Then there exists a type graph T such that $P \triangleright T$ and a consistent mapping σ such that for every $1 \leq i \leq |\tilde{t}|$ it holds that $\sigma(\lfloor \chi_T \rfloor_i) = \Gamma(\lfloor \tilde{t} \rfloor_i)$.

4.2   Concurrent Object-Oriented Programming

We now show how to model a concurrent object-oriented system by graph rewriting and then present a type system. In our model, several objects may compete in order to receive a message, and several messages might be waiting at the same object. Typically, type systems in object-oriented programming are there to ensure that an object that receives a message is able to process it.

Definition 8. (Concurrent object-oriented rewrite system) Let $(\mathcal{C}, <:)$ be a lattice of classes with a top class[6] $\top$ and a bottom class $\bot$. We denote classes by the letters $A, B, C, \ldots$. Furthermore let $\mathcal{M}$ be a set of method names. The function $ar : \mathcal{C} \cup \mathcal{M} \to \mathbb{N} \backslash \{0\}$ assigns an arity to every class or method name.

An object graph G is a hypergraph with a duplicate-free string of external nodes, labelled with elements of $\mathcal{C} \backslash \{\bot\} \cup \mathcal{M}$ where for every edge e it holds that $ar(e) = ar(l_G(e))$. A concurrent object-oriented rewrite system (specifying the semantics) consists of a set of rules $\mathcal{R}$ satisfying the following conditions:

- the left-hand side of a rule always has the form shown in Figure (C) below (where $A \in \mathcal{C} \backslash \{\bot\}$, $ar(A) = n$, $m \in \mathcal{M}$, $ar(m) = k + 1$).

The right-hand side is again an object graph of arity $n + k$. If a left-hand side $R_{A,m}$ exists, we say that A understands m.

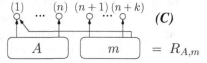

- If $A <: B$, $A \neq \bot$ and B understands m, then A also understands m.
- For all $m \in \mathcal{M}$, either $\{A \mid A$ understands $m\}$ is empty or it contains a greatest element.

An object graph G contains a "message not understood"-error if G contains a subgraph $R_{A,m}$, but A does not understand m.

Thus the predicate X for this section is defined as follows: $X(G)$ if and only if G does not contain a "message not understood"-error.

In contrast to the previous section, we now use annotated type graphs: the annotation mapping $\mathcal{A}$ assigns a lattice $(\{a : V_H \to \mathcal{C} \times \mathcal{C}\}, \leq)$ to every hypergraph H. The partial order is defined as follows: $a_1 \leq a_2 \iff \forall v : (a_1(v) = (A_1, B_1) \wedge a_2(v) = (A_2, B_2) \Rightarrow A_1 <: A_2 \wedge B_1 :> B_2)$, i.e. we have covariance in the first and contravariance in the second position. If a node v is labelled (A, B), this has the following intuitive meaning: we can accept at least as many messages as an object of class A on this node *and* we can send at most as many messages as an object of class B can accept.

---

[6] This corresponds to the class Object in Java

Furthermore we define $\mathcal{A}_\phi(a)(v') = \bigvee_{\phi(v)=v'} a(v)$ where $\phi : H \to H'$, a is an element of $\mathcal{A}(H)$ and $v' \in V_{H'}$.

We now define the operator f: let $T[a]$ be a type graph of arity n where it holds for all nodes v that $a(v) = (A, B)$ implies $A <: B$ (otherwise f is undefined). Then f reduces the graph to its string of external nodes, i.e $f(T[a]) = \mathbf{n}[b]$ where $b(\lfloor \chi_{\mathbf{n}} \rfloor i) = a(\lfloor \chi_T \rfloor i)$.

The linear mapping t determines the type of a class or method. It is necessary to choose a linear mapping that preserves the interface of left-hand and right-hand sides, i.e. we can use any t that satisfies condition (9) and the following two conditions below for $A \in \mathcal{C} \backslash \{\bot\}$ and $m \in \mathcal{M}$:

$$t([A]_n) = [A]_n[a] \text{ where } a(\lfloor \chi_{[A]_n} \rfloor 1) \geq (A, \top)$$
$$t([m]_n) = [m]_n[a] \text{ where } a(\lfloor \chi_{[m]_n} \rfloor n) \geq (\bot, \max\{B \mid B \text{ understands } m\})$$

Proposition 5. *The annotation mapping $\mathcal{A}$, the mappings f and t and the predicate X defined above satisfy conditions (6)–(9) of Theorem 1. Thus if $G \triangleright T$, then G will never produce a "message not understood"-error during reduction.*

In this case we do not prove that this type systems corresponds to an object-oriented type system, but rather present a semi-formal argument: we give the syntax and a type system for a small object calculus, and furthermore an encoding into hypergraphs, without really defining the semantics. For the formal semantics of object calculi see [20,9], among others.

An expression e in the object calculus either has the form $new\ A(e_1, \ldots, e_n)$ where $A \in \mathcal{C} \backslash \{\bot\}$ and $ar(A) = n + 1$ or $e.m(e_1, \ldots, e_n)$ where $m \in \mathcal{M}$ and $ar(m) = n + 2$. The e_i are again expressions. Every class A is assigned an $(ar(A) - 1)$-tuple of classes defining the type of the fields of A ($A : (A_1, \ldots, A_n)$) and every method m with $ar(m) = n + 2$ defined in class B is assigned a type $B.m : C_1, \ldots, C_n \to C$. If a method is overwritten in a subclass it is required to have the same type. A simple type systems looks as follows:

$$\frac{e : A,\ A <: B}{e : B} \qquad \frac{A : (A_1, \ldots, A_n),\ e_i : A_i}{new\ A(e_1, \ldots, e_n) : A} \qquad \frac{e : B,\ B.m : C_1, \ldots, C_n \to C,\ e_i : C_i}{e.m(e_1, \ldots, e_n) : C}$$

Now an encoding $[\![\cdot]\!]$ can be defined as shown in the figure to the right. We introduce the convention that the penultimate node of a message can be used to access the result after the rewriting step.

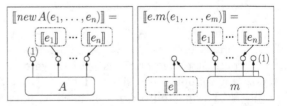

If $A : (A_1, \ldots, A_n)$ we define t in such a way that the $n + 1$ external nodes of $t([A]_{n+1})$ are annotated by $(A, \top)$, $(\bot, A_1)$, $\ldots$, $(\bot, A_n)$. And if $B.m : C_1, \ldots, C_n \to C$ (where B is the maximal class which understands method m), we annotate the external nodes of $t([m]_{n+2})$ by $(\bot, C_1)$, $\ldots$, $(\bot, C_n)$, $(C, \top)$, $(\bot, B)$. Now we can show by induction on the typing rules that if $e : A$, then there exists a type graph $T[a]$ such that $[\![e]\!] \triangleright T[a]$ and $a(\lfloor \chi_T \rfloor 1) = (A, \top)$.

5   Conclusion and Comparison to Related Work

This is a first tentative approach aimed at developing a general framework for the static analysis of graph rewriting in the context of type systems. It is obvious that there are many type systems which do not fit well into our proposal. But since we are able to capture the essence of two important type systems, we assume to be on the right track.

Types are often used to make the connection of components and the flow of information through a system explicit (see e.g. the type system for the π-calculus, where the type trees indicate which tuple of channels is sent via which channel). Since connections are already explicit in graphs, we can use them both as type and as the expression to be typed. Via morphisms we can establish a clear connection between an expression and its type. Graphs are furthermore useful since we can easily add an extra layer of annotation.

Work that is very close in spirit to ours is [8] by Honda which also presents a general framework for type systems. The underlying model is closer to standard process algebras and the main focus is on the characterisation and classification of type systems.

The idea of composing graphs in such a way that they satisfy a certain property was already presented by Lafont in [14] where it is used to obtain deadlock-free nets.

In graph rewriting there already exists a concept of typed graphs [1], related to ours, but nevertheless different. In that work, a type graph is fixed a priori and there is only one type graph for every set of productions. Graphs are considered valid only if they can be mapped into the type graph by a graph morphism (this is similar to our proposal). In our case, we compute the type graphs a posteriori and it is a crucial point in the design of every type system to distinguish as many graphs as possible by assigning different type graphs to them.

This paper is a continuation of the work presented in [10] where the idea of generic type systems for process graphs (as defined in section 4.1) was introduced, but no proof of the equivalence of our type system to the standard type system for the π-calculus was given. The ideas presented there are now extended to general graph rewriting systems.

Further work will consist in better understanding the underlying mechanism of the type system. An interesting question in this context is the following: given a set of rewrite rules, is it possible to automatically derive mappings f and t satisfying the conditions of Theorem 1?

Acknowledgements: I would like to thank Reiko Heckel and Andrea Corradini for their comments on drafts of this paper, and Tobias Nipkow for his advice. I am also grateful to the anonymous referees for their valuable comments.

References

1. A. Corradini, U. Montanari, and F. Rossi. Graph processes. *Fundamenta Informaticae*, 26(3/4):241–265, 1996.

2. H. Ehrig. Introduction to the algebraic theory of graphs. In *Proc. 1st International Workshop on Graph Grammars*, pages 1–69. Springer-Verlag, 1979. LNCS 73.
3. H. Ehrig, H.-J. Kreowski, U. Montanari, and G. Rozenberg, editors. *Handbook of Graph Grammars and Computing by Graph Transformation, Vol.3: Concurrency, Parallellism, and Distribution*. World Scientific, 1999.
4. F. Gadducci and U. Montanari. Comparing logics for rewriting: Rewriting logic, action calculi and tile logic. *Theoretical Computer Science*, 2000. to appear.
5. Philippa Gardner. Closed action calculi. *Theoretical Computer Science (in association with the conference on Mathematical Foundations in Programming Semantics)*, 1998.
6. Annegret Habel. *Hyperedge Replacement: Grammars and Languages*. Springer-Verlag, 1992. LNCS 643.
7. Masahito Hasegawa. *Models of Sharing Graphs (A Categorical Semantics of Let and Letrec)*. PhD thesis, University of Edingburgh, 1997. available in Springer Distinguished Dissertation Series.
8. Kohei Honda. Composing processes. In *Proc. of POPL'96*, pages 344–357. ACM, 1996.
9. Atsushi Igarashi, Benjamin Pierce, and Philip Wadler. Featherweight Java: A core calculus for Java and GJ. In *Proc. of OOPSLA 1999*, 1999.
10. Barbara König. Generating type systems for process graphs. In *Proc. of CONCUR '99*, pages 352–367. Springer-Verlag, 1999. LNCS 1664.
11. Barbara König. A general framework for types in graph rewriting. Technical Report TUM-I0014, Technische Universität München, 2000.
12. Barbara König. A graph rewriting semantics for the polyadic pi-calculus. In *Workshop on Graph Transformation and Visual Modeling Techniques (Geneva, Switzerland), ICALP Workshops 2000*, pages 451–458. Carleton Scientific, 2000.
13. Barbara König. Hypergraph construction and its application to the compositional modelling of concurrency. In *GRATRA 2000: Joint APPLIGRAPH/GETGRATS Workshop on Graph Transformation Systems*, 2000.
14. Yves Lafont. Interaction nets. In *Proc. of POPL '90*, pages 95–108. ACM Press, 1990.
15. James J. Leifer and Robin Milner. Deriving bisimulation congruences for reactive systems. In *Proc. of CONCUR 2000*, 2000. LNCS 1877.
16. José Meseguer. Rewriting logic as a semantic framework for concurrency: A progress report. In *Concurrency Theory*, pages 331–372. Springer-Verlag, 1996. LNCS 1119.
17. Robin Milner. The polyadic π-calculus: a tutorial. In F. L. Hamer, W. Brauer, and H. Schwichtenberg, editors, *Logic and Algebra of Specification*. Springer-Verlag, Heidelberg, 1993.
18. Robin Milner. Calculi for interaction. *Acta Informatica*, 33(8):707–737, 1996.
19. Benjamin Pierce and Davide Sangiorgi. Typing and subtyping for mobile processes. In *Proc. of LICS '93*, pages 376–385, 1993.
20. David Walker. Objects in the π-calculus. *Information and Computation*, 116:253–271, 1995.
21. Nobuko Yoshida. Graph notation for concurrent combinators. In *Proc. of TPPP '94*. Springer-Verlag, 1994. LNCS 907.

The Ground Congruence for Chi Calculus

Yuxi Fu* and Zhenrong Yang

Department of Computer Science
Shanghai Jiaotong University, Shanghai 200030, China

Abstract. The definition of open bisimilarity on the χ-processes does not give rise to a sensible relation on the χ-processes with the mismatch operator. The paper proposes ground open congruence as a principal open congruence on the χ-processes with the mismatch operator. The algebraic properties of the ground congruence is studied. The paper also takes a close look at barbed congruence. This relation is similar to the ground congruence. The precise relationship between the two is worked out. It is pointed out that the sound and complete system for the ground congruence can be obtained by removing one tau law from the complete system for the barbed congruence.

1 Introduction and χ-Calculus with Mismatch

The π-calculus ([6]) is a powerful process calculus. The expressiveness is partly supported by input processes of the form $a(x).P$ and output processes of the form $\bar{a}x.P$. The former may receive a name at channel name a before evolving as P with x replaced by the received name. The latter can emit x at a and then continues as P. The expressiveness is also supported by processes of the form $(x)P$. The localization operator (x) encapsulates the name x in P. In χ-calculus ([1,2,3,4]) the input and output processes are unified as $\alpha[x].P$, in which α stands for either a name or a coname.

Formally χ-processes are defined by the following abstract syntax:

$$P := \mathbf{0} \mid \alpha[x].P \mid P|P \mid (x)P \mid [x{=}y]P \mid P{+}P$$

where $\alpha \in \mathcal{N} \cup \overline{\mathcal{N}}$. Here $\mathcal{N}$ is the set of names ranged over by small case letters. The set $\{\bar{x} \mid x \in \mathcal{N}\}$ of conames is denoted by $\overline{\mathcal{N}}$. The name x in $(x)P$ is local. A name is global in P if it is not local in P. The global names, the local names and the names of a syntactical object, as well as the notations $gn(\_)$, $ln(\_)$ and $n(\_)$, are defined with their standard meanings. We adopt the α-convention widely used in the literature on process algebra. We do not consider replication or recursion operator since it does not affect the results of this paper.

---
* The author is funded by NNSFC (69873032) and 863 Hi-Tech Project (863-306-ZT06-02-2). He is also supported by BASICS, Center of Basic Studies in Computing Science, sponsored by Shanghai Education Committee. BASICS is affiliated to the Department of Computer Science at Shanghai Jiaotong University.

S. Kapoor and S. Prasad (Eds.): FST TCS 2000, LNCS 1974, pp. 385–396, 2000.

The following labeled transition system defines the operational semantics of χ-calculus, in which symmetric rules are systematically omitted. In the following rules the letter γ ranges over the set $\{\alpha(x), \alpha[x] \mid \alpha \in \mathcal{N} \cup \overline{\mathcal{N}}, x \in \mathcal{N}\} \cup \{\tau\}$ and the letter λ over the set $\{\alpha(x), \alpha[x], [y/x] \mid \alpha \in \mathcal{N} \cup \overline{\mathcal{N}}, x, y \in \mathcal{N}\} \cup \{\tau\}$. The symbols $\alpha(x), \alpha[x], [y/x]$ represent restricted action, free action and update action respectively. The x in the label $\alpha(x)$ is local.

Sequentialization

$$\frac{}{\alpha[x].P \xrightarrow{\alpha[x]} P} Sqn$$

Composition

$$\frac{P \xrightarrow{\gamma} P' \quad ln(\gamma) \cap gn(Q) = \emptyset}{P|Q \xrightarrow{\gamma} P'|Q} Cmp_0 \qquad \frac{P \xrightarrow{[y/x]} P'}{P|Q \xrightarrow{[y/x]} P'|Q[y/x]} Cmp_1$$

Communication

$$\frac{P \xrightarrow{\alpha(x)} P' \quad Q \xrightarrow{\overline{\alpha}[y]} Q'}{P|Q \xrightarrow{\tau} P'[y/x]|Q'} Cmm_0 \qquad \frac{P \xrightarrow{\alpha(x)} P' \quad Q \xrightarrow{\overline{\alpha}(x)} Q'}{P|Q \xrightarrow{\tau} (x)(P'|Q')} Cmm_1$$

$$\frac{P \xrightarrow{\alpha[x]} P' \quad Q \xrightarrow{\overline{\alpha}[y]} Q' \quad x \neq y}{P|Q \xrightarrow{[y/x]} P'[y/x]|Q'[y/x]} Cmm_2 \qquad \frac{P \xrightarrow{\alpha[x]} P' \quad Q \xrightarrow{\overline{\alpha}[x]} Q'}{P|Q \xrightarrow{\tau} P'|Q'} Cmm_3$$

Localization

$$\frac{P \xrightarrow{\lambda} P' \quad x \notin n(\lambda)}{(x)P \xrightarrow{\lambda} (x)P'} Loc_0 \qquad \frac{P \xrightarrow{\alpha[x]} P' \quad x \notin \{\alpha, \overline{\alpha}\}}{(x)P \xrightarrow{\alpha(x)} P'} Loc_1 \qquad \frac{P \xrightarrow{[y/x]} P'}{(x)P \xrightarrow{\tau} P'} Loc_2$$

Condition

$$\frac{P \xrightarrow{\lambda} P'}{[x=x]P \xrightarrow{\lambda} P'} Mtch$$

Summation

$$\frac{P \xrightarrow{\lambda} P'}{P+Q \xrightarrow{\lambda} P'} Sum$$

A substitution is a function from $\mathcal{N}$ to $\mathcal{N}$ that is identical on all but a finite number of names. Substitutions are usually denoted by $\sigma, \sigma', \ldots$. The notations $\Longrightarrow$ and $\overset{\widehat{\lambda}}{\Longrightarrow}$ are used in their standard meanings.

We will use two induced prefix operators, tau and update prefixes, defined as follows: $[y|x].P \overset{\text{def}}{=} (a)(\overline{a}[y]|a[x].P)$ and $\tau.P \overset{\text{def}}{=} (b)[b|b].P$ where a, b are fresh.

The subject language of this paper is $\chi^{\neq}$-calculus, the χ-calculus with the mismatch operator. The operational semantics of the mismatch combinator is defined as follows:

$$\frac{P \xrightarrow{\lambda} P' \quad x \neq y}{[x \neq y]P \xrightarrow{\lambda} P'} Mismtch$$

The set of $\chi^{\neq}$-processes is denoted by $\mathcal{C}$. Suppose Y is a finite set $\{y_1, \ldots, y_n\}$ of names. The notation $[y \notin Y]P$ will stand for $[y \neq y_1] \ldots [y \neq y_n]P$, where the order of mismatch operators is immaterial. We will write ϕ and ψ, called conditions, to stand for sequences of match and mismatch combinators concatenated one after another, μ for a sequence of match operators, and δ for a sequence of mismatch operators. Consequently we write ψP, μP and δP. When the length of ψ (μ, δ) is zero, ψP (μP, δP) is just P. The notation $\phi \Rightarrow \psi$ says that ϕ logically implies ψ and $\phi \Leftrightarrow \psi$ that ϕ and ψ are logically equivalent. A substitution σ agrees with ψ, and ψ agrees with σ, when $\psi \Rightarrow x{=}y$ if and only if $\sigma(x){=}\sigma(y)$.

Bisimulation equivalence relations on mobile processes are a lot more complex than those on CCS processes. The complication is mainly due to the dynamic aspect of mobile processes. The names in a process are subject to updates during the evolution of the process. These updates could be caused either by actions in which the process participates or by changes incurred by environments. A sensible observational equivalence for mobile processes must take that into account. To illustrate what kind of relations one would obtain if s/he ignored the mobility, we introduce the following definition for χ-calculus:

Definition 1. *Let $\mathcal{R}$ be a symmetric binary relation on the set of χ-processes. It is called a naked bisimulation if whenever $P\mathcal{R}Q$ and $P \xrightarrow{\lambda} P'$ then some Q' exists such that $Q \overset{\widehat{\lambda}}{\Longrightarrow} Q'\mathcal{R}P'$. The naked bisimilarity $\approx$ is the largest naked bisimulation.*

It is obvious that the definition of $\approx$ is simply a reiteration of the weak bisimilarity of CCS in terms of the operational semantic of χ-calculus. However the naked bisimilarity is not a good equivalence relation since it is not closed under the parallel composition. For instance one has $a[x]|\bar{b}[y] \approx a[x].\bar{b}[y]+\bar{b}[y].a[x]$ but not $(a[x]|\bar{b}[y])|(c[a]|\bar{c}[b]) \approx (a[x].\bar{b}[y]+\bar{b}[y].a[x])|(c[a]|\bar{c}[b])$. Process equivalence is observational equivalence. One of the defining properties for an observational equivalence is that the equivalence should be closed under parallel composition. In [1,2,3,4], it has been argued that bisimulation equivalences for χ-calculus are closed under substitution. This suggests to introduce the following definition:

Let $\mathcal{R}$ be a symmetric binary relation on the set of χ-processes that is closed under substitution. It is called an open bisimulation if whenever $P\mathcal{R}Q$ and $P \xrightarrow{\lambda} P'$ then some Q' exists such that $Q \overset{\widehat{\lambda}}{\Longrightarrow} Q'\mathcal{R}P'$. The open bisimilarity $\approx_o$ is the largest open bisimulation.

The open bisimilarity $\approx_o$ has been studied in [1,2,3,4] in both the symmetric and the asymmetric frameworks. It must be pointed out that the investigations carried out in [1,2,3,4] are for the χ-calculus *without* the mismatch combinator. For the χ-calculus *with* the mismatch operator, one should ask the question whether the open bisimilarity $\approx_o$ is a sensible equivalence. In [5] the present authors have given a negative answer to the question. As it turned out the open bisimilarity defined above is not closed under parallel composition in $\chi^{\neq}$-calculus! One has $[x{\neq}y]a[x].P + a[x].[x{\neq}y]\tau.P \approx_o a[x].[x{\neq}y]\tau.P$ but it is clear that $\overline{a}[y]|([x\neq y]a[x].P + a[x].[x\neq y]\tau.P) \not\approx_o \overline{a}[y]|(a[x].[x\neq y]\tau.P)$. This is a serious problem because closure under parallel composition is an intrinsic property of observational equivalence. In [5] we have studied the problem and introduced two modified open congruences. These are early open congruence and late open congruence. Their relationship strongly recalls that between the weak early equivalence and the weak late equivalence ([6]). It should be said however that both the early open congruence and the late open congruence are the obvious modifications with motivation from π-calculus. They are not *the* open congruence for the χ-calculus with the mismatch operator. What is then the principal open congruence for χ-calculus with the mismatch combinator? We will give our answer to the question in this paper. The way to arrive to the definition of the open congruence is via a particular naked bisimulation. In order to define this relation we need the notion of contexts defined as follows: (i) $[]$ is a context; (ii) If $C[]$ is a context then $\alpha[x].C[]$, $C[]|P$, $P|C[]$, $(x)C[]$ and $[x=y]C[]$ are contexts.

Definition 2. *The ground bisimilarity $\approx_g$ is the largest naked bisimulation that is closed under context.*

In the above definition the requirement of closure under the prefix operator is reasonable since it is equivalent to that of closure under substitution. We will give an equivalent characterization of $\approx_g$ in the style of open semantics, which we argue is the principal open bisimilarity.

As it turns out the equivalence $\approx_g$ is very similar to the barbed bisimilarity of the χ-calculus with the mismatch operator. The difference is very subtle. The barbed bisimilarity also has an equivalent open characterization. The similarity and the difference between the ground bisimilarity and the barbed bisimilarity are revealed through their open characterizations.

This paper continues the work of [5] by studying the ground congruence and the barbed congruence for the $\chi^{\neq}$-calculus. The main contributions of this paper are as follows:

- We give an alternative characterization of the weak barbed bisimilarity. This characterization points out the complex nature of the weak barbed bisimilarity. Many unknown equalities are discovered. A complete system for the weak barbed congruence is provided. The new tau laws used to establish the completeness result are surprisingly complex.
- We study what we call ground open bisimilarity. A complete system for the ground open congruence is given. The relationship between the ground open congruence and the weak barbed congruence is revealed.

Due to space limitation, all proofs have been omitted.

2   Barbed Congruence

The barbed equivalence is often quoted as a universal equivalence relation for process algebras. For a specific process calculus barbed equivalence immediately gives rise to an observational equivalence. For two process calculi barbed equivalence can be used to compare the semantics of the two models. Despite the universal nature, barbed equivalence can have quite different displays in different process calculi. The barbed equivalence for the χ-calculus has brought some new insight into the calculi of mobile processes. In this section we demonstrate that the barbed equivalence for the $\chi^{\neq}$-calculus is even more different. A characterization theorem for the barbed bisimilarity on $\chi^{\neq}$-calculus is given. Some illustrating pairs of barbed equivalent processes are given. First we introduce the notion of barbedness.

Definition 3. *A process P is strongly barbed at a, notation $P{\downarrow}a$, if $P \xrightarrow{\alpha(x)} P'$ or $P \xrightarrow{\alpha[x]} P'$ for some P' such that $a \in \{\alpha, \overline{\alpha}\}$. P is barbed at a, written $P{\Downarrow}a$, if some P' exists such that $P \Longrightarrow P'{\downarrow}a$. A binary relation $\mathcal{R}$ is barbed if $\forall a \in \mathcal{N}.P{\Downarrow}a \Leftrightarrow Q{\Downarrow}a$ whenever $P\mathcal{R}Q$.*

From the point of view of barbed equivalence an observer can not see the content of a communication. What an observer can detect is the ability of a process to communicate at particular channels. Two processes are identified if they can simulate each other in terms of this ability.

Definition 4. *Let $\mathcal{R}$ be a barbed symmetric relation on $\mathcal{C}$ closed under context. The relation $\mathcal{R}$ is a barbed bisimulation if whenever $P\mathcal{R}Q$ and $P \xrightarrow{\tau} P'$ then $Q \Longrightarrow Q'\mathcal{R}P'$ for some Q'. The barbed bisimilarity $\approx_b$ is the largest barbed bisimulation.*

The trade-off of the simplicity of the above definition is that it provides little intuition about equivalent processes. We know that it is weaker than most bisimulation equivalences. But we want to know how much weaker it is. We first give some examples of barbed equivalent processes. To make the examples more readable, we will write $A \stackrel{\text{def}}{=} P\mathcal{R}(A+Q)$ for $P\mathcal{R}(P+Q)$, where $\mathcal{R}$ is a binary relation on processes. The first example of an equivalent pair is this:

$$A_1 \stackrel{\text{def}}{=} \alpha[x].(P_1+[x{=}y_1]\tau.Q)+\alpha[x].(P_2+[x{\neq}y_1]\tau.Q) \approx_b A_1 + \alpha[x].Q$$

If $\alpha[x].Q$ on the right hand side is involved in a communication in which x is replaced by y_1 then $\alpha[x].(P_1+[x{=}y_1]\tau.Q)$ can simulate the action. Otherwise $\alpha[x].(P_2+[x{\neq}y_1]\tau.Q)$ would do the job. The second example is more interesting:

$$A_2 \stackrel{\text{def}}{=} (z)\alpha[z].(P_1+[z{=}y_2][z|x].Q)+\alpha[x].(P_2+[x{\neq}y_2]\tau.Q[x/z])$$
$$\approx_b A_2 + \alpha[x].Q[x/z]$$

The communication $\overline{\alpha}[y_2]|(x)(A_2+\alpha[x].Q[x/z]) \xrightarrow{\tau} \mathbf{0}|Q[x/z][y_2/x]$ for instance can be matched up by $\overline{\alpha}[y_2]|(x)A_2 \xrightarrow{\tau} \mathbf{0}|(x)(P_1[y_2/z]+[y_2{=}y_2][y_2|x].Q[y_2/z]) \xrightarrow{\tau} \mathbf{0}|Q[y_2/z][y_2/x]$. The third example is unusual:

$$A_3 \stackrel{\text{def}}{=} \alpha[y_3].(P_1+[y_3|x].Q)+\alpha[x].(P_2+[x{\neq}y_3]\tau.Q) \approx_b A_3 + \alpha[x].Q$$

If $\alpha[x].Q$ participates in a communication in which x exchanges for y_3 then its role can be simulated by $\alpha[y_3].(P_1+[y_3|x].Q)$. The fourth is similar:

$$A_4 \stackrel{\text{def}}{=} [y_4|x].(P_1+\alpha[y_4].Q)+\alpha[x].(P_2+[x{\neq}y_4]\tau.Q) \approx_b A_4 + \alpha[x].Q$$

If $(y_4)((A_4+\alpha[x].Q)|\overline{\alpha}[y_4].O) \stackrel{\tau}{\longrightarrow} Q[x/y_4]|O[x/y_4]$ then the simulation is:

$$(y_4)(A_4|\overline{\alpha}[y_4].O) \stackrel{\tau}{\longrightarrow} (P_1[x/y_4]+\alpha[x].Q[x/y_4])|\overline{\alpha}[x].O[x/y_4] \stackrel{\tau}{\longrightarrow} Q[x/y_4]|O[x/y_4]$$

The fifth example is the combination of the fourth and the second:

$$A_5 \stackrel{\text{def}}{=} [y_5|x].(P_1+(z)\alpha[z].(P_1'+[z{=}y_5]\tau.Q))+\alpha[x].(P_2+[x{\neq}y_5]\tau.Q[x/z])$$
$$\approx_b A_5 + \alpha[x].Q[x/z]$$

Notice that the component $[y_5|x].(P_1+(z)\alpha[z].(P_1'+[z{=}y_5]\tau.Q))$ is operationally the same as the process $[y_5|x].(P_1+(z)\alpha[z].(P_1'+[z{=}y_5][z|x].Q))$.

In the above examples, all the explicit mismatch operators contain the name x. In general there could be other conditions. The treatment of match operator is easy. The mismatch operator is however nontrivial. Suppose δ is a sequence of mismacth operators such that all names in δ are different from both x and z. An example more gerneral than A_1 is this:

$$A_1' \stackrel{\text{def}}{=} \alpha[x].(P_1+\delta[x{=}y_1]\tau.Q)+\alpha[x].(P_2+\delta[x{\neq}y_1]\tau.Q) \approx_b A_1' + [x{\notin}n(\delta)]\delta\alpha[x].Q$$

We need to explain the mismatch sequence in $[x{\notin}n(\delta)]\delta\alpha[x].Q$. The δ before $\alpha[x].Q$ is necessary for otherwise an action of $([x{\notin}n(\delta)]\alpha[x].Q)\sigma$ may not be simulated by any action from $A_1'\sigma$ when σ invalidates δ. The $[x{\notin}n(\delta)]$ is necessary because otherwise it would not be closed under substitution. A counter example is the pair $\alpha[x].[y{\neq}z][x{=}y_1]\tau.Q+\alpha[x].[y{\neq}z][x{\neq}y_1]\tau.Q+[y{\neq}z]\alpha[x].Q$ and $\alpha[x].[y{\neq}z][x{=}y_1]\tau.Q+\alpha[x].[y{\neq}z][x{\neq}y_1]\tau.Q$. If we substitute x for z in the two processes we get two processes that are not barbed bisimilar. Similarly the example A_2 can be generalized to the following:

$$A_2' \stackrel{\text{def}}{=} (z)\alpha[z].(P_1+[x{\notin}n(\delta)]\delta[z{=}y_2][z|x].Q)+\alpha[x].(P_2+\delta[x{\neq}y_2]\tau.Q[x/z])$$
$$\approx_b A_2' + [x{\notin}n(\delta)]\delta\alpha[x].Q[x/z]$$

The general form of A_3 is more delicate:

$$A_3' \stackrel{\text{def}}{=} [x{\neq}y_3]\alpha[y_3].(P_1+[x{\notin}n(\delta)]\delta[y_3|x].Q)+\alpha[x].(P_2+\delta[x{\neq}y_3]\tau.Q)$$
$$\approx_b A_3' + [x{\neq}y_3][x{\notin}n(\delta)]\delta\alpha[x].Q$$

In both $[x{\neq}y_3]\alpha[y_3].(P_1+[x{\notin}n(\delta)]\delta[y_3|x].Q)$ and $[x{\neq}y_3][x{\notin}n(\delta)]\delta\alpha[x].Q$ there is the mismatch $[x{\neq}y_3]$. If this operator is removed from A_3' one has

$$B_3' \stackrel{\text{def}}{=} \alpha[y_3].(P_1+[x{\notin}n(\delta)]\delta[y_3|x].Q)+\alpha[x].(P_2+\delta[x{\neq}y_3]\tau.Q)$$
$$\not\approx_b B_3' + [x{\notin}n(\delta)]\delta\alpha[x].Q$$

The inequality is clearer if one substitutes x for y_3 in the above:

$$C_3' \stackrel{\text{def}}{=} \alpha[x].(P_1+[x\notin n(\delta)]\delta[x|x].Q)+\alpha[x].(P_2+\delta[x\neq x]\tau.Q)$$
$$\not\approx_b C_3' + [x\notin n(\delta)]\delta\alpha[x].Q$$

The component $[x\notin n(\delta)]\delta\alpha[x].Q$ may be involved in a communication in which x is replaced by a name in δ. This action can not be simulated by C_3'. The general forms of A_4 and A_5 are as follows:

$$A_4' \stackrel{\text{def}}{=} [y_4|x].(P_1+\delta\alpha[y_4].Q)+\alpha[x].(P_2+\delta[x\neq y_4]\tau.Q) \approx_b A_4' + [x\notin n(\delta)]\delta\alpha[x].Q$$
$$A_5' \stackrel{\text{def}}{=} [y_5|x].(P_1+(z)\alpha[z].(P_1'+\delta[z=y_5]\tau.Q))+\alpha[x].(P_2+\delta[x\neq y_5]\tau.Q[x/z])$$
$$\approx_b A_5' + [x\notin n(\delta)]\delta\alpha[x].Q[x/z]$$

If we replace the second summand $\alpha[x].(P_2+\delta[x\neq y_1]\tau.Q)$ of A_1' by $(z)\alpha[z].(P_2+[x\notin n(\delta)]\delta[z\neq y_1][z|x].Q)$ and Q by $Q[x/z]$, we get an interesting variant of A_1' as follows:

$$A_1'' \stackrel{\text{def}}{=} \alpha[x].(P_1+\delta[x=y_1]\tau.Q[x/z])+(z)\alpha[z].(P_2+[x\notin n(\delta)]\delta[z\neq y_1][z|x].Q)$$
$$\approx_b A_1'' + [x\notin n(\delta)]\delta\alpha[x].Q[x/z]$$

The bisimilar pairs A_2' through A_5' have similar variants:

$$A_2'' \stackrel{\text{def}}{=} (z)\alpha[z].(P_1+[x\notin n(\delta)]\delta[z=y_2][z|x].Q)+O_2$$
$$\approx_b A_2'' + [x\notin n(\delta)]\delta\alpha[x].Q[x/z]$$
$$A_3'' \stackrel{\text{def}}{=} [x\neq y_3]\alpha[y_3].(P_1+[x\notin n(\delta)]\delta[y_3|x].Q[x/z])+O_3$$
$$\approx_b A_3'' + [x\neq y_3][x\notin n(\delta)]\delta\alpha[x].Q[x/z]$$
$$A_4'' \stackrel{\text{def}}{=} [y_4|x].(P_1+\delta\alpha[y_4].Q[x/z])+O_4$$
$$\approx_b A_4'' + [x\notin n(\delta)]\delta\alpha[x].Q[x/z]$$
$$A_5'' \stackrel{\text{def}}{=} [y_5|x].(P_1+(z)\alpha[z].(P_1'+\delta[z=y_5]\tau.Q[z/x]))+O_5$$
$$\approx_b A_5'' + [x\notin n(\delta)]\delta\alpha[x].Q[x/z]$$

where O_i is $(z)\alpha[z].(P_2+[x\notin n(\delta)]\delta[z\neq y_i][z|x].Q)$ for $i \in \{2,3,4,5\}$. The most complicated situation arises when all the five possibilities as described by A_1'' through A_5'' happen at one go:

$$A \stackrel{\text{def}}{=} (z)\alpha[z].(P_2+[x\notin n(\delta)]\delta[z\notin\{y_1,y_2,y_3,y_4,y_5\}][z|x].Q)$$
$$+\alpha[x].(P_1+\delta[x=y_1]\tau.Q[x/z])$$
$$+(z)\alpha[z].(P_1+[x\notin n(\delta)]\delta[z=y_2][z|x].Q)$$
$$+[x\neq y_3]\alpha[y_3].(P_1+[x\notin n(\delta)]\delta[y_3|x].Q[x/z])$$
$$+[y_4|x].(P_1+\delta\alpha[y_4].Q[x/z])$$
$$+[y_5|x].(P_1+(z)\alpha[z].(P_1'+\delta[z=y_5]\tau.Q[z/x]))$$
$$\approx_b A + [x\neq y_3][x\notin n(\delta)]\delta\alpha[x].Q[x/z]$$

Similarly the examples A'_1 through A'_5 can be combined into one as follows:

$$A' \stackrel{def}{=} \alpha[x].(P_2+\delta[x\notin\{y_1,y_2,y_3,y_4,y_5\}]\tau.Q[x/z])$$
$$+\alpha[x].(P_1+\delta[x=y_1]\tau.Q[x/z])$$
$$+(z)\alpha[z].(P_1+[x\notin n(\delta)]\delta[z=y_2][z|x].Q)$$
$$+[x\neq y_3]\alpha[y_3].(P_1+[x\notin n(\delta)]\delta[y_3|x].Q[x/z])$$
$$+[y_4|x].(P_1+\delta\alpha[y_4].Q[x/z])$$
$$+[y_5|x].(P_1+(z)\alpha[z].(P'_1+\delta[z=y_5]\tau.Q[z/x]))$$
$$\approx_b A' + [x\neq y_3][x\notin n(\delta)]\delta\alpha[x].Q[x/z]$$

Having seen so many bisimilar pairs of processes, the reader might wonder how we have discovered them. As a matter of fact these examples are all motivated by an alternative characterization of the barbed bisimilarity. This characterization is given by an open bisimilarity as defined below.

Definition 5. *Let $\mathcal{R}$ be a binary symmetric relation on $\mathcal{C}$ closed under substitution. The relation $\mathcal{R}$ is a barbed open bisimulation if the following properties hold for P and Q whenever $P\mathcal{R}Q$:*

(i) If λ is an update or a tau and $P \stackrel{\lambda}{\longrightarrow} P'$ then Q' exists such that $Q \stackrel{\widehat{\lambda}}{\Longrightarrow} Q'\mathcal{R}P'$.

(ii) If $P \stackrel{\alpha[x]}{\longrightarrow} P'$ then one of the following properties holds:

- *Q' exists such that $Q \stackrel{\alpha[x]}{\Longrightarrow} Q'\mathcal{R}P'$;*
- *Q' and Q'' exist such that $Q \Longrightarrow \stackrel{\alpha(z)}{\longrightarrow} Q''$ and $Q''[x/z] \Longrightarrow Q'\mathcal{R}P'$;*

and, for each y different from x, one of the following properties holds:

- *Q' and Q'' exist such that $Q \Longrightarrow \stackrel{\alpha[x]}{\longrightarrow} Q''$ and $Q''[y/x] \Longrightarrow Q'\mathcal{R}P'[y/x]$;*
- *Q' and Q'' exist such that $Q \Longrightarrow \stackrel{\alpha(z)}{\longrightarrow} Q''$ and $Q''[y/z] \stackrel{[y/x]}{\Longrightarrow} Q'\mathcal{R}P'[y/x]$;*
- *Q' exists such that $Q \stackrel{\alpha[y]}{\Longrightarrow}\stackrel{[y/x]}{\longrightarrow} Q'\mathcal{R}P'[y/x]$;*
- *Q' exists such that $Q \stackrel{[y/x]}{\Longrightarrow}\stackrel{\alpha[y]}{\longrightarrow} Q'\mathcal{R}P'[y/x]$;*
- *Q' and Q'' exist such that $Q \stackrel{[y/x]}{\Longrightarrow}\stackrel{\alpha(z)}{\longrightarrow} Q''$ and $Q''[y/z]\Longrightarrow Q'\mathcal{R}P'[y/x]$.*

(iii) If $P \stackrel{\alpha(x)}{\longrightarrow} P'$ then, for each y, one of the following properties holds:

- *Q' and Q'' exist such that $Q \Longrightarrow\stackrel{\alpha(x)}{\longrightarrow} Q''$ and $Q''[y/x]\Longrightarrow Q'\mathcal{R}P'[y/x]$;*
- *Q' exists such that $Q \stackrel{\alpha[y]}{\Longrightarrow} Q'\mathcal{R}P'[y/x]$.*

The barbed open bisimilarity $\approx^b_{open}$ is the largest barbed open bisimulation.

With a definition as complex as Definition 5, it is not very clear if the relation it introduces is well behaved. The next lemma gives one some confidence on the barbed open bisimilarity.

Lemma 6. $\approx^b_{open}$ *is closed under localization and composition.*

Since $\approx^b_{open}$ is closed under substitution, it must also be closed under prefix operation. It is also clear that $\approx^b_{open}$ is closed under match operation. However the relation is closed neither under the mismatch operation nor under the summation operation. For instance $[x{\neq}y]P \approx^b_{open} [x{\neq}y]\tau.P$ does not hold. To obtain the largest congruence contained in $\approx^b_{open}$ we use the standard approach.

Definition 7. *Two processes P and Q are barbed congruent, notation $P \simeq_b Q$, if $P \approx^b_{open} Q$ and for each substitution σ whenever $P\sigma \xrightarrow{\tau} P'$ then Q' exists such that $Q\sigma \Longrightarrow Q' \approx^b_{open} P'$ and vice versa.*

The notation $\simeq_b$ is not confusing because it is also the largest congruence contained in $\approx_b$. This is guaranteed by the next theorem.

Theorem 8. $\approx^b_{open}$ *and* $\approx_b$ *coincide.*

3   Axiomatic System

In this section we give a complete system for the barbed congruence on the finite $\chi^{\neq}$-processes. In order to prove the completeness theorem, we need some auxiliary definitions.

Definition 9. *Let V be a finite set of names. We say that ψ is complete on V if $n(\psi) \subseteq V$ and for each pair x, y of names in V it holds that either $\psi \Rightarrow x{=}y$ or $\psi \Rightarrow x{\neq}y$. A substitution σ is induced by ψ, and ψ induces σ, if σ agrees with ψ and $\sigma\sigma = \sigma$.*

We now begin to describe a system complete for the barbed congruence. Let AS denote the system consisting of the rules and laws in Figure 2 plus the following expansion law:

$$P|Q = \sum_i \phi_i(\tilde{x})\pi_i.(P_i|Q) + \sum_{\substack{\pi_i=a_i[x_i] \\ \gamma_j=\overline{b_j}[y_j]}} \phi_i\psi_j(\tilde{x})(\tilde{y})[a_i{=}b_j][x_i|y_j].(P_i|Q_j) +$$

$$\sum_j \psi_j(\tilde{y})\gamma_j.(P|Q_j) + \sum_{\substack{\pi_i=\overline{a_i}[x_i] \\ \gamma_j=b_j[y_j]}} \phi_i\psi_j(\tilde{x})(\tilde{y})[a_i{=}b_j][x_i|y_j].(P_i|Q_j)$$

where P is $\sum_i \phi_i(\tilde{x})\pi_i.P_i$ and Q is $\sum_j \psi_j(\tilde{y})\gamma_j.Q_j$, π_i and γ_j range over $\{\alpha[x] \mid \alpha \in \mathcal{N} \cup \overline{\mathcal{N}}, x \in \mathcal{N}\}$.

Using axioms in AS, a process can be converted to a process that contains no occurrence of composition operator, the latter process is of special form as defined below.

Definition 10. *A process P is in normal form on $V \supseteq fn(P)$ if P is of the form $\sum_{i \in I_1} \phi_i\alpha_i[x_i].P_i + \sum_{i \in I_2} \phi_i(x)\alpha_i[x].P_i + \sum_{i \in I_3} \phi_i[z_i|y_i].P_i$ such that x does not appear in P, ϕ_i is complete on V for each $i \in I_1 \cup I_2 \cup I_3$, P_i is in normal form on V for $i \in I_1 \cup I_3$ and is in normal form on $V \cup \{x\}$ for $i \in I_2$. Here I_1, I_2 and I_3 are pairwise disjoint finite indexing sets.*

T1	$\lambda.\tau.P = \lambda.P$	
T2	$P+\tau.P = \tau.P$	
T3	$\lambda.(P+\tau.Q) = \lambda.(P+\tau.Q)+\lambda.Q$	
T4	$\tau.P = \tau.(P+\psi\tau.P)$	
T5	$[y|x].(P+\delta\tau.Q) = [y|x].(P+\delta\tau.Q)+\psi\delta[y|x].Q$	$C(\psi,\delta)$
T6	$FF = FF+[x\notin Y_3][x\notin n(\delta)]\delta\alpha[x].Q[x/z]$	$z\notin n(\delta)$
T7	$FR = FR+[x\notin Y_3][x\notin n(\delta)]\delta\alpha[x].Q[x/z]$	$z\notin n(\delta)$
TD1	$RO = RO+\delta(x)\alpha[x].Q$	$x\notin n(\delta)$

Fig. 1. Tau Laws

The depth of a process measures the maximal length of nested prefixes in the process. The structural definition goes as follows: (i) $d(\mathbf{0}) = 0$; (ii) $d(\alpha[x].P) = 1+d(P)$; (iii) $d(P|Q) = d(P)+d(Q)$; (iv) $d((x)P) = d(P)$; (v) $d([x=y]P) = d(P)$, $d([x\neq y]P) = d(P)$; (vi) $d(P+Q) = max\{d(P), d(Q)\}$.

Lemma 11. *For a process P and a finite set V of names such that $fn(P) \subseteq V$ there is a normal form Q on V such that $d(Q) \leq d(P)$ and $AS \vdash Q = P$.*

In order to obtain a complete system for the barbed congruence, we need some tau laws, some of which are new and complex. Figure 1 contains seven tau laws used in this paper. T4, introduced by the first author in previous publication, is a necessary law for open congruences. T5 holds under the condition $C(\psi,\delta)$:

If $\delta \Rightarrow [u\neq v]$ then either $\psi \Rightarrow [x=u][y\neq v]$ or $\psi \Rightarrow [x=v][y\neq u]$ or $\psi \Rightarrow [y=u][x\neq v]$ or $\psi \Rightarrow [y=v][x\neq u]$ or $\psi \Rightarrow [x\neq u][x\neq v][y\neq u][y\neq v]$.

This law was used for the first time in [5]. The laws T6 and T7 are equational formalization of the examples given in Section 2 in a more general form. In these axioms, FF (respectively FR) stands for

$\alpha[x].(P+\delta[x\notin Y_1 \cup \ldots \cup Y_5]\tau.Q[x/z])$

(respectively $(z)\alpha[z].(P+[x\notin n(\delta)]\delta[z\notin Y_1 \cup \ldots \cup Y_5][z|x].Q))$

$+\Sigma_{y\in Y_1}\alpha[x].(P_y+\delta[x=y]\tau.Q[x/z])+\Sigma_{y\in Y_2}(z)\alpha[z].(P_y+[x\notin n(\delta)]\delta[z=y][z|x].Q)$

$+\Sigma_{y\in Y_3}[x\neq y]\alpha[y].(P_y+[x\notin n(\delta)]\delta[y|x].Q[x/z])$

$+\Sigma_{y\in Y_4}[y|x].(P_y+\delta\alpha[y].(P'_y+\delta\tau.Q[x/z]))$

$+\Sigma_{y\in Y_5}[y|x].(P_y+\delta(z)\alpha[z].(P'_y+\delta[z=y]\tau.Q))$

These two laws are new. In TD1, which is derivable from T6, RO is

$\Sigma_{y\in Y_1}\alpha[y].(P_y+\delta\tau.Q[y/x]) + \Sigma_{y\in Y_2}(x)\alpha[x].(P_y+\delta[x=y]\tau.Q)$
$+ (x)\alpha[x].(P+\delta[x\notin Y_1 \cup Y_2]\tau.Q)$

Let $AS \cup \{T1, T2, T3, T4, T5, T6, T7\}$ denote AS_o^b. Without further ado, we state the main result of this section.

Theorem 12. *AS_o^b is sound and complete for $\simeq_b$.*

E1	$P = P$	
E2	$P = Q$	if $Q = P$
E3	$P = R$	if $P = Q$ and $Q = R$
C1	$\alpha[x].P = \alpha[x].Q$	if $P = Q$
C2	$(x)P = (x)Q$	if $P = Q$
C3a	$[x{=}y]P = [x{=}y]Q$	if $P = Q$
C3b	$[x{\neq}y]P = [x{\neq}y]Q$	if $P = Q$
C4	$P{+}R = Q{+}R$	if $P = Q$
C5	$P_0|P_1 = Q_0|Q_1$	if $P_0 = Q_0$ and $P_1 = Q_1$
L1	$(x)0 = 0$	
L2	$(x)\alpha[y].P = 0$	$x \in \{\alpha, \overline{\alpha}\}$
L3	$(x)\alpha[y].P = \alpha[y].(x)P$	$x \notin \{y, \alpha, \overline{\alpha}\}$
L4	$(x)(y)P = (y)(x)P$	
L5	$(x)[y{=}z]P = [y{=}z](x)P$	$x \notin \{y, z\}$
L6	$(x)[x{=}y]P = 0$	$x{\neq}y$
L7	$(x)(P{+}Q) = (x)P{+}(x)Q$	
L8	$(x)[y|z].P = [y|z].(x)P$	$x \notin \{y, z\}$
L9	$(x)[y|x].P = \tau.P[y/x]$	$y \neq x$
L10	$(x)[x|x].P = \tau.(x)P$	
M1	$\phi P = \psi P$	if $\phi \Leftrightarrow \psi$
M2	$[x{=}y]P = [x{=}y]P[y/x]$	
M3a	$[x{=}y](P{+}Q) = [x{=}y]P{+}[x{=}y]Q$	
M3b	$[x{\neq}y](P{+}Q) = [x{\neq}y]P{+}[x{\neq}y]Q$	
M4	$P = [x{=}y]P{+}[x{\neq}y]P$	
M5	$[x{\neq}x]P = 0$	
S1	$P{+}0 = P$	
S2	$P{+}Q = Q{+}P$	
S3	$P{+}(Q{+}R) = (P{+}Q){+}R$	
S4	$P{+}P = P$	
U1	$[y|x].P = [x|y].P$	
U2	$[y|x].P = [y|x].[x{=}y]P$	
U3	$[x|x].P = \tau.P$	

Fig. 2. Axiomatic System AS

LD1	$(x)[x|x].P = [y|y].(x)P$	U3 and L8
LD2	$(x)[y{\neq}z]P = [y{\neq}z](x)P$	L5, L7 and M4
LD3	$(x)[x{\neq}y]P = (x)P$	L6, L7 and M4
MD1	$[x{=}y].0 = 0$	S1, S4 and M4
MD2	$[x{=}x].P = P$	M1
MD3	$\phi P = \phi(P\sigma)$ where σ is induced by ϕ	M2
SD1	$\phi P{+}P = P$	S-rules and M4
UD1	$[y|x].P = [y|x].P[y/x]$	U2 and M2

Fig. 3. Some Laws Derivable from AS

4   Ground Congruence

In this section we sketch some main properties about $\approx_g$. First of all the ground bisimilarity can be characterized by an open bisimilarity called ground open bisimilarity, notation $\approx^g_{open}$. The definition of the ground open bisimilarity is obtained from Defintion 5 by replacing clause (ii) by

(ii') If $P \xrightarrow{\alpha[x]} P'$ then Q' exists such that $Q \xRightarrow{\alpha[x]} Q'\mathcal{R}P'$.

It is easy to prove that $\approx^g_{open}$ is closed under localization and composition and that $\approx^g_{open}$ coincides with $\approx_g$. By definition the ground open bisimilarity is contained in the barbed one. The inclusion is strict because T7 is not valid for $\approx^g_{open}$.

Let $\simeq_g$ be the largest congruence contained in $\approx^g_{open}$. Its formal definition is completely similar to that of $\simeq_b$. Let AS^g_o stand for $AS \cup \{T1, T2, T3, T4, T5, T6\}$. It can be similarly proved that AS^g_o is sound and complete for $\simeq_g$.

5   Remark

Parrow and Victor have studied fusion calculus ([7]). It is a polyadic version of $\chi^{\neq}$-calculus. The main observational equivalence they have studied is what they call weak hyperequivalence. The weak hyperequivalence is essentially a polyadic version of the open bisimilarity $\approx_o$ we have defined in the introduction. Since $\chi^{\neq}$-calculus is a monadic version of the fusion calculus and therefore is a subcalculus of the latter, the counter example given in the introduction is valid in fusion calculus as well. One of the motivations of the ground bisimilarity is to rectify the weak hyperequivalence. Apart from its theoretical interest, the barbed bisimilarity is introduced partly to study the ground bisimilarity.

References

1. Y. Fu. A Proof Theoretical Approach to Communications. *ICALP'97*, Lecture Notes in Computer Science 1256, Springer, 325–335, 1997.
2. Y. Fu. Bisimulation Lattice of Chi Processes. *ASIAN'98*, Lecture Notes in Computer Science 1538, Springer, 245–262, 1998.
3. Y. Fu. Variations on Mobile Processes. *Theoretical Computer Science*, **221**: 327-368, 1999.
4. Y. Fu. Open Bisimulations of Chi Processes. *CONCUR'99*, Eindhoven, The Netherlands, August 24-27, Lecture Notes in Computer Science 1664, Springer, 304-319, 1999.
5. Y. Fu, Z. Yang. Chi Calculus with Mismatch. *CONCUR 2000*, Pennsylvania, USA, August 22-25, Lecture Notes in Computer Science 1877, Springer, 2000.
6. R. Milner, J. Parrow, D. Walker. A Calculus of Mobile Processes. *Information and Computation*, **100**: 1-40 (Part I), 41-77 (Part II), Academic Press.
7. J. Parrow, B. Victor. The Fusion Calculus: Expressiveness and Symmetry in Mobile Processes.
8. J. Parrow, D. Sangiorgi. Algebraic Theories for Name-Passing Calculi. *Journal of Information and Computation*, **120**: 174-197, 1995.

Inheritance in the Join Calculus

(Extended Abstract)

Cédric Fournet[1], Cosimo Laneve[2], Luc Maranget[3], and Didier Rémy[3]

[1] Microsoft Research, 1 Guildhall Street, Cambridge, U.K.
[2] Dipartimento di Scienze dell'Informazione, Università di Bologna,
Mura Anteo Zamboni 7, 40127 Bologna, Italy
[3] INRIA Rocquencourt, BP 105, 78153 Le Chesnay Cedex France

Abstract. We propose an object-oriented calculus with internal concurrency and class-based inheritance that is built upon the join calculus. Method calls, locks, and states are handled in a uniform manner, using labeled messages. Classes are partial message definitions that can be combined and transformed. We design operators for behavioral and synchronization inheritance. Our model is compatible with the JoCaml implementation of the join calculus.

1 Introduction

Object-oriented programming has long been praised as favoring abstraction, incremental development, and code reuse. Objects can be created by instantiating definition patterns called *classes*, and in turn complex classes can be built from simpler ones. To make this approach effective, the assembly of classes should rely on a small set of operators with a clear semantics and should support modular proof techniques. In a concurrency setting, such promises can be rather hard to achieve.

The design and the implementation of concurrent object-oriented languages, *e.g.* [2, 20, 1, 4], has recently prompted the investigation of the theoretical foundations of concurrent objects. Several works provide encodings of objects in process calculi [19, 18, 12, 5] or, alternatively, supplement objects with concurrent primitives [16, 3, 11]. Those works promote a unified framework for reasoning about objects and processes, but they do not address the composition of object definitions or its typechecking.

In this work, we model concurrent objects in a simple process calculus—a variant of the *join calculus* [7, 6], we design operators for behavioral and synchronization inheritance, and we give a type system that statically enforces basic safety properties.

The join calculus is a simple name-passing calculus, related to the pi calculus but with a functional flavor. It is the core of a distributed programming language, currently implemented as an extension of ML [8, 13]. In the join calculus, communication channels are statically scoped: when channels are created, their definition provides a set of *reaction rules* that specify, once for all, how

S. Kapoor and S. Prasad (Eds.): FST TCS 2000, LNCS 1974, pp. 397–408, 2000.
© Springer-Verlag Berlin Heidelberg 2000

messages sent on these names will be synchronized and processed. Although the join calculus does not have a primitive notion of object, definitions encapsulate the details of synchronization, much as concurrent objects.

Applying the well-known objects-as-records paradigm to the join calculus, we obtain a simple language of objects with asynchronous message passing. Method calls, locks, and states are handled in a uniform manner, using labeled messages. There is no primitive notion of functions, calling sequences, or threads (they can all be encoded using continuation messages). Our language—the *objective join calculus*—allows fine-grain internal concurrency, as each object may send and receive several messages in parallel.

For every object of our language, message synchronization is defined and compiled as a whole. This allows an efficient compilation of message delivery into automata [14] and simplifies reasoning on objects. However, the static definition of behavior can be overly restrictive for the programmer. This suggests some compile-time mechanism for assembling partial definitions. To this end, we promote partial definitions into classes. Classes can be combined and transformed to form new classes. They can also be closed to create objects.

The class language is layered on top of the core objective calculus, with a semantics that reduces classes into plain object definitions. We thus retain strong static properties for all objects at run-time. Some operators are imported from sequential languages and adapted to a concurrent setting. For instance, multiple inheritance is expressed as a disjunction of join definitions, but some disjunctions have no counterpart in a sequential language. In addition, we propose a new operator, called *selective refinement*. Selective refinement applies to a parent class, and rewrites the parent reaction rules according to their synchronization patterns. Selective refinement treats synchronization concretely, but it handles the parent processes abstractly. Our approach is compatible with the JoCaml implementation of the join calculus [13], which already singles out synchronization patterns using concrete compile-time representation, and, on the contrary, compiles behaviors into functional closures.

Our design of the class language follows from common programming patterns in the join calculus. We motivate it further by coding some standard problematic examples that mix synchronization and inheritance.

The language is equipped with a polymorphic type system, in the style of [9]; in addition to basic safety properties, the type system also enforces privacy. The formal presentation of both dynamic and static semantics, the soundness results, and their proofs are omitted from this extended abstract. They can be found in the full paper [10].

The paper is organized as follows. In section 2, we present the objective join calculus and develop a few examples. In section 3, we supplement the language with classes. In section 4, we provide more involved examples of inheritance and concurrency. In section 5, we discuss related works and possible extensions.

2   The Objective Join Calculus

Getting Started. The basic operation of our calculus is asynchronous message passing in object style. For instance, the process *out.print_int*(*n*) sends a message with label *print_int* and with content *n* to an object named *out*, meant to print integers on the terminal.

Accordingly, the definition of an object describes how messages received on some labels can trigger processes. For instance,

obj *continuation* = *reply*(*x*) ▷ *out.print_int*(*x*)

defines an object that reacts to messages on *reply* by printing their content on the terminal. Another example is the rendez-vous, or synchronous buffer:

obj *sbuffer* = *get*(*r*) & *put*(*a,s*) ▷ *r.reply*(*a*) & *s.reply*()

The object *sbuffer* has two labels *get* and *put*; it reacts to the *simultaneous presence* of one message on each of these labels by passing a message to the continuation *r*, with label *reply* and contents *a*, and passing an empty message to *s*. (Object *r* may be the previously-defined *continuation*; object *s* is another continuation taking no argument on *reply*.) As regards the syntax, concurrent execution and message synchronization are expressed in a symmetric manner using the same infix operator &. Also, the calculus is polyadic, *i.e.*, messages carry tuples of values.

Some labels may convey messages representing the internal state of an object, rather than an external method call. This is the case of label *Some* in the following unbounded, unordered, asynchronous buffer:

obj *abuffer* = self(*z*)
    put(*a,r*) ▷ *r.reply*() & *z.Some*(*a*)
or *get*(*r*) & *Some*(*a*) ▷ *r.reply* (*a*)

The object *abuffer* can react in two different ways: a message (*a, r*) on *put* may be consumed by storing the value *a* in a self-inflicted message on *Some*; alternatively, a message on *get* and a message on *Some* may be jointly consumed, and then the value stored on *Some* is sent to the continuation received on *get*. The indirection through *Some* makes *abuffer* behave asynchronously: messages on *put* are never blocked, even if no message is ever sent on *get*. As regards the syntax, the prefix self(*z*) explicitly binds the name *z* to the defined object.

In the example above, the messages on label *Some* encode the state of *abuffer*. The following definition illustrates a tighter management of state that implements a one-place buffer:

obj *buffer* = self(*z*)
    put(*a,r*) & *Empty*() ▷ *r.reply*() & *z.Some*(*a*)
or *get*(*r*) & *Some*(*a*) ▷ *r.reply*(*a*) & *z.Empty*()
init *buffer.Empty*()

Such a buffer can either be empty or contain one element. The state is encoded as a message pending on *Empty* or *Some*, respectively. Object *buffer* is created empty, by sending a first message on *Empty* in the (optional) init part of the

Fig. 1. Syntax for the core object calculus

$P ::=$		**Processes**
	0	null process
	$x.M$	message sending
	P_1 & P_2	parallel composition
	obj $x = \mathsf{self}(z)\, D\; \mathsf{init}\; P_1\; \mathsf{in}\; P_2$	object definition
$D ::=$		**Definitions**
	$M \rhd P$	reaction rule
	$D_1\; \mathsf{or}\; D_2$	disjunction
$M ::=$		**Patterns**
	$\ell(\widetilde{u})$	message
	M_1 & M_2	synchronization

obj construct. As opposed to *abuffer* above, a *put* message is blocked when the buffer is not empty.

To keep the *buffer* object consistent, there should be a single message pending on either *Empty* or *Some*. This invariant holds as long as external users cannot send messages on these labels directly. In the full paper [10], we describe a refined semantics and a type system that distinguishes private labels such as *Empty* and *Some* from public labels, and restrict access to private labels. In the examples, private labels conventionally bear an initial capital letter.

Once private labels are hidden, each variant of our buffer provides the same interface to the outside world (two methods labeled *get* and *put*) but their concurrent behaviors are very different.

Syntax. We use two disjoint countable sets of identifiers for object names $x, z, u \in \mathcal{O}$ and for labels $\ell \in \mathcal{L}$. Tuples of names are written $x_i{}^{i \in I}$ or simply $\widetilde{x}$. The grammar of the *objective join calculus* (without classes) is given in Figure 1; it has syntactic categories for processes P, definitions D, and patterns M. We abbreviate obj $x = \mathsf{self}(z)\, D\; \mathsf{init}\; P\; \mathsf{in}\; Q$ by omitting $\mathsf{self}(z)$ when z does not occur free in D and omitting init P when P is 0.

A reaction rule $M \rhd P$ associates a pattern M with a guarded process P. Every message pattern $\ell(\widetilde{u})$ in M binds the object names $\widetilde{u}$ with scope P. We require that every pattern M guarding a reaction rule be linear; that is, each label and each name appears at most once in M. This will be enforced by typing. In addition, an object definition obj $x = \mathsf{self}(z)\, D\; \mathsf{init}\; P_1\; \mathsf{in}\; P_2$ binds two names x and z to D. The scope of x is the processes P_1 and P_2; the scope of z is every guarded process in D. Free names in processes and definitions, noted $\mathsf{fn}(\cdot)$, are defined according to these binders. Terms are taken modulo renaming of bound names (or α-conversion).

The reduction relation on processes is defined using a reflexive chemical abstract machine; it appears in the full paper.

3   Inheritance and Concurrency

We now extend the calculus of concurrent objects with classes and inheritance. The behavior of objects in the join calculus is statically defined: once an object is created, it cannot be extended with new labels or with new reaction rules synchronizing existing labels. Instead, we provide this flexibility at the level of classes. Our operators on classes can express various object paradigms, such as method overriding (with late binding) or method extension. As regards concurrency, these operators are also suitable to define synchronization policies in a modular manner.

Deriving a Concurrent Class. We introduce the syntax for classes in a series of simple examples. We begin with a class *buffer* behaving as the one-place buffer of Section 2:

$$\text{class } \textit{buffer} = \mathsf{self}(z)$$
$$\textit{get}(r) \ \& \ \textit{Some}(a) \rhd r.\textit{reply}(a) \ \& \ z.\textit{Empty}()$$
$$\text{or } \textit{put}(a,r) \ \& \ \textit{Empty}() \rhd r.\textit{reply}() \ \& \ z.\textit{Some}(a)$$

The class *buffer* can be used to create objects:

$$\text{obj } b = \textit{buffer} \text{ init } b.\textit{Empty}()$$

Assume that, for debugging purposes, we want to log the buffer content on the terminal. Our first solution uses an explicit *log* label.

$$\text{class } \textit{logged\_buffer} = \mathsf{self}(z)$$
$$\textit{buffer}$$
$$\text{or } \textit{log}() \ \& \ \textit{Some}(a) \rhd \textit{out.print\_int}(a) \ \& \ z.\textit{Some}(a)$$
$$\text{or } \textit{log}() \ \& \ \textit{Empty}() \rhd \textit{out.print\_string}(\text{"Empty"}) \ \& \ z.\textit{Empty}()$$

The class body above is a disjunction of an inherited class and of additional reaction rules. The intended meaning of disjunction is that reaction rules are cumulated, yielding competing behaviors for messages on labels that appear in several disjuncts. The order of the disjuncts does not matter. The programmer that writes *logged_buffer* must have some knowledge of the parent class *buffer*, namely the use of private labels *Some* and *Empty* for representing the state.

Other possible debugging information is the synchronous log of all messages that are consumed on *put*. This is done by selecting the patterns in which *put* occurs and adding a printing message to the corresponding guarded processes:

$$\text{class } \textit{logged\_buffer\_bis} =$$
$$\text{match } \textit{buffer} \text{ with}$$
$$\textit{put}(a,r) \Rightarrow \textit{put}(a,r) \rhd \textit{out.print\_int}(a)$$
$$\text{end}$$

The match construct can be understood by analogy with pattern matching *à la* ML, applied to the reaction rules of the parent class. Here, any reaction rule from the parent *buffer* whose pattern contains the label *put* is replaced in the derived *logged_buffer_bis* by a rule with the same pattern (*put* appears on both sides of ⇒) and with the original guarded process in parallel with a printing message (the parent action and the & are left implicit in the match syntax). Any other parent rule remains unchanged. Hence, the above definition behaves as the following definition:

> class *logged_buffer_bis* =
>   *get(r)* & *Some(a)* ▷ *r.reply(a)* & *z.Empty()*
> or *put(a,r)* & *Empty()* ▷ *r.reply()* & *z.Some(a)* & *out.print_int(a)*

Yet another kind of debugging information is a log of *put* attempts:

> class *logged_buffer_ter* = self(*z*)
>   match *buffer* with
>   *put(a,r)* ⇒ *Parent_put(a,r)* ▷ 0
>   end
> or *put(a,r)* ▷ *out.print_int(a)* & *z.Parent_put(a,r)*

In this case, the match construct performs a renaming of *put* into *Parent_put* in the patterns of selected rules, without affecting their guarded processes. The net effect is similar to parent method overriding, with the new *put* calling the parent one and a late-binding semantics for guarded processes.

The examples above illustrate that the very idea of class refinement is less abstract in a concurrent setting than in a sequential one. In the first *logged_buffer* example, logging the buffer state requires knowledge of how this state is encoded; otherwise, some states might be forgotten or logging might lead the buffer into deadlock. The other two examples expose another subtlety: in a sequential language, the distinction between logging *put* attempts and *put* successes is irrelevant. Thinking in terms of sequential object invocations, one may be unaware of the concurrent behavior of the object, and thus write *logged_buffer_ter* instead of *logged_buffer_bis*.

Syntax. The language with classes extends the core calculus of section 2; its grammar is given in Figure 2. Classes are taken up to the associative-commutative laws for disjunction or. We use two additional sets of identifiers for class names $c \in C$ and for sets of labels $L \in 2^{\mathcal{L}}$. Such sets L are used to represent abstract classes that declare the labels in L but do not define them.

Join patterns J generalize the syntactic category of patterns M given in Figure 1 with an or operator that represents alternative synchronization patterns. Join patterns are taken up to simple equivalence laws: & and or are associative-commutative, and & distributes over or. Hence, every join pattern J can be written as a non-empty alternative of patterns $\text{or}_{i \in I} M_i$, and the reaction rule $(\text{or}_{i \in I} M_i) \triangleright P$ behaves as $\text{or}_{i \in I}(M_i \triangleright P)$.

Selection patterns K are either join patterns or the empty pattern 0. Their normal forms are of the form above, except that I can be empty. We always assume that patterns J and K meet the following well-formed conditions. Free names $fn(K)$ are defined in the obvious way. As usual, ⊎ means disjoint union.

Fig. 2. Syntax for classes

$P ::=$	**Processes**
$\quad \dots$	(as in figure 1)
$\quad$ obj $x = \mathsf{self}(z)\, C$ init P_1 in P_2	object definition
$\quad$ class $c = \mathsf{self}(z)\, C$ in P	class definition
$C ::=$	**Classes**
$\quad c$	class variable
$\quad L$	abstract class
$\quad J \triangleright P$	reaction rule
$\quad C_1$ or C_2	disjunction
$\quad \mathsf{self}(x)\, C$	self binding
$\quad$ match C with S end	selective refinement
$S ::=$	**Refinement clauses**
$\quad (K_1 \Rightarrow K_2 \triangleright P) \mid S$	refinement sequence
$\quad \emptyset$	empty refinement
$J ::=$	**Join patterns**
$\quad \ell(\tilde{u})$	message
$\quad J_1 \,\&\, J_2$	synchronization
$\quad J_1$ or J_2	alternative
$K ::=$	**Selection patterns**
$\quad 0$	empty pattern
$\quad J$	join pattern

1. All conjuncts M_i in the normal form of K are linear (as defined in section 2) and bind the same names. By extension, we say that K binds the names $\mathit{fn}(M_i)$ bound in every M_i, and write $\mathit{fn}(K)$ for these names.
2. In a refinement clause $K_1 \Rightarrow K_2 \triangleright P$, the pattern K_1 is either M or 0, and the pattern K_2 binds at least the names of K_1 ($\mathit{fn}(K_1) \subseteq \mathit{fn}(K_2)$).

As described in section 2, binders for object names include object definitions (binding the defined object to name x and self name z) and patterns (binding the received names). In a reaction rule $J \triangleright P$, the join pattern J binds $\mathit{fn}(J)$ with scope P. In a refinement clause $K_1 \Rightarrow K_2 \triangleright P$, the selection pattern K_1 binds $\mathit{fn}(K_1)$ with scope K_2 and P; the modification pattern K_2 binds $\mathit{fn}(K_2) \setminus \mathit{fn}(K_1)$ with scope P. Finally, the self renaming $\mathsf{self}(x)\, C$ binds the object name x with scope C. Class definitions class $c = C$ in P are the only binders for class names c, with scope P. Processes, classes, and reaction rules are taken up to α-conversion.

Labels don't have scopes. Join patterns J declare the labels appearing in their message. Classes C declare the labels of their reaction rules. Abstract classes trivially declare their labels. Finally, selective refinements declare labels appearing either in the parent class or in a refinement clause.

Class expressions are simplified by means of a reduction semantics, that allows to obtain processes in the core calculus without classes. These reduction sementics (see the full paper [10]) has been designed to support separate compilation of classes.

4   Inheritance Anomaly

As remarked by many authors, the classical point of view on class abstraction—method names and signatures are known, method bodies are abstract—does not mix well with concurrency. More specifically, concurrency and class-based inheritance are not orthogonal. This well-known problem is often referred to as the *inheritance anomaly*. Unfortunately, inheritance anomaly is not defined formally, but by means of examples as in [15], where three patterns of inheritance anomaly are given.

The examples in [15] demonstrate that extending a base class by new capacities has an impact on the (desirable) concurrent behavior of the capacities that are inherited from the base class. Straightforward extensions to concurrency of sequential languages, such as implementing synchronization in method bodies or providing simple locking policies (cf. **synchronized** from Java) prove unsufficient here. The former because synchronization code is not accessible, the latter because the provided synchronization policies are not expressive enough. [15] (partially) solve inheritance anomalies by making concrete some parts of the parent class (such as "concurrency-control"). It is to be noticed that they do so by considering a new extension for each anomaly.

Our approach is different: starting from a concurrent language we are more concerned with the expressive power of our inheritance operators. Solving the three categories of inheritance anomaly, as we do, appears to be a valuable test.

To this aim, we consider the same running example as Matsuoka and Yonezawa: a FIFO buffer with two methods *put* and *get* to store and retrieve items. We also adopt their taxonomy of inheritance anomaly: inheritance induces desirable modifications of "acceptable states" [of objects], and a solution is a way to express these modifications.

We extend our programming language with booleans and integers, with usual primitive operations. Arrays are created by create(n), that gives an uninitialized array of size n. The size of array A is retrieved by $A.size$. Finally $A[i] \leftarrow v$ is array A where the i-th item has been replaced with v.

```
class buff = self (z)
    put(v,r) & (Empty(A, i, n) or Some(A, i, n)) ▷
        r.reply() & z.Check(A[(i+n) mod A.size] ← v, i, n+1)
or get(r) & (Full(A, i, n) or Some(A, i, n)) ▷
        r.reply(A[i]) & z.Check(A, (i+1) mod A.size, n−1)
or Check(A,i,n) ▷
        if n = A.size then z.Full(A, i, A.size)
        else  if n = 0 then z.Empty(A, 0, 0)
```

else $z.Some(A, i, n)$
or $Init(size) \;\triangleright\; z.Empty(\mathsf{create}(size),0,0)$

The state of the buffer is encoded as a message with label *Empty*, *Some*, or *Full*. The buffer may react to messages on *put* when non-full, and to messages on *get* when non-empty; this is expressed in a concise manner using the or operator in patterns. Once a request is accepted, the state of the buffer is recomputed by sending an internal message on *Check*. As *Check* appears alone in a join pattern, message sending on *Check* acts like a function call, which can be in-lined by an optimizing compiler.

Partitioning of acceptable states. The class *buff2* supplements *buff* with a new method *get2* that atomically retrieves two items from the buffer. For simplicity, we assume here $size > 2$.

Since *get2* succeeds when the buffer contains two elements or more, the buffer state needs to be refined. Furthermore, since for instance, a successful *get2* may disable *get* or enable *put*, the addition of *get2* has an impact on the "acceptable states" of *get* and *put*, which are inherited from the parent *buff*. Therefore, label *Some* is no longer pertinent and is replaced with two labels *One* and *Many*. *One* models the state when the buffer holds exactly one item; *Many* defines a state with two items or more in the buffer.

class $buff2 = \mathsf{self}(z)$
   $get2(r)$ & $(Full(A,i,n)$ or $Many(A, i, n)) \;\triangleright$
         $r.reply(A[i], A[(i{+}1)$ mod $A.size])$
      & $z.Check(A, (i{+}2)$ mod $A.size, n{-}2)$
   or match $buff$ with
      $Some(A, i, n) \Rightarrow (One(A, i, n)$ or $Many(A, i, n)) \;\triangleright\; 0$ end
   or $Some(A, i, n) \;\triangleright$
         if $n > 1$ then $z.Many(A, i, n)$ else $z.One(A, i, n)$

In the program above, a new method *get2* is defined, with its own synchronization condition. The new reaction rule is cumulated with those of *buff*, using a selective refinement that substitutes "$One(\ldots)$ or $Many(\ldots)$" for every occurrence of "$Some(\ldots)$" in a join pattern. The refinement eliminates *Some* from any inherited pattern, but it does not affect occurrences of *Some* in inherited guarded processes: the parent code is handled abstractly, so it cannot be modified. Instead, the new class provides an adapter rule that consumes any message on *Some* and issues a message on either *One* or *Many*, depending on the value of n.

History-dependent acceptable states. The class *gget_buff* alters *buff* as follows: the new method *gget* returns one item from the buffer (like *get*), except that a request on *gget* can be served only immediately after serving a request on *put*. More precisely, a *put* transition enables *gget*, while *get* and *gget* transitions disable it. This condition is reflected in the code by introducing two labels *After-Put* and *NotAfterPut*. Then, messages on *gget* are synchronized with messages on *AfterPut*.

```
class gget_buff = self (z)
    gget(r) & AfterPut()
            & (Full(A, i, n) or Some(A, i, n)) ▷
        r.reply(A[i]) & z.NotAfterPut()
        & z.Check(A, (i+1) mod A.size, n−1)
or match buff with
    Init(size) ⇒ Init(size ) ▷ z.NotAfterPut()
  | put(v, r) ⇒
        put(v, r) & (AfterPut() or NotAfterPut()) ▷
        z.AfterPut()
  | get(r) ⇒
        get(r) & (AfterPut() or NotAfterPut()) ▷
        z.NotAfterPut()
end
```

The first clause in the match construct refines initialization, which now also issues a message on *NotAfterPut*. The two other clauses refine the existing methods *put* and *get*, which now consume any message on *AfterPut* or *NotAfterPut* and produce a message on *AfterPut* or *NotAfterPut*, respectively.

Modification of acceptable states. A general-purpose lock may be defined as:

```
class locker = self (z)
    suspend(r) & Free() ▷ r.reply() & z.Locked()
or resume(r) & Locked() ▷ r.reply() & z.Free()
```

The class *locker* can be used to create locks, but it can also be combined with some other class c to control message processing on the labels of c. To this end, a simple disjunction of c and *locker* is not enough and some refinement of the parent c is required.

```
class locked_buff = self (z)
    locker
or match buff with
      Init(size) ⇒ Init(size ) ▷ z.Free()
    | 0 ⇒ Free () ▷ z.Free()
    end
```

The first clause in the match construct supplements the initialization of *buff* with an initial *Free* message for the lock. The second clause matches every other rule of *buff*, and requires that the refined clause consume and produce a message on *Free*. (The semantics of clause selection follows the textual priority scheme of ML pattern-matching, where a clause applies to all reaction rules that are not selected by previous clauses, and where the empty selection pattern acts as a default case.)

As a consequence of these changes, parent rules are blocked between a call to *suspend* and the next call to *resume*, and parent rules leave the state of the lock unchanged. In contrast with previous examples, the code above is quite general; it applies to any class following the same convention as *buff* for initialization.

5   Related and Future Works

There are many works on supplementing object calculi with concurrent primitives [16, 3, 11], or on supplementing process calculi with objects (usually by the mean of an encoding in the original calculus) [19, 18, 12, 5]. Our work follows the latter tradition. However, to our knowledge, it is the only one to address safe object composition in a process calculus.

In [17], Odersky proposes an object-oriented extension of a language based on the join calculus. His proposal amounts to adding some record structure to join definitions, as we do in section 2. However, Odersky does not consider the problem of inheritance and refinement of synchronization. A small technical difference is that Odersky's calculus is designed with pattern matching on values: values in Odersky's calculus are not only object names but also constructed values (such as strings, integers, lists, etc.); then, the shape of values can be taken into account during synchronization. For instance, a rule of the form $\ell(h :: t) \triangleright P$ is allowed, and will only fire when a message on ℓ carries a list that contains at least one cell. The extension of our calculus to structured values is easy (see [9]). However, we believe that synchronization should only concern names, and not depend on the shape of values itself. In particular, this allows a simpler semantics and an efficient compilation of synchronization into automata [14].

Since our type system abstracts from the shape of synchronization patterns in classes, it is blind to a number of relevant properties of concurrency, such as the presence of race conditions or deadlock freedom. The design of a sophisticated analyzer that is sensitive to synchronizations is a promising research direction.

6   Conclusions

We have proposed a simple, object-based variant of the join calculus. Every object is defined as a fixed set of reaction rules that describe its synchronization behavior. The expressiveness of the language is significantly increased by adding classes—a form of open definitions that can be incrementally assembled before object instantiation. Thereby, we partially recover the ability of the pi calculus to dynamically define the receptive behavior for messages. Our layered design confines this capability to classes. From a programming-language viewpoint, this seems a good compromise between flexibility and simplicity of the model. Indeed, our proposal still enables efficient compilation of synchronization and type inference.

References

[1] G. Agha, P. Wegner, and A. Yonezawa. *Research Directions in Concurrent Object-Oriented Programming.* MIT Press, 1993.

[2] P. America. Issues in the design of a parallel object-oriented language. *Formal Aspects of Computing*, 1(4):366–411, 1989.

[3] P. D. Blasio and K. Fisher. A calculus for concurrent objects. In U. Montanari and V. Sassone, editors, *Proceedings of the 7th International Conference on Concurrency Theory (CONCUR '96)*, LNCS 1119, pages 406–421, 1996.

[4] L. Cardelli. Obliq A language with distributed scope. SRC Research Report 122, Digital Equipment, June 1994.

[5] S. Dal-Zilio. Quiet and bouncing objects: Two migration abstractions in a simple distributed blue calculus. In H. Hüttel and U. Nestmann, editors, *Proceedings of the Worshop on Semantics of Objects as Proceedings (SOAP '98)*, pages 35–42, June 1998.

[6] C. Fournet. *The Join-Calculus: a Calculus for Distributed Mobile Programming*. PhD thesis, Ecole Polytechnique, Palaiseau, Nov. 1998.

[7] C. Fournet and G. Gonthier. The reflexive chemical abstract machine and the join-calculus. In *Proceedings of POPL '96*, pages 372–385, Jan. 1996.

[8] C. Fournet, G. Gonthier, J.-J. Lévy, L. Maranget, and D. Rémy. A calculus of mobile agents. In U. Montanari and V. Sassone, editors, *Proceedings of the 7th International Conference on Concurrency Theory (CONCUR '96)*, LNCS 1119, pages 406–421, 1996.

[9] C. Fournet, C. Laneve, L. Maranget, and D. Rémy. Implicit typing à la ML for the join-calculus. In A. Mazurkiewicz and J. Winkowski, editors, *Proceedings of the 8th International Conference on Concurrency Theory*, LNCS 1243, pages 196–212, 1997.

[10] C. Fournet, C. Laneve, L. Maranget, and D. Rémy. Inheritance in the join-calculus. Full version. Available electronically at `http://cristal.inria.fr/~remy/work/ojoin.ps.gz`, June 2000.

[11] A. D. Gordon and P. D. Hankin. A concurrent object calculus: reduction and typing. In U. Nestmann and B. C. Pierce, editors, *HLCL '98: High-Level Concurrent Languages*, volume 16(3) of *entcs*, Nice, France, Sept. 1998.

[12] J. Kleist and D. Sangiorgi. Imperative objects and mobile processes. June 1998.

[13] F. Le Fessant. The JoCAML system prototype. Software and documentation available from `http://pauillac.inria.fr/jocaml`, 1998.

[14] F. Le Fessant and L. Maranget. Compiling join-patterns. *Electronic Notes in Computer Science*, 16(2), 1998.

[15] S. Matsuoka and A. Yonezawa. Analysis of inheritance anomaly in object-oriented concurrent programming languages. In G. Agha, P. Wegner, and A. Yonezawa, editors, *Research Directions in Concurrent Object-Oriented Programming*, chapter 4, pages 107–150. The MIT Press, 1993.

[16] O. Nierstrasz. Towards an object calculus. In O. N. M. Tokoro and P. Wegner, editors, *Proceedings of the ECOOP'91 Workshop on Object-Based Concurrent Computing*, LNCS 612, pages 1–20, 1992.

[17] M. Odersky. Functional nets. In *European Symposium on Programming 2000*, LNCS. Springer Verlag, 2000.

[18] D. Sangiorgi. An interpretation of typed objects into typed π-calculus. *Information and Computation*, 143(1):34–73, 1998.

[19] D. J. Walker. Objects in the pi-calculus. *Information and Computation*, 116(2):253–271, 1995.

[20] A. Yonezawa, J.-P. Briot, and E. Shibayama. Object-oriented concurrent programming in ABCL/1. *ACM SIGPLAN Notices*, 21(11):258–268, Nov. 1986. Proceedings of OOPSLA '86.

Approximation Algorithms for Bandwidth and Storage Allocation Problems under Real Time Constraints*

Stefano Leonardi, Alberto Marchetti-Spaccamela, and Andrea Vitaletti**

Dipartimento di Informatica Sistemistica, Università di Roma "La Sapienza",
via Salaria 113, 00198-Roma, Italia.
{leon,marchetti,vitale}@dis.uniroma1.it

Abstract. The problem we consider is motivated by allocating bandwidth slots to communication requests on a satellite channel under real time constraints. Accepted requests must be scheduled on non-intersecting rectangles in the time/bandwidth Cartesian space with the goal of maximizing the benefit obtained from accepted requests. This problem turns out to be equal to the maximization version of the well known Dynamic Storage Allocation problem when storage size is limited and requests must be accommodated within a prescribed time interval.
We present constant approximation algorithms for the problem introduced in this paper using as a basic step the solution of a fractional Linear Programming formulation.
This problem has been independently studied by Bar-Noy et al [BNBYF+00] with different techniques. Our approach gives an improved approximation ratio for the problem.

1   Introduction

The problem we study in this paper has been encountered in the context of the EU research project Euromednet on scheduling requests for remote medical consulting on a shared satellite channel [Eme]. Every request asks for a number of contiguous bandwidth slots to provide every end user involved in the consulting with a TCP/IP satellite channel. Bandwidth is assigned in slots of 64 kb/sec. The number of slots per end user depends on the type of service desired (typical values are 64 kb/sec for common internet services – 384 Kb/sec for audio/video.) At most 48 slots of 64 Kb/sec are available on the channel in this specific application. Requests also specify a duration of the consulting (typical values are from 1/2 hour to 2 hours) to be allocated within a time interval specified in the request. Requests are typically issued a few days in advance. The service manager will

---

* Partially supported by the IST Programme of the EU under contract number IST-1999-14186 (ALCOM-FT), IST-1999-10440 (BRAHMS) and IST-2000-14084 (APPOL), and by the Italian Research Project MURST 40% "Resource Allocation in Computer Networks".
** Partially supported by Etnoteam (Italy).

reply soon with a positive or a negative answer on the basis of the pending requests and of the requests already accepted. Every accepted request is allocated starting from a base bandwidth for a contiguous number of slots along a time duration within the indicated time interval. The total bandwidth assigned to a single request must be contiguous due to the constraints imposed from FDMA (Frequency Division Multiple Access) technology. Other details regarding this specific application are available to [Eme].

The problem encountered in this application is a natural interesting combinatorial problem: if we consider a Cartesian space with the bandwidth on the ordinate and the time on the abscissa then every accepted request is a rectangle of basis equal to the duration and height equal to the bandwidth requested. Accepted requests must observe the *packing constraint*, that is any two rectangles are non-overlapping. A benefit is also assigned to every request, modeling its relevance or the economic revenue from its acceptance. The objective is to maximize the overall benefit of accepted requests. In the sequel of the paper we denote this problem as the Rectangle Packing (*RP*) problem.

This problem is related to a number of well studied combinatorial problems. Consider the machine scheduling problem with real time constraints in which every job asks to be scheduled for a given duration between a release time and a deadline. Only one job can be scheduled at any time on every single machine. A benefit is associated with every job with the goal of maximizing the benefit obtained from scheduled jobs. This is an old NP-hard scheduling problem [GJ79]. Very recently the first constant approximation algorithms has been proposed for this problem [BNGNS99] both on single and parallel machines.

A second related problem is the Dynamic Storage Allocation problem (DSA). DSA concerns the dynamic allocation of contiguous areas in a storage device. In DSA a set of requests for a contiguous area of memory along a specified time duration has to be accommodated while minimizing the storage space required. DSA is a classical problem in computer science [Knu73] whose study backs to the sixties. The rectangle packing problem is a maximization version of DSA where we have to allocate bandwidth rather than storage space. In *RP* storage space is limited. On the other end we insert real time constraints on the temporal allocation of the process. We believe that this version of DSA is of relevance in many practical settings. DSA has been shown to be tightly related to interval graph coloring. Kierstead and Slusarek [Kie91, Slu89] proposed 3-approximation algorithms for aligned DSA, where the storage space required is a power of 2. More recently Gergov [Ger96] proposed a 5/2 approximation algorithm for aligned DSA that implies a 5 approximation for DSA and claimed a 3 approximation algorithm [Ger99].

A third closely related problem is the call control problem on linear networks [GGK97, BNCK+95]. This problem has been typically considered in its on-line version. At any step a request for establishing a connection between two vertices on the line network with a given bandwidth is presented. The algorithm has to accept or reject the request without knowledge of the requests presented in the future. If the request is accepted, a given benefit is obtained. The objective is to maximize the obtained benefit without violating the bandwidth constraint on any link of the network. In the call control problem every request must be assigned on a fixed path in the line network, while in the RP problem some slackness in the time allocation may be allowed. However, the major difference from RP is that in call control we only require the bandwidth constraint, imposing that the overall bandwidth allocated on a link does not exceed the capacity of the link, rather than the stronger packing constraint of the RP problem.

Results of the paper. We present a 12 approximation algorithm for the RP problem. As a basic step of the algorithm we solve a fractional LP problem in which we only enforce the bandwidth constraint and requests can be fractionally accepted. We then show with a novel rounding technique that the optimal fractional solution is a convex combination of a set of integral solutions with a specific property that we call stability, of which we select that with highest benefit. The selected solution may still contain intersecting rectangles. However it can be partitioned into three feasible solutions of which we select the best one as final solution of the algorithm. The approximation ratio we obtain is 6 if the bandwidth requested is a power of 2, 12 in the general case. The proposed solution runs in pseudopolynomial time. It can be transformed into a fully polynomial time algorithm at the expenses of a small increase in the approximation ratio. We also show a combinatorial algorithm with approximation ratio arbitrarily close to $26 + \epsilon$. This algorithm uses as a basic step the combinatorial algorithm devised in Bar-Noy et al. [BNBYF$^+$00].

Independently from our paper, Bar-Noy et al. [BNBYF$^+$00] proposed a 35 approximation for our problem that they call Benefit DSA. Their approach is to solve a version of the problem where requests are either accepted or rejected in an integral sense, while the packing constraint is relaxed to the milder bandwidth constraint. A solution of this problem is then combined with an algorithm for the DSA problem. In a later version of their paper they improve the result to a $6\gamma - 1$ combinatorial approximation and to a $6\gamma - 3$ LP-based approximation, where γ is the approximation ratio for DSA. If we consider the 5-approximation for DSA of [Ger96] this yields respectively a 29 combinatorial and a 27 LP-based approximation for the problem. If we consider the 3-approximation claimed in [Ger99], this yields a 17 combinatorial and a 15 LP-based approximation for the problem.

We finally show how to extend our algorithm to the multiple channel case for bandwidth allocation or, equivalently, to the multiple storage devices case in the DSA problem.

Structure of the paper. In Section 2 we formally describe the RP problem. In Section 3 we describe the LP based approximation algorithm for the RP problem. In section 4 we show how the algorithm is turned into a fully polynomial time algorithm. In Section 5 we present a combinatorial version of the algorithm. Finally, in Section 6 we describe the extension to multiple channels.

2 The RP Problem

Given an input set of n requests $\{< r_i, d_i, b_i, l_i, \omega_i >\}_1^n$, where $r_i, d_i, b_i, l_i, \omega_i$ are integers, the generic request asks for a bandwidth interval of size b_i in $[0, B]$ along a continuous time interval of length l_i contained in $[r_i, d_i]$. A request can be either accepted or rejected. A request that is accepted is scheduled on a bandwidth interval $[f(i), f(i) + b_i]$ and a time interval $[t(i), t(i) + l_i]$ and a benefit ω_i is accrued. An accepted request is represented with a rectangle of basis l_i and height b_i on a Cartesian space having the time on the abscissa and the bandwidth on the ordinate. The schedule must observe the constraint that any two rectangles are non-overlapping. In the following we denote by *packing constraint* the constraint that two rectangles are non-overlapping. The packing constraint is stronger than the *bandwidth constraint* imposing that the overall bandwidth allocated at time t cannot exceed B. In the following we assume $b_i \leq 1$ and $B = 1$. In the aligned version of the RP every bandwidth request is a power of $1/2$. The objective of the algorithm is to maximise the overall profit obtained from accepted requests.

3 A LP Based Approximation Algorithm

We present an LP based approximation algorithm for RP. We first round all the bandwidth requests to the nearest higher power of 2. As a basic step of the algorithm we solve a fractional LP problem in which we only enforce the bandwidth constraint. We then show that the optimal solution to the fractional RP problem is a convex combination of a set of integral solutions holding a property that we will call *stability*. We select the best among these stable solutions that has benefit at least $1/2$ the optimum to the LP problem. The selected solution can contain intersecting intervals since the packing constraint has not been imposed. In the final step of the algorithm we show that the selected solution can be decomposed into three feasible solution of which we select that with highest benefit that will form the final solution to the problem. The obtained solution is a 6 approximation for the aligned version and a 12 approximation for the original problem.

Thus the algorithm consists of three main steps:

1. Solve the LP formulation;
2. Find a stable solution;
3. Obtain a feasible solution.

3.1 The LP Formulation

In this section we present the LP formulation we use as a basic step for the solution of the RP problem.

We first round every bandwidth request to the lowest higher power of $1/2$, namely $\overline{b}_i = \min_k \left\{ \frac{1}{2^k} : \frac{1}{2^k} \geq b_i \right\}$.

Variables x_{it}, $t = r_i, .., d_i - l_i$, are associated with request i. Variable x_{it} is ranging in $[0,1]$, and denotes the schedule of request i with $t(i) = t$. Every request can be fractionally scheduled along a set of intervals for an overall value not exceeding one, thus we impose $\sum_{t=r_i}^{d_i - l_i} x_{it} \leq 1$. Denote by T the latest deadline of a request. In the LP formulation we also impose the bandwidth constraint at any time $t \in 1, .., T$, namely that the overall bandwidth assigned to the requests fractionally scheduled at time t is at most 1.

$$\max \sum_{i=1}^{n} \sum_{t=r_i}^{d_i - l_i} \omega_i x_{it}$$

$$\sum_{i,t':t\in[t',t'+l_i)} \bar{b}_i x_{it'} \leq 1, \forall t$$

$$\sum_{t=r_i}^{d_i - l_i} x_{it} \leq 1, \forall i$$

$$x_{it} \in [0,1], \forall t, i$$

$$x_{it} = 0, \forall i, t \notin [r_i, d_i - l_i]$$

We will denote with x_{it} both a variable and its value. The optimum of the LP problem is related to the optimum of the RP problem by the following Lemma:

Lemma 1. *For the RP problem it holds $OPT(LP) \geq OPT(RP)/2$. For the aligned version of RP it holds $OPT(LP) \geq OPT(RP)$.*

Proof: Consider a new formulation LP_1 obtained from LP by replacing in the bandwidth constraint the $\bar{b}_i$'s with the original b_i's.

$$\sum_{i,t':t\in[t',t'+l_i)} b_i x_{it'} \leq 1, \forall t, \tag{1}$$

and by imposing the integrality constraints $x_{it} \in \{0,1\}$. Observe that any solution to RP is also a solution to LP_1 for which $OPT(LP_1) \geq OPT(RP)$. Since $\bar{b}_i \leq 2b_i$, any solution to LP_1 with values $\frac{x_{it}}{2}$ is also a solution to LP, with a benefit at least $\frac{1}{2}$ of the benefit of LP_1. Therefore, from a solution to LP_1 we obtain a solution to LP of value at least $\frac{1}{2}$ of the value of the solution to LP_1. Then $OPT(LP) \geq OPT(LP_1)/2 \geq OPT(RP)/2$.

For the aligned case, we simply obtain $OPT(LP) \geq OPT(RP)$. ∎

3.2   The Algorithm for Obtaining a Stable Solution

In this section we present the LP based algorithm for the RP problem. We denote by i^t the request i scheduled at time t.

Definition 1. *Given a schedule of requests, the support at time t', denoted by $support(t')$, is the maximum value such that there exists a set of j non-overlapping requests $i_1, i_2, .., i_j$ scheduled at time t' for which $f(i_1) = 0$, $f(i_k) = f(i_{k-1}) + b_{i_{k-1}}$, $k = 2, .., j$, $f(i_j) = support(t')$.*

Request i^t is (h, t') stable if $h = support(t') = max_{t'' \in [t, t+l_i)} support(t'')$.

A schedule of requests is stable if every request i in the schedule is (h_i, t_i) stable for some h_i and t_i.

The geometrical interpretation of a request i (h, t) stable is a rectangle placed on the top of a pile of non-overlapping rectangles of total bandwidth h (see Figure 1). We will say that the rectangles in the pile form the support of i.

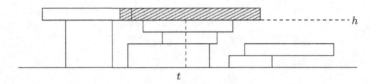

Figure 1. Request i is the filled rectangle in the figure. Request i is (h, t) stable. Observe that 2 requests in a stable schedule can overlap.

Let x_{it} be the value of a variable in the optimal LP solution. Given an optimal LP solution we denote by α the largest rational such that x_{it} is an integer multiple of α.

The algorithm for obtaining a good stable solution first finds at most $\frac{2}{\alpha}$ integral stable solutions and then chooses that one with highest benefit. Denote by s the number of solutions constructed at a generic step of the algorithm. The algorithm is composed of the following steps:

1. Order the non-zero x_{it} by non increasing $\bar{b}_i$.
2. Replicate $\frac{x_{it}}{\alpha}$ times request i^t.
3. Assign every replication of i^t to a solution with the following algorithm:
   (a) Select those solutions $S_1, ..., S_m$ not containing a copy of i, out of the s constructed solutions.
   (b) Merge the m solution $S_1, ..., S_m$ of bandwidth 1 into a single solution S of bandwidth m.(The relative order of the solution is not relevant for the algorithm.)
   (c) Let the replication of i^t be (h, t') stable in S.
   (d) If $h \leq m$, then assign the copy of i^t to solution $S_{\lfloor h \rfloor + 1}$ with $f(i^t) = h \mod 1$; If $h = m$, then construct a new solution having i^t assigned with $f(i^t) = 0$.
4. Select the solution with highest benefit that we call S_{best}.

The simpler alternative would just place every i^t in the first solution where it fits, i.e. where i^t is (h, t) stable with $h \leq 1 - \bar{b}_i$, if any. However this alternative fails to place all the replications into at most $2/\alpha$ solutions as we are able to show for our algorithm.

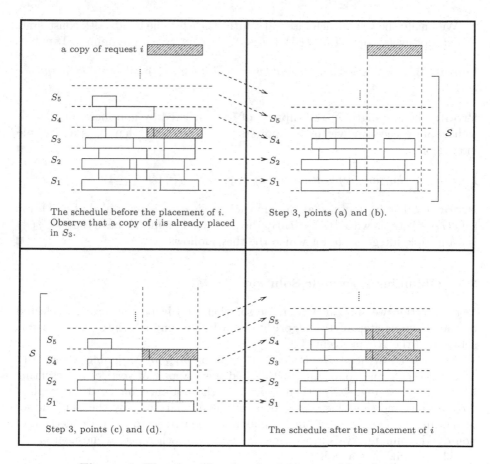

Figure 2. The algorithm for obtaining a stable solution.

Lemma 2. *Every copy of i^t is placed in a solution $S_j \in \mathcal{S}$, $j \leq m$, if $s = \frac{2}{\alpha}$ solutions have been constructed.*

Proof: We prove that i^t is (h, t') stable for a value $h < m$. Assume by contradiction $h \geq m$. At most $1/\alpha$ distinct copies of i^t need to be placed for every request i. Since $2/\alpha$ solutions are available, at least $m \geq 1/\alpha + 1$ solutions $S_1, .., S_m$ not containing a copy of i are available for a single i^t. It follows that at time t' the whole bandwidth has been assigned on all the m solutions, namely for any S_j,

$$\sum_{i^t \in S_j : t' \in [t, t+l_i)} \overline{b}_i = 1.$$

From the packing constraint in LP, at time t' we have : $\alpha \displaystyle\sum_{i, t : t' \in [t, t+l_i)} \frac{\overline{b}_i x_{it}}{\alpha} \leq 1.$

It follows that at time t':

$$1 \geq \alpha \sum_{i, t : t' \in [t, t+l_i)} \frac{x_{it} \overline{b}_i}{\alpha} = \alpha \sum_{S_j} \sum_{i^t \in S_j : t' \in [t, t+l_i)} \overline{b}_i \geq m\alpha \geq 1 + \alpha,$$

thus a contradiction.   ■

We finally show that each integral solution satisfies the bandwidth constraint, namely for every copy of i^t, $f(i^t) + \bar{b}_i \leq 1$. We first give a preliminary Lemma:

Lemma 3. *For every request i and every solution S, $f(i) = k\bar{b}_i$ for some integer k.*

Proof: The rectangles in the support of i are ordered by non increasing bandwidth. Since the bandwidths are powers of $1/2$, we have that h_i is a multiple integer of $\bar{b}_i$. ∎

Lemma 4. *For any replication of any request i^t, $f(i^t) + \bar{b}_i \leq 1$.*

Proof: By definition of the algorithm each replication is scheduled at $f(i^t)$ if it is $(f(i^t), t')$ stable with $f(i^t) < 1$. By the previous Lemma we have that $1 - f(i^t)$ is a multiple integer of $\bar{b}_i$ for which the thesis follows. ∎

3.3   Obtaining a Feasible Solution

In this section we show how a stable solution, and in particular S_{best} selected at the previous step of the algorithm, can be decomposed into three feasible solution of the RP problem.

We first construct the intersection graph of S_{best} by assigning a vertex to every rectangle and connecting with an edge every pair of vertices representing intersecting rectangles. We then show that the obtained intersection graph is 3-colourable and that this can be done in polynomial time. By choosing the set of rectangles of same color having maximum benefit we obtain a collection of non-overlapping intervals in which every request is scheduled at most once.

The algorithm is as follows:

1. Construct the intersection graph of solution S_{best};
2. Colour the intersection graph with three colours with the following algorithm:

   (i.) Consider the requests in order of non increasing bandwidth $\bar{b}_i$;
   (ii.) Colour the requests with same $\bar{b}_i$ and $f(i)$, in order of starting point, assigning greedily the 3 available colours.

3. Accept those requests coloured with same colour having total highest benefit;
4. Bring every rectangle's height $\bar{b}_i$ to the original original b_i.

In order to prove that the algorithm gives a legal 3-coloring of the graph we state a set of properties of a stable schedule. We first give a direct corollary of Lemma 3.

Corollary 1. *Consider two requests i and j with $\bar{b}_i \geq \bar{b}_j$, respectively (h_i, t_i) and (h_j, t_j) stable. Request i intersects with request j only if $h_i \leq h_j < h_i + \bar{b}_i$.*

The following Lemma is used to prove the two following Lemmas.

Lemma 5. *Consider a schedule with two intersecting requests i and j that are respectively (h_i, t_i) and (h_j, t_j) stable. Therefore $t_i, t_j \notin [t(i), t(i) + l_i) \cap [t(j), t(j) + l_j)$*

Proof: The proof is by contradiction. Assume request i placed before j. If $t_i \in [t(i), t(i) + l_i) \cap [t(j), t(j) + l_j)$ then i is part of the support of j, $h_j \geq h_i + \bar{b}_i$, a contradiction since the two requests are intersecting.

Assume $t_j \in [t(i), t(i) + l_i) \cap [t(j), t(j) + l_j)$. It must be $h_j \geq h_i$, otherwise by Lemma 3, $h_j + \bar{b}_j \leq h_i$, a contradiction to the intersection of the two rectangles. Since we are considering the aligned case, at least one rectangle of the support of j in t_j, say h, will be scheduled between $h_i - \bar{b}_h$ and h_i. Therefore, i is part of the support of j, a contradiction. ∎

The next Lemma states that if i and j are intersecting the two associated time intervals are not nested.

Lemma 6. *For any two intersecting requests i, j, it never holds $[t(i), t(i) + l_i) \subseteq [t(j), t(j) + l_j)$.*

Proof: The proof is by contradiction. If i and j are overlapping and $[t(i), t(i) + l_i) \subseteq [t(j), t(j) + l_j)$ then for the support of i it holds $t_i \in [t(i), t(i) + l_i) \cap [t(j), t(j) + l_j)$, a contradiction to Lemma 5. ∎

Lemma 7. *The maximum clique size of the intersection graph is 2.*

Proof: Assume by contradiction that requests i, j and k form a clique of size 3 and that k is assigned to a solution after i and j. Assume i (h_i, t_i) stable, j (h_j, t_j) stable and $t_i < t_j$. Request k must be completely contained in the interval (t_i, t_j), otherwise k is either $(h_i + \bar{b}_i, t_i)$ stable or $(h_j + \bar{b}_j, t_j)$ stable, thus it does not intersect with i or j.

Therefore $[t(k), t(k) + l_k)$ is completely contained in (t_i, t_j) that leads to the fact that either $t_k \in (t(i), t(i) + l_i)$ or $t_k \in (t(i), t(i) + l_i)$. By Lemma 5 this is a contradiction to the assumption that k intersects both i and j. ∎

We finally prove that the algorithm produces a legal 3 colouring of the intersection graph.

Theorem 1. *The algorithm colours the intersection graph with 3 colours.*

Proof: By Corollary 1, requests with same $\bar{b}_i$ and different $f(i)$ are non intersecting. Therefore they can be coloured independently. Concentrate on a set of requests with same $\bar{b}_i$ and $f(i)$. They are coloured greedily in order of starting point, i.e. from left to right.

Consider one such request i. By Lemma 6, every request intersecting i can intersect either $t(i)$ or $t(i) + l_i - 1$, but not both endpoints. If more than 1 request intersect an endpoint, by Corollary 1, these all intersect in that point thus creating a clique of size at least 3, by Lemma 7 a contradiction. Therefore at most 1 request intersects each endpoint of i, at most 2 requests intersecting i are already coloured, leaving one colour available for i. ∎

We finally show the approximation ratio we obtain.

Theorem 2. *The algorithm for the RP problem is 12-approximated in the general case and 6-approximated in the aligned case.*

Proof: The algorithm selects a solution S_{best} whose benefit is at least $OPT(LP)/2$ as it follows from: $OPT(LP) = \sum_{S_l} \sum_{i^t \in S_l} \alpha \omega_i \leq \frac{2}{\alpha} \sum_{i^t \in S_{Best}} \alpha \omega_i \leq 2 \sum_{i^t \in S_{Best}} \omega_i$.

By Lemma 1 $OPT(RP) \geq OPT(LP)/2$ in the general case for which the benefit of S_{best} is at least $1/4$ of $OPT(RP)$, while in the aligned case we have $OPT(RP) \geq OPT(LP)/2$. Moreover we colour the requests of S_{best} with 3 colours and select the set of intervals with same colour of highest benefit, for which the final solution has a benefit of at least $1/3$ of the benefit of S_{best}. Altogether we obtain an approximation ratio of 12 for the general case and of 6 for the aligned case. ∎

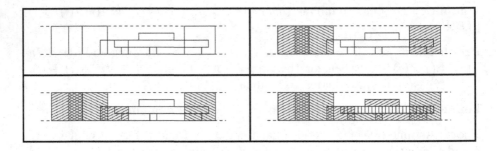

Figure 3. Requests are coloured by non-increasing bandwidth size.

4 A Fully Polynomial Algorithm

The number of constraints in the LP formulation is $O(nT)$, thus leading to a pseudopolynomial algorithm. Bar-Noy et al.[BNGNS99] showed how to reduce the number of time slots to a polynomial in n in a LP formulation for the maximum throughput scheduling problem under real-time constraints. The application of their technique to our case allows to express the LP solution as a convex combination of $3/\alpha$ rather than $2/\alpha$ integral solutions, therefore leading to an integral solution with a benefit at least $1/3$ of the optimum LP. This results in a higher approximation ratio of 18 for the general version and of 9 for the aligned version.

5 A Combinatorial Algorithm

In this section we sketch how to replace the basic step of the approximation algorithm based on the solution of a fractional LP formulation with a combinatorial algorithm that delivers a constant approximation solution to the LP problem.

We partition the requests into *wide* requests, that ask at least $1/2$ of the available bandwidth, and *narrow* requests whose bandwidth requirement is less than $1/2$. We solve the RP problem separately for wide requests and narrow requests and we choose the best solution. If all requests are wide then RP is equivalent to interval scheduling for which a 2 approximation algorithm is known [Spi99].

For narrow requests we replace the basic step of the algorithm based on solving the LP formulation with a combinatorial algorithm. We divide every request in k identical requests each one with a fraction $1/k$ of the bandwidth and of the profit of the original request. We then apply the combinatorial algorithm by [BNBYF⁺00] for finding an approximate integral solution to the problem in which the only bandwidth constraint is imposed. Lemma 3.2 of [BNBYF⁺00] states the following:

Lemma 8. *For each integer k there exists a combinatorial algorithm that finds a $2 + 1/k$ approximate solution to the LP formulation.*

Therefore the combinatorial algorithm gives a solution that is away form the optimal LP solution for at most a $2 + \frac{1}{k}$ factor thus leading to a $12(2 + \frac{1}{k})$ approximate solution for narrow requests. Combined with the 2 approximation for wide requests we obtain:

Theorem 3. *For every k there exists a $26 + 1/k$ combinatorial approximation algorithm for the RP problem.*

6   The Multiple Channel Case

In this section we assume that m channels, each one with a bandwidth $B_j \leq 1$, are available. For the sake of simplicity we assume the B_j's to be powers of $1/2$. We briefly sketch the extension of known techniques [BNGNS99], to obtain a $c+1$ throughput maximization approximation algorithm for m parallel unrelated machines provided a c algorithm for a single machine. We consider a Linear Programming formulation with variables x_{ijt} indicating the allocation of request i at time t on machine j. We set $x_{ijt} = 0$ for those machines j where $\bar{b}_i > B_j$. We then solve the LP problem and apply our rounding algorithm in order from channel 1 to channel m while we disregard on channel j requests already accepted on a previous channel. We then conclude with the following theorem:

Theorem 4. *Provided a c approximation algorithm for the RP problem on a single channel, there exists a $c+1$ approximation algorithm for the RP problem on multiple channels.*

7   Conclusions

In this paper we present constant approximation algorithms for the RP problem, a throughput version of bandwidth and storage allocation problems when real time constraints are imposed. Our algorithm uses as a basis a solution of a Linear

Programming formulation and partitions it into a convex combination of integral solutions with a novel rounding technique. We improve the approximation results found independently from our work in [BNBYF+00].

An interesting open problem is to study the problem in the on-line model in which requests for bandwidth allocation are presented over time. An interesting model is also to not allow rejection of the requests if enough bandwidth is available. We finally mention the improvement of the approximability of the problem, in particular by exploiting some of the ideas behind the recent approximation for DSA of [Ger96, Ger99]

References

[BNBYF+00] Amotz Bar-Noy, Reuven Bar-Yehuda, Ari Freund, Joseph (Seffi) Naor, and Baruch Schieber. A unified approach to approximating resource allocation and scheduling. In *Proceedings of the Thirty-Second Annual ACM Symposium on the Theory of Computing*, Las Vegas, Nevada, 2000.

[BNCK+95] Amotz Bar-Noy, Ran Canetti, Shay Kutten, Yishay Mansour, and Baruch Schieber. Bandwidth allocation with preemption. In *Proceedings of the Twenty-Seventh Annual ACM Symposium on the Theory of Computing*, pages 616–625, Las Vegas, Nevada, 29 May–1 June 1995.

[BNGNS99] Amotz Bar-Noy, Sudipto Guha, Joseph (Seffi) Naor, and Baruch Schieber. Approximating the throughput of multiple machines under real-time scheduling. In *Proceedings of the thirty-first annual ACM symposium on Theory of computing*, pages 622–631, Atlanta, GA USA, 1–4 May 1999.

[Eme] http://www.estec.esa.nl/artes3/projects/12telbios/telbios.htm.

[Ger96] Jordan Gergov. Approximation algorithms for dynamic storage allocation. In *European Symposium on Algorithms (ESA '96)*, volume 1136 of *Lecture Notes in Computer Science*, pages 52–61. Springer, 1996.

[Ger99] Jordan Gergov. Algorithms for compile-time memory optimization. In *Proc. of the 10th ACM-SIAM Symposium on Discrete Algorithms*, pages 907–908, 1999.

[GGK97] Juan A. Garay, Inder S. Gopal, and Shay Kutten. Efficient on-line call control algorithms. *Journal of Algorithms*, 23(1):180–194, April 1997.

[GJ79] M. R. Garey and D. S. Johnson. *Computers and Intractability – A Guide to the Theory of NP-Completeness*. Freeman, San Francisco, 1979.

[Kie91] H. A. Kierstead. A polynomial time approximation algorithm for dynamic storage allocation. *Disccrete Mathematics*, 88:231–237, 1991.

[Knu73] D.E. Knuth. *The Art of Computer Programming, Vol. 1: Fundamental Algorithms, 2nd Edition*. Addison-Wesley, 1973.

[Slu89] M. Slusarek. A coloring algorithm for interval graphs. In *Proc. of the 14th Mathematical Foundations of Computer Science*, pages 471–480, 1989.

[Spi99] F.C.R. Spieksma. On the approximability of an interval scheduling problem. *Journal of Scheduling*, 2:215–2227, 1999.

Dynamic Spectrum Allocation:
The Impotency of Duration Notification[*]

Bala Kalyanasundaram[1] and Kirk Pruhs[2]

[1] Dept. of Computer Science
Georgetown University
Washington D.C. 20057 USA
kalyan@cs.georgetown.edu
[2] Dept. of Computer Science
University of Pittsburgh
Pittsburgh, PA. 15260 USA
kirk@cs.pitt.edu
http://www.cs.pitt.edu/∼kirk

Abstract. For the classic dynamic storage/spectrum allocation problem, we show that knowledge of the durations of the requests is of no great use to an online algorithm in the worst case. This answers an open question posed by Naor, Orda, and Petruschka [9]. More precisely, we show that the competitive ratio of every randomized algorithm against an oblivious adversary is $\Omega(\frac{\log x}{\log \log x})$, where x may be any of several different parameters used in the literature. It is known that First Fit, which does not require knowledge of the durations of the task, is logarithmically competitive in these parameters.

1 Introduction

The dynamic storage/spectrum allocation (DSA) problem is a classic combinatorial optimization problem in the computer science literature (for surveys see [2] or [13]).

Dynamic Storage/Spectrum Allocation Problem Statement: An online algorithm is equipped with a linear resource, for example memory or radio spectrum, that it must use efficiently to satisfy a sequence of n requests for this resource. The ith request R_i is a pair (s_i, d_i) that is revealed to the online algorithm at some release time r_i. The parameter s_i denotes the bandwidth of the request R_i, and d_i denotes the duration of R_i. We number the requests by increasing order of release times, that is, $r_i \leq r_j$ for $i < j$. In response to request R_i , the online algorithm must allocate s_i units of contiguous resource to R_i during the time interval $[r_i, r_i + d_i]$. We say that R_i is *active* during $[r_i, r_i + d_i]$. Importantly, at no time may any two active requests share a common unit of resource. The objective function is to minimize the total resource size required to satisfy all of the given requests.

[*] Supported in part by NSF grant CCR-9734927 and by ASOSR grant F49620010011.

S. Kapoor and S. Prasad (Eds.): FST TCS 2000, LNCS 1974, pp. 421–428, 2000.
© Springer-Verlag Berlin Heidelberg 2000

In essentially all programming systems, dynamic memory managers do not, and more generally can not, know the exact duration of dynamically allocated objects. Hence, if the linear resource is memory, the standard assumption is that the online scheduler does not learn d_i at time r_i [13]. However, in many wireless data transmission applications, where the linear resource is frequency spectrum, the online scheduler does learn d_i at time r_i. For example in the testing of aircraft, d_i is known because the test is predetermined [6]. More generally, the value of d_i will be known if the information to be transmitted is known a priori. In this paper, we consider the online version of DSA with known durations.

The standard measure used to compare online algorithms is competitiveness [1]. In the context of DSA, an online randomized algorithm A is c-competitive if for all input sequences I, the expected value of the bandwidth $A(I)$ used by A on input I is at most c times $\text{OPT}(I)$, the optimal bandwidth to satisfy I [1].

Most of the literature on online dynamic storage allocation deals with the unknown duration case, that is when the online algorithm does not learn the duration of a request R_i until R_i leaves the system at time $r_i + d_i$. There is no constant competitive algorithm in the unknown duration case [10,11]. Every deterministic algorithm is $\Omega(\log \beta)$ competitive, where β is the ratio of the largest bandwidth of a request to the shortest bandwidth of a request [10]. Every randomized algorithm is $\Omega(\min(\frac{\log \beta}{\log \log \beta}, \frac{\log \alpha}{\log \log \alpha}))$ competitive against an oblivious adversary [8], here α is the maximum number of active requests at any one time. An *oblivious* adversary must specify all of the requests before the online algorithm begins execution [1]. The competitive ratio of First Fit, which places each request in the lowest feasible location in the linear resource, was observed to have a competitive ratio of $\Theta(\log(2 + \alpha\beta/M))$ in [4], where M is the maximum amount of occupied linear resource.

In [9], the dynamic storage allocation problem with known durations is considered. They show that in the case of known durations that an online algorithm may be logarithmically competitive in other parameters besides α and β. More precisely, an online algorithm may be $O(\log \Delta)$ competitive, here Δ is the ratio of the largest duration of a request to the smallest duration of a request, and may be $O(\log \tau)$ competitive, here τ is the number of distinct durations. In [9] the fundamental problem of determining whether knowledge of the durations of the requests is of significant benefit to the online algorithm, i.e. whether it is possible for an online algorithm to be constant competitive, is left open. Here we prove the following theorem.

Theorem 1. *Assuming an oblivious adversary, the competitive ratio of every randomized online algorithm for the dynamic storage allocation with known durations is* $\Omega(\frac{\log x}{\log \log x})$, *where x may be any one of n, α, β, τ or $\log \Delta$.*

We feel that the proper interpretation of these results is that knowledge of the durations does not greatly benefit the online allocator in the worst case. Although these results do leave open the possibility that knowing the durations may logarithmically improve the competitive ratio achievable by an online algorithm when the competitive ratio is measured in terms of Δ.

The offline version of DSA, where the algorithm has complete knowledge of future requests, is NP-hard (see [3], which credits this result to a personal communication from Stockmeyer), and a 3-approximation polynomial time algorithm is known [5]. Several average case analyses of various algorithms are known, for more information see [13]. Some early papers on DSA are [7,12,14].

For convenience, we will refer to the linear resource as spectrum for the rest of the paper.

2   The Lower Bound Construction

In this section we prove Theorem 1 using the following statement of Yao's principle for online cost minimization problems [1].

Theorem 2. *Suppose that there exists a input distribution on the inputs I such that for all deterministic algorithms A it is the case that*

$$\liminf_{n \to \infty} \frac{E\left[A(I)\right]}{E\left[\text{OPT}(I)\right]} \geq c$$

and

$$\limsup_{n \to \infty} E\left[\text{OPT}(I)\right] = \infty$$

where the expectation is over the marginal distribution on inputs of size n, Then the competitive ratio of every randomized algorithm is at least c.

In order to apply Theorem 2, fix a deterministic online algorithm A, and fix some $c > 2$. Let $W = 12(12c)^{47c}$. Note that $c = \Theta(\frac{\log W}{\log \log W})$. We will give a request distribution that forces $E[A(I)] = \Theta(cW)$ while $E[\text{OPT}(I)] = \Theta(W)$.

Request Distribution: We partition the requests into $24c$ rounds. The bandwidth of each request in round i, $1 \leq i \leq 24c$, will be $w(i) = (12c)^{2i-2}$. Further, we partition round i, $1 \leq i \leq 24c$, into $\frac{W}{b(i)}$ stages, where $b(i) = (12c)^{2i-1}$. The number $\ell_{i,j}$ of requests in stage j, $1 \leq j \leq \frac{W}{b(i)}$, of round i, will be selected uniformly at random over the range $[1, 2j\frac{b(i)}{w(i)}]$. Note that we have set the value of W so that $\frac{W}{b(i)} \geq 12$ for any $1 \leq i \leq 24c$.

We associate an interval $I_{i,j} = [a_{i,j}, b_{i,j}]$ with the stage j of round i. Initially, $I_{1,1} = [a_{1,1}, b_{1,1}] = [0, 1]$. Each request in stage j of round i is released at time $a_{i,j}$.

The duration of the kth request, $1 \leq k \leq \ell_{i,j}$, in stage j of round i is then $(b_{i,j} - a_{i,j})(1 - 1/2^k)$. Notice that the durations increase throughout a stage. We call the last request in each stage a *fang*.

Finally, the interval associated with the next stage (whether it is in the same round or a different round is irrelevant) is $[a_{i,j} + (b_{i,j} - a_{i,j})(1 - 1/2^{\ell_{i,j}-1}), a_{i,j} + (b_{i,j} - a_{i,j})(1 - 1/2^{\ell_{i,j}})]$. That is, the interval for the next stage is between the end of the penultimate request in the previous stage and the end of the fang in the previous stage. Thus all fangs released during a stage will stay present till the arrival of the last request of the last round, while requests that are not fangs leave the system before requests in the next stage appear. For a graphical depiction of the construction see Figure 1.

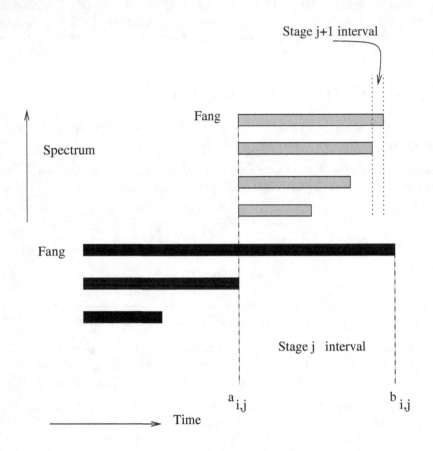

Fig. 1. Lower Bound Construction

Intuitively, the adversary may allocate fangs consecutively in the spectrum. However, since the online algorithm can not identify a fang when it arrives, it will not be able to allocate fangs to one part of the spectrum, and thus will have a more fragmented resource when it proceeds to the next stage. We will first prove that $\text{OPT} = \Theta(W)$, and then show that the expected bandwidth used by A is $\Omega(cW)$.

Lemma 1. *With probability one,* $\text{OPT}(I) \leq 4W$

Proof. One possible strategy to achieve $4W$ is to divide the spectrum into two pieces N and F, each of size $2W$. All non-fangs are allocated using to N using First Fit, and and all fangs are allocated to F using First Fit.

The total bandwidth of the non-fang requests in a stage j in round i is at most the product of the bandwidth $w(i)$ of the requests in round i, and the number of requests $2j\frac{b(i)}{w(i)}$. Since j is at most $W/b(i)$, the total bandwidth of the non-fang request in stage j of round i is at most

$$w(i) \cdot 2 \cdot \frac{W}{b(i)} \cdot \frac{b(i)}{w(i)} = 2W$$

Hence, all the non-fangs within any stage can be fit into N. Observe that any two non-fang requests not from the same stage do not overlap in time. Therefore, the same spectrum N can be reused for all of the stages.

We now calculate required bandwidth to satisfy fang requests. For round i, there are $\frac{W}{b(i)}$ fang requests of width $w(i)$ each. Therefore, the total bandwidth of the fangs is at most

$$\sum_{i=1}^{24c} \frac{W}{b(i)} w(i) = \sum_{i=1}^{24c} \frac{W}{12c} = 2W$$

Hence, all the fangs can fit into F. ∎

It order to analyze the performance of A it is convenient to grant A additional powers. At the end of each round, say round i, we allow A to reorganize its spectrum in the following manner. Assume that there are two fangs R_j and R_k such that the space between R_j and R_k is less than $w(i+1)$, the bandwidths of the requests in the next round. Notice that the space used by R_j and R_k, and the space between R_j and R_k, is of no use to A in future rounds. The online algorithm A may then delete the space used by R_j, the space used by R_k, and the space between these two requests from its spectrum, and meld the remaining two pieces of spectrum. We say that the remaining contiguous portion of the melded spectrum is *available* spectrum for the next round. Note that this is strictly to A's benefit since any feasible assignment on the original spectrum is also a feasible assignment in the modified spectrum.

We now wish to define what we call a fragmented round. Intuitively, a fragmented round is one in which A assigns many fangs to different parts of the available spectrum. Let i be some round. We partition A's available spectrum into contiguous *i-blocks* $B_{i,k}$ of size $b(i)$. We say that a fang from round i is assigned to an *i*-block $B_{i,k}$ if either the fang is contained entirely within $B_{i,k}$ or the fang crosses the boundary between $B_{i,k}$ and $B_{i,k+1}$. Let $f(i)$ be the number distinct *i*-blocks that have at least one fang assigned to them. We say that a round i is *fragmented* if $f(i) \geq \frac{W}{2b(i)}$.

We will now argue that:

- the expected number of fragmented rounds is at least one half of the total number of rounds, and
- the available spectrum decreases by $\Theta(W)$ after each fragmented round provided that A does not use too much spectrum.

Lemma 2. *The expected number of fragmented rounds is at least* $12c$.

Proof. Consider stage j of round i. Let $f_j(i)$ be the number of distinct i-blocks assigned at least one fang from stages 1 through j in round i. By our construction and the definition of fang requests, we have $f_{j-1}(i) \leq j - 1$. Therefore, at most $(j-1)\frac{b(i)}{w(i)}$ of the requests in stage j of round i can occupy i-blocks that have already been assigned a fang in round i. Since the number of requests in stage j of round i is uniformly at random selected from the range $[1, 2j\frac{b(i)}{w(i)}]$, the probability that the fang of stage j of round i is assigned to an i-block that does not already contain a fang is at least $1/2$. Hence, the probability that half of the fangs in round i are assigned to an i-block that did not previously contain a fang from round i is at least $\frac{1}{2}$. The result then follows since there are $24c$ rounds. ∎

Lemma 3. *During every fragmented round, either A deleted at least $\frac{W}{12}$ units of spectrum at the end of this round, or A used at least cW units of spectrum during this round.*

Proof. Fix a fragmented round i, and consider the assignment of the fangs from round i. Within each i-block $B_{i,k}$ that contains a fang from round i, arbitrary pick a *canonical* fang from among those fangs assigned to $B_{i,k}$ during round i. Recall that there are at least $\frac{W}{2b(i)}$ canonical fangs since i is a fragmented round. We sort these canonical fangs according to their position in A's spectrum, and group them into $\frac{W}{6b(i)}$ groups, each containing 3 consecutive canonical fangs.

We say that a group is *sparse* if the gap between any two consecutive canonical fangs in the group is at least $w(i + 1)$. Recall that we set W such that $\frac{W}{12b(i)} \geq 1$ for all $1 \leq i \leq 24c$. First, suppose that $\frac{W}{12b(i)}$ of the groups are sparse. Then the total spectrum used by A must be at least $\frac{W}{12b(i)}w(i + 1)$, which by substitution is cW

On the other hand, suppose that $\frac{W}{12b(i)}$ of the groups are not sparse. Hence, the spectrum used by the canonical fangs in this group, as well as the spectrum in the gaps between these fangs, will be deleted by A at the end of the round. Since the gap between first and third requests of each group is at least $b(i)$, the total spectrum deleted by A during this round is at least $\frac{W}{12}$. ∎

We are now ready to prove Theorem 1.

Proof. First, the fact that OPT $= \Theta(W)$ follows from lemma 1. By lemma 2 and lemma 3, the expected spectrum used by A is at least cW. Note that the expected spectrum used by OPT goes to ∞ as n goes to ∞. Hence, we get a lower bound of $\Omega(\frac{\log W}{\log \log W})$ on the competitive ratio.

Observe that we get a lower bound of $\Omega(\frac{\log \beta}{\log \log \beta})$ on the competitive ratio since $\beta \leq W$.

Observe that n is no smaller than α and τ. So in order to show a lower bound of $\Omega(\frac{\log x}{\log \log x})$ on the competitive ratio, for x equal to n, α, and τ, it is sufficient to show that $n = O(W^2)$. Since the number of stages in round i is $\frac{W}{b(i)}$, and the number of requests in stage j of round i is at most $2j\frac{b(i)}{w(i)}$, it follows that the number requests in round i is at most

$$2\frac{b(i)}{w(i)}\sum_{j=1}^{\frac{W}{b(i)}} j$$

$$= \frac{b(i)}{w(i)}\frac{W}{b(i)}\left(\frac{W}{b(i)}+1\right)$$

$$\leq \frac{b(i)}{w(i)}\frac{W}{b(i)}\left(2\frac{W}{b(i)}\right)$$

$$= \frac{2W^2}{w(i)\cdot b(i)}$$

Since $b(i)w(i) = (12c)^{4i-3} \geq 12c \geq 4$, the number of requests per round is going down by more than a factor of two each round. Hence, the total number of requests is at most

$$2\frac{2W^2}{w(1)\cdot b(1)} \leq 2W^2$$

Finally, that the competitive ratio is $\Omega(\frac{\log\log\Delta}{\log\log\log\Delta})$ follows from the fact that $\Delta = 2^n$ by construction.   ∎

References

1. A. Borodin, and R. El-Yaniv, *Online Computation and Competitive Analysis*, Cambridge University Press, 1998.
2. E. Coffman, "An introduction to combinatorial models of dynamic storage allocation", *SIAM Review*, **25**, 311 – 325, 1999.
3. M. Garey and D. Johnson, Computers and Intractability: A Guide to NP-completeness, W.H. Freeman and Company, 1979.
4. J. Gergov, "Approximation algorithms for dynamic storage allocation", European Symposium on Algorithms, 52 – 61, 1996.
5. J. Gergov, "Algorithms for compile-time memory allocation", ACM/SIAM Symposium on Discrete Algorithms, S907 – S908, 1999.
6. Proceedings of the 2000 *Symposium of the International Test and Evaluation Association*, http://www.edwards.af.mil/itea/.
7. S. Krogdahl, "A dynamic storage allocation problem", *Information Processing Letters*, **2**(4), 96 – 99, 1973.
8. M. Luby, J. Naor, A. Orda, "Tight Bounds for Dynamic Storage Allocation" *SIAM Journal of Discrete Mathematics*, **9**, 156 – 166, 1996.
9. J. Naor, A. Orda, and Y. Petruschka, "Dynamic storage allocation with known durations", European Symposium on Algorithms, 378 – 387, 1997.
10. J. Robson, "An estimate of the store size necessary for dynamic storage allocation", *Journal of the ACM*, **18**, 416 – 423, 1971.
11. J. Robson, "Bounds on some functions concerning dynamic storage allocation", *Journal of the ACM*, **21**, 491 – 499, 1974.

12. J. Robson, "Worst case fragmentation of first fit and best fit storage allocation strategies", *Computer Journal*, **20**, 242 – 244, 1977.
13. P. Wilson, M. Johnstone, M. Neely, and D. Boles, "Dynamic storage allocation: a survey and critical review", International Workshop on Memory Management, *Lecture Notes in Computer Science*, **986**, 1 – 116, 1995.
14. D. Woodall, "The bay restaurant — a linear storage problem", *American Mathematical Monthly*, **81**, 240 – 246, 1974.

The Fine Structure of Game Lambda Models

Pietro Di Gianantonio and Gianluca Franco

Dipartimento di Matematica e Informatica
Università di Udine
Via delle Scienze, 206 - 33100 Udine - Italy
Tel. +39 0432 558469
Fax +39 0432 558499
pietro,gfranco@dimi.uniud.it

Abstract. We study models of the untyped lambda calculus in the setting of game semantics. In particular, we show that, in the category of games $\mathcal{G}$, introduced by Abramsky, Jagadeesan and Malacaria, all categorical λ-models can be partitioned in three disjoint classes, and each model in a class induces the same theory (*i.e.* the set of equations between terms), that are the theory $\mathcal{H}^*$, the theory which identifies two terms iff they have the same Böhm tree and the theory which identifies all the terms which have the same Lévy-Longo tree.

Introduction

In this work we explore the methodology for giving denotational semantics based on games, introduced by Abramsky – Jagadeesan – Malacaria, Hyland – Ong and Nickau (see [AJM96, HO00, Nic94]). We use game semantics to build models of the untyped λ-calculus, focusing on which λ-theories can be modeled. λ-theories are congruences over λ-terms, which extend pure β-conversion. Their interest lies in the fact that they correspond to the possible *operational* (*observational*) semantics of the λ-calculus. Brute force, purely syntactical techniques are usually extremely difficult to use in the study of λ-theories. Therefore, since the seminal work of Dana Scott on D_∞ in 1969 [Sco72], semantical tools have been extensively investigated [HR92, HL95, Ber].

This paper is the completion of the work initiated in [DGFH99] and gives a complete characterization of the theories induced by general game models of the λ-calculus. In [DGFH99] we considered just models which validate the η-rule. In order to obtain our new results, new proof techniques have been introduced.

We show that the theory induced by each categorical model of the λ-calculus in the Cartesian closed category $K_!(\mathcal{G})$ of games and history-free strategies is either: the theory $\mathcal{H}^*$ (the maximal sensible theory), the theory $\mathcal{B}$ which equates two terms if and only if they have the same Böhm tree or the theory $\mathcal{L}$ which equates two terms if and only if they have the same Lévy-Longo tree.

This result suggests that there exists a strong connection between a *strategy* which interprets a term in the game semantics setting and the tree form of the term. The existence of relations between strategies and some syntactical normal

S. Kapoor and S. Prasad (Eds.): FST TCS 2000, LNCS 1974, pp. 429–441, 2000.

form of terms (like trees) is described in many works on game semantics. In all these works, a relation is established to study the fine structure of a particular game model and prove that is fully abstract. In our work we give a somewhat stronger result and prove that such a relation exists for all the game models of the λ-calculus.

Works related to ours, in the slight different game semantics paradigm of Hyland and Ong, are [KNO99, KNO00]. There, two particular game models for the λ-calculus are built and it is proved, using techniques quite different from ours, that the two models induce respectively the theories $\mathcal{H}^*$ and $\mathfrak{B}$.

The present paper is organized as follows. In section 1 we remind the basic definitions of game semantics and introduce the extensions $\mathcal{G}^s$ and $K_!(\mathcal{G}^s)$ of the classical categories of games $\mathcal{G}$ and $K_!(\mathcal{G})$. In Section 2 we introduce the main tool for the study of the fine structure of the models built in the categories of Section 1, that is the *approximating strategies* whose intended meaning is to give a finite approximation of the interpretation of a λ-term. Section 3 is devoted to the study of the models previously introduced and to the proof of the main theorem of this work, *i.e.* the characterization of all the λ-theories induced by the models of the untyped λ-calculus in the category $K_!(\mathcal{G})$.

We assume the reader familiar with the basic notions and definitions of λ-calculus, see e.g. [Bar84]. For lack of space, all the proofs have been omitted. A complete version of this paper is [DGF].

Acknowledgements. We wish to thank Fabio Alessi, Corrado Böhm, Furio Honsell and Luke Ong for useful discussions during the period of which this work generated. We also thank the anonymous referees whose comments have contributed to improve this work.

1   Categories of Games

Throughout this paper, we shall make use of the well-known category $\mathcal{G}$ of games and history-free strategies and its Cartesian closed companion $K_!(\mathcal{G})$, and of the categories $\mathcal{G}^s$ of games and history-sensitive strategies and $K_!(\mathcal{G}^s)$ that are new. $\mathcal{G}^s$ is a straightforward extension (super-category) of $\mathcal{G}$ and it has been introduced for technical reasons. We briefly remind the basic definitions of game semantics [AJM96] and introduce the new categories we shall utilize.

Definition 1 (Games). *A game has two participants: the Player and the Opponent. A game A is a quadruple $(M_A, \lambda_A, P_A, \approx_A)$ where*

- *M_A is the set of moves of the game.*
- *$\lambda_A : M_A \to \{O, P\} \times \{Q, A\}$ is the labeling function: it tells us if a move is made by the Opponent or by the Player, and if it is a Question or an Answer. We can decompose λ_A into $\lambda_A^{OP} : M_A \to \{O, P\}$ and $\lambda_A^{QA} : M_A \to \{Q, A\}$ and put $\lambda_A = \langle \lambda_A^{OP}, \lambda_A^{QA} \rangle$. We denote by $^-$ the function which exchanges Player and Opponent, i.e. $\overline{O} = P$ and $\overline{P} = O$. We also denote with $\overline{\lambda_A^{OP}}$ the function defined by $\overline{\lambda_A^{OP}}(a) = \overline{\lambda_A^{OP}(a)}$. Finally, we denote with $\overline{\lambda_A}$ the function $\langle \overline{\lambda_A^{OP}}, \lambda_A^{QA} \rangle$.*

- P_A is a non-empty and prefix-closed subset of the set $M_A^\circledast$ (which will be written as $P_A \subseteq^{nepref} M_A^\circledast$), where $M_A^\circledast$ is the set of all sequences of moves which satisfy the following conditions:
  - $s = at \Rightarrow \lambda_A(a) = OQ$
  - $\forall i : 1 \leq i < |s| . \lambda_A^{OP}(s_{i+1}) = \overline{\lambda_A^{OP}(s_i)}$
  - $\forall t \sqsubseteq s . |t \!\upharpoonright\! M_A^A| \leq |t \!\upharpoonright\! M_A^Q|$

  where M_A^A and M_A^Q denote the subsets of game moves labeled respectively as Answers and as Questions, $s \!\upharpoonright\! M$ denotes the set of moves of M which appear in s and $\sqsubseteq$ is the substring relation. P_A is called the set of positions of the game A. Questions and answers behave in a parenthesis-like fashion, that is each question waits for a corresponding answer to appear and each answer in a sequence corresponds to the last pending question in that sequence.
- $\approx_A$ is an equivalence relation on P_A which satisfies the following properties:
  - $s \approx_A s' \Rightarrow |s| = |s'|$
  - $sa \approx_A s'a' \Rightarrow s \approx_A s'$
  - $s \approx_A s' \wedge sa \in P_A \Rightarrow \exists a' . sa \approx_A s'a'$

In the above s, s', t and t' range over sequences of moves, while a, a', b and b' range over moves. The empty sequence is written ϵ.

Definition 2 (Tensor product). Given games A and B, the tensor product $A \otimes B$ is the game defined as follows:

- $M_{A \otimes B} = M_A + M_B$
- $\lambda_{A \otimes B} = [\lambda_A, \lambda_B]$
- $P_{A \otimes B} \subseteq M_{A \otimes B}^\circledast$ is the set of positions, s, which satisfy the following:
  - the projections on each component (written as $s \!\upharpoonright\! A$ or $s \!\upharpoonright\! B$) are positions for the games A and B respectively;
  - every answer in s must be in the same component game as the corresponding question.
- $s \approx_{A \otimes B} s' \Leftrightarrow s \!\upharpoonright\! A \approx_A s' \!\upharpoonright\! A, s \!\upharpoonright\! B \approx_B s' \!\upharpoonright\! B, \forall i . s_i \in M_A \Leftrightarrow s'_i \in M_A$

Here $+$ denotes disjoint union of sets, that is $A + B = \{in_l(a) \mid a \in A\} \cup \{in_r(b) \mid b \in B\}$, and $[-, -]$ is the usual (unique) decomposition of a function defined on disjoint unions.

It is easy to see that in such a game only the Opponent can switch component.

Definition 3 (Unit). The unit element for the tensor product is given by the empty game $I = (\varnothing, \varnothing, \{\epsilon\}, \{(\epsilon, \epsilon)\})$.

Definition 4 (Linear implication). Given games A and B, the compound game $A \multimap B$ is defined as the tensor product but for the condition $\lambda_{A \multimap B} = [\overline{\lambda_A}, \lambda_B]$.

It is easy to see that in such a game only the Player can switch component.

Definition 5 (Exponential). *Given a game A, the game $!A$ is defined by:*

- $M_{!A} = \omega \times M_A = \sum_{i \in \omega} M_A$
- $\lambda_{!A}(\langle i, a \rangle) = \lambda_A(a)$
- $P_{!A} \subseteq M_{!A}^{\circledast}$ *is the set of positions, s, which satisfy the following:*
    - $\forall i \in \omega \;.\; s \upharpoonright \langle i, A \rangle \in P_{\langle i, A \rangle};$
    - *every answer in s is in the same index as the corresponding question.*
- $s \approx_{!A} s' \Leftrightarrow \exists$ *a permutation of indexes $\alpha \in S(\omega)$ such that:*
    - $\pi_1^*(s) = \alpha^*(\pi_1^*(s'))$
    - $\forall i \in \omega \;.\; \pi_2^*(s \upharpoonright \alpha(i)) \approx \pi_2^*(s \upharpoonright i)$

where π_1 and π_2 are the projections of $\omega \times M_A$, π_1^* and π_2^* are the (unique) extensions of π_1 and π_2 to sequences of moves and $s \upharpoonright i$ is an abbreviation of $s \upharpoonright \langle i, A \rangle$.

Definition 6 (Strategies). *A strategy for the Player in a game A is a non-empty set $\sigma \subseteq P_A^{even}$ of positions of even length such that $\overline{\sigma} = \sigma \cup dom(\sigma)$ is prefix-closed, where $dom(\sigma) = \{t \in P_A^{odd} \mid \exists !a \;.\; ta \in \sigma\}$, and P_A^{odd} and P_A^{even} denote the sets of positions of odd and even length respectively. A strategy can be seen as a set of rules which tells (in some position) the Player which move to make after the last move by the Opponent.*

The equivalence relation on positions $\approx_A$ can be extended to strategies in the following way.

Definition 7 (Equivalence of strategies). *Let σ, τ be strategies, $\sigma \approx \tau$ if and only if*

- $sab \in \sigma, s'a'b' \in \tau, sa \approx_A s'a' \Rightarrow sab \approx_A s'a'b'$
- $s \in \sigma, s' \in \tau, sa \approx_A s'a' \Rightarrow ((\exists b \;.\; sab \in \sigma) \Leftrightarrow (\exists b' \;.\; s'a'b' \in \tau))$

Such an extension is not in general an equivalence relation since it might lack reflexivity. If σ is a strategy for a game A such that $\sigma \approx \sigma$, we write $\sigma : A$ and denote with $[\sigma]$ the equivalence class containing σ.

Definition 8 (History-free strategies). *A strategy σ for a game A is history-free if it satisfies the following properties:*

- $sab, tac \in \sigma \Rightarrow b = c$
- $sab, t \in \sigma, ta \in P_A \Rightarrow tab \in \sigma$

Definition 9 (The category of games $\mathcal{G}$). *The category $\mathcal{G}$ has as objects games and as morphisms, between games A and B, the equivalence class, w.r.t. the relation $\approx_{A \multimap B}$, of the history-free strategies for the game $A \multimap B$. The identity, for each game A, is given by the (equivalence class) of the copy-cat strategy $id_A = \{s \in P_{A' \multimap A''} \mid \forall t \sqsubseteq s \;.\; \mathsf{even}(|t|) \Rightarrow t \upharpoonright A' = t \upharpoonright A''\}$ where $\mathsf{even}(-)$ is the obvious predicate and the superscripts are introduced to distinguish between the two different occurrences of the game A. Composition is given by the extension on equivalence classes of the following composition of strategies. Given strategies $\sigma : A \multimap B$ and $\tau : B \multimap C$, $\tau \circ \sigma : A \multimap C$ is defined by*

$$\tau \circ \sigma = \{s \upharpoonright (A, C) \mid s \in (M_A + M_B + M_C)^* \;\wedge\; s \upharpoonright (A, B) \in \overline{\sigma}, s \upharpoonright (B, C) \in \overline{\tau}\}^{even}$$

It is not difficult to check that the above definitions are well posed and that the constructions introduced in Definitions 2, 4 and 5 can be made functorial. Notice that there is a natural isomorphism in the category of sets between $(M_A + M_B) + M_C$ and $M_A + (M_B + M_C)$ which induces a natural transformation $\Lambda^l_{A,B,C} : \hom(A \otimes B, C) \to \hom(A, B \multimap C)$ in $\mathcal{G}$, that is the category $\mathcal{G}$ is monoidal closed. If we define, for each pair of games B and C of $\mathcal{G}$, the strategy $ev^l_{B,C}$ as the set $\{s \in P_{((A' \multimap B') \otimes A'') \multimap B''} \mid \forall t \sqsubseteq s \,.\, \mathsf{even}(|t|) \Rightarrow t \restriction A' = t \restriction A''$ $\&\ t \restriction B' = t \restriction B''\}$ we have, for each strategy $\sigma : A \otimes B \multimap C$, the identity $[\sigma] = [ev^l_{B,C}] \circ (\Lambda^l_{A,B,C}([\sigma]) \otimes [id_B])$. However $\mathcal{G}$ is not Cartesian.

Definition 10 (The Cartesian closed category of games $K_!(\mathcal{G})$). *The category $K_!(\mathcal{G})$ is the category obtained by taking the co-Kleisli category over $\mathcal{G}$ over the co-monad $(!, \mathsf{der}, \delta)$ [AJM96], where, for each game A, the (history-free) strategies $\mathsf{der}_A : \,!A \multimap A$ and $\delta_A : \,!A \multimap !!A$ are defined as follows:*
$$\mathsf{der}_A = \{s \in P_{!A \multimap A} \mid \forall t \sqsubseteq s \,.\, \mathsf{even}(|t|) \Rightarrow t \restriction \langle 0, A \rangle = t \restriction A\}$$
$$\delta_A = \{s \in P_{!A \multimap !!A} \mid \forall t \sqsubseteq s \,.\, \mathsf{even}(|t|) \Rightarrow t \restriction \langle p(i, j), A \rangle = t \restriction \langle j, \langle i, A \rangle\rangle\}$$

where $p : \mathbb{N} \times \mathbb{N} \to \mathbb{N}$ is a pairing function. By the above definition the category $K_!(\mathcal{G})$ has as objects games and as morphisms between games A and B the equivalence classes of the history-free strategies for the game $!A \multimap B$. Moreover, $K_!(\mathcal{G})$ is Cartesian.

Definition 11 (Cartesian product). *The Cartesian product $A \times B$ of two games A and B is defined by:*
$$M_{A \times B} = M_A + M_B \qquad \lambda_{A \times B} = [\lambda_A, \lambda_B]$$
$$P_{A \times B} = P_A + P_B \qquad \approx_{A \times B} = \,\approx_A + \approx_B$$

The projection morphism $\pi^{A,B}_A : A \times B \to A$ is defined as
$$[\{s \in P_{A' \times B \multimap A''} \mid \forall t \sqsubseteq s \,.\, \mathsf{even}(|t|) \Rightarrow t \restriction A' = t \restriction A''\} \circ \mathsf{der}_{A \times B}]$$

From the isomorphisms $!(A \times B) \cong \,!A \otimes !B$ and $!I \cong I$ it follows easily that $K_!(\mathcal{G})$ is Cartesian closed [AJM96].

Definition 12 (Exponent). *The exponent game $A \Rightarrow B$ is the game $!A \multimap B$. The natural transformation $\Lambda_{A,B,C} : \hom(A \times B, C) \to \hom(A, B \Rightarrow C)$ is $\Lambda^l_{!A,!B,C}$, and $ev_{B,C} = ev^l_{!B,C} \circ (\mathsf{der}_{!B \multimap C} \times id_{!B})$.*

In order to carry out the proofs of our main theorem, we need to introduce the category $\mathcal{G}^s$ having as morphisms all the strategies (not only the history-free ones). This because we shall use *approximations* of history-free strategies that are not, in general, history-free. We call these morphisms history-sensitive strategies. It is worth noting that almost all the definitions in the categories $\mathcal{G}$ and $\mathcal{G}^s$ coincide.

Definition 13 (The category of games $\mathcal{G}^s$). *The category $\mathcal{G}^s$ has as objects games and as morphisms, between games A and B, the equivalence classes, w.r.t. the relation $\approx_{A \multimap B}$, of the strategies $\sigma : A \multimap B$. The identity, for each game A, is given by the (equivalence class) of the copy-cat strategy id_A. Composition is given as in $\mathcal{G}$.*

$K_!(\mathcal{G}^s)$ is obtained like $K_!(\mathcal{G})$, since $(!,[\mathbf{der}],[\delta])$ is a co-monad also over $\mathcal{G}^s$. Together with the category $\mathcal{G}^s$, we need to introduce a new relation on strategies of $\mathcal{G}^s$ which induces a partial order on equivalence classes of strategies. This notion can be easily proved equivalent to the standard one.

Definition 14 (Partial order relation on strategies). *Given a game A and strategies $\sigma : A$ and $\tau : A$ (hence such that $\sigma \approx \sigma$ and $\tau \approx \tau$) we define $\sigma \sqsubseteq \tau \Leftrightarrow \forall s \in \sigma . \exists t \in \tau . s \approx t$ and then $[\sigma] \sqsubseteq [\tau] \Leftrightarrow \sigma \sqsubseteq \tau$.*

2 Approximating Strategies

The argument of this section is the general concept of *approximating strategy*, which can be seen as a finite approximation of a strategy. It will be used to prove that the interpretation of a term is the least upper bound of the interpretations of its "approximate normal forms".

Definition 15. *1. Let D be a game. A sub-game D' of D (written $D' \trianglelefteq D$) is a game such that $M_{D'} \subseteq M_D$, $\lambda_{D'} = \lambda_D \upharpoonright M_{D'}$, $P_{D'} \subseteq P_D$ and $\approx_{D'} = \approx_D \upharpoonright P_{D'} \times P_{D'}$.*
2. Let D be a game. We indicate with D^n the sub-game of D in which $P_{D^n} = \{s \in P_D \mid |s| \leq n\}$.
3. Let A' be a sub-game of A and let σ be a strategy for the game A. We write $\sigma | A'$ for the strategy $\{s \in \sigma \mid s \in P_{A'}\}$.
4. Let $\sigma : A \multimap B$ be a strategy. We indicate with σ^n the history-sensitive strategy $\sigma | A \multimap B^n$ and with $[\sigma]^n$ the equivalence class $[\sigma^n]$.

Observe that if $\sigma \approx \tau$ then $\sigma^n \approx \tau^n$, since equivalent positions have the same length. Thus we can write $[\sigma]^n$ with no ambiguity. In general the strategy σ^n can be history-sensitive also if the strategy σ is history-free. This is because σ^n can reply to a move a of the Opponent in some position and does not reply in some others. In order to accommodate and freely use the strategies σ^n we introduce the category $\mathcal{G}^s$ of games and history-sensitive strategies. The strategies σ^n can be seen as a finite approximation of the strategy σ, and they will be use to prove an approximation theorem along the same line of the works [Hyl76, Wad78]. In these works the approximation of a semantical point is obtained through a series of projection functions. Here we use a different approach that, in the context of games, is simpler and more direct. We need to state a series of properties enjoyed by the approximating strategies. The basic ones are the following.

Proposition 1. *For each pair of games A and B and strategy $\sigma : A \multimap B$, the following properties hold:*

1. $\sigma^0 = \{\epsilon\}$ 2. $\sigma^n \subseteq \sigma^{n+1}$
3. $\bigcup_{n \in \omega}\{\sigma^n\} = \sigma$ 4. $(\sigma^n)^m = \sigma^{min\{m,n\}}$

Lemma 1. *For each pair of games A and B we have:*

1. $(A \multimap B)^{n+1} \trianglelefteq A^n \multimap B^{n+1}$
2. $ev_{A,B}^l | (A^n \multimap B^m) \otimes A \multimap B = ev_{A,B}^l | (A \multimap B) \otimes A^n \multimap B^m$

3   The Fine Structure of the Game Models

In this section the study of the λ-theory (*i.e.* the set of equations between λ-terms) supported by the models built in $K_l(\mathcal{G})$ is carried out. The theory induced by a model is also known as its *fine structure*. The equations on terms are described by means of the equality of some tree of the terms. The trees we consider are the Lévy-Longo trees [Lév75, Lon83] and the Böhm trees [Bar84, Hyl76]. We remind briefly the definitions.

Definition 16 (Trees). *Let* $\Sigma^1 = \{\lambda x_1 \ldots x_n.\bot \mid n \in \omega\} \cup \{T\} \cup \{\lambda x_1 \ldots x_n.y \mid n \in \omega\}$, *let* $\Sigma^2 = \{\bot\} \cup \{\lambda x_1 \ldots x_n.y \mid n \in \omega\}$, *let* $x_1, \ldots x_n, y$ *be variables and let* $M \in \Lambda$ *be a term. If* M *is solvable it is intended to have principal head normal form* $\lambda x_1 \ldots x_n.y M_1 \ldots M_m$.

1. *The Lévy-Longo tree of* M, $LLT(M)$ *is a* Σ^1*-labelled infinitary tree defined informally as follows:*

$$LLT(M) = \qquad T \qquad \text{if } M \text{ is unsolvable of order } \infty$$
$$LLT(M) = \qquad \lambda x_1 \ldots x_n.\bot \qquad \text{if } M \text{ is unsolvable of order } n$$
$$LLT(M) = \qquad \lambda x_1 \ldots x_n.y \qquad \text{if } M \text{ is solvable}$$

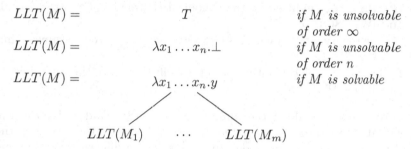

2. *The Böhm tree of* M, $BT(M)$ *is a* Σ^2*-labelled tree defined informally as follows:*

$$BT(M) = \qquad \bot \qquad \text{if } M \text{ is unsolvable}$$
$$BT(M) = \qquad \lambda x_1 \ldots x_n.y \qquad \text{if } M \text{ is solvable}$$

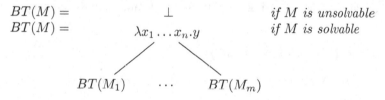

On Lévy-Longo trees (Böhm trees) there is a natural order relation defined by $LLT(M) \subseteq LLT(N)$ iff $LLT(N)$ is obtained by $LLT(M)$ by replacing $\bot$ in some leaves of $LLT(M)$ by Lévy-Longo trees of λ-terms or by replacing some $\lambda x_1 \ldots x_n.\bot$ by T ($BT(M) \subseteq BT(N)$ iff $BT(N)$ is obtained by $BT(M)$ by replacing $\bot$ in some leaves of $BT(M)$ by Böhm trees of λ-terms).

In this work we are interested in categorical models of the untyped λ-calculus, that is, reflexive objects in a Cartesian closed category.

Definition 17 (Categorical λ-model).

1. *Let* $\mathcal{C}$ *be a category and* $A, B \in Obj(\mathcal{C})$. B *is a* retract *of* A *if there exists a pair of morphisms* $f : A \to B$ *and* $g : B \to A$, *such that* $f \circ g = id_B$. *We write* $(B \lhd A, f, g)$ *to indicate that* B *is a retract of* A *via* f *and* g.

2. *A reflexive object is a retract* $(D \Rightarrow D \lhd D, f, g)$, *between an object D and its exponent $D \Rightarrow D$. We write $\langle D, (f, g) \rangle$ to indicate that D is a reflexive object via morphisms f and g, and call it* categorical λ-model.

Definition 18 (Classes of models). *Let $\mathcal{D}$ be the class of all the categorical λ-models $\langle D, ([\varphi], [\psi]) \rangle$, with $D \neq I$, in the category $K_!(\mathcal{G})$. We partition $\mathcal{D}$ in the following subclasses:*

1. $\mathcal{D}^{\mathcal{E}} = \{ \langle D, ([\varphi], [\psi]) \rangle \in \mathcal{D} \mid \psi \circ \varphi \approx id_D \}$
2. $\mathcal{D}^{\mathcal{B}} = \{ \langle D, ([\varphi], [\psi]) \rangle \in \mathcal{D} \mid \psi \circ \epsilon_{I \Rightarrow (D \Rightarrow D)} = \epsilon_{I \Rightarrow D} \text{ and } \psi \circ \varphi \not\approx id_D \}$
3. $\mathcal{D}^{\mathcal{L}} = \{ \langle D, ([\varphi], [\psi]) \rangle \in \mathcal{D} \mid \psi \circ \epsilon_{I \Rightarrow (D \Rightarrow D)} \neq \epsilon_{I \Rightarrow D} \}$

The main result of this paper states that, given a categorical λ-model $\mathsf{D} \in \mathcal{D}$, the theory it induces is either

1. $\mathcal{H}^*$, the theory induced by the canonical D_∞ model of Scott [Sco72, Bar84] and [Wad78], if $\mathsf{D} \in \mathcal{D}^{\mathcal{E}}$;
2. $\mathcal{B}$, the theory which identifies two terms iff they have the same Böhm tree, if $\mathsf{D} \in \mathcal{D}^{\mathcal{B}}$;
3. $\mathcal{L}$ the theory which identifies two terms iff they have the same Lévy-Longo tree if $\mathsf{D} \in \mathcal{D}^{\mathcal{L}}$.

The proof proceeds along the same lines of [Bar84, Wad78, Hyl76]. First we show that if two terms are equated in one of the above theories then they are equal in the corresponding model. In order to prove this, we state an important property satisfied by all the models. The *approximation theorem* says that the interpretation of a term is the least upper bound of the interpretations of its approximants. The following definitions and lemmata are necessary to state this result.

Definition 19 (Indexed terms).

1. *The set of $\lambda\Omega$-terms, $\Lambda(\Omega)(\ni M)$ is defined from a set of variables $Var(\ni x)$ as $M ::= x \mid MM \mid \lambda x.M \mid \Omega$.*
2. *The set of (possibly) indexed terms $\Lambda(\Omega)^{\mathbb{N}}(\ni M)$ is the superset of $\Lambda(\Omega)$ defined as $M ::= x \mid MM \mid \lambda x.M \mid \Omega \mid M^n$.*
3. *A term is* truly indexed *if it is of the shape M^n. A term is* completely indexed *if all its subterms of the shape variable, abstraction, and application are immediate subterms of truly indexed terms.*

Notice that in a truly indexed term the constant Ω does not need to be indexed. The reduction rules are extended to indexed terms as follows.

Definition 20 (Approximate reduction).

1. *The following reduction rules are definable on $\Lambda(\Omega)$:*
   $(\Omega_1) \quad \lambda x.\Omega \to \Omega \qquad (\Omega_2) \quad \Omega M \to \Omega$

2. *The following reduction rules are definable on indexed terms of $\Lambda(\Omega)^{\mathbb{N}}$:*

(Ω^n) $\quad \Omega^n \to \Omega$ $\qquad\qquad\qquad\qquad$ (Ω^0) $\quad M^0 \to \Omega$

(β_I) $\quad ((\lambda x.P^n)^{m+1}Q^p)^h \to (P[Q^a/x])^b$ $\quad (\beta_{i,j})$ $\quad (M^i)^j \to M^{min\{i,j\}}$

where $b = min\{n, m+1, h\}, a = min\{m, p\}$

Lemma 2. *A completely indexed term Q is $\Omega^n \Omega^0 \beta_I \beta_{i,j}$-normalizing.*

Denotational semantics is readily defined. The denotation of a pure λ-term $M \in \Lambda$ is defined along the usual categorical definition. To accommodate indexed terms we need to introduce two new rules and use the larger categories of games and history-sensitive strategies.

Definition 21. *Let $\mathsf{D} \in \mathcal{D}$ be a categorical λ-model. The interpretation of a term $M \in \Lambda(\Omega)^{\mathbb{N}}$ (whose free variables are among the list $\Gamma = \{x_1, \ldots, x_k\}$) in D, $[\![M]\!]_\Gamma^{\mathsf{D}} : \mathsf{D}^{|\Gamma|} \Rightarrow \mathsf{D}$ is the strategy inductively defined as follows:*

$[\![x]\!]_\Gamma^{\mathsf{D}} = \pi_x^\Gamma;$

$[\![MN]\!]_\Gamma^{\mathsf{D}} = [\![M]\!]_\Gamma^{\mathsf{D}} \cdot [\![N]\!]_\Gamma^{\mathsf{D}} = ev \circ \langle \varphi \circ [\![M]\!]_\Gamma^{\mathsf{D}}, [\![N]\!]_\Gamma^{\mathsf{D}} \rangle;$

$[\![\lambda x.M]\!]_\Gamma^{\mathsf{D}} = \psi \circ \Lambda([\![M]\!]_{\Gamma,x}^{\mathsf{D}});$

$[\![M^n]\!]_\Gamma^{\mathsf{D}} = ([\![M]\!]_\Gamma^{\mathsf{D}})^n$

$[\![\Omega]\!]_\Gamma^{\mathsf{D}} = \epsilon_{\mathsf{D}^{|\Gamma|} \Rightarrow \mathsf{D}}.$

It is immediate to observe that for each term with no indexes $M \in \Lambda$, the strategy $[\![M]\!]_\Gamma^{\mathsf{D}}$ is history-free.

Proposition 2. *Let A be a game, $\langle D, ([\varphi], [\psi]) \rangle$ be a reflexive object in the Cartesian closed category of games $K_!(\mathcal{G})$ and $\sigma, \tau : A \Rightarrow D$ be two strategies. Let $\epsilon_{A \Rightarrow D} : A \Rightarrow D = \{\epsilon\}$ be the empty strategy. Then we have*

1. $\sigma^0 \cdot \tau = \epsilon_{A \Rightarrow D}$
2. $\sigma^{n+1} \cdot \tau \sqsubseteq (\sigma \cdot \tau^n)^{n+1}$

Theorem 1 (Validity of indexed reduction). *Rules (Ω_2), (Ω^n), (Ω^0), (β_I) and $(\beta_{i,j})$ are valid in each categorical λ-model $\mathsf{D} \in \mathcal{D}$; the rule Ω_1 is valid in each categorical λ-model $\mathsf{D} \in \mathcal{D}^{\mathcal{E}} \cup \mathcal{D}^{\mathcal{B}}$. The validity of a rule γ is intended in the following sense: for each $P, Q \in \Lambda(\Omega)^{\mathbb{N}}$ if $(P \to_\gamma Q)$ then $[\![P]\!]_\Gamma^{\mathsf{D}} \sqsubseteq [\![Q]\!]_\Gamma^{\mathsf{D}}$.*

Each λ-term M can be approximated by a "partially evaluated" term $A \in \Lambda(\Omega)$ which is called an *approximant*. Different notions of approximants arise for the different classes of models.

Definition 22. *For each term $M \in \Lambda$ the sets of approximants are defined by:*

1. $\mathcal{A}^{\mathcal{E}}(M) = \{A \in \Lambda(\Omega) \mid BT(A[\Delta\Delta/\Omega]) \subseteq BT(M)$ *and A is in* $\beta\eta\Omega_1\Omega_2\text{-nf}\,\}$
2. $\mathcal{A}^{\mathcal{B}}(M) = \{A \in \Lambda(\Omega) \mid BT(A[\Delta\Delta/\Omega]) \subseteq BT(M)$ *and A is in* $\beta\Omega_1\Omega_2\text{-nf}\,\}$
3. $\mathcal{A}^{\mathcal{L}}(M) = \{A \in \Lambda(\Omega) \mid LLT(A[\Delta\Delta/\Omega]) \subseteq LLT(M)$ *and A is in* $\beta\Omega_2\text{-nf}\,\}$

Lemma 3. *For each categorical λ-model $\mathsf{D} \in \mathcal{D}$, λ-term M and approximant $A \in \mathcal{A}^\star(M)$ with $\star \in \{\mathcal{E}, \mathcal{B}, \mathcal{L}\}$, $[\![A]\!]_\Gamma^\mathsf{D} \sqsubseteq [\![M]\!]_\Gamma^\mathsf{D}$.*

Definition 23 (Erasing function). *The erasing function $\mathcal{R} : \Lambda(\Omega)^\mathbb{N} \to \Lambda(\Omega)$ is inductively defined as follows:*

1. $\mathcal{R}(x) = x; \ \mathcal{R}(\Omega) = \Omega$ 2. $\mathcal{R}(PQ) = \mathcal{R}(P)\mathcal{R}(Q)$
3. $\mathcal{R}(\lambda x.P) = \lambda x.\mathcal{R}(P)$ 4. $\mathcal{R}(M^n) = \mathcal{R}(M)$

Lemma 4. *For each categorical λ-model $\mathsf{D} \in \mathcal{D}^\star$, and for each completely indexed term $M \in \Lambda(\Omega)^\mathbb{N}$ there exists a term $N \in \Lambda(\Omega)^\mathbb{N}$ such that $[\![M]\!]_\Gamma^\mathsf{D} \sqsubseteq [\![N]\!]_\Gamma^\mathsf{D}$ and $\mathcal{R}(N) \in \mathcal{A}^\star(\mathcal{R}(M))$ with $\star \in \{\mathcal{E}, \mathcal{B}, \mathcal{L}\}$.*

Lemma 5. *For each categorical λ-model $\mathsf{D} \in \mathcal{D}$, λ-term M and $n \in \omega$ there exists a completely indexed term M^* such that: $[\![M^n]\!]_\Gamma^\mathsf{D} = [\![M^*]\!]_\Gamma^\mathsf{D}$.*

At last we are ready to state the following.

Theorem 2 (Approximation theorem). *For each categorical λ-model $\mathsf{D} \in \mathcal{D}^\star$ and each λ-term M, $[\![M]\!]_\Gamma^\mathsf{D} = \bigsqcup\{[\![A]\!]_\Gamma^\mathsf{D} \mid A \in \mathcal{A}^\star(M)\}$ with $\star \in \{\mathcal{E}, \mathcal{B}, \mathcal{L}\}$.*

From Theorem 2 we can readily conclude that if two terms have the same tree they also have the same interpretation in the different game models, that is:

Proposition 3. *For each categorical λ-model $\mathsf{D} \in \mathcal{D}$, λ-terms M, N we have:*

1. *if $\mathsf{D} \in \mathcal{D}^\mathcal{L}$ and $LLT(M) = LLT(N)$ then $[\![M]\!]_\Gamma^\mathsf{D} = [\![N]\!]_\Gamma^\mathsf{D}$;*
2. *if $\mathsf{D} \in \mathcal{D}^\mathcal{B}$ and $BT(M) = BT(N)$ then $[\![M]\!]_\Gamma^\mathsf{D} = [\![N]\!]_\Gamma^\mathsf{D}$;*
3. *if $\mathsf{D} \in \mathcal{D}^\mathcal{E}$ then $M =_{\mathcal{H}^*} N \Leftrightarrow [\![M]\!]_\Gamma^\mathsf{D} = [\![N]\!]_\Gamma^\mathsf{D}$.*

In the following part of the section we shall prove that if two terms have different Lévy-Longo trees or different Böhm trees they also have different interpretation in corresponding game models of the λ-calculus. This will characterize completely the theories induced by game models and will substantiate the intuitive impression that the strategy which interprets a term is strongly connected with the tree of the term. The following definition and Lemma 6 are standard (see for instance [Bar84]).

Definition 24 (Similar terms). *Given two terms $M, N \in \Lambda$, we say that M and N are similar and we write $M \sim N$ if both M and N are unsolvable or they are solvable with principal head normal forms respectively $\lambda x_1 \ldots x_n.yM_1 \ldots M_m$ and $\lambda x_1 \ldots x_{n'}.y'N_1 \ldots N_{m'}$ in which $y \equiv y'$ and $m - n = m' - n'$.*

Lemma 6. *For each compositional non-trivial model of the λ-calculus D, for each pair of λ-terms M, N if $M \not\sim N$ then $[\![M]\!]_\Gamma^\mathsf{D} \neq [\![N]\!]_\Gamma^\mathsf{D}$.*

The following properties do not necessarily hold for any compositional λ-model.

Lemma 7. *For each categorical λ-model $\mathsf{D} \in \mathcal{D}$, for each sequence of terms $M, N, M_1, \ldots, M_m, N_1, \ldots, N_m$, for each sequence of variables $x, y, x_1, \ldots, x_n$, the following properties hold:*

1. *if $[\![xM_1 \ldots M_m]\!]^{\mathsf{D}}_\Gamma \approx [\![xN_1 \ldots N_m]\!]^{\mathsf{D}}_\Gamma$ then $\forall 1 \leq i \leq m$. $[\![M_i]\!]^{\mathsf{D}}_\Gamma \approx [\![N_i]\!]^{\mathsf{D}}_\Gamma$;*
2. *if $\mathsf{D} \in \mathcal{D}^{\mathcal{B}} \cup \mathcal{D}^{\mathcal{L}}$ then $[\![x]\!]^{\mathsf{D}}_\Gamma \not\approx [\![\lambda y.M]\!]^{\mathsf{D}}_\Gamma$;*
3. *if $\mathsf{D} \in \mathcal{D}^{\mathcal{B}} \cup \mathcal{D}^{\mathcal{L}}$ and $n' < n$ then*
   $$[\![\lambda x_1 \ldots x_n.yM_1 \ldots M_m]\!]^{\mathsf{D}}_\Gamma \not\approx [\![\lambda x_1 \ldots x_{n'}.yN_1 \ldots N_{m'}]\!]^{\mathsf{D}}_\Gamma;$$
4. *if $\mathsf{D} \in \mathcal{D}^{\mathcal{L}}$ and M and N are both unsolvable but of different order then $[\![M]\!]^{\mathsf{D}}_\Gamma \not\approx [\![N]\!]^{\mathsf{D}}_\Gamma$.*

Theorem 3. *Let $\mathsf{D} \in \mathcal{D}$ be a categorical λ-model and let M and N be two untyped λ-terms. If $[\![M]\!]^{\mathsf{D}}_\Gamma = [\![N]\!]^{\mathsf{D}}_\Gamma$ then we have:*

1. $LLT(M) = LLT(N)$ *if $\mathsf{D} \in \mathcal{D}^{\mathcal{L}}$;*
2. $BT(M) = BT(N)$ *if $\mathsf{D} \in \mathcal{D}^{\mathcal{B}}$.*

4 Conclusions

In the present paper we have studied the λ-theories induced by the game models without performing the extensional collapse. Through the extensional collapse it is possible to identify strategies that have the same observational behavior. In general, the extensional collapse is fundamental in order to obtain fully abstract game models of programming languages. Therefore it is still possible to use game models to capture λ-theories that are strictly coarser than the three considered in this paper. An example of such a theory can be found in [AM95] where, through the extensional collapse of a model D in $\mathcal{D}^{\mathcal{L}}$, a fully abstract model of the lazy λ-calculus is obtained. However, in general, models obtained through the extensional collapse are more difficult to study, *e.g.* the equivalence between strategies is not decidable also in the finite case. Our main theorem defines precisely those theories that can be obtained using simple (not collapsed) game models, and hence it implies also that the theories obtained through the extensional collapse lie only in between the theories $\mathcal{L}$ and $\mathcal{H}^*$.

A second consideration concerns the class of the game models we consider in this work. We have focused on games and history-free strategies mainly for historical reasons. We claim that the paper can be easily reformulated in order to prove the same results for the category of games and innocent strategies [HO00]. We can substantiate our claim by observing that the main tools used in the proofs — history-sensitive strategies, approximating strategies, Lemma 7 — are not peculiar to the history-free strategies and can be reformulated and applied in the context of innocent strategies.

A final point concerns the construction of game models. In this paper we do not build any example of game model for the λ-calculus; however in [DGFH99] a general method to obtain non-initial solutions of recursive equations is presented.

It is then quite simple to find extensional game models: several examples are presented there. Non-extensional game models can be obtained through the standard tricks used in the setting of the cpo models. For example a non-extensional model whose theory is $\mathfrak{B}$ can be obtained by taking the initial solution of the recursive equation $D = (D \Rightarrow D) \times A$ while a model whose theory is $\mathcal{L}$ can be obtained by taking the initial solution of the equation $D = (D \Rightarrow D)_\perp \times A$, where, in both equations, A is an arbitrary game.

References

[AJM96]   S. Abramsky, R. Jagadeesan, and P. Malacaria. Full abstraction for PCF. Accepted for pubblication, *Information and Computation*, 1996.

[AM95]   S. Abramsky and G. McCusker. Games and Full Abstraction for the Lazy Lambda-Calculus. In D. Kozen, editor, *Proceedings of the Tenth Annual Symposium on Logic in Computer Science*, pages 234–243. IEEE Computer Society Press, June 1995.

[Bar84]   H. Barendregt. *The Lambda Calculus: Its Syntax and Semantics*, volume 103 of *Studies in Logic and the Foundations of Mathematics*. North-Holland, 1984. Revised edition.

[Ber]   C. Berline. From Computation to Foundations via Functions and Application: the λ-calculus and its Webbed Models. To appear in Theoretical Computer Science.

[DGF]   P. Di Gianantonio and G. Franco. The Fine Structure of Game Lambda Models. Technical report, Department of Mathematics and Computer Science, University of Udine, 2000 Electronically available: `http://www.dimi.uniud.it/~pietro/Papers`.

[DGFH99] P. Di Gianantonio, G. Franco, and F. Honsell. Game semantics for the untyped λβη-calculus. In *Proceedings of the International Conference on Typed Lambda Calculi and Applications 1999*, volume 1591 of *Lecture Notes in Computer Science*, pages 114–128. Springer-Verlag, 1999.

[HL95]   F. Honsell and M. Lenisa. Final semantics for untyped λ-calculus. In M. Dezani, editor, *Proceedings of the International Conference on Typed Lambda Calculi and Applications 1995*, volume 902 of *Lecture Notes in Computer Science*, pages 249–265. Springer-Verlag, 1995.

[HO00]   J. M. E. Hyland and C. H. L. Ong. On full abstraction for PCF:I. Models, observables and the full abstraction problem, II. Dialogue games and innocent strategies, III. A fully abstract and universal game model. To appear in *Information and Computation*, 2000.

[HR92]   F. Honsell and S. Ronchi Della Rocca. An Approximation Theorem for Topological Lambda Models and the Topological Incompleteness of Lambda Calculus. *Journal of Computer and System Sciences*, 45:49–75, 1992.

[Hyl76]   J. M. E. Hyland. A syntatic characterization of the equality in some models of the λ-calculus. *Journal of London Mathematical Society*, 12(2):361–370, 1976.

[KNO99]   A. D. Ker, H. Nickau, and C. H. L. Ong. A universal innocent game model for the Böhm tree lambda theory. In *Computer Science Logic: Proceedings of the 8th Annual Conference of the EACSL Madrid, Spain*, volume 1683 of *Lecture Notes in Computer Science*, pages 405–419. Springer-Verlag, September 1999.

[KNO00]   A. D. Ker, H. Nickau, and C. H. L. Ong. Innocent game models of untyped lambda calculus. To appear in *Theoretical Computer Science*, 2000.

[Lév75]   J.J. Lévy. An algebraic interpretation of λ-calculus and a labelled λ-calculus. In C. Böhm, editor, *Lambda Calculus and Computer Science*, volume 37 of *Lecture Notes in Computer Science*, pages 147–165. Springer-Verlag, 1975.

[Lon83]   G. Longo. Set-theoretical models of λ-calculus: theories, expansions and isomorphisms. *Annals of Pure and Applied Logic*, 24:153–188, 1983.

[Nic94]   H. Nickau. Hereditarily sequential functionals. In *Proceedings of the Symposium on Logical Foundations of Computer Science: Logic at St. Petersburg*, Lecture Notes in Computer Science. Springer-Verlag, 1994.

[Sco72]   D. Scott. Continuous lattices. In *Toposes, Algebraic Geometry and Logic*, volume 274 of *Lecture Notes in Mathematics*. Springer-Verlag, 1972.

[Wad78]   C. P. Wadsworth. Approximate Reduction and Lambda Calculus Models. *SIAM Journal of Computing*, 7(3):337–356, August 1978.

Strong Normalization of Second Order Symmetric λ-Calculus

Michel Parigot

Equipe de Logique Mathématique
case 7012, Université Paris 7
2 place Jussieu, 75251 Paris cedex 05, France

Abstract. Typed symmetric λ-calculus is a simple computational interpretation of classical logic with an involutive negation. Its main distinguishing feature is to be a true non-confluent computational interpretation of classical logic. Its non-confluence reflects the computational freedom of classical logic (as compared to intuitionistic logic).
Barbanera and Berardi proved in [1,2] that first order typed symmetric λ-calculus enjoys the strong normalization property and showed in [3] that it can be used to derive symmetric programs.
In this paper we prove strong normalization for second order typed symmetric λ-calculus.

1 Introduction

The quest for computational interpretations of classical logic, started 10 years ago from the work of Felleisen [4,5] and Griffin [8]. It has been shown that classical natural deduction allows to model imperative features added to functional languages like Scheme, Common Lisp or ML. Two particular systems, λ_C-calculus [4,5] and $\lambda\mu$-calculus [12], have been intensively studied and the relation between features of languages, rules of natural deduction, machines and semantics seems to be well understood [9,10,15,16].

In the context of sequent calculus, several other computational interpretations of classical logic have been constructed following the spirit of Girard's linear logic [6]. It is often claimed in this context that computational interpretations of negation in classical logic should be involutive, that is, $\neg\neg A = A$ should be realized at the computational level. It is even sometimes claimed that this is the distinguishing feature of classical logic. But the real computational effect of the involutive character is not clear.

Systems coming from a natural deduction setting, like λ_C-calculus or $\lambda\mu$-calculus, don't have an involutive negation.

The symmetric λ-calculus of Barbanera and Berardi [1,2] is a simple computational interpretation of classical logic which is explicitly based on an involutive negation. Contrary to $\lambda\mu$-calculus, symmetric λ-calculus is non-confluent. But this non-confluence is an essential non-confluence which is supposed to reflect the computational freedom given by classical logical (compared to intuitionistic

S. Kapoor and S. Prasad (Eds.): FST TCS 2000, LNCS 1974, pp. 442–453, 2000.

logic). In [3] it is shown that it can be used to derive symmetric programs, which cannot be derived in the usual confluent systems.

In [1,2] Barbanera and Berardi proved that first order typed symmetric λ-calculus satisfies the strong normalization property. The proof is based on an original construction of reducibility candidates using fixed points.

In this paper we push one step further the understanding of symmetric λ-calculus: we prove that second order typed symmetric λ-calculus satisfies also the strong normalization property. Note that the second order setting gives a complete kernel of a typed programming language where data types can be defined internally. Moreover, from our strong normalization result, it can be easily deduced that one can also extract correct programs from proofs in this setting.

The proof mixes ingredients from the proof of Barbanera and Berardi and from our proof of strong normalization of second order typed λμ-calculus [13].

The section 2 is devoted to the definition of symmetric λ-calculus and typed symmetric λ-calculus of Barbanera and Berardi [1,2]. We show in section 2.3 how to extend this calculus with second order types.

The section 3 is devoted to the proof of strong normalization. Because we have a second order type system we need a notion of reducibility candidate defined independently of the notion of type. Because we have an involutive negation we also need to define reducibility candidates using fixed points. Our notion of reducibility candidates is defined in section 3.1 and its fundamental properties are proved in section 3.2: it is proved in particular that reducibility candidates are sets of strongly normalizable terms. In section 3.3, we define an interpretation of types by reducibility candidates. We finish the proof by showing in section 3.4 that, if a term has a certain type, then it belongs to the interpretation of that type and thus is strongly normalizable.

In the sequel types are designated by letters A, B, C etc., while atomic types are designated by P, Q, R, etc.

2   The Symmetric λ-Calculus of Barbanera and Berardi

The symmetric λ-calculus, introduced by Barbanera and Berardi [1,2], originated from a computational interpretation of classical logic with an involutive negation. It is basically a λ-calculus with a symmetric application.

In the following, the symmetric application is denoted by $*$, as in [1,2], and the abstraction by μ instead of λ, because it corresponds to negation and not to implication (see [14] for a discussion of this point).

2.1   Pure Symmetric λ-Calculus

Terms Symmetric λ-Calculus

Let Var be an infinite set of variables (denoted $x, x_1, x_2, ...$). Terms of symmetric λ-calculus are defined by:

$$t := x \mid \mu x.t \mid t * t \mid \langle t, t \rangle \mid \sigma_1(t) \mid \sigma_2(t)$$

where x ranges over variables.

Terms are denoted by letters t, u, v, w. The set of terms is denoted by $\mathcal{T}$.

Reduction Rules of Symmetric λ-Calculus

(β) $\mu x.u * v \rhd^c u[v/x]$

$(\beta^\perp)$ $u * \mu x.v \rhd^c v[u/x]$

(π) $\langle u_1, u_2 \rangle * \sigma_i(v_i) \rhd^c u_i * v_i$ for $i \in \{1, 2\}$

$(\pi^\perp)$ $\sigma_i(v_i) * \langle u_1, u_2 \rangle \rhd^c v_i * u_i$ for $i \in \{1, 2\}$

The one-step reduction relation between terms u and v is defined from teh previous rules as follows: $u \rhd_1 v$ iff v is obtained from u by replacing a subterm u_1 by v_1 with $u_1 \rhd^c v_1$.

The reduction relation $\rhd$ is defined as the reflexive and transitive closure of the one-step reduction relation $\rhd_1$.

A term u is *strongly normalizable* if there is no infinite reduction sequence, i.e. no infinite sequence $(u_i)_{i<\omega}$ such that $u_0 = u$ and $u_i \rhd_1 u_{i+1}$, for all $i < \omega$. The set of strongly normalizable terms is denoted by $\mathcal{N}$.

Comment. Due to the symmetric character of the rules, symmetric λ-calculus is obviously not confluent.

2.2   Typed Symmetric λ-Calculus

Types

The *m-types* of the system are defined by:

$$A := P \mid \neg P \mid A \wedge A \mid A \vee A$$

where P ranges over atomic types.

The *types* of the system are either m-types or the special type $\perp$.

An involutive negation on m-types is defined as follows:

$\neg(P) = \neg P$

$\neg(\neg P) = P$

$\neg(A \wedge B) = \neg(A) \vee \neg(B)$

$\neg(A \vee B) = \neg(A) \wedge \neg(B)$

In the following we freely use $\neg A$ instead of $\neg(A)$.

Comment. The fact that the type $\perp$ is not among the set of atomic types is necessary to have a strong normalization result. It is shown in [14] that if $\neg$ is involutive and $\perp$ is among the set of atomic types, then normalization fails.

Typing Rules of Symmetric λ-Calculus

$$\overline{\Gamma, x : A \vdash x : A} \quad axiom$$

$$\frac{\Gamma \vdash u : A \quad \Gamma \vdash v : B}{\Gamma \vdash \langle u, v \rangle : A \wedge B} \quad \wedge\text{-}intro \qquad\qquad \frac{\Gamma \vdash u_i : A_i}{\Gamma \vdash \sigma_i(u_i) : A_1 \vee A_2} \quad \vee\text{-}intro \quad (i{=}1,2)$$

$$\frac{\Gamma, x : A \vdash u : \bot}{\Gamma \vdash \mu x.u : \neg A} \quad \neg\text{-}intro \qquad\qquad \frac{\Gamma \vdash u : \neg A \quad \Gamma \vdash v : A}{\Gamma \vdash u * v : \bot} \quad \neg\text{-}elim$$

In the previous rules, Γ denotes an arbitrary context of the form $x_1 : A_1, ..., x_n : A_n$. As usual in typed λ-calculi, we adopt in these rules an implicit management of contraction and weakening: weakening is obtained by allowing an arbitrary context in axioms and contraction by merging contexts in rules with two premises.

Barbanera and Berardi proved in [1,2] that this system satisfies the strong normalization property i.e.:

if $\Gamma \vdash u : A$ is derivable, then u is strongly normalizable.

2.3   Second Order Typed Symmetric λ-Calculus

We extend the previous typed symmetric λ-calculus to second order.

Terms

We first extend the definition of terms with two constructions which reflect the presence of the two quantifiers ∀ and ∃.
Terms are defined by:

$$t := x \mid \mu x.t \mid t * t \mid \langle t, t \rangle \mid \sigma_1(t) \mid \sigma_2(t) \mid a.t \mid e.t$$

where x ranges over variables.

Types

We start from an infinite set of type variables (denoted $X, Y, ...$).
The *m-types* of the system are defined by:

$$A := X \mid \neg X \mid A \wedge A \mid A \vee A \mid \forall X A \mid \exists X A$$

where X ranges over type variables.

The *types* of the system are either m-types or the special type $\bot$.

An involutive negation on m-types is defined as follows:

$\neg(X) = \neg X$
$\neg(\neg X) = X \quad \neg(A \wedge B) = \neg(A) \vee \neg(B)$

$$\neg(A \vee B) = \neg(A) \wedge \neg(B)$$
$$\neg(\forall X A) = \exists X \neg(A)$$
$$\neg(\exists X A) = \forall X \neg(A)$$

Reduction Rules

We add two symmetric reduction rules for eliminating a and e.

$$(q) \qquad a.u * e.v \; \triangleright^c \; u * v$$
$$(q^\perp) \qquad e.v * a.u \; \triangleright^c \; v * u$$

The notions of reduction and strongly normalizable term are extended in the obvious way to this rules.

Typing Rules

We add two introduction rules for the quantifiers $\forall$ and $\exists$.

$$\frac{\Gamma \vdash u : A[Y/X]}{\Gamma \vdash a.u : \forall X A} \; \forall\text{--}intro \; (*) \qquad\qquad \frac{\Gamma \vdash u : A[B/X]}{\Gamma \vdash e.u : \exists X A} \; \exists\text{--}intro$$

$(*)$ Y is not free in $\Gamma, \forall X A$

Comment. The constructions a and e are trivial witnesses of the quantifiers at the level of terms. They have no real computational effect. Contrary to the case of second order typed λ-calculus or $\lambda\mu$-calculus, such witnesses are needed for second order typed symmetric λ-calculus.

If one takes instead the rules:

$$\frac{\Gamma \vdash u : A[Y/X]}{\Gamma \vdash u : \forall X A} \; \forall\text{--}intro \; (*) \qquad\qquad \frac{\Gamma \vdash u : A[B/X]}{\Gamma \vdash u : \exists X A} \; \exists\text{--}intro$$

then reduction doesn't preserve typing of the system. The crucial situation is the following:

$$\frac{\dfrac{\cdots\cdots\cdots\cdots\cdots}{\dfrac{\Gamma, x : \neg A[Y/X] \vdash t : \bot}{\dfrac{\Gamma \vdash \mu x.t : A[Y/X]}{\Gamma \vdash \mu x.t : \forall X A}}} \quad \dfrac{\cdots\cdots\cdots\cdots}{\dfrac{\Gamma, y : \forall X A \vdash s : \bot}{\Gamma \vdash \mu y.s : \exists X \neg A}}}{\Gamma \vdash \mu x.t * \mu y.s : \bot}$$

2.4 Extensions

In § 3 we prove strong normalization of second order typed symmetric λ-calculus presented in § 2.3. The result easily extends in two directions: one can add simplification rules and other basic connectives.

Simplification Rules

Symmetric λ-calculus of Barbanera and Berardi has in addition to the reduction rules presented in § 2.1, other reduction rules, that we call simplification rules:

$$\mu x.(u * x) \rhd^c u$$

$$\mu x.(x * u) \rhd^c u$$

$$E[u * v] * w \rhd^c u * v$$

$$w * E[u * v] \rhd^c u * v$$

These rules are subject to the following restrictions: in the first two rules x has no free occurrence in u; in the last two rules, $E[\ ]$ is a context which doesn't bind any free variable of $u * v$.

Strong normalization for the reduction with simplification rules deduces from strong normalization for the reduction without simplification rules. It is sufficient to remark that:

1) there is no infinite sequence of reduction using only simplification rules, because each application of a simplication rule strictly decreases the length of the term;

2) in reduction sequences, one can always push applications of the original rules before applications of simplification rules.

Additional Connectives

In § 3 we prove strong normalization of typed symmetric λ-calculus based on the connectives ∧ and ∨. The proof extends in a straightforward manner to other pairs of dual connectives. One interesting case is the calculus based on → and its dual −, which is easier ot relate to typed λ-calculus and λμ-calculus than the original one.

The typing rules for → and − are the following:

$$\frac{\Gamma, \ x : A \vdash u : B}{\Gamma \vdash \lambda x.u : A \to B} \qquad\qquad \frac{\Gamma \vdash u : A \quad \Gamma \vdash v : \neg B}{\Gamma \vdash (u, v) : A - B}$$

The corresponding reduction rules are:

$$(u, v) * \lambda x.t \rhd^c t[u/x] * v$$

$$\lambda x.t * (u, v) \rhd^c t[u/x] * v$$

3   Proof of Strong Normalization

We prove the strong normalization using the reducibility method: each type is interpreted by a set of terms. In section 3.1 we define the set of possible interpretations of types, called reducibility candidates. In section 3.2 we prove that each reducibility candidate is a set of strongly normalisable terms. In section 3.3 we define the notion interpretation such that each type is interpreted by a reducibility candidate. In section 3.4 we prove that each term of type A belongs to the interpretation of A and therefore is strongly normalisable.

3.1   Reducibility Candidates

For $C, D \in \mathcal{P}(\mathcal{T})$ and $\mathcal{S} \subseteq \mathcal{P}(\mathcal{T})$, one defines the following constructions:

$$C \times D = \{\langle u, v \rangle; u \in C, v \in D\}$$
$$C + D = \{\sigma_1(u); u \in C\} \cup \{\sigma_2(u); u \in D\}$$
$$\neg(C) = \{\mu x.u; \text{ for all } v \in C, \ u[v/x] \in \mathcal{N}\}$$
$$\bigcap \mathcal{S} = \{a.t; \text{ for all } C \in \mathcal{S}, \ t \in C\}$$
$$\bigcup \mathcal{S} = \{e.t; \text{ there exists } C \in \mathcal{S}, \ t \in C\}$$

If $F : \mathcal{P}(\mathcal{T}) \to \mathcal{P}(\mathcal{T})$ is an increasing function with respect to set-theoretic inclusion, then F has a smallest fixed point denoted by $\mu X.F(X)$.
For $C, D \in \mathcal{P}(\mathcal{T})$, one defines

$$\text{Neg}_D(C) = \text{Var} \cup D \cup \neg(C)$$

For each $D \in \mathcal{P}(\mathcal{T})$, Neg_D is a decreasing function from $\mathcal{P}(\mathcal{T})$ to $\mathcal{P}(\mathcal{T})$. Thus, for each $D, D' \in \mathcal{P}(\mathcal{T})$, $\text{Neg}_D \circ \text{Neg}_{D'}$ is an increasing function which has a fixed point, $\mu X.\text{Neg}_D(\text{Neg}_{D'}(X))$.
For $\mathcal{F} \subseteq \mathcal{P}(\mathcal{T}) \times \mathcal{P}(\mathcal{T})$, we define $p_1\mathcal{F} = \{C; \text{there exists } C', (C, C') \in \mathcal{F}\}$ and $p_2\mathcal{F} = \{C'; \text{there exists } C, (C, C') \in \mathcal{F}\}$.

Definition 1. *The set $\mathcal{R}$ of reducibility pairs is the smallest subset of $\mathcal{P}(\mathcal{T}) \times \mathcal{P}(\mathcal{T})$ such that:*

1) $(\mu X.\text{Neg}_\emptyset(\text{Neg}_\emptyset(X)), \ \text{Neg}_\emptyset(\mu X.\text{Neg}_\emptyset(\text{Neg}_\emptyset(X)))) \in \mathcal{R}$;

2) If $(C, C') \in \mathcal{R}$ and $(D, D') \in \mathcal{R}$, then
$(\mu X.\text{Neg}_{C \times D}(\text{Neg}_{C'+D'}(X)), \ \text{Neg}_{C'+D'}(\mu X.\text{Neg}_{C \times D}(\text{Neg}_{C'+D'}(X)))) \in \mathcal{R}$;

3) If $\emptyset \neq \mathcal{F} \subseteq \mathcal{R}$, $\mathcal{S} = p_1\mathcal{F}$ and $\mathcal{S}' = p_2\mathcal{F}$, then
$(\mu X.\text{Neg}_{\cap\mathcal{S}}(\text{Neg}_{\cup\mathcal{S}'}(X)), \ \text{Neg}_{\cup\mathcal{S}'}(\mu X.\text{Neg}_{\cap\mathcal{S}}(\text{Neg}_{\cup\mathcal{S}'}(X)))) \in \mathcal{R}$

4) If $(C, C') \in \mathcal{R}$, then $(C', C) \in \mathcal{R}$.
The set $\mathcal{R}_0$ of reducibility candidates is $\mathcal{R}_0 = p_1\mathcal{R} = p_2\mathcal{R}$.

Comment. Because we have an involutive negation, $\neg\neg A$ and A need to have the same interpetation. This is achieved by constructing reducibility candidates which are fixed points with respect to double negation. Reducibility pairs correspond intuitively to interpretations of pairs of formulas $(A, \neg A)$.

3.2   Properties of Reducibility Candidates

Lemma 1. *If $(C, C') \in \mathcal{R}$, then one of the following cases holds:*

1) $C = \text{Neg}_\emptyset(C')$ and $C' = \text{Neg}_\emptyset(C)$;

2) $C = \text{Neg}_{D_1 \times D_2}(C')$ and $C' = \text{Neg}_{D_1' + D_2'}(C)$ with $(D_i, D_i') \in \mathcal{R}$ for $i = 1, 2$;

3) $C = \text{Neg}_{D_1 + D_2}(C')$ and $C' = \text{Neg}_{D_1' \times D_2'}(C)$ with $(D_i, D_i') \in \mathcal{R}$ for $i = 1, 2$;

4) $C = \text{Neg}_{\cap\mathcal{S}}(C')$ and $C' = \text{Neg}_{\cup\mathcal{S}'}(C)$ with $\mathcal{S} = p_1\mathcal{F}$, $\mathcal{S}' = p_2\mathcal{F}$ and $\mathcal{F} \subseteq \mathcal{R}$;

5) $C = \text{Neg}_{\cup\mathcal{S}}(C')$ and $C' = \text{Neg}_{\cap\mathcal{S}'}(C)$ with $\mathcal{S} = p_1\mathcal{F}$, $\mathcal{S}' = p_2\mathcal{F}$ and $\mathcal{F} \subseteq \mathcal{R}$.

Proof. We prove the result by induction on the construction of (C, C').

1) $C = \mu X.\mathrm{Neg}_{\emptyset}(\mathrm{Neg}_{\emptyset}(X))$ and $C' = \mathrm{Neg}_{\emptyset}(\mu X.\mathrm{Neg}_{\emptyset}(\mathrm{Neg}_{\emptyset}(X)))$.
We have $C' = \mathrm{Neg}_{\emptyset}(C)$ and $C = \mu X.\mathrm{Neg}_{\emptyset}(\mathrm{Neg}_{\emptyset}(X))$; by definition of the fixed point, we have $C = \mathrm{Neg}_{\emptyset}(\mathrm{Neg}_{\emptyset}(C))$ and therefore $C = \mathrm{Neg}_{\emptyset}(C')$.

2) $C = \mu X.\mathrm{Neg}_{D_1 \times D_2}(\mathrm{Neg}_{D'_1 + D'_2}(X))$
and $C' = \mathrm{Neg}_{D'_1 + D'_2}(\mu X.\mathrm{Neg}_{D_1 \times D_2}(\mathrm{Neg}_{D'_1 + D'_2}(X)))$.
We have $C' = \mathrm{Neg}_{D'_1 + D'_2}(C)$ and $C = \mu X.\mathrm{Neg}_{D_1 \times D_2}(\mathrm{Neg}_{D'_1 + D'_2}(X))$; by definition of the fixed point, we have $C = \mathrm{Neg}_{D_1 \times D_2}(\mathrm{Neg}_{D'_1 + D'_2}(C))$ and therefore $C = \mathrm{Neg}_{D_1 \times D_2}(C')$.

3) $C = \mu X.\mathrm{Neg}_{\cap S}(\mathrm{Neg}_{\cup S'}(X))$ and $C' = \mathrm{Neg}_{\cup S'}(\mu X.\mathrm{Neg}_{\cap S}(\mathrm{Neg}_{\cup S'}(X)))$.
We have $C' = \mathrm{Neg}_{\cup S'}(C)$ and $C = \mu X.\mathrm{Neg}_{\cap S}(\mathrm{Neg}_{\cup S'}(X))$; by definition of the fixed point, we have $C = \mathrm{Neg}_{\cap S}(\mathrm{Neg}_{\cup S'}(C))$ and therefore $C = \mathrm{Neg}_{\cap S}(C')$.

4) If (C, C') is not obtained by clauses 1), 2) or 3) of definition 1, then (C', C) is obtained by one of these clauses and we are in case 1), 3) or 5) of lemma 1.

Lemma 2. *Let $(C, C') \in \mathcal{R}$ and $u \in \mathcal{T}$.*
Then $\mu x.u \in C$ iff $\mu x.u \in \neg(C')$
$\qquad\qquad$ *iff for all $v \in C'$, $u[v/x] \in \mathcal{N}$.*

Proof. Let $(C, C') \in \mathcal{R}$ and $u \in \mathcal{T}$. By lemma 1, we have $C = \mathrm{Neg}_D(C') = \mathrm{Var} \cup D \cup \neg(C')$, with D being $\emptyset$, $E \times F$, $E + F$, $\cap S$ or $\cup S$, with $E, F \in \mathcal{R}_0$ and $S \subseteq \mathcal{R}_0$. Because $\mathrm{Var} \cup D$ doesn't contain terms starting with a μ, we have $\mu x.u \in C$ iff $\mu x.u \in \neg(C')$.

Lemma 3. *If $C \in \mathcal{R}_0$, then $\mathrm{Var} \subseteq C \subseteq \mathcal{N}$.*

Proof. First remark that, for each $C \in \mathcal{R}_0$, we have by lemma 1, $C = \mathrm{Neg}_D(C') = \mathrm{Var} \cup D \cup \neg(C')$ and therefore $\mathrm{Var} \subseteq C$.
Let $C \in \mathcal{R}_0$ and $t \in C$. We prove $t \in \mathcal{N}$. By lemma 1, we have $C = \mathrm{Neg}_D(C') = \mathrm{Var} \cup D \cup \neg(C')$, with $C' \in \mathcal{R}_0$ and D being $\emptyset$, $E \times F$, $E + F$, $\cap S$ or $\cup S$, with $E, F \in \mathcal{R}_0$ and $S \subseteq \mathcal{R}_0$. Therefore one of the following cases holds:
1) $t \in \mathrm{Var}$. In this case, $t \in \mathcal{N}$.
2) $t \in \neg(C')$. In this case, $t = \mu x.u$ and for all $v \in C'$, $u[v/x] \in \mathcal{N}$; because $\mathrm{Var} \subseteq C'$, we have $x \in C'$ and $u \in \mathcal{N}$; therefore $\mu x.u \in \mathcal{N}$.
3) $t \in D$. One considers the possibilities for D given by lemma 1.
a) $D = E \times F$, with $E, F \in \mathcal{R}_0$.
In this case $t = \langle u, v \rangle$ with $u \in C$ and $v \in D$; by induction hypothesis, $u, v \in \mathcal{N}$ and therefore $t \in \mathcal{N}$.
b) $D = E + F$ with $E, F \in \mathcal{R}_0$.
In this case $t = \sigma_1(u)$ with $u \in E$ or $t = \sigma_2(v)$ with $v \in F$; by induction hypothesis, $u, v \in \mathcal{N}$ and therefore $t \in \mathcal{N}$.
c) $D = \cap S$ with $S \subseteq \mathcal{R}_0$.
In this case $t = a.u$ with $u \in C$ for all $C \in S$; by induction hypothesis, $u \in \mathcal{N}$ and therefore $t \in \mathcal{N}$.

d) $D = \bigcup \mathcal{S}$ with $\mathcal{S} \subseteq \mathcal{R}_0$.

In this case $t = e.u$ with $u \in C$ and $C \in \mathcal{S}$; by induction hypothesis, $u \in \mathcal{N}$ and therefore $t \in \mathcal{N}$.

Lemma 4. *Let $(C, C') \in \mathcal{R}$ and $u, u' \in \mathcal{T}$.*
If $u \in C$ and $u \vartriangleright_1 u'$, then $u' \in C$.

Proof. Let $(C, C') \in \mathcal{R}$ and $u, u' \in \mathcal{T}$ such that $u \in C$ and $u \vartriangleright_1 u'$. One proves $u' \in C$ by induction on the construction of (C, C').

By lemma 1, we have $C = \mathrm{Neg}_D(C') = \mathrm{Var} \cup D \cup \neg(C')$, with D being $\emptyset$, $C_1 \times C_2$, $C_1 + C_2$, $\bigcap \mathcal{S}$ or $\bigcup \mathcal{S}$, with $C_1, C_2 \in \mathcal{R}_0$ and $\mathcal{S} \subseteq \mathcal{R}_0$.

One considers the different possibilities for u.

If $u \in \mathrm{Var}$, the result is trivial.

Suppose $u \in \neg(C')$. Then $u = \mu x.t$ with $t[v/x] \in \mathcal{N}$, for all $v \in C'$, and $u' = \mu x.t'$ with $t \vartriangleright_1 t'$. Let $v \in C'$; since $t[v/x] \in \mathcal{N}$ and $t \vartriangleright_1 t'$ we have $t'[v/x] \in \mathcal{N}$. Therefore $\mu x.t' \in C$, i.e. $u' \in C$.

Suppose $u \in D$. One considers the possibilities for D given by lemma 1.

1) $D = C_1 \times C_2$ with $C_1, C_2 \in \mathcal{R}_0$. In this case $u = \langle u_1, u_2 \rangle$ with $u_1 \in C_1$ and $u_2 \in C_2$. There are two possibilities for u': either $u' = \langle u_1', u_2 \rangle$ with $u_1 \vartriangleright_1 u_1'$ or $u' = \langle u_1, u_2' \rangle$ with $u_2 \vartriangleright_1 u_2'$. By induction hypothesis we have $u_i' \in C_i$ and therefore $u' \in C$.

2) $D = C_1 + C_2$ with $C_1, C_2 \in \mathcal{R}_0$.

In this case there exists $i \in \{1, 2\}$ such that $u = \sigma_i u_i$ with $u_i \in C_i$. Since $u \vartriangleright_1 u'$, we have $u' = \sigma_i u_i'$ with $u_i \vartriangleright_1 u_i'$. By induction hypothesis we have $u_i' \in C_i$ and therefore $u' \in C$.

3) $D = \bigcap \mathcal{S}$ with $\mathcal{S} \subseteq \mathcal{R}_0$.

In this case $u = a.t$ with $t \in E$ for each $E \in \mathcal{S}$. Since $u \vartriangleright_1 u'$, we have $u' = a.t'$ with $t \vartriangleright_1 t'$. By induction hypothesis we have $t' \in E$ for each $E \in \mathcal{S}$ and therefore $u' \in C$.

4) $D = \bigcup \mathcal{S}$ with $\mathcal{S} \subseteq \mathcal{R}_0$.

In this case there exist $E \in \mathcal{S}$ and $t \in E$ such that $u = a.t$. Since $u \vartriangleright_1 u'$, we have $u' = a.t'$ with $t \vartriangleright_1 t'$. By induction hypothesis we have $t' \in E$ and therefore $u' \in C$.

Lemma 5. *Let $(C, C') \in \mathcal{R}$ and $u, u' \in \mathcal{T}$.*
If $u \in C$ and $u' \in C'$, then $u * u' \in \mathcal{N}$.

Proof. Let $(C, C') \in \mathcal{R}$, $u \in C$ and $u' \in C'$. By lemma 3, we have $u \in \mathcal{N}$ and $u' \in \mathcal{N}$. Let $N(u)$ (resp. $N(u')$) be the sum of the lengths of the reduction sequences of u (resp. u'). We prove $u * u' \in \mathcal{N}$ by a double induction on the construction of (C, C') and $N(u) + N(u')$.

In order to prove $u * u' \in \mathcal{N}$ we prove: for all $w \in \mathcal{T}$, if $u * u' \vartriangleright_1 w$ then $w \in \mathcal{N}$. We consider the different possibilities for w.

1) $w = t[u'/x]$ with $u = \mu x.t$.

By lemma 2, we have $\mu x.t \in \neg(C')$ and therefore $t[u'/x] \in \mathcal{N}$.

2) $w = t'[u/x]$ with $u' = \mu x.t'$.

The proof is analogous to that of case 1).

3) $w = u_i * u_i'$ with $u = \langle u_1, u_2 \rangle$, $u' = \sigma_i(u_i')$ and $i \in \{1, 2\}$.

By lemma 1, we have $C = Neg_{C_1 \times C_2}(C')$, $C' = Neg_{C_1' + C_2'}(C)$ with $(C_i, C_i') \in \mathcal{R}$, $u_i \in C_i$ and $u_i' \in C_i'$. Since $u_i \in C_i$ and $u_i' \in C_i'$, we have by induction hypothesis $u_i * u_i' \in \mathcal{N}$.

4) $w = u_i * u_i'$ with $u = \sigma_i(u_i)$, $u' = \langle u_1', u_2' \rangle$ and $i \in \{1, 2\}$.

The proof is analogous to that of case 3).

5) $w = t * t'$ with $u = a.t$ and $u' = e.t'$.

By lemma 1, we have $C = Neg_{\cap S}(C')$, $C' = Neg_{\cup S'}(C)$ with
$S = \{D;$ there exists $D', (D, D') \in \mathcal{F}\}$, $S' = \{D';$ there exists $D, (D, D') \in \mathcal{F}\}$,
$\mathcal{F} \subseteq \mathcal{R}$, $t \in D$ for all $D \in S$ and $t' \in D_0'$ for a certain $D_0' \in S'$. Let D_0 such that $(D_0, D_0') \in \mathcal{F}$; we have $t \in D_0$ and $t' \in D_0'$; by induction hypothesis, it follows $t * t' \in \mathcal{N}$.

6) $w = t * t'$ with $u = e.t$ and $u' = a.t'$.

The proof is analogous to that of case 5).

7) $w = u_1 * u'$ with $u \rhd_1 u_1$.

By lemma 4, we have $u_1 \in C$. Because $N(u_1) < N(u)$, we have by induction hypothesis $u_1 * u' \in \mathcal{N}$.

8) $w = u * u_1'$ with $u' \rhd_1 u_1'$.

The proof is analogous to that of case 7).

3.3 Interpretation of Formulas

Let Δ be the set of type variables and negated type variables.

Definition 2. *A valuation α is a function from Δ to $\mathcal{R}_0$ such that for each type variable X, $(\alpha(X), \alpha(\neg X)) \in \mathcal{R}$. For $U \in \Delta$ and $C \in \mathcal{R}_0$, we denote by $\alpha[C/U]$, the valuation α' defined by $\alpha'(U) = C$ and $\alpha'(V) = \alpha(V)$ for $V \neq U$.*

The value $\|A\|^\alpha$ of an m-type A for a valuation α is defined inductively as follows:

$\|X\|^\alpha = \alpha(X)$, *for X a type variable;*

$\|\neg X\|^\alpha = \alpha(\neg X)$, *for X a type variable;*

$\|A \wedge B\|^\alpha = \mu X.Neg_{\|A\|^\alpha \times \|B\|^\alpha}(Neg_{\|\neg A\|^\alpha + \|\neg B\|^\alpha}(X))$

$\|A \vee B\|^\alpha = Neg_{\|A\|^\alpha + \|B\|^\alpha}(\|\neg A \wedge \neg B\|^\alpha)$

$\|\forall X A\|^\alpha = \mu X.Neg_{\cap \{\|A\|^{\alpha[C/X, C'/\neg X]}; (C,C') \in \mathcal{R}\}}$
$$(Neg_{\cup \{\|\neg A\|^{\alpha[C/X, C'/\neg X]}; (C,C') \in \mathcal{R}\}}(X))$$

$\|\exists X A\|^\alpha = Neg_{\cup \{\|A\|^{\alpha[C/X, C'/\neg X]}; (C,C') \in \mathcal{R}\}}(\|\forall X \neg A\|^\alpha)$

The definition is extended to types by $\| \perp \|^\alpha = \mathcal{N}$.

Lemma 6. *For each valuation α and each m-type A, $(\|A\|^\alpha, \|\neg A\|^\alpha) \in \mathcal{R}$.*

Proof. Easy induction on A. The case where A is a type variable is given by the definition of valuation.

Lemma 7. *Let A, B be m-types and α, α' valuations.*

(1) Suppose that $\alpha(X) = \alpha'(X)$ and $\alpha(\neg X) = \alpha'(\neg X)$, for each type variable X free in A. Then $\|A\|^{\alpha} = \|A\|^{\alpha'}$.

(2) $\|A[B/X]\|^{\alpha} = \|A\|^{\alpha[\|B\|^{\alpha}/X,\, \|\neg B\|^{\alpha}/\neg X]}$.

Proof. Easy but tedious inspection (this lemma says only that our notion of value is correctly defined).

3.4 Proof of Strong Normalization

Lemma 8. *Let $A_1, ..., A_n$ be m-types and C a type.*
If $x_1 : A_1, ..., x_n : A_n \vdash t : C$, then for all $u_1 \in \|A_1\|^{\alpha}, ..., u_n \in \|A_n\|^{\alpha}$,

$$t[u_1/x_1, ..., u_n/x_n] \in \|C\|^{\alpha}.$$

Proof. By induction on the derivation of $x_1 : A_1, ..., x_n : A_n \vdash t : C$. One considers the different possibilities for t.

1) $t = x_i$ and $C = A_i$. In this case, we have $t[u_1/x_1, ..., u_n/x_n] = u_i \in \|C\|^{\alpha}$.

2) $t = \langle t_1, t_2 \rangle$ and $C = C_1 \wedge C_2$. By induction hypothesis we have $t_1[u_1/x_1, ..., u_n/x_n] \in \|C_1\|^{\alpha}$ and $t_2[u_1/x_1, ..., u_n/x_n] \in \|C_2\|^{\alpha}$. Therefore $t[u_1/x_1, ..., u_n/x_n] \in \|C_1\|^{\alpha} \times \|C_2\|^{\alpha}$ and $t[u_1/x_1, ..., u_n/x_n] \in \|C_1 \wedge C_2\|^{\alpha}$.

3) $t = \sigma_i(t_i)$ with $i \in \{1, 2\}$ and $C = C_1 \vee C_2$. By induction hypothesis we have $t_i[u_1/x_1, ..., u_n/x_n] \in \|C_i\|^{\alpha}$. Therefore $t[u_1/x_1, ..., u_n/x_n] \in \|C_1\|^{\alpha} + \|C_2\|^{\alpha}$ and $t[u_1/x_1, ..., u_n/x_n] \in \|C_1 \vee C_2\|^{\alpha}$.

4) $t = t_1 * t_2$ and $C = \perp$. In this case $\perp$ is obtained from C_1 and $\neg C_1$. By induction hypothesis we have $t_1[u_1/x_1, ..., u_n/x_n] \in \|C_1\|^{\alpha}$ and $t_2[u_1/x_1, ..., u_n/x_n] \in \|\neg C_1\|^{\alpha}$. Therefore by lemmas 6 and 5, $t[u_1/x_1, ..., u_n/x_n] \in \mathcal{N}$ i.e. $t[u_1/x_1, ..., u_n/x_n] \in \| \perp \|^{\alpha}$.

5) $t = \mu x.s$ and $C = \neg A$. In this case $t[u_1/x_1, ..., u_n/x_n] = \mu x.s[u_1/x_1, ..., u_n/x_n]$. By induction hypothesis, we have $s[u_1/x_1, ..., u_n/x_n, v/x] \in \| \perp \|^{\alpha} = \mathcal{N}$, for all $v \in \|A\|^{\alpha}$. Therefore by lemma 2, $\mu x.s[u_1/x_1, ..., u_n/x_n] \in \|\neg A\|^{\alpha}$ i.e $t[u_1/x_1, ..., u_n/x_n] \in \|C\|^{\alpha}$.

6) $t = a.s$ and $C = \forall X A$. In this case $\forall X A$ is deduced from $A[Y/X]$ with Y not free in $A_1, ..., A_n, \forall X A$. We have to show $a.s[u_1/x_1, ..., u_n/x_n] \in \|\forall X A\|^{\alpha}$. By definition of $\|\forall X A\|^{\alpha}$, it suffices to show $s[u_1/x_1, ..., u_n/x_n] \in \|A\|^{\alpha[C/X, C'/\neg X]}$, for all $(C, C') \in \mathcal{R}$. Let $(C, C') \in \mathcal{R}$. Because Y is not free in $A_1, ..., A_n$, we have by lemma 7, $\|A_i\|^{\alpha[C/Y, C'/\neg Y]} = \|A_i\|^{\alpha}$ and therefore $u_i \in \|A_i\|^{\alpha[C/Y, C'/\neg Y]}$, for each $i \in \{1, ..., n\}$. By induction hypothesis, $s[u_1/x_1, ..., u_n/x_n] \in \|A[Y/X]\|^{\alpha[C/Y, C'/\neg Y]}$. Because Y is not free in $\forall X A$, $\|A[Y/X]\|^{\alpha[C/Y, C'/\neg Y]} = \|A\|^{\alpha[C/X, C'/\neg X]}$ and therefore $s[u_1/x_1, ..., u_n/x_n] \in \|A\|^{\alpha[C/X, C'/\neg X]}$.

7) $t = e.s$ and $C = \exists X A$. In this case $\exists X A$ is deduced from $A[B/X]$, for a certain type B. We have to show $e.s[u_1/x_1, ..., u_n/x_n] \in \|\exists X A\|^{\alpha}$. By definition of $\|\exists X A\|^{\alpha}$, it suffices to show that there exists $(C, C') \in \mathcal{R}$ such that $s[u_1/x_1, ..., u_n/x_n] \in \|A\|^{\alpha[C/X, C'/\neg X]}$. Let $(C, C') = (\|B\|^{\alpha}, \|\neg B\|^{\alpha})$. By induction hypothesis we have $s[u_1/x_1, ..., u_n/x_n] \in \|A[B/X]\|^{\alpha}$. By lemma 7, we

have $\|A[B/X]\|^\alpha = \|A\|^{\alpha[\|B\|^\alpha/X, \|\neg B\|^\alpha/\neg X]}$ and therefore $s[u_1/x_1, ..., u_n/x_n] \in \|A\|^{\alpha[C/X, C'/\neg X]}$.

Theorem 1. *If* $x_1 : A_1, ..., x_n : A_n \vdash t : C$, *then* t *is strongly normalizable.*

Proof. Suppose $x_1 : A_1, ..., x_n : A_n \vdash t : C$. For each $i \in \{1, ..., n\}$, we have $\|A_i\|^\alpha \in \mathcal{R}_0$ by lemma 6 and $x_i \in \|A_i\|^\alpha$ by lemma 3. Therefore by lemma 8, $t \in \|C\|^\alpha$. If $C = \perp$, then $\|C\|^\alpha = \mathcal{N}$ and $t \in \mathcal{N}$; otherwise by lemma 6, $\|C\|^\alpha \in \mathcal{R}_0$ and therefore by lemma 3, $t \in \mathcal{N}$.

References

1. F. Barbanera, S. Berardi : A symmetric lambda-calculus for classical program extraction. Proceedings TACS'94, Springer LNCS **789** (1994).
2. F. Barbanera, S. Berardi : A symmetric lambda-calculus for classical program extraction. Information and Computation **125** (1996) 103-117.
3. F. Barbanera, S. Berardi, M. Schivalocchi : "Classical" programming-with-proofs in lambda-sym: an analysis of a non-confluence. Proc. TACS'97.
4. M. Felleisen, D.P. Friedman, E. Kohlbecker, B. Duba : A syntactic theory of sequential control. Theoretical Computer Science **52** (1987) 205-237.
5. M. Felleisen, R. Hieb : The revised report on the syntactic theory of sequential control and state. Theoretical Computer Science **102** (1994) 235-271.
6. J.Y. Girard : Linear logic. Theoretical Computer Science. **50** (1987) 1-102.
7. J.Y. Girard, Y. Lafont, and P. Taylor : Proofs and Types. Cambridge University Press, 1989.
8. T. Griffin : A formulae-as-types notion of control. Proc. POPL'90 (1990) 47-58.
9. M. Hofmann, T. Streicher : Continuation models are universal for λμ-calculus. Proc. LICS'97 (1997) 387-397.
10. M. Hofmann, T. Streicher : Completeness of continuation models for λμ-calculus. Information and Computation (to appear).
11. M. Parigot : Free Deduction: an Analysis of "Computations" in Classical Logic. Proc. Russian Conference on Logic Programming, 1991, Springer LNCS **592** 361-380.
12. M. Parigot : λμ-calculus: an Algorithmic Interpretation of Classical Natural Deduction. Proc. LPAR'92, Springer LNCS **624** (1992) 190-201.
13. M. Parigot : Strong normalization for second order classical natural deduction, Proc. LICS'93 (1993) 39-46.
14. M. Parigot : On the computational interpretation of negation, Proc. CSL'2000, Springer LNCS **1862** (2000) 472-484.
15. C.H.L. Ong, C.A. Stewart : A Curry-Howard foundation for functional computation with control. Proc. POPL'97 (1997)
16. P. Selinger : Control categories and duality: on the categorical semantics of lambda-mu calculus, Mathematical Structures in Computer Science (to appear).

Scheduling to Minimize the Average Completion Time of Dedicated Tasks*

Foto Afrati[1], Evripidis Bampis[2], Aleksei V. Fishkin[3], Klaus Jansen[3], and Claire Kenyon[4]

[1] National Technical University of Athens,
Heroon Polytechniou 9, 15773, Athens, Greece,
`afrati@softlab.ece.ntua.gr`
[2] LaMI, Université d'Evry, Boulevard François Mitterand, 91025 Evry Cedex, France,
`bampis@lami.univ-evry.fr`
[3] Institut für Informatik und praktische Mathematik, Universität Kiel,
Olshausenstrasse 40, 24098 Kiel, Germany,
`{avf,kj}@informatik.uni-kiel.de`
[4] LRI, Bât 490, Université Paris-Sud, 91405 Orsay Cedex, France,
`Claire.Kenyon@lri.fr`

Abstract. We propose a polynomial time approximation scheme for scheduling a set of dedicated tasks on a constant number m of processors in order to minimize the sum of completion times $Pm|\text{fix}_j| \sum C_j$. In addition we give a polynomial time approximation scheme for the weighted preemptive problem with release dates, $Pm|\text{fix}_j, pmtn, r_j| \sum w_j C_j$.

1 Introduction

In the last few years, an important amount of work is devoted to the study of scheduling problems in which the objective is to minimize the sum of completion times. In [1], the authors presented the first *polynomial-time-approximation-schemes* (PTASs) for scheduling to minimize the average weighted completion time (in the presence of release dates) in various machine models including one, identical parallel, unrelated parallel machines, with and without preemption. In all these models each task is processed on at most one machine at a time. On the contrary, no PTAS was known for scheduling problems, in which the objective is to minimize the average completion time, involving *multiprocessor* tasks i.e. tasks that may require more than one processors at a time.

In this paper, we propose the first PTASs for the *dedicated multiprocessor task model* in which the objectives are the minimization of the average completion time in the non-preemptive case, and the average weighted completion time in the

---

* This research was partially supported by the ASP "Approximabilité et Recherche Locale" of the French Ministry of Education, Research and Technology (MENRT), by the DFG - Graduiertenkolleg "Effiziente Algorithmen und Mehrskalenmethoden" and by the EU project APPOL, "Approximation and on-line algorithms", IST-1999-14084.

S. Kapoor and S. Prasad (Eds.): FST TCS 2000, LNCS 1974, pp. 454–464, 2000.

preemptive case in the presence of release dates. For the dedicated multiprocessor model, the only known PTAS was for the case where the objective is to minimize the makespan [2].

Using the standard three field notation [5], the problem of scheduling dedicated tasks on a set of processors in order to minimize the sum of task completion times is denoted as $P|\text{fix}_j| \sum C_j$. This problem was studied for the first time by Hoogeveen et al. in [6], where it was shown to be $\mathcal{NP}$-hard in the strong sense, even in the case where all the tasks have unit execution times. Cai, Lee and Li [4] proved that the problem is also strongly $\mathcal{NP}$-hard, even in the case where there are just 2 processors. On the other hand, in [3], Brucker and Krämer proved that the problem is polynomial in the case where the tasks have unit execution times and the number of processors is a fixed constant, i.e. for $Pm|\text{fix}_j, p_j = 1| \sum C_j$, as well as when in addition the tasks have release dates. In terms of approximation algorithms, in [4], a 2-approximation is given for the 2 processor problem $P2|fix_j| \sum C_j$. In this paper an approximation scheme is given for the m processor problem $Pm|fix_j| \sum C_j$ with m constant, via a reduction to the preemptive version of the problem.

When the tasks can be preempted, the problem is a bit easier. In fact the 2 processor case $P2|fix_j, pmtn| \sum C_j$ is polynomial [4]. We present an approximation scheme for the generalization where the tasks have weights and release dates, and the goal is to minimize the weighted sum of completion times, $Pm|r_j, pmtn| \sum w_j C_j$. Note that Labetoulle et al. [7] proved that the single processor version of this problem, $P1|r_j, pmtn| \sum w_j C_j$, is already strongly $\mathcal{NP}$-hard.

In section 2, we present a reduction from the non-preemptive problem $Pm|fix_j| \sum C_j$ to the preemptive problem $Pm|fix_j, pmtn| \sum C_j$. In section 3, we present an approximation scheme for the (more general) preemptive problem $Pm|fix_j, pmtn, r_j| \sum_j w_j C_j$.

2   A Reduction from Non-preemptive to Preemptive

Formulation of the problem. We are given a set of n tasks $\mathcal{T} = \{1, 2, \ldots, n\}$ and a set of m processors M. The tasks are dedicated: each task j requires for its execution the simultaneous availability of a prespecified subset of processors $\tau_j \subseteq M$ for p_j units of time. The set τ_j is called the *type* of task j. We denote by S_j and C_j the starting and completion times of task j. The problem is to design a schedule which minimizes the sum of task completion times, $\sum_{j=1}^n C_j$.

2.1   The Algorithm

Let UB denote the upper bound to the optimal cost obtained by processing all the tasks sequentially in order of non-decreasing processing times. Note that this upper bound is within a constant factor m of optimal. The lemma below shows how to separate tasks into "long" tasks and "short" tasks, with a few (negligible) "medium" tasks in between. Its proof is a simple algebraic manipulation.

Lemma 1. *Let L_i denote the set of tasks j with processing time $p_j \geq \epsilon^{5i+5}UB$, M_i denote the set of tasks j with processing time $\epsilon^{5i+10}UB \leq p_j < \epsilon^{5i+5}UB$, and $S_i = \mathcal{T} \setminus L_i \cup M_i$ denote the remaining tasks. If $L_1 \neq \emptyset$, then there exists an $i < \log_{1+\epsilon}(1/\epsilon^{10})$ such that $OPT(L_i \cup M_i) \leq (1 + \epsilon^2)OPT(L_i)$.*

At a high level, the algorithm is as follows.

1. For every $i < \log_{1+\epsilon}(1/\epsilon^{10})$, partition the tasks into $\mathcal{T} = L_i \cup M_i \cup S_i$ and construct a schedule of $\mathcal{T}$ as follows.
2. Construct a non-preemptive schedule of S_i as follows.
   (a) Solve the preemptive problem for S_i with relative error ϵ.
   (b) For each subset τ of the machines, consider all the time intervals during which the schedule executes tasks of type τ, and reorder the tasks of type τ in these time intervals by order of increasing processing time.
   (c) Stretch time by a factor of $(1 + \epsilon)$, so that each task in the schedule corresponds to an interval or a set of intervals (if preemption occurred) of total measure $(1+\epsilon)p_j$; leave the first ϵp_j section idle and process task j during the last p_j part of the interval or set of intervals.
   (d) Modify the schedule to reduce the number of preemptions in the following way. In each interval $I = [(1 + \epsilon)^k, (1 + \epsilon)^{k+1}]$, look at the schedule during that interval, erasing the names of the tasks and only remembering the task types ; at every instant a certain set of task types are being processed; call this a configuration. We reorder the schedule inside I so that identical configurations are contiguous; put the tasks back in, in order of increasing processing time. Note that in this modified schedule, a task can only be preempted when a configuration ends, thus the number of preemptions in the interval is at most m times the number A of configurations, which is $O(1)$ since the number m of processors is bounded and configurations, which are just partitions of $\{1, 2, \dots, m\}$ into task types, are also in constant number.
   (e) Modify the schedule to make it non-preemptive in the following way: for each preempted task j, if its completion time t is greater than Ap_j/ϵ^2, then finish executing j there, otherwise remove task j and insert it in the gap at time $(1 + \epsilon)^i$ such that $(1 + \epsilon)^{i-1} < mt/\epsilon^2 \leq (1 + \epsilon)^i$.
   (f) Remove all times during which all processors are idle from the resulting schedule.
3. Construct an optimal non-preemptive schedule of $L_i \cup M_i$ by exhaustive search.
4. Concatenate the schedule of S_i and the schedule of $L_i \cup M_i$.
5. Output the best resulting schedule, over all choices of i.

2.2   Analysis of Running Time

There are only $O(1)$ possibilities for i. For each choice of i, we partition the tasks in $O(n)$, run the preemptive approximation scheme once, reorder the tasks of the same type in $O(n \log n)$, perform the rest of step 2 in $O(n)$, and construct an optimal schedule of $L_i \cup M_i$ in time $O(|L_i \cup M_i|!)$. Since $L_i \cup M_i$

consists of tasks with processing time greater than $\epsilon^{5i+10} \sum p_k$, there can be at most ϵ^{-5i-10} such tasks. By Lemma 1, $i < \log_{1+\epsilon}(1/\epsilon^{10})$ and so $|L_i \cup M_i| \leq (1/\epsilon)^{O(1/\epsilon^{10})}$, so that $|L_i \cup M_i|! = O(1)$. Thus the overall running time is $O(n \log n) + O(\text{preemptive approximation scheme})$.

2.3   Analysis of the Sum of Completion Times

At the end of step 2a, the schedule of S_i has cost at most $(1 + \epsilon)$ times the optimal preemptive schedule cost, which is at least as good as the non-preemptive schedule, hence the schedule of S_i has cost as most $(1+\epsilon)OPT(S_i)$. Step 2b can only decrease the cost. Step 2c increases the cost by a factor of $(1 + \epsilon)$. Since step 2d only modifies the completion times inside the intervals, it also increases the cost by a factor of $(1 + \epsilon)$ at most.

Step 2e is more difficult to analyze. First, finishing all the short preempted tasks of interval $I_k = [(1 + \epsilon)^k, (1 + \epsilon)^{k+1}]$ creates a delay of at most $p_j < \epsilon^2(1+\epsilon)^{k+1}/A$ for each of the A configurations, adding up to a delay of at most $\epsilon^2(1+\epsilon)^{k+1}$ due to interval I_k. Thus a task j completed in interval I_l is delayed by intervals $I_1, I_2, \ldots, I_{l-1}$, for a total delay of at most

$$\epsilon^2 + \epsilon^2(1 + \epsilon) + \ldots + \epsilon^2(1 + \epsilon)^{l-1} \leq \epsilon(1 + \epsilon)^l < \epsilon C_j.$$

Thus these delays increase the cost by a factor of $(1 + \epsilon)$ at most.

Secondly, the long tasks displaced in step 2e may also cause further delays. A gap at time $t' = (1+\epsilon)^l$ receives only tasks previously scheduled preemptively to complete at time $t \leq t'\epsilon^2/m$. These tasks use up a space of at most $t'\epsilon^2$ in the gap at t', which again sums to a negligible delay in the schedule, a factor of at most $(1 + \epsilon)$.

Thirdly, the long tasks displaced see their own completion times greatly increased. Call $\mathcal{D}$ the set of such displaced tasks. They were displaced because their completion time in the preemptive schedule was smaller than Ap_j/ϵ^2, and the displacement increased their completion time by a factor of m/ϵ^2, thus the sum of their new completion times is at most $\sum_{j \in \mathcal{D}} mAp_j/\epsilon^4$. But their processing times sum to at most m times the makespan of the schedule $\mathcal{S}$.

Let $p_{\max}(S_i)$ be the maximum processing time of S_i and $\mathcal{M}$ the makespan of $\mathcal{S}$. Considering that at least $\mathcal{M}/(2p_{\max})$ tasks will be executed during the last $\mathcal{M}/2$ steps of $\mathcal{S}$ and hence have completion time greater than $\mathcal{M}/2$, we obtain that the cost of $\mathcal{S}$ is at least $\mathcal{M}^2/(4p_{\max})$, hence

$$\mathcal{M} \leq 2\sqrt{p_{\max}COST(\mathcal{S})} \leq 2\sqrt{\epsilon^{10}OPT(\mathcal{T})(1 + \epsilon)^5 OPT(S_i)/m}.$$

Thus the new completion times of the tasks of $\mathcal{D}$ sum to at most $2A\epsilon\sqrt{m}OPT(\mathcal{T})$.

Finally, at the end of step 2, the non-preemptive schedule $\mathcal{X}$ of S_i has cost at most $(1 + \epsilon)^5 OPT(S_i) + 2A\sqrt{m}\epsilon OPT(\mathcal{T})$ and makespan at most $\mathcal{M}(\mathcal{X}) \leq 2\sqrt{p_{\max}(S_i)COST(\mathcal{X})}$.

Step 3 constructs an optimal schedule $\mathcal{Y}$ of $L_i \cup M_i$ of cost $OPT(L_i \cup M_i)$ which is at most $(1 + \epsilon)^2 OPT(L_i)$ by Lemma 1.

Step 4 concatenates the two schedules, for a cost of $COST(\mathcal{X})+COST(\mathcal{Y})+$ $|L_i \cup M_i|\mathcal{M}(\mathcal{X})$ which we now need to analyze. L_i satisfies $OPT(L_i) > |L_i|^2 p_{\min}(L_i)/(2m^2)$, thus

$$(L_i\mathcal{M}(\mathcal{X}))^2 < 2m^2 \frac{OPT(L_i)}{p_{\min}(L_i)} 4p_{\max}(S_i)COST(\mathcal{X}) < 8m^2(1+\epsilon)^5\epsilon^5 OPT(\mathcal{T})^2$$

since the processing times in L_i and S_i differ by a factor of ϵ^5 at least.

Moreover, M_i satisfies $OPT(M_i) > |M_i|^2 p_{\min}(M_i)/(2m^2)$, and moreover $OPT(M_i) < OPT(L_i \cup M_i) - OPT(L_i) < \epsilon^2 OPT(\mathcal{T})$, thus

$$(|M_i|\mathcal{M}(\mathcal{X}))^2 < 2m^2\epsilon^2 \frac{OPT(\mathcal{T})}{p_{\min}(M_i)} 4p_{\max}(S_i)COST(\mathcal{X}) \leq 8m^2\epsilon^2(1+\epsilon)^5 OPT(\mathcal{T})^2.$$

Thus the concatenated schedule has overall cost

$$(1+\epsilon)^5 OPT(S_i) + 2\sqrt{m}A\epsilon OPT(\mathcal{T}) + (1+\epsilon^2)OPT(L_i)+$$

$$2\sqrt{2}m(1+\epsilon)^{2.5}\epsilon^{2.5}OPT(\mathcal{T}) + 2\sqrt{2}m\epsilon(1+\epsilon)^5 OPT(\mathcal{T}).$$

Since $OPT(S_i) + OPT(L_i) \leq OPT(\mathcal{T})$, we obtain that the cost of the schedule is $(1+O(\epsilon))OPT(\mathcal{T})$.

3 Solving the Preemptive Problem

In this section we present a PTAS for $Pm|fix_j, r_j, pmtn| \sum w_j C_j$ with release dates r_j and weights w_j for each task. First, using ideas in [1] and new ideas we simplify the problem instance. Then, we apply a dynamic programming technique to find an approximative schedule. Inside of the dynamic program we use an optimal algorithm of Amoura et al. [2] for $Pm|fix_j, pmtn|C_{max}$ (makespan optimization) to test whether tasks can be processed in an interval or not. In total, we prove the following result:

Theorem 1. There is a PTAS for $Pm|fix_j, r_j, pmtn| \sum w_j C_j$ that constructs a $1 + \epsilon$-approximation in $O(n \log n)$ time (with m and $\epsilon > 0$ constant).

As in section 2 we partition the time $(0, \infty)$ into disjoint intervals of the form $I_x := [R_x, R_{x+1})$ with $R_{x+1} = R_x(1+\epsilon)$ and $I_x = \epsilon R_x$ (we use I_x to refer to both $|I_x|$ and I_x). Let $\mathcal{T}^\tau \subseteq \mathcal{T}$ denotes the set of tasks with the same type $\tau \subseteq M$; $\mathcal{T}_x^\tau$ is the set of tasks in $\mathcal{T}^\tau$ that are released at I_x. Let C_j be the completion and S_j be the starting time of task j. The values $x(j)$ and $z(j)$ denote the indices of the intervals $I_{x(j)}$ and $I_{z(j)}$ where task j is released and completed, respectively.

First, we simplify the problem instance. With at most $1 + \epsilon$ loss in the objective function, we can assume that all release dates r_j and processing times p_j are integer powers of $1 + \epsilon$; $r_j \geq \epsilon p_j$ and $p_j \geq 1$ [1]. As consequence, the processing time of each task j is at most $\frac{1}{\epsilon^2}$ times more than the length of the interval where this task is released (i.e. $p_j \leq \frac{I_{x(j)}}{\epsilon^2}$). This means that every task crosses at most a constant number of intervals.

Furthermore, we can assume that all quotients p_j/w_j are different. Let $p_{j_1}/w_{j_1} \le p_{j_2}/w_{j_2} \le \ldots \le p_{j_n}/w_{j_n}$. Suppose that some tasks have the same values p_j/w_j. In this case, we can increase the weights such that $w_j < w_j' \le (1+\epsilon)w_j$ such that all quotients p_j/w_j are different. The objective value of a schedule with the new weights $\sum w_j' C_j$ is bounded by $(1+\epsilon)\sum w_j C_j$. Finally, we can rearrange tasks inside an interval and consider $\sum w_j R_{z(j)}$ instead of our original objective function $\sum w_j C_j$.

Now we introduce two types of tasks. Using the assumption above, every release date r_j is the left endpoint of an interval $I_{x(j)}$. A task j is *large*, if the processing time p_j is larger than $\frac{\epsilon^2 I_{x(j)}}{2m}$, and is *small* otherwise. Let $\mathcal{LT}_x^\tau$ be the set of large tasks and $\mathcal{ST}_x^\tau$ the set of small tasks in $\mathcal{T}_x^\tau$. We may assume that $\epsilon \le \frac{1}{2m}$ and $\log_{1+\epsilon} \frac{1}{\epsilon}, \frac{1}{\epsilon}$ are integral.

For an optimal schedule and tasks in $\mathcal{T}_x^\tau$ we can assume that each task is processed completely before another one starts to be processed. In other words for $t, \ell \in \mathcal{T}_x^\tau$ we have $C_t \le S_\ell$ or $C_\ell \le S_t$.

Lemma 2. *With at most $1 + O(\epsilon)$ loss, we can assume that each task in $\mathcal{ST}_x^\tau$ is processed completely in one interval.*

Proof. Fix an interval I_y where this property does not hold. Consider one processor type $\tau \subseteq M$. Using the oberservation above it follows that there is at most one small task j_x released in interval I_x with $x \le y$, such that j_x is started in I_y with $x \le y$ but finished later.

The goal is to complete all these small tasks (among all previous intervals I_x) in I_y. The total processing time of these tasks can be bounded as follows:

$$\sum_{x \le y} p_{j_x} \le \sum_{x \le y} \frac{\epsilon^2 I_x}{2m} \le \sum_{t \ge 0} \frac{\epsilon^2 I_y}{2m(1+\epsilon)^t} \le \frac{\epsilon^2 I_y}{2m} \sum_{t \ge 0} \frac{1}{(1+\epsilon)^t} \le \frac{\epsilon(1+\epsilon)I_y}{2m} \le \frac{2\epsilon}{2m} I_y.$$

Adding the bound for all types $\tau \subseteq M$, the total time to complete all these small tasks is at most $2\epsilon I_y$. To create $2\epsilon I_y$ idle time, we shift the entire schedule two intervals forward. This increase the objective function by at most $1 + (1 + \epsilon)^2$. Using these idle times and preemptions we are able to reschedule and to complete the small tasks within I_y. ∎

The tasks in $\mathcal{ST}_x^\tau$ are *scheduled by Smith's rule* if they are scheduled in order of increasing $\frac{p_j}{w_j}$. We say that two tasks $t \in \mathcal{ST}_{x(t)}^\tau$ and $t' \in \mathcal{ST}_{x(t')}^\tau$ (with release indices $x(t') \le x(t)$) are *scheduled by Smith's rule* if one of two following conditions holds:

(1) t' is completed before t is released,
(2) t' is not started before t is released and
  (2.1) if $\frac{p_t}{w_t} < \frac{p_{t'}}{w_{t'}}$ then t' starts only after t is completed,
  (2.2) if $\frac{p_t}{w_t} > \frac{p_{t'}}{w_{t'}}$ then t starts only after t' is completed.

The following lemma gives us a powerful tool to handle small tasks.

Lemma 3. *With at most $1 + O(\epsilon)$ loss, for each processor set τ we can assume that all small tasks in T^τ are scheduled by Smith's rule.*

Proof. Consider an optimal schedule where no small task crosses an interval and where the quotients p_j/w_j are different. Let S^τ be the set of all small tasks with processor set τ, and let $S_x^\tau \subseteq S^\tau$ be the subset of small tasks that are executed in I_x. Then, define $p(I_x, \tau) = \sum_{j \in S_x^\tau} p_j$ as the total time to process small tasks in I_x. Furthermore, $x(j)$ denotes the index of the interval where task j is released and L denotes the index of the last interval. For any type $\tau \subseteq M$, we study the following linear program:

Minimize $\sum_{j \in S_\tau} w_j \sum_{i=x(j)}^{L} y_{j,i} R_i$ s.t.

(1) $\sum_{i=x(j)}^{L} y_{j,i} = 1, \quad \forall j \in S^\tau,$

(2) $\sum_{j \,:\, j \in S_\tau, \, x(j) \leq i} y_{j,i} p_j \leq p(I_i, \tau), \quad \forall I_i$ and $\tau \subset M,$

(3) $y_{j,i} \geq 0, \quad \forall j \in S^\tau, \, i = x(j), \dots, L.$

First, the objective value of the linear program is not larger than the weighted average completion time for the small tasks and the fact that the fractional assignment gives only a smaller value. In other words, the value of an optimal fractional solution is a lower bound of the weighted completion time. Consider an optimal solution $(y_{j,i}^*)$ of the linear program. Suppose that two tasks t and t' are scheduled not by Smith's rule. Without loss of generality we suppose that $y_{t,i_t}^* > 0$, $y_{t',i_{t'}}^* > 0$, $x(t') \leq x(t) \leq i_t < i_{t'}$ and $\frac{p_{t'}}{w_{t'}} < \frac{p_t}{w_t}$.

Then, there exist values z_t and $z_{t'}$ such that $0 < z_t \leq y_{t,i_t}^*$, $0 < z_{t'} \leq y_{t',i_{t'}}^*$ and $z_t p_t = z_{t'} p_{t'}$. Now we exchange parts of the variables:

$$y_{t,i_t}' = y_{t,i_t}^* - z_t \quad y_{t,i_{t'}}' = y_{t,i_{t'}}^* + z_t$$
$$y_{t',i_{t'}}' = y_{t',i_{t'}}^* - z_{t'} \quad y_{t',i_t}' = y_{t',i_t}^* + z_{t'}$$

and $y_{j,i}' = y_{j,i}^*$ for the remaining variables. The new solution $(y_{j,i}')$ is feasible and the objective value $\sum_{j \in S^\tau} w_j \sum_{i=x(j)}^{L} y_{j,i}' R_i$ is equal to $\sum_{j \in S^\tau} w_j \sum_{i=x(j)}^{L} y_{j,i}^* R_i + R_{t,t'}$ where $R_{t,t'} = (R_{i_t} - R_{i_{t'}})(w_{t'} z_{t'} - z_t w_t)$. Using $z_{t'} = z_t \frac{p_t}{p_{t'}}$, $\frac{p_{t'}}{w_{t'}} < \frac{p_t}{w_t}$ and $z_t > 0$, the second factor $(w_{t'} z_{t'} - z_t w_t) = z_t(w_{t'} \frac{p_t}{p_{t'}} - w_t)$ is larger than 0. The inequality $i_t < i_{t'}$ implies $R_{i_t} < R_{i_{t'}}$ and $R_{t,t'} < 0$. In other words, the new solution $(y_{j,i}')$ has a lower objective value and gives us a contradiction. This means that the two tasks t and t' are scheduled by Smith's rule.

Now we use some properties about the optimal solution of the linear program above. There is an optimal solution such that for each interval I_i we have at most one task $j \in S^\tau$ with $x_{j,i} \in (0, 1)$ and that is assigned for the first time. Otherwise we can use the same argument as above (and the fact that the quotients $\frac{p_j}{w_j}$ are different) to improve the objective value. To turn the fractional solution into an integral, we need only to increase the values $p(I_i, \tau)$ by at most $\frac{\epsilon I_i}{2^m}$ (because all tasks are small). Thus for all $\tau \subseteq M$ we have to create at most $2\epsilon I_i$ idle time. Then we shift the schedule two intervals forwards and use the created idle time to reschedule small tasks by Smith's rule. ∎

Let $p(T')$ be the processing time of all tasks in $T' \subset T$. By delaying of tasks to later intervals we can bound the number of long tasks and the total length of small tasks for each interval I_x:

Lemma 4. *With $1 + O(\epsilon)$ loss, each instance I of $Pm|fix, r_j, pmtn| \sum w_j C_j$ can be transformed in $O(n \log n)$ time into an instance I' such that for each type $\tau \subseteq M$ and release date R_x:*

- $|\mathcal{LT}_x^\tau| \leq K := \frac{2^m k}{\epsilon^2}$, *where* $k = 5 \log_{1+\epsilon} \frac{1}{\epsilon}$,
- $p(\mathcal{ST}_x^\tau) \leq 2I_x$.

Proof. Consider an interval I_x. Using Lemma 3 we order $\mathcal{ST}_x^\tau$ by Smith's rule. In the interval I_x the total available time to schedule tasks from T_x^τ is I_x. Thus we select tasks from $\mathcal{ST}_x^\tau$ until the total processing time of selected tasks is greater or equal to I_x. Since $p_j \leq \frac{\epsilon^2}{2^m} I_x$ for each job j in $\mathcal{ST}_x^\tau$ we have at most $2I_x$ for selected tasks. Within large tasks in $\mathcal{LT}_x^\tau$ of the same size we select at most $\frac{2^m}{\epsilon^2}$ tasks in order of decreasing weights (only they can be started in I_x). We have at most $\frac{2^m k}{\epsilon^2}$ selected large jobs. After that we increase the release time of not selected tasks.

∎

The difficult part in the dynamic programming is to show that it is sufficient to maintain informations for a small number of tasks. To do this we introduce a compact representation of small and long tasks. We start with the small tasks and assume that the small tasks in $\mathcal{ST}_x^\tau$ are ordered by Smith's rule (i.e. in increasing order of p_j/w_j). Then we select the tasks one by one and create sets $\mathcal{ST}_{x,i}^\tau \subseteq \mathcal{ST}_x^\tau$, $1 \leq i \leq 2^{m+3}/\epsilon^2$ (the last sets may be empty) of lengths roughly equal to $\frac{\epsilon^2 I_x}{2^{m+1}}$ (but not greater than $\frac{\epsilon^2 I_x}{2^m}$). We always create a new set $\mathcal{ST}_{x,i+1}^\tau$ when the total processing time of tasks in $\mathcal{ST}_{x,i}^\tau$ and the last selected task is greater than $\frac{\epsilon^2 I_x}{2^{m+1}}$. This last selected task is placed into $\mathcal{ST}_{x,i+1}^\tau$. The following Lemma shows how the small tasks can be scheduled without increasing the objective function too much.

Lemma 5. *With $1 + O(\epsilon)$ loss we can assume that in each interval I_y, $y \geq x$ for all subsets $\tau \subseteq M$, either*

- *a consecutive sequence of task sets (at least one set) $\mathcal{ST}_{x,a_y}^\tau$, $\mathcal{ST}_{x,a_y+1}^\tau$, ..., $\mathcal{ST}_{x,b_y}^\tau$ is scheduled in I_y, or*
- *all tasks in $\mathcal{ST}_x^\tau$ have already been scheduled.*

Proof. Fix one processor set $\tau \subseteq M$ and consider all small tasks that require τ. Using Lemma 3, these small tasks are scheduled by Smith's rule. Next consider the first interval I_y, $y \geq x$ where the properties in the Lemma above does not hold. Then there is one set $\mathcal{ST}_{x,b_y}^\tau$ that is not completely scheduled in I_y (or there is no task from $\mathcal{ST}_x^\tau$). If we increase the processing time $p(I_y, \tau)$ by an amount of $\frac{\epsilon^2 I_x}{2^m}$ (for each such I_x) then the sets ST^τ_{x,b_y} can be scheduled completely in I_y. The total enlargement for all I_x, $x \leq y$, is bounded by

$$\sum_{x \leq y} p(ST^\tau_{x,b_x}) \leq \sum_{x \leq y} \frac{\epsilon^2 I_x}{2^m} \leq \sum_{t \geq 0} \frac{\epsilon^2 I_y}{2^m (1+\epsilon)^t} \leq \frac{\epsilon(1+\epsilon) I_y}{2^m} \leq \frac{2\epsilon}{2^m} I_y.$$

This implies that we have to increase the processing times $p(I_y, \tau)$ by at most $\frac{2\epsilon}{2^m} I_y$. For all processor sets $\tau \subseteq M$ we have to create at most $2\epsilon I_y$ idle time to complete the small tasks from previous intervals in I_y. Again we create $2\epsilon I_y$ idle time by shifting the schedule two intervals forward.

∎

The schedule type above allows us now to represent the set ST^τ_x in a more compact way by a set $\widehat{ST}^\tau_x$ with at most $\bar{T} = \frac{2^{m+3}}{\epsilon^2}$ new created small tasks. Each task $T_{x,i} \in \widehat{ST}^\tau_x$ represents the corresponding set $ST^\tau_{x,i}$. The processing time of $T_{x,i}$ is equal to $p(ST^\tau_{x,i})$ and the weight $w_{x,i}$ of $T_{x,i}$ is equal to $\sum_{j \in ST^\tau_{x,i}} w_j$. Finally the new tasks have to be processed in the total order $T_{x,1}, T_{x,2}, \ldots, T_{x,\bar{T}}$.

Finally we use a similar idea for the large tasks in $\mathcal{LT}^\tau_x$. We notice that $\mathcal{LT}^\tau_x$ contains at most a constant number of at most $K = \frac{2^m k}{\epsilon^2}$ large tasks where $k = 5 \log_{1+\epsilon} \frac{1}{\epsilon}$ and the processing time p_j of each large task j in $\mathcal{LT}^\tau_x$ is bounded by $\frac{I_x}{\epsilon^2}$.

Lemma 6. *With $1 + O(\epsilon)$ loss, we can assume that in each interval I_y, $y \geq x$ for all subsets $\tau \subseteq M$, the partial time $p_{j,y}$ that is used in I_y to process a large task $j \in \mathcal{LT}^\tau_x$, either*

- *is equal to $\ell_{j,y} \Delta_x$, where $\Delta_x = \frac{\epsilon^2 I_x}{2^m K} = \frac{\epsilon^4 I_x}{k 2^{2m}}$, $\ell_{j,y} \in \{1, \ldots, H\}$, $H := \frac{I_x}{\epsilon^2 \Delta_x} = \frac{k 2^{2m}}{\epsilon^6}$ and $k = 5 \log_{1+\epsilon} \frac{1}{\epsilon}$, or*
- *j has been already completed.*

Proof. Notice that the total enlargement needed in I_y for tasks from $\mathcal{LT}^\tau_x$, $x \leq y$ is at most $\frac{\epsilon^2 I_x}{2^m}$, since there are at most $K = \frac{2^m k}{\epsilon^2}$ large tasks in $\mathcal{LT}^\tau_x$. The rest follows as in Lemma 5.

∎

Corollary 1. *All tasks in $\mathcal{T}_x = \cup_{\tau \subseteq M} \mathcal{T}^\tau_x$ are scheduled completely within the next $O(s) := \max\{\bar{T}, H\}$ intervals following I_x.*

For the dynamic programming, we introduce now a block structure on the time line. The basic idea is to decompose the time line into a sequence of *blocks*. Let $\mathcal{A} = \{a_1, \ldots, a_r\}$ be the indices of release dates $R_{a_1}, \ldots, R_{a_r}$ with $a_1 < a_2 < \ldots < a_r$. Corollary 1 implies that if $a_{i+1} - a_i > O(s)$ then all tasks that are released at R_{a_i} can be scheduled in the intervals $I_{a_i}, \ldots, I_{a_i + O(s)}$ (further we consider only them). Thus, to find an optimal restricted schedule we have to consider only $nO(s)$ intervals.

Let $\mathcal{B} = \{a_1, a_1 + 1, \ldots, a_{r+1}\}$ be the indices of the corresponding intervals (at most $nO(s)$), where $a_{r+1} = a_r + O(s)$. We partition the set $\mathcal{B}$ into a sequence of blocks $\mathcal{B}_1, \ldots, \mathcal{B}_r$ where r is at most n and each block $\mathcal{B}_i$, $i = 1, \ldots, r$ consists of $O(s)$ intervals with indices from $\mathcal{B}$. Notice that the tasks that are released in the intervals of block $\mathcal{B}_i$ either finish in $\mathcal{B}_i$ or in $\mathcal{B}_{i+1}$, this set of tasks is denoted by $\mathcal{BT}_i$. Furthermore, there is at most a constant number $\mu := 2^m \bar{T} O(s)$ of small and a constant number $\nu := 2^m K O(s)$ of large tasks in $\mathcal{BT}_i$.

Let $\mathcal{G}_i$ be the different ways how the tasks from $\mathcal{BT}_i$ are scheduled in the intervals of $\mathcal{B}_i \cup \mathcal{B}_{i+1}$. Each small task is processed completely in one of the $\bar{T}$ intervals after its release. Each large task can be splitted in at most H intervals with sizes $\ell_{j,y} \Delta_x$ where $\ell_{j,y} \in \{1, \ldots, H\}$. This gives at most H^H possibilities for a large task. The total number of different ways $G_i \in \mathcal{G}_i$ to schedule these tasks is bounded by a constant $q := (1 + O(s))^\mu (O(s)^{O(s)})^\nu$. Now we can describe our objective function as follows:

$$\sum w_j R_{z(j)} = \sum_{i=1}^{r} \sum_{z(j) \in \mathcal{B}_i} w_j R_{z(j)} = \sum_{i=1}^{r} W(i, G_i, G_{i-1}),$$

where $W(i, G_i, G_{i-1})$ is the total weighted completion time of tasks that complete in block $\mathcal{B}_i$ corresponding to ways G_i and G_{i-1} (we use the fact that in block $\mathcal{B}_i$ only tasks that are released in $\mathcal{B}_i$ and $\mathcal{B}_{i-1}$ can be scheduled).

The dynamic programming table entry $\mathcal{O}(i, G_i)$ stores the minimum weighted completion time among all restricted schedules, where:

(1) G_i represents the way in which the tasks released in $\mathcal{B}_i$ are scheduled in the intervals of block $\mathcal{B}_i$ and $\mathcal{B}_{i+1}$, and

(2) all tasks that are released before block $\mathcal{B}_i$ are completely finished.

To compute the table $\mathcal{O}_i$ we use the following recursive equation:

$$\mathcal{O}(i, G_i) = \begin{cases} W(1, G_1, -), \; G_1 \in \mathcal{G}_1 \text{ for } i = 1; \\ \min_{G_{i-1} \in \mathcal{G}_{i-1}}[\mathcal{O}(i-1, G_{i-1}) + W(i, G_i, G_{i-1})], \\ \quad G_i \in \mathcal{G}_i \text{ for } i = 2, \ldots, r. \end{cases}$$

Lemma 7. *The time to compute the table $\mathcal{O}(i)$ for all $1 \leq i \leq r$ can be bounded by $O(n) \cdot T(W, n)$ where $T(W, n)$ denotes the maximal time to compute the function W for one triple (i, G_i, G_{i-1}).*

In the following we describe the procedure to test the feasibility for the set of tasks defined by G_i and G_{i-1} to be scheduled in $\mathcal{B}_i$ and to compute the value $W(i, G_i, G_{i-1})$. First there are $O(s)$ intervals in $\mathcal{B}_i$. Using the information from G_i and G_{i-1} we know precisely the finishing interval $I_{z(j)}$ of each task $j \in \mathcal{BT}_{i-1} \cup \mathcal{BT}_i$. Thus if G_i and G_{i-1} give us a feasible schedule then we can compute the value W directly. To test the feasibility we use the following idea. Consider the intervals in $\mathcal{B}_i$. For each interval $I_{i,t}$, $t = 1, \ldots, O(s)$ in $\mathcal{B}_i$ we compute the set V_t of tasks that are processed in $I_{i,t}$ and the set P_{V_t} of processing times that are used to process these tasks. Then we have to verify whether the set V_t with P_{V_t} can be scheduled in $I_{i,t}$ for each $t = 1, \ldots, O(s)$. In total, the problem of testing is equivalent to a sequence (of constant length $O(s)$) of problem instances of $Pm|fix_j, pmtn|C_{\max}$. Each such problem can be solved in linear time optimally with respect to the number of tasks in the instance [2]. Since the number of tasks in each set V_t, $t = 1, \ldots, O(s)$ is at most the number of tasks defined by G_i and G_{i-1} (this number is constant because there are only $O(1)$ tasks released in $\mathcal{B}_i$ and $\mathcal{B}_{i-1}$), we have obtained the following result:

Lemma 8. *Given G_{i-1} and G_i, the feasibility test and the computation of the value $W(i, G_i, G_{i-1})$ can be done in $O(1)$ time.*

This Lemma implies that the time to compute the entire table is bounded by $O(n)$ and that our algorithm for the preemptive variant runs in $O(n \log n)$ time.

References

1. F. Afrati, E. Bampis, C. Chekuri, D. Karger, C. Kenyon, S. Khanna, I. Milis, M. Queyranne, M. Skutella, C. Stein, M. Sviridenko, *Approximation schemes for minimizing average weighted completion time with release dates*, 40th Annual Symposium on Foundations of Computer Science (FOCS'99), 1999, 32-43.
2. A.K. Amoura, E. Bampis, C. Kenyon, Y. Manoussakis, *How to schedule independent multiprocessor tasks*, Proceedings of the 5th European Symposium on Algorithms (ESA'97), LNCS 1284, 1-12.
3. P. Brucker, A. Krämer, *Polynomial algorithms for resource constrained and multiprocessor task scheduling problems*, European Journal of Operational Research, **90** (1996) 214-226.
4. X. Cai, C.-Y. Lee, C.-L. Li, *Minimizing total completion time in two-processor task systems with prespecified processor allocations*, Naval Research Logistics, **45** (1998) 231-242.
5. R. L. Graham, E. L. Lawler, J. K. Lenstra, K. Rinnooy Kan, *Optimization and Approximation in Deterministic Scheduling: A Survey*, Annals Discrete Mathematics, **5** (1979) 287-326.
6. J.A. Hoogeveen, S.L. van de Velde, B. Veltman, *Complexity of scheduling multiprocessor tasks with prespecified processor allocations*, Discrete Applied Mathematics, **55** (1994) 259-272.
7. J. Labetoulle, E. L. Lawler, J. K. Lenstra, A. H. G. Rinnooy Kan, *Preemptive scheduling of uniform machines subject to release dates*, in: W. R. Pulleyblank (ed.), *Progress in Combinatorial Optimization*, Academic Press, New York, 1984, 245-261.

Hunting for Functionally Analogous Genes

Michael T. Hallett[1] and Jens Lagergren[2]

[1] McGill Centre for Bioinformatics
McGill University, Montreal, Canada
hallett@cs.mcgill.ca
[2] Stockholm Bioinformatics Center and
Dept. of Numerical Analysis of Computing Science, KTH, Stockholm, Sweden
jensl@nada.kth.se

Abstract. Evidence indicates that members of many gene families in the genome of an organism tend to have homologues both within their own genome and in the genomes of other organisms. Amongst these homologues, typically only one or a few per genome perform an analogous function in their genome. Finding subsets of these genes which show evidence of performing a common function is an important first step towards, for instance, the creation of phylogenetic trees, multiple sequence alignments and secondary structure predictions.

Given a collection of taxa $P = \{P_1, P_2, \ldots, P_k\}$ where P_i contains genes $\{p_{i,1}, p_{i,2}, \ldots, p_{i,n_i}\}$, we ask to choose one gene from each of the taxa P_i such that these chosen vertices *most agree*. We define *most agreeing* in three distinct ways: most tree-like, pairwise closest, and pairwise most similar.

We show these problems to be computationally *hard* from almost every angle via classical, parameterized and approximation complexity theory. However, on the positive side, we give *randomized approximation* algorithms following ideas from [GGR98] for the *pairwise closest* and *pairwise most similar* variants.

1 Introduction

Given a new nucleo- or peptide sequence, the standard "first step" of any inquiry into the determination of the evolution, chemical properties, and (ultimately) function of this biomolecule is to align it against every entry in a large molecular dataset such as EMBL[S99] or SwissProt[BA]. Since properties such as *function* are extremely complex and still largely unknown, no simple search of a dataset can answer these questions directly. The standard alignment tools [AGMWL90, PL88] only return entries which show statistically significant signs of *pairwise* evolutionary relationships. The end result is that many of the returned sequences will belong to gene families other than the family of our new sequence.

There are many reasons why this is the case such as *partial domain agreement*, *long distance homology* and *parology* via gene duplications and losses. We refer readers to [B99], [BDDEHY98], [GCMRM79], [KTG98], [MMS95], [P98], [PC97],

S. Kapoor and S. Prasad (Eds.): FST TCS 2000, LNCS 1974, pp. 465–476, 2000.

[S99], [SM98], [TKL97], [YEVB98] and the authors' paper [HL99a] for a more thorough treatment of the problems, models, and experimental results.

In any study of evolution, chemical properties, and function, care must be taken to use sequences that are *all* pairwise homologous (all related by a common evolutionary ancestor) and that all perform an analogous function[1] in their respective genome. When such care is not taken in the selection of sequences, gene trees will not reflect the true evolutionary relationships of the species, multiple sequence alignments will not display regions of conservation and change, and predictions of secondary structure will be inaccurate [B92, BDDEHY98, F88].

We introduce the following model of the above selection problem. A collection of sets $P = \{P_1, P_2, \ldots, P_k\}$ is given where P_i corresponds to taxa i and contains the homologues $\{p_{1,i}, p_{2,i}, \ldots, p_{n_i,i}\}$ found in the genome of taxa i. The goal is to choose one gene from each of the P_i such that these genes *agree the most*. Such a subset is refered to as a *core* of the weighted k-partite graph. We introduce three distinct definitions of *most agreeing*: most tree-like, pairwise closest, and pairwise most similar.

MOST TREE LIKE IN A k-PARTITE GRAPH (CORE-TREE)

input: A complete k-partite graph $G = (P_1, P_2, \ldots, P_k, E)$, edge weights $w : E \to \mathbb{R}$.

output: A set $P' = \{p_1, p_2, \ldots, p_k\}$ where $p_i \in P_i$ such that $||D(P') - A(D(P'))||_z$ is minimized where $D(P')$ is the distance matrix formed in the obvious way from P' and $A(D(P'))$ is the closest additive approximation to $D(P')$ under the L_z norm for some $z \in \{1, 2, \ldots, \infty\}$.

That is, one vertex (one gene) is selected from each partition (each genome) such that the distance matrix formed from the pairwise comparisons of the genes is as close to additive (as close to "tree-like") as possible. The assumption behind this optimization criteria is that genes, which have a different function (hence, a significantly different underlying sequence) than the gene family, should introduce non-additivity when placed into a distance matrix consisting of genes from the gene family.

Assume we are given a set of homologous genes where some are functionally analogous and others are effectively functionally inactive. The functionally inactive copies of the gene should drift in a random direction through the amino acid "space" whilst the functionally active genes in our family should mutate relatively slowly. Therefore the genes performing analogous function should be identifiable by being mutually similiar or closer in distance than any other homologues. Furthermore, sequences which have domains foreign to the gene family will also induce distance measures significantly greater than pairwise measurements between members of the gene family. We arrive at our second and third notions of *most agreeing*:

---

[1] We say *analogous function* here and not simply *function* to stress that the role a specific gene in a family plays is almost never exactly the same between organisms.

MINIMUM WEIGHT CLIQUE IN A k-PARTITE GRAPHS (CORE-CLIQUE)
input: A complete k-partite graph $G = (P_1, P_2, \ldots, P_k, E)$, edge weights
   $w : E \to \mathbb{R}$.
output: A set $P' = \{p_1, p_2, \ldots, p_k\}$ such that $p_i \in P_i$ and $\Sigma_{1 \le i < j \le k} w(p_i, p_j)$
   is minimum.

Note that the edges between vertices in different partitions could correspond to either (1) an estimate of the distance between the two genes, or (2) a statistical measure of similarity (eg. a maximum likelihood score). The first variant induces a minimization problem whilst the second variant induces a maximization problem. In most cases, the behaviour of either problem is the same and thus we focus attention on the former. Note also that the gene family is not assumed to have any sort of nice "tree-like" behavior. This problem may be particularly suited to studying microbial taxa as it is becoming clear that gene and species phylogenies are often tentative at best.

In the remainder of this paper we show that choosing cores under any of these optimization criteria is hard from the classical, parameterized and approximation complexity frameworks. That is, the general versions of these problems are NP-complete and hard for complexity class $W[1]$ for versions of the problem when the number of partitions, the size of each partition, the maximum weight of an edge, or the overall weight of the core are parameters. We also show that all of these problems are not approximable within a polynomial function of n in polynomial time. On the positive side, we give a *randomized approximation* algorithm using ideas from [GGR98, RS96] for these last two problems. For a confidence parameter δ and a accuracy parameter ϵ, this algorithm will correctly find a core-clique of weight $opt + \epsilon\sigma \cdot k^2$ with probability $1 - \delta/2$, where opt is the optimal weight core clique in the input graph, k is the number of partitions and σ is the maximum difference between the weight of two edges adjacent to the same vertex.

2   Background

Trees and Graphs A *phylogenetic tree* $T = (V, E)$ is a binary connected acyclic graph. A *leaf* in T has degree 1 and L_T is used to denote the subset of V which contain the leaves of T. For $S \subseteq T$, we let $T[S]$ represent the subtree of T induced by S. A *weighted phylogenetic tree* is a phylogenetic tree with a weight function associated with the edges, $T = (V, E, w)$ where $w : E_T \to [0, \infty)$. A complete k-partite graph is $(k+1)$-tuple $P = (P_1, P_2, \ldots, P_k, E)$ where P_i contains vertices $\{p_{i,1}, p_{i,2}, \ldots, p_{i,n_i}\}$ for some n_i where $P_i \cap P_j = \emptyset$, and where E, the edge set, contains edges between every two vertices in two different partitions P_i and P_j. Weighted k-partite graphs are defined similarly. A *clique* of size t in a graph G is a set of t distinct vertices which are mutually adjacent. The weight of an edge is written $w(x, y)$ as a short hand for $w((x, y))$ for some edge (x, y).
Distance/Similarity Matrices A *distance matrix* D is a 0 diagonal, symmetric, nonnegative matrix, indexed by the set of taxa L_T for a phylogenetic tree

T where the entry D_{ij} is the distance (an estimated distance) between taxa i and taxa j. An $n \times n$ distance matrix D is *additive*, if there exists a weighted phylogenetic tree T with n leaves such that entry D_{ij} equals to the sum of the edge weights in the tree along the path connecting i and j. A *similarity matrix* S is the same as a distance matrix except that diagonal elements have value ∞ and entry S_{ij} is a similarity score between taxa i and j.

Theorem 1 ([B71]). *A matrix D is additive if and only if for all i, j, k, l (not necessarily distinct), the maximum of $D_{ij} + D_{kl}, D_{ik} + D_{jl}, D_{il} + D_{jk}$ is not unique. The edge weighted tree (with positive weights on internal edges and non-negative weights on leaf edges) representing the additive distance matrix is unique among the trees without vertices of degree two.*

Error Measurements The L_k *norm* between distance matrices D and D', written $||D - D'||_k$, is defined as $||D - D'||_k = \left(\Sigma_{i<j} \left(|D_{ij} - D'_{ij}| \right)^k \right)^{\frac{1}{k}}$ for $k \geq 1$. For $k = \infty$, the L_∞ *norm* is defined as $||D - D'||_\infty = max_{i<j} |D_{ij} - D'_{ij}|$

Approximation Ratios An approximation algorithm is said to achieve an *approximation ratio* of α for a maximization problem Π if for each input x, it computes a solution y of cost at least OPT/α, where OPT is the cost of the optimum. For a minimization problem, the algorithm must return a solution y of cost at most $\alpha \cdot OPT$. Note that $\alpha \geq 1$.

We refer the reader to [DF99] for a complete description of parameterized complexity. The following three items are the main ingredients of this tool.

FPT, Completeness, Reductions (1) For a parameterized language L, $L \subseteq \Sigma^* \times \Sigma^* : \langle x, k \rangle$, ($k$ is the parameter) we say that L is *(uniformly) fixed parameter tractable (FPT)* if there exists a constant α and an algorithm Φ such that Φ decides if $\langle x, k \rangle \in L$ in time $f(k)|x|^\alpha$ where $f : \mathbb{N} \to \mathbb{N}$ is an arbitrary function. (2) We say that L reduces to L' by a standard parameterized m–reduction if there is an algorithm Φ which transforms $\langle x, k \rangle$ into $\langle x', g(k) \rangle$ in time $f(k)|x|^\alpha$, where $f, g : \mathbb{N} \to \mathbb{N}$ are arbitrary functions and α is a constant independent of k, so that $\langle x, k \rangle \in L$ if and only if $\langle x', g(k) \rangle \in L'$. (3) k-CLIQUE, parameterized by the clique set size k, is complete for complexity class $W[1]$. That is, k-CLIQUE is not in *FPT* unless problems like k-STEP TURING MACHINE and many other problems whose best known algorithms run in time $\Sigma(n^k)$, can be solved in *FPT* $(f(k) \cdot n^\alpha)$ time.

3   Complexity Results

3.1 Core-Clique. The decision version of this problem takes as input a parameter $r \in \mathbb{R}$ and answers "yes" iff the core-clique has weight $\leq r$. Theorem 2 below states that even when the number of candidate genes per genome is bounded by 3, an extremely simple weighting function is used, and a bound of 0 is placed on the size of the core-clique, the problem remains NP-complete. Theorem 3 states that a modified (easier to approximate) version of CORE-CLIQUE cannot be approximated within any function of n (the number of vertices of the input graph) in polynomial time. Both these theorems follows easily from the following lemma.

Lemma 1. *Let $f(n)$ be a function such that $f(n) > 0$ for all $n \geq 1$, then* CORE-CLIQUE *restricted to partitions of size 3 and with a weighting function w which assigns an edge either 0 or $f(n)$, and $r = 0$ is NP-complete, where n is the size of the input graph.*

Proof. The problem is in NP. To show hardness, we reduce from 3SAT which accepts as input a formula Φ in 3-CNF over a set of variables $X = \{x_1, x_2, \ldots, x_t\}$, and asks if there is a truth assignment to X such that each clause of Φ has at least one true literal.

Let $X = \{x_1, x_2, \ldots, x_t\}$ be the set of variables and $C = \{C_1, C_2, \ldots, C_k\}$ be the set of clauses of an arbitrary instance of this problem. To construct an instance of the CORE-CLIQUE problem (G, w, r), we create k partitions $P_1, P_2, \ldots, P_k$ and associate P_i with clause C_i. The 3 vertices in P_i are labeled by the literals in C_i. The weight of an edge between two vertices in different partitions corresponding to two negated literals x_j and $\bar{x}_j$ is $f(n)$. Otherwise, the weight is 0.

Claim G has a weight 0 core-clique if and only if Φ is satisfiable.

$(\Rightarrow)$ Let $p^1, p^2, \ldots, p^k$ be the set of vertices which induce a core-clique of weight 0. Now there can be no weight $f(n)$ edges between any p^i and p^j which implies that it is never the case that p^i is some literal x whilst p^j is the negated literal $\bar{x}$. Hence, we may set the literal p^i to be true. Since we may do this for all k of the partitions, we have a truth assignment for Φ with at least one true literal in each clause.

$(\Leftarrow)$ Let $T : X \to \{true, false\}$ be a truth assignment to Φ such that at least one literal x in each clause C_i is true. Consider any two distinct such literals x_i and x_j which are true in clauses C_i and C_j. Then the vertex labelled x_i in P_i and the vertex lapelled x_j in P_j have no weight $f(n)$ edge between them, since T is a satisfying assignment for Φ and there is an edge of weight $f(n)$ only if two literals are negations of each other. Hence, we may place x_i and x_j in the core-clique.

Theorem 2. CORE-CLIQUE *restricted to partitions of size 3 and with a weighting function w which assigns an edge either 0 or 1, and $r = 0$ is NP-complete.*

No minimization problem for which it is NP-complete to distinguish between instances with 0 minimum cost and instances with cost $c > 0$ can be approximated within any ratio in polynomial time. Since this comment applies to the CORE-CLIQUE problem, we formulate a slightly modified version of the optimization form (MODIFIED-CORE-CLIQUE) of the problem which asks for the P' which minimizes $1 + \Sigma_{1 \leq i < j \leq k} w(p_i, p_j)$, for which non-trivial non-approximability results can be proved.

Theorem 3. *If $P \neq NP$, then* MODIFIED-CORE-CLIQUE *is not approximable within any function of n in polynomial time, where n is the size of the input graph.*

Proof. Assume that MODIFIED CORE-CLIQUE can be approximated in polynomial time approximated to within a function $g(n)$. It follows immediately that $g(n) \geq 1$ for all $n \geq 1$. By Lemma 1, it is NP-hard to distinguish between instances of MODIFIED CORE-CLIQUE with a minimum of 1 and those with a minimum of $1 + g(n)$. However, using the assumed approximation algorithm it is possible to distinguish between such instances. From this contradiction the theorem follows.

Next we examine the CORE-CLIQUE problem from the perspective of parameterized complexity (see § 2 and [DF99]). The main principle here is that, although the general form of the problem is NP-complete, our reduction does not disclose exactly where the source of intractability lies. We see at least the following four possible parameterizations of the problem: (1) $m = max_{\forall i}|P_i|$, the maximum size of a partition, (2) k, the number of partitions, (3) r, the total weight of the core-tree, and (4) ω, the maximum weight of a distance between two vertices. Note, Theorem 2 shows that any subset of parameters 1, 3 and 4 are not enough as the problem remains NP-complete. Our next theorem rules out the possibility of an FPT algorithm for any subset of parameters 2, 3, and 4.

Theorem 4. $2, 3, 4$-CORE-CLIQUE *is hard for* $W[1]$.

Proof. Let $(C = (V, E), K)$ be an instance of the K-CLIQUE PROBLEM. We construct an instance of the CORE-CLIQUE problem $(G = (P_1, P_2, \ldots, P_k, E), w, r)$, where r, ω, and k are functions depending only on K and show that (C, K) is a "yes" instance if and only if (G, w, r) is a "yes" instance.

Let the vertices in V_C be labeled by $1, 2, \ldots, |V_C| = m$. Let $r = \binom{K}{2}$. We create partitions $P_1, P_2, \ldots, P_{K=k}$ and include vertices labeled $p_{i,j}$ for $1 \leq j \leq m$ in partition P_i. We place an edge between all vertices in G which are not in the same partition: for all i, j, $1 \leq i < j \leq k$, and for all q, q', $1 \leq q < q' \leq m$, $(p_{i,q}, p_{j,q'}) \in E_G$. If $(u, v) \notin E_C$, then $w(p_{i,u}, p_{j,v}) = c$ for all $1 \leq i < j \leq k$. c is an arbitrarily large constant at least as big as $\binom{K}{2} + 1$. If $(u, v) \in E_C$, then $w(p_{i,u}, p_{j,v}) = 1$ for all $1 \leq i < j \leq k$. For all edges of the form $(p_{i,u}, p_{j,u}) \in E_G$, let $w(p_{i,u}, p_{j,u}) = c$.

We omit the remainder of the (straightforward) argument due to space limitations.

Observe that $1, 2$-CORE-CLIQUE is fixed parameter tractable with an algorithm running in time $O(m^k)$. We simply try all $O(m^k)$ possible ortho-sets.

Theorem 2 shows that the problem remains hard for partition size 3 with constant edge weight functions and a constant bound on the core-clique. Our next theorem shows that restricted to partition size 2 and constant edge weight functions it still stays hard.

Theorem 5. $1, 4$-CORE-CLIQUE *is NP-complete even when the number of vertices in each partition is at most 2 and the edges are assigned a weight of either 0 or 1.*

Proof. Reduction from the MAXIMUM 2SAT problem omitted.

3.2 Most Tree Like. We restrict our attention to the L_∞ norm throughout the following analysis, but note that our reductions also work for other norms. Clearly, the decision version of the CORE-TREE problem, which asks if there is a P' such that $||D(P') - A(D(P'))||_\infty \le \Delta$ for input parameter $\Delta \in \mathbb{R}$, is NP-complete since NUMERICAL TAXONOMY [ABFNPT96] [2] is simply a restricted version (specifically, all partitions having size 1) of it. We begin our analysis with a sub-version of the problem where we ask if there exists a choice of one leaf from each partition in the input graph that induces an additive tree. Furthermore, we are given the unweighted topology of the tree, so the problem reduces to just choosing one vertex per partition so that the pairwise distances fit to the tree. This problem, when each partition just has a single vertex, is not NP-complete [F88].

EXACT TREE IN A k-PARTITE GRAPH (EXACT-CORE-TREE)
input: As with CORE-TREE but also an unweighted leaf-labeled tree T with each leaf receiving a distinct label from $\{P_1, P_2, \ldots, P_k\}$.
question: Does there exist a set $P' = \{p_1, p_2, \ldots, p_k\}$ where $p_i \in P_i$ such that $D(P')$ is additive, where $D(P')$ is the distance matrix formed from P', and such that the corresponding tree $T(D(P'))$ is isomorphic to T and for $u \in T(D(P'))$, $u \in P_i$, the corresponding leaf in T has label P_i.

Again, we analyze this problem from the perspective of parameterized complexity. Our parameters remain the same: (1) $m = max_{\forall i} |P_i|$, the maximum size of a partition, (2) k, the number of partitions, (3) r, the total weight of the core-tree, and (4) ω, the maximum weight of a distance between two leaves. Our first theorem shows that no FPT algorithms are possible for any subset of parameters 2, 3, or 4, unless $W[1] = FPT$.

Theorem 6. $2, 3, 4$-EXACT-CORE-TREE *is hard for* $W[1]$.

Proof. Given an instance of the K-CLIQUE PROBLEM $(C = (V, E), K)$, we create an instance of the $2, 3, 4$-EXACT-CORE-TREE problem (G, T) and show that (C, K) is a "yes" instance if and only if (G, w, T) is a "yes" instance.

We construct $K + 4(= k)$ partitions $\{A, B, C, D, P_1, P_2, \ldots, P_K\}$. Partition A contains one vertex a, B contains b, C contains c, and D contains d. Each partition P_i contains $|V_C| = m$ vertices labeled $p_{i,1}, p_{i,2}, \ldots, p_{i,m}$. Our tree T is created as in Figure 1: the caterpillar with (A, B) and (C, D) as its "head" and "tail". That is, our tree has internal vertices $\{h, t, n_1, \ldots, n_K\}$ with edges $\{(h, A), (h, B), (t, C), (t, D), (h, n_1), (t, n_K)\}$ and $\{(n_i, n_{i+1}) : 1 \le i < K\}$.

Let $D_{a,b} = D_{c,d} = 2$, $D_{a,c} = D_{a,d} = D_{b,c} = D_{b,d} = 4 + (K - 1)$. Let $D_{x,p_{i,j}} = 2 + i$ for $x = \{a, b\}$, $1 \le i \le K$ and $1 \le j \le m$. Let $D_{y,p_{i,j}} = 2 + (K - i + 1)$ for $y = \{c, d\}$, $1 \le i \le K$ and $1 \le j \le m$. Let $D_{p_{i,j},p_{i',j}} = 3K + 10$ for all $1 \le i \ne i' \le K$ and $1 \le j \le m$. If $(u, v) \notin E_C$, then $D_{p_{i,u},p_{i',v}} = 3K + 10$ for all $1 \le i \ne i' \le K$. If $(u, v) \in E_C$, then for $1 \le i < j \le K$, $D_{p_{i,u},p_{j,v}} = 2 + j - i$.

---

[2] NUMERICAL TAXONOMY. **input:** An $n \times n$ distance matrix D, a bound $\Delta \in \mathbb{R}$.
question: Is $||A(D) - D||_\infty \le \Delta$?

($\Rightarrow$) Let $V' = \{v_1, v_2, \ldots, v_K\}$ where $v_i \in V_C$ a clique in C. We show how to choose one vertex from each of the P_i in G such that the distance matrix formed from these vertices alongside with a, b, c and d is additive. Note that we must choose a, b, c and d, and that the distance matrix these four vertices induce is additive (see Theorem 1) and agrees with the topology T.

Now consider the set of vertices $\{p_{1,v_1}, p_{2,v_2}, \ldots, p_{K,v_K}\} = P'$ in G. From the construction, $D_{p_{i,v_i}, p_{j,v_j}} = 2 + j - i$ as any two distinct vertices p_{i,v_i}, p_{j,v_j} from this set are mutually adjacent. We must show how weights can be applied to the edges of T such that the distances in T between p_{i,v_i} and p_{j,v_j}, $d(p_{i,v_i}, p_{j,v_j})$ are equal to the entries $D_{p_{i,v_i}, p_{j,v_j}}$. This can be accomplished by assigning 1 to every edge on the path between p_{i,v_i} and p_{j,v_j} in T. It is easy to verify that $d_T(x, p_{i,v_i}) = D_{x, p_{i,v_i}}$, for $x \in \{a, b, c, d\}$ and that the matrix can be realized as a tree.

($\Leftarrow$) Let $P' = \{a, b, c, d, p_1, p_2, \ldots, p_K\}$ be the set of vertices from G which induces a tree with topology T. By Theorem 1, the underlying distance matrix D is additive. For a leaf vertex x, let $n(x)$ be the unique neighbour of x in T. Focus on the four vertices $\{a, b, c, d\}$. By Theorem 1, the edge weights in this subtree must be 1 for edges of the form $(x, n(x))$ where $x \in \{a, b, c, d\}$. The weight of the path between (a, b) and (c, d) receives weight $2 + (K - 1)$. We now analyze the "choice" of vertices $\{p_1, p_2, \ldots, p_K\}$.

Claim: [No Fit] P' does not contain two vertices $p_{i,j}$ and $p_{i',j}$, $i \neq i'$.

(By contradiction) Suppose there exist $p_{i,j}, p_{i',j} \in P'$ simultaneously (w.l.o.g. $i < i'$). Then, by the construction, $D_{p_{i,j}, a} = 2 + i$, $D_{p_{i,j}, c} = 2 + (K - i + 1)$, $D_{a,b} = 2$ and $D_{p_{i,j}, p_{i',j}} = 3K + 10$. Focus on the quartet formed by $\{a, b, p_{i,j}, c\}$. It is easy to verify that the edge $(p_{i,j}, n(p_{i,j}))$ must have weight 1. Furthermore, the path from vertex (AB) to $n(p_{i,j})$ must have total weight i and the path from vertex (CD) to $n(p_{i,j})$ must have weight $K - i + 1$. The same argument holds for the edge weights in the quartet $\{a, b, p_{i',j}, c\}$, that is, the edge weight of $(p_{i',j}, n(p_{i',j}))$ is also 1. Allowing $n(x)$ to denote the unique neighbor of a leaf vertex x in T, it is easy to verify that the weight of the path from $n(p_{i,j})$ to $n(p_{i',j})$ must be $i' - i$. Since $i' - i + 2 < 3K + 10$, we reach a contradiction since we can not assign edge weights to T so that they agree with the distance matrix induced by $\{a, b, c, d, p_{i,j}, p_{i',j}\}$. Hence, by Theorem 1, this matrix is not additive.

Claim. P' does not contain two vertices $p_{i,j}$ and $p_{i',j'}$, $i < i'$, $j \neq j$, such that $(v_j, v_{j'}) \notin E_C$.

This claim can be proved in the same way as Claim *No Fit* above. Simply note we assigned $D_{p_{i,j}, p_{i',j'}}$ to be $3K + 10$ when $(v_j, v_{j'}) \notin E_C$.

The previous two claims establish the fact that we must include K distinct vertices in G which correspond to pairwise adjacent vertices in C. Hardness for $W[1]$ follows from the fact that our construction required only $K + 4$ partitions, all edge weights are a function only of K and the overall weight of the clique-tree is also a function only of K.

Our second theorem shows that this problem is NP-complete even when the number of candidate homologous genes per genome is at most 3.

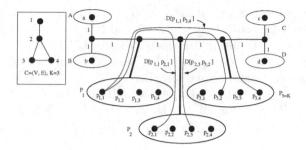

Fig. 1. Construction for the 2, 3, 4-EXACT-CORE-TREE.

Theorem 7. 1-EXACT-CORE-TREE *restricted to partitions of size* 3 *is NP-complete.*

Proof. Proof omitted due to space limitations but uses many of the same ideas from the proof above. We reduce for 3SAT and map each clause to a partition with extra partitions A, B, C, D as above.

Parameterizing on both the number of partitions k and the size of each partition m leads to a trivial FPT algorithm for 1, 2-CORE-TREE with a running time of $O(m^k)$.

Consider the relaxation of EXACT-CORE-TREE to the optimization version which asks for the core-set P' which best fits to the topology T and we modify this optimization criteria so that it is always > 0, we can prove the following non-approximation results via Theorem 7:

Theorem 8. *The always positive, optimization version of* EXACT-CORE-TREE *is not approximable within any function of n in polynomial time, where n is the size of the graph G, unless $P = NP$.*

Proof. Similar to Theorem 3.

4   A Randomized Approximation Algorithm for the CORE-CLIQUE Problem

Following [GGR98], we will now give a randomized approximation algorithm for the CORE-CLIQUE problem. The algorithm runs in linear time if each P_i has size bounded by a constant m, and polynomial time in the general case. Let $\sigma(G, w)$ denote the maximum difference between the weights of two edges adjacent to a vertex v, over all vertices v of G and its adjacent edges.

Theorem 9. *For any $\epsilon, \delta \in (0, 1)$, there is a randomized algorithm for the* CORE-CLIQUE *problem that for a given instance G, w with probability $\geq 1 - \delta/2$ in polynomial time finds a solution of cost $\leq c^* + \epsilon\sigma(G, w)k^2$, where c^* is the cost of the minimum cost core-clique.*

Consider a given CORE-CLIQUE instance G, w and let $\sigma = \sigma(G, w)$. Let ϵ, our *distance parameter*, be such that $0 < \epsilon < 1$ and δ, our *confidence parameter*, be such that $0 < \delta < 1$. We use $[k]$ to denote the set $\{1, 2, \ldots, k\}$.

Let $l = \lceil 8/\epsilon \rceil$ and $t = \Theta(\frac{1}{\epsilon^2} \log \frac{1}{\epsilon \delta})$. Consider a partition of $[k]$ into l sets $A_1, \ldots, A_l$ of approximately equal size. Let $V_j = \cup_{i \in A_j} P_i$ and $W_j = V(G) \setminus V_j$. For $U = U_1, \ldots, U_l$ where $U_j \subseteq [k] \setminus A_j$, let $X(U_j)$ be the family of all $X \subseteq W_j$ such that $|X \cap P_i| = 1$ for all $i \in U_j$ and $X \cap P_i = \emptyset$ for $i \notin U_j$, and let $X(U) = \{(X_1, \ldots, X_l) : X_j \in X(U_i), i = 1 \ldots l\}$.

ALGORITHM RANDOMIZED A

1. Choose $U = U_1, \ldots, U_l$ where U_j has size t and is chosen uniformly in $[k] \setminus A_j$.
2. For each $X \in X(U)$
3. Let

$$O^X = \{\mathrm{argmin}_{v \in P_i} w(v, X_j) : 1 \le j \le l, i \in A_j\}.$$

4. Output the core-clique O^X which has minimum weight over all $X \in X(U)$.

We will denote the minimum cost core-clique by O^*.

Lemma 2. *With probability $1 - \delta/2$ over the choice of U there is an $X \in X(U)$ such that $w(O^X) \le w(O^*) + \epsilon \sigma k^2/2$.*

Proof. The proof is omitted due to space limitations, but very similar proofs can be found in [GGR98].

ALGORITHM RANDOMIZED B

1. Choose $U = U_1, \ldots, U_l$ where U_j has size t and is chosen uniformly in $[k] \setminus A_j$.
2. Uniformly chose a subset $C = \{c_1, \ldots, c_r\}$ of even size $\Theta(\frac{lt \log m + \log(1/\delta)}{\epsilon^2})$ from $[k]$.
3. For each $X \in X(U)$
4. For each $i \in C$, let

$$v_i^X = \mathrm{argmin}_{v \in P_i, 1 \le j \le l, i \in A_j} w(v, X_j).$$

5. Output the tuple X which minimize

$$\sum_{i=1}^{r/2} w(v_{2i-1}^X, v_{2i}^X)$$

over all $X \in X(U)$.

The final version of our algorithm does the following. It computes a tuple X using Algorithm Randomized B and then outputs the core-clique $O = \{\operatorname{argmin}_{v \in P_i} w(v, X_j) : 1 \le j \le l, i \in A_j\}$. Since

$$2 \sum_{i=1}^{r/2} w(v_{2i-1}^X, v_{2i}^X)/r$$

has expected value $w(O^X)/k^2$, it follows that

$$\operatorname{Pr}_C[|2 \sum_{i=1}^{r/2} w(v_{2i-1}^X, v_{2i}^X)/r - w(O^X)/k^2| > \epsilon\sigma/4] \le e^{-\Theta(\epsilon^2 r)} \le O(\delta m^{-lt}).$$

Since $|X(U)| \le m^{lt}$, it follows that

$$\operatorname{Pr}_C[\forall X \in X(U), |2 \sum_{i=1}^{r/2} w(v_{2i-1}^X, v_{2i}^X)/r - w(O^X)/k^2| < \epsilon\sigma/4] \ge 1 - \delta/2.$$

Discussion Our experimental results for a family of suspected Rubredoxin proteins suggest that the CORE-CLIQUE optimization critera *does* in fact allow us to distinguish families of analogously functioning genes.

All of the algorithms mentioned in this paper have been implemented and tested. We note that our randomized approximation algorithm performs best when the input graph is quite large. We have also tried a number of greedy and randomized greedy heuristics for these problems and we have found that these simple heuristics tend to out-perform our randomized approximation algorithm in practice. There are a number of ways that ideas in the approximation algorithms can be used to derive more advanced heuristics (dominating the simpler ones) and possibly more practical algorithms with proven performance bounds. This is certainly a very challenging line of research that needs further consideration.

References

[ABFNPT96] Agarwala, R. et. al. (1996) On the approximability of numerical taxonomy. In: *Proceedings of the Seventh Annual ACM-SIAM Symposium on Discrete Algorithms*, 365–372.

[AGMWL90] Altschul, S. F. et. al. (1990) Basic local alignment search tool. *J. Mol. Biol.*, 215, 403-410.

[BA] Bairoch, A. and Apweiler, R.(1999) The SWISS-PROT protein sequence data bank and its supplement TrEMBL in 1999. *Nuc. Acids Res.*, 27, 49-54.

[BSFDHW00] an Beilen, J. B., Smits, T., Franchini, A., Disch, T., Hallett, M., Witholt, B. (2000) Two types of rubredoxins involved in alkane oxidation. To be submitted to *Gene*. ETH Zürich.

[B92] Benner, S. A. (1992) Predicting de novo the folded structure of proteins. *Current Opinion in Structural Biology*, 2:402–412.

[B99] Benner, S. A. (1998) Personal communication.

[BDDEHY98] Bork, P. et. al. (1998) Predicting function: from genes to genomes and back. *J. Mol. Biol.* 283, 707–725.

[B71] Buneman, P. (1971) The recovery of trees from measures of dissimilarity. In: *Mathematics in the Archaeological and Historical Sciences*, F. R. Hodson, D. G. Kendall, P. Tauto, eds.: Edinburgh University Press, Edinburgh, 387–395.

[DF99] Downey, R. and Fellows, M. R. (1999) Parameterized Complexity. *Springer Verlag*, New York.

[F88] Felsenstein, J. (1988) Phylogenies from molecular sequences: inference and reliability. *Annual Revue of Genetics*, 22, 521-565.

[GGR98] Goldreich et. al. (1998) Property testing and its connection to learning and approximation. *J. of the ACM*, 45:4, 653–750.

[GCMRM79] Goodman, M. et. al. (1979) Fitting the Gene Lineage into its Species Lineage: A parsimony strategy illustrated by cladograms constructed from globin sequences, *Syst.Zool.*, 28.

[GMS96] Guigó, R. et. al. (1996) Reconstruction of Ancient Molecular Phylogeny. *Molec. Phylogenet. and Evol.*, 6(2), pp. 189–213, 1996.

[HL99a] Hallett, M. T. and Lagergren. J. (1999) Hunting for Functionally Analogous Genes: Cores of Partite Graphs (Full Paper). (1999) Tech. Report ETH Zürich, No. 327.

[HL99b] Hallett, M. T. and Lagergren. J. (2000) New Algorithms for the Duplication-Loss Model. *RECOMB '00*, Tokyo, Japan, p. 136–148.

[H63] W. Hoeffding. (1963) Probability inequalities for sums of bounded random variables. *Amer. Statist. Assoc. J.*, 58, 13–30.

[KTG98] Koonin, E. V. et. al. (1998) Beyond complete genomes: from sequence to structure and function. *Curr Opin Struct Biol*, 8(3), 355-63.

[MMS95] Mirkin, B. et. al. (1995) A biologically consistent model for comparing molecular phylogenies. *Journal of computational biology*, 2(4), 493–507.

[P98] Page, R. (1998) GeneTree: comparing gene and species phylogenies using reconciled trees. *Bioinformatics*, 14(9), 819–820.

[PC97] Page, R. and M. Charleston, M. (1997) From Gene to organismal phylogeny: reconciled trees and the gene tree/species tree problem. *Molec. Phyl. and Evol.* 7, 231–240.

[PL88] Pearson, W. R. and Lipman, D. J. (1988) Improved tools for biological sequence comparison. *Proc. Natl. Acad. Sci.*, 85:2444-2448.

[RS96] Rubinfeld, R. and Sudan, M. (1996) Robust characterization of polynomials with applications to program testing. *SIAM J. Comput.* 25, 2, 252-271.

[SN87] Saitou, N. and Nei, N. (1987) The neighbour-joining method: a new method for reconstructing phylogenetic trees. *Mol. Biol. Evol.*, 4, pp. 406–425, 1987.

[SM98] Slonimski et. al. (1998) The first law of genomics. *Abstract "Microbial Genomes II"*, Hilton Head, January.

[S99] Stoesser, G. et. al. (1999) The EMBL Nucleotide Sequence Database. *Nuc. Acids Res.*, 27(1), 18-24.

[TKL97] Tatusov, R. L. et. al. (1997) A genomic perspective on protein families. *Science*, 278(5338), 631-7.

[YEVB98] Yuan, Y. P. et. al. (1998) Towards detection of orthologues in sequence databases. *Bioinformatics*, 14(3), 285–289.

Keeping Track of the Latest Gossip in Shared Memory Systems

Bharat Adsul, Aranyak Mehta, and Milind Sohoni

Dept of Computer Science & Engineering,
Indian Institute of Technology, Mumbai 400 076, India
{abharat, aranyak, sohoni}@cse.iitb.ernet.in

Abstract. In this paper we present a solution to the 'Latest Gossip Problem' for a shared memory distributed system. The Latest Gossip Problem is essentially one of bounded timestamping in which processes must locally keep track of the 'latest' information, direct or indirect, about all other processes. A solution to the Latest Gossip Problem is fundamental to the understanding of information flow in a distributed computation, and has applications to problems such as global state detection and mutual exclusion. Our solution is along the lines of that for message passing systems in [6], and for synchronously communicating systems [8].

Our algorithm uses a modified version of the *consume* and *update* protocols of Dwork and Waarts [3], where these were introduced to construct a 'Bounded Concurrent Timestamping System (BCTS)'. As applications of our Gossip Protocol, we also indicate another construction of a BCTS and a solution to the global state detection problem, which, we believe, are improvements over older solutions.

1 Introduction

Consider a distributed system of processes which communicate through protocols utilizing shared memory. The *online latest gossip problem* for this system is the following:

> Whenever a process q reads the information written by another process p, q should be able decide which of p and q has more recent information, direct or indirect, about r, for every other process r in the system.

Once q makes this decision, it can systematically collate and update this information to maintain, on-line, its 'latest gossip' about every other process. The latest gossip provides crucial information to each process about the unfolding of the global computation.

Note that there is a trivial solution to the latest gossip problem if we allow unbounded labels. For example, each process could label its writes from the set of integers, labeling its i-th write operation with the number i. This unique labeling of the writes also reflects their temporal order. Whence the process q, on reading p, may compare the r-label that p holds, with the r-label that q

S. Kapoor and S. Prasad (Eds.): FST TCS 2000, LNCS 1974, pp. 477–488, 2000.

holds and update its latest information on r. Unfortunately, as time progresses, the labels increase in size without bounds. To eliminate this problem we need an unambiguous labeling of the writes from a finite set of labels. This requires a careful reuse of 'old' labels and a method of comparing 'latest information' using these labels. In our solution to the problem, with each write operation, processes attach a bounded 'gossip' information. The 'gossip' at each process is a selection of a finite number of operations in its recent past arranged in a suitable structure, and constitutes the bounded timestamp.

Bounded timestamping protocols to solve the latest gossip problem have been exhibited for other distributed models. In [8] the authors introduce the latest gossip problem and present a solution in a distributed system where processes use handshake communication to synchronize and exchange information. This solution has also been extended to the message-passing model in which processes exchange information through messages sent from process to process [6]. However, there the underlying computation is restricted to that in which, at every instant, the number of 'unacknowledged messages' is uniformly bounded. As shown there, this requirement may be implemented by each process by waiting for acknowledgments to come in. It is argued in [7] that for applications which require robust solutions such as *mutual exclusion*, such wait is essential.

With the same restriction on computations, the solution in [6] may be adapted to the shared memory model. However, in this model, the results of [7] do not hold: there are many situations in which computations with unbounded unacknowledged writes, are essential. We present here, protocols for reading and writing shared memory such that gossip information may be maintained in the general case, without any restrictions on the underlying computation. In other words, the underlying computation may run 'wait-free', and maintain gossip at the same time. However, this is not without cost: Each request by the computation for an operation on the shared memory unfolds into a *sequence* of the atomic read and write operations of the underlying system. Furthermore, a few additional synchronization variables are required to implement the label. All the same, the original computation proceeds wait-free and with identical semantics.

The protocols by which a computation may read or write into shared memory are called READ and WRITE respectively. These protocols are simple adaptations of those presented by Dwork and Waarts [3] in their construction of a Bounded Concurrent Timestamping System (BCTS). In return, our solution indicates another construction of a BCTS in which we permit any number of hops (indirection) in the flow of information. The solution to the BCTS as presented in [3] also has the following drawback. Processes label their writes from a pool of available timestamps. Once this pool is exhausted, the process performs a 'garbage collection' operation to replenish its pool of usable timestamps. Thus a process must stop work every once in a while and perform garbage collection. Our modification eliminates the need for garbage collection. Processes can keep track of usable timestamps 'on-line' during their normal operation in such a way that the pool of available timestamps is never exhausted.

2   The Model

We begin by describing the distributed system as an *abstract machine*. Let $\mathcal{P} = \{p_1, p_2, \ldots, p_N\}$ be a set of N processes which collaborate via shared memory. The shared memory is segmented into N parts, with the part M_p associated with the process p. The 'public' memory M_p may be read by any process, while it may be written into only by process p. Besides this, each process also has some *private* memory L_p which is operable only by p. The memories L_p and M_p will be called p's local memory, and may be further organized into a sequence of variables, say, $M_p = \{p : \mathtt{X}_1, \ldots, p : \mathtt{X}_r\}$ and $L_p = \{p : Y_1, \ldots, p : Y_s\}$. We use the typewriter font for public variables, and the *italic* font for private variables.

The processes are allowed to manipulate their public memory via two atomic operations – read and write. The read operation can be performed by a process p to read some public variable of another process q. The value gets written into some private variable of p. The write operation enables a process p to copy the contents of a local private variable to a local public variable. An operation of the machine is the occurrence of an operation in one of the processes. A computation is a (temporal) sequence of such operations, with obvious semantics. For any computation, there is the *causality* partial order on the operations in the computation. For operations e, f in a computation, we say that $e \leq f$ if the outcome of e may affect the outcome of f. The locality of each operation to just one or two processes usually makes causality coarser than the total order.

Each process p runs a program written in a suitable programming language. This program may use its local private memory L_p as it chooses. However, the local public memory M_p may be accessed by two prescribed protocols: (i) WRITE$(p : \mathtt{X}, p : Y)$, which acts to copy the contents of $p : Y \in L_p$ into $p : \mathtt{X} \in M_p$ and (ii) READ$(q : \mathtt{X})$, which acts to copy the public variable $q : \mathtt{X}$ of another process q into the local private memory of p. Any execution of the distributed program must result in a computation. In particular, the above protocols must be 'compiled' into the elementary operations allowed by our machine.

Our task is to supply protocols for READ and WRITE which will enable any computation arising from any program to correctly compute the latest information, online. The simplest possibility is to replace READ and WRITE by the atomic read and write operations of the machine. However, in this case, unambiguous timestamping from a bounded set becomes impossible, *e.g.*, consider the situation in which a process p performs only WRITEs and no READs. Since p will never know which WRITEs have been READ by other processes, it must label each successive WRITE with a distinct label, eventually exhausting its finite set of labels. This points to the need for some elaborate handshaking protocols for READ and WRITE.

3   The Protocol

We first present our adaptations of the *update* and *consume* protocols of [3]. Next, we describe the protocols, READ and WRITE, which use the above consume and update. Henceforth, for simplicity, we assume that the distributed program requires only one public variable for each process, *viz.*, $p : \mathsf{X}$.

3.1   Update and Consume Protocols

Each process p maintains, for every other process q, the following *additional* public variables: (i) two public bit-variables: $p : \mathsf{demand}_q$ and $p : \mathsf{supply}_q$, (ii) two public variables $p : \mathsf{B1}_{pq}$ and $p : \mathsf{B2}_{pq}$, and (iii) a public variable $p : \mathsf{A}_{pq}$. These variables are protocol variables used only in the protocol and may not be used by the distributed program.

For a process p, the protocol **update**$(p : \mathsf{X}, p : Y)$ writes the value of a private variable $p : Y$ in the public variable $p : \mathsf{X}$. On the other hand, the protocol **consume**$(q : \mathsf{X})$ reads q's public variable $q : \mathsf{X}$ and copies it into p's private memory (into the private variable $p : temp$). Roughly speaking, the bit **demand** is used to indicate a desire of one process to read the public memory of another, and **supply**, its satisfaction. During an update, if p discovers unsatisfied **demand**, it proceeds to *set aside* a copy of its public variable. In the consume protocol, the process p first raises **demand**, and then proceeds to read $q : \mathsf{X}$ successively. If it notices a stable value, then the consume exits, declaring this stable value as the contents of $q : \mathsf{X}$. If this fails, then p reads the value set aside by q, and declares this as the contents of $q : \mathsf{X}$. The detailed protocols are described below.

---

consume$(q : \mathsf{X})$

Perform handshake
1. Read $q : \mathsf{supply}_p$
2. Write $p : \mathsf{demand}_q = \neg q : \mathsf{supply}_p$

Remainder of the first Read-Write-Read (RWR)
3. Read $q : \mathsf{X}$
4. Write $p : \mathsf{B1}_{pq} := q : \mathsf{X}$
5. Read $q : \mathsf{X}$
6. If (the label of) $q : \mathsf{X}$ is unchanged since Line 3 then $p : temp = p : \mathsf{B1}_{pq}$, else

Remainder of the second Read-Write-Read (RWR)
7. Write $p : \mathsf{B2}_{pq} := q : \mathsf{X}$
8. Read $q : \mathsf{X}$
9. If (the label of) $q : \mathsf{X}$ is unchanged since Line 5 then $p : temp = p : \mathsf{B2}_{pq}$ else

Read Set-Aside
10. Read $q : \mathsf{A}_{qp}$
11. $p : temp = q : \mathsf{A}_{qp}$

---

update$(p : \mathsf{X}, p : Y)$

1. For each $q \neq p$, read $q : \mathsf{demand}_p$
2. For each $q \neq p$, if $q : \mathsf{demand}_p \neq p : \mathsf{supply}_q$ then $p : \mathsf{A}_{pq} = p : \mathsf{X}$
3. Atomically write $p : \mathsf{X} = p : Y$ and for each q, $p : \mathsf{supply}_q = q : \mathsf{demand}_p$

Consume and update operations performed by a process p are called p-consume and p-update respectively. The *label* in statement 6 (resp. 9) of the p-consume protocol above refers to labels read in statements 3 and 5 (resp. 5 and 8) and which were attached to the quantity $q : X$ by process q during its q-updates. This *labeling* is the central construction of this paper. For the moment, we assume that if any two of the $q : X$ read in statements 3, 5 and 8 were written in distinct q-updates then their labels would also be distinct. Next, note that a consume $\mathcal{C}$ may have had a 'successful' RWR on the information written by an update $\mathcal{U}$, or it may have read the information set aside by a later $\mathcal{U}'$ but of an earlier $\mathcal{U}$. In either case the consume $\mathcal{C}$ is said to have *succeeded* on the update $\mathcal{U}$.

3.2 The READ and WRITE Protocols

The protocols for READ and WRITE appear below. Note that each WRITE contains exactly one update operation. Whence, we may define the label of a WRITE to be that of the update within. Also, note that the consume and the update operations are *wait-free* - processes can consume and update independent of the state of other processes. Whence, the READ and WRITE are also *wait-free*.

READ($q : X$)	WRITE($p : X, p : Y$)

1. **consume**($q : X$)	1. For each process $q \neq p$,
	     Read $q : \text{B1}_{qp}$ and $q : \text{B2}_{qp}$
	     consume($q : X$)
	 2. **update**($p : X, p : Y$)

3.3 Causal Order

Suppose, now, that each process executes a (distributed) program which, besides accessing its local private variables, manipulates its public memory by the READ and WRITE operations. Any finite (partial) computation of this program will result in a sequence of READ/WRITE operations. These operations will henceforth be referred as *program events*, or simply *events*. The events at the process p will be called p-events.
For any two events e and e', we say that $e \sqsubset e'$ iff

- Either: e is a WRITE q-event and e' is a READ p-event and $\mathcal{C}'$ succeeds on $\mathcal{U}$, where $\mathcal{U}$ is the update in e and $\mathcal{C}'$ is the consume operation in e'. In this case e' is called an external successor of e. We also say that e' succeeds on e.
- Or: e is a WRITE q-event and e' is a WRITE p-event and $\mathcal{C}'$ succeeds on $\mathcal{U}$, where $\mathcal{U}$ is the update in e and $\mathcal{C}'$ is the consume($q : X$) operation in e'. In this case also e' is called an external successor of e. We also say that e' succeeds on e.
- Or: both e and e' are p-events, and e' is the very next p-event after e. In this case e' is called an internal successor of e.

Define $e \sqsubseteq e'$ if $e = e'$ or $e \sqsubset e'$. Define $\sqsubseteq^*$ as the reflexive transitive closure of $\sqsubset$. The partial order $\sqsubseteq^*$, henceforth referred as *causal order*, records the information we require about causality and independence between events. For any event e, we denote by $e{\downarrow}$ the (partially ordered) set of events e' such that $e' \sqsubseteq^* e$. For a process p and a p-event e_p, $e_p{\downarrow}$ is the 'local view' of p at e_p.

Note that the READ and WRITE calls can be 'compiled' into the elementary `read` and `write` operations using the protocols mentioned above. Hence any execution of the distributed program will result in a computation of our abstract machine. The causal order $\sqsubseteq^*$ is really a coarsening of the order $\leq$ on the atomic operations of the abstract machine. Indeed, the order $\sqsubseteq^*$ between program events may be 'observed' as the relation $\leq$ between specific critical atomic operations encountered during the execution of those events. We state here two interesting properties satisfied by the causal order:

Regularity If a p-event e succeeds on a q-event e' then there is no WRITE q-event e'' after e' which finishes before the **consume**$(q : X)$ in e starts. In a sense this means that the READ/WRITE events do not read very old information.

Monotonicity If two p-events e and e' succeed on two q-events f and f' respectively, then if e' occurs after e then f' can not occur before f.

These properties follow from the corresponding properties of the consume and update operations [3].

Let p, q be two processes and let e be a p-event. We define $latest_{p \to q}(e)$ as the $\sqsubseteq^*$-maximum q-event f such that $f \sqsubseteq^* e$. This is the latest information that p has about q after e. Note that $latest_{p \to q}(e)$ need not exist if there is no q-event in $e{\downarrow}$. We set $latest_{p \to p}(e)$ to e itself. Observe that, for $q \neq p$, $latest_{p \to q}(e)$ is always a WRITE q-event.

At any point during a computation, let e_p be the last p-event. Similarly let e_q be the last q-event. Let e_1, e_1' denote $latest_{p \to r}(e_p)$ and $latest_{q \to r}(e_q)$ respectively. Suppose that $e_1 \sqsubseteq^* e_1'$. Consider a path $e_1 \sqsubset e_2 \cdots \sqsubset e_k \cdots \sqsubset e_p$ from e_1 to e_p. Let the next WRITE operation in the path after e_1 be e_k, an s-event. Then we have the following lemma. Refer [1] for the proofs.

Lemma 1 *If e_1 is maximal (w.r.t. $\sqsubseteq^*$) in $e_p{\downarrow} \cap e_q{\downarrow}$, then e_k finishes after the consume(s : X) in e_1' starts.*

4   The Gossip Algorithm

The Gossip Algorithm requires each process to maintain a pre-specified set of events in its recent past. For each process p, besides the events $latest_{p \to q}$, for all q, this set contains some additional auxiliary 'unacknowledged' events. These events and the $\sqsubseteq^*$-relationship between them constitute the *primary graph*. Let us suppose that, at the end of a certain computation, the events within the primary graphs of each process have been labeled distinctly. If the next event is, say an event in which p READs q, then (i) p may correctly compare its latest gossip with that of q's, and if it is, say a WRITE event of p, then (ii) provided p can discover a suitable label, it may update its primary graph, maintaining the distinctness invariant. The hunt for a suitable label is aided by *secondary information*.

4.1   Primary Information

Let e be a p-event. We denote by $latest(e) = \cup_{q \in \mathcal{P}} latest_{p \to q}(e)$, the latest information p has about all processes, at e.

To maintain and update the latest information of processes, we need to expand the set of events that each process keeps track of. This expanded set is the *primary information*. The primary information of a process contains not only its latest information about every other process but also information about unacknowledged WRITEs.

Unacknowledgment Recall that when a process q performs a consume$(p : \mathtt{X})$ operation $\mathcal{C}$, it writes twice a temporary value (with a label) in its public memory (this is the 'W' of the RWRs of the consume). This information is written in the public variables $\mathtt{B1}_{qp}$ and $\mathtt{B2}_{qp}$. These are the variables that p reads in its WRITE events. Also recall that a process p may, during an update, set aside the previous event for another process q in the variable $p : \mathtt{A}_{pq}$. Now given a fixed run of READ and WRITE events we have, for every pair of processes p and q, functions $\mathcal{F}_{p,q,\mathtt{B1}}, \mathcal{F}_{p,q,\mathtt{B2}}$ and $\mathcal{F}_{p,q,\mathtt{A}}$ as follows:

- For $\mathtt{i} = 1, 2$, $\mathcal{F}_{p,q,\mathtt{Bi}}$ is a function from the set of all WRITE p-events to itself. For any WRITE p-event e, $\mathcal{F}_{p,q,\mathtt{Bi}}(e)$ is the WRITE p-event whose label was written by q during some q-consume in its $q : \mathtt{Bi}_{qp}$ variable and read by p during e from the $q : \mathtt{Bi}_{qp}$ variable of q.
- $\mathcal{F}_{p,q,\mathtt{A}}$ is a function from the set of all WRITE p-events to itself. $\mathcal{F}_{p,q,\mathtt{A}}(e)$ is the event which was last set aside by p for q at or before e.

For a process p and for any WRITE p-event e, we define the set $unacklist(e)$ as the set of the following events:

- For each $q \neq p$, $\mathcal{F}_{p,q,\mathtt{B1}}(e)$, $\mathcal{F}_{p,q,\mathtt{B2}}(e)$ and $\mathcal{F}_{p,q,\mathtt{A}}(e)$,
- The previous WRITE p-event, and
- The event e itself.

Note that the size of $unacklist(e)$ is bounded above by $3N$, where N is the number of processes.

For a p-event e, we define $unack_r(e)$ to be the set $unacklist(latest_{p \to r}(e))$ for $r \neq p$. Define $unack_p(e) = unacklist(e')$ where e' is the $\sqsubseteq^*$-maximum WRITE p-event in $e \downarrow$, that is, the last WRITE p-event before e. We set $unack(e) = \cup_{r \in \mathcal{P}} unack_r(e)$. Note that $unack(e)$ is defined for WRITE as well as READ events. The following lemma relates the causal order with unacklists.

Lemma 2 *Let e_1 and e_2 be WRITE r-events and e_s be a WRITE s-event for some two processes r and s. Suppose $e_1 \sqsubseteq^* e_2$ and $e_1 \sqsubset e_s$. If it is not that $e_s \sqsubseteq^* e_2$ then $e_1 \in unacklist(e_2)$.*

The *primary information* of a process p after an event e consists of $latest(e)$ and $unack(e)$. Events in the primary information are called *primary events* at e, and the set is denoted by $primary(e)$. Note that the size of $primary(e)$ is bounded above by $3N^2+N$, $3N^2$ elements in $unack(e)$ and N elements in $latest(e)$.

To compare and update primary information, processes also need to remember how their primary events are ordered by $\sqsubseteq^*$.

Primary graph The *primary graph* of a process p after an event e, denoted by $primary\text{-}graph(e)$, is the directed graph (V, E) where:

- $V = $ the set of primary events at e.
- For $v_1, v_2 \in V$, let e_1 and e_2 be the corresponding primary events. Then, $(v_1, v_2) \in E$ iff $e_1 \sqsubseteq^* e_2$.

The primary graph is the basic structure which is recursively maintained by each process during a computation. We assume that, at any point during a computation, the events in the primary graphs of the processes have received distinct labels. In other words, if e_p and e_q are two $\sqsubseteq^*$-maximal events, and $e, f \in primary(e_p) \cup primary(e_q)$, then $e \neq f$ iff $label(e) \neq label(f)$. Each process p maintains this graph as one on labels and writes it into the public variable $p : \mathtt{X}$ during every WRITE event, thus making it available to other processes.

4.2   Comparing Primary Information

Let e_q be a q-event and e_p a WRITE p-event, such that $e_p \sqsubset e_q$, that is, e_q is an external successor of e_p. Let e'_q be the q-event which immediately precedes e_q. Let $I_p = e_p\!\downarrow$ and $I_q = e'_q\!\downarrow$. In general, before the occurrence of e_q, the processes p and q will have incomparable information. The events known to both p and q lie in $I_p \cap I_q$. Events lying 'above' $I_p \cap I_q$ are known to only one of the processes.

Let us assume that p has computed $primary\text{-}graph(e_p)$ at the end of e_p, and that q has computed $primary\text{-}graph(e'_q)$ at the end of e'_q. When q successfully reads $primary\text{-}graph(e_p)$, it will have to compare that latest information with the latest information it itself knows and has kept in $primary\text{-}graph(e'_q)$.

Now if $e_p = latest_{q \to p}(e'_q)$, then there is nothing to do, as no new information has reached q from p in e_q. So the interesting case is when e_q succeeds on a newer p-event. Our first observation is that if q knows both $primary\text{-}graph(e_p)$ and $primary\text{-}graph(e'_q)$, it can 'determine' $I_p \cap I_q$, the events in I_p which q already knew, before e_q.

Lemma 3 *For each maximal event e ($\neq e_p, e'_q$) in $I_p \cap I_q$, either $e \in latest(e_p) \cap latest(e'_q)$ or $e \in latest(e_p) \cap unack(e'_q)$ or $e \in unack(e_p) \cap latest(e'_q)$.*

Thus, when q reads p's primary graph, q can collect together in a set M all the events that lie in $latest(e_p) \cap latest(e'_q)$, $latest(e_p) \cap unack(e'_q)$ and $unack(e_p) \cap latest(e'_q)$. By the preceding lemma, the events in M subsume the maximal events in the intersection $I_p \cap I_q$. (It is easy to see that those events in M which are not actually maximal still lie within the intersection.)

Process q can use M to check whether a primary event $e \in primary(e_p) \cup primary(e'_q)$ lies inside or outside the intersection — e lies inside the intersection iff it lies below one of the elements in M. These comparisons can be made using the edge information in the graphs $primary\text{-}graph(e_p)$ and $primary\text{-}graph(e_q)$.

Now, it is easy for q to compare the events in $latest(e_p)$ with those in $latest(e'_q)$ to determine which of p and q have more recent information about every other process r.

Lemma 4 Let $e = latest_{p \to r}(e_p)$ and $f = latest_{q \to r}(e'_q)$ such that $e \neq f$. Then, $e \sqsubseteq^+ f$ iff $f \in I_q - I_p$.

Once q has compared all events of the form $latest_{p \to r}(e_p)$ and $latest_{q \to r}(e'_q)$, it can easily update its sets $unack_r(e'_q)$. The process which has better information about r also has better information about r's unacklist. In other words, q inherits the set $unack_r(e_p)$ for every process r such that $latest_{p \to r}(e_p)$ is more recent than $latest_{q \to r}(e'_q)$. On the other hand, if $latest_{p \to r}(e_p)$ is older than $latest_{q \to r}(e'_q)$, then q ignores p's set $unack_r(e_p)$ since it already has better information about these events. Furthermore, q has the latest $unack_q(e_q)$ with it, and does not need to update this set.

At this stage, q has updated its primary information and formed $primary(e_q)$ using the information in $primary\text{-}graph(e_p)$ and $primary\text{-}graph(e'_q)$. We now need to extend this set to the graph $primary\text{-}graph(e_q)$.

Let $f_1, f_2 \in primary(e_q)$. If both f_1 and f_2 came from $primary(e_p)$, then we add an edge from f_1 to f_2 in $primary\text{-}graph(e_q)$ iff a corresponding edge existed in $primary\text{-}graph(e_p)$. A symmetric situation applies if both f_1 and f_2 were contributed by $primary(e_q)$. So, the only interesting case is when f_1 and f_2 originally came from different processes. Without loss of generality, suppose that f_1 came from $primary(e_p)$ and f_2 from $primary(e_q)$. Then, from the method which we used to compare events, we know that f_1 must have been in $I_p - I_q$ and f_2 must have been in $I_q - I_p$. So, it is clear that f_1 and f_2 are unordered and there is therefore no edge between them in $primary\text{-}graph(e_q)$.

We now have the proposition:

Proposition 5 Let e_q be a q-event and e_p a p-event, such that e_q succeeds on e_p (that is, e_q is an external successor of e_p). Let e'_q be the q-event just preceding e_q. Then, q can construct $primary\text{-}graph(e_q)$ from the graphs $primary\text{-}graph(e_p)$ and $primary\text{-}graph(e'_q)$.

Notice that the procedure for updating primary graphs only checks the labels of events which actually lie in the primary graphs. Call a q-event e 'current' if e belongs to $primary(e_p)$ for the last ($\sqsubseteq^*$-maximum) event e_p of some process p. Recall that N is the number of processes in the system. We know that there are at most $3N^2+N$ distinct events in $primary(e_p)$ for process p. So, at any given time, the number of events across the system which are current is bounded by $N(3N^2+N)$.

Each event begins by being current. Meanwhile, as the computation progresses, this event may get added to the primary information of other processes. However, it gradually recedes into the past, until it drops out of the primary information of *all* processes. At this time, the label assigned to this event can be reused, since the old event with the same label can *never* become current again.

A process can keep track of which of its events in the system are current by keeping track of one additional level of events, called secondary information.

4.3   Secondary Information

Consider a p-event e for some process p. The *secondary information* of p at e is the collection of sets $primary(f)$ for each event f in $primary(e)$. This collection of sets is denoted $secondary(e)$.

The following lemma says that the only p-events which can be current in the system are those which occur in p's secondary information.

Lemma 6 *Let, at any time, e_p be the last p-event, and e_q the last q-event, for some processes p and q. Let $I_p = e_p\!\downarrow$ and $I_q = e_q\!\downarrow$. If e is a p-event which belongs to $primary(e_q)$, then $e \in secondary(e_p)$.*

We will use the preceding result in the following form.

Corollary 7 *Let e be a p-event such that $e \notin secondary(e_p)$. Then e does not belong to $primary(e_q)$ for the last event e_q of any $q \in \mathcal{P}$.*

As long as all processes which refer to the same label in their primary information are actually pointing to the same event, reusing labels should cause no confusion. Therefore, if p knows that no p-event labeled ℓ is currently part of the primary information of any process in the system, it can safely use ℓ to timestamp the next WRITE which it performs.

Secondary information can be updated in a straightforward manner when we update primary information — if q inherits an event e from p's primary information, it also inherits the secondary information $primary(e)$ associated with e. Notice that it suffices to maintain secondary information as an indexed set — we do not need to maintain secondary *graphs* in a manner similar to primary graphs. Note that the number of events in the secondary information is less than $10N^4$.

4.4   Labeling from a Bounded Set

In this subsection we describe precisely how bounded timestamping is performed using the results of this section. For each process p let $\mathcal{L}_p$ be a finite set of labels such that $|\mathcal{L}_p| > 10N^4$. Process p uses the set $\mathcal{L}_p$ to label its WRITEs. It maintains two primary graphs, the *public primary graph* in its public memory M_p and the *private primary graph* in its private memory L_p. It also maintains a *public secondary information* in M_p and a *private secondary information* in L_p. The public primary graph and the public secondary information of process p are available to other processes through $p : \mathbf{X}$.

At a READ p-event, only the private primary graph and the private secondary information are updated. While at a WRITE p-event, both the primary graphs and both the secondary informations are updated. We describe next what steps p has to take at every READ and WRITE.

When p performs a READ : Process p will read q's public primary graph and public secondary information and compare it with its own private primary graph and private secondary information.

 – Extract the label ℓ of the $\sqsubseteq^*$-maximum q-event in q's public primary graph.
 – If a new event has been read then update the private primary graph and private secondary information as described earlier in this section.

When p performs a WRITE : Let e denote this new WRITE event and e' the previous WRITE p-event. Recall that a WRITE consists of $N-1$ consumes (for each $q \neq p$, **consume**$(q : X)$) and an update (**update**$(p : X, p : Y)$).

 – On the **consume**$(q : X)$: The steps are the same as when it performs a READ.
 – On the **update**$(p : X, p : Y)$: Process p does the following:
   • Choose a label ℓ for e, from $\mathcal{L}_p$ which does not appear in the private secondary information.
   • Replace e' by e in the *latest* component of the private primary graph.
   • Replace *unacklist*(e') in the private primary graph by *unacklist*(e), consisting of the events $B1_{*p}$, $B2_{*p}$ and A_{p*} read during the present WRITE event e, as well as e and e'. The ordering among these is available from the discarded unacklist.
   • Update the private secondary information in accordance with the change in the private primary graph.
   • Copy *new* private primary graph and private secondary information into the public primary graph and public secondary information respectively.

Putting together all the results we have proved so far, we can state the following theorem.

Theorem 8 *The algorithm described above solves the latest gossip problem in a shared memory system for computations consisting of* READs *and* WRITEs, *with only a bounded amount of additional information being attached to each* WRITE.

5   Discussion

In this paper, we have presented a solution to the latest gossip problem in a shared memory system. The gossip construction is extremely powerful and immediately leads to the effective construction of the latest operator, denoted Latest, with the expected semantics. This operator may be used to define and maintain auxiliary variables such as $p : \text{Latest}_{p \to q}(q : X)$ which is a local private variable of p, but which tracks the latest contents of $q : X$. The latest operator may even be composed and (causally) compared, *e.g.*, the program at process p, may check whether $p : \text{Latest}_{p \to q}(\text{Latest}_{q \to p}(p : \text{flag}))$ refers to the current contents of $p : \text{flag}$. Such variables (and causality comparisons) should prove useful in writing parallel programs which meet desired behavioural specifications. Refer [5], for an algorithm for mutual exclusion which uses such auxiliary variables. As opposed to this, *e.g.*, in [4], for the same mutual exclusion problem, even with a BCTS, the bakery algorithm requires an intricate manipulation of another set of variables p : choosing, one for each process p, which is not part of the behavioural specification of the mutual exclusion problem.

The gossip construction immediately leads to a BCTS: The *scan* operation of a BCTS translates to a sequence of $N-1$ READs, one for every other process. The *label* operation translates to a WRITE. The output of the *scan* operation is the sequence $\{\text{Latest}_{p \rightarrow q} \mid q \in \mathcal{P}\}$ ordered by $\sqsubseteq^*$. The solution of the global state detection problem as posed in [2] is even simpler: the primary information at each process p provides a global state. Furthermore, global states are always current and may be maintained online without requiring additional communications.

The gossip problem was originally motivated by problems of logical specification and verification of distributed systems. The solution for synchronization systems, as in [8], was crucially used in [9] for the effective construction of a trace based-extension of linear temporal logic to reason about synchronization protocols. We believe that the shared-memory solution presented here, besides being useful in protocol synthesis, will also be useful in developing an automata-theoretic framework for protocol verification and logics.

Acknowledgments We have benefited greatly from discussions with Madhavan Mukund and K. Narayan Kumar. The work of Bharat Adsul was supported by an INFOSYS FELLOWSHIP.

References

1. B. Adsul, A. Mehta and M. Sohoni: Keeping Track of the Latest Gossip in Shared Memory Systems, *Technical Report*, Dept of CSE, IIT Bombay. Electronic version available @ http://www.cse.iitb.ernet.in/~abharat/gossip.html
2. K. M. Chandy and L. Lamport: Distributed Snapshots: Determining Global States of Distributed Systems, *ACM Transactions on Computer Systems* **3**(1) (1985) 63-75.
3. C. Dwork and O. Waarts: Bounded Concurrent Timestamp Systems are Comprehensible!, *Proc. ACM STOC* (1992) 655-666.
4. N. A. Lynch: Distributed Algorithms, *Morgan Kaufmann* (1996).
5. A. Mehta: Keeping Track of the Latest Gossip in Shared Memory Systems, *BTech Project Report*, Dept of CSE, IIT Bombay. Electronic version available @ http://www.cse.iitb.ernet.in/~abharat/gossip.html
6. M. Mukund, K. Narayan Kumar and M. Sohoni: Keeping Track of the Latest Gossip in Message-Passing Systems, *Proc. Structures in Concurrency Theory (STRICT)*, Berlin 1995, Workshops in Computing Series, Springer-Verlag (1995) 249-263.
7. M. Mukund, K. Narayan Kumar, J. Radhakrishnan and M. Sohoni: Robust Asynchronous Protocols are Finite-State, *Proc. 25th ICALP*, Springer LNCS 1443 (1998) 188-199.
8. M. Mukund and M. Sohoni: Keeping Track of the Latest Gossip in a Distributed System, *Distributed Computing*, **10**(3) (1997) 137-148.
9. P. S. Thiagarajan: TrPTL: A Trace Based Extension of Linear Time Temporal Logic, *Proc. 9th IEEE LICS* (1994) 438-447.

Concurrent Knowledge and Logical Clock Abstractions

Ajay D. Kshemkalyani

Dept. of EECS, University of Illinois at Chicago, Chicago, IL 60607-7053, USA
ajayk@eecs.uic.edu

Abstract. Vector and matrix clocks are extensively used in asynchronous distributed systems. This paper asks, "how does the clock abstraction generalize?" and casts the problem in terms of concurrent knowledge. To this end, the paper motivates and proposes logical clocks of arbitrary dimensions. It then identifies and explores the conceptual link between such clocks and knowledge. It establishes the necessary and sufficient conditions on the size and dimension of clocks required to declare k-level concurrent knowledge about the most recent global facts for which this is possible without using control messages. It then gives algorithms to compute the latest global fact about which a specified level of knowledge is attainable in a given state, and to compute the earliest state in which a specified level of knowledge about a given global fact is attainable.

1 Introduction

1.1 Motivation

A large number of application areas in asynchronous distributed message-passing systems use vector clocks and matrix clocks. Some example areas that use vector clocks [6,14] are checkpointing, garbage collection, causal memory, maintaining consistency of replicated files, taking efficient snapshots of a system, global time approximation, termination detection, bounded multiwriter construction of shared variables, mutual exclusion, debugging, and defining concurrency measures. Some example areas that use matrix clocks are designing fault-tolerant protocols and distributed database protocols [9,20], including protocols to discard obsolete information in distributed databases [18], and protocols to solve the replicated log and replicated dictionary problems [20].

Vector clocks can be thought of as imparting knowledge to a process: when $V[i] = x$ at process h, process h knows that process i has executed at least x events. Matrix clocks give one more level of knowledge: when $M[i, j] = x$ at process h, process h knows that process i knows that process j has executed at least x events. Vector and matrix clocks are convenient as they are updated without sending additional messages; knowledge is imparted via the inhibition-free *ambient message passing* that (i) *eliminates control messages* by using piggybacking, and (ii) *diffuses knowledge using* only *computation messages, whenever sent*.

S. Kapoor and S. Prasad (Eds.): FST TCS 2000, LNCS 1974, pp. 489–502, 2000.

This paper asks the question: *"how does this clock abstraction generalize?"* The problem is cast in terms of concurrent knowledge ("everybody knows on consistent cuts"), which is a form of knowledge appropriate for (time-free) asynchronous distributed systems [16] — all the applications mentioned above implicitly use concurrent knowledge that is *not* common knowledge in their clock algorithms, although this has never been formally studied as such.

1.2   Background

A distributed system can be modeled by a network (N, L), where N is the set of processes that communicate by message passing over L, the set of logical links. We assume an asynchronous distributed (message passing) system, i.e., there is no global clock or shared memory, relative process speeds are independent, and message delivery times are finite but unbounded [2,16]. *Common knowledge*, which has been proposed as a definition of agreement in distributed systems, is defined as follows [8]. A process i that knows a fact ϕ is said to have knowledge $K_i(\phi)$, and if "every process in the system knows ϕ", then the system exhibits knowledge $E^1(\phi) = \bigwedge_{i \in N} K_i(\phi)$. A knowledge level of $E^2(\phi)$ indicates that every process knows $E^1(\phi)$, i.e., $E^2(\phi) = E(E^1(\phi))$. Inductively, a hierarchy of levels of knowledge $E^j(\phi)$ $(j > 0)$ gets defined, where $E^{k+1}(\phi) \Longrightarrow E^k(\phi)$. Common knowledge of ϕ, denoted as $C(\phi)$, is defined as the knowledge X which is the greatest fixed point of $E(\phi \wedge X)$ and is equivalent to $\bigwedge_{j \in Z*} E^j(\phi)$, where $Z*$ is used to denote the set of whole numbers. Common knowledge requires simultaneous action for its achievement and is therefore unattainable in asynchronous distributed systems [5,7,8,17].

Panangaden and Taylor proposed *concurrent common knowledge* (CCK) which is required to be attained simultaneously in *logical time* based on *causality* [13], and is attainable in asynchronous distributed systems [16]. Specifically, CCK can be attained at a "consistent cut" or possible global state [1] in the system execution. To define concurrent common knowledge, [16] first defines $P_i(\phi)$ to represent the statement "there is some consistent global state of the current execution that includes i's local state, in which ϕ is true." $E^C(\phi) = \bigwedge_{i \in N} K_i P_i(\phi)$ and is attainable by the processes at a consistent global state. Likewise, higher levels of knowledge $(E^C)^k(\phi)$, for $k > 1$, are attainable by the processes at a consistent global state. Concurrent common knowledge of ϕ, denoted by $C^C(\phi)$, is defined as the knowledge X which is the greatest fixed point of $E^C(\phi \wedge X)$ and is equivalent to $\bigwedge_{j \in Z*} (E^C)^j(\phi)$. This form of knowledge underlies many existing protocols involving processes reaching agreement about some property of a consistent global state, defined using logical time and causality.

$C^C(\phi) \Longrightarrow (E^C)^j(\phi)(j \in Z*)$. Several applications (see Section 1.1) need only lower levels of such knowledge. Vector clocks [6,14] provide specific knowledge of a global fact/state ϕ, equivalent to concurrent knowledge $(E^C)^0(\phi)$, in the application domain. However, vector clocks are not sufficient for other applications, for which it is necessary to use matrix clocks. Matrix clocks provide concurrent knowledge $(E^C)^1(\phi)$ about facts ϕ in the application domain. Thus, although levels of concurrent knowledge (besides CCK) have not received formal attention besides [16], they are implicitly used in a wide range of applications. Hence, studying levels of concurrent knowledge is important.

Two important and desirable characteristics of the clock protocols used to achieve $(E^C)^0(\phi)$ and $(E^C)^1(\phi)$ are that they *do not use any control messages* and *they diffuse knowledge on a continual basis*, using piggybacked timestamps on the application messages as and when they are sent (see Section 1.1). These clock protocols are not full-information protocols [5], yet for each of the above applications, they suffice to provide the required degree of concurrent knowledge because the clocks are defined so as to capture the property of interest.

All other known protocols to attain concurrent common knowledge and levels of concurrent knowledge are variants of the global state recording algorithm [1,4,16]. Such global state protocols require (i) $O(min(k \cdot |N|, |L|))$ messages to attain $(E^C)^k(\phi)$ and $O(|L|)$ messages to attain $C^C(\phi)$; (ii) $O(d)$ communication time steps, where d is the network diameter. In the above, the $|L|$ factor in the message complexity can be reduced to $|N|$ if inhibitory protocols are used, at the cost of inhibitory time delays [4]. All protocols based on global state recording require control messages for a one-time knowledge attainment of each fact, and may additionally use freezing/inhibition. Hence, they are not considered further.

In Theorem 1, we show that for the class of facts we consider and which includes the applications listed in Section 1.1, $(E^C)^k(\phi)$ is equivalent to $E^k(\phi)$. So we also refer to $(E^C)^k(\phi)$ as just $E^k(\phi)$.

1.3   Objectives

This paper examines the feasibility of and mechanisms for achieving levels of knowledge $E^k(\phi)$ ($k > 0, 1$) using ambient message passing, i.e.,

1. No control messages can be used. Control information may be piggybacked on computation messages. Also, no freezing/inhibition is allowed.
2. The latest knowledge about the (past) computation should be diffused as much as possible, using only the computation messages, whenever sent.

As justified in Section 1.2, we focus only on clock-based protocols. We now formalize the objectives. The *full-information protocol (FIP)* which attains common knowledge (in a synchronous system) has been defined such that "at each step (after each local event), a process broadcasts (via control messages) to other processes its local state (which captures everything it knows)" [5]. The FIP is very expensive, and does not meet our criteria. We now define a "no control messages" protocol, the *full-information piggybacking protocol (FIPP)*, to be one in which on each computation message of the application, the local state information is piggybacked by the sender. This protocol meets the criteria but is expensive in terms of the information piggybacked. We define the k-bounded-information piggybacking protocol to overcome this drawback of the FIPP.

Definition 1. *The* k-bounded-information piggybacking protocol (KIPP) *is such that on each computation message of the application, k-bounded state information[1] is piggybacked by the sender j, where k-bounded information is in-*

---

[1] Presumably about some property of interest.

formation of the form: $K_j(K_{i_1}(K_{i_2} \ldots (K_{i_k}(\phi)) \ldots))$, *where* $i_1, i_2, \ldots i_k \in N$, *for any fact* ϕ *on the system state.*

Facts about the property of interest, which are a function of the system state, are *represented by the timestamp of that system state* in the applications of Section 1.1. *Similarly, 0- and 1-bounded information about these facts are also represented as timestamps* in these applications. Therefore, we will assume that appropriate timestamps can represent facts relevant to the application, and k-bounded information about them. The type of facts considered in Section 1.1 and which we will consider satisfy *monotonicity*. Informally, ϕ is a *monotonic fact* in a run if ϕ holds in some global state, and for every later global state, some ψ holds and $\psi \Longrightarrow \phi$ in that later state (see Definition 5). Monotonic facts are also stable. The paper answers the following questions.

Problem 1. In a system using the KIPP protocol, what are the necessary and sufficient conditions on the timestamp information required to achieve and declare $E^k(\phi)$, where ϕ is the greatest possible monotonic fact (most recent possible system state) about which $E^k(\phi)$ is possible to be declared in the current state?

Problem 2. For any global monotonic fact ϕ on the system state, what is the earliest global state in which $E^k(\phi)$ is attained using the KIPP protocol?

Problem 3. Given a timestamp of a system state, what is the maximum possible monotonic fact ϕ (most recent possible system state) about which $E^k(\phi)$ can be declared in the given state in a system using the KIPP protocol?

Section 2 describes the system model and existing clock systems. Section 3 defines monotonic facts. Section 4 proposes α-dimensional clocks. Section 5 analyzes the levels of knowledge that can be inferred using α-dimensional clocks and answers the above problems. Section 6 concludes. See [12] for the full paper.

2   Preliminaries

2.1   System Model

We assume an asynchronous distributed system (see the first paragraph of Section 1.2). The notion of the local state of a process is primitive. An event e at process i is denoted e_i. An event causes a local state transition. The local history of process i, denoted h_i, is a possibly infinite sequence of alternating local states (beginning with a distinguished initial state) and events [16]. It is equivalently described by the initial state and the sequence of local events.

Formally, an asynchronous distributed system consists of (i) a network (N, L), (ii) a set $\mathcal{H}_i$ of possible local histories for each process i, (iii) a set $\mathcal{A}$ of *asynchronous runs* or *executions*, or *computations*, each of which is a vector of local histories, one per process, and (iv) a set of messages sent in any possible asynchronous run. The system follows the KIPP protocol (Definition 1).

A given run of a distributed system has a poset event structure model as in [13]. Let $(H, \prec)$ represent the set of events H in a system run that are related by the causality relation $\prec$, an irreflexive partial order [13]. H is partitioned into local executions, one per process. Each local execution defines the local history. We assume the initial state of each process is common knowledge.

A *global state* (or *cut*) of run a is a n-vector of prefixes of local histories of a, one prefix per process. It can be viewed equivalently as the union of the events in prefixes of the local histories of a, one prefix per process. A consistent global state (*consistent cut*) is a global state such that if the receipt of a message is recorded, then the sending of that message is also recorded [1]. It can be viewed equivalently as a downward-closed subset of H. Let $H^\perp$ denote the empty cut.

For a given run, the set of all cuts, $Cuts$, forms a lattice ordered by "$\subset$" (subset); the set of downward-closed cuts is its sublattice [14]. The seq. of states in actual time is a chain in this sublattice. The sublattice is not visible to any process, but gives the possible consistent cuts which could have occurred and are "valid" views of the run. Our results implicitly deal with such a run.

We define $F(Cut)$ to be the set consisting of the latest event at each process in cut Cut. $F(Cut)$ denotes the "front" of cut Cut.

Definition 2. $F(Cut) =_{def} \{e_i \in Cut \mid \forall e_i' \in Cut, e_i' \preceq e_i\}$

Given a cut Cut, its projection $F_i(Cut)$ is the element of $F(Cut)$ at process i.

Define $\downarrow e$ as $\downarrow e =_{def} \{e' \mid e' \preceq e\}$. The cut $\downarrow e$ has a unique maximal event e and is downward-closed in $(H, \prec)$. As the set of all downward-closed cuts forms a lattice, therefore $\bigcap_{x \in X} \downarrow x$ and $\bigcup_{x \in X} \downarrow x$, also denoted as $\cap_\Downarrow X$ and $\cup_\Downarrow X$, resp., are downward-closed cuts for any set of events X. These cuts are used to prove Theorems 6 and 7. Based on the definition of $\downarrow e$, we can assert as follows.

Proposition 1. $e \in \cap_\Downarrow X \iff \forall x \in X, e \preceq x$

$\cap_\Downarrow X$, the largest set of events that causally precede every $x \in X$, represents the largest execution prefix with the following property: any fact in this execution prefix can be known in the local state of each process after event $x \in X$ [11].

A Kripkean interpretation of knowledge modality requires the identification of an appropriate set of possible worlds – in the system model, the possible worlds are the (consistent) cuts of the set of possible asynchronous runs [7,8]. (a, c) denotes cut c in run a. Standard definitions for the modal operators K_i and P_i, and for various forms of knowledge are used. The formal semantics are given by the satisfaction relation $\models$ and are the same as in [16]. Proposition 1 can now be expressed in this logic. Assuming that adequate knowledge about local histories is propagated, for any cut X, $(a, X) \models E(\cap_\Downarrow F(X))$, i.e., all the processes know $\cap_\Downarrow F(X)$ after execution of X.

2.2 Logical Clocks

Logical clocks track causality which determines the extent of the past computation that could possibly be known at any state/event. A clock is a function that maps cuts in a run to elements in the time domain $\mathcal{T}$. Thus, $Clk : Cuts \mapsto \mathcal{T}$. Clocks provide a quantitative identifier for cuts. For any run, the timestamp of a cut (which is the union of a prefix of the local history of each process), is defined using timestamps of cuts of the form $\downarrow e$. When we say that an event e is assigned a clock value/timestamp, more formally we mean that the cut $\downarrow e$ is assigned that clock value/timestamp. Also, a subscripted timestamp T_i denotes a timestamp of an event at process i, and $|N|$ is also denoted as n.

Scalar clocks [13], vector clocks [6,14], and matrix clocks [9,18,20] are the only clocks proposed in the literature. A canonical clock updates the local component of the clock by one at each local event. Henceforth, we assume canonical clocks. A canonical vector clock assigns timestamps to an event as follows.

Definition 3. $T(e) =_{def} (i \in N)\ T(e)[i] = |\{e_i \mid e_i \preceq e \}|$, i.e., $T(e)[i]$ is the *number of events on process i that causally precede or equal e.*

For any run, vector clocks of size n track the progress at each process (and are needed to capture concurrency; see discussion on dimension of $(H, \prec)$ [3]). For cut Cut, we define its timestamp $T(Cut)$ such that its ith component is the ith component of the timestamp of event $F_i(Cut)$ [11].

Definition 4. $T(Cut) =_{def} (i \in N)\ T(Cut)[i] = T(F_i(Cut))[i]$

The vector timestamp of a cut identifies the number of events at each process in the cut. For any run, there is an isomorphism between $Cuts$ and $\mathcal{T}^1$, the set of canonical vector timestamps such that $(T \in \mathcal{T}^1)\ T[i] \leq |h_i|$ in that run.

Proposition 2. *For a run $(H, \prec)$, $(Cuts, \subset)$ is isomorphic to $(\mathcal{T}^1, <)$.*

Lemma 1. *The timestamp of cut $\bigcap_{x \in X} \downarrow x$, denoted $T(\cap_{\Downarrow} X)$, is expressed as a function of the timestamps of the members of X as follows. $(i \in N)\ T(\cap_{\Downarrow} X)[i] = min_{x \in X}(T(x)[i])$.*

In Lemma 1, X can be an arbitrary set of events, also termed a nonatomic event [10,11]. Lemma 1 will be shown to have a counterpart Lemma 3 that is based on higher dimensional clocks, and which is used in the proof of Theorem 7.

For any run $(H, \prec)$, observe from Definition 2 that there is a bijection from the set containing each cut Cut to the set containing each front of a cut $F(Cut)$. So, the timestamp of $F(Cut)$ is defined to be the timestamp of Cut.

3   Monotonic Facts

We now define monotonic facts – such facts capture the relevant properties of the applications in Section 1.1, and it is this class of facts that we consider. Examples of such facts are "computation has progressed at least up to global state *state_vector*", and "all logs upto global state *state_vector* can be discarded". As in the applications in Section 1.1, we assume facts of interest are related by a semantic inclusion relation "$\sqsubseteq$" (if $\phi \sqsubseteq \psi$, then ψ semantically includes ϕ).

Definition 5. *For a given run a, any fact ϕ is monotonic iff for every cut c at which $(a, c) \models \phi$, and for every cut c' such that $c \prec c'$, there exists some fact ψ such that $(a, c') \models \psi$ and $(a, c') \models (\phi \sqsubseteq \psi)$.*

Monotonic facts are also stable; however, not all stable facts are monotonic.

Lemma 2. *For a monotonic fact ϕ, the following are all stable facts: ϕ, $K_i(\phi)$, $K_iP_i(\phi)$, $E(\phi)$, and $E^k(\phi)$.*

Let ψ be any of ϕ, $K_i(\phi)$, $K_iP_i(\phi)$, $E(\phi)$, and $E^k(\phi)$, where ϕ is a monotonic fact. When process m receives a message with ψ piggybacked on it from process j at event e_m^y resulting in local state s_m^y, we have $s_m^y \models K_mP_mK_j(\psi)$. Using Lemma 2, we can show that "the P_m operator can be safely removed", and hence $s_m^y \models K_mK_j(\psi)$ (and also $s_m^y \models K_m(\psi)$). Similarly, $\bigwedge_i K_iP_i(\psi)$ is equivalent to $\bigwedge_i K_i(\psi)$. Developing this idea further leads to Theorem 1 that allows us to replace concurrent knowledge with the equivalent normal knowledge.

Theorem 1. *In a system following the KIPP protocol, the greatest possible monotonic fact ϕ about which $(E^C)^k(\phi)$ is possible to be declared in a given state is the greatest possible monotonic fact ϕ' about which $E^k(\phi')$ is possible to be declared in the given state.*

As in the applications of Sect. 1.1, we assume that for any run, the set of monotonic facts ordered by $\sqsubseteq$ is a lattice, there is a semantically greatest fact in each state, and that there is an (iso/homo)morphism from $(Cuts, \subset)$ to $(\mathcal{M}, \sqsubseteq)$, where $\mathcal{M}$ is the set that contains the greatest monotonic fact (of interest) at each cut in $(Cuts, \subset)$. Combining this (iso/homo)morphism with Prop. 2 (and restricting to consistent states for semantic integrity) leads to Prop. 3.

Proposition 3. *The (semantically greatest) monotonic fact of interest in a global state, whose truth value is a function of that global state, will be uniquely identified by the timestamp of that global state.*

4   Clocks of Arbitrary Dimensions

Definition 6. *An α-dimensional clock Clk^α defines the mapping $Clk^\alpha : Cuts \mapsto (Z*)^{n^\alpha}$ (i.e., Clk^α is an α-dimensional array of integers, where each dimension is of size n), satisfying the following properties.*

SP1. *The local clock component at process j, $Clk_j^\alpha[j, j, \ldots, j]$, is common knowledge in the initial system state, i.e., $(a, H^\perp) \models C(Clk_j^\alpha[j, j, \ldots, j])$.*

SP2. *The local clock component at process j, $Clk_j^\alpha[j, j, \ldots, j]$, must be incremented by a natural number when a computation event occurs at j.*

SP3. *Any element $Clk^\alpha(e_j)[i_1, i_2, \ldots, i_\alpha]$ is the maximum scalar clock value ϕ_{i_α} $= Clk_{i_\alpha}[i_\alpha, i_\alpha, \ldots, i_\alpha]$ at i_α such that $K_j(K_{i_1}(K_{i_2}(K_{i_3}(\ldots K_{i_\alpha}(\phi_{i_\alpha})\ldots))))$.*

R0. (Initial state:) $Clk_i^\alpha = \alpha$ dimensional 0-vector

R1. (Internal event:) Before process i executes the event, $Clk_i^\alpha[i, i, \ldots, i] = Clk_i^\alpha[i, i, \ldots, i] + d \ (d > 0)$

R2. (Send event:) Before process i executes the event, $Clk_i^\alpha[i, i, \ldots, i] = Clk_i^\alpha[i, i, \ldots, i] + d \ (d > 0)$. Send message timestamped with Clk_i^α.

R3. (Receive event:) When process j receives a message with timestamp T^α from process i,

　1. **for** $\beta = 1$ **to** $\alpha - 1$ **do**

$$\forall q_1 \in N \setminus \{j\}, \forall q_2, q_3, \ldots, q_\beta \in N,$$

$$Clk_j^\alpha[\underbrace{j, \ldots, j}_{\alpha - \beta \ times}, \underbrace{q_1, q_2, \ldots, q_\beta}_{\beta \ entries}] =$$

$$max(Clk_j^\alpha[\underbrace{j, \ldots, j}_{\alpha - \beta \ times}, \underbrace{q_1, q_2, \ldots, q_\beta}_{\beta \ entries}], T^\alpha[\underbrace{i, \ldots, i}_{\alpha - \beta \ times}, \underbrace{q_1, q_2, \ldots, q_\beta}_{\beta \ entries}])$$

　2. $\forall q_1 \in N \setminus \{j\}, \forall q_2, \ldots, q_\alpha \in N,$
$$Clk_j^\alpha[q_1, q_2, \ldots, q_\alpha] = max(Clk_j^\alpha[q_1, q_2, \ldots, q_\alpha], T^\alpha[q_1, q_2, \ldots, q_\alpha])$$

　3. $Clk_j^\alpha[j, j, \ldots, j] = Clk_j^\alpha[j, j, \ldots, j] + d \ (d > 0)$

　4. Deliver the message.

Fig. 1. Protocol to operate α-dimension clocks

With canonical clocks, $d = 1$ and $Clk^\alpha(e)[i_1, i_2, \ldots, i_\alpha] = |F_{i_\alpha}(\ldots \downarrow F_{i_3}(\downarrow F_{i_2}(\downarrow F_{i_1}(\downarrow e)))\ldots)|$. The value of Clk^α assigned as a timestamp is denoted T^α. $T^\alpha[i]$, also represented as $T^\alpha[i, \cdot]$, is a timestamp of dimension $(\alpha - 1)$ and is derived from T^α by instantiating the first dimension variable i_1 by i. $T^\alpha(e_p)[i, \cdot]$ is the $(\alpha - 1)$ dimensional timestamp of the most recent event at process i, as known to process p after event e_p. Moreover, this most recent event at process i has a scalar timestamp $T^\alpha(e_p)[i, i, \ldots, i]$. In terms of knowledge, $T^\alpha(e_p^x)[i, \cdot]$ represents the knowledge $s_p^x \models K_p(K_i K_{i_2} K_{i_3} \ldots K_{i_\alpha}(\phi))$, where only i_1 is instantiated by i in $K_p(K_{i_1} K_{i_2} K_{i_3} \ldots K_{i_\alpha}(\phi))$, for all $i_1, i_2, \ldots i_\alpha \in N$. Analogously, $T^\alpha[\underbrace{a, b, \ldots, f}_{\beta \ entries, \ \beta \geq 0}, \cdot]$ is a timestamp of dimension $(\alpha - \beta)$. $T^\alpha(e_p^x)[a, b, \ldots, f, \cdot]$ represents the knowledge $s_p^x \models K_p(K_a K_b \ldots K_f K_{i_{\beta+1}} K_{i_{\beta+2}} \ldots K_{i_\alpha}(\phi))$, where the first β dimension variables $i_1, \ldots, i_\beta$ are instantiated in $K_p(K_{i_1} K_{i_2} K_{i_3} \ldots K_{i_\alpha}(\phi))$, for all $i_1, i_2, \ldots i_\alpha \in N$. When $p = (a = b = \ldots = f)$, $T^\alpha(e_p^x)[a, b, \ldots, f, \cdot]$ is effectively a $(\alpha - \beta)$ dimensional timestamp of e_p^x.

Theorem 2. *The protocol in Fig. 1 implements the α-dimensional clock specification of Definition 6.*

The protocol in Fig. 1 has a space and time complexity of $\Theta(n^\alpha)$. Rules (R3.1 and R3.2) can be simplified using simple observations, as shown in [12]. The size of each clock of dimension α is n^α integers. This clock/timestamp size may be reduced by using information such as the message pattern, logical network topology, and the partial order $(H, \prec)$, using analysis such as in [15,19,20], or by using approximations to the true clock, using schemes such as in [9,20].

The α-dimensional timestamp of a cut is defined using the $(\alpha-1)$-dimensional timestamp of the latest event at each process in that cut.

Definition 7. $T^\alpha(Cut) =_{def} (i \in N) \; T^\alpha(Cut)[i, \cdot] = T^\alpha(F_i(Cut))[i, \cdot]$

Lemma 3. *The timestamp of cut* $\bigcap_{x \in X} \downarrow x$, *denoted* $T^\alpha(\cap_\Downarrow X)$, *is expressed as a function of the timestamps of the members of* X *as follows.* $(i \in N) T^\alpha(\cap_\Downarrow X)[i, \cdot]$ *is the* $(\alpha - 1)$-*dimensional timestamp* $T^\alpha(x')[i, \cdot]$, *where* $T^\alpha(x')[i, i, \ldots, i] = min_{x \in X}(T^\alpha(x)[i, i, \ldots, i])$.

Lemma 3 gives a way to implement the test for Proposition 1. It will be used in Theorem 7 to identify the maximum computation prefix ϕ (cut) about which knowledge $E^k(\phi)$ has been attained at a given cut Cut.

Recall that by Proposition 3, the problem of identifying the minimum possible computation prefix (cut) c such that $(a, c) \models E^k(\phi)$ for a given ϕ (Problem 2) is equivalent to the problem of identifying the minimum possible computation prefix c such that $(a, c) \models E^k(Cut)$, where Cut is the cut in which ϕ is true. Likewise, the problem of identifying the maximum possible fact ϕ such that $(a, c) \models E^k(\phi)$ at a given cut c (Problems 1 and 3), is equivalent to the problem of identifying the maximum possible computation prefix Cut such that $(a, c) \models E^k(Cut)$. We now give the main results linking clocks and knowledge.

5   Attaining Knowledge Using Clocks

At process i, k-bounded knowledge (of global facts about a property of interest) is of the form $K_i(K_{i_1} K_{i_2} K_{i_3} \ldots K_{i_k}(\phi))$. The number of unique permutations of the K_{i_j} operators that represent k-bounded knowledge is computed as follows. $i_1 \neq i$, and $\forall j \in [2, k], i_j \neq i_{j-1}$. Thus, $\forall j \in [1, k], i_j$ can take one of $n-1$ values, giving $(n-1)^k$ permutations; each permutation denotes a global fact about which k-bounded knowledge exists at process i. Each global fact is represented by a cut and requires a vector (n integers). Thus, the space for k-bounded knowledge at i is $n \cdot (n - 1)^k$ integers. The space requirement for all levels of knowledge upto k at process i is $n \cdot \sum_{j=0}^{k} (n - 1)^j$ integers.

Lemma 4. *Representation of k-bounded knowledge (of global facts about a property of interest) needs* $n \cdot \sum_{j=0}^{k} (n - 1)^j$ *integers.*

From the inequality $n^k <> n \cdot \sum_{j=0}^{k}(n-1)^j < n^{k+1}$, we now get Theorem 3.

Theorem 3. *k-bounded knowledge (of global facts about a property of interest) cannot be represented by a k-dimensional clock system, but can be represented by a $(k + 1)$-dimensional clock system.*

By definition, $E^k(\phi) = \bigwedge_i K_i P_i(E^{k-1}(\phi))$, i.e., each process knows $E^{k-1}(\phi)$ along some (consistent) global state. To identify the bounds on space complexity to determine $E^k(\phi)$ knowledge for the latest possible ϕ in a system

(1) **Problem Inputs:**
(1a) **array of int** T_ϕ^1;   //vector timestamp of earliest state in which ϕ is true
(1b) **int** k;   //level of knowledge $E^k(\phi)$ to be attained
(2) **Problem Output:**
(2a). **array of int** $TS^1 = Compute\_State(T_\phi^1, k)$.
(2b)   //vector timestamp of earliest state in which $E^k(\phi)$ is attained

(3) *function Compute_State*(**array of int** T_ϕ^1; **int** k) **returns** TS^1
(4)   **for** $lvl = 1$ **to** $k + 1$ **do**
(5) $\forall p \in N$ **do**
(6)   identify earliest event $e_p \mid T^1(e_p) \geq TS^1$;
(7) $T'^1[p] = T^1(e_p)[p]$;
(8) $\forall p \in N$ **do**
(9) $TS^1[p] = max(T^1(e_1)[p], T^1(e_2)[p], \ldots, T^1(e_n)[p])$;
(10)   // $(a, TS^1) \models E^{lvl}(T_\phi^1) \bigwedge \not\exists TS'^1 \mid (TS'^1 < TS^1 \wedge (a, TS'^1) \models E^{lvl}(T_\phi^1))$
(11) **return**(TS^1).

Fig. 2. Given ϕ, protocol to compute earliest system state in which $E^k(\phi)$ is achievable

following the KIPP protocol (Problem 1), it can be shown that $\forall i_1, i_2, \ldots i_k$, $K_i(K_{i_1} K_{i_2} \ldots, K_{i_k}(\psi_{i_1, i_2, \ldots, i_k}))$ must be available at each process i, where $\psi_{i_1, i_2, \ldots, i_k}$ is the max. execution prefix about which the corresponding knowledge is available, i.e., "i knows i_1 knows i_2 knows $\ldots i_k$ knows $\psi_{i_1, i_2, \ldots, i_k}$". The max. execution prefix ϕ about which $E^k(\phi)$ is attained is given by $\bigcap_{i_1, i_2, \ldots, i_k \in N} \psi_{i_1, i_2, \ldots, i_k}$.

Theorem 4. *In a system following the KIPP protocol, k-bounded knowledge at each process is required to attain and declare $E^k(\phi)$, where ϕ is the maximum possible monotonic fact (most recent possible system state) about which $E^k(\phi)$ is possible to be declared in the current state.*

Theorem 5 ($=$ Thms. 3 $+$ 4) and Theorem 6 answer Problems 1 and 2, resp..

Theorem 5. *In a system following the KIPP protocol, a $(k+1)$-dim clock system is sufficient but a k-dim clock system is not sufficient to attain and declare $E^k(\phi)$, where ϕ is the maximum possible monotonic fact (most recent possible system state) about which $E^k(\phi)$ is possible to be declared in the current state.*

Theorem 6. *Given a global monotonic fact ϕ, the earliest global state in which $E^k(\phi)$ is attained in a system following the KIPP protocol is given by the protocol in Fig. 2.*

Given the earliest cut where ϕ becomes true, specified by T_ϕ^1, (which by Proposition 3 captures fact ϕ), Fig. 2 gives a protocol to determine the earliest global state at which $E^k(\phi)$ can be attained. The protocol is iterative. Function *Compute_State* uses two inputs: (i) T_ϕ^1, the vector timestamp of the earliest state in which ϕ holds, and (ii) k, the level of knowledge $E^k(\phi)$ to be attained. The output is TS^1, the vector timestamp of the earliest state in which assertion $E^k(\phi)$ can be made. The protocol is proved correct by showing that the invariant in line (10) holds after each iteration. Note that in each iteration, T'^1 (line (7)) identifies a global state that may not be consistent; hence a consistent global state TS^1 (line (9)) is computed.

Complexity: Time complexity is (# send and receive events in $(H, \prec)$ after the cut at which ϕ is defined). Space complexity is that of a vector clock system, and also requires each process to store a trace of the timestamps of its send and receive events beyond the cut at which ϕ is defined.

To answer Problem 3, "Given a timestamp $T^{\beta+1}$ of a state, what is the maximum ϕ such that $(a, T^{\beta+1}) \models E^\beta(\phi)$?" we can apply the function min to the n^β 1-dimensional timestamps of size n in the given $T^{\beta+1}$. This requires $n \cdot n^\beta$ comparisons. Theorem 7 gives a solution of $\Theta(\beta \cdot (n^2 + n))$ time complexity.

Theorem 7. *Given a timestamp T^β, the maximum possible monotonic fact ϕ (most recent possible system state) about which $E^k(\phi)$, where $k \leq \beta - 1$, can be declared at the given state T^β in a system following the KIPP protocol is given by the protocol in Fig. 3.*

The proof is by construction. Fig. 3 gives a protocol to derive the max. computation prefix ϕ about which the processes have knowledge $E^k(\phi)$, given the timestamp Ob_T^β, where $\beta > k$. $Compute\_Phi$ has inputs (i) T^α, the (variable dim.) timestamp of the maximum cut about which knowledge E^{atn} is attained in Ob_T^β, (ii) m, the level of knowledge that is yet to be attained, and (iii) atn, the level of knowledge already attained. The output is the timestamp ϕ of the max. cut about which E^k knowledge is attained in the given state Ob_T^β.

$Compute\_Phi$ is invoked as $Compute\_Phi(Ob\_T^\beta, k, 0)$ and is tail-recursive. T^α is progressively decreased at each recursion level to add another level of knowledge to what is known of T^α at cut Ob_T^β. So at each additional recursion level, T^α therein converges towards ϕ. Each recursion level behaves as follows.

- Given $T^\alpha(Cut)$, the loop in lines (5)-(6) computes the $(\alpha - 1)$-dimensional timestamp of $F_p(Cut)$ which is the latest event of the cut Cut at process p, $(p \in N)$. $T^{(\alpha-1)}(F_p(Cut))$ is simply $T^\alpha[p, \cdot]$.
- Let X denote the events $F(Cut)$ identified in line (6). The loop in lines (7)-(9) applies Lemma 3 to X to compute $\cap_\Downarrow X$. By doing so, it identifies the timestamps $T^{(\alpha-2)}(F_p(\cap_\Downarrow X))$ for each process p. Then $T^{(\alpha-1)}(\cap_\Downarrow X)$ is simply the aggregation of the n timestamps $T^{(\alpha-2)}(F_p(\cap_\Downarrow X))$, as shown in line (10). By Proposition 1, $T^{(\alpha-1)}$ is the timestamp of the maximal prefix about which all the processes have knowledge at $X = F(Cut)$ and this can be asserted only at or after $F(Cut)$. Thus, $E(T^{(\alpha-1)})$ holds in the state with timestamp T^α in this recursion level and we assert the invariant on line (11).
- The above steps also add a level of knowledge to that at the given initial state Ob_T^β; we assert this in the invariant on line (13). If this is the desired level of knowledge, then we have the terminating case for the recursion and the value of $T^{(\alpha-1)}$ is returned (lines (14)-(16)), else $Compute\_Phi$ is recursively

(1) **Problem Inputs:**
(1a) β-dim. array of int Ob_T^β;   // timestamp of observation state
(1b)   int k, where $\beta > k \geq 1$;   // level of knowledge to be attained
(2) **Problem Output:**
(2a) $(\beta - k)$ **dim. array of int** $\phi = Compute\_Phi(Ob\_T^\beta, k, 0)$.
(2b)   //timestamp of maximum possible state such that $(a, Ob\_T^\beta) \models E^k(\phi)$

(3) *function Compute_Phi*(**var dim. array of int** T^α; **int** m, atn) **returns** ϕ
(4a) // T^α is timestamp of the max. possible state such that $(a, Ob\_T^\beta) \models E^{atn}(T^\alpha)$
(4b) // m is the level of knowledge yet to be attained
(4c) // atn is the level of knowledge already attained. $atn = k - m$.
(5) $\forall p \in N$ **do**
(6) $T_p^{\alpha-1} = T^\alpha[p, \cdot]$;
(7) $\forall p \in N$ **do**
(8)   let r be such that $T_r^{\alpha-1}[p, p, \ldots, p] = min_{q \in N}(T_q^{\alpha-1}[p, p, \ldots, p])$;
(9) $T_p^{\alpha-2} = T_r^{\alpha-1}[p, \cdot]$;
(10) $T^{\alpha-1} = [T_1^{\alpha-2}, T_2^{\alpha-2}, \ldots, T_n^{\alpha-2}]$;
(11) // $(a, T^\alpha) \models E^1(T^{\alpha-1}) \wedge \not\exists T'^{\alpha-1} \mid (T'^{\alpha-1} > T^{\alpha-1} \wedge (a, T^\alpha) \models E^1(T'^{\alpha-1}))$
(12) $atn = atn + 1$; $m = m - 1$;
(13)// $(a, Ob\_T^\beta) \models E^{atn}(T^{\alpha-1}) \wedge \not\exists T'^{\alpha-1} \mid (T'^{\alpha-1} > T^{\alpha-1} \wedge (a, Ob\_T^\beta) \models E^{atn}(T'^{\alpha-1}))$
(14) **if** $m = 0$ **then**
(15) $\phi = T^{\alpha-1}$;
(16)   **return**(ϕ);
(17) **else**
(18) $\phi = Compute\_Phi(T^{\alpha-1}, m, atn)$;
(19)   // $(a, T^{\alpha-1}) \models E^m(\phi) \wedge \not\exists \phi' \mid (\phi' > \phi \wedge (a, T^{\alpha-1}) \models E^m(\phi'))$
(20)   // $(a, T^\alpha) \models E^{m+1}(\phi) \wedge \not\exists \phi' \mid (\phi' > \phi \wedge (a, T^\alpha) \models E^{m+1}(\phi'))$
(21)   // $(a, Ob\_T^\beta) \models E^{atn+m}(\phi) \wedge \not\exists \phi' \mid (\phi' > \phi \wedge (a, Ob\_T^\beta) \models E^{atn+m}(\phi'))$
(22)   **return**(ϕ).

Fig. 3. Protocol to compute latest ϕ for which $E^k(\phi)$ holds in a state with timestamp T^β, where $\beta > k$

invoked to determine the greatest ϕ that is known at $T^{(\alpha-1)}$ for the remaining m levels of knowledge to be attained (lines (17)-(18)).

The invariants on lines (11,13,19,20,21) are seen to hold. Hence, ϕ is the max prefix such that $(a, Ob\_T^\beta) \models E^k(\phi)$, derived from the recursive use of Lemma 3. **Complexity:** The time complexity is $\Theta(k \cdot (n^2 + n))$. The space complexity is that of β-dimensional clocks, which is $\Theta(n^\beta)$ integers and meets the tight bound established by Theorem 5. The time complexity is less than the space complexity because information is selectively accessed dynamically.
Necessary and sufficient conditions required to declare $E^k(\phi)$ using the KIPP: Lemma 4 and Theorem 4 together give the conditions on the exact size of clocks, whereas Theorem 5 gave the conditions on the dimension of clocks.

6   Concluding Remarks

So far, concurrent knowledge has been studied much less than normal knowledge although asynchronous systems are much more prevalent than synchronous ones. This paper made significant contributions to the theory of concurrent common knowledge and proposed logical clock systems of arbitrary dimensions. Specifically, it made the following contributions. (i) It motivated and proposed logical clocks of arbitrary dimensions, and also formalized the KIPP protocol for knowledge transfer used by such clock systems. (ii) It showed that there exists a tight relation between the dimension of logical clocks and the level of concurrent knowledge attainable, and established some complexity bounds. Here it identified and explored an important conceptual link. (iii) It proposed algorithms to compute the latest global fact about which a specified level of knowledge is attainable in a given state, and to compute the earliest state in which a specified level of knowledge about a given global fact is attainable.

Acknowledgements: This work was supported by the U.S. National Science Foundation grant CCR-9875617.

References

1. M. Chandy, L. Lamport, Finding global states of a distributed system, *ACM Transactions on Computer Systems*, 3(1): 63-75, 1985.
2. M. Chandy, J. Misra, How processes learn, *Distributed Computing*, 1: 40-52, 1986.
3. B. Charron-Bost, Concerning the size of clocks in distributed systems, *Information Processing Letters*, 39: 11-16, 1991.
4. C. Critchlow, K. Taylor, The inhibition spectrum and the achievement of causal consistency, *Distributed Computing*, 10(1): 11-27, 1996.
5. R. Fagin, J. Halpern, Y. Moses, M. Vardi, Reasoning about Knowledge, MIT Press, 1995.
6. C. Fidge, Timestamps in message-passing systems that preserve partial ordering, *Australian Computer Science Communications*, 10(1): 56-66, Feb. 1988.
7. J. Halpern, R. Fagin, Modeling knowledge and action in distributed systems, *Distributed Computing*, 3(4): 139-179, 1989.
8. J. Halpern, Y. Moses, Knowledge and common knowledge in a distributed environment, *Journal of the ACM*, 37(3): 549-587, 1990.
9. N. Krishnakumar, A. Bernstein, Bounded ignorance in replicated systems, *Proc. ACM Symp. on Principles of Database Systems*, 1991.
10. A. Kshemkalyani, Temporal interactions of intervals in distributed systems, *Journal of Computer and System Sciences*, 52(2): 287-298, Apr. 1996.
11. A. Kshemkalyani, Causality and atomicity in distributed computations, *Distributed Computing*, 11(4): 169-189, Oct. 1998.
12. A. Kshemkalyani, On continuously attaining levels of concurrent knowledge without control messages, Tech. Rep. UIC-EECS-98-6, Univ. Illinois at Chicago, 1998.
13. L. Lamport, Time, clocks, and the ordering of events in a distributed system, *Communications of the ACM*, 21(7): 558-565, July 1978.
14. F. Mattern, Virtual time and global states of distributed systems, *Parallel and Distributed Algorithms*, North-Holland, pp 215-226, 1989.

15. S. Meldal, S. Sankar, J. Vera, Exploiting locality in maintaining potential causality, *Proc. 10th ACM Symp. on Principles of Distributed Computing*, 231-239, 1991.
16. P. Panangaden, K. Taylor, Concurrent common knowledge: Defining agreement for asynchronous systems, *Distributed Computing*, 6(2): 73-94, Sept. 1992.
17. R. Parikh, P. Krasucki, Levels of knowledge in distributed computing, *Sadhana Journal*, 17(1): 167-191, 1992.
18. S. Sarin, N. Lynch, Discarding obsolete information in a distributed database system, *IEEE Transactions on Software Engineering*, 13(1): 39-46, 1987.
19. M. Singhal, A. Kshemkalyani, Efficient implementation of vector clocks, *Information Processing Letters*, 43, 47-52, Aug. 1992.
20. G. Wuu, A. Bernstein, Efficient solutions to the replicated log and dictionary problems, *Proc. 3rd ACM Symp. on Principles of Distributed Computing*, 232-242, 1984.

Decidable Hierarchies of Starfree Languages

Christian Glaßer[*] and Heinz Schmitz[**]

Theoretische Informatik, Universität Würzburg, 97074 Würzburg, Germany
{glasser,schmitz}@informatik.uni-wuerzburg.de

Abstract. We introduce a strict hierarchy $\{\mathbb{L}_n^{\mathcal{B}}\}$ of language classes which exhausts the class of starfree regular languages. It is shown for all $n \geq 0$ that the classes $\mathbb{L}_n^{\mathcal{B}}$ have decidable membership problems. As the main result, we prove that our hierarchy is levelwise comparable by inclusion to the dot-depth hierarchy, more precisely, $\mathbb{L}_n^{\mathcal{B}}$ contains all languages having dot-depth $n+1/2$. This yields a lower bound algorithm for the dot-depth of a given language. The same results hold for a hierarchy $\{\mathbb{L}_n^{\mathcal{C}}\}$ and the Straubing-Thérien hierarchy.

1 Introduction

We contribute to the study of starfree regular languages (SF, for short) which are constructed from alphabet letters using Boolean operations together with concatenation. To determine for a given language the minimal number of alternations between these two kinds of operations is known as the dot-depth problem, recently considered as one of the most important open questions on regular languages [9]. For an overview we refer to [8].

We deal with the dot-depth hierarchy [3] and the Straubing-Thérien hierarchy [12,15,13], which both formalize the dot-depth problem in terms of the membership problems of their hierarchy classes. Fix some finite alphabet A with $|A| \geq 2$. For a class $\mathcal{C}$ of languages let $\mathrm{Pol}(\mathcal{C})$ be its polynomial closure, i.e., the closure under finite union and concatenation, and denote by $\mathrm{BC}(\mathcal{C})$ its Boolean closure. The classes $\mathcal{B}_{n/2}$ of the dot-depth hierarchy (DDH) and the classes $\mathcal{L}_{n/2}$ of the Straubing-Thérien hierarchy (STH) can be defined as follows.

$$\mathcal{B}_{1/2} := \mathrm{Pol}(\{\{a\} : a \in A\} \cup \{A^+\}) \qquad \mathcal{L}_{1/2} := \mathrm{Pol}(\{A^*aA^* : a \in A\})$$

$$\mathcal{B}_{n+1} := \mathrm{BC}(\mathcal{B}_{n+1/2}) \quad \text{for } n \geq 0 \qquad \mathcal{L}_{n+1} := \mathrm{BC}(\mathcal{L}_{n+1/2}) \quad \text{for } n \geq 0$$

$$\mathcal{B}_{n+3/2} := \mathrm{Pol}(\mathcal{B}_{n+1}) \quad \text{for } n \geq 0 \qquad \mathcal{L}_{n+3/2} := \mathrm{Pol}(\mathcal{L}_{n+1}) \quad \text{for } n \geq 0$$

By definition, all these classes are closed under union and it is known, that they are also closed under intersection and under taking residuals [1,10]. Up to now, levels 1/2, 1 and 3/2 of both hierarchies are known to be decidable [11,7,1,10,6] while the question is open for any other level. Partial results are known for level 2 of the STH which is decidable if a two-letter alphabet is considered [14].

---

[*] Supported by the Studienstiftung des Deutschen Volkes.
[**] Supported by the Deutsche Forschungsgemeinschaft (DFG), grant Wa 847/4-1.

S. Kapoor and S. Prasad (Eds.): FST TCS 2000, LNCS 1974, pp. 503–515, 2000.
© Springer-Verlag Berlin Heidelberg 2000

We take up the discussion started in [6] and look at known results of the type *"L belongs to the class C if and only if the accepting automaton does not have subgraph S in its transition graph"*. Such a forbidden pattern characterization implies decidablility of the membership problem of C, and even more, it reflects the effect of language operations in the structure of automata. The present paper continues this approach in a natural way.

More precisely, we observe how the forbidden pattern characterizing $\mathcal{L}_{1/2}$ acts as a building block in the forbidden pattern which characterizes $\mathcal{L}_{3/2}$ [10]. Surprisingly, we find this observation confirmed, if we compare the pattern for $\mathcal{B}_{1/2}$ [10] with the characterization of $\mathcal{B}_{3/2}$ [6]. Note from the definition above that in both hierarchies we get with the same operations from one level to the next. Together, this motivates the introduction of an iteration rule IT on patterns, which continues the just observed formation procedure.

In general, starting from an initial class of patterns $\mathcal{I}$, our iteration rule generates for $n \geq 0$ classes of patterns $\mathbb{P}_n^{\mathcal{I}}$ which in turn define language classes $\mathbb{L}_n^{\mathcal{I}}$ by prohibiting the patterns $\mathbb{P}_n^{\mathcal{I}}$ in the transition graphs of deterministic finite automata. We prove that $\mathbb{L}_n^{\mathcal{I}} \cup \mathrm{co}\mathbb{L}_n^{\mathcal{I}} \subseteq \mathbb{L}_{n+1}^{\mathcal{I}} \cap \mathrm{co}\mathbb{L}_{n+1}^{\mathcal{I}}$ and, as the main technical result, that $\mathrm{Pol}(\mathrm{co}\mathbb{L}_n^{\mathcal{I}}) \subseteq \mathbb{L}_{n+1}^{\mathcal{I}}$ holds (cf. Theorem 1). With the latter we relate in a very general way Boolean operations and concatenation to the structural complexity of transition graphs.

Then we apply our results to particular initial classes of patterns $\mathcal{B}$ and $\mathcal{L}$ corresponding to the DDH and STH, respectively. As a consequence, we obtain decidable hierarchies of classes $\mathbb{L}_n^{\mathcal{B}}$ and $\mathbb{L}_n^{\mathcal{L}}$ which exhaust the class of starfree languages and for which it holds that:

$$\begin{array}{ll}
\mathcal{B}_{1/2} = \mathbb{L}_0^{\mathcal{B}} & \mathcal{L}_{1/2} = \mathbb{L}_0^{\mathcal{L}} \\
\mathcal{B}_{3/2} = \mathbb{L}_1^{\mathcal{B}} & \mathcal{L}_{3/2} = \mathbb{L}_1^{\mathcal{L}} \\
\mathcal{B}_{n+1/2} \subseteq \mathbb{L}_n^{\mathcal{B}} & \mathcal{L}_{n+1/2} \subseteq \mathbb{L}_n^{\mathcal{L}}
\end{array}$$

These inclusions imply in particular a lower bound algorithm for the dot-depth of a given language L. One just has to determine the class $\mathbb{L}_n^{\mathcal{B}}$ or $\mathbb{L}_n^{\mathcal{L}}$ for minimal n to which L belongs and it follows that L has at least dot-depth n (for another lower bound result see [17]). However, it remains to argue that the forbidden pattern classes are not too large, e.g., if they all equal SF nothing is won. For this end, we provide more structural similarities between the DDH and STH and the forbidden pattern classes: All hierarchies show the same inclusion structure (see Fig. 4) and, interestingly, the typical languages that separate the levels of the DDH and STH also separate levelwise our forbidden pattern classes. In particular, it holds that $\mathbb{L}_n^{\mathcal{L}}$ (just as $\mathcal{L}_{n+1/2}$) does not capture $\mathcal{B}_{n+1/2}$.

2   Preliminaries

All definitions of language classes will be made w.r.t. the fixed alphabet A. The empty word is denoted by ε, the set of all non-empty words over A is denoted by A^+. We consider all languages as subsets of A^+. For a class C of languages the set of complements is denoted by $\mathrm{co}\mathcal{C} := \{ A^+ \setminus L \mid L \in \mathcal{C} \}$. For a word $w \in A^*$ denote by $|w|$ its number of letters. A deterministic finite automaton (dfa) F is given by $F = (A, S, \delta, s_0, S')$, where A is its input alphabet, S is its set of states, $\delta : A \times S \to S$ is its total transition function, $s_0 \in S$ is the starting state and $S' \subseteq S$ is the set of accepting states. We denote by $L(F)$ the language accepted by F. As usual, we extend transition functions to input words, and we denote by $|F|$ the number of states of F. We say that a state $s \in S$ has a loop $v \in A^*$ (has a v-loop, for short) if and only if $\delta(s, v) = s$. If a dfa F is fixed we write $s_1 \xrightarrow{w} s_2$ instead of "$\delta(s_1, w) = s_2$", and $s_1 \longrightarrow s_2$ instead of "$\delta(s_1, w) = s_2$ for some $w \in A^*$". Every $w \in A^*$ induces a total mapping $\delta^w : S \to S$ with $\delta^w(s) := \delta(s, w)$. We define that a total mapping $\delta' : S \to S$ leads to a certain structure in a dfa (for instance a v-loop) if and only if for all $s \in S$ the state $\delta'(s)$ has this structure (has a v-loop). We will also say that $w \in A^*$ leads to a certain structure in a dfa if δ^w does so. An obvious property of dfa's is that they run into loops after a small number of successive words in the input.

Proposition 1. *Let F be a dfa, $w \in A^*$, $r \geq |F|$. Then w^r leads to a $w^{r!}$-loop.*

The following inclusion relations in each hierarchy are easy to see from the definitions. We can compare the hierarchies to each other in both directions.

Proposition 2. *It holds that $\mathcal{B}_{n+1/2} \cup \mathrm{co}\mathcal{B}_{n+1/2} \subseteq \mathcal{B}_{n+1} \subseteq \mathcal{B}_{n+3/2} \cap \mathrm{co}\mathcal{B}_{n+3/2}$ and $\mathcal{L}_{n+1/2} \cup \mathrm{co}\mathcal{L}_{n+1/2} \subseteq \mathcal{L}_{n+1} \subseteq \mathcal{L}_{n+3/2} \cap \mathrm{co}\mathcal{L}_{n+3/2}$ for $n \geq 0$.*

Proposition 3. *For $n \geq 1$ the following holds.*

1. $\mathcal{L}_{n-1/2} \subseteq \mathcal{B}_{n-1/2} \subseteq \mathcal{L}_{n+1/2}$
2. $\mathrm{co}\mathcal{L}_{n-1/2} \subseteq \mathrm{co}\mathcal{B}_{n-1/2} \subseteq \mathrm{co}\mathcal{L}_{n+1/2}$
3. $\mathcal{L}_n \subseteq \mathcal{B}_n \subseteq \mathcal{L}_{n+1}$

By [4] we have $\bigcup_{n \geq 1} \mathcal{L}_{n/2} = \bigcup_{n \geq 1} \mathcal{B}_{n/2} = SF$. By [5] for $n \geq 1$ it holds that $\mathcal{L}_{n+1/2} = \mathrm{Pol}(\mathrm{co}\mathcal{L}_{n-1/2})$ and $\mathcal{B}_{n+1/2} = \mathrm{Pol}(\mathrm{co}\mathcal{B}_{n-1/2})$.

3   A Theory of Forbidden Patterns

We consequently take up the idea of forbidden pattern characterizations and develop a general method for a uniform definition of hierarchies via iterated patterns in transition graphs. Such a definition starts with an initial pattern which determines the first level of the corresponding hierarchy. Using an iteration rule we obtain more complicated patterns which define the higher levels. Theorem 1 states that a complementation followed by a polynomial closure operation on the language side is captured by our iteration rule on the forbidden pattern side.

3.1   Hierarchies of Iterated Patterns

The known forbidden pattern characterizations for the levels 1/2 and 3/2 of the DDH and STH are of the following form: There appear two states s_1, s_2 and a word z such that $s_1 \xrightarrow{z} +$, $s_2 \xrightarrow{z} -$ and we find a certain structure between s_1 and s_2. Since in the following we consider only patterns of this form it suffices to describe the structures that occur between s_1 and s_2. Usually, both states s_1 and s_2 have a loop of the same structure in the dfa, the *loop-structure*. This structure in turn determines the subgraph we need to find between s_1 and s_2, together with the loop-structure we call this the *bridge-structure* (cf. Fig. 1). Let us first define what we mean by an initial class of patterns.

Definition 1. *We define an initial pattern $\mathcal{I}$ to be a subset of $A^* \times A^*$ such that for all $r \geq 1$ and $v, w \in A^*$ it holds that $(v, w) \in \mathcal{I} \Longrightarrow (v, v), (v^r, w \cdot v^r) \in \mathcal{I}$. For $p = (v, w) \in \mathcal{I}$ and given states s, s_1, s_2 of some dfa F we say:*

- *p appears at s $\xleftrightarrow{\text{def}}$ s has a v-loop, and*
- *s_1, s_2 are connected via p (in symbols $s_1 \overset{p}{\rightsquigarrow} s_2$) $\xleftrightarrow{\text{def}}$ p appears at s_1 and s_2, and $s_1 \xrightarrow{w} s_2$.*

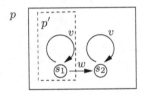

Fig. 1. The pattern for $\mathcal{B}_{1/2}$ from [10] can be written as the initial pattern $\mathcal{B} = A^+ \times A^+$. Here $p = (v, w) \in \mathcal{B}$ with loop-structure p' and bridge-structure p.

Consider the initial pattern $\mathcal{B} = A^+ \times A^+$ and some $p = (v, w) \in \mathcal{B}$. We interpret p as the structure shown in Fig. 1. In the initial case the loop-structure of p is simply a v-loop (cf. p' in Fig. 1), whereas its bridge-structure requires two states s_1, s_2 both having a v-loop such that $s_1 \xrightarrow{w} s_2$ (cf. p in Fig. 1). We say that p *appears at* some state s if we find the loop-structure of p at this state. In contrast, if two states s_1, s_2 are *connected via* p, then we find the bridge-structure of p between them.

As a next step, we observe how the patterns for the levels 1/2 are used in those for levels 3/2. In case of the STH the reader may consider the results from [10] (with an appropriate rewriting of patterns), for the DDH we refer to Fig. 1 and 2. They show that (i) the loop-structures p_i' of $\mathcal{B}_{1/2}$ patterns appear on the path $s_1 \longrightarrow s_2$ in the $\mathcal{B}_{3/2}$ pattern and (ii) the bridge-structures p_i of $\mathcal{B}_{1/2}$ patterns appear as building blocks in the loop-structures of the $\mathcal{B}_{3/2}$ pattern. We formalize this observation as the following iteration rule.

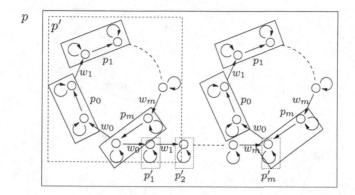

Fig. 2. The pattern for $\mathcal{B}_{3/2}$ from [6] can be written as $\mathbb{P}_1^{\mathcal{B}} = \mathrm{IT}(\mathcal{B})$ with $\mathcal{B} = A^+ \times A^+$. Here $p = (w_0, p_0, w_1, p_1, \dots, w_m, p_m) \in \mathbb{P}_1^{\mathcal{B}}$ with loop-structure p' and bridge-structure p. Moreover, $p_i \in \mathcal{B}$ has loop-structure p_i' and bridge-structure p_i.

Definition 2. *For sets* $\mathbb{P}$ *let* $\mathrm{IT}(\mathbb{P}) := \{(w_0, p_0, \dots, w_m, p_m) : p_i \in \mathbb{P}, w_i \in A^+\}$.

In the following we start with an initial pattern $\mathcal{I}$ and generate classes of iterated patterns by repeated applications of IT.

Definition 3. *For an initial pattern* $\mathcal{I}$ *we define* $\mathbb{P}_0^{\mathcal{I}} := \mathcal{I}$ *and* $\mathbb{P}_{n+1}^{\mathcal{I}} := \mathrm{IT}(\mathbb{P}_n^{\mathcal{I}})$ *for* $n \geq 0$. *For some* $p = (w_0, p_0, \dots, w_m, p_m) \in \mathrm{IT}(\mathbb{P}_n^{\mathcal{I}})$ *and given states* s, s_1, s_2 *of some dfa* F *we say:*

- p *appears at* s $\overset{\text{def}}{\Longleftrightarrow}$ *there exist states* $q_0, r_0, \dots, q_m, r_m$ *such that*
  $$s \xrightarrow{w_0} q_0 \overset{p_0}{\rightsquigarrow} r_0 \xrightarrow{w_1} q_1 \overset{p_1}{\rightsquigarrow} r_1 \xrightarrow{w_2} \dots \xrightarrow{w_m} q_m \overset{p_m}{\rightsquigarrow} r_m = s$$
- s_1, s_2 *are connected via* p *(in symbols* $s_1 \overset{p}{\rightsquigarrow} s_2$*)* $\overset{\text{def}}{\Longleftrightarrow}$ p *appears at* s_1 *and* s_2, *there exist states* $q_0, \dots, q_m$ *such that* $s_1 \xrightarrow{w_0} q_0 \xrightarrow{w_1} q_1 \xrightarrow{w_2} \dots \xrightarrow{w_m} q_m = s_2$ *and* p_i *appears at state* q_i *for* $0 \leq i \leq m$

Again, let us comment on this definition and see how we can understand it with the known results at hand (cf. Fig. 1 and 2). Consider the initial pattern $\mathcal{B} = A^+ \times A^+$ and some $p \in \mathbb{P}_1^{\mathcal{B}}$. This means that $p = (w_0, p_0, \dots, w_m, p_m)$ for words w_i and elements $p_i \in \mathbb{P}_0^{\mathcal{B}} = \mathcal{B}$. The loop-structure described by p is a loop with factors of words $w_0, w_1, \dots, w_m$ in this ordering such that between each w_i, w_{i+1} we find the bridge-structure of p_i. Here we see how elements of $\mathbb{P}_0^{\mathcal{B}}$ appear as building blocks in the loop-structure of elements of $\mathbb{P}_1^{\mathcal{B}}$. The bridge-structure of p connects two states s_1, s_2 such that we find the loop-structure of p at both of them. Additionally, it holds that $s_1 \xrightarrow{w_1 \cdots w_m} s_2$ and after each prefix $w_0 \cdots w_i$ we reach a state at which the loop-structure of p_i appears ($= p_i'$ in Fig. 2). An example of the next iteration step for initial pattern $\mathcal{B}$ is given in Fig. 3.

As mentioned at the beginning of this subsection we use the just defined patterns in the following way as forbidden patterns in dfa's.

p

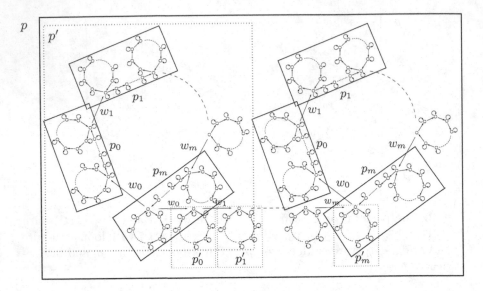

Fig. 3. Pattern $\mathbb{P}_2^{\mathcal{B}} = \mathrm{IT}(\mathbb{P}_1^{\mathcal{B}})$. Here $p = (w_0, p_0, w_1, p_1, \ldots, w_m, p_m) \in \mathbb{P}_2^{\mathcal{B}}$ with loop-structure p' and bridge-structure p. Moreover, $p_i \in \mathbb{P}_1^{\mathcal{B}}$ has loop-structure p'_i and bridge-structure p_i.

Definition 4. *For a dfa $F = (A, S, \delta, s_0, S')$, an initial pattern $\mathcal{I}$ and $n \geq 0$ we say F has pattern $\mathbb{P}_n^{\mathcal{I}}$ if and only if there exist $s_1, s_2 \in S$, $u, z \in A^*$, $p \in \mathbb{P}_n^{\mathcal{I}}$ such that $\delta(s_0, u) = s_1$, $\delta(s_1, z) \in S'$, $\delta(s_2, z) \notin S'$ and $s_1 \overset{p}{\leadsto} s_2$.*

3.2   Auxiliary Results

To handle patterns p in a better way, we define a word $\overline{p}^{\circ}$ obtained from the loop-structure of p (call this the loop-word) and a word $\overline{p}$ which is derived from the bridge-structure of p (bridge-word).

Definition 5. *Let $\mathcal{I}$ be an initial pattern. For $p = (v, w) \in \mathbb{P}_0^{\mathcal{I}}$ let $\overline{p} := w$ and $\overline{p}^{\circ} := v$. For $n \geq 0$ and $p = (w_0, p_0, \ldots, w_m, p_m) \in \mathbb{P}_{n+1}^{\mathcal{I}}$ let $\overline{p} := w_0 \cdots w_m$ and $\overline{p}^{\circ} := w_0 \overline{p_0} \cdots w_m \overline{p_m}$.*

The following is clear by definition. If p appears at some state s, then this state has a $\overline{p}^{\circ}$-loop, and if s_1 and s_2 are connected via p then the bridge-word $\overline{p}$ leads from s_1 to s_2. Moreover, for $n \geq 1$ and $p = (w_0, p_0, \ldots, w_m, p_m) \in \mathbb{P}_n^{\mathcal{I}}$ we have $\overline{p}, \overline{p}^{\circ} \in A^+$, and if p appears at some state then also p_m appears there.

In order to establish a relation between the polynomial closure operation and the iteration rule, we isolate the main argument of the proof of Theorem 1 in Lemma 3 below, for which the following two constructions are needed. First, for every $p \in \mathbb{P}_n^{\mathcal{I}}$ some $\lambda(p) \in \mathbb{P}_n^{\mathcal{I}}$ can be defined such that if p appears at some state s then s, s are connected via $\lambda(p)$ (cf. Definition 6 and Lemma 1). Secondly, in

Definition 7 and Lemma 2 we pump up the loop-structure of p to construct for given $r \geq 3$ some $\pi(p,r) \in \mathbb{P}_n^{\mathcal{I}}$ such that for every dfa F we have

(i) if two states are connected via p, then they are connected via $\pi(p,r)$ and
(ii) if $|F| \leq r$ then $\overline{\pi(p,r)}$ and $\overline{\pi(p,r)}^{\circ}$ lead to states where $\pi(p,r)$ appears.

Definition 6. *Let $\mathcal{I}$ be an initial pattern. For $p = (v, w) \in \mathbb{P}_0^{\mathcal{I}}$ let $\lambda(p) := (v, v)$. For $n \geq 1$ and $p = (w_0, p_0, \dots , w_m, p_m) \in \mathbb{P}_n^{\mathcal{I}}$ let $\lambda(p) := (\overline{p}, \lambda(p_m))$.*

The following lemma is easy to see by an induction on n.

Lemma 1. *For every initial pattern $\mathcal{I}$, $n \geq 0$ and $p \in \mathbb{P}_n^{\mathcal{I}}$ we have $\lambda(p) \in \mathbb{P}_n^{\mathcal{I}}$. Moreover, if p appears at state s of some dfa, then s, s are connected via $\lambda(p)$.*

Definition 7. *Let $\mathcal{I}$ be an initial pattern and $r \geq 3$. For $p = (v, w) \in \mathbb{P}_0^{\mathcal{I}}$ let $\pi(p,r) := (v^{r!}, w \cdot v^{r!})$. For $n \geq 1$ and $p = (w_0, p_0, \dots , w_m, p_m) \in \mathbb{P}_n^{\mathcal{I}}$ we define:*

$$p_i' := \pi(p_i, r)$$
$$w := w_0 \cdot \overline{p_0'}^{\circ} \cdot \overline{p_0'}^{\circ} \cdots w_m \cdot \overline{p_m'}^{\circ} \cdot \overline{p_m'}$$
$$\pi(p,r) := (w_0 \cdot \overline{p_0'}^{\circ}, p_0', \dots , w_m \cdot \overline{p_m'}^{\circ}, p_m', \underbrace{w, \lambda(p_m'), \dots , w, \lambda(p_m')}_{(r! - 1) \ \text{times} \ \text{“} w, \lambda(p_m') \text{”}})$$

Again, it is immediate by definition that $\pi(p,r) \in \mathbb{P}_n^{\mathcal{I}}$. Furthermore, a short observation makes clear that (i) if p appears at state s of some dfa, then also $\pi(p,r)$ appears there and (ii) if the states s_1, s_2 of some dfa are connected via p, then these states are also connected via $\pi(p,r)$. In addition, we prove the following lemma.

Lemma 2. *Let $\mathcal{I}$ be an initial pattern, $r \geq 3$, $n \geq 0$, $p \in \mathbb{P}_n^{\mathcal{I}}$, and let F be a dfa with $|F| \leq r$.*

1. *$\overline{\pi(p,r)}^{\circ}$ leads to states in F where $\pi(p,r)$ appears.*
2. *$\overline{\pi(p,r)}$ leads to states in F where $\pi(p,r)$ appears.*
3. *$\overline{\pi(p,r)}^{\circ}, \overline{\pi(p,r)}^{\circ} \overline{\pi(p,r)}$ lead to states in F which are connected via $\pi(p,r)$.*

Proof. We prove the lemma by induction on n. For $n = 0$ we have $p = (v, w)$ and $\pi(p,r) = (v^{r!}, w \cdot v^{r!})$. Since $v^{r!}$ leads to $v^{r!}$-loops in F, we obtain that $\overline{\pi(p,r)}^{\circ} = v^{r!}$ and $\overline{\pi(p,r)} = w \cdot v^{r!}$ lead to states where $\pi(p,r)$ appears. Hence $\overline{\pi(p,r)}^{\circ}, \overline{\pi(p,r)}^{\circ} \cdot \overline{\pi(p,r)}$ lead to states which are connected via $\pi(p,r)$.

For the induction step let $n = l + 1$, $p = (w_0, p_0, \dots , w_m, p_m) \in \mathbb{P}_{l+1}^{\mathcal{I}}$, and let w, p_i' as in Definition 7. First of all we show the following claim.

Claim 1: $w^{r!-1}$ leads to states in F where $\pi(p,r)$ appears.

Observe that $w^{r!-1}$ leads to a $w^{r!}$-loop in F. So let s be a state in F that has a $w^{r!}$-loop, we will show that $\pi(p,r)$ appears at s. Define the witnessing states:

$$q_0 := \delta(s, w_0 \cdot \overline{p_0'}^{\circ}) \qquad r_0 := \delta(q_0, \overline{p_0'})$$
$$q_i := \delta(r_{i-1}, w_i \cdot \overline{p_i'}^{\circ}) \qquad r_i := \delta(q_i, \overline{p_i'}) \qquad \text{for } 1 \leq i \leq m$$
$$q_{m+j} := \delta(r_m, w^j) \qquad r_{m+j} := q_{m+j} \qquad \text{for } 1 \leq j \leq r! - 1$$

So we have the following situation where $m' := m + r! - 1$.

$$s \xrightarrow{w_0 \cdot \overline{p_0'}^\circ} q_0 \xrightarrow{\overline{p_0'}} r_0 \xrightarrow{w_1 \cdot \overline{p_1'}^\circ} q_1 \xrightarrow{\overline{p_1'}} r_1 \xrightarrow{w_2 \cdot \overline{p_2'}^\circ} \cdots \xrightarrow{w_m \cdot \overline{p_m'}^\circ} q_m \xrightarrow{\overline{p_m'}} r_m$$
$$r_m \xrightarrow{w} q_{m+1} = r_{m+1} \xrightarrow{w} q_{m+2} = r_{m+2} \xrightarrow{w} \cdots \xrightarrow{w} q_{m'} = r_{m'}$$

Therefore, from induction hypothesis it follows that q_i, r_i are connected via p_i' for $0 \le i \le m$. Moreover, the hypothesis also shows that p_m' appears at q_j for $m + 1 \le j \le m'$, since $\overline{p_m'}$ is a suffix of w. From Lemma 1 it follows that q_j, r_j are connected via $\lambda(p_m')$. Finally, by the definition of w we have $r_m = \delta(s, w)$ and $r_{m'} = \delta(s, w^{r!}) = s$. Hence we have shown the following.

$$s \xrightarrow{w_0 \cdot \overline{p_0'}^\circ} q_0 \overset{p_0'}{\rightsquigarrow} r_0 \xrightarrow{w_1 \cdot \overline{p_1'}^\circ} q_1 \overset{p_1'}{\rightsquigarrow} r_1 \xrightarrow{w_2 \cdot \overline{p_2'}^\circ} \cdots \xrightarrow{w_m \cdot \overline{p_m'}^\circ} q_m \overset{p_m'}{\rightsquigarrow} r_m$$
$$r_m \xrightarrow{w} q_{m+1} \overset{\lambda(p_m')}{\rightsquigarrow} r_{m+1} \xrightarrow{w} q_{m+2} \overset{\lambda(p_m')}{\rightsquigarrow} r_{m+2} \xrightarrow{w} \cdots \xrightarrow{w} q_{m'} \overset{\lambda(p_m')}{\rightsquigarrow} r_{m'} = s$$

So $\pi(p, r)$ appears at s which shows our claim.

Since $\overline{p_m'}$ is a suffix of w, it follows that w leads to states where p_m' appears (induction hypothesis). From Lemma 1 we obtain that w leads to a $\lambda(p_m')$-loop in F. Hence Claim 1 also holds for $\left(w \cdot \lambda(p_m')\right)^{r!-1}$. Now observe the following.

$$\overline{\pi(p, r)} = w_0 \cdot \overline{p_0'}^\circ \cdots w_m \cdot \overline{p_m'}^\circ \cdot w^{r!-1}$$
$$\overline{\pi(p, r)}^\circ = w_0 \cdot \overline{p_0'}^\circ \cdot \overline{p_0'}^\circ \cdots w_m \cdot \overline{p_m'}^\circ \cdot \overline{p_m'}^\circ \cdot \left(w \cdot \overline{\lambda(p_m')}\right)^{r!-1}$$

It follows that $\overline{\pi(p, r)}$ and $\overline{\pi(p, r)}^\circ$ lead to states in F where $\pi(p, r)$ appears. This shows the statements 1 and 2 of the lemma.

Let us turn to statement 3 and choose an arbitrary state s of F. For $s_1 := \delta(s, \overline{\pi(p, r)}^\circ)$ and $s_2 := \delta(s, \overline{\pi(p, r)}^\circ \cdot \overline{\pi(p, r)})$ we show that s_1, s_2 are connected via $\pi(p, r)$. Let $m' := m + r! - 1$ and define the following witnessing states.

$$q_0 := \delta(s_1, w_0 \cdot \overline{p_0'}^\circ)$$
$$q_{i+1} := \delta(q_i, w_{i+1} \cdot \overline{p_{i+1}'}^\circ) \quad \text{for } 0 \le i < m$$
$$q_{j+1} := \delta(q_j, w) \quad \text{for } m \le j < m'$$

We have already seen that $\pi(p, r)$ appears at s_1 and at s_2. Observe that $q_{m'} = \delta(s_1, \overline{\pi(p, r)}) = s_2$. So it remains to show that (i) p_i' appears at q_i for $0 \le i \le m$ and (ii) $\lambda(p_m')$ appears at q_j for $m + 1 \le j \le m'$.

By induction hypothesis, $\overline{p_i'}^\circ$ leads to states in F where p_i' appears. Hence p_i' appears at state q_i for $0 \le i \le m$. Note that $\overline{p_m'}$ is a suffix of w. So the induction hypothesis shows that p_m' appears at q_j for all j with $m + 1 \le j \le m'$. From Lemma 1 it follows that q_j, q_j are connected via $\lambda(p_m')$. Particularly, $\lambda(p_m')$ appears at state q_j. This proves the lemma.

Now we isolate the main argument of the proof of Theorem 1. The following lemma says that under certain assumptions we can replace bridge-words by their respective loop-words without leaving the language of some dfa.

Lemma 3. *Let $\mathcal{I}$ be an initial pattern, $r \geq 3$, $n \geq 0$, $p \in \mathbb{P}_{n+1}^{\mathcal{I}}$ and let F be a dfa with $|F| \leq r$ which does not have pattern $\mathbb{P}_n^{\mathcal{I}}$. Then for all $u, z \in A^*$ we have*

$$u\overline{\pi(p,r)}z \in \mathrm{L}(F) \implies u\overline{\pi(p,r)}^{\circ}z \in \mathrm{L}(F).$$

Proof. Suppose $F = (A, S, \delta, s_0, S')$ and $p = (w_0, p_0, \ldots, w_m, p_m)$ for suitable $m \geq 0$, $w_i \in A^+$ and $p_i \in \mathbb{P}_n^{\mathcal{I}}$. Let $u, z \in A^*$ such that $u\overline{\pi(p,r)}z \in \mathrm{L}(F)$, and let p_i' and w as in Definition 7. Compare the following factorizations.

$$\overline{\pi(p,r)}^{\circ} = w_0 \cdot \overline{p_0'}^{\circ} \cdot \overline{p_0'}^{\circ} \cdots w_m \cdot \overline{p_m'}^{\circ} \cdot \overline{p_m'}^{\circ} \cdot \left(w \cdot \lambda(p_m')\right)^{r!-1} \tag{1}$$

$$\overline{\pi(p,r)} = w_0 \cdot \overline{p_0'}^{\circ} \cdot \quad \cdots w_m \cdot \overline{p_m'}^{\circ} \cdot \quad \left(w \quad\quad\right)^{r!-1} \tag{2}$$

We already know that (i) $\overline{p_m'}$ leads to a state in s where p_m' appears and that (ii) such a state has a connection with itself via $\lambda(p_m')$. It follows that $\overline{p_m'}$ leads to an $\overline{\lambda(p_m')}$-loop in F, which in turn implies that also w leads to such a loop. So taking (1) and (2) into account, it remains to show the following for all $u', z' \in A^*$.

$$u'\overline{p_i'}^{\circ}z' \in \mathrm{L}(F) \implies u'\overline{p_i'}^{\circ} \cdot \overline{p_i'} \cdot z' \in \mathrm{L}(F)$$

Suppose the contrary and let $s_1 := \delta(s_0, u'\overline{p_i'}^{\circ})$, $s_2 := \delta(s_0, u'\overline{p_i'}^{\circ} \cdot \overline{p_i'})$. By Lemma 2.3 we know that s_1, s_2 are connected via p_i'. Since $\delta(s_1, z')$ is accepting and $\delta(s_2, z')$ is rejecting, we have found pattern $\mathbb{P}_n^{\mathcal{I}}$ in F, a contradiction.

3.3 Pattern Iterator versus Polynomial Closure

The proof the following theorem can be carried out with Lemma 3. It says that pattern iteration captures complementation followed by polynomial closure.

Definition 8. *Let $\mathcal{I}$ be an initial pattern. For $n \geq 0$ we define the class of languages corresponding to $\mathbb{P}_n^{\mathcal{I}}$ as*

$$\mathbb{L}_n^{\mathcal{I}} := \left\{L \subseteq A^+ : L \text{ is accepted by some dfa } F \text{ which does not have } \mathbb{P}_n^{\mathcal{I}}\right\}.$$

With Lemma 2 one can show that this is well-defined.

Theorem 1. *Let $\mathcal{I}$ be an initial pattern and $n \geq 0$. Then $\mathrm{Pol}(\mathrm{co}\mathbb{L}_n^{\mathcal{I}}) \subseteq \mathbb{L}_{n+1}^{\mathcal{I}}$.*

Proof. We assume that there exists an $L \in \mathrm{Pol}(\mathrm{co}\mathbb{L}_n^{\mathcal{I}}) \setminus \mathbb{L}_{n+1}^{\mathcal{I}}$, this will lead to a contradiction. Let $F = (A, S, \delta, s_0, S')$ be some dfa with $\mathrm{L}(F) = L$. Since $L \in \mathrm{Pol}(\mathrm{co}\mathbb{L}_n^{\mathcal{I}})$, we have

$$L = \bigcup_{i=1}^{k} L_{i,0}L_{i,1} \cdots L_{i,k_i}$$

for languages $L_{i,j} \in \mathrm{co}\mathbb{L}_n^{\mathcal{I}}$. Choose $r \geq 1$ sufficiently large, i.e., larger than k, k_i and the size of some dfa's accepting L, $L_{i,j}$ and the complement of $L_{i,j}$.

Since $L \notin \mathbb{L}_{n+1}^{\mathcal{I}}$ there exist states $s_1, s_2 \in S$ and words $u, z \in A^*$ such that s_1, s_2 are connected via some $p \in \mathbb{P}_{n+1}^{\mathcal{I}}$, $\delta(s_0, u) = s_1$, $\delta(s_1, z)$ is accepting and $\delta(s_2, z)$ is rejecting. It follows that s_1, s_2 are also connected via $\pi(p, r)$ and $u \left(\overline{\pi(p, r)}^\circ \right)^r z \in L_{i', 0} L_{i', 1} \cdots L_{i', k_{i'}}$ for some $1 \leq i' \leq k$. Since $r > k_{i'}$, the latter word can be factorized as $u'' u' \overline{\pi(p, r)}^\circ z' z''$ such that $u'' \in L_{i', 0} \cdots L_{i', j'-1}$, $u' \overline{\pi(p, r)}^\circ z' \in L_{i', j'}$ and $z'' \in L_{i', j'+1} \cdots L_{i', k_{i'}}$ for some $j' \leq k_{i'}$. Because there is a dfa of size $\leq r$ accepting $L_{i', j'}$, the word $\overline{\pi(p, r)}^\circ$ leads to a $\overline{\pi(p, r)}^\circ$-loop in this dfa (Lemma 2.1). Hence for all $i \geq 1$ we obtain

$$u' \left(\overline{\pi(p, r)}^\circ \right)^i z' \in L_{i', j'}. \tag{3}$$

Moreover, $u' \overline{\pi(p, r)}^\circ \pi(p, r) z' \notin L_{i', j'}$, otherwise we would obtain

$$u \left(\overline{\pi(p, r)}^\circ \right)^j \overline{\pi(p, r)} \left(\overline{\pi(p, r)}^\circ \right)^{r-j} z \in L$$

for some $j \leq r$, which in turn implies the contradiction $u \overline{\pi(p, r)} z \in L$ (recall that s_1, s_2 are connected via $\pi(p, r)$ in F). Observe that some dfa accepting the complement of $L_{i', j'}$ is of size $\leq r$ and does not have pattern $\mathbb{P}_n^{\mathcal{I}}$. From Lemma 3 it follows that $u' \overline{\pi(p, r)}^\circ \overline{\pi(p, r)}^\circ z' \notin L_{i', j'}$. This contradicts (3).

3.4   Hierarchies, Decidability, and Starfreeness

Let $\mathcal{I}, \mathcal{J}$ be initial patterns and $i, j \geq 0$. We say that any pattern from $\mathbb{P}_j^{\mathcal{J}}$ can be interpreted as a pattern from $\mathbb{P}_i^{\mathcal{I}}$ if and only if for every $p \in \mathbb{P}_j^{\mathcal{J}}$ there exists a $p' \in \mathbb{P}_i^{\mathcal{I}}$ such that (i) if p appears at state s of some dfa, then also p' appears at this state and (ii) if the states s_1, s_2 of some dfa are connected via p, then they are also connected via p'. An easy induction shows that if any pattern from $\mathbb{P}_j^{\mathcal{J}}$ can be interpreted as a pattern from $\mathbb{P}_i^{\mathcal{I}}$, then $\mathbb{L}_{i+n}^{\mathcal{I}} \subseteq \mathbb{L}_{j+n}^{\mathcal{J}}$ for all $n \geq 0$. Particularly, if any pattern from $\mathbb{P}_1^{\mathcal{I}}$ can be interpreted as a pattern from $\mathbb{P}_0^{\mathcal{I}}$ (which is a weak assumption), then we obtain $\mathbb{L}_n^{\mathcal{I}} \subseteq \mathbb{L}_{n+1}^{\mathcal{I}}$ for $n \geq 0$. Together with Theorem 1 this yields $\mathbb{L}_n^{\mathcal{I}} \cup \operatorname{co}\mathbb{L}_n^{\mathcal{I}} \subseteq \mathbb{L}_{n+1}^{\mathcal{I}} \cap \operatorname{co}\mathbb{L}_{n+1}^{\mathcal{I}}$.

Let us turn to the decidability of pattern classes. It is reasonable to consider initial patterns $\mathcal{I}$ such that for every $k \geq 1$ there exists an algorithm $\mathcal{A}_k$ which does the following in nondeterministic logspace NL: On input F, k states of F and k pairs of states of F it decides whether there is some $p \in \mathcal{I}$ appearing at each of the given single states and connecting each of the given pairs. For $n \geq 0$ this leads by induction to an NL-algorithm for the membership problem of $\mathbb{L}_n^{\mathcal{I}}$.

The pattern iterator IT can be considered as a starfree iterator. Let $\mathcal{I}$ be an arbitrary initial pattern and recall that SF denotes the class of starfree languages. One can show that for $n \geq 1$ it holds that $\mathbb{L}_n^{\mathcal{I}} \subseteq SF$ if and only if $\bigcup_{i \geq 0} \mathbb{L}_i^{\mathcal{I}} \subseteq SF$ (actually this does not hold for $n = 0$).

4   Consequences for Concatenation Hierarchies

From now on we consider two special initial patterns. With $\mathcal{L} := \{\varepsilon\} \times A^*$ and $\mathcal{B} := A^+ \times A^+$ we meet the known forbidden pattern characterizations for $\mathcal{L}_{1/2}$ and $\mathcal{B}_{1/2}$ from [10]. Furthermore, if we compare the characterizations for $\mathcal{L}_{3/2}$ and $\mathcal{B}_{3/2}$ from [10] and [6], respectively, we observe that $\mathcal{L}_{3/2} = \mathbb{L}_1^{\mathcal{L}}$ and $\mathcal{B}_{3/2} = \mathbb{L}_1^{\mathcal{B}}$. From the results of the previous section we obtain for the pattern classes the same inclusion structure as it is known for the concatenation hierarchies in question (see Propositions 2 and 3). Moreover, it follows from Theorem 1 that the pattern classes contain the respective levels of the concatenation hierarchies (cf. Fig. 4).

Theorem 2. *For $n \geq 0$ the following holds.*

1. $\mathbb{L}_n^{\mathcal{L}} \cup \mathrm{co}\mathbb{L}_n^{\mathcal{L}} \subseteq \mathbb{L}_{n+1}^{\mathcal{L}} \cap \mathrm{co}\mathbb{L}_{n+1}^{\mathcal{L}}$
2. $\mathbb{L}_n^{\mathcal{B}} \cup \mathrm{co}\mathbb{L}_n^{\mathcal{B}} \subseteq \mathbb{L}_{n+1}^{\mathcal{B}} \cap \mathrm{co}\mathbb{L}_{n+1}^{\mathcal{B}}$
3. $\mathbb{L}_n^{\mathcal{L}} \subseteq \mathbb{L}_n^{\mathcal{B}} \subseteq \mathbb{L}_{n+1}^{\mathcal{L}}$

Theorem 3. *For $n \geq 0$ it holds that $\mathcal{L}_{n+1/2} \subseteq \mathbb{L}_n^{\mathcal{L}}$ and $\mathcal{B}_{n+1/2} \subseteq \mathbb{L}_n^{\mathcal{B}}$.*

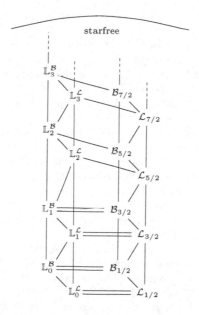

Fig. 4. Concatenation hierarchies and forbidden pattern classes. Inclusions hold from bottom to top, doubled lines stand for equality

The pattern hierarchies even exhaust the class of starfree languages.

Theorem 4. *It holds that $\bigcup_{n \geq 0} \mathbb{L}_n^{\mathcal{B}} = \bigcup_{n \geq 0} \mathbb{L}_n^{\mathcal{L}} = SF$.*

Next, we want to show the strictness of $\{\mathbb{L}_n^{\mathcal{B}}\}$ and $\{\mathbb{L}_n^{\mathcal{L}}\}$ in a certain way, namely we take witnessing languages from [16] (see also [2]) that were used there to separate the classes of the DDH. As remarked in [16], these languages can also be used to show that the STH is strict. W.l.o.g. we assume that $A = \{a, b\}$. Let us recall the definition of a particular family of languages of A^+ from [16]. Denote for $w \in A^+$ by $|w|_a$ the number of occurrences of the letter a in w. Now define for $n \geq 1$ the language L_n to be the set of words $w \in A^+$ such that $|w|_a - |w|_b = n$ and for every prefix v of w it holds that $0 \leq (|v|_a - |v|_b) \leq n$. It was shown in [16] that $L_n \in \mathcal{B}_n \backslash \mathcal{B}_{n-1}$. So with $\mathcal{B}_n \subseteq \mathcal{B}_{n+1/2}$ we get from Theorem 3 that $L_n \in \mathbb{L}_n^{\mathcal{B}}$. We can prove that the minimal dfa F_n with $L_n = L(F_n)$ has pattern $\mathbb{P}_n^{\mathcal{L}}$. It follows that $L_n \in \mathbb{L}_n^{\mathcal{B}} \setminus \mathbb{L}_n^{\mathcal{L}}$.

Theorem 5. *Let $n \geq 1$. Then the following holds.*

1. $\mathbb{L}_{n-1}^{\mathcal{B}} \subsetneq \mathbb{L}_n^{\mathcal{B}}$ and $\mathcal{B}_{n+1/2} \not\subseteq \mathbb{L}_{n-1}^{\mathcal{B}}$.
2. $\mathbb{L}_{n-1}^{\mathcal{L}} \subsetneq \mathbb{L}_n^{\mathcal{L}}$ and $\mathcal{L}_{n+1/2} \not\subseteq \mathbb{L}_{n-1}^{\mathcal{L}}$.

We adapt the well-known algorithm that solves the graph accessibility problem to show that the initial patterns $\mathcal{L}$ and $\mathcal{B}$ allow algorithms $\mathcal{A}_k$ as mentioned in subsection 3.4, so the membership problems for $\mathbb{L}_n^{\mathcal{B}}$ and $\mathbb{L}_n^{\mathcal{L}}$ are decidable in NL.

References

1. M. Arfi. Opérations polynomiales et hiérarchies de concaténation. *Theoretical Computer Science*, 91:71–84, 1991.
2. J. A. Brzozowski and R. Knast. The dot-depth hierarchy of star-free languages is infinite. *Journal of Computer and System Sciences*, 16:37–55, 1978.
3. R. S. Cohen and J. A. Brzozowski. Dot-depth of star-free events. *Journal of Computer and System Sciences*, 5:1–16, 1971.
4. S. Eilenberg. *Automata, languages and machines*, volume B. Academic Press, New York, 1976.
5. C. Glaßer. A normal form for classes of concatenation hierarchies. Technical Report 216, Inst. für Informatik, Univ. Würzburg, 1998.
6. C. Glaßer and H. Schmitz. Languages of dot-depth 3/2. In *Proceedings 17th STACS*, volume 1770 of *LNCS*, pages 555–566. Springer Verlag, 2000.
7. R. Knast. A semigroup characterization of dot-depth one languages. *RAIRO Inform. Théor.*, 17:321–330, 1983.
8. J.-E. Pin. Syntactic semigroups. In G.Rozenberg and A.Salomaa, editors, *Handbook of formal languages*, volume I, pages 679–746. Springer, 1996.
9. J.-E. Pin. Bridges for concatenation hierarchies. In *Proceedings 25th ICALP*, volume 1443 of *LNCS*, pages 431–442. Springer Verlag, 1998.
10. J.-E. Pin and P. Weil. Polynomial closure and unambiguous product. *Theory of computing systems*, 30:383–422, 1997.
11. I. Simon. Piecewise testable events. In *Proceedings 2nd GI Conference*, volume 33 of *Lecture Notes in Computer Science*, pages 214–222. Springer-Verlag, 1975.
12. H. Straubing. A generalization of the Schützenberger product of finite monoids. *Theoretical Computer Science*, 13:137–150, 1981.
13. H. Straubing. Finite semigroup varieties of the form V * D. *J.Pure Appl.Algebra*, 36:53–94, 1985.

14. H. Straubing. Semigroups and languages of dot-depth two. *Theoretical Computer Science*, 58:361–378, 1988.
15. D. Thérien. Classification of finite monoids: the language approach. *Theoretical Computer Science*, 14:195–208, 1981.
16. W. Thomas. An application of the Ehrenfeucht–Fraïssé game in formal language theory. *Société Mathématique de France, mémoire 16*, 2:11–21, 1984.
17. P. Weil. Some results on the dot-depth hierarchy. *Semigroup Forum*, 46:352–370, 1993.

Prefix Languages of Church-Rosser Languages

Jens R. Woinowski

TU Darmstadt
Fachbereich Informatik
Fax:+49-(0)6151-16-6185
woinowski@iti.informatik.tu-darmstadt.de

Abstract. Church-Rosser languages are mainly based on confluent length reducing string rewriting systems. In general, the prefix language of a Church-Rosser language may not be describable by such a system, too. In this paper it is shown that under certain conditions it is possible to give a construction for a system defining the prefix language and to prove its correctness. The construction also gives a completion of prefixes to full words in the original language. This is an interesting property for practical applications, as it shows potential for error recovery strategies in parsers.

1 Introduction

The Church-Rosser languages (CRL) are a relatively new class of languages. Basically they are defined by confluent and length reducing string-rewriting systems with a distinction between terminals and nonterminals and the possibility to mark word ends. They were defined by McNaughton, Narendran, and Otto in [MNO88] and are the deterministic variant of the growing context sensitive languages defined by Dahlhaus and Warmuth in [DW86]. This was proved by Niemann and Otto in [NO98]. On the one hand, they have some nice properties, e.g. solvability of the word problem in deterministic linear time, closure against the mirror operation, and they are a superset of the deterministic context free languages (detCF). On the other hand, they are a basis of the recursively enumerable languages. That means, given an alphabet Σ ($\mathtt{c}, \sharp \notin \Sigma$) and any r.e. language $L \subseteq \Sigma^*$ there is a CRL $L' \subseteq \Sigma^* \cdot \{\mathtt{c}\} \cdot \{\sharp\}^*$ so that deleting the letters $\mathtt{c}$ and $\sharp$ with a homomorphism h which leaves letters of Σ unchanged leads to $h(L') = L$ (see also [OKK97]). Therefore there are prefix languages of CRL's which are not CRL's themselves. So a natural question is, given a rewriting system for a CRL, under which conditions a new CRL system can be constructed and proved to be correct that delivers exactly the prefix language. In this article it is shown that under some restrictions for the CRL system this is possible.

There also is a practical aspect. Although the theory of parsing programming languages seems to be fully elaborated (with [Knu65] being a turning point), its main coverage is the parsing of correct programs. Even relatively new compiler generators like CUP [App98] do not provide much help to produce useful output

S. Kapoor and S. Prasad (Eds.): FST TCS 2000, LNCS 1974, pp. 516–529, 2000.
© Springer-Verlag Berlin Heidelberg 2000

in the error case. Most error recovery strategies are either relatively simple or they are very complicated and time consuming [SSS90], [App98]. The prefix construction for CRL given in this paper offers some new potential. Whenever the sufficient conditions are fulfilled, the construction does not only produce a CRL system for the prefix language—it also gives a method to deduce correct completions to a word of the language.[1] This is more than the prefix closure proof of detCF (see for example [Har78]) delivers.

This paper is organised in the following way: The next section gives the necessary basic definitions and a technical result about a normal form for CRLS's. Also, the construction for socalled prefix systems is introduced. The third section contains some examples for the effects of the construction and how prefixes are accepted. The fourth section goes into the details of the correctness problem in a slightly informal way and adds the construction of an enriched version of prefix systems. In the fifth section the main result is stated and proved.

Because of the limited space this text can only contain basic ideas. It tries to give an overview which allows to assess the theoretical value of the results. Proofs of the theorems are rather technical and lengthy. Because of this they are omitted or only briefly sketched. Full proofs can be found in the technical report [Woi00b].

2   Basic Definitions and Ideas

The following definitions are mostly necessary to identify the notations used throughout this paper. See also [Har78], [Jan88], [BO93], and [BO98].

Let Σ be a finite alphabet, Σ^* denotes the free monoid over Σ, for the empty word we write $\square$. A subset $L \subseteq \Sigma^*$ is called a language. If w is a word of length n, we write $|w| := n$. To address single letters of w we use $w = a_1 \cdots a_i \cdots a_n, a_i \in \Sigma (1 \leq i \leq n)$. Then $\mathrm{Pref}(w)$ ($\mathrm{Suff}(w)$) is the set of prefixes (suffixes) of w. With L being a language, $\mathrm{Pref}(L) := \bigcup_{w \in L} \mathrm{Pref}(w)$ and $\mathrm{Suff}(L) := \bigcup_{w \in L} \mathrm{Suff}(w)$.

A *string-rewriting system (or simply rewriting system)* R on Σ is a subset of $\Sigma^* \times \Sigma^*$. For $(u, v) \in R$ we also write $(u \to v) \in R$ and call (u, v) a rule. The *rewriting relation* between words in Σ^* is defined as $\xrightarrow{R} := \{(sut, svt)|s, t \in \Sigma^*, (u, v) \in R\}$. The reflexive and transitive closure is denoted with $\xrightarrow{R}^*$. A word $w \in \Sigma^*$ is called *irreducible modulo R* if there exists no w' with $w \xrightarrow{R} w'$. The set of all irreducible words of R is denoted with $\mathrm{IRR}(R)$.

A *weight function* is a function $f : \Sigma \to \mathbb{N}$. It is recursively extended to a function on Σ^* by $f(wx) := f(w) + f(x)$ and $f(\square) := 0$ with $w \in \Sigma^*, x \in \Sigma$. An example for a weight function is the *length function* with $f(x) := 1$ for all $x \in \Sigma$, then $f(w) = |w|$. A string-rewriting system R will be called a *weight reducing system*, if there exists a weight function f so that $f(u) > f(v)$ for all $(u, v) \in R$.

---

[1] This already has been implemented in a CRL development system [Rot00].

A string-rewriting system R is confluent if for all w, w_1, w_2, $w \xrightarrow{R}^* w_1$ and $w \xrightarrow{R}^* w_2$ there exists a $w_3 \in \text{IRR}(R)$ so that $w_1 \xrightarrow{R}^* w_3$ and $w_2 \xrightarrow{R}^* w_3$. If so, and if R is a weight reducing system[2], w_3 is unique and is called *irreducible normal form of* w (and, of course, w_1 and w_2). For a word w we denote its irreducible normal form with $[w]_R$.

Definition 1. *A* Church-Rosser language system (CRLS) *is a 5-tuple* $C = (\Gamma, \Sigma, R, k_l, k_r, y)$ *with finite alphabet* Γ, *terminal alphabet* $\Sigma \subset \Gamma$ ($\Gamma \setminus \Sigma$ *is the alphabet of nonterminals), finite confluent weight reducing system* $R \subseteq \Gamma^* \times \Gamma^*$, *left and right end marker words* $k_l, k_r \in (\Gamma \setminus \Sigma)^* \cap \text{IRR}(R)$, *and accepting letter* $y \in \Gamma \cap \text{IRR}(R)$ *The language defined by* C *is defined as:* $L_C := \{w \in \Sigma^* | k_l \cdot w \cdot k_r \rightarrow^*_R y\}$

A language L *is called a* Church-Rosser language (CRL) *if there exists a CRLS* C *with* $L_C = L$. *We say that a word* w *is* accepted *by* C *if* $w \in L_C$, *thus stressing the fact that a CRL is defined by a reduction process.*

To address the rewriting system R *of a CRLS* C, *we use* $\text{REW}(C)$.

The definition of Church-Rosser languages is due to McNaughton, Narendran, and Otto [MNO88]. The definition of Church-Rosser language systems given here is a convenient notation for their definition. Nieman and Otto proved in [NO98] that the expressive power of Church-Rosser languages is not enhanced by allowing arbitrary weight functions instead of the length function, so this fact is used, too.

In order to be able to give a prefix construction for CRLS's, some restrictions will be made. The first observation is that detecting the left and right end of a word is relatively difficult because of the arbitrary end marker words k_l and k_r. In consequence, only single letters will be used: $k_l = ¢$ and $k_r = \$$. Furthermore it will be required that these are not changed, removed, or added throughout the reduction process. Only if the word is accepted they will be deleted. Secondly, we will require all rules to be of a limited form which makes some operations easier. This form is inspired by the shift-reduce automatons for detCF languages. Furthermore it is very similar to well known forms of context-sensitive grammars.

Definition 2. *A CRLS* $C = (\Gamma, \Sigma, R, ¢, \$, y)$ *is* prefix splittable (C *is a psCRLS) if* $¢, \$, y \in \text{IRR}(R) \cap \Gamma \setminus \Sigma$ *(let the* inner alphabet *be* $\Gamma_{inner} := \Gamma \setminus \{¢, \$, y\}$) *and for any rule* $r \in R$ *there exists a splitting* (u, v, w, x) *with:*

1. $r = (uvw, uxw)$
2. v *is non-empty.*
3. *uvw may contain at most one* ¢ *and if so at its beginning. Also it can have at most one* \$ *which only may appear at the end. All other letters of uvw have to be from the inner alphabet* Γ_{inner}.

---

[2] Or otherwise terminating, but this will not be considered in this text. Also, in case of terminating rewriting systems, local confluence implies confluence.

4. x is a single letter not equal to ¢ or \$ or it is the empty word.
5. If v contains a ¢ or \$, then $x = y$, u and w are empty, and v is of the form
   $¢ \cdot \Gamma_{inner}^* \cdot \$$.
6. If $x = y$, then u and w are empty, and v is of the form $¢ \cdot \Gamma_{inner}^* \cdot \$$.

The splitting (u, v, w, x) of a rule r allowed by this is called a potential prefix splitting, u and w are called the left and right context, respectively.

Also this definition seems to be rather restrictive, it is possible to show that it is a normal form for CRLS's. We only state the result because the proof for this is not in the scope of this text:

Theorem 1. Let $C = (\Gamma, \Sigma, R, k_l, k_r, y)$ be a CRLS (without restriction of generality let R be length reducing [NO98]) with language L_C. Then there exists a psCRLS C' with $L_{C'} = L_C$.

Note on proof. The main idea is to use a compression technique. That is, an alphabet of non-terminals which can store more than one letter (of the input or of intermediate reduction results). The biggest problem is to ensure that the new system is weight reducing. This can be done by *spreading weights* over more than one of these compression letters. In order to achieve this it is necessary to simulate single rules by chains of rules that are linked to each other in a way delivering confluence. This also requires a system where at any time the place of the next possible reduction can be uniquely identified. Working on a compression alphabet, the restriction of the end marker words to special letters is a very simple problem. For more details see [Woi00a].

Remark 1. A rule may have several different prefix splittings. For our investigation it will not matter which prefix splitting we choose. Because of this situation we just choose a prefix splitting arbitrarily (see also [Woi00b]).

The idea of constructing a prefix CRLS to a psCRLS is very basic: Simply cut off suffixes of rules. To be precise, some efforts are necessary to handle the right end of words. Given any unique definition of prefix splittings, a prefix system is defined as follows:

Definition 3. Let C be a psCRLS, $r \in \text{REW}(C)$ with prefix splitting (u, v, w, x), $v = a_1 \cdots a_i \cdots a_{|v|}$, $a_i \in \Gamma$, $w = b_1 \cdots b_i \cdots b_{|w|}$, $b_i \in \Gamma$.
The prefix rules of r (PREF(r)) are defined as:

$$\text{PREF}(r) := (\ \{(uvb_1 \cdots b_j\$, uxb_1 \cdots b_j\$)|0 \le j < |w|\}$$
$$\cup \begin{cases} \{(a_1 \cdots a_j\$, y)|0 < j < |v|\} & x = y \\ \{(ua_1 \cdots a_j\$, ux\$)|0 < j < |v|\} & else \end{cases}$$
$$)$$
$$\setminus \{(w, w)|w \in \Gamma^*\}$$

Define the following rewriting system R':

$$R' = R \cup \bigcup_{r \in R} \text{PREF}(r).$$

If R' is weight reducing and confluent (which is decidable) then the CRLS $C' = (\Gamma, \Sigma, R', \text{¢}, \$, y)$ *is called* prefix system of C, $C' = \text{PREF}(C)$.

We also use $R' = \text{PREF}(R)$ in that case and call R the origin *of R'. If R' is not confluent or not weight reducing $\text{PREF}(R)$ and $\text{PREF}(C)$ are not defined. The reason for this will be discussed later. The process of building $\text{PREF}(R)$ is called* prefix construction.

3   Some Examples

The first example will be used throughout the rest of this paper to show the effects of the prefix construction:

Example 1. The psCRLS C is defined in the following way: Let $\Sigma = \{a, b, c, d, e\}$, $\Gamma = \Sigma \cup \{\$, \text{¢}, y\}$ with left and right end markers ¢ and \$ and accepting symbol y. Let the rewriting system R be defined as follows. We mark a prefix splitting[3] with concatenation dots '·', these dots do not belong to the rules:

$$\cdot abc \cdot \rightarrow \cdot b \cdot$$
$$\cdot abb \cdot bc \rightarrow \cdot a \cdot bc$$
$$\cdot bb \cdot \$ \rightarrow \cdot b \cdot \$$$
$$\cdot db \cdot \rightarrow \cdot b \cdot$$
$$\cdot e \cdot c \rightarrow \cdot b \cdot c$$
$$\cdot \text{¢} b \$ \cdot \rightarrow \cdot y \cdot$$

In figure 1 the prefix system R' of $C' = \text{PREF}(C)$ is given. The parts appearing in brackets and the double rules may be ignored at this point. They contain information about the parts cut off which will be used later.

Since it is easily verified that R' is confluent and weight reducing we omit this here. Now have a look at a prefix of the word $adabbdecc \in L_C$ and its acceptance in C'. Observe the mixture of rules already in R and those new rules from $R' \setminus R$:

$$\text{¢} adabbde\$ \underset{r_{12}}{\rightarrow} \text{¢} adabbdb\$ \underset{r_9}{\rightarrow} \text{¢} adabbb\$ \underset{r_5}{\rightarrow} \text{¢} adab\$ \underset{r_2}{\rightarrow} \text{¢} adb\$ \underset{r_9}{\rightarrow} \text{¢} ab\$ \underset{r_2}{\rightarrow} \text{¢} b\$ \underset{r_{13}}{\rightarrow} y$$

This seems to work fine, but we cannot always be sure that a prefix system is doing what we would expect it to do. Regard these three examples:

---

[3] Note that this choice is not necessarily optimal, but here, this will not be discussed further.

$$
\begin{array}{llll}
r_1 & abc \rightarrow b & [\square/\square] \\
r_2 & ab\$ \rightarrow b\$ & [c/\square] \\
r_3 & a\$ \rightarrow b\$ & [bc/\square] \\
r_4 & abbbc \rightarrow abc & [\square/\square] \\
r_5 & abbb\$ \rightarrow ab\$ & [\square/c] \\
r_6 & abb\$ \rightarrow a\$ & [\square/bc] \\
r_7 & ab\$ \rightarrow a\$ & [b/bc] \\
r_8 & bb\$ \rightarrow b\$ & [\square/\$] \\
r_9 & db \rightarrow b & [\square/\square] \\
r_{10} & d\$ \rightarrow b\$ & [b/\square] \\
r_{11} & ec \rightarrow bc & [\square/\square] \\
r_{12} & e\$ \rightarrow b\$ & [\square/c] \\
r_{13} & ¢b\$ \rightarrow y & [\square/\$] \\
r_{14} & ¢\$ \rightarrow y & [b/\$]
\end{array}
$$

Fig. 1. The prefix system R' for example 1.

Let the psCRLSs be C_1, C_2, C_3. Let $\Sigma = \{a, b, c\}$ and $\Gamma = \{a, b, c, D, ¢, \$, y\}$ be the common alphabets of the three systems. Build prefix rewritings systems R'_1, R'_2, R'_3 with the prefix construction.

Example 2. $R_1 = \{(a, b), (ba, a), (bb, a), (¢a\$, y)\}$
$R'_1 = R_1 \cup \{(b\$, a\$), (¢\$, y)\}$
R'_1 is not weight reducing: $a\$ \xrightarrow[R'_1]{} b\$ \xrightarrow[R'_1]{} a\$$

Example 3. $R_2 = \{abb\$, a\$), (bbc, d), (¢a\$, y)$
$R'_2 = R_2 \cup \{ab\$, a\$), (bb\$, d\$), (b\$, d\$), (¢\$, y)\}$
R'_2 is not confluent:
$¢abb\$ \xrightarrow[R'_2]{}^* y \in \mathrm{IRR}(R'_2)$ and $¢abb\$ \xrightarrow[R'_2]{}^* ¢ad\$ \in \mathrm{IRR}(R'_2)$

Example 4. $R_3 = \{(¢D\$, y)\}$
$R'_3 = R_3 \cup \{(¢\$, y)\}$
$L_{C_3} = \emptyset$ but $L_{C'_3} = \{\square\}$.

These examples lead to the following definition and a result stated as remark. They also answer the question why we required R' to be confluent and weight reducing in the definition of $\mathrm{PREF}(C)$.

Definition 4. *Let C be a psCRLS. Then $\mathrm{PREF}(C)$ is correct if and only if it is defined and $L_{\mathrm{PREF}(C)} = \mathrm{Pref}(L_C)$.*

Remark 2. There are psCRLS C so that $\mathrm{PREF}(C)$ is not defined or not correct.

On the other hand we have a positive result:

Theorem 2. *Let C be a psCRLS and $C' = \text{PREF}(C)$ the prefix system with R' as string-rewriting system of C'. Then $\text{Pref}(L_C) \subseteq L_{C'}$.*
The proof is a simple induction over the length of reductions in C.

4   What Is Happening in Prefix Reductions?

A closer look at the rules of psCRLS reveals the different roles of the parts of the prefix splittings. Let r be a rule with prefix splitting (u, v, w, x). Then in the prefix construction the left context is never deleted. In contrast to this, parts v and w are deleted. Still those two differ in their meaning, because by applying r the information of v is always lost (resp. substituted by x), whereas w remains unchanged. Obviously, this is important for accepting correct prefixes. In order to use this, we now define *expanded* prefix CRLSs which have 4-tuples instead of pairs as rules. In these, the first two components have the same meaning as in usual rewriting systems, including the relation $\rightarrow$. In the other two we store what has been cut off during the prefix construction:

Definition 5. *Let R be a rewriting system defined on the alphabet Γ, and $\alpha \in \Gamma$. Then $R \backslash \alpha$ is the subsystem of R that is obtained by removing all rules containing the letter α.*

Definition 6. *Let C be a psCRLS with rewriting system R and $r \in R$ with prefix splitting (u, v, w, x), $v = a_1 \cdot a_2 \cdots a_i \cdots a_{|v|}$, $a_i \in \Gamma$, $w = b_1 \cdot b_2 \cdots b_i \cdots b_{|w|}$, $b_i \in \Gamma$. If $R \backslash \$$ is confluent and $\text{PREF}(C)$ is defined, we define the set $\text{CPREF}(r)$ of* completion prefix rules *of r (or, shorter,* completion rules*):*

$$\text{CPREF}(r) = (\quad \{(uvb_1 \cdots b_j\$, [uxb_1 \cdots b_j\$]_{R \backslash \$}, \square, b_{j+1} \cdots b_{|w|}) | 0 \le j < |w|\}$$

$$\cup \begin{cases} \{(a_1 \cdots a_j\$, y, a_{j+1} \cdots a_{|v|-1}, \$) | 0 < j < |v|\} & x = y \\ \{(ua_1 \cdots a_j\$, [ux\$]_{R \backslash \$}, a_{j+1} \cdots a_{|v|}, w) | 0 < j < |v|\} & else \end{cases}$$

$$\cup \{(uvw, uxw, \square, \square) | vw \notin \Gamma^* \cdot \$\}^4$$
$$)$$
$$\backslash \{(w', w', u', v') | w', u', v' \in \Gamma^*\}$$

- If $R \backslash \$$ is not confluent, $\text{CPREF}(r)$ is not defined.[5]
- Reducing the second component with all rules of R that do not work at the right end of words (which $R \backslash \$$ means) does not change the defined language.

---

[4] This set is empty or a singleton.
[5] It should be possible to avoid this restriction but this is not in the scope of this investigation.

- The third component of the completion rules is called *consumed completion*.
- The fourth component is called *unconsumed completion*
- The system R' is called *completion prefix system of* R with $R' = \bigcup_{r \in R} \text{CPREF}(r)$. With $\text{REW}(R')$ we denote the extraction of the first two components, which has a rewriting system as result. Note: If $\text{PREF}(C)$ is defined then $\text{REW}(R') = \text{REW}(\text{PREF}(C))$.
- In the same manner, CPREF will be used for the completion prefix CRLS given by the alphabets and accepting letter of C and R' iff $\text{PREF}(C)$ is defined.
- $R'_{\text{pref}} := R' \setminus \{(u, v, \square, \square) | (u, v) \in \Sigma^* \times \Sigma^*\}$ is the set of rules from R' with nonempty completions.
- The set of all (old) rules that do not work at the right end can be identified with $R' \setminus \$:= R' \setminus R'_{\text{pref}}$.
- C'_{pref} is the completion prefix which is given by the expanded rewriting system R'_{pref} and the remaining parts being identical to those of C'.
- $C' \setminus \$$ is defined accordingly.

Remark 3. By using R'_{pref} we can speak of all *newly generated* rules. On the other hand $R' \setminus \$$ are all *old* rules that do *not* work at the *right end* of words. Those old rules that *do* work at the right end of words will have a representant in R'_{pref}.

Now we can explain the brackets in figure 1. The words left of the slashes are consumed completions. They will be of no importance for further reductions. Right of the slashes are the unconsumed completions. They play an important role for the correctness of prefix systems: they link applications of prefix rules. To explain this, we have a closer look at a part of the above reduction. Below the $\rightarrow$ we write the bracket expression of the respective rules:

$$\text{¢} adabbde\$ \xrightarrow[[\square/c]]{} \text{¢} adabbdb\$ \xrightarrow[[\square/\square]]{} \text{¢} adabbb\$ \xrightarrow[[\square/c]]{} \text{¢} adab\$ \xrightarrow[[c/\square]]{} \text{¢} adb\$$$

The application of rule r_{12} means: "guess that the next letter would be a c and that it belonged to the unchanged right context." Then rule r_9 is used. Since this is an old rule, it does not change the guess of a completion. After that, rule r_5 is used. It again assumes an unchanged c to be the next letter of a possible completion to a correct word. The application of rule r_5 fits to that of r_{12} which assumed the same. Now rule r_2 is used. Here the c is still the same letter, but is part of the consumed completion. This means in other words: "we guessed the next letter to be a c, this guess was correct, and now it is completely used, so we do not need to consider it further."

In contrast to this, consider using rule r_8 twice instead of r_5. This rule guesses the end of the word. So its unconsumed completion will not fit to the completion of r_2. This is of major importance. Because of r_8, $L_{C'}$ would also contain $abbbb$ which clearly is no prefix of a word in L_C.

This has two consequences: (a) we have to find a method to check such cases and (b) an extension of remark 2:

Remark 4. There are psCRLS C without true nonterminals, i.e. $\Gamma = \Sigma \cup \{\mathfrak{c}, \$, y\}$, so that $\text{PREF}(C)$ is defined but not correct, which means $L_{\text{PREF}(C)} \neq \text{Pref}(L_C)$. So, a natural question is, if it is possible to determine whether or not the system $\text{PREF}(C)$ is correct. This will be discussed in the next section.

5   Main Result

Our first step is to introduce a way to store information about the interaction between old rules and new rules and the involved completions.

Definition 7. *Let C be a psCRLS and an expanded CRLS $C' = \text{CPREF}(C)$ with expanded rewriting system R'. A set K of 4-tuples is called* candidate set *of C', if for all $u = (u_1, u_2, u_3, u_4) \in K$ the following holds:*

1. *u_1 ends with $\$$: $u_1 \in \Gamma^* \cdot \$$*
2. *u_2 is the accepting letter or ends with $\$$: $u_2 \in (\{y\} \cup (\Gamma^* \cdot \$))$*
3. *u_2 is irreducible w.r.t. all old rules that do not work at the right end: $u_2 \in \text{IRR}(R' \setminus \$)$*
4. *u_3 is built from the inner alphabet: $u_3 \in (\Gamma \setminus \{\mathfrak{c}, \$, y\})^*$*
5. *u_4 is mainly from the inner alphabet, yet it may have a $\$$ at its end: $u_4 \in (\Gamma \setminus \{\mathfrak{c}, \$, y\})^* \cdot \{\Box, \$\}$*
6. *a) u_1 can be reduced to u_2 using only old rules that do not work at the word end: $u_1 \xrightarrow[R' \setminus \$]{*} u_2$*
   or
   b) u itself is a new rule: $u \in R'_{pref}$.

The elements of candidate sets are called candidates*. In the above case 6a u is a representative of a reduction with $R' \setminus \$$. We also call u a reduction candidate. In case 6b we call u a rule candidate from R'_{pref}.*

Now we want to know if two chains of reductions on partial words, each represented by a candidate, can happen after each other:

Definition 8. *Let C be a psCRLS, $C' = \text{CPREF}(C)$, K a candidate set of C' and $u = (u_1, u_2, u_3, u_4), v = (v_1, v_2, v_3, v_4) \in K$*
We say u allows v in K with rest w, $u \vdash_{K,w} v$, if one of the following conditions holds:

(i) *u_2 and v_1 overlap so that their right ends are matched together, then w is empty:*
   $w = \Box$ and $v \in C'_{pref}$ and $u_2 \in \text{Suff}(v_1) \vee v_1 \in \text{Suff}(u_2)$

(ii) u_2 *and* v_1 *do not overlap in that way but one of the old rules not working at the right end of words can be padded at the right end of its first component with a word* w *so that an indirect overlap (via reduction) is possible. Furthermore the first component of this old rule has an overlap with* u_2 *that reaches into the part of the word which is changed by the reduction represented by the candidate* u:

$$w \neq \square \text{ and } \exists(v_1', v_2') \in \text{REW}(C \setminus \$), v' \in \Gamma^* \text{ so that}$$
$$|v_1'| - |v'| > \text{lcp}(u_1, u_2) \text{ and } v_1 = v_1' w = v' u_2 \text{ and } v_2 = [v_2' w]_{\text{REW}(C' \setminus \$)}.$$

With $\text{lcp}(u_1, u_2)$ *we denote the length of the longest common prefix of* u_1 *and* u_2.

In figure 2 the second alternative of the definition, which is more complicated, is illustrated. Irreducible parts are set in boldface.

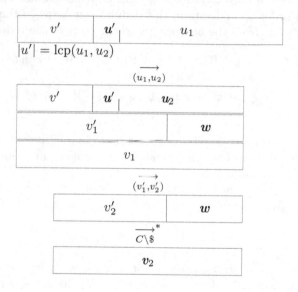

Fig. 2. The allows relation $\vdash_{K,w}$ with $w \neq \square$.

The last step is to check if the completions appearing in these reductions fit together.

Definition 9. *Let* C *be a psCRLS, let* $C' = \text{CPREF}(C)$, K *a candidate set of* C', *and* $u = (u_1, u_2, u_3, u_4), v = (v_1, v_2, v_3, v_4) \in K$.

We say u *allows* v *in* K *and with rest* w *and correct completion   if* $u_4 \in \text{Pref}(v_3 v_4)$ *and* $u \vdash_{K,w} v$.

That means the reduction represented by v *may safely be applied after the one represented by* u *since the rest of the completion left by* u *fits to the completion of* v.

Notation: $u \succ_{K,w} v$. *Also, if we are not interested in* w: $u \succ_K v$, *then* $\succ_K^+$ *and* $(\succ_K^*)$ *denote the transitive (reflexive and transitive) closure.*

If $u \succ_K^+ v$ *and* u, v *are rule candidates from* R'_{pref},
and if no u' *exists which is a rule candidate from* R'_{pref}, *so that* $u \not\succ_{K \setminus \{u'\}}^+ v$
we say u *directly allows* v *in* K *with correct completion, short* $u \succ_K^{\oplus} v$.

The definition of $u \succ_K^+ v$ does not distinguish how many prefix rules are used between u and v. By $u \succ_K^{\oplus} v$ we can make sure that v can be reached from u by using exactly one prefix rule.

Example 5. (for definition 9) There exists no candidate set K so that $r_8 \succ_K^+ r_2$. There exists a K so that $r_5 \succ_K^{\oplus} r_2$.

Now we can define sets of correctly working right end reductions. This definition is inductive. In order to understand this definition it helps to think of the reductions *backwards*, i.e. from the accepting y to the start of the reductions.

Definition 10. *Let* C *be a psCRLS,* $C' = \mathrm{CPREF}(C)$, W *a candidate set of* C'.
W *is a* working set *of* C' *if for all* $u = (u_1, u_2, u_3, u_4) \in W$ *one of the following conditions holds:*

(i) $u_4 = \square$ *(Fully consumed completion means that the next reduction(s) may be applied without regard of the completion; of course only until a new unconsumed completion appears.)*

(ii) $u_4 = \$ \wedge u_2 = y$ *(After an accepting rule where only the word end marker is left as completion no further harm can be done. The reduction is finished ...)*

(iii) *for all* $v = (v_1, v_2, v_3, v_4) \in W, w \in \Gamma^*$ *with* $u \vdash_{W,w} v$
exists $v' = (v_1, v_2, v_3', v_4') \in W$ *with* $u \succ_{W,w} v'$
(Whenever a reduction with v *after* u *can take place, there is a variant* v' *of* v *with fitting next completion that does exactly the same w.r.t. to the rewriting relation.)*

The set of all working sets of C' *is denoted by* $\mathrm{WORKING}(C')$. *Since the definition is inductive, an algorithm can be given to compute or at least enumerate working sets. For details, see [Woi00b].*

The following lemma shows that working sets can be used to check the correctness of prefix systems:

Lemma 1. *Let* C *be a psCRLS, with* $\Gamma = \Sigma \cup \{\mathrm{c}, \$, y\}$, *and* $C' = \{(u_1, u_2', u_3, u_4) | (u_1, u_2, u_3, u_4) \in \mathrm{CPREF}(C), u_2' = [u_2]_{C \setminus \$}\}$. *Let* $W \in \mathrm{WORKING}(C')$ *and* C'' *be a subset of* C' *so that for all rules* $r \in \mathrm{REW}(C'')$ *either* $r \in \mathrm{REW}(C \setminus \$)$ *or* $r \in \mathrm{REW}(C') \cap \mathrm{REW}(W)$. *Let* $w \in \Sigma^*$ *be accepted by* C'': $\mathrm{c}w\$ \xrightarrow[C'']{*} y$. *Then there exists a* $\tilde{w} \in \Sigma^*$ *so that* $\mathrm{c}w\tilde{w}\$ \xrightarrow[C]{*} y$.

Proof. Since for any reduction in a confluent weight reducing system there exists a right canonical reduction and because of the properties of working sets there is an $n \geq 1$ and a reduction of the form:

$$w \xrightarrow[C \backslash \$]{*} w_1 \xrightarrow[r_1 \in \text{REW}(C''_{\text{pref}})]{} w'_1 \xrightarrow[C \backslash \$]{*} w_2 \xrightarrow[r_2 \in \text{REW}(C''_{\text{pref}})]{} w'_2$$

$$\xrightarrow{*} \cdots \xrightarrow[r_i \in \text{REW}(C''_{\text{pref}})]{} w'_i \xrightarrow{*} \cdots$$

$$\xrightarrow[r_{n-1} \in \text{REW}(C''_{\text{pref}})]{} w'_{n-1} \xrightarrow[C \backslash \$]{*} w_n \xrightarrow[r_n \in \text{REW}(C''_{\text{pref}})]{} y$$

With $r_i \succ_W^{\oplus} r_{i+1}$ or $r_i = (u, v, s, \square)$ for all $1 \leq i < n$.

We show the lemma with an induction over n.

Basis. $n = 1$: Then rule r_n is accepting with consumed completion s (possibly empty) and unconsumed completion \$. So $\tilde{w} = s$ is the correct completion.

Claim. Let the lemma hold for $n \geq 1$

Induction step. $(n \to n + 1)$

We can find w', w''_1, and w'''_1 so that $w = \text{¢}w'\$$, $w_1 = \text{¢}w''_1\$$, and $w'_1 = \text{¢}w'''_1\$$. There are three cases, we will only show the proof for the first one, the other two cases are similar:

Case 1: $r_1 = (u\$, v\$, s, \square)$ With induction claim there always exists a completion t with
$\text{¢}w'''_1 t\$ \xrightarrow[C]{*} y$. There exists $r'_1 \in \text{REW}(C)$ with $r'_1 = (us, v)$. This leads to $\text{¢}w' \xrightarrow[C]{*} \text{¢}w''_1$ and in consequence $\text{¢}w's \xrightarrow[C]{*} \text{¢}w''_1 s \xrightarrow[r'_1]{} \text{¢}w'''_1$. So, $\text{¢}w'st \xrightarrow[C]{*} \text{¢}w''_1 st \xrightarrow[r'_1]{} \text{¢}w'''_1 t$. Now, $\tilde{w} = st$ is the correct completion: $\text{¢}w'st\$ \xrightarrow[C]{*} y$.

Case 2: $r_1 \succ_W r_2$ and $r_1 = (u, v, s, s')$ with $s, s' \in (\Gamma \backslash \{\$\})^*$ (analogous to case 1, because of $r_1 \succ_W^{\oplus} r_2$ we know that $s' \in \text{Pref}(t)$ and therefore st is the correct completion).

Case 3: $r_1 \succ_W r_2$ and $r_1 = (u, v, s, s'\$)$ with $s, s' \in (\Gamma \backslash \{\$\})^*$(analogous to case 2, st is the correct completion). $\square$

This directly gives the following result:

Theorem 3. *Let C be a psCRLS, with $\Gamma = \Sigma \cup \{\text{¢}, \$, y\}$, let $C' = \text{CPREF}(C)$ (which is equivalent to $\text{PREF}(C)$ w.r.t. to the defined words), and $W \in \text{WORKING}(C')$. If for all $r = (u, v) \in \text{REW}(C'_{\text{pref}})$ the condition $(u, v) \in \text{REW}(W)$ holds, then $\text{CPREF}(C)$ is correct, and therefore also $\text{PREF}(C)$ is correct.*

Final remark. Obviously, these results can be used for a similar suffix construction. The problem of true nonterminals ($\Gamma_{\text{inner}} \neq \Sigma$) cannot be discussed here for lack of space.

6   Conclusion

We have shown that under certain conditions it is possible to give an effective construction for CRLS's defining the prefix language of a CRL. Due to the fact that prefix splitable systems are a normal form this syntactical restriction is not a hinderance to this on its own account. But because the CRL's are a basis for the r.e. languages there still will be languages whose prefix systems cannot be correct. To be precise, chances are high that the correctness of prefix systems is undecidable. Furthermore, there could be CRL's whose prefix languages are CRL's but for which the construction given here fails on any system defining them. Another problem arises from true nonterminals, that is such letters in Γ that are neither end markers nor the accepting letters. These questions show a line of further research in the theoretical aspects of Church-Rosser prefix languages.

From the practical point of view, one might ask if more than "toy" languages are possible. In [Rot00] this is answered in the positive. In this diploma thesis a substantial subset (i.e., covering main syntactical problems) of JAVA syntax has been described with a CRLS for which the prefix construction gives a correct system. One possibility to incorporate this into usable software tools would be to design something like "hybrid" $LR(k)$/CRL-compilers.

Altogether one can conclude that under theoretical as well as under practical aspects prefix languages of CRL's are worth future investigation.

Acknowledgements

The author wishes to thank the anonymous referees for their helpful comments, especially for detecting a (correctable) error in figure 1.

References

[App98]   A. W. Appel. *Modern Compiler Implementation in Java.* Cambridge University Press, 1998.

[BO93]   R.V. Book and F. Otto. *String-Rewriting Systems.* Springer-Verlag, New York, 1993.

[BO98]   G. Buntrock and F. Otto. Growing context-sensitive languages and Church-Rosser languages. *Information and Computation,* 141:1–36, 1998.

[DW86]   E. Dahlhaus and M.K. Warmuth. Membership for growing context-sensitive grammars is polynomial. *Journal of Computer and System Sciences,* 33:456–472, 1986.

[Har78]   M. A. Harrison. *Introduction to Formal Language Theory.* Addison-Wesley, Reading, Mass., 1978.

[Jan88]   M. Jantzen. *Confluent String Rewriting.* Springer-Verlag, 1988.

[Knu65]   D. E. Knuth. On the translation of languages from left to right. *Information and Control,* 8:607–639, 1965.

[MNO88]   R. McNaughton, P. Narendran, and F. Otto. Church-Rosser Thue systems
          and formal languages. *Journal Association Computing Machinery*, 35:324–
          344, 1988.

[NO98]   G. Niemann and F. Otto. The Church-Rosser languages are the deterministic
          tic variants of the growing context-sensitive languages. In M. Nivat, editor,
          *Foundations of Software Sscience and Computation Structures, Proceedings
          FoSSaCS'98*, volume 1378 of *LNCS*, pages 243–257, Berlin, 1998. Springer-
          Verlag.

[OKK97]   F. Otto, M. Katsura, and Y. Kobayashi. Cross-sections for finitely presented
          monoids with decidable word problems. In H. Comon, editor, *Rewriting
          Techniques and Applications*, volume 1232 of *LNCS*, pages 53–67, Berlin,
          1997. Springer-Verlag.

[Rot00]   T. Rottschäfer.   Eine Entwicklungsumgebung für Church-Rosser
          Präfixparser. Diplomarbeit, TU-Darmstadt, February 2000.

[SSS90]   S. Sippu and E. Soisalon-Soininen. *Parsing Theory. Volume II: LR(k) and
          LL(k) Parsing*. Springer-Verlag, Berlin, 1990.

[Woi00a] J. R. Woinowski. A normal form for Church-Rosser language systems. Re-
          port, TU-Darmstadt, www.iti.tu-darmstadt.de/~woinowsk/, June 2000.

[Woi00b] J. R. Woinowski. Prefixes of Church-Rosser languages. Report TI-2/00,
          TU-Darmstadt, www.iti.tu-darmstadt.de/~woinowsk/, February 2000.

Author Index

Lecture Notes in Computer Science

For information about Vols. 1–1893
please contact your bookseller or Springer-Verlag

Vol. 1927: P. Thomas, H.W. Gellersen, (Eds.), Handheld and Ubiquitous Computing. Proceedings, 2000. X, 249 pages. 2000.

Vol. 1928: U. Brandes, D. Wagner (Eds.), Graph-Theoretic Concepts in Computer Science. Proceedings, 2000. X, 315 pages. 2000.

Vol. 1929: R. Laurini (Ed.), Advances in Visual Information Systems. Proceedings, 2000. XII, 542 pages. 2000.

Vol. 1931: E. Horlait (Ed.), Mobile Agents for Telecommunication Applications. Proceedings, 2000. IX, 271 pages. 2000.

Vol. 1658: J. Baumann, Mobile Agents: Control Algorithms. XIX, 161 pages. 2000.

Vol. 1756: G. Ruhe, F. Bomarius (Eds.), Learning Software Organization. Proceedings, 1999. VIII, 226 pages. 2000.

Vol. 1766: M. Jazayeri, R.G.K. Loos, D.R. Musser (Eds.), Generic Programming. Proceedings, 1998. X, 269 pages. 2000.

Vol. 1791: D. Fensel, Problem-Solving Methods. XII, 153 pages. 2000. (Subseries LNAI).

Vol. 1799: K. Czarnecki, U.W. Eisenecker, Generative and Component-Based Software Engineering. Proceedings, 1999. VIII, 225 pages. 2000.

Vol. 1812: J. Wyatt, J. Demiris (Eds.), Advances in Robot Learning. Proceedings, 1999. VII, 165 pages. 2000. (Subseries LNAI).

Vol. 1932: Z.W. Raś, S. Ohsuga (Eds.), Foundations of Intelligent Systems. Proceedings, 2000. XII, 646 pages. (Subseries LNAI).

Vol. 1933: R.W. Brause, E. Hanisch (Eds.), Medical Data Analysis. Proceedings, 2000. XI, 316 pages. 2000.

Vol. 1934: J.S. White (Ed.), Envisioning Machine Translation in the Information Future. Proceedings, 2000. XV, 254 pages. 2000. (Subseries LNAI).

Vol. 1935: S.L. Delp, A.M. DiGioia, B. Jaramaz (Eds.), Medical Image Computing and Computer-Assisted Intervention – MICCAI 2000. Proceedings, 2000. XXV, 1250 pages. 2000.

Vol. 1937: R. Dieng, O. Corby (Eds.), Knowledge Engineering and Knowledge Management. Proceedings, 2000. XIII, 457 pages. 2000. (Subseries LNAI).

Vol. 1938: S. Rao, K.I. Sletta (Eds.), Next Generation Networks. Proceedings, 2000. XI, 392 pages. 2000.

Vol. 1939: A. Evans, S. Kent, B. Selic (Eds.), «UML» – The Unified Modeling Language. Proceedings, 2000. XIV, 572 pages. 2000.

Vol. 1940: M. Valero, K. Joe, M. Kitsuregawa, H. Tanaka (Eds.), High Performance Computing. Proceedings, 2000. XV, 595 pages. 2000.

Vol. 1941: A.K. Chhabra, D. Dori (Eds.), Graphics Recognition. Proceedings, 1999. XI, 346 pages. 2000.

Vol. 1942: H. Yasuda (Ed.), Active Networks. Proceedings, 2000. XI, 424 pages. 2000.

Vol. 1943: F. Koornneef, M. van der Meulen (Eds.), Computer Safety, Reliability and Security. Proceedings, 2000. X, 432 pages. 2000.

Vol. 1945: W. Grieskamp, T. Santen, B. Stoddart (Eds.), Integrated Formal Methods. Proceedings, 2000. X, 441 pages. 2000.

Vol. 1948: T. Tan, Y. Shi, W. Gao (Eds.), Advances in Multimodal Interfaces – ICMI 2000. Proceedings, 2000. XVI, 678 pages. 2000.

Vol. 1952: M.C. Monard, J. Simão Sichman (Eds.), Advances in Artificial Intelligence. Proceedings, 2000. XV, 498 pages. 2000. (Subseries LNAI).

Vol. 1953: G. Borgefors, I. Nyström, G. Sanniti di Baja (Eds.), Discrete Geometry for Computer Imagery. Proceedings, 2000. XI, 544 pages. 2000.

Vol. 1954: W.A. Hunt, Jr., S.D. Johnson (Eds.), Formal Methods in Computer-Aided Design. Proceedings, 2000. XI, 539 pages. 2000.

Vol. 1955: M. Parigot, A. Voronkov (Eds.), Logic for Programming and Automated Reasoning. Proceedings, 2000. XIII, 487 pages. 2000. (Subseries LNAI).

Vol. 1960: A. Ambler, S.B. Calo, G. Kar (Eds.), Services Management in Intelligent Networks. Proceedings, 2000. X, 259 pages. 2000.

Vol. 1961: J. He, M. Sato (Eds.), Advances in Computing Science – ASIAN 2000. Proceedings, 2000. X, 267 pages. 2000.

Vol. 1963: V. Hlaváč, K.G. Jeffery, J. Wiedermann (Eds.), SOFSEM 2000: Theory and Practice of Informatics. Proceedings, 2000. XI, 460 pages. 2000.

Vol. 1966: S. Bhalla (Ed.), Databases in Networked Information Systems. Proceedings, 2000. VIII, 247 pages. 2000.

Vol. 1967: S. Arikawa, S. Morishita (Eds.), Discovery Science. Proceedings, 2000. XII, 332 pages. 2000. (Subseries LNAI).

Vol. 1968: H. Arimura, S. Jain, A. Sharma (Eds.), Algorithmic Learning Theory. Proceedings, 2000. XI, 335 pages. 2000. (Subseries LNAI).

Vol. 1969: D.T. Lee, S.-H. Teng (Eds.), Algorithms and Computation. Proceedings, 2000. XIV, 578 pages. 2000.

Vol. 1970: M. Valero, V.K. Prasanna, S. Vajapeyam (Eds.), High Performance Computing – HiPC 2000. Proceedings, 2000. XVIII, 568 pages. 2000.

Vol. 1971: R. Buyya, M. Baker (Eds.), Grid Computing – GRID 2000. Proceedings, 2000. XIV, 229 pages. 2000.

Vol. 1974: S. Kapoor, S. Prasad (Eds.), FST TCS 2000: Foundations of Software Technology and Theoretical Computer Science. Proceedings, 2000. XIII, 532 pages. 2000.

Vol. 1975: J. Pieprzyk, E. Okamoto, J. Seberry (Eds.), Information Security. Proceedings, 2000. X, 323 pages. 2000.

Vol. 1976: T. Okamoto (Ed.), Advances in Cryptology – ASIACRYPT 2000. Proceedings, 2000. XII, 630 pages. 2000.

Vol. 1977: B. Roy, E. Okamoto (Eds.), Progress in Cryptology – INDOCRYPT 2000. Proceedings, 2000. X, 295 pages. 2000.

Vol. 1983: K.S. Leung, L.-w. Chan, H. Meng (Eds.), Intelligent Data Engineering and Automated Learning – IDEAL 2000. Proceedings, 2000. XVI, 573 pages. 2000.

Vol. 1987: K.-L. Tan, M.J. Franklin, J. C.-S. Lui (Eds.), Mobile Data Management. Proceedings, 2001. XIII, 290 pages. 2001.